D0138581

# Modern Physics

# Modern Physics

Jeremy Bernstein

Paul M. Fishbane
*University of Virginia*

Stephen Gasiorowicz
*University of Minnesota*

PRENTICE HALL
Upper Saddle River, NJ 07458

**Library of Congress Cataloging-in-Publication Data**

Bernstein, Jeremy
     Modern physics/Jeremy Bernstein, Paul M. Fishbane, Stephen Gasiorowicz
          p. cm.
     Includes bibliographical references and index.
     ISBN 0-13-955311-8
     1. Physics. I. Fishbane, Paul M. II. Gasiorowicz, Stephen. III. Title
     QC21.2.B478 2000
     539—dc21                             99-086230
                                               CIP

Executive Editor: Alison Reeves
Editorial Director: Paul F. Corey
Development Editor: David Chelton
Editor in Chief, Development: Ray Mullaney
Production Editor: Joanne Hakim
Assistant Vice President of Production & Manufacturing: David W. Riccardi
Executive Managing Editor: Kathleen Schiaparelli
Assistant Managing Editor: Lisa Kinne
Marketing Manager: Erik Fahlgren
Creative Director: Paul Belfanti
Art Director: Joseph Sengotta
Interior Design: Maureen Eide
Cover Design: Maureen Eide
Art Manager: Gus Vibal
Art Editor: Karen Branson
Manufacturing Manager: Trudy Pisciotti
Manufacturing Buyer: Michael Bell
Photo Editor: Beth Boyd
Photo Researcher: Yvonne Gerin
Photo Coordinator: Zina Arabia
Illustrations: Scientific Illustrators
Copy Editor: Brian Baker
Editorial Assistants: Christian Botting and Nancy Bauer
Text Composition/Prepress: Prepare Inc.
Cover Image: An artistic view of the atomic-scale topography of Si (114)
          as revealed by Scanning Tunneling Microscopy (STM).
          Naval Research Laboratory, Washington, D.C.

© 2000 by Prentice-Hall, Inc.
Upper Saddle River, NJ 07458

All rights reserved. No part of this book may be
reproduced, in any form or by any means,
without permission in writing from the publisher.

Printed in the United States of America

10 9 8 7 6 5 4 3 2 1

**ISBN 0-13-955311-8**

Prentice-Hall International (UK) Limited, *London*
Prentice-Hall of Australia Pty. Limited, *Sydney*
Prentice-Hall Canada Inc., *Toronto*
Prentice-Hall Hispanoamericana, S.A., *Mexico*
Prentice-Hall of India Private Limited, *New Delhi*
Prentice-Hall of Japan, Inc., *Tokyo*
Pearson Education Asia Pte. Ltd.
Editora Prentice-Hall do Brasil, Ltda., *Rio de Janeiro*

# Brief Contents

# Contents

# 12 Statistical Physics 326

# 13 Decays, Radiation from Atoms, and Lasers 370

# 14 Conductors, Semiconductors, and Superconductors 394

# 15 The Atomic Nucleus 441

# About the Authors

### Jeremy Bernstein

Jeremy Bernstein has had a dual career in physics and writing. He was on the staff of the *New Yorker* from 1963 to 1993 and was a Professor of Physics at the Stevens Institute of Technology from 1968 until his retirement in 1993, when he became Professor emeritus. He has won several awards for his writing about science and mountain travel. He has also published widely in both technical and non-technical journals. Some of his recent books are: *An Introduction to Cosmology, Albert Einstein and the Frontiers of Physics, A Theory for Everything, In the Himalayas,* and *Dawning of the Raj.* He has held visiting appointments at The Rockefeller University, The University of Islamabad, The Ecolé Polytechnique, CERN laboratory, Princeton University, and Oxford. This photograph of Jeremy was taken on a bicycle trip in northern California. The thumb, which is on the grounds of the Clos Pegase art gallery and winery in Calistoga, was the work of the French artist Cesar Baldachini. Bernstein has bicycled in many countries including Bali and Crete. He makes his home in New York City and Aspen, Colorado.

### Paul M. Fishbane

Paul Fishbane has been teaching undergraduate courses at the University of Virginia, where he is Professor of Physics, for some 25 years. He received his doctoral degree from Princeton University in 1967 and has published some 100 papers in his field, theoretical high energy physics. He is co-author of *Physics for Scientists and Engineers* with Stephen Gasiorowicz and Stephen Thornton. Paul has held visiting appointments at the State University of New York at Stony Brook, Los Alamos Scientific Laboratory, CERN laboratory in Switzerland, Amsterdam's NIKHEF laboratory, France's Institut de Physique Nucleaire, the University of Paris-Sud, and the Ecolé Polytechnique. He has been active for many years at the Aspen Center for Physics, where current issues in physics are discussed with an international group of participants. His other interests include biking, music, and the physics of the kitchen. All of the rest of his time is spent trying to keep up with his family, especially his youngest son Nicholas.

### Stephen Gasiorowicz

Stephen Gasiorowicz was born in Poland and received his Ph.D. in physics at the University of California, Los Angeles in 1952. After spending 8 years at the Lawrence Radiation Laboratory in Berkeley, California, he joined the faculty of the University of Minnesota, where his field of research is theoretical high energy physics. As a visiting professor, he has traveled to the Niels Bohr Institute, NORDITA in Copenhagen, the Max Planck Institute for Physics and Astrophysics in Munich, DESY in Hamburg, Fermilab in Batavia, and the Universities of Marseille and Tokyo. He has been a frequent visitor and an officer of the Aspen Center for Physics. Steve is co-author of *Physics for Scientists and Engineers* with Paul Fishbane and Stephen Thornton and has written books on elementary particle physics and quantum physics. A relatively new occupation is that of grandfather, which still leaves some time for reading (history), biking, canoeing, and skiing.

# Preface

Knowledge of the revolutions of 20th-century physics is an indispensable part of the training of any engineer and physical scientist. That is because virtually all of today's technology is based, at least in part, on this knowledge. The basic subject material of what is called modern physics is very nearly 100 years old, so that it is hardly modern at at all. Yet just as Newton's laws, today 300 years old, Maxwell's equations, today nearly 150 years old, and the laws of classical statistical physics, more than 100 years old, remain applicable and essential in their respective domains of physical law, so too do the two major developments of the first half of this century: relativity and quantum mechanics. These fundamental subjects underlie a vast scope of application that continues its inventive course today. Moreover, research on fundamental physics has not stopped with relativity and quantum mechanics, and working scientists still face questions as interesting as any that have been answered in the past.

Both relativity and quantum mechanics require the student to make difficult changes in how he or she thinks the physical world works. The subjects violate prejudices that have been built up by everyday experience. For this reason, precision and clarity of explanation are, for us, the first and most important part of the material. We have made every effort to avoid the "it can be shown" approach and to present modern physics in a way that makes its interconnectedness, as well as its connection to classical physics, evident.

Throughout this text, we have built in a historical approach—a discussion of how a subject developed and the thinking that led to its maturation. Often this historical perspective is interwoven with the material; at other times it would interrupt an efficient and compact presentation, and then we present it on the side, as it were. We feel that this approach is useful in that it stresses that the roots of the revolutionary advances lie in experiment; it also makes the text more fun to read.

The book forms the basis of a traditional course in the subject. It contains, in a mathematical language that we have deliberately kept at a level we felt students would be comfortable with, descriptions of special relativity and of the laws of quantum mechanics. It describes applications of these fundamental ideas to both technological and scientific issues. Finally, it describes the subject matter that is of fundamental interest today. All this material is too much to cover in one semester, the usual length of time for such courses, so a more detailed explanation of what we do is in order. This will allow the instructor to make a reasonable choice of what to cover and provide guidance to the reader for the use of the material in the book.

We have broken the material into several parts, even if the boundary between the coverage of the different parts is not always perfectly sharp. The first chapter replaces what would otherwise be a steady set of footnotes referring the reader to an introductory calculus-based textbook.[1] In other words, Chapter 1

---

[1] For example, see Fishbane, Gasiorowicz, and Thornton, *Physics for Scientists and Engineers*, 2d ed. (Englewood Cliffs, Prentice Hall, 1996).

is a place to remind students of things that, ideally, they should have fully absorbed in their introductory courses. While the chapter cannot replace such a textbook, it can be a convenient road map to the introductory material. It also constitutes a type of formulary of classical physics. But we urge the student to keep his or her introductory text and to consult it when necessary. The chapter contains no examples or problems, and it is not meant to be assigned as normal course material.

Part 1 consists of two chapters on special relativity. These are divided more or less according to traditional lines, with a discussion of space and time in one chapter and momentum and energy in the next. Our approach is to extract length contraction from the Michelson–Morley result and use it as a jumping-off point for the other effects of special relativity, including the Lorentz transformations. This tack differs somewhat in detail, although not in spirit, from the approach that abstracts special relativity from a moving light-clock. We believe that the formal approach starting early with the full set of Lorentz transformations, less suitable for physicists and engineers than mathematicians, can miss the physics of the subject. We also save our discussion of general relativity for a much later chapter. Even though the origins of general relativity are old, there is much exciting current material to cover.

Part 2 is a treatment of the fundamental laws of quantum mechanics. This is a subject with a fascinating yet complex history, but we feel that the number of missteps in the development of quantum mechanics speak against a full historical interweaving of the material with the rest of the text. Thus a separate historical introduction is presented. Chapter 4 describes the experimental data that could not be encompassed by classical physics and examines the daring ideas that opened the gateway to the development of quantum mechanics. Bohr's approach to the structure of the hydrogen atom provided the critical breakthrough, and it merits a chapter on its own, Chapter 5. Extended to circular orbits for other central forces, that approach leads to the quantum nature of rotational and vibrational motion, and it also provides a useful tool for the dependence of energy levels on the relevant physical parameters.

Chapter 6 introduces the Schrödinger equation. Here the problem is to find a way to present this material without getting too heavily into mathematics. One common approach is via wave packets, but they are something with which many students using this text may feel uncomfortable. Instead, we motivate the Schrödinger equation by using classical parallels and the physical meaning of a wave function to argue the form of the Schrödinger equation. This involves bringing in the probability interpretation of the wave function in what we feel is its proper place: right at the beginning. We have kept the mathematics involved in actually solving the Schrödinger equation low, treating just the infinite well here. Only in Chapter 7 do we go into the addition of plane waves with easily managed distributions to get at the concepts of wave packets and of probabilities for measurements of momentum. In that way we can understand the particlelike behavior of a superposition of waves. This material also allows us to introduce the uncertainty relations. We show how they "shield" quantum mechanics from contradictions, and we illustrate their utility in making estimates of ground-state energies.

Starting with Chapter 8 we are in position to see what the Schrödinger equation has to say about some interesting potentials, namely barriers and wells. A good deal of useful physics about scattering and bound states can be conveyed for these mathematically simple situations. We pay particular attention to the physics of tunneling, relating it to internal reflection and to a demonstration that can actually be done in class and describing where it is relevant to physical phenomena. Chapter 9 is a discussion of the Schrödinger equation in the context of the coulomb potential. It is in this chapter that we treat angular momenta, even if we do not employ much in the way of mathematical rigor, and our discussion of the hydrogen atom is concentrated in this chapter, along with the Zeeman effect and the concept of spin. In Chapter 10 we conclude our discussion of the principles of quantum mechanics with the treatment of many-body systems and the symmetry of the wave function for identical particles. This subject is indispensable for an understanding of solids and other material systems, and by putting the exclusion principle here we are prepared for its applications in many domains.

Part 3 of the book is labeled "applications," and it contains discussions of those areas, both in nature and in technology, that cannot be understood without quantum mechanics. The instructor can easily pick and choose among the chapters in this part of the text if he or she is pressed for time. Still, one needs to be aware that there are constraints in some cases; for example, it would be difficult to teach the physics of semiconductors without having first seen the Fermi–Dirac distribution.

Part 3 begins (in Chapter 11) with a discussion of complex atoms and of molecules. We are primarily interested in the quantum mechanical basis of the periodic table, in the way that minima in energy are associated with the mechanisms by which atoms can form molecules, and in simple molecular spectra. The next chapter is a treatment of thermal systems, and because some of the students who take this course may not have had a good background in that material we begin with a simple treatment of classical statistical mechanics, an extremely useful subject for any future engineer or scientist. A treatment of specific heats allows us to understand why one needs a discussion of statistical quantum mechanics. The Boltzmann distribution, a major target, is not only extremely important on its own, it also provides a guide for the development of the quantum mechanical distributions for identical particles. In each case, very simple arguments based on the idea of thermal equilibrium are used. We can also make the connection back to the blackbody distribution first described in Chapter 4, closing a circle.

In Chapter 13 we describe how one can think about unstable systems in quantum mechanics, a topic relevant to atoms in excited states and, by extension, to lasers, whose operation and use form a major part of the chapter. Chapter 14 describes applications to the solid state, a topic so large that we have been forced to make some restrictive choices. We have tried in part to choose according to topics of the greatest current interest to engineers. Accordingly, we have begun with a treatment of how electricity is conducted in materials. When this is coupled with the essential description of band structure, we are led in a natural way to the behavior of semiconductors, a subject with exceptionally rich, diverse applications. We nevertheless restrict ourselves to the more comprehensible topics, leaving out a detailed treatment of many of the more complicated ones—the many varieties of transistors, for example. We also take the opportunity to describe what we think are the most interesting and physically significant aspects of superconductivity. The last chapter in this part, Chapter 15, contains a selection of topics in nuclear physics. The subject is a complex one, and we have chosen on the basis of what we think will illuminate best its various facets; the applications that we examine are equally diverse.

Part 4 of the text contains a discussion of topics that are, at least in part, at the forefront of the unknown. We think it important that students—even students who are going to work in highly applied areas—be exposed to this sort of material. It helps to dispel the notion that the subject is a closed one in which all one has to do is know how to plug things into formulas, and it emphasizes the overarching role that simple scientific curiosity plays. The three chapters of this part treat, respectively, elementary particle physics, general relativity, and cosmology. While general relativity *per se* is old, it is deeply implicated in our understanding of cosmological issues, and its reconciliation with quantum mechanics represents one of the great unanswered questions. Particle physics, too, is an important piece of the puzzle that cosmologists are attempting to assemble.

Chapter 16, on particle physics, addresses the unanswered questions of just what are the underlying laws that govern all the other aspects of matter we have described in this book. It is a highly qualitative and descriptive chapter, but it is also a modern one, concentrating on those issues that are actively addressed today. In addition to covering the older topics, the chapter on general relativity contains a deeper and more physical discussion of such issues as black holes and gravitational radiation than is usual. We think these issues are of great interest to students. The chapter on cosmology speaks to the question of the nature of the universe; this chapter also contains a detailed discussion of the motivation and evidence for the big bang. The discussion of the evolution of the universe from a big bang brings in many of the topics we have discussed throughout and, we hope, will convey the fundamental unity of physics to the reader.

We would like to offer thanks for the considerable help we were given in the process of writing this book. In addition to the many scientific colleagues who clarified issues we did not understand well enough, we want to thank our editor Alison Reeves, our developmental editor David Chelton, and our production editor Joanne Hakim. Many others at Prentice Hall have helped us, too. In particular, we want to thank Yvonne Gerin and Ray Mullaney. We would also like to acknowledge the following reviewers, who provided valuable feedback.

Albert Altman
*University of Massachusetts, Lowell*

David Curott
*University of North Alabama*

Luther Frommhold
*University of Texas, Austin*

Richard T. Hammond
*Rensselaer Polytechnic Institute*

Roger J. Hanson
*University of Northern Iowa*

Edward Hart
*University of Tennessee, Knoxville*

Gary G. Ihas
*University of Florida*

Rondo Jeffery
*Weber State University*

John Kenny
*Bradley University*

Sanford Kern
*Colorado State University*

John M. Knox
*Idaho State University*

Arthur Z. Kovacs
*Rochester Institute of Technology*

Curt Larson
*University of Wisconsin, River Falls*

Paul L. Lee
*California State University, Northbridge*

Nathaniel P. Longley
*Colorado College*

Wolfgang Lorenzon
*University of Michigan*

Thomas Moses
*Knox College*

Joseph F. Owens III
*Florida State University, Tallahassee*

Stephen Pate
*New Mexico State University*

Joseph Priest
*Miami University*

Robert Ross
*University of Detroit, Mercy*

Weidian C. Shen
*Eastern Michigan University*

Paul Sokol
*Pennsylvania State University*

Takamasa Takahashi
*St. Norbert College*

Frank C. Taylor
*Furman University (now retired)*

Larry H. Toburen
*East Carolina University*

Jack Tuszynski
*University of Alberta*

C. Wesley Walter
*Denison University*

Jeffrey L. Wragg
*College of Charleston*

# Modern Physics

# A Review

S cience is an edifice, with the latest brick almost always laid on an earlier one. The physics described in this book depends on classical concepts and techniques described in an earlier course in physics. You have taken such a course, one that uses calculus, but while you have not forgotten everything, in all likelihood you have not perfectly retained all the material presented in that course. Our aim in this chapter is to provide you with a kind of road map of the most important material from your introductory course. We emphasize those things that are important in a course in modern physics. We also remind you of some of the history of this material, because we think it is useful for you to be aware of the way things have developed, not just the results.

We do assume that you have retained *some* things. In particular, we do not review the mathematics that was necessary for your first physics course. Thus, you are expected to know that vectors are not the same as scalars, that the acceleration is a vector given by the second derivative of the displacement, and that partial derivatives may enter into some equations, as well as what the area integral in Gauss' law means.

Our review is no substitute for a first course. You learn physics by understanding the material well enough to solve problems, and we make no attempt to do examples or present problems in this chapter. That will come later. The material we present here is so brief that it is at best a reminder; if you want details, consult the text from your first physics course, which is a far better reference.

## 1–1 Newton's Laws

Isaac Newton was one of the singular geniuses in all human history. However, he had predecessors in the development of physics—Copernicus, Galileo, and Kepler come to mind—and contemporaries, of whom Christian Huygens and Newton's great rival Robert Hooke were perhaps the most illustrious. Gottfried Leibniz, also a contemporary of Newton, independently invented the differential calculus, something that Newton would never admit. While Newton was aware of such earlier work, it was rare that he acknowledged it. But he transformed it in such a way that it would have been all but unrecognizable to his predecessors. Newton created what we think of as theoretical physics—the idea of describing nature in terms of equations. The gravitational force was the inspiration for this process: In order to solve the problem of gravitation, Newton invented the differential and integral calculus and introduced the three basic laws of mechanics that we know today as Newton's laws. These three laws are true foundation stones of physics.

**Newton's second law** describes the motion of an object of mass $m$ that is acted on by a set of forces that add *vectorially* to a net force:

$$\vec{\mathbf{F}}_{net} = m\vec{\mathbf{a}}. \tag{1--1}$$

Here, $\vec{\mathbf{a}}$ is the acceleration of the object. For a given net force, $\vec{\mathbf{a}}$ is inversely proportional to the mass—the smaller the mass, the larger is the acceleration and vice versa. The meaning of this mass, more properly termed the **inertial mass**, is provided by Newton's second law itself. In classical physics, mass is conserved, although it can be spread around when an object breaks up or conjoined when an object is formed by amalgamation. We sometimes see the second law written in the form

$$\vec{\mathbf{F}}_{net} = \frac{d\vec{\mathbf{p}}}{dt}, \tag{1--2}$$

where the **momentum** of the object is $\vec{\mathbf{p}} = m\vec{\mathbf{v}}$. This form allows us to account for an object with a changing mass.

In order for Newton's second law to be useful, we must know something about the forces produced in different circumstances; that is, we must have **force laws**. When we do know about the forces, Newton's second law becomes a dynamical equation—an equation of motion—that allows us to predict the motion of the object in question. Some of the forces you will have studied are gravity, electrical and magnetic forces, and contact forces such as friction and the tension in a rope. An important example is the spring force, governed by the law which states that, when a spring is stretched or compressed by an amount $x$ from its equilibrium position, it exerts a force given by $-kx$ on a mass attached to its end. Here, $k$ is the spring constant, a quantity that varies from spring to spring. This particular force law is known as **Hooke's law**. A great deal of information is packed into Hooke's law when it is taken in conjunction with Newton's second law. The minus sign in Hooke's law tells us that the force is a restoring force, tending to move the mass back to the equilibrium position of the spring. When an object of a given mass is attached to the end of a spring and no other forces act, then the second law becomes a differential equation for the position of the mass, relating the position to its second derivative, the acceleration. In this context, we refer to the second law as the *equation of motion* of the spring, and the equation can be solved—meaning that we can find the function of time that represents the position. We will actually solve an equation of motion given by Hooke's law shortly. The importance of this particular case is that it applies to nearly any situation in which there is stable equilibrium—that is, a situation in which a slight displacement of an object from an equilibrium position produces a net force that pulls the object back to that position. But no matter how important this example is, it is only an example, one that fits into the framework of Newton's second law along with all the other examples we can imagine.

**Newton's first law** is superficially[1] a special case of the second law:

$$\text{If } \vec{\mathbf{F}}_{net} = \vec{\mathbf{0}}, \text{ the motion is uniform.} \tag{1--3}$$

---

[1] We use this word because the first law is really more than a special case—it has to do with the existence of inertial frames—reference frames in which all forces have identifiable sources.

Here, when we say the motion is uniform, we mean that the acceleration is zero, or that the velocity is constant. This includes the case where the object in question is at rest. The novelty in Newton's first law is that it asserts that the role of a force is only to *accelerate* an object. Galileo's predecessors—especially Aristotle and all those who continued to quote him—assumed that it took a force to keep an object in motion. They could not imagine a frictionless world in which an object would simply keep going once it got started. This law is often used to deduce the presence, magnitude, and direction of an unknown force. For example, if a falling object reaches a constant terminal velocity, then we can deduce the magnitude of the drag force on it if we know that of the force of gravity that acts on it. Of greater significance, the first law tells us that there is no way of distinguishing between a "uniformly moving" and a "stationary" observer.

▌ The lack of distinction between observers moving relative to each other with constant velocities will play a crucial role in the development of the theory of relativity; see Chapters 2 and 3.

## Gravity

We have said that when a given net force acts on an object, the acceleration of the object is inversely proportional to its mass. But we seem to have one conspicuous exception: the case in which a body falls free of all influence but that of gravity. One of the common errors of pre-Galilean thinking was the idea that heavier (more massive) objects fall more quickly (have larger accelerations) than lighter ones—something that contradicts the more careful observation that *all* objects falling under the (sole) influence of gravitation at the surface of the Earth do so with the same acceleration. Galileo claimed to have dropped objects from the Leaning Tower of Pisa to determine whether they all had the same acceleration. If he had actually done that, he would have found out that air drag spoils the experiment. As you might have learned in your introductory course, you can eliminate air drag by creating a laboratory vacuum in which a feather falls as rapidly as a penny. What, then, has happened to Newton's second law, which seems to imply that, because $\vec{a} = \vec{F}/m$, the acceleration *does* depend on the mass? The only way that the second law can be consistent is for the force law for gravitation to be proportional to the mass itself. In this way, the mass cancels from both sides of Newton's second law, and the acceleration is independent of the mass.

▌ When we discuss Einstein's theory of gravitation at the end of the book, we will see that this cancellation involves some very deep physics.

## Hooke's Law

Let us examine how the equation expressing Newton's second law can be solved by reviewing an important example: the spring force, $F = -kx$ (•Fig. 1–1a). Since we are working in one dimension, vectorial aspects do not matter, and we can dispense with vector notation in the equation of motion:

$$-kx = m\,\frac{d^2x}{dt^2}. \tag{1–4}$$

This equation relates a position, a function of time, to its second derivative. Not just any function can satisfy the equation for all time. Finding the function or functions that can do this is what we mean by solving the differential equation. The first derivative of the sine function is the cosine, and the derivative of the cosine is the sine again, but with a minus sign. Thus, the solution takes the form

$$x(t) = A \sin(\omega t + \varphi). \tag{1–5}$$

The quantities $A$, the *amplitude* of the motion, and $\varphi$, the *phase*, are determined by initial conditions; for example, specifying the initial position and velocity would determine $A$ and $\varphi$. The quantity $\omega$, on the other hand, is determined by

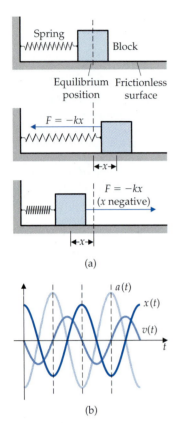

(a)

(b)

• **Figure 1–1** (a) The force due to an idealized spring, which is proportional to the deviation of the length of the spring from its equilibrium length, leads to simple harmonic motion. (b) Plots of displacement, velocity, and acceleration as functions of time of a mass at the end of a spring.

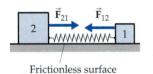

Frictionless surface

• **Figure 1–2** The force $\vec{\mathbf{F}}_{12}$ exerted on block 1 by block 2 is equal and opposite to the force $\vec{\mathbf{F}}_{21}$ exerted on block 2 by block 1.

Eq. (1–4) itself; we say that the dynamics determines $\omega$. We find $\omega$ simply by insisting that Eq. (1–5) satisfy Eq. (1–4), with the result that

$$\omega = \sqrt{k/m}. \tag{1–6}$$

The quantity $\omega$, which is termed the *angular frequency*, is closely related to the repeat time, or *period T*, of the solution. The period and the *frequency f* of the repeating motion are related by

$$T = \frac{1}{f}, \tag{1–7}$$

where $f = \omega/2\pi$. Note, finally, that by taking the first and second time derivatives of the position, Eq. (1–5), we find the velocity and acceleration, respectively (•Fig. 1–1b).

The last of Newton's three laws states that forces act between pairs of objects. That is, a force $\vec{\mathbf{F}}_{21}$ acts on object 2 *due to object 1*, then a force $\vec{\mathbf{F}}_{12}$ acts on object 1 *due to object 2*. For example, when a hand exerts a force on a piano key, the piano key exerts a force on the hand, or when Earth exerts a force on a tennis ball, the tennis ball exerts a force on Earth. **Newton's third law** states that these pairs of forces are represented by equal and opposite vectors (•Fig. 1–2):

$$\vec{\mathbf{F}}_{12} = -\vec{\mathbf{F}}_{21}. \tag{1–8}$$

This law is sometimes called the law of equal action and reaction. Of course, that is somewhat of a misnomer if you think of the word "reaction" as a description of the motion of the objects on which forces act. When a less massive object interacts with a more massive one, the forces may be equal and opposite, but Newton's second law ensures us that the motions of the two objects will be quite different. The tennis ball reacts visibly when the force due to Earth acts on it, but the motion of Earth due to the tennis ball is not visible even to our best instruments.

The third law may be restated in terms of momentum. Imagine that you have two objects labeled 1 and 2, isolated from the outside world and interacting with each other. From Eq. (1–2), the force on object 1 is the rate of change of its momentum, $\vec{\mathbf{p}}_1$; the force on object 2 is the rate of change of $\vec{\mathbf{p}}_2$. Then Eq. (1–8) becomes

$$\frac{d\vec{\mathbf{p}}_1}{dt} = -\frac{d\vec{\mathbf{p}}_2}{dt}, \qquad \text{or} \qquad \frac{d}{dt}\left(\vec{\mathbf{p}}_1 + \vec{\mathbf{p}}_2\right) = \vec{\mathbf{0}}.$$

Thus, the vector $\vec{\mathbf{p}}_1 + \vec{\mathbf{p}}_2$ is constant in time, a great simplification when Newton's laws are applied to what may otherwise be a complicated interaction between two objects. This result extends to more than one interacting object, taking the form

$$\vec{\mathbf{P}}_{tot} = \text{a constant vector}, \tag{1–9}$$

where $\vec{\mathbf{P}}_{tot}$ is the total momentum. We refer to Eq. (1–9) as the law of **conservation of momentum**. It may be taken as an alternative to Newton's third law. Conservation laws play a central role in modern ideas about the operation of physical laws.

## 1–2 Work, Energy, and the Conservation of Energy

Energy is an extremely useful and fundamental quantity. The concept of energy was developed by early 19th-century engineers studying the operation of steam engines. Today, we can solve Newton's law numerically with computers if we know the forces involved. However, this was not true in the 18th and 19th centuries, and the search for simplifications led to a whole world of ideas encompassed under the general term "energy."

Consider a constant net force $F$ acting in one dimension on an object of mass $m$. In this case the motion is entirely determined, with the velocity $v$ changing linearly with time; that is $v \propto t$. If we eliminate the time and find the speed as a function of the distance moved, $x$, we obtain the relation

$$F \times (x_f - x_i) = \frac{1}{2} m v_f^2 - \frac{1}{2} m v_i^2, \tag{1–10}$$

where the initial point of the motion is labeled with the subscript $i$ and the final point with the subscript $f$. It is customary to call the quantity $mv^2/2$ the **kinetic energy** $K$. Then the right side of Eq. (1–10) is the *change* in kinetic energy of the object. The left side is referred to as the **work** $W$ done by the force on the object. When many forces act, each can perform work, and it is the work done by the net force, or the net work, that enters into Eq. (1–10). The work can be positive or negative, according to whether the kinetic energy increases or decreases. The sign is implicit in the definition of work and is determined by whether the force lies parallel [$F$ and $x_f - x_i$ have the same sign] or antiparallel [$F$ and $x_f - x_i$ have the opposite sign] to the displacement in this one-dimensional example.

Of course, not every force is constant, and not every force acts only in one dimension. The more general form of Eq. (1–10) consists in finding the definition of work suitable for the more general form of the net force. As far as forces that vary in space are concerned, we note that for a sufficiently small interval, the force can be regarded as constant, and we can sum the constant forces over all the small intervals. This is in fact, the definition of an integral, so that, for a one-dimensional *non*constant force (a force whose magnitude might vary), the work is defined as

$$W = \int_{x_i}^{x_f} F(x)\, dx. \tag{1–11}$$

For a force that acts in more than one dimension, or, what is the same thing, for motion in more than one dimension, only the component of the force that lies along the motion changes the speed, and hence the kinetic energy, and we pick out this component through the dot, or scalar, product of the force vector and the displacement vector. The result of including both effects is that a suitable definition of the work done on an object as it moves from position $\vec{r}_i$ to $\vec{r}_f$ is

$$W = \int_{\vec{r}_i}^{\vec{r}_f} \vec{F} \cdot d\vec{r}. \tag{1–12}$$

With these definitions, Eq. (1–10) remains

$$W_{\text{net}} = \Delta K = K_f - K_i. \tag{1–13}$$

This equation is known as the **work–energy theorem**.

The work–energy theorem is useful if you are not interested in the time dependence of the motion. The equation can be applied to any force, although whether it is easy to use depends on whether the integration in Eq. (1–11) or Eq. (1–12) can be performed analytically. However, the work–energy theorem can be recast into another form for so-called conservative forces. Then it takes on a significance beyond convenience. A conservative force is a force for which the integration that expresses the work, Eq. (1–12), is independent of the path between the initial and final points. In •Fig. 1–3, we show initial and final points

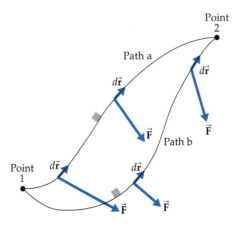

• **Figure 1–3** When the work done by a force that acts on an object which moves from point 1 to a second point 2 is the same for any two paths between the points, the force is said to be conservative.

together with two paths, A and B, between them. A **conservative force** is a force for which values of the work done over the two paths—and indeed, over any paths connecting the two points—are *identical*. For such forces, the work depends only on the two endpoints:

$$\int_{\vec{r}_i}^{\vec{r}_f} \vec{F} \cdot d\vec{r} = f(\vec{r}_f) - f(\vec{r}_i).$$

We get the simplest form of the work–energy theorem if, instead of the function $f(\vec{r})$, we use its negative,

$$-\int_{\vec{r}_i}^{\vec{r}_f} \vec{F} \cdot d\vec{r} = U(\vec{r}_f) - U(\vec{r}_i). \tag{1–14}$$

With this notation, we can recast the work–energy theorem as

$$U(\vec{r}_i) + K_i = U(\vec{r}_f) + K_f. \tag{1–15}$$

This equation is in the form of a conservation law for the **total energy**, defined by

$$E \equiv U(\vec{r}) + K, \tag{1–16}$$

where the **potential energy**—also called the energy of position—is defined by

$$U(\vec{r}) = -\int_{\vec{r}_0}^{\vec{r}} \vec{F} \cdot d\vec{r} + U(\vec{r}_0). \tag{1–17}$$

$U(\vec{r}_0)$ is a constant that drops out of Eq. (1–14). The potential energy is thus defined only up to this constant, which plays no role in the connection between the force and the potential energy, viz., $\vec{F} = -\vec{\nabla}U(\vec{r})$, a relation discussed shortly.

A conservation principle such as the law of conservation of energy is both a central principle of physical laws and a powerful problem-solving tool. Energy conservation states that there is no change in the numerical value of $E$ in the course of time. This principle is not affected by the presence of an arbitrary constant in the definition of $U$, namely, its value at an arbitrary initial point $\vec{r}_0$ in the integral. That point and the value of $U$ there can be chosen for convenience. We usually express this arbitrariness by setting $U$ equal to zero at some convenient point; we will see some examples later. Another thing to keep in mind is that energy is a scalar, not a vector, quantity. Of course, scalar quantities have signs. Thus, while $K$ is positive, $U$ and hence $E$ can perfectly well be negative; again, only *changes* in conserved quantities matter.

What forces are conservative? In other words, what forces are associated with a potential energy, so that we can express the work–energy theorem as the law of conservation of energy? The answer for forces that act in one dimension is simple: If the force is a function of the single position variable only, *and not the time*, it is conservative. In three dimensions, the answer is not quite so simple, but an important and often relevant case is provided by any *central* force, a force that depends only on the distance between the two objects between which the force acts and that is directed along the line between the objects.

Three of the most important examples of conservative forces are the force of gravity (the gravitational force in the immediate vicinity of Earth's surface), the spring force, and the gravitational force. The potential energies for these forces are as follows (•Fig. 1–4):

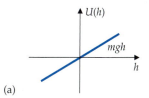

(a)

$$\text{Local gravity:} \quad U(h) = mgh. \tag{1–18}$$

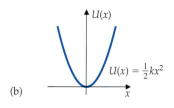

(b)

$$\text{Hooke's law:} \quad U(x) = \frac{1}{2}kx^2. \tag{1–19}$$

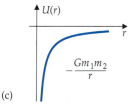

(c)

$$\text{Gravitation:} \quad U(r) = -\frac{Gm_1 m_2}{r}. \tag{1–20}$$

In these three cases, the zero of the potential energy is at $h = 0$, $x = 0$, and $r = \infty$, respectively.

The primary example of a force that is *not* conservative is friction, which looks superficially like a constant force. However, its *direction* depends on the way the object it acts on is moving—the force of friction acts to oppose the motion—so implicitly, there is a dependence on velocity as well as position. Thus, there is no potential energy associated with friction. We say such forces are dissipative: Energy is superficially "lost" when they act. The "lost" energy goes into heat. Of course, the work–energy theorem still holds for frictional forces.

But given the potential energy, what can we say about the associated force? Because the potential energy is an integral of the force, the force is a derivative of the potential energy. In three dimensions, as was mentioned earlier,

$$\vec{\mathbf{F}}(\vec{\mathbf{r}}) = -\vec{\nabla}U = -\left(\frac{\partial U}{\partial x}, \frac{\partial U}{\partial y}, \frac{\partial U}{\partial z}\right), \tag{1–21}$$

which, for a potential energy that depends on only one spatial variable, reduces to

$$F(x) = -\frac{dU}{dx}. \tag{1–22}$$

• **Figure 1–4**   Plots of the potential energy as a function of appropriate position variables. (a) Local gravity as a function of height; (b) Hooke's law as a function of the displacement of the mass from its equilibrium position; (c) Universal gravitation as a function of the distance between two masses in gravitational interaction.

## 1–3   Rotations and the Center of Mass

In classical physics, all of rotational kinematics and dynamics can be derived from Newton's laws. Moreover, the structure of both the kinematics and equations of rotational motion have a great deal of formal similarity to the kinematics and equations of linear motion.

Rotations occur when (rigid or nonrigid) extended objects are acted on by forces, depending on just where the forces act. (These forces also cause linear motion, described in our discussion of Newton's laws.) We restrict ourselves here to rotations of a rigid body about an axis, and we measure the amount of rotation by an angle, $\theta$; indeed, the fact that an object is rigid is what allows us to use the single variable $\theta$ for the description of the object's rotation. The rate at which the angle $\theta$ changes is the **angular velocity $\vec{\omega}$**, a vector whose direction is given by a right-hand rule for rotation about the axis (•Fig. 1–5). The rate at which the angular velocity $\vec{\omega}$ changes is the **angular acceleration** vector $\vec{\alpha}$. The dynamical equations for rotations involve the angular acceleration, just as the

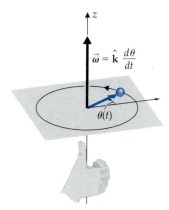

• **Figure 1–5**   The angular velocity vector describes the rate of rotation. Its direction is specified by a right-hand rule.

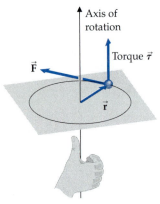

Axis of
rotation

Torque $\vec{\tau}$

$\vec{F}$

$\vec{r}$

• **Figure 1–6** The cause of rotation is torque, specified by a cross product (vector product) of the position vector of the point of application of a force with the force itself. Here, too, a right-hand rule is involved.

equations for linear motion—Newton's second law—involve the linear acceleration. There is an equivalent to mass, the **rotational inertia** $I$, and an equivalent to force, the **torque** $\vec{\tau}$, and the relation is the same, namely,

$$\vec{\tau} = I\vec{\alpha} = \frac{d\vec{L}}{dt}, \tag{1–23}$$

where $\vec{L}$ is the **angular momentum**. The torque is expressed with respect to an axis and is given by

$$\vec{\tau} \equiv \vec{r} \times \vec{F}, \tag{1–24}$$

where $\vec{r}$ is a vector running perpendicularly from the axis of rotation to the point where the force is applied (•Fig. 1–6). The rotational inertia $I$ is associated with the way the mass of the body is distributed. Like all the other rotational quantities that appear here, it is defined with respect to the axis of rotation. If we imagine breaking up the object in question, whose total mass is $M$, into many discrete portions of mass $\Delta m_i$, then

$$I \equiv \sum_i (\Delta m_i) r_i^2, \tag{1–25}$$

where $r_i$ is the distance from the mass point labeled $i$ to the axis of rotation. In the limit of a continuously distributed mass, the sum becomes an integration.

The work–energy theorem for rotational motion takes a form that extends the parallels we see in the dynamical equations. The work done in rotating a rigid body through an angle $\Delta\theta = \theta_f - \theta_i$ is

$$W = \int_{\theta_i}^{\theta_f} \tau \, d\theta, \tag{1–26}$$

and the kinetic energy of an object rotating with angular speed $\omega$ is

$$K = \frac{1}{2} I\omega^2 \tag{1–27}$$

The relation between these quantities is the familiar work–energy theorem, Eq. (1–13).

There is, in fact, nothing special about the axis of rotation: No physical effect depends on the choice of axis, any more than any physical effect depends on choice of origin for Newton's laws. Any axis can be employed as a reference, and the preceding relations remain true as long as the same axis is used for all of the quantities employed.

The quantities just described can be extended to nonrigid objects. To do so, we need only sum over them for a set of point objects that make up a suitable description of the entire system. We can also describe the rotational quantities with respect to an origin, not an axis. If $\vec{r}$ describes the vector from this origin to a given point object, then the angular momentum of the object is

$$\vec{L} = \vec{r} \times \vec{p}, \tag{1–28}$$

and the torque $\vec{\tau}$ remains as in Eq. (1–24) and describes the rate of change of $\vec{L}$. The angular momentum of a system is the sum of the angular momenta of the system's individual mass components. In particular, if there is no external torque acting on the system, and if the internal forces are central—directed along the lines between the internal masses—then Newton's third law assures us that *the angular momentum of the system is conserved.*

The study of rotations is a study of extended objects—that is, of mass distributions, rigid or otherwise. This study reveals the presence of a location with special properties within a mass distribution, the **center of mass**. If our object has a total mass $M$, the center of mass is given by

$$\vec{\mathbf{R}} = \frac{\sum_i m_i \vec{\mathbf{r}}_i}{M}. \tag{1–29}$$

Note the important property that when a net external force acts on an extended object, the object's center of mass moves as a point mass with mass $M$ according to Newton's second law:

$$\vec{\mathbf{F}}_{\text{net, external}} = M\vec{\mathbf{A}} \tag{1–30}$$

where $\vec{\mathbf{A}} = d^2\vec{\mathbf{R}}/dt^2$. This is a great simplification.

## 1–4  Elastic Media and Waves

The collections of interacting masses that make up matter are capable of a type of collective motion that we refer to as **waves**. Waves are some of the most obvious features of our physical environment; they are also pervasive, if less obvious, in light, in sound, and within all types of matter. Because light waves have some special features—in particular, they do not require matter in which to propagate—in this section we refer to all types of waves *except* light or other electromagnetic waves. (See Section 1–8.)

Internal forces that resemble the force in Hooke's law lead to wave motion within materials. This is not so unlikely as it may sound: It really means only that there is a stable equilibrium and that the internal forces tend to lead back to that equilibrium. For example, the pressure in air is a stable quantity. A fluctuation within the air such that the pressure varies from its average is subject to forces that tend to bring it back to the average, and a *small* deformation within many materials is subject to forces that tend to remove the deformation. Let us imagine that there is some quantity $h$ that represents a variation from equilibrium. This quantity could be the pressure in a gas, or a displacement from an equilibrium position along a string or within a material. The quantity $h$ varies throughout the medium, as well as with time, so that $h = h(x, t)$. Here, we have let the position within the medium be represented by the single variable $x$, as if the medium were one dimensional, but the position could more generally be a multidimensional position vector $\vec{\mathbf{r}}$. The variable $h$ satisfies what is called the wave equation,

$$\frac{\partial^2 h}{\partial x^2} = \frac{1}{v^2}\frac{\partial^2 h}{\partial t^2}. \tag{1–31}$$

This equation is a direct result of applying Newton's second law to the elastic medium. The equation has the important property that it is **linear**: If we have two solutions to the equation, then the sum of those two solutions is also a solution. This is sometimes referred to as the **superposition principle**. The quantity $v$, whose interpretation is the **wave speed**, is a function of the internal restoring forces and of an inertial factor that describes how nimbly the medium responds to the forces:

$$v = \sqrt{\frac{\text{restoring force factor}}{\text{inertial factor}}}. \tag{1–32}$$

For example, the speed of sound waves in a solid of mass density $\rho$ is given by the expression $\sqrt{Y/\rho}$, where $Y$ is Young's modulus, a quantity that determines how the solid responds to a stretching force. Analogously, the speed of a wave moving along a string of mass per unit length $\mu$ under a tension $T$ is $\sqrt{T/\mu}$.

The solution of Eq. (1–31) is given by $h(x, t) = g(x - vt)$, where $g$ is *any* function of the combination $x - vt$. We call the wave denoted by $h$ a **traveling wave**. It represents some shape described by the function $g$ moving in the $x$-direction with (positive or negative) velocity $v$. The wave could be a pulse of some kind; alternatively, it could repeat regularly. In particular, if $g$ takes a sinusoidal form, $h$ is known as a **harmonic wave**:

$$h(x, t) = H \sin[k(x - vt)] = H \sin(kx - \omega t). \tag{1–33}$$

The parameters that appear here have simple interpretations. The **amplitude** of the wave—its maximum departure from equilibrium—is $H$, while $k$ and $\omega$ specify the repeat length and repeat time, respectively, for the wave. In particular, if the repeat length is the **wavelength** $\lambda$, the repeat time is the period $T$, and the repeat frequency is $f = 1/T$, then $k$, termed the **wave number**, and $\omega$, termed the **angular frequency**, are related to $\lambda$ and $T$ by

$$\omega = 2\pi f = \frac{2\pi}{T} \quad \text{and} \quad k = \frac{2\pi}{\lambda}. \tag{1–34}$$

The wave speed links all these quantities:

$$v = \lambda f. \tag{1–35}$$

•Fig. 1–7 illustrates some of these points. Harmonic waves are of special importance because of Fourier's theorem, which states that any wave can be broken down into (or constructed using) harmonic waves of different frequencies. In some media, the wave speed is a function of frequency. Such media are said to be **dispersive**; the treatment of waves in dispersive media is more complicated.

The wave equation has a second type of solution, corresponding to **standing waves**. This type of solution separates the time and space dependence:

$$h(x, t) = H \sin(kx + \varphi)\cos(\omega t + \delta). \tag{1–36}$$

These waves oscillate in place, in contrast to traveling waves, which represent a moving disturbance (•Fig. 1–8). Note the presence of **nodes**, places where there is no disturbance. Here, the frequency $f = \omega/2\pi$ is determined by the wave equation. But what determines the wavelength? The answer is that the wavelength is determined by boundary conditions. For example, a common boundary condition is that the endpoints of a string of length $L$ on which the

• **Figure 1–7**   A harmonic wave has a repetition length called the *wavelength* $\lambda$ and a maximum value of the variable it describes called the *amplitude*. The time necessary for one wavelength to pass a given point $x$ is the *period*. The wavelength and period are connected by the wave speed.

• **Figure 1–8**   In harmonic standing waves, the changing variable oscillates in place. Points at which the variable vanishes are fixed in space at all times and are called *nodes*.

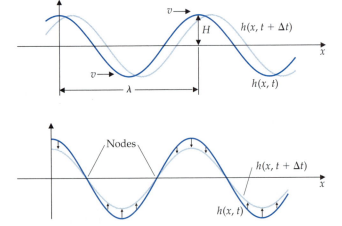

waves are set up must be fixed; that is, the quantity $h$ must vanish at the two ends, $x = 0$ and $x = L$. In this case, $\varphi = 0$ and $kL = n\pi$; that is, $\lambda = 2L/n$, where $n$ is a positive integer. These conditions correspond to a half-integer number of wavelengths fitting the string, as in •Fig. 1–9. The appearance of integers is often cited as unique to quantum physics; here, we see that it is perfectly possible in a classical context. It is actually typical of wavelike phenomena, and these phenomena are in many ways common to classical and quantum physics.

We have not been very specific about the characteristic wave variable $h$, which is generally a displacement from equilibrium. That is because a large variety of physical variables satisfy wave equations, depending on the medium. There are two main categories of waves: **transverse** and **longitudinal**. This distinction depends on whether the displacement is perpendicular to, or along, the direction of propagation of the wave itself. As an example of a medium that supports both types of wave, we could take a long spring such as a Slinky™. If the spring is stretched along the $x$-direction, then an initial pulse formed by moving one end of the spring in the $y$-direction will form a propagating transverse wave, as in •Fig. 1–10a. If, instead, an initial pulse is formed by pushing the spring along its length, in the $x$-direction, then this pulse will propagate as a longitudinal wave, as in •Fig. 1–10b. Perhaps the most familiar longitudinal wave is that of **sound**, which is a wave in which the density of the air varies about its equilibrium value in the direction of the propagation of the sound wave.

▌ We discuss quantum numbers throughout this book.

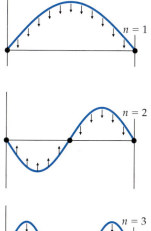

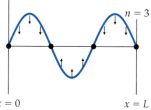

• **Figure 1–9**   For a harmonic standing wave on a string with boundary conditions that fix the ends, only certain wavelengths are allowed. This is a "quantization" phenomenon.

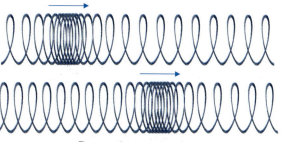

Propagating transverse wave pulse

Propagating compression

• **Figure 1–10**   Waves can be characterized according to whether the disturbance is (a) perpendicular to the direction of the propagation, in which case the wave is transverse, or (b) along the direction of the propagation, in which case the wave is longitudinal. Sometimes a combination of the two occurs. The two types of waves are illustrated here by a Slinky™.

### Power and Energy in Waves

In a wave, a disturbance propagates from one place to another; there is no over-all motion of the medium. Nevertheless, a traveling wave can transport the energy that is put into the formation of a pulse or into a periodic wave from one end of a medium to another. The harmonic wave of Eq. (1–33) propagating on a string of mass density $\mu$ provides the essential features: The power—that is, the rate at which energy is delivered through a unit area perpendicular to the direction of wave propagation—in this wave is

$$P = \mu v \omega^2 H^2 \cos^2(kx - \omega t).$$

The energy density in the harmonic wave is the power divided by the wave speed $v$. The power and energy density are themselves traveling waves, and each of these quantities is proportional to the *amplitude squared* and the *frequency squared*.

### Reflection and Refraction

Waves that reach boundaries between different media reflect back into the medium upon which they were incident and refract into the medium on the far side of the boundary. (We also refer to the refracted wave as a **transmitted** wave.) For waves on one-dimensional systems such as strings, we need only note the possibility of a phase change at the boundary. Thus, a wave that has reflected from a boundary at which the string is fixed undergoes a phase change of 180°; that is, the wave is inverted (•Fig. 1–11a). If the string is free at the end, the wave is reflected upright, without a phase change (•Fig. 1–11b). In these cases no wave is transmitted, but we can arrange for one by connecting the string to another string of different mass density. In a two- or three-dimensional system (such as light), we have angles of reflection and refraction to deal with. In any case, the amplitudes of reflected and refracted waves are constrained by the requirement that energy be conserved.

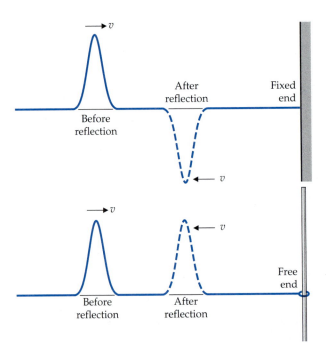

• **Figure 1–11** How waves reflect depends on boundary conditions. (a) If one end of a string is fixed, an incoming pulse will be inverted on reflection. (b) If the reflecting end is free, the reflected pulse will remain upright.

### Coherence, Interference, and Diffraction

The superposition principle tells us that any sum of harmonic waves is also an acceptable solution of the wave equation. We have already remarked that this principle relates pulses to harmonic waves. In addition, a standing wave of the type in Eq. (1–36) is a superposition of two traveling waves like those of Eq. (1–33), one moving to the right and one moving to the left. Now, when two waves superpose, interesting things can happen. If, at a certain point in space and time, there are two waves with $h$-values that are equal and opposite, then at that point the net displacement cancels; we say the waves interfere destructively. If, in contrast, we have two waves with $h$-values that are equal and of the same sign, then at that point in space the two waves interfere constructively. An interference pattern between two or more waves can be set up throughout space and time, and this pattern can be regular if the two waves are themselves regular. We say that harmonic waves are **coherent** if there is a definite relation between their frequencies and phases. The possibility of coherent waves can be realized in various ways. Let us enumerate several of the typical patterns that result.

- *Beats.* Imagine two interfering waves of slightly different angular frequencies $\omega_1$ and $\omega_2$ propagating in the same medium, so that they have the same wave speed. (For sound, this can easily be arranged with two slightly different tuning forks.) Then if they each have the same amplitude, their algebraic sum takes the form

$$H \sin(k_1 x - \omega_1 t) + H \sin(k_2 x - \omega_2 t)$$
$$= 2H \sin(Kx - \Omega t)\cos\left(\frac{\delta k}{2} x - \frac{\delta \omega}{2} t\right),$$

where $K$ is the average of $k_1$ and $k_2$, $\Omega$ is the average of $\omega_1$ and $\omega_2$, $\delta k$ is the difference of $k_1$ and $k_2$, and $\delta \omega$ is the difference of $\omega_1$ and $\omega_2$. The interfering waves produce the product of two waves, one with a very small wave number $\delta k/2$—that is, a very long wavelength—and a very small angular frequency. The part with the small frequency is termed the **beat**.

- *Interference patterns in space with two sources.* •Figure 1–12 shows one way to set up an interference pattern in space. A single wave is sent through two slits that act as two sources of the same frequency. A point $P$ is located at different distances from the two slits, and hence there is a definite phase difference between the waves at that point. The two waves have the same amplitude and wavelength $\lambda$, but because one has to travel farther than the other to get to point $P$, they arrive at $P$ with different phases. Depending on the difference in path length, $\Delta L$, the two waves could interfere constructively or destructively. The condition for constructive interference is that $\Delta L = n\lambda$, while for destructive interference, the condition is $\Delta L = (n + 1/2)\lambda$. In each case, $n$ is either zero or a positive or negative integer. To find how these conditions correspond to a particular set of points in space, an exercise in geometry is necessary. For instance, •Fig. 1–12 shows a situation in which the distance $D$ from the two sources to an observing screen is much larger than the distance $d$ between the sources. In that case, the light rays leaving from $S_1$ and $S_2$ are nearly parallel, and we have $\theta \approx \theta'$. Thus, the difference between the two path lengths is given approximately by $d \sin\theta$, and the condition for fully constructive interference becomes $d \sin\theta = n\lambda$, where $n$ is any integer, positive or negative. A pattern of maxima and minima is the result. Another way to get two coherent sources is through the use of split

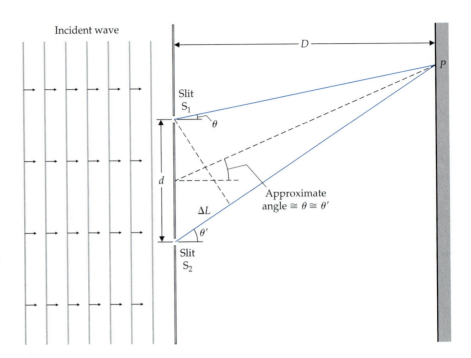

• **Figure 1–12**   An arrangement for producing an interference pattern. The two apertures produce two coherent waves whose disturbances systematically add or subtract on a screen, here assumed to be distant compared with the separation between the sources. The waves could be sound, light, or water waves in a ripple tank. For $L \gg d$, $\theta' \cong \theta$.

beams that are partially reflected and partially transmitted and then are rejoined by means of mirrors.

• *Gratings* involve spatial interference patterns similar to those produced by two sources. However, in a grating, $N$ sources are used, and if $N$ is very large, it has the effect of greatly sharpening the interference pattern. Gratings are particularly important when the wave involved is light. (See Section 1–8.)

Christian Huygens demonstrated how a wave front is due to interference. He pictured the propagation of a wave front as a continual regeneration of "wavelets" along the front. The straight-line propagation of the front is due to the constructive interference of the wavelets all along the front. If, on the other hand, the front is broken by, say, the presence of a barrier, then the constructive interference is no longer present on the far side of the barrier, and the wave front will bend around the edge of the barrier. This phenomenon is known as **diffraction**.

### The Doppler Shift

This phenomenon describes how the movement of the receiver or emitter of a wave or the medium itself affects the frequency or wavelength of the wave. The Doppler shift plays an important role in relativity; accordingly, we shall reserve treatment of it in some detail in that discussion.

## 1–5  Thermal Phenomena

Thermal phenomena are a familiar aspect of everyday experience. The first principles in the description of thermal systems—a container of a gas such as air is an example to keep in mind—are that such systems have a **temperature** and that **thermal contact** allows two thermal systems to reach a common temperature. These systems are then said to be in **thermal equilibrium**. Other parameters of thermal systems—thermodynamic variables—are closely tied to temperature; two important ones are the **volume** and the **pressure** of a gas.

(The pressure is the force per unit area exerted on any surface in contact with the gas.) Temperature scales can be defined in a variety of ways, but by far the most important scale, the *absolute* or *Kelvin* scale, is based on a proportionality to pressure:

$$T = \lim_{p \to 0} \frac{p}{p_{tp}} \, (273.16 \text{ K}). \qquad (1\text{–}37)$$

Here, the subscript "tp" indicates the triple point of water—the set of values for thermodynamic variables at which ice, steam, and liquid water can coexist in equilibrium.

The limiting procedure is used in Eq. (1–37) because, at least in classical physics (quantum gases need a different treatment), all gases behave alike when they are sufficiently dilute, and it is not necessary to specify a particular gas in the definition of temperature. This behavior is specified in the **equation of state**, a relation between thermodynamic variables that holds for all gases. When gases are dilute, they satisfy the **ideal-gas** equation of state,

$$pV = nRT = NkT, \qquad (1\text{–}38)$$

where the quantity $n$ is the number of moles of gas, while $N$ is the number of individual gas molecules. The two are related by $n = N \times N_A$, with $N_A$ being **Avogadro's number**, roughly $6 \times 10^{23}$. The quantities $R$ and $k$ are the **universal gas constant** and **Boltzmann's constant**, respectively. Their values are given in Appendix A.

Thermodynamic transformations occur when the thermodynamic variables change under outside influences, subject to the constraints of the equation of state. The variables can change in a controlled way, as when a piston is pushed into a cylinder held at a fixed temperature and the pressure of the gas within increases while the volume decreases. In this case, we have a **reversible transformation**. In contrast, **irreversible transformations** often are spontaneous, as when a gas freely escapes from a container into empty space. A reversible transformation can be traced on a *p–V* diagram; •Fig. 1–13 shows the path of a gas undergoing three important types of reversible transformation: isothermal, isobaric, and isochoric, in which the temperature, the pressure, and the volume, respectively, of the gas remain fixed. These transformations can be arranged easily enough; for example, to make sure the pressure is fixed, we need only

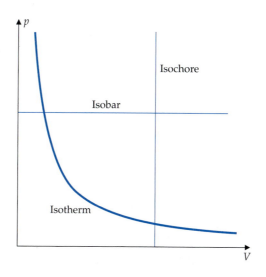

• **Figure 1–13**   Some possible transformations for a gas. The thermodynamic variables $p$, $V$, and $T$ are all related by an *equation of state* for a fixed number of moles of gas, so that two variables suffice to label the state of the system. This state can be changed by external means; if the change is controlled (reversible), it can be plotted as here on a *p–V* diagram. Changes (*tranformations*) at constant temperature lie on curves called *isotherms*, those at constant pressure follow *isobars*, and those at constant volume follow *isochores*.

exert a fixed force against the exterior of the piston containing the gas within a cylinder.

The fact that force must be exerted against a piston to make it move into the cylinder suggests that work is done when thermodynamic transformations occur. Let us think about the force exerted *by* the gas *on* the piston, a force that will make the piston move. The infinitesimal work done by the gas when the piston moves an infinitesimal distance $dx$ is $dW = F\,dx = pA\,dx$, where $A$ is the area of the piston's face. The quantity $A\,dx$ is the change in volume, $dV$, of the gas, so that the work done is $dW = p\,dV$. The net work when the system changes its volume from $V_i$ to $V_f$ is then

$$W = \int_{V_i}^{V_f} p\,dV. \tag{1–39}$$

This is the area under the curve of the transformation on a $p$–$V$ diagram. Evidently, the work is a function of the particular transformation—it is a path-dependent quantity.

For all but isothermal transformations, the temperature of a thermal system changes when work is done on it. There is, of course, another way to change the temperature, and that is to hold a match under the system or, more generally, to put the system in thermal contact with another thermal system of a different temperature. When the temperature is changed this way, we say that heat has been added to the system, and the amount of heat $dQ$ that has been added is proportional to the temperature change $dT$, so that

$$dQ = C\,dT. \tag{1–40}$$

Here, $C$ is the **heat capacity**, a quantity that varies from substance to substance. $C$ depends on *how* the temperature change has been made, and this is indicated with a subscript; for example, $C_V$ means the temperature change has been effected while the system remains at a constant volume. Thus, heat, like work, is a path-dependent quantity. It is also proportional to the amount of material present, and one uses the specific heat $c$, the heat capacity of one gram of material, or the molar heat capacity $c'$, the heat capacity of one mole of the material. One other item is important here: If no work is done when heat is transferred between two thermal systems, the heat is conserved—as much is taken from one system as is added to the other.

Once we have established the idea that heat transfers between systems occur through thermal contact, we can also imagine a transformation that involves no heat transfer, because the system that is transformed is held in thermal isolation. Such a transformation is called **adiabatic**.

One of the handful of truly important discoveries of the 19th century clarified the idea of energy and of its conservation in a conclusive way. This work was announced in 1847 by James Joule, who was the son of a wealthy English brewer and who devoted his life to independent scientific research. By a series of experiments, Joule found that when a thermal system underwent a change of temperature, *there was no way to tell* by looking at the state of the system whether the change was due to work having been done on it or to heat having been transferred to it. Joule also found that a given quantity of heat added to the system corresponded precisely to performing a given amount of work on the system. This is called the **mechanical equivalent of heat**, and it forms the foundation for the extension of the principle of the conservation of energy to thermal systems.

The connection to energy arises because of the work–energy theorem, which is a consequence of Newton's laws. The extension of the concept of energy to

**thermal energy**—also known as internal energy—is made by noting that when positive work is done on the thermal system, energy is added to it. Let us suppose the system has a thermal energy $U$. Then, recalling that we have defined $dW$ as the work done by the system, we write $dU = -dW$ for the change in thermal energy, assuming that no heat is added to or removed from the system. If, in addition, an amount of heat $dQ$ is going *to* the system, then the mechanical equivalent of heat suggests that the work–energy theorem generalizes for infinitesimal changes to

$$dU = đQ - đW. \tag{1–41}$$

This form of the principle of conservation of energy is the **first law of thermodynamics**. Note that we have written the change in energy as the difference of two differences. Both the work done, $đW$, and the change in the heat absorbed or emitted, $đQ$, depend on the path taken between the initial and final states of the system. That is why we have used the symbol $đ$ to express these changes. But the change in energy is a function of only the thermodynamic variables of the system, sometimes called a function of state. Thus, in contrast to the heat transfer and the work done, *the change in energy is not dependent on the path taken.* Accordingly, even though changes in both $Q$ and $W$ are path dependent, the change in their difference is not!

Just how the internal energy depends on the thermodynamic variables is an interesting question. Joule is also responsible for showing that the internal energy of an ideal gas is a function of temperature only; that is, $U = U(T)$. In the case of constant-volume transformations, during which no work is done, we find that

$$U(T) = \int^{T} C_V \, dT. \tag{1–42}$$

If we recognize that $C_V$ is constant over wide ranges of temperature, this equation simplifies still further, to

$$U(T) = C_V T. \tag{1–43}$$

An important application of thermodynamics occurs when thermal systems operate in cycles, meaning that they follow a closed path on a $p$–$V$ diagram (•Fig. 1–14). In a given cycle, there is no change in the energy, since the path ends in the same state at which it started; on the other hand the work done is the area contained within the (clockwise) cycle. Thus, there must have been heat transferred from some sources outside the thermal system. This cycle neatly describes the operation of an engine, which takes in heat in the form of burning fuel and does work as a result. The reverse cycle has work done on it and absorbs heat from outside the system; this cycle describes the operation of a refrigerator. According to the first law of thermodynamics, it is possible to run an engine with perfect efficiency, transforming all the heat put into it to useful work. However, the **second law of thermodynamics** states that in the real world no engine can be perfectly efficient. An engine operates by burning fuel or through contact with a hot thermal system, taking the thermal energy (or positive heat flow) it gets and turning it into work. The efficiency is the ratio of the work done in one cycle to the positive heat flow provided to the engine. The impossibility of perfect efficiency means that there is some leftover heat which must be rejected to a second, cooler reservoir. Indeed, the second law specifies the maximum efficiency possible for an engine whose cycle runs between two extreme temperatures $T_h$ and $T_c$:

$$\text{maximum efficiency} = 1 - T_c/T_h. \tag{1–44}$$

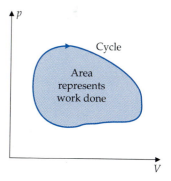

• **Figure 1–14**  A closed path on a $p$–$V$ diagram describes the working of a cyclical system—for example, an engine. Such a system can turn thermal energy into mechanical work. The area enclosed by the closed path in the $p$–$V$ diagram shown is the work done during the cycle, which is positive if the path follows a clockwise direction.

In particular, when the cycle includes irreversible portions, the efficiencies decrease below that of Eq. (1–44).

Irreversibility provides a link to still another function of state called the **entropy**. In an isolated system, entropy has the property that it increases when irreversible processes occur. We can always create isolated systems by making them ever larger, until the ultimate isolated system is the entire physical world. Referring to the increase in entropy is, in fact, another way to state the second law of thermodynamics: For an isolated system, the stable equilibrium state is the state of maximum entropy.

### Kinetic Theory

Temperature is most easily understood in terms of the **kinetic theory of gases**, which treats a gas in a container as a very large number of molecules moving through space, occasionally colliding elastically with one another and with the walls. In this picture, the internal energy of a system of pointlike molecules is the sum of the kinetic energies of all the molecules, which in turn can be expressed as the average kinetic energy per molecule times the number of molecules ($U = NK_{av}$; if the molecules have structure, then energy can be associated with their rotations or vibrations.) Similarly, the origin of pressure is the transfer of momentum imparted to the walls when the molecules bounce off of them.

Recall how pressure is calculated in this picture. As a simplifying starting point, we make the assumption that *all* the molecules have a speed $v$. Even with uniform speeds, however, the molecules move in a variety of directions, subject to the fact that there are as many moving to the left as to the right, and so forth. •Figure 1–15 shows a single molecule with initial velocity $\vec{v}$ bouncing elastically from a section of the wall with area $A$ lying in the $yz$-plane. The transfer of momentum to the wall as a result of the molecule's colliding with it is $2mv_x$. Now we count the number of molecules that impart momentum to the wall. First, think about those molecules that have velocities between $\vec{v} = (v_x, v_y, v_z)$ and $\vec{v} + d\vec{v} = (v_x + dv_x, v_y + dv_y, v_z + dv_z)$—remember, this is just a question of direction. The number of molecules with these velocities are specified by a **distribution function** $n(\vec{v})$; $n(\vec{v}) d^3\vec{v}$ is the number of molecules

> ▌ Quantum mechanics is necessary to deal correctly with the problem of the internal motion of molecules, as we shall see in Chapter 4.

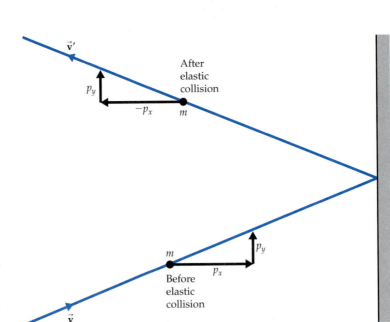

• **Figure 1–15**   The path of a molecule of momentum $\vec{p} = m\vec{v}$ colliding with a wall forming the $yz$-plane. The component of the momentum along the direction perpendicular to the wall changes sign in an elastic collision. The components that lie in the plane of the wall are unchanged.

per unit volume in the range of velocities we are considering. All the molecules in this range that are in a cylinder of height $v_x \, dt$ and area $A$ in the vicinity of the wall will strike the wall. This adds up to the number per unit volume times the volume $v_x \, dt \, A$ of the cylinder, namely, $n(\vec{v}) d^3\vec{v}(v_x \, dt \, A)$ molecules. These molecules impart to the wall a transfer of momentum

$$(2mv_x)(n(\vec{v}) d^3\vec{v})(v_x \, dt \, A) = 2m \, dt \, A \, v_x^2 \, n(\vec{v}) d^3\vec{v}.$$

Finally, the pressure $dp$ due to these molecules is the transfer of momentum per unit time per unit area; we divide by $dt$ and by $A$ to obtain

$$dp = 2m \, v_x^2 \, n(\vec{v}) d^3\vec{v}.$$

To find the total pressure $p$, we must sum over all the different velocity ranges that can occur. This is equivalent to integrating our result over all $v_z$ and all $v_y$, but only over positive $v_x$, since the molecules must be headed for the wall in order to collide with it. On the other hand, there are as many molecules with positive $v_x$ as with negative $v_x$, so we can simply integrate over all $v_x$ and divide by 2:

$$p = 2m \, \frac{1}{2} \int v_x'^2 n(\vec{v}') d^3\vec{v}'.$$

We have labeled the integration variable with a prime. All that is left at this point is to apply the assumptions we have made about the molecules' distribution. First, all directions are equally probable, meaning that $n(\vec{v}')$ depends only on speed, and not direction; that is, $n(\vec{v}') = n(v')$. Moreover, we can replace $v_x'^2$ in the integrand with $(v_x'^2 + v_y'^2 + v_z'^2)/3 = v'^2/3$. (We have used the fact that since there is no preferred direction, we have, on average, $v_z'^2 = v_x'^2 = v_y'^2$.) Finally, we note that we have assumed that only one speed, $v$, occurs, so that the integration just picks out the value $v^2/3$, while the integration over $n$ gives us the total number of molecules per unit volume, $N/V$. (As we shall see later, $\int v^2 \, n(\vec{v}) d^3\vec{v}$ is $N/V$ times the average of the square of the speed, so our assumption that only one speed occurs is just the assumption that the square of the speed for each molecule is the same as the average value of the square of the speed.) Thus, the pressure is

$$p = \left(\frac{mv^2}{3}\right)\left(\frac{N}{V}\right) = \left(\frac{2}{3}\right)\left(\frac{mv^2}{2}\right)\left(\frac{N}{V}\right) = \left(\frac{2}{3}\right)K_{av}\left(\frac{N}{V}\right). \qquad (1\text{–}45)$$

This important relation in turn tells us, through the relation $U = N K_{av}$, that the internal energy and the pressure are related by

$$pV = \left(\frac{2}{3}\right)U. \qquad (1\text{–}46)$$

According to the ideal-gas equation of state, Eq. (1–38), we can replace $pV$ by the factor $NkT$, so that

$$U = \left(\frac{3}{2}\right)NkT. \qquad (1\text{–}47)$$

Comparing Eq. (1–47) with Eq. (1–43), we find that $C_V = (3/2)Nk$. What is more, if we recall that $U$ is $N$ times the average kinetic energy per molecule, we have

$$kT = \left(\frac{2}{3}\right)K_{av}. \qquad (1\text{–}48)$$

The atomic picture thus gives a simple interpretation of temperature: It measures the average kinetic energy of the gas molecules.

■ The statistical treatment of matter at different temperatures is the subject of Chapter 12.

The entropy also has an interpretation in terms of the microscopic view of the thermal system: It is a measure of disorder in the system, and all the ramifications of the second law can be derived by treating entropy as disorder. The entropy is thus closely linked to statistical ideas.

## 1–6 The Atomic Structure of Matter

In Section 1–5, we referred to the atomic picture of matter. The modern atomic view originated with the English Quaker schoolteacher John Dalton through his chemical investigations. The chemical elements, of which more than 100 are currently known, consist of atoms of a given type, and each chemically distinct compound consists of a set of distinct molecules, themselves constructed of combinations of atoms. Early in the 19th century, the remarkable Englishman Thomas Young, who, among other things, made important contributions to the deciphering of Egyptian hieroglyphics, had measured atomic diameters to be around $10^{-10}$ m by simple techniques, such as letting a drop of oil spread on water until it forms a floating layer a single molecule thick. More accurate determinations had to wait until the invention of statistical mechanics later in the century.

Bulk matter comes in three states: gaseous, liquid, and solid. These states correspond to different ways of organizing the atoms (or molecules) that make up matter. In a gas, the atoms are separated by distances many times their sizes. In liquids, the atoms are closely bunched. In the solid form, the atoms are closely bunched and, in addition, are ordered in their arrangement. In fact, there is often more than one way to order the atoms in solids, leading to different crystalline forms.

What are the forces that cause atoms to group in different ways and to form different types of molecules? These are electromagnetic forces, the subject of Section 1–7. At this point, we simply refer to the presence of **electric charge**, which is the property a particle must have in order to undergo or be the cause of these forces. Electric charge comes in two types: positive and negative. Like charges repel, while unlike charges attract. Moreover, the force between two charges depends on the distance between them. A great variety of forces thus exists among collections of charges, a fact closely associated with the variety of ways in which bulk matter can form. The formation of bulk matter and of molecules also depends on the way atoms themselves are constructed.

The New Zealand–born physicist Ernest Rutherford was responsible for a set of experiments performed early this century that established the general structure of atoms (•Fig. 1–16). While the atom as a whole is electrically neutral—that is, it has no net charge—it does have constituents with individual electric charges of opposite sign. In classical terms, an atom resembles a miniature solar system, with negatively charged electrons circling a positively charged nucleus. The electron, which had been observed earlier by the English physicist J. J. Thomson, has a negative electric charge conventionally labeled[2] $-e$ and a mass much less than the mass of a hydrogen atom. Thus, the nucleus carries most of the mass of the atom. The differences between the different kinds of atoms lie only in the number of electrons circling an increasingly heavy nucleus. The atomic number in Mendeleev's periodic table of the elements is just the number of electrons an atom possesses, and the increasing mass of the nucleus is due in part simply to the necessarily increasing number of positive charge carriers in the nucleus.

---

[2] By definition, the quantity $e$ is positive.

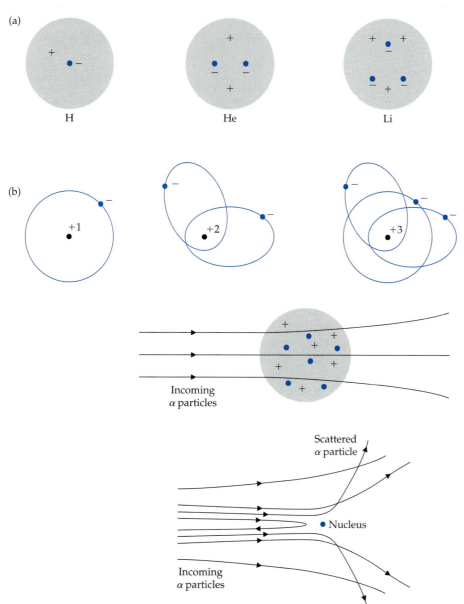

(a)

H          He          Li

(b)

+1          +2          +3

Incoming
α particles

Scattered
α particle

Nucleus

Incoming
α particles

• **Figure 1–16**   (a) In the Thomson model of the atom, negative point charges (electrons) are embedded in a uniform continuum of positive charge. (b) In the Rutherford model of the atom, one or more electrons travel in Keplerian orbits about a pointlike nucleus carrying the positive charge necessary to make the atoms electrically neutral.

• **Figure 1–17**   The deflection of incident alpha particles in the Thomson and Rutherford models, respectively. In the Thomson atom, the pointlike electrons, whose mass is 1/8,000 that of the alpha particles, do not deflect them, while the smooth distribution of positive charge can deflect the alpha particles by an angle that is at most as large as that given in Eq. (1–50). For the Rutherford atom, the electrons again have little effect, but the pointlike massive nucleus can deflect the alpha particles through large angles.

In 1911, Rutherford performed an experiment that established the planetary picture we have described (•Fig. 1–17). Let us look closely at Rutherford's experiment, which consisted of the firing of **alpha particles**—particles with charge $+2e$ that later were shown to be helium nuclei—at a thin foil. According to what was then the favorite atomic model—Thomson's, in which light-weight electrons are scattered throughout a positive uniform background charge that fills the entire atomic volume and has almost the entire atomic mass—the alpha particles would simply pass through this diffuse structure and hardly be scattered at all. Some few would be repelled by the positive charge distribution and be deflected to some small degree. (Of course, the alpha particles would also scatter from the electrons, but the electrons are so much lighter than the alpha particles, that the consequent deflection would be even smaller.) We can estimate the deflection quite readily. The maximum force on the alpha particle occurs when it grazes the atom—that is, when it passes at a distance $R$—to the atomic radius—from the center of the atom. There, the force due to a positive charge $Ze$ is the so-called coulomb force,[3]

$$F = \frac{(2e)(Ze)}{4\pi\varepsilon_0 R^2}.$$  (1–49)

The time that the alpha particle spends near the charge $Ze$ is approximately $2R/v$, where $v$ is the velocity of the alpha particle. Thus, we may estimate the deflection angle as

$$\frac{\Delta p}{p} = \frac{(F)(2R/v)}{M_\alpha v} = \frac{4Ze^2/4\pi\varepsilon_0 R}{M_\alpha v^2}.$$  (1–50)

For gold, with $Z = 79$, and a 5-MeV alpha particle, typical of Rutherford's experiments, Eq. (1–50) gives a deflection on the order of $10^{-3}$ radian. In this picture, deflections as large as 90° can occur only through the superposition of multiple small deflections, and since the direction of these is random, the chance of getting a 90° deflection is very small, on the order of $10^{-3,500}$ for the conditions cited. But Rutherford found that one alpha particle in 8,000 scattered through 90°!

Rutherford's explanation of this finding was that the radius of the positive charge distribution must be on the order of $10^4$ times smaller than the $10^{-10}$ m proposed by Thomson. In other words, the positive charge and virtually all the atomic mass would be packed into a very tiny nucleus. Since the atomic size had to be $10^{-10}$ m, the discrepancy could be accounted for by assuming that the electrons travel in planetary orbits of size $10^{-10}$ m about the central nucleus. The atom could then be viewed as a tiny solar system, with the nucleus as the central star about which the planetary electrons orbit. Since, like the gravitational force, the coulomb force varies as $1/r^2$, the orbits have the same structure as in planetary physics.

As we shall describe in detail in Chapter 5, the Rutherford model had basic difficulties. It could not explain the frequency spectrum of radiation emitted by the atom. Nor could it explain why all atoms of an element are the same, since, in the planetary picture, the radius, the energy, and so on depend on arbitrary initial conditions. Finally, because the orbiting electrons are constantly being accelerated, the atom would radiate all the time and would ultimately collapse into a point. Applications of classical electricity and magnetism show that this collapse would occur within about $10^{-10}$ s!

> ▌ Explanation of these and other features of atomic and bulk phenomena requires the use of quantum mechanics, a subject that will make up much of this book.

## 1–7  Electricity and Magnetism

The story of the discovery of the laws of electromagnetism, in the nineteenth century, is one of the glorious tales of science. A full recounting of this history would take us too far afield; but if you are not familiar with it, we cannot recommend too strongly that you take the trouble to read about it.[4]

**Electrical forces** occur between stationary electric charges; **magnetic forces** occur between moving electric charges, or, equivalently, **electric currents**. If one maps the forces acting on a tiny test charge in the presence of a given distribution of charges, moving or otherwise, and then divides out the effects of the test charge itself, one is left with something that depends on the distribu-

---

[3] We discuss electric forces in more detail in the next section.

[4] See, for example, Charles Singer, *A Short History of Scientific Ideas to 1900* (Oxford, U.K.: Oxford University Press, 1959).

tion. As the test charge moves through space, it traces the **electric and magnetic fields** that are due to the original distribution. These fields are vectors—the forces have directions—that have a value at every point in space. We label the electric field $\vec{\mathbf{E}}(\vec{\mathbf{r}})$ and the magnetic field $\vec{\mathbf{B}}(\vec{\mathbf{r}})$. As a first approximation, electric charges are associated with electric fields, and electric currents are associated with magnetic fields. (The direction of positive current is, by definition, the direction of the movement of positive charges. We emphasize, however, that you can have a current in an electrically neutral situation, as long as equal amounts of positive and negative charge are present—all that is necessary is that *one* of the two charge components be moving. This is the situation in real wires. In addition, note that a current moving to the right could correspond to positive charges moving to the right or negative charges moving to the left.) As soon as a dependence on time is included, the sources of electric and magnetic forces become more complicated.

The electric field, at least as it is set up by electric charges (•Fig. 1–18), is associated with conservative forces. Therefore, there is a potential energy, a function of position, for these electric forces. This potential energy is that of a (test) charge in the presence of some given distribution of charge that sets up the field, and by dividing out the test charge, we are left with something that depends only on the charge distribution, just as the electric field does. We refer to this quantity as the **electric potential**, and it is given by

$$V(\vec{\mathbf{r}}) = -\int_{\vec{\mathbf{r}}_0}^{\vec{\mathbf{r}}} \vec{\mathbf{E}} \cdot d\vec{\mathbf{r}} + V(\vec{\mathbf{r}}_0). \qquad (1\text{–}51)$$

As with potential energy, we are free to choose the location $\vec{\mathbf{r}}_0$ where the potential is zero, typically at infinity. Generally speaking, the potential is positive near positive charges and negative near negative charges.

The fields represent more than a simple way to summarize the forces: The fundamental laws of electricity and magnetism can be formulated in terms of them. The most important of these laws are known as **Maxwell's equations**, after James Clerk Maxwell, who formulated them in 1867. The four Maxwell equations describe the behavior of electric and magnetic fields, which are associated with electric charges and electric currents. The equations are as follows:

**1. Gauss' law for electric fields:**

$$\iint\limits_{\text{closed surface}} \vec{\mathbf{E}} \cdot d\vec{\mathbf{A}} = \frac{Q}{\varepsilon_0}. \qquad (1\text{–}52)$$

Here, $Q$ is the net electric charge (the algebraic sum of the positive and negative charges) contained within the closed surface over which the electric field is integrated, while $\varepsilon_0$ is a constant called the **permittivity of free space**. This constant is associated with the units of charge. The meaning of the surface integration is as follows: $d\vec{\mathbf{A}}$ is a surface element that is oriented outward from the closed surface, perpendicular to it. The integral is known as the **electric flux** through the surface; although the flux is in this case over a closed surface, it is a concept that has meaning even when the surface is not closed. While Gauss' law holds even when a dependence on time is present, it is useful to note that in static situations it is equivalent to **Coulomb's law** for the force between charges, which, in the context of the electric field and for a point charge $q$, takes the form

$$\vec{\mathbf{E}} = \frac{q}{4\pi\varepsilon_0 r^2}\,\hat{\mathbf{r}}. \qquad (1\text{–}53)$$

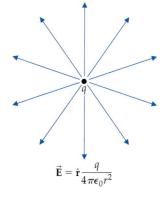

$$\vec{\mathbf{E}} = \hat{\mathbf{r}}\frac{q}{4\pi\epsilon_0 r^2}$$

• **Figure 1–18**   The electric field due to a positive point charge points away from the charge and falls as $1/r^2$, where $r$ is the distance from the charge.

Here, $r$ is the distance from the point charge. (Recall that the force on a test charge is the electric field times the magnitude of the charge.) From Coulomb's law, we note that electric fields point away from positive charges and toward negative charges.

2. **Gauss' law for magnetic fields:**

$$\iint\limits_{\text{closed surface}} \vec{\mathbf{B}} \cdot d\vec{\mathbf{A}} = 0. \tag{1-54}$$

In this case, it is the **magnetic flux** through a closed surface that is involved. The fact that it is zero means that magnetic fields trace out closed loops. As far as we know, there are no analogues of electric charges for magnetic fields. Although we know of no deep reason that forbids such analogues, called magnetic monopoles, they simply have never been detected.

3. The **generalized Ampère's law** relates magnetic fields to the currents and changing electric fields that produce them by the equation

$$\oint \vec{\mathbf{B}} \cdot d\vec{\mathbf{s}} = \mu_0 I + \mu_0 \varepsilon_0 \frac{d}{dt} \iint\limits_{\text{surface}} \vec{\mathbf{E}} \cdot d\vec{\mathbf{A}}. \tag{1-55}$$

The left side of this equation is a line integral around a closed loop (•Fig. 1–19). If one breaks the closed loop into tiny segments, then the line element $d\vec{\mathbf{s}}$ is a vector of length $ds$ pointing in the direction followed by the loop. This direction, clockwise or counterclockwise, sets information for the quantities on the right-hand side of the equation. The quantity $I$ is the electric current—the rate at which charge passes—through the closed loop. The second term on the right contains a surface integral that is the electric flux through the surface—any surface—that spans the loop. This term is known as the **Maxwell displacement current** term, after Maxwell, who noted that the meaning of the electric current passing through the loop is ambiguous without this second term. The direction of positive current, or of positive

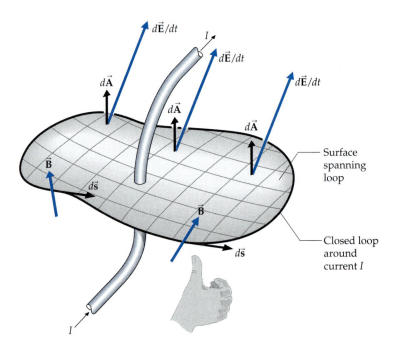

• **Figure 1–19** Ampère's law relates the integral of the magnetic field around a closed loop to the current through the loop and the rate of change of the electric flux crossing a surface spanning the loop. A right-hand rule is involved.

flux (meaning positive area elements $d\vec{A}$ in the surface integral), is defined by a right-hand rule: Curl the fingers of the right hand in the direction in which the loop is followed, and the right thumb defines the positive direction. Finally, a new constant, $\mu_0$, appears here, the **permeability of free space**. This constant is associated with units of current and the strength of magnetic forces. The generalized Ampère's law pinpoints the origins of magnetic fields: currents or changing electric fields.

4. **Faraday's law:**

$$\oint \vec{E} \cdot d\vec{s} = -\frac{d}{dt} \iint\limits_{\text{surface}} \vec{B} \cdot d\vec{A}. \tag{1–56}$$

Here, as in the generalized Ampère's law, a loop integral occurs, this time over the electric field; conventions about directions are the same as for Ampère's law. On the right-hand side, we see the rate of change of magnetic flux for any surface spanning the loop that appears on the left-hand side. Faraday's law describes how a changing magnetic flux can generate an electric field, as do electric charges (by Gauss' law). This electric field is quite different from the field generated by charges, however, because it can form closed loops, while the electric field generated by charges must begin or end on those charges. The electric field described by Eq. (1–54) generates its own secondary magnetic field by the generalized Ampère's law. The minus sign that appears in Eq. (1–56) implies that the secondary magnetic field has a magnetic flux that tends to oppose the original change in the magnetic flux; in this form, the law is also known as **Lenz's law**. Note the similarity between Faraday's law and the generalized Ampère's law. If magnetic monopoles existed, then one might expect a term in Faraday's law analogous to the electric charge in Ampère's law, representing the movement of magnetic charges.

To the four Maxwell equations, we add the force laws that describe how electric and magnetic fields affect charges and currents. The force on a charge $q$ moving with velocity $\vec{v}$ in an electric field $\vec{E}$ and a magnetic field $\vec{B}$ is given by the **Lorentz force law**:

$$\vec{F} = q(\vec{E} + \vec{v} \times \vec{B}). \tag{1–57}$$

Recognizing that a current is no more than moving charges, we can sometimes write the velocity-dependent part of this expression as a force on a length of wire $d\vec{\ell}$ carrying a current element in a magnetic field. We then have

$$d\vec{F} = Id\vec{\ell} \times \vec{B}. \tag{1–58}$$

The net magnetic force on a wire of finite length is found by integrating elements like those of Eq. (1–58).

By adding (integrating) the magnetic force on a succession of wire elements, we can find forces on finite wire segments. A particularly important example is the case of a current loop (•Fig. 1–20). If such a loop is placed in a uniform magnetic field $\vec{B}$, there is no net force, but there is a torque that tends to twist the loop. This torque is given by

$$\vec{\tau} = \vec{\mu} \times \vec{B}. \tag{1–59}$$

The magnetic moment $\vec{\mu}$ has magnitude $IA$, where $I$ is the current carried in the loop and $A$ is the area of the loop. The direction of $\vec{\mu}$ is defined by a right-hand rule, as in •Fig. 1–19. Thus, the torque tends to rotate the loop so that

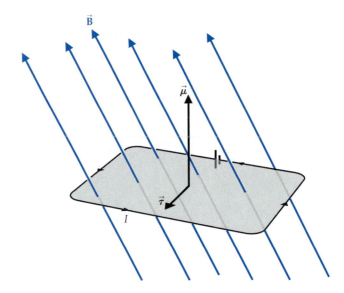

• **Figure 1–20**  A current loop placed in a magnetic field undergoes a torque. In the figure, the loop is placed at an angle to a constant magnetic field. The magnetic moment points in a direction perpendicular to the loop, and the torque points in a direction that lies in the plane and is perpendicular to the magnetic field.

the magnetic moment is parallel to $\vec{\mathbf{B}}$. When it is placed in the field, the loop has a potential energy that is a minimum when $\vec{\boldsymbol{\mu}}$ and $\vec{\mathbf{B}}$ are aligned, namely,

$$U = -\vec{\boldsymbol{\mu}} \cdot \vec{\mathbf{B}}. \tag{1–60}$$

The fact that electric and magnetic fields are associated with forces means that they are also associated with energy. In other words, *the fields themselves contain energy*. In a region of space with permittivity $\varepsilon_0$ and permeability $\mu_0$ where the electric field has magnitude $E$ and the magnetic field has magnitude $B$, the energy density, or energy per unit volume, is

$$u = \frac{1}{2\mu_0} B^2 + \frac{1}{2} \varepsilon_0 E^2. \tag{1–61}$$

The presence of matter has an effect on Maxwell's equations that can be accounted for in a direct way. The mixtures of moving charges that constitute matter react to the presence of fields and modify those fields. How they do so depends on how the atoms form that matter. For our purposes, we can divide materials into several classes. **Insulators** have the property that electric charges—either those supplied from within the atoms that make up the material or those introduced from the outside—cannot move through them very easily. At most, the charges within individual atoms can move from side to side within the atom. **Conductors** have the property that, in effect, electrons move freely within them. These materials are bulk metals; the electrons, which are also known as *valence electrons*, come from the very atoms that make up the conductor. **Semiconductors** lie somewhere between these two groups.

For insulators in the presence of an external electric field, the atoms align so that the electric field within the material is diminished (•Fig. 1–21). The effects of insulators, also known as dielectrics, can be summarized by replacing $\varepsilon_0$ by $\varepsilon = \kappa\varepsilon_0$, where $\kappa$, the *dielectric constant*, is greater than unity. Conductors react much more dramatically: The charges within them move until the electric field within is canceled, at least if the field is not dependent on time. Materials also can be classified according to their magnetic properties. The constant $\mu_0$ is replaced by a modified constant $\mu$ in magnetic materials. The most important class of magnetic materials is known as **ferromagnetic**; these materials pro-

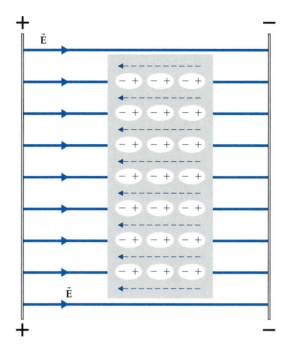

• **Figure 1–21** When an insulator is placed in an external electric field, dipoles align to produce a field that partially cancels the external field, so that the net field within the insulator is less than the external field.

duce a magnetic field on their own. The source of this field is the alignment of atoms, which, by virtue of the current loops formed by the circulating charge of the atomic electrons, themselves act like tiny magnets.

Other electric properties of materials are of importance in understanding electric circuits. *Ohm's law* describes how currents in a piece of conducting material are formed when an electric field is maintained across the material by a battery or its equivalent:

$$V = IR. \tag{1-62}$$

Here, $V$ is the potential difference from one end of the wire to another. The proportionality constant $R$ is the resistance, and when it is independent of $V$, we say that Ohm's law applies, or that the material is ohmic.

## 1–8 Electromagnetic Waves and Light

Taken together, Maxwell's equations imply two connected wave equations, one for the electric field and one for the magnetic field. As we shall see shortly, Maxwell recognized that the waves described by these equations had the right speed to represent light, and this understanding, together with the German physicist Heinrich Hertz's generation and detection of these waves in a direct electromagnetic context, closed a major chapter in the history of science.

The 17th century was rich in optical discoveries, beginning with those of Willebrord Snell, who quantified **refraction**, the bending of light as it propagates from one medium to another. Somewhat later, Francisco Grimaldi observed—and named—the phenomenon of diffraction, the bending of light as it goes past a barrier. Diffraction is manifested in the fact that light going through a narrow slit is spread on the other side and that a shadow has no perfect edge. This phenomenon had also been noted by Robert Hooke, who suspected that it might indicate the wave nature of light. In the second half of the century, Olaf Roemer made the first quantitative determination of the speed of light. He compared the calculated and the observed times of the eclipses of Jupiter's satellites.

The difference between these two times was due to the fact that light does not propagate with an infinite speed. We see the light after an event—the eclipse in this case—has taken place.

By the first half of the 17th century, Pierre de Fermat had conjectured the correct law of light propagation, namely, that light propagates between two points along the path that minimizes the time it takes to make the trip. Some decades later, Newton and Huygens each elaborated rather different ideas about the nature of light, although Newton's ideas were more subtle and complex than his followers and successors acknowledged. They assumed that Newton had in mind a simple "particle" theory of light, in which indivisible "atoms" of light moved in straight lines and were subject to the effects of gravitation. But Newton also investigated the colors seen in thin layers, such as those of soap bubbles. He noted that if one placed a slightly convex lens on a flat piece of glass such that the surfaces were not quite in contact, brilliantly colored rings, called *Newton's rings*, would be observed. This did not seem explicable by the propagation of particles, and indeed, Newton referred to light propagation in this effect as being in "fits" of reflection and transmission, which one might take to describe the motion of a longitudinal wave like a sound wave. Newton also knew about certain polarization phenomena—the fact that certain crystals will transmit light with various intensities, depending on how their axes are oriented—but he attributed these phenomena to the notion that his atoms of light had particular shapes.

At about the same time that Newton was doing his work, Christian Huygens developed a wave picture of light. The full generality of his construction was exploited only in the beginning of the 19th century, but Huygens was able to explain refraction using it. The idea is that when the leading edge of the wave hits, say, the denser medium, it is slowed down, and the wave is turned around this point as the faster moving portion of it catches up. This is a phenomenon you can observe as water waves hit a beach.

During the next century, the particle theory became the dominant theory of light propagation. Nonetheless, by the beginning of the 19th century, the next great period in the history of optical discovery, the particle theory of light had been swept aside in favor of the wave theory, chiefly due to the work of Thomas Young and Augustin Fresnel. The latter used the Huygens construction to give a complete theory of diffraction, including the discovery that in the center of any circular shadow there is a bright spot.

In Young's most famous experiment, early in the 19th century, the distance between two pinholes that supply coherent light that can interfere on a screen was set to be about a millimeter. The screen was about a meter away from the pinholes. As the pinholes are made smaller, the two patches of light they make on the screen get larger due to diffraction, and eventually they will overlap. At this point, an array of light and dark bands—the interference pattern of the two sources—appears. Apparently, these phenomena could be explained only if light was described by a wave. Using the conditions for constructive interference described in Section 1–4, Young could even determine the wavelength of the light. Such wavelengths $\lambda$ are quite small by ordinary standards, on the order of several hundred nanometers.

This historical diversion brings us back to Maxwell's equations. For the electric field, the presence of a set of charges oscillating in the $x$-direction allows us to generate a wave equation of the form

$$\frac{\partial^2}{\partial z^2} E_x = \mu_0 \varepsilon_0 \frac{\partial^2}{\partial t^2} E_x. \tag{1–63}$$

As in the discussion of Eq. (1–31), Eq. (1–63) represents a wave propagating in the $z$-direction with speed $v = 1/\sqrt{\varepsilon_0\mu_0}$. (On purely dimensional grounds, $1/\mu_0\varepsilon_0$ has the dimensions of velocity squared.) If we put in the numbers for $\mu_0$ and $\varepsilon_0$, as Maxwell did, we find that

$$\frac{1}{\sqrt{\mu_0\varepsilon_0}} = c, \tag{1–64}$$

the observed speed of light. Maxwell concluded immediately that light was an example of these waves. The same conditions that give Eq. (1–63) also give an equation for the magnetic field, which is in the $y$-direction:

$$\frac{\partial^2}{\partial z^2} B_y = \mu_0\varepsilon_0 \frac{\partial^2}{\partial t^2} B_y. \tag{1–65}$$

Equations (1–63) and (1–65) have general solutions of the form

$$E_x = E_0 \cos(kz - \omega t + \varphi) \quad \text{and} \quad B_y = B_0 \cos(kz - \omega t + \varphi). \tag{1–66}$$

The phase $\varphi$ is the same in these solutions because of a further consequence of Maxwell's equations that requires this condition. As with the waves of general form in Section 1–4, the wavelength $\lambda = 2\pi/k$ and the frequency $f = \omega/2\pi$ are related according to $\lambda f = c$.

Waves produced in this way have some interesting properties (•Fig. 1–22):

- Nothing in the wave equation itself fixes the wavelength—or, equivalently, the frequency—although the initial conditions, here the frequency at which the original charges oscillate, will do so. Thus, although visible light corresponds to a certain limited range of wavelengths in the wave equation, an infinite range is possible. We call this range the **electromagnetic spectrum**. As •Fig. 1–23 shows, the spectrum encompasses a huge variety of phenomena.

- Accelerating charges are necessary to get electromagnetic waves started, but once this has happened, the wave equation does not require that any further charges be present. Nor, for that matter, does it require a propagating medium. Electromagnetic waves can, however, propagate in dielectric media.

▌ The fact that Maxwell's equations require no medium for electromagnetic waves to propagate plays an important role in relativity, as we shall see in Chapter 2.

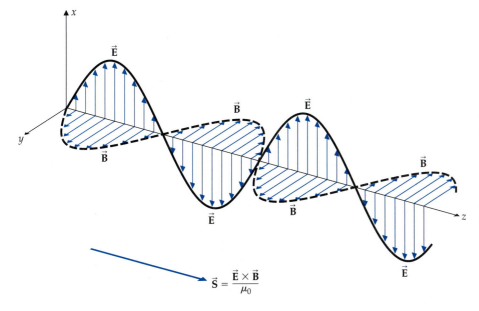

• **Figure 1–22** The electric and magnetic fields in a traveling harmonic electromagnetic wave are transverse to the direction of propagation and perpendicular to each other. They are also in phase with one another: When one is a maximum, so is the other. The Poynting vector $\vec{S}$ [see Eq. (1–68)] describes the momentum density of the wave and points in the direction of its propagation.

$$\vec{S} = \frac{\vec{E} \times \vec{B}}{\mu_0}$$

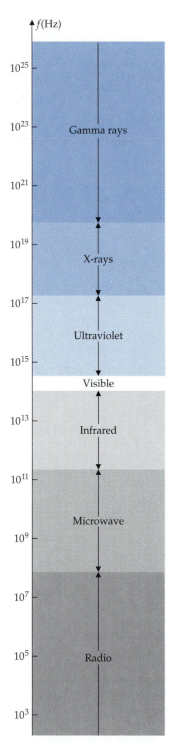

f(Hz)

$10^{25}$

Gamma rays

$10^{23}$

$10^{21}$

$10^{19}$

X-rays

$10^{17}$

Ultraviolet

$10^{15}$

Visible

$10^{13}$

Infrared

$10^{11}$

Microwave

$10^9$

$10^7$

$10^5$   Radio

$10^3$

• **Figure 1–23**   The electromagnetic spectrum. All these waves have the same structure, determined by Maxwell's equations. Only the wavelengths differ.

If there are no mechanisms in these media that destroy the waves, then the media are transparent, and in line with what we stated in Section 1–7, the only change is that $\varepsilon_0$ is replaced by $\varepsilon$ and $\mu_0$ is replaced by $\mu$. This means that the speed of light changes to $c' = c/n$, where the quantity $n$ is the **index of refraction** of the material. Transparent materials have $\mu \approx \mu_0$, so that $n \approx \sqrt{\kappa}$, where $\kappa$ is the dielectric constant, a quantity greater than unity. The reduced speed of waves in such media neatly explains refractive effects.

• The fact that the phase of the magnetic and electric fields in these waves is the same is a general feature. When the electric field is a maximum, so is the magnetic field; when one field is a minimum, so is the other.

• An additional restriction on the relative size of $E_0$ and $B_0$, and hence on the size of $E$ and $B$, in electromagnetic waves is that $E = cB$.

• The electric and magnetic fields are perpendicular to the direction of propagation of the wave and to each other, so that electromagnetic waves are *transverse*. This is also a general feature.

### Energy and Momentum Transport

Equation (1–61) shows that there is energy in electromagnetic waves, and the energy density in a region where they propagate is

$$u = \frac{1}{2} \varepsilon_0 (E^2 + c^2 B^2) = \varepsilon_0 E^2. \qquad (1\text{–}67)$$

Here we have used the relation between $E$ and $B$ in the wave, a relation that tells us that the energy density is equally split between the electric and the magnetic components of the wave. Moreover, the energy density is itself a propagating wave—it contains the factor $\cos^2(kz - \omega t + \varphi)$, which corresponds to something propagating with speed $c$. In other words, electromagnetic waves transport energy, and the speed of transport is the speed of light. The rate at which this energy arrives at a surface perpendicular to the direction of propagation of the waves is the **energy flux**, and it is given by the product $cu$. The energy flux can be more fully characterized by a vector that is along the direction of propagation—that is, proportional to $\vec{E} \times \vec{B}$—and of magnitude $cu$. This vector is known as the **Poynting vector**,

$$\vec{S} = (\vec{E} \times \vec{B})/\mu_0. \qquad (1\text{–}68)$$

Another quantity used to measure energy transport is the **intensity** $I$, defined as the time average of the energy flux. Because the average of the cosine squared is $1/2$, $I = S/2$.

The fields of an electromagnetic wave exert a net force in the direction of propagation on a charge of any sign. This shows that the wave also transports momentum to the charge when the wave hits it. The momentum per unit volume, or momentum density, carried by the wave is $\vec{S}/c^2$. One consequence of this fact is a pressure is exerted on a material when an electromagnetic wave is incident on it. This *radiation pressure* is given by $2u$ for a perfectly reflecting surface and by $u$ for a perfectly absorbing surface.

### Polarization

Although the electric field lies in a plane perpendicular to the direction of propagation of an electromagnetic wave (•Fig. 1–22), there is nothing that picks a direction in that plane. The particular direction chosen specifies the **polarization** of the wave. It is possible to measure this direction, as well as to "filter" the waves so that only certain directions are passed out of a beam that consists of a mixture of directions. A material filter consists of a kind of grating that projects only the component of the electric field in a certain direction. Since the intensity of the light is proportional to $E^2$ in the wave, incident polarized light of intensity $I_0$ whose polarization direction makes an angle $\theta$ with the direction of the polarizer will have an intensity $I_0 \cos^2 \theta$ after it passes through the polarizer. This relationship is known as *Malus's law*. In addition to being generated by material filters, polarized light can be produced by scattering and by reflection.

## CONCLUSION

The scientific achievements described in this chapter are, by any measure, monumental. Yet classical physics is applicable only over a limited domain: the domain of speeds much less than that of light, of energies much less than the energies associated with mass itself, and of distances that are large on the atomic scale. The story that we tell in this book is about the truly revolutionary extensions of our understanding beyond these realms.

# Relativity

The laws of classical mechanics, as formulated by Newton and developed right through to the end of the 19th century, are, it turns out, only approximations. These approximations are extremely good at speeds much less than the speed of light, $c$, but become extremely bad as the speeds of the objects in question approach $c$. Nineteenth-century physicists dealt with material objects—planets, billiard balls, and the like—that move much more slowly than the speed of light, so they were quite unaware that the mechanical world view they had constructed was only an approximate one. The equations of electromagnetism were known, thanks to Maxwell, but their implications for classical physics were not fully appreciated. It took the genius of Albert Einstein to realize that a crisis in classical physics having to do with the propagation of light waves could be resolved only within an entirely new framework for the fundamental laws of physics. This new framework required us to modify the very concepts of space and time. Einstein's contribution to this modification includes as a first step what became known as the special theory of relativity. Its mathematics is not very difficult; its ideas, however, are subtle and require close attention.

This chapter explores the concepts of space and time, and how coupling a careful understanding of these concepts with Maxwell's equations—particularly with the description of the propagation of light—leads to some surprising features. The next chapter deals with consequences of our modern understanding of space and time for energy and momentum.

# The Basics of Relativity

In giving us special relativity, Albert Einstein clarified and, indeed, changed our ideas about space and time. In this chapter, we present those ideas and the path by which Einstein arrived at them. When one deals with unaccelerated motion, as we mainly do here, one is led to *special* relativity. General relativity, which we treat in Chapter 17, carries these ideas into the realm of accelerated motion.

## 2–1 Some Historical Background

### The Ether

Early in the 19th century, the French physicist Augustin Fresnel and his collaborator Dominique F. J. Arago made the important discovery that two beams of light with different polarizations cannot be made to interfere. From this observation, Thomas Young concluded that light waves had to be, within his experimental accuracy, wholly transverse; that is, they vibrated only in directions at right angles to their direction of propagation. (See •Fig. 1–22a.) This was a very novel idea, because, up to that time, waves that had been studied in material media, such as sound waves in air, were all longitudinal, or at least always had a longitudinal component. That is because material waves involve some pushing and pulling—compressing and decompressing—of the medium. In fact, the notion that there could be a wave motion in the absence of a medium—a material that oscillated—had never occurred to anyone.

This, naturally raised the question of what it was that the light waves were oscillating in. Since light waves come to us from the stars through what seemed to be otherwise empty space, a novel space-filling medium had to be invented and its properties investigated. This medium was dubbed the *ether*, or, in the language of the day, the *luminiferous aether*. Later, as a result of Maxwell's work, it was realized that this hypothetical medium had to be very peculiar because the electromagnetic fields in Maxwell's waves were purely transverse, confirming and interpreting conclusions that Young had drawn much earlier in the 19th century. But in that case, the medium would have to be infinitely rigid, so that it would not undergo the compression and rarefaction characteristic of longitudinal waves. Nonetheless, the Earth and the planets appeared to travel through the ether as if it were not there.

Nineteenth-century physicists thought—reasonably at the time—that light must propagate in the ether analogously to the way that sound propagates in a material medium such as air. The speed of sound in air depends on properties of the air. Most important, the speed of sound has a characteristic value *relative to the rest frame of the air*: The speed of sound an observer measures depends on the observer's motion relative to the air. If the ether carried light the way air carries sound, then an observer moving relative to the ether would measure the speed of a light wave so that it would vary according to the observer's speed. The observer could "catch up with" or "fall back from" a propagating disturbance. Certainly, one can catch up with a sound wave; indeed, this happens in supersonic airplanes—that is what "supersonic" means. We illustrate some experimental consequences of this idea in Example 2–1.

**Example 2–1**    Imagine a coordinate system $S'$ fixed to a train moving with speed $v$ on a track while a second coordinate system $S$ is fixed to the track (•Fig. 2–1). We assume that system $S$ is at rest relative to the ether. Suppose that at time $t = 0$ lightning strikes at points $x = -L$ and $+L$ on the track. At this instant the train has its center, $x' = 0$, at the origin of the $S$-system, $x = 0$. Then an observer on the ground at $x = 0$ will see the lightning flashes arrive from either side at a time $t = L/c$; that is, the flashes will arrive at this vantage point "simultaneously." Will this be true for an observer stationed at the center of the train? If not, what is the time interval between the arrival of the flashes?

**Solution**    In the time interval $L/c$, the train will have moved a distance $vL/c$. The path length of the light will be shortened by this amount for the light coming from where the train is going and lengthened by the same amount for the light coming from where the train has been. But according to the reasoning described in the text, the observed speed of light is either $c + v$ or $c - v$, depending on whether one is moving toward or away from the light. Putting these two effects together, we find that the time at which the lightning flash arrives from where the train has been will be

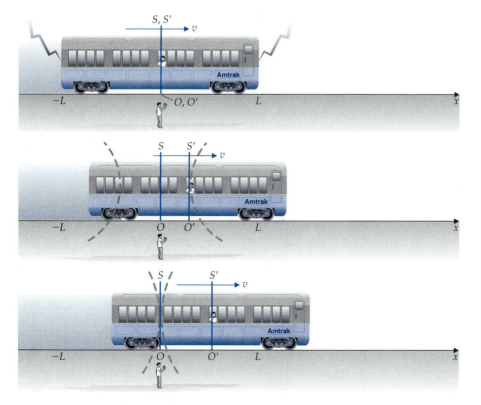

• **Figure 2–1**    (a) At $t = 0$, when lightning strikes at $x = L$ and $-L$, the coordinate frame $S$ attached to the track and the coordinate frame $S'$ attached to the train coincide. (b) At a later time, the pulse of radiation from the front of the train reaches the midpoint of the train $O'$ before the pulse from the rear reaches it. (c) At time $t = L/c$, the two pulses reach the point $O$ simultaneously. The pulse from the left will not reach the center of the train till later.

$$\frac{\left[L + (vL/c)\right]}{(c - v)} = \frac{(L/c)(c + v)}{(c - v)}.$$

The time at which the light arrives from where the train is going is

$$\frac{\left[L - (vL/c)\right]}{(c + v)} = \frac{(L/c)(c - v)}{(c + v)}.$$

Hence, the observer on the train will not see the lightning flashes arrive simultaneously, and the difference in their arrival time is

$$\Delta t = \frac{L}{c}\left(\frac{c + v}{c - v} - \frac{c - v}{c + v}\right) = \frac{4Lv}{c^2}\frac{1}{1 - v^2/c^2}.$$

Incidentally, Example 2–1 contains an assumption that will be very important throughout the chapter: that an observer in a particular frame can measure the time that "something" happens in his or her frame. (We refer to this "something" as an "event.") Just how can we justify this assumption? First we can ask, How does our observer find the place at which the event happens? The answer is that the observer lays out a grid, with spacing in whatever units he or she likes, and marks the space coordinates of every point on the grid. If the grid is made fine enough, the event will occur arbitrarily close to a grid point, and the observer uses the labeling of the point to locate the event.

Now we turn to the time of the event. The observer does something quite similar to laying out a grid. He or she places a clock at every grid point, and once all the millions of clocks are synchronized, the observer need only have the clock that is located where the event occurs register the time of the event. That way, there is no time delay between when the event occurs and a clock's "seeing" the event. But how is the synchronization accomplished? We start with a master clock at some point—for the example, the origin of our reference frame. We then send signals from the master clock to other points. These signals need only have the property that in the frame of the observer they move with a universal unchanging speed $v_0$. Consider a clock at a particular grid point a distance $D$ from the master clock. A signal is sent out at $t = 0$ according to the master clock. When the signal arrives at the grid point, the clock there will be set to $t = D/v_0$. In this way, all the clocks throughout the observer's frame are synchronized: We now have a clock and a grid label at every point and can say unambiguously what we mean by the time and place of an event according to our observer.

Let us return now to questions of the ether. In the 1870s, Maxwell asked whether the velocity of Earth relative to the ether might affect the observed speed of light. Earth traces out an elliptical orbit as it moves around the Sun. This motion can be well approximated by uniform motion in a straight line with a speed $v$ over time intervals substantially shorter than a year. But how big is $v$? We can approximate Earth's orbit by a circle with a radius of about $1.5 \times 10^8$ km, around whose circumference our planet moves uniformly. With one year about equal to $3 \times 10^7$ s, $v$ is approximately 30 km/s. The speed of light, $c$, is approximately $3 \times 10^5$ km/s, so that $v/c \approx 10^{-4}$. Thus, fractional variations in the speed of light of around $v/c = 10^{-4}$ would have to be observed.

### How to Measure Earth's Speed through the Ether

Maxwell had a simple idea for measuring the speed of Earth relative to the ether, which we illustrate in •Fig. 2–2. Call the distance from the candle to the mirror $L$. Then if Earth is not moving in the ether, the round-trip time for the

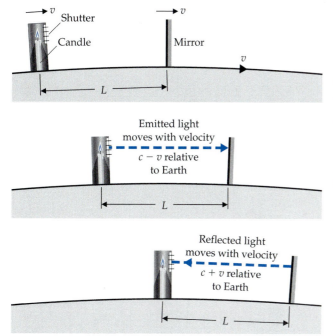

• **Figure 2–2**  Measurement of the time required for a flash of light emitted from a candle in a shuttered enclosure to reach a mirror and be reflected back to the enclosure. The light source and the mirror are attached to Earth, which moves through the stationary ether with speed $v$. The speed of light in the fixed ether is $c$. The idea was impractical at the time because $L/c$ is so short.

• **Figure 2–3**  Albert Michelson was born in Strelno in what is now Poland in 1852. He emigrated to this country with his parents when he was four. The family followed the gold rush to California and then Nevada. Michelson was sent to a boarding school in San Francisco for his high school education. A teacher who had aroused his interest in science suggested that he might apply to enter the U.S. Naval Academy in Annapolis, Maryland, where he could get a free education with a large scientific component. When he did not get one of the appointments from Nevada, he made the trip to Washington to try to persuade President Grant to give him a presidential appointment. Michelson learned that Grant came out of the White House to take evening walks. He waited for Grant to come out and then presented his case. Grant was sufficiently impressed to give Michelson an appointment. After graduation, Michelson went to sea and then returned to the academy as a physics instructor. It was at this time that Michelson became interested in the properties of light.

light—what we shall call the "resting time"—is $T = (2L)/c$. (All the actual methods used to measure the speed of light at that time used one version or another of this round-trip idea. To measure the speed of light using a one-way trip between two clocks required a then impossibly accurate synchronization of the clocks.) Now suppose the candle and the mirror, which are attached to Earth, move with a speed $v$ in the ether in, say, the $x$-direction. The light always moves with a speed $c$ in the ether, since, in this picture, the speed of light depends only on the properties of the ether. As the light moves from the candle to the mirror, one is catching up with it, so that the speed of light observed from Earth would be $c - v$. Coming back from the mirror, the light has a speed $c + v$. The total time for the round-trip—the "moving time"—is

$$\frac{L}{c-v} + \frac{L}{c+v} = L\left(\frac{c+v}{(c-v)(c+v)} + \frac{c-v}{(c+v)(c-v)}\right) = \frac{2Lc}{c^2 - v^2}$$

$$= \frac{2L}{c}\frac{1}{1 - v^2/c^2} \approx \frac{2L}{c}\left(1 + v^2/c^2\right). \qquad (2\text{–}1)$$

In the last step, we have taken advantage of the fact that $v^2/c^2 \cong 10^{-8}$, so that the small-$x$ approximation $1/(1 - x) \cong (1 + x)$ applies. The resting time and the moving time differ as a function of Earth's speed through the ether. However, as Eq. (2–1) shows, the difference between the resting time and the moving time is of order $(v/c)^2$—that is, one part in a hundred million!

### The Michelson–Morley Experiment

Maxwell did not think that experiments which required this kind of accuracy were possible. This is what he wrote in an 1879 letter that found its way to the young American Albert Michelson, who had already embarked on experiments to measure the speed of light. Michelson (•Fig. 2–3) took Maxwell's letter as a challenge. He became determined to measure the speed of Earth through the ether.

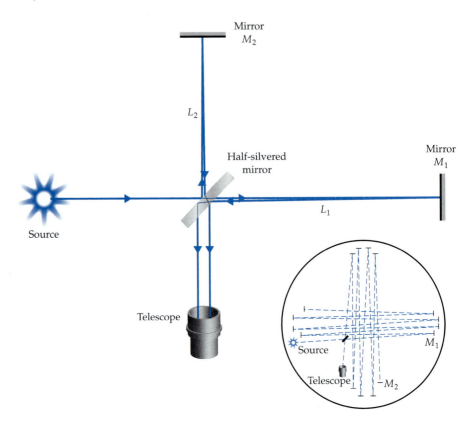

• **Figure 2–4** A schematic drawing of the Michelson interferometer. The reflection of the incident beam and the beam reflected from mirror $M_1$ takes place at the front face of the partially silvered mirror. The partial reflection of the beam returning from mirror $M_2$ is not sketched here, since it is irrelevant to the discussion. The insert shows the use of multiple reflections from mirrors to make the effective lengths approximately 11 m long.

In the winter of 1880–81, Michelson went to Germany to study in the laboratory of the physicist Hermann Helmholtz. Some of the best work on experimental optics was being done in Helmholtz's laboratory. During the course of this visit, Michelson invented the so-called *Michelson interferometer*, which became the essential tool in his subsequent research on the speed of light. •Figure 2–4 is a simplified drawing of the interferometer. At the left of the figure is a light source. (In the actual experiments, light from a sodium flame with a wavelength $\lambda = 590$ nm was used.) The light impinges on a half-silvered mirror that splits the beam. Half the light is transmitted to the mirror $M_1$ and half to the mirror $M_2$. Each of these two beams is reflected back to the central mirror, and a single, combined beam passes to a telescope at which an observer is stationed. There is no practical way of making the lengths of the paths to the mirrors identical, a problem that, as we shall see, Michelson found an ingenious way to deal with. We shall call the path lengths $L_1$ and $L_2$, respectively. If the interferometer is at rest in the ether, then the difference in time between the two round-trips is given by

$$\Delta T_{\text{rest}} = \frac{2L_1}{c} - \frac{2L_2}{c}, \qquad (2\text{–}2)$$

where the subscript "rest" refers to the fact that the apparatus is at rest in the ether. In this picture, $c$ always refers to the speed of light that an observer at rest in the ether would measure. If we could measure $L_1$ and $L_2$ with unlimited precision, we could make use of the time difference to predict what the interference pattern would look like when the two reflected beams come together after making their round-trips through the interferometer. If the beams started in phase, then they would be slightly out of phase when they came back, because the time it takes for the two round-trips is slightly different. That would produce an interference pattern[1] (which is why Michelson called his apparatus an interferometer). We cannot measure the two lengths to arbitrary accuracy, but before we see how Michelson dealt with this problem, let us repeat our calculation of what happens when the interferometer moves through the ether.

Suppose that the motion of the apparatus through the ether occurs at a constant speed $v$ in what we can call the $x$-direction. We then have two different travel times to consider, namely $T_1$, the back-and-forth time in the direction of motion, and $T_2$, the travel time for the path at right angles to the motion. Earlier, in our discussion of Maxwell's scheme for finding the motion of Earth through the ether, we in fact computed $T_1$ [Eq. (2–1)]:

$$T_1 = \frac{2L_1}{c} \frac{1}{1 - v^2/c^2} \approx \frac{2L_1}{c} \left( 1 + \frac{v^2}{c^2} \right). \tag{2–3}$$

To compute $T_2$, we use •Fig. 2–5. This figure shows that the beam makes a longer trip than the direct transverse distance would give. Since $T_2$ represents the time it takes the light to make the entire trip over the distance $2d$, the quantity $T_2/2$ represents the time it takes for half the trip. During this time, the observer will see Earth move a distance $vT_2/2$. The Pythagorean theorem allows us to equate the square of the diagonal distance $d = cT_2/2$ to the square of the transverse distance $L_2$ plus the square of the distance Earth moves; that is,

$$\left( \frac{cT_2}{2} \right)^2 = L_2^2 + \left( \frac{vT_2}{2} \right)^2.$$

We then solve this equation for $T_2$:

$$T_2 = 2L_2 / \sqrt{c^2 - v^2}.$$

The result is that an earthbound observer sees an effective speed of light for this beam that is reduced from $c$ to

$$c' = \sqrt{c^2 - v^2}. \tag{2–4}$$

If we remove a factor of $c$ from the square root, we are left with

$$T_2 = \frac{2L_2}{c} \frac{1}{\sqrt{1 - v^2/c^2}} \approx \frac{2L_2}{c} \left( 1 + \frac{v^2}{2c^2} \right). \tag{2–5}$$

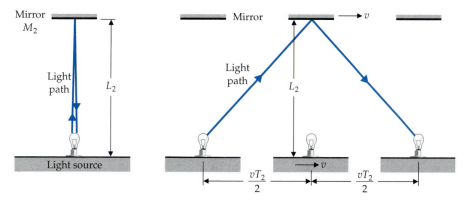

• **Figure 2–5**   (a) Light from a source is reflected by a mirror; here neither source nor mirror is moving with respect to the ether. The time taken is $L_2/c$. (b) In this case, both the light source and the mirror are moving with speed $v$ with respect to the ether. The computation of the time $T_2$ for the beam to make a round-trip must take into account the fact that the mirror will have moved a distance $vT_2/2$ during the time it takes for the light ray to reach it.

---

[1] Why a "pattern"? From the discussion, you would expect constructive or destructive interference across the whole area of the beam. But the insertion of a wedge across the beam will make one side of the beam a little longer than the other side, and then a set of stripes corresponding to successive constructive and destructive interference will be introduced. Any predicted phase shift is then observable as a sideways translation of the pattern of stripes.

In the last expression, we have again used the fact that $v/c$ is small to make an approximation. Next, as we did for the resting case, we find the difference in time for the two beams, which also describes the phase difference of the two beams and hence their interference. Let us define $\Delta T(0) \equiv T_1 - T_2$. (The reason for the zero will become clear shortly.) Then, from Eqs. (2–3) and (2–5),

$$\Delta T(0) \approx \Delta T_{\text{rest}} + \frac{v^2}{c^2} \frac{2L_1 - L_2}{c}, \tag{2–6}$$

where $\Delta T_{\text{rest}} = 2(L_1 - L_2)/c$. Note that the two lengths $L_1$ and $L_2$ still appear in $\Delta T_{\text{rest}}$. To eliminate the necessity of measuring these two lengths very precisely, Michelson employed an ingenious idea. Suppose we build the interferometer so that we can physically rotate the arms. If we rotate the arms by $90° = \pi/2$ rad, we interchange the roles of $L_1$ and $L_2$. Then, if we redo the calculation we have just done for this new configuration, we find that

$$\Delta T(\pi/2) \approx \Delta T_{\text{rest}} + \frac{v^2}{c^2} \frac{L_1 - 2L_2}{c}. \tag{2–7}$$

Since there is a time difference for each of these configurations, the observer would observe interference fringes in each case. But, as Eqs. (2–6) and (2–7) show, the fringes would be different. And if we take the *difference* of these differences, the quantity $\Delta T_{\text{rest}}$ cancels out. Indeed,

$$\Delta T \equiv \Delta T(0) - \Delta T(\pi/2) = \frac{v^2}{c^2} \frac{L_1 + L_2}{c}. \tag{2–8}$$

This time difference describes how the interference fringes *shift* when the apparatus is turned by $90°$. Moreover, $\Delta T$ is very much less sensitive to the precise knowledge of $L_1$ and $L_2$ than is either $\Delta T(0)$ or $\Delta T(\pi/2)$ by itself, because $\Delta T$ depends on the *sum* $L_1 + L_2$ rather than each of the addends alone. Thus, by measuring the shift in the interference fringes as the interferometer is rotated, Michelson had high expectations that he would be able to meet Maxwell's challenge.

In 1885, Michelson took a position at what is now Case–Western Reserve University in Cleveland. There, he met the chemist Edward Morley, and the two of them began working on what became known as the Michelson–Morley experiment. It was carried out during a five-day period in July of 1887. For this purpose, the two men had constructed an interferometer on a sandstone slab floating on a mercury pool, so that the arms could turn. Each arm was some 11 meters long.[2] If we put the numbers into Eq. (2–8), using $v/c \approx 10^{-4}$, we find that $\Delta T = \Delta T(0) - \Delta T(\pi/2) \approx 0.3 \times 10^{-15}$ s. What does this mean in terms of the experiment? The frequency $(\Delta T)^{-1} = 3 \times 10^{15}$ s$^{-1}$ corresponds to a wavelength of about 100 nm. But the sodium light Michelson and Morley were using corresponds, as we have noted, to a wavelength of 590 nm. Hence, the effect they were looking for amounted to a shift comparable to a substantial fraction of a wavelength. This was something that was absolutely measurable; indeed, it could not be missed. *But when they did the experiment, they found nothing!* There was no fringe shift, indicating no phase shift between the waves and hence no time difference $\Delta T$. They repeated the experiment later in the year, when Earth

---

[2] Not literally. The arms had an effective length greater than their physical length through multiple reflections from partially silvered mirrors; the total distance traveled by light along an arm is multiplied if the light bounces back and forth from mirrors at the ends of the arms. (See •Fig. 2–4.)

was in a different position in its orbit, just in case the planet had been accidentally at rest in the ether the first time. Still they found nothing; there was no shift at all.

We might summarize this result by saying that the speed of light is the same, no matter how the frame in which its speed is measured moves, at least for the kind of approximately uniform motion an earthbound instrument undergoes. But the first reaction to the Michelson–Morley experiment was not this simplest interpretation; rather, it was something quite different.

### The Lorentz–Fitzgerald Length Contraction

To the end of his life, in 1931, Michelson regarded his result as a failure.[3] Michelson was not the only one who was disturbed by his discovery. To both the Irish physicist George Francis Fitzgerald and the Dutch physicist Hendrik Antoon Lorentz, the idea that there was no medium in which light waves propagated seemed unthinkable, and they decided that drastic measures were needed to save the idea of the ether. To see what they proposed, let us write out the $\Delta T$ of Eq. (2–8) without making any approximations. Using $T_1$ and $T_2$ in their unapproximated form, we find that

$$\Delta T = \frac{2L_1}{c} \frac{1}{1 - v^2/c^2} - \frac{2L_2}{c} \frac{1}{\sqrt{1 - v^2/c^2}} - \frac{2L_1}{c} \frac{1}{\sqrt{1 - v^2/c^2}} + \frac{2L_2}{c} \frac{1}{1 - v^2/c^2}.$$

$$(2–9)$$

Recall that in the first term of the expression, the arm corresponding to the length $L_1$ is in the direction of the motion, while in the fourth term in the expression, it is the arm of length $L_2$ that is in the direction of motion. To make everything cancel out so that $\Delta T = 0$, one had to suppose, Lorentz and Fitzgerald independently noted, that the arm in the direction of motion *shrinks—contracts*—by a factor of $\sqrt{1 - v^2/c^2}$! More generally, suppose we have a one-dimensional object whose length is measured to be $L(0)$ when it is not moving. (Indeed, prior to Lorentz and Fitzgerald, no one would have thought it necessary to specify the motion!) Then according to Lorentz and Fitzgerald, if the object begins moving uniformly in the direction of its length with a speed $v$, its length changes to

$$L(v) = L(0)\sqrt{1 - v^2/c^2}. \qquad (2–10)$$

This equation became known as the **Lorentz–Fitzgerald length contraction**, or just **length contraction** for short.

At first sight, the Lorentz–Fitzgerald length contraction seems like an entirely crazy idea: In our common experience, we do not observe any such effect when we measure the lengths of moving objects. But to be fair, we should note two things. In the first place, $v/c$ is tiny for objects with which we are familiar, and the contraction would be minuscule. For example, for an object moving with the speed of Earth, the fractional contraction would be only

$$\left| \frac{L(v) - L(0)}{L(0)} \right| \approx \frac{1}{2} \frac{v^2}{c^2} \approx 10^{-9}. \qquad (2–11)$$

It would take a highly precise optical experiment to measure such a minute change in length—indeed, just the sort of experiment that Michelson was the

---

[3] He was so embarrassed by it that when he won the Nobel prize in physics in 1907—the first American to do so in the sciences—he did not even mention it in his Nobel prize lecture. Late in life, he was still referring to the "beloved old ether (which is now abandoned, though I personally still cling a little to it)."

first person to perform. The second point is that Lorentz, who was one of the greatest theoretical physicists of his era, was actually able to produce electromagnetic models of matter in which such contractions appear. In retrospect, this is not too surprising: These models turned out to be consistent with Einstein's theory of relativity, in which the Lorentz–Fitzgerald contraction emerges naturally. Any model of matter consistent with the theory of relativity would exhibit such a contraction. So why is relativity thought of as Einstein's theory and not Lorentz's? The reason is that Lorentz's results were based on specific and very limited models of matter. These models have long since lost any utility they once had, while Einstein's approach involved profound insights into the nature of space and time—insights that are with us to the present day.

## 2–2 Classical Relativity

Relativity in mechanics—although not by that name—was something that had been familiar to physicists for nearly three centuries prior to Einstein. The first scientist to describe a relativity principle was Galileo, in the first quarter of the 17th century. Galileo was persuaded that Earth was in motion around the Sun. To most of his contemporaries, that idea seemed to violate common sense; it was argued, for example, that birds leaving the ground would be left behind by the speeding Earth. Galileo reasoned instead that if a ship moved uniformly on the sea, a sailor could not distinguish between the situation in which the ship is at rest and the sea is in motion and the situation in which the ship is in motion and the sea at rest. Thus, according to Galileo, a weight dropped from the top of a uniformly moving ship should land at the base of the mast, not behind it as might be the case if the boat moved from beneath the falling weight. In Galileo's view—a position he claimed was upheld by experiment—the only motion that is measurable is the *relative* motion between the sea and the ship; hence the term **relativity**. One could say that the fact that the only motion that is measurable is relative motion forms a principle of relativity.

It is unclear whether Galileo actually carried out his experiment or simply wrote about it as if the result had already been demonstrated. If he had done it, presumably the weight would not have landed exactly at the base of the mast, partly due to effects of the wind and partly due to the fact that an actual moving ship is unlikely to move uniformly. Relativity, at least in its simplest form, applies only to uniform motions, something that, incidentally, Galileo was aware of. But we know that the ideal experiment would have shown Galileo to be right.

### Coordinate Transformations

The sort of mechanical relativity just discussed, which in the 20th century came to be called *Galilean*, or *Newtonian*, *relativity*, can be established in terms of Newton's equations of motions. This is an instructive exercise to carry out because it prepares us for some features of Einstein's theory. We imagine two *frames of reference S* and *S'* moving with respect to one another, as shown in •Fig. 2–6. Each frame has observers equipped to measure position and motion in the coordinates of their respective frames. When an observer within a given frame sees no acceleration of an object without definite, identifiable forces responsible, we say the frame is an *inertial frame*. If *S* is an inertial frame and the relative speed

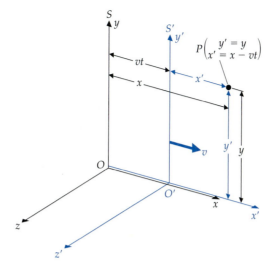

• **Figure 2–6** Two frames of reference, $S$ and $S'$. $S'$ moves in the $x$- (and $x'$-) direction with speed $v$ relative to the frame $S$. The other axes ($y$ and $y'$, and $z$ and $z'$) remain coincident with each other.

$v$ is a constant, as it is assumed to be throughout Chapters 2 and 3, then $S'$ is also an inertial frame.[4]

Suppose that $S'$ moves relative to the frame $S$ in the $x$-direction with constant speed $v$. Furthermore, suppose that at time $t = 0$ the frames $S$ and $S'$ coincide. To locate a point $P$ in space—the location of an object, for example—we must specify the coordinates of that point in some specific reference frame. If $P$ is specified by the coordinates $(x, y, z)$ in the $S$ frame, then in the $S'$ frame the same point in space will be specified at time $t$ by the coordinates $(x - vt, y, z)$. We can write this connection between the location of a point in the two frames—the *Galilean coordinate transformation*—as

$$x' = x - vt;$$

$$y' = y;$$

$$z' = z;$$

$$t' = t. \tag{2-12}$$

We have written the last equation $t' = t$ to emphasize that something has been assumed about time in the two systems, namely, that it is "absolute." Clocks in all systems moving with uniform speeds with respect to each other have been assumed to go at the same rate. This notion was taken to be self-evident by all physicists from Newton up until Einstein.

Now suppose the point $P$ is itself in motion along the $x$-axis with a speed[5] $dx/dt \equiv u$. Then if we differentiate the first of the transformation equations (2–12) with respect to time, we find that

$$u' = u - v, \tag{2-13}$$

---

[4] Nonuniform motion introduces effects that cannot be distinguished from the effects of forces. Einstein's general theory of relativity (Chapter 17) includes the effects of accelerations; such effects turn out to describe gravitation!

[5] Not to be confused with the speed $v$ with which the two frames move with respect to one another!

where $u'$ is the speed of the point $P$ in the $S'$ frame. This result, which is known as the *Galilean addition theorem* for velocities, describes the relation between the velocities of a moving object as measured by observers in the two frames. The ether physicists' assumption that the speed of light obeyed Eq. (2–13) is what led to the incorrect prediction for the Michelson–Morley experiment. It also leads to an incorrect prediction for the Doppler shift, as we shall see next. But, as with the Michelson–Morley experiment, this breakdown in classical optics shows up only when experiments are performed of higher accuracy than $v/c$.

### The Classical Doppler Shift

The Doppler shift for sound is readily observable on a daily basis. If you have studied the derivation of the equation, you know that there are two different Doppler shifts, one for the case in which the observer moves while the source is at rest and another for the case in which the source moves and the observer is at rest. The situations are not symmetric. Why is this? The reason is that for sound there is an "ether," the air, a medium whose rest frame is a privileged frame. That is the source of the asymmetry. By carefully measuring the Doppler shift, you can tell whether you are at rest or moving in the "ether." The "ether" might be totally invisible to you, yet an accurate measurement of a Doppler shift could tell whether you were moving through the "ether" rather than it moving past you. This measurement would violate the principle of relativity as stated previously.

Let us review the classical Doppler effect, the one an ether theorist might derive. We begin by assuming that the resting source emits a light wave with period $T$ and wave length $\lambda$. We know that $\lambda$ and $T$ are connected by the equation $\lambda = cT$, where $c$ is the speed of light in the ether. But now suppose, as in •Fig. 2–7(a), that the observer is moving toward the source with a speed $v$. Someone who believed that the Galilean addition theorem for velocities ap-

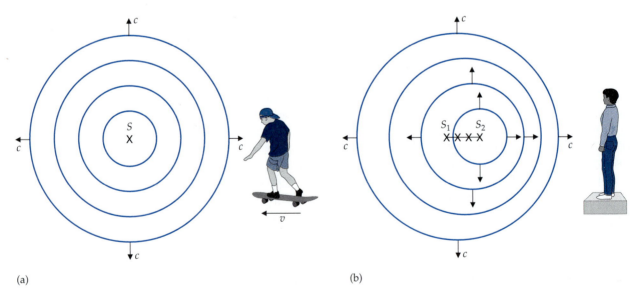

(a)                                                              (b)

• **Figure 2–7** (a) Waves with a fixed wavelength emitted by a source that is stationary with respect to the ether. An observer $O$ moving toward the source sees the waves approaching with speed $c + v$. (b) A source moving with respect to the ether emits waves that are detected by an observer at rest with respect to the ether. The gap between the wave crests—the wavelength—emitted from points $S_1$ and $S_2$ is smaller to an observer toward whom the source moves than it would be to an observer who is stationary with respect to the ether.

plied to light would then argue that the moving observer would find the speed of light to be $c' = c + v$. Hence, such an observer would find that it takes less time for a wavelength to pass. The new period is given by

$$T' = \frac{\lambda}{c + v} = \frac{T}{1 + v/c} \approx T\left(1 - \frac{v}{c} + \frac{v^2}{c^2} - \cdots\right). \qquad (2\text{–}14)$$

As usual, the approximation corresponds to an expansion in the small quantity $v/c$. Equation (2–14) gives the Doppler shift for the case at hand. (One sometimes expresses the Doppler shift in terms of the modification of the frequency $f = 1/T$ rather than that of the period $T$.)

Now suppose that the source moves toward the observer, rather than vice versa [•Fig. 2–7(b)]. In that situation, the distance between wave maxima is given by $\lambda' = \lambda - vT$. Thus,

$$T' = \frac{\lambda'}{c} = \frac{\lambda - vT}{c} = T\left(1 - \frac{v}{c}\right). \qquad (2\text{–}15)$$

Equation (2–15) gives the Doppler shift for *this* case. As we anticipated, Eqs. (2–14) and (2–15) differ; however, note that they are the same if we neglect terms of order $(v/c)^2$ and higher. The fact that the two equations differ is what we meant when we said that the ether theory of light violates the principle of relativity.

This result was well known to Einstein. It was one of the reasons he thought that the ether theory of light propagation was nonsense. He simply did not believe that there was any way, using optical experiments such as the one hypothesized here, that you could tell if you were at rest or moving uniformly with respect to an otherwise invisible ether. Einstein also knew about Maxwell's wave solution of the equations of electrodynamics for empty space, although, as it turned out, Einstein had to teach it to himself, since it was not part of the regular physics curriculum in Switzerland, where he was studying. We have already recalled in Chapter 1 that Maxwell's wave equation gives a speed $c$ for light that depends only on static constants of electricity and magnetism. The speed is independent of any motion of the source (accelerating charges) and does not seem to involve any reference frame. This independence is something that also made a deep impression on Einstein. With these preliminaries out of the way, we are now ready to turn to Einstein's theory of relativity.

## 2–3  Einstein's Special Theory of Relativity

Albert Einstein (•Fig. 2–8) was born in Ulm, Germany, in 1879. He was not precocious; quite the contrary, he was such a late talker that the family worried that there might be something wrong with him. He took an interest in science and mathematics at an early age. When he was very young, his father showed him a compass, and Einstein became fascinated with the question of how it "knew" to point north. Einstein was always a good student, but he acquired a reputation, even in high school, of being "disrespectful" to his teachers. He could never stand authority of any kind. By the age of 15, Einstein had decided that he would like to enter the Swiss Technical High School, the ETH—one of the best places to study science and engineering in Europe—located in Zürich. Although he was somewhat underage, he took the qualifying examination. He did very well in science and mathematics, but failed the parts on languages. But he made a sufficiently favorable impression on his examiners that they

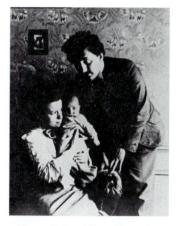

• **Figure 2–8**   Albert Einstein (1879–1955). This photo, which was taken in 1904, shows Einstein with his wife Mileva and their young son Hans Albert. We devote more space to the life of Albert Einstein and to the history of his ideas than we do to most other historical figures in this book. The reason is that Einstein almost single-handedly created what we now call "modern physics" and so occupies a special place in the subject. It is likely that relativity and quantum theory would have been discovered sooner or later by other people, just as it is likely that Newtonian mechanics would have been discovered even if Newton had never lived. But that is not how it happened; Newtonian physics depends on Newton, and relativity and the quantum theory depend on Einstein.

proposed that he spend a year in a progressive high school in Aarau near Zürich and then reapply to the ETH. Einstein thrived at Aarau.

Many years later, Einstein recalled that the first thoughts which eventually led him to relativity occurred during that year. They involved what he saw as problems with the ether theory of light. It bothered him—or at least, this is what he later recalled—that Newtonian physics would allow an observer who boosted himself to the speed of light to catch up to a light wave, which, to the observer, would merely become electric and magnetic fields of constant amplitude. The observer could then tell that he was traveling at the speed of light in the ether. As with the Doppler shift, this determination was a violation of the relativity principle according to which one could not distinguish between rest and uniform motion. Einstein also recalled another, similar puzzle that had occurred to him when he was 16. This one involved trying to look at himself in a mirror. If the mirror moved at the speed of light, the light would never catch up to it, and he would not be able to see his reflection. At least, that is what the ether theory seemed to predict, again a violation of the relativity principle. There was a conflict between Newtonian mechanics, which allowed one to travel with the speed of light or faster, and the relativity principle, which did not allow one to distinguish between rest and uniform motion.

In 1900 Einstein completed his studies at the ETH, graduating with good, but not outstanding, grades. For two years he was unable to get a real job, partly because he had not made an entirely favorable impression on his professors at the ETH—again, it was a matter of "disrespect." But in 1902 he became a patent examiner at the Swiss National Patent Office in Bern. All of his physics was done in his spare time. While he did not have access to a good physics library or to fellow physicists with whom he could discuss his ideas, this did have one advantage: His work was completely original. His paper on relativity, which was the last of the three great papers he wrote in 1905, when he was 26—the papers that founded modern physics—contains no reference to any other physics paper.

### Einstein's Postulates

Einstein's theory begins with two postulates, or axioms:

I. The principle of relativity. **The test of a physical law by any experiment carried out in a uniformly moving frame of reference does not depend on the speed of that frame relative to any other frame moving uniformly with respect to it.**

We have said enough about relativity so that this principle should be familiar by now. In essence, it says that *there is no experiment one can perform in a uniformly moving system in order to tell whether one is at rest or in a state of uniform motion.* The new feature of Einstein's formulation is that the principle is taken to be exact and to apply to all phenomena—not merely those of mechanics. It implies, for example, that the classical Doppler shift formula cannot be right for light.

II. The principle of constancy. **There exists a frame of reference S (call it the rest frame) with respect to which the speed of light is c. The speed is then also c in every other frame of reference moving uniformly with respect to S. This implies, as a corollary, that the speed of light is independent of the motion of the source.**[6]

---

[6] Actually, the theory assumes only that there is a maximum speed of propagation that is the same in all frames. There are good reasons for taking this maximum speed to be that of light.

While relativity had a certain familiarity, this assumed property of light propagation was completely novel. No other speed that we are familiar with obeys such a law. When asked late in life why he had made this assumption, Einstein answered that he believed in Maxwell's equations. By 1905, it was clear to Einstein that the ether was unnecessary in explaining the physics of light. Maxwell's equations gave a perfectly good description of the phenomena with no reference to an ether. Indeed, the only mention of the ether in Einstein's paper on relativity is the famous sentence, "The introduction of a 'luminiferous ether' will prove to be superfluous in as much as the view here to be developed will not require an 'absolutely stationary space' provided with special properties...." A century of work on the ether thus became irrelevant.

Given Einstein's two postulates there are a number of ways one can develop the theory. We choose not to follow Einstein's original discussion, which is logically correct, but somewhat abstract. Instead, we consider the Michelson–Morley experiment from the point of view of Einstein's theory.[7] This approach will serve to introduce many of the basic ideas. In particular, we are going to use the Lorentz–Fitzgerald contraction, something that can be derived directly from Einstein's original approach.

### Another Look at the Michelson–Morley Result: Time Dilation

Let us consider an idealized Michelson–Morley experiment from the point of view of an observer moving uniformly with a speed $v$ with respect to Earth. Let us call this system $S$ and the system in which Earth is at rest $S'$. The Michelson–Morley apparatus is at rest in $S'$. To simplify the discussion we assume that the lengths of the two arms—shown in •Fig. 2–4—as measured in $S'$, are *exactly* the same. (Recall that the ability to rotate the apparatus allows us to correct for this assumption.) We label these lengths $L'$. Incidentally, $L'$—the length of the arm *in its own rest frame*—is called the **proper length** of the arm. We shall see that the concept of proper length plays an important role throughout the discussion that follows.

As measured in $S'$, the round-trip times $T'_1$ and $T'_2$ are equal and are given by

$$T'_1 = \frac{2L'}{c} = T'_2, \tag{2–16}$$

where $c$ is the speed of light as measured in *any* uniformly moving frame of reference. If the principle of relativity is to be valid, the round-trip times $T_1$ and $T_2$—the corresponding times as measured in $S$—must also be identical, because otherwise the two uniformly moving observers would disagree about whether there was or was not an interference pattern characteristic of motion through an ether. It is in seeing how these times agree that the subtleties of Einstein's relativity begin to emerge.

Let us first analyze the time $T_2$ corresponding to the light passage at right angles to the motion (•Fig. 2–5). As viewed in $S$, the light travels along the triangular path shown in the figure. The distance to the mirror $M_2$ is the same in the two systems, since it is at right angles to the motion, so we do not expect it to be affected by the Lorentz–Fitzgerald contraction—a matter to be discussed in detail later. This distance is $L'$. Thus, by the Pythagorean theorem, we have

---

[7] Einstein was always careful to point out that the Michelson–Morley experiment had essentially no influence on his thinking. Indeed, he was not sure that he had ever heard of it. He once commented that if he had, he would surely have mentioned it in his paper. Einstein was certain that the principle of relativity was *exact*. So as far as he was concerned, he did not learn anything new from the experiment when he finally heard of it.

$$d = \sqrt{L'^2 + \frac{v^2 T_2^2}{4}}. \tag{2–17}$$

Hence, using the principle of constancy—that light propagates with the same speed $c$ in both frames—we obtain

$$\frac{2d}{c} = T_2 = \frac{2}{c}\sqrt{L'^2 + \frac{v^2 T_2^2}{4}}. \tag{2–18}$$

We can solve this equation for $T_2$:

$$T_2 = \frac{2L'}{c}\frac{1}{\sqrt{1 - v^2/c^2}} = T'_2\frac{1}{\sqrt{1 - v^2/c^2}}. \tag{2–19}$$

Since $1/\sqrt{1 - v^2/c^2} \geq 1$, we have $T_2 \geq T'_2$. This is an example of the important idea of **time dilation.** We will return to it shortly, but here we note that the setup acts as a kind of clock. We can call it a "light clock." The period of this clock in a frame in which it is resting—something known as the **proper time—** is given by $2L'/c$. In this case, the rest frame is the Earth frame. But if the period of the same clock is measured when the clock is in motion with a uniform velocity, then the period will be longer. Thus, *a clock moving uniformly with respect to a given observer moves more slowly.* In our example, the observer moving with respect to Earth sees the clock moving past him or her, and we have seen that *that* observer measures a longer "tick." We shall see shortly that time dilation is a statement about the nature of time and not about the mechanics of some special kind of clock: It holds for *every* clock.[8]

The fact that time dilation holds for any clock means that it is very readily observed, as the next example shows. This example involves a particular kind of unstable particle, one that decays into other particles after a certain characteristic timescale known as its lifetime. We'll discuss the lifetime in much greater detail throughout the book, because it has an important physical significance, but for now, you can think of it as the "tick" of a kind of internal clock, one that undergoes time dilation, like all clocks.

**Example 2–2**   A beam of *muons*—unstable particles with a proper lifetime of $2.2 \times 10^{-6}$ s—are measured to move with a speed of $0.99c$. (a) Without the time dilation effect, how far would muons travel before they decay? (b) Include the time dilation effect, and redo your calculation. (Use the description of the lifetime just given—later we will see that what we are calculating is not how far each muon will travel, but rather an average distance traveled before decay.)

**Solution**   (a) Denote the proper lifetime of muons as $\tau$. Then the muons travel a distance $x = v\tau$ if there is no time dilation. Numerically, this gives

$$x = (0.99c)(2.2 \times 10^{-6}\,\text{s}) = 0.99 \times 3.0 \times 10^8\,\text{m/s} \times 2.2 \times 10^{-6}\,\text{s} = 650\,\text{m}.$$

(b) According to the experimenter who observes the muons moving with a speed of $0.99c$, the muons' "clock" slows by the factor

$$1/\sqrt{1 - v^2/c^2} = 1/\sqrt{1 - (0.99)^2} = 1/0.14 = 7.1$$

in other words, the lifetime is not $\tau$, but has increased to $7.1 \times \tau = 15.6 \times 10^{-6}$ s. And correspondingly, the distance the muons travel before decaying is increased by a factor of 7.1, to 4.62 km!

---

[8] If not, one could use the measured difference in times to tell which of two inertial frames is moving and which is at rest.

It is no trouble at all, these days, to accelerate a muon so that they travel at 99% of the speed of light. The effect we have calculated here cannot be missed.

---

**Example 2–3** In his 1905 paper on relativity, Einstein made a prediction. He imagined taking two identical clocks and putting one at the equator and one at the north pole (•Fig. 2–9). He concluded that the clock "at the equator must go more slowly, by a very small amount, than a precisely similar clock situated at one of the poles under otherwise identical conditions." How did Einstein reach this conclusion? Give a rough estimate of the magnitude of the effect.

**Solution** Following Einstein, we treat this as a problem involving only uniformly moving systems: For short periods, the motion of a point on the equator is simply motion in a straight line with a constant speed. (The pole is at rest.) Once we know this speed, we see that the moving clock runs slowly according to the time dilation relation, Eq. (2–19). To estimate the average speed, which is actually the difference in speed between a point on the pole and a point on the equator, only the rotational speed of Earth about its axis comes into play. This angular speed is $\omega = 360°/\text{day}$, or

$$(2\pi/\text{day})(1 \text{ day}/24 \text{ hr})(1 \text{ hr}/3{,}600 \text{ s}) \approx 7 \times 10^{-5} \text{ rad/sec}.$$

Then if Earth's radius is $R_E \approx 6{,}378$ km, the equatorial speed is

$$v \approx R_E\omega = (6{,}378 \text{ km})(7 \times 10^{-5} \text{ rad/s}) \approx 0.46 \text{ km/s}.$$

Therefore,

$$(v/c)^2 \approx \left[(0.46 \text{ km/s})/(3 \times 10^5 \text{ km/s})\right]^2 = 2 \times 10^{-12}.$$

According to an observer on the pole, whose clock ticks to give a time $t$, the equatorial clock reads

$$t' = t\sqrt{1 - v^2/c^2} = t - \left(1 - \sqrt{1 - v^2/c^2}\right)t,$$

or, if we expand the square root for small values of $v^2/c^2$,

$$(t - t')/t \approx (1/2)\left(v^2/c^2\right) \approx 10^{-12}.$$

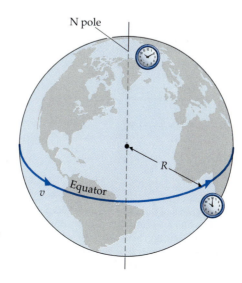

N pole

Equator

$v$

$R$

• **Figure 2–9** If we assume that the effects of acceleration due to the circular motion of a clock at the equator can be neglected, the clock on the equator will run slowly compared to the clock at the pole, because it moves relative to the clock at the pole.

This expression gives us a measure of the degree by which the equatorial clock runs slowly compared with the polar clock. A difference of about one part in $10^{12}$ would be quite visible with today's clocks.[9]

We still have not completed our job with the Michelson–Morley experiment. We must now examine $T_1$, the time the moving observer, in $S$, measures for the signal to propagate back and forth in the direction of motion. Thus our calculation involves the Lorentz–Fitzgerald length contraction, because all motion is parallel. In the theory of relativity one often finds time dilation and length contraction entering into a situation in a complementary fashion to reconcile different points of view. The time $T_1$ is the sum of two times: $T_{out}$, which is the time it takes for the light to get to $M_1$, and $T_{back}$, which is the time it takes for the light to return to the origin of $S$. These two times are not the same. To compute, say, $T_{out}$, we focus on the distance traveled by the light (•Fig. 2–10). During the time $T_{out}$, as an observer in $S$ views the situation, the $S'$ frame has moved a distance $vT_{out}$. Light starting from the origins of $S$ and $S'$ when they overlapped at the common time $t = t' = 0$ would, according to the observer in $S$, have to travel this additional distance to reach the mirror $M_1$. But there is a second effect: The observer in $S$ measures a contracted distance $L'$ to the mirror $M_1$. If we add these two effects, the total distance traveled by the light to the mirror is, according to the observer in $S$,

$$d_1 = vT_{out} + L'\sqrt{1 - v^2/c^2}. \tag{2-20}$$

If we divide this equation by $c$, we have

$$T_{out} = \frac{v}{c} T_{out} + \frac{L'}{c} \sqrt{1 - v^2/c^2}, \tag{2-21}$$

or, solving for $T_{out}$, we obtain

$$T_{out} = \frac{L'}{c} \frac{\sqrt{1 - v^2/c^2}}{1 - v/c}. \tag{2-22}$$

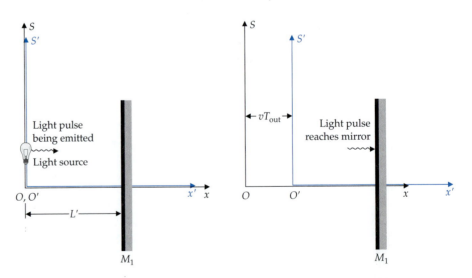

• **Figure 2–10**  During the time $T_{out}$, the frame $S'$ and the mirror $M_1$ with it will have moved a distance $vT_{out}$.

[9] It turns out that this is the one incorrect prediction in Einstein's paper. It is not that he made a mistake in arithmetic; rather, it is that special relativity alone does not fully explain this situation. There is also a connection between gravitation and time—something not included in the 1905 paper. It took Einstein another decade of work before he understood how to include the effect of gravity. As we shall see in Chapter 17, gravity precisely cancels the special-relativity effect in this instance.

To find the return time $T_{\text{back}}$, we replace $v$ by $-v$. Thus,

$$T_{\text{back}} = \frac{L'}{c} \frac{\sqrt{1 - v^2/c^2}}{1 + v/c}. \tag{2–23}$$

The total round-trip time is then

$$T_1 = T_{\text{out}} + T_{\text{back}} = \frac{L'}{c} \sqrt{1 - \frac{v^2}{c^2}} \left\{ \frac{1}{1 + v/c} + \frac{1}{1 - v/c} \right\}$$

$$= \frac{2L'}{c} \frac{1}{\sqrt{1 - v^2/c^2}}. \tag{2–24}$$

Comparing Eqs. (2–19) and (2–24), we see that we have $T_1 = T_2$. Thus, the moving observer agrees with the earthbound observer that there is no Michelson–Morley effect, as is required by the relativity principle.

## 2–4  The Lorentz Transformations

We may use the set of ideas described in the previous section to "derive" what are known as the **Lorentz transformations**. These are the transformations that take the four coordinates $(x, y, z, t)$ that describe an "event"—something that occurs at a definite point in space and at a definite time, an explosion for example, or the arrival of a light ray at a mirror—in the frame $S$ into the four coordinates $(x', y', z', t')$ that describe the same event in a system $S'$ that is moving uniformly with a velocity $\vec{v}$ with respect to $S$. For convenience, we take this motion to be in the $x$-direction. We have put "derive" in quotation marks because, in our derivation, we shall *assume* the Lorentz–Fitzgerald contraction, as we did in our discussion of the Michelson–Morley experiment. Indeed, our derivation will be in the spirit of that discussion. This is not what Einstein did. He derived the Lorentz transformations from very general, albeit rather abstract, principles and then showed how the Lorentz–Fitzgerald contraction and time dilation followed from these transformations. We present this more abstract derivation of the transformations in an appendix to this chapter. Our own way of doing things may make the physics a little clearer.

Incidentally, you may wonder why the transformations are named after Lorentz. They are named after Lorentz since he discovered them first. In his efforts to save the ether, Lorentz investigated how forces might produce the contraction. In this process, he discovered, at least for small $v/c$, that "his" transformations left Maxwell's equations in the same form they had in the rest frame when they were expressed in the new, transformed coordinates. He regarded these transformations as a sort of mathematical convenience, useful for solving the equations, and did not realize their deeper significance. Lorentz published his most detailed paper on the topic in 1904, but by all accounts, Einstein discovered the Lorentz transformations independently.

To proceed, we start with the frames $S$ and $S'$ (•Fig. 2–6) and recall the Galilean (classical) transformations between their coordinates, Eq. (2–12). We shall see later that the transverse coordinate transformations—the ones involving the $y$ and $z$ coordinates—remain as they are in the Galilean form. The transformation $t' = t$ is what we have referred to as the "absoluteness" of time. We now know enough about relativity to be sure that transformation will be modified. Finally, having seen the Lorentz contraction, we can also be sure that the transformation $x' = x - vt$ will be modified.

Implicit in the classical transformation is the assumption that the length scales in $S$ and $S'$ are the same. We must be careful about this. In this transformation equation, what is called $x'$ is the distance from the origin of the $S'$ coordinate system to the point marked $x'$ in that system. In classical physics, observers in both $S$ and an $S'$ would measure the same distance for this variable. But in relativity, this length, according to $S$, is contracted by the factor $\sqrt{1 - v^2/c^2}$. Thus, the relativity-corrected version of $x' = x - vt$ is $x'\sqrt{1 - v^2/c^2} = x - vt$. This equation yields the spatial Lorentz transformation

$$x' = \frac{x - vt}{\sqrt{1 - v^2/c^2}}. \tag{2--25a}$$

The factor $1/\sqrt{1 - v^2/c^2}$ occurs so frequently in this and in other contexts, that we follow custom and give it a name:

$$\gamma(v) \equiv \frac{1}{\sqrt{1 - v^2/c^2}}. \tag{2--26}$$

We have made the fact that $\gamma$ depends on $v$ explicit here by writing $\gamma(v)$, but in keeping with standard usage, we shall drop the argument and write, more simply, "$\gamma$," unless there may be confusion in situations involving more than one speed. Keep in mind that when $\gamma$ appears, you must know to what speed it refers; here, it is the relative speed of the two frames. We have thus shown the spatial Lorentz transformation,

$$x' = \gamma(x - vt). \tag{2--25b}$$

To find the time transformation, we employ an argument that is similar in style to the one we used to analyze the Michelson–Morley experiment. From •Fig. 2–11, we see that at time $t$, as measured in $S$, the light beam will have reached the point $x = ct$. However, we have already argued that

$$x = ct = vt + x'\sqrt{1 - \frac{v^2}{c^2}} = \frac{v}{c}x + x'\sqrt{1 - \frac{v^2}{c^2}}.$$

But in $S'$, it takes a time $t' = x'/c$ to reach the point $x'$. When we substitute $x' = ct'$, we find that

$$ct = \frac{v}{c}x + ct'\sqrt{1 - \frac{v^2}{c^2}}.$$

This equation can be solved for $t'$ to find the relation between the time $t$ for the light to reach $x$ in $S$ and the time $t'$ for the light to reach $x'$ in $S'$:

$$t' = \gamma\left[t - \left(\frac{v}{c^2}\right)x\right]. \tag{2--27}$$

Equation (2–27) is the Lorentz transformation for time; it follows from the principle of constancy and the Lorentz–Fitzgerald contraction. Compare it with the transformation $t' = t$ in the Galilean transformations!

Equations (2–25) and (2–27) tell us how to find the position and time in the $S'$ system in terms of the coordinates of the $S$ system. How can we go backwards to find the position and time in the $S$ system in terms of the coordinates of the $S'$ system—that is, how do we find the **inverse transformation**? Here, we can appeal to the symmetry implicit in the postulates of relativity: We simply change the sign of $v$, as well as the labels of the primed and unprimed coordinates:

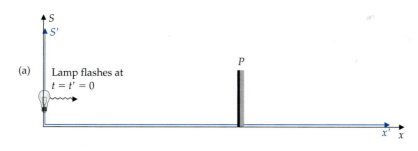

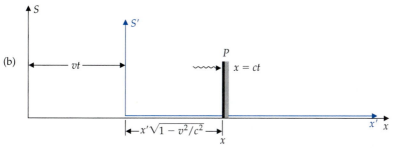

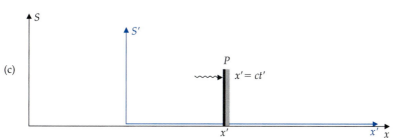

• **Figure 2–11** (a) A lamp flashes at $t = t' = 0$, when the origins of $S$ and $S'$ coincide. (b) The arrival of the light pulse at $P$, as described in frame $S$. The pulse arrives at a time given by $x = ct$. The location of $P$ in the $S$ frame is at $x$, which can be expressed in terms of $x'$, $v$, and $t$. (c) The arrival of the light pulse arrival as described in the $S'$ frame. The pulse arrives at a time given by $x' = ct'$.

$$t = \gamma\left[t' + \left(\frac{v}{c^2}\right)x'\right] \qquad \text{and} \qquad x = \gamma(x' + vt'). \qquad (2\text{--}28)$$

[Note that $\gamma$ depends only on the magnitude of the velocity, not its sign, so it is the same here as it is in Eqs. (2–25) and (2–27).] Of course, you could also explicitly invert Eqs. (2–25) and (2–27); you would get the same result.

**Example 2–4** Consider a spaceship of proper length 100 m that moves along the positive $x$-axis at $0.9c$ with respect to the ground (•Fig. 2–12). If $S$ is a coordinate frame fixed to the ground and $S'$ is a coordinate frame fixed to the ship, then the origins are set so that at $t = t' = 0$ the front of the ship is at $x = x' = 0$. Where is the back of the ship at $t = 0$, according to an observer on the ground, and at $t = 0$ what time $T'$ will a clock fixed to the back of the ship read, according to a ground-based observer? Do the problem (a) classically and (b) according to special relativity.

**Solution** (a) The answers to our questions are "obvious" classically: At $t = 0$, the back of the ship is at $x = x' = -100$ m, while the clock-based ship reads the same as the Earth-based clock, namely, $T' = 0$.

(b) To answer this problem in special relativity, we specify "events"—space–time locations—and then use the Lorentz transformations to relate the coordinates of these events to coordinates in other frames. *(This is an extremely good way to approach relativity problems.)* The problem specifies the following events: In the $S$-frame, there is one event, the location of the front of the ship at $x = 0$, $t = 0$. In the $S'$-frame, there are two events—first, the location of the front of the ship at $x' = 0$, $t' = 0$, and second, the location of the back of the ship at $x' \equiv X' = -100$ m, $t' = 0$. We are asked to find, first, the $S$-coordinate $X$

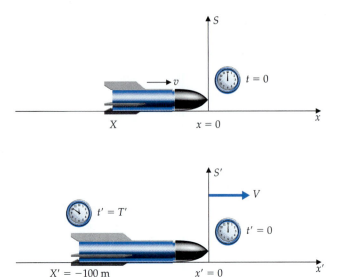

• **Figure 2–12** The relativistic description of a moving rocket. (a) The front and rear ends of the rocket as described in reference frame $S$ fixed to the ground. (b) The front and rear ends of the rocket as described in reference frame $S'$ attached to the rocket. The frame $S'$ moves with speed $v$ relative to the frame $S$. The rocket is at rest in reference frame $S'$.

of the back of the ship at $t = 0$. We can find it by using the Lorentz transformation, Eq. (2–25b), for $x' = X'$, $t = 0$, and unknown $x = X$, namely,

$$X' = \gamma(X - v \times 0) = \gamma X, \text{ or } X = X'\sqrt{1 - \frac{v^2}{c^2}} = (-100 \text{ m})\sqrt{1 - (0.9)^2} = -43.6 \text{ m}.$$

This is the by-now familiar Lorentz contraction.

Next, we must find the reading of the onboard clock at the rear of the ship when the $S$-coordinates are $t = 0$ and $x = X = -43.6$ m. Here, the Lorentz transformation of Eq. (2–27) applies, namely, $T' = \gamma[t - (v/c^2)x] = \gamma[0 - (v/c^2)X]$. While you might be tempted to insert numbers at this point, it is useful to use the previous result for $X$ with the explicit form of $\gamma$ inserted:

$$T' = \frac{-(v/c^2)(X'\sqrt{1 - v^2/c^2})}{\sqrt{1 - v^2/c^2}} = -X'\left(\frac{v}{c^2}\right) = (-100 \text{ m})(0.9/c) = -3 \times 10^{-7} \text{ s.}$$

The square-root factor has canceled out. The clock at the back of the spaceship does not read zero, but instead shows a slightly earlier time.

The preceding example illustrates a point of great importance: The arrival of the front of the ship at the spatial coordinate $x = 0$ and that of the back of the ship at $X$ are simultaneous events according to the $S$-based observer—they both occur at $t = 0$. But these same events are not simultaneous to the $S'$-based observer! We have already discussed classical simultaneity in Example 2–1; we shall go more deeply into the relativity of simultaneity in Section 2–5.

### The Transverse Lorentz Transformations

The transformations $y' = y$ and $z' = z$ apply to the coordinates perpendicular to the direction of relative motion. A simple physical argument makes these transformations easy to understand. In particular, we shall show that a rod aligned in the $y$-direction does not change length when it moves along the $x$-axis. We take such a rod, with proper length 1 m, with a pencil at the end such that it can mark its passage along a wall (•Fig. 2–13). The wall has a long blue line drawn on it at a height $y = 1$ m; we could even draw that line with the slow passage of the rod. Now we imagine that the rod passes the wall at a rel-

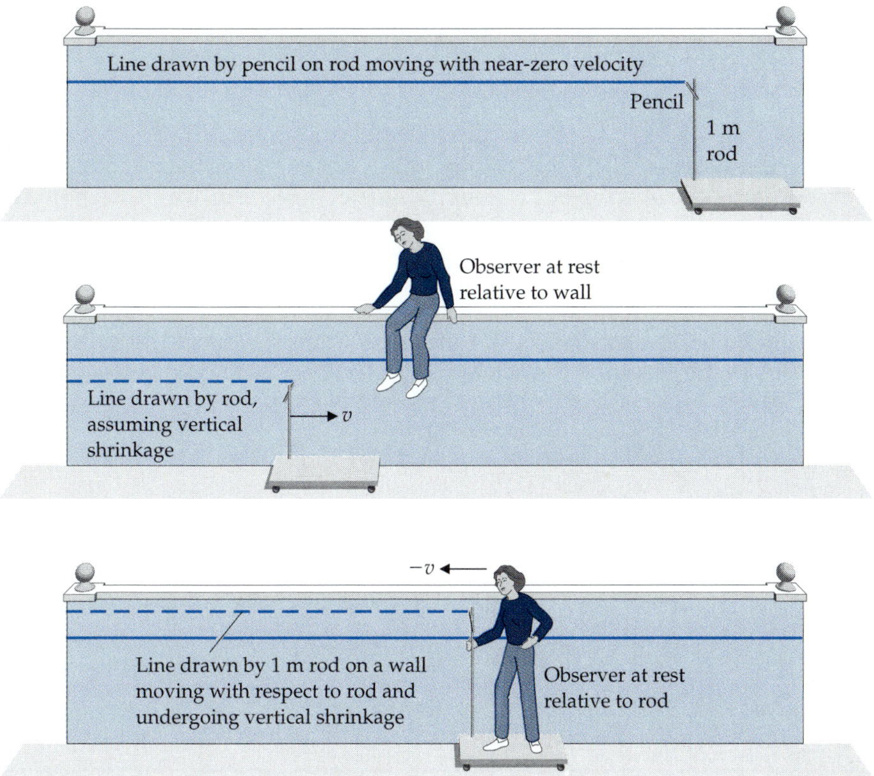
Line drawn by pencil on rod moving with near-zero velocity
Pencil
1 m
rod
Observer at rest
relative to wall
Line drawn by rod,
assuming vertical
shrinkage
v
−v
Line drawn by 1 m rod on a wall
moving with respect to rod and
undergoing vertical shrinkage
Observer at rest
relative to rod

• **Figure 2–13** Trace made on a wall by a pencil attached to a transversely oriented rod. (a) Line made by a rod moving with an infinitesimally small velocity. (b) Observer at rest relative to the wall sees a shrunken rod, which makes a line below the line (a). (c) Observer at rest relative to the rod sees the wall moving with speed $v$ and thus sees it and the original line (a) lower than the line drawn by the pencil.

ativistic speed. If an observer on the wall sees the rod shrink, then, to that observer the pencil on the rod will make a line on the wall *below* the red line. But if the relativity principle remains valid, an observer fixed to the rod will see the height of the red line shrink and will see the pencil draw a line *above* the red line. Now, however, the observers can come together and see where the pencil line really is! The same argument can be made if there is elongation. In either case there is an inconsistency with the relativity principle that can be resolved only if there is no contraction or elongation effect whatsoever. Hence, we have the transformation laws $y' = y$ and $z' = z$.

With both the longitudinal and transverse Lorentz transformations at hand, we can ask, How does an angle $\theta$ appear to an observer moving uniformly with respect to it? For this purpose, we consider a reference frame $S'$ moving with speed $v$ with respect to another frame $S$; the two frames have coincident $x$-axes. The angle $\theta$ is defined as in •Fig. 2–14, namely, with the point $P \equiv (x, y)$ such that $\tan\theta = y/x$. The moving observer will see the point $P$ at the coordinates $(x', y')$, with $x'$ and $y'$ related to $x$ and $y$, respectively, according to our Lorentz transformations. The $y'$-value is the same as the $y$-value, while the $x'$-value is Lorentz contracted. Thus, the angle $\theta'$ is given by

$$\tan\theta' = y'/x' = \gamma/[x/\gamma] = \gamma\tan\theta.$$

This result says that as the speed of $S'$ approaches that of light, any angle oriented as we have hypothesized approaches 90° as viewed in $S'$, no matter what the $S$-based observer claims the angle to be. This is because the $x'$-component which is part of the definition of the angle shrinks to zero.

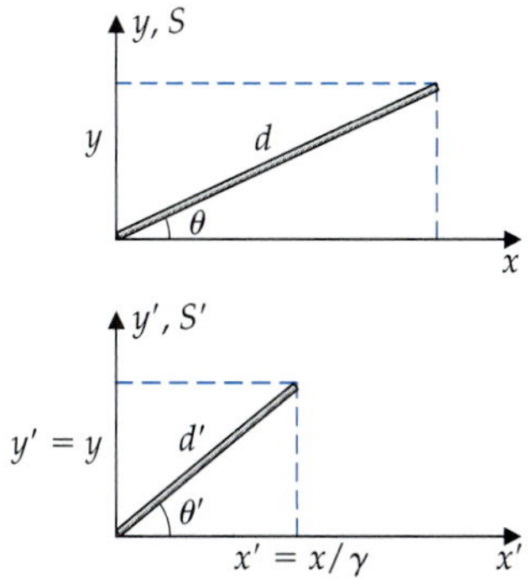
y, S
y
d
θ
x
y′, S′
y′ = y
d′
θ′
x′ = x/γ
x′

• **Figure 2–14** A rod inclined at an angle $\theta$ with respect to the $x$-axis has a different length and a different angle of inclination as seen from a frame moving with velocity $v$ along the $x$-axis.

**Example 2–5**    Consider a rod of proper length $d$. What is its length, as seen by an observer moving with velocity $\mathbf{v}$ relative to the rod, when the direction of the velocity makes an angle $\theta$ with the direction of the rod? Check that your result has the proper limits when $\theta$ is zero and 90°.

**Solution**    As we see in •Fig. 2–12, if we orient the velocity $\vec{v}$ along the $x$-direction, then the angles are as described in the previous paragraph. In this case, we are interested in the length of the rod, not the angle it makes as seen by the observer in the $S'$ system. We can place one endpoint of the ruler at the origin and the other at the points $x = d \sin\theta$ and $y = d \cos\theta$ in the ruler's rest frame. In that frame, the ruler's length is $d$. The observer in $S'$ can measure the length of the rod when the first endpoint is also at his origin and the second endpoint has components $x'$ and $y'$, the sum of whose squares add up to the square of the ruler's length as seen by $S'$, namely, $d'$ (•Fig. 2–12). From the Lorentz transformations, we have $y = y' = d \sin\theta$ (the transverse direction is unaffected), while in the $x$, $x'$ direction there is a Lorentz contraction, so that $x' = x\sqrt{1 - v^2/c^2} = d \cos\theta\sqrt{1 - v^2/c^2}$. To find $d'$, we use the Pythagorean theorem to obtain

$$d'^2 = x'^2 + y'^2 = d^2\left[1 - \left(\frac{v^2}{c^2}\right)\cos^2\theta\right],$$

or

$$d' = d\sqrt{1 - \left(\frac{v^2}{c^2}\right)\cos^2\theta}.$$

This result reduces to the usual Lorentz contraction when $\theta = 0$, and at $\theta = \pi/2$ we see that $d = d'$, which is what we expect.

## Invariant Quantities

By thinking about the propagation of light in different frames, we learn that some quantities have the same value in different inertial frames. Once again, we consider frames $S$ and $S'$, moving with a relative velocity $\vec{v}$ along the $x$-axis. Suppose that at $t = t' = 0$ the origins of these two frames coincide. Just at that instant, a light source at the common origin emits a pulse of light that moves out in all directions. The observer in system $S$—we'll just refer to him or her as "$S$"—will claim to be at rest and will see the light propagating in a spherical wave whose radius at time $t$ is $ct$. The equation of the sphere is

$$x^2 + y^2 + z^2 = c^2t^2. \tag{2–29}$$

But $S'$ will also claim to be at rest and will also claim an equation for the light wave of the same form, viz.,

$$x'^2 + y'^2 + z'^2 = c^2t'^2. \tag{2–30}$$

Are these equations compatible? The answer lies in the Lorentz transformations. Since $y' = y$ and $z' = z$, it is the $x$ and $t$ transformations that matter. To see that Eq. (2–30) implies Eq. (2–29), we simply use the Lorentz transformations for $x'$ and $t'$ in the former equation; that is, we substitute the Lorentz transformation into both sides of Eq. (2–30). If we do this, we are led to the relation

$$\frac{(x - vt)^2}{1 - v^2/c^2} + y^2 + z^2 = \frac{c^2(t - xv/c^2)^2}{1 - v^2/c^2}. \tag{2–31}$$

If we square out the factors, we find that this equation reduces to Eq. (2–29), so that the Lorentz transformations do indeed leave the relation $x^2 + y^2 + z^2 = c^2t^2$ unchanged. Each observer in fact sees a spherical wave. We say that $x^2 + y^2 + z^2 - c^2t^2$ is an *invariant*.

The quantity $x^2 + y^2 + z^2 - c^2t^2$ refers particularly to the front made by a light pulse. The same algebra, however, (see Problem 13 at the end of the chapter), leaves the more general quantity

$$s^2 = x^2 + y^2 + z^2 - c^2t^2 \tag{2-32}$$

invariant. In relativity, the four coordinates $x$, $y$, $z$, and $t$ define what is called an "event." We then define a "distance" between two events—say, $(x_1, y_1, z_1, t_1)$ and $(x_2, y_2, z_2, t_2)$—by

$$s^2 = (x_1 - x_2)^2 + (y_1 - y_2)^2 + (z_1 - z_2)^2 - c^2(t_1 - t_2)^2. \tag{2-33}$$

The square of this "distance" can be positive or negative, depending on the relations between the spatial and temporal variables. One can approach the derivation of the Lorentz transformations in a completely different way by showing that under certain general and simple assumptions, they are the set of transformations that leave Eq. (2–33) invariant.

Relations such as Eq. (2–32), in which the quantity $ct$ appears in a role completely analogous to that of the ordinary spatial coordinates, suggest that it is often useful to think of $ct$ as just one of four coordinates that describe an event. This event is said to occur in a four dimensional space that one calls "space–time." In the next chapter, we shall encounter another four-dimensional space, "momentum–energy," that is also useful.

## 2–5  Simultaneity and Relativity

The concept of simultaneity is normally taken for granted. Yet, as we shall discuss in more detail in this section, it was a deep examination of simultaneity and its relation to time that led Einstein to formulate the special theory of relativity. Even in classical physics, simultaneity is relative. (Recall Example 2–1.) In this section, however, we deal with the issue in terms of Einstein's relativity. Here, the speed of light will be the same for all observers; in Example 2–1 it is not.

At the end of the previous section, we saw that observers in $S$ and $S'$ would, in terms of their own coordinates, each describe a spherical wave emanating from a common origin at $t = t' = 0$. This must be the case, since each observer can claim to be at rest, with the *other* observer in motion. But now we raise the following question: The observer in $S'$ will find that the spherical wave has arrived at the same instant—$t'$—at all points on the surface of the sphere whose radius is $ct'$. The arrival of the light wave at the various points on this surface is a *simultaneous* set of events in the frame of reference $S'$. But what about the observer in $S$? Will he or she perceive these events to be simultaneous? To answer this question, it is best to use the Lorentz transformations.

We make our point by considering two positions on the surface of radius $ct'$. Let us take a point on the positive $y$-axis given by $x' = z' = 0$ and $y' = ct'$, along with a point on the positive $x$-axis given by $y' = z' = 0$ and $x' = ct'$. As indicated, these points are specified at the common time $t'$. We may now use the Lorentz transformation that takes us from $S'$ back to $S$ to find the times—let us call them $t_x$ and $t_y$—in the $S$ frame corresponding to the two spatial points in $S'$. Substituting into Eq. (2–28), we find at once that

$$t_y = \gamma t', \tag{2-34}$$

while, from the fact that $x' = ct'$, and using the explicit form of $\gamma$, we have

$$t_x = t' \frac{1 + v/c}{\sqrt{1 - v^2/c^2}} = t' \sqrt{\frac{1 + v/c}{1 - v/c}}. \tag{2-35}$$

It is clear from Eqs. (2–34) and (2–35) that $t_x > t_y$; from the point of view of $S$, the two events in question do *not* happen simultaneously. This means that *simultaneity is not an absolute concept*: Two events that are simultaneous in $S'$ will not necessarily be simultaneous in $S$ and vice versa. This is not at all at odds with the relativity principle, according to which there is no way to tell which observer is moving and which is at rest. The only thing relativity requires is that the lack of simultaneity be symmetric. The observer in $S'$ could make exactly the same set of arguments to show that, from his or her point of view, the arrival of the light at all points of the sphere of radius $ct$ in $S$ is not simultaneous.

The issue of simultaneity was the key that unlocked the relativity of time for Einstein. Before explaining this remark, let us give a different example—one that Einstein was fond of—which shows that the idea was already apparent in classical physics, even if no one paid any attention to it. Suppose a train moves with velocity $v$ along a platform. The platform is our system $S$, while the train itself is our system $S'$. Suppose further that when the front of the train passes the common origin of $S$ and $S'$, two lightning bolts strike at positions $\pm L$ at time $t = 0$ as measured in $S$. An observer in $S$ knows that this has happened because at time $t = L/c$, as measured in $S$, the light from the two bolts arrives at the origin of $S$. Thus, from the simultaneous observation of the bolts and the fact that they hit at positions $\pm L$, the observer in $S$ states that the bolts struck simultaneously.

But what about $S'$—the system of train? Without invoking the Lorentz transformations or anything else, it is quite clear that as the train is moving in the positive $x$-direction, the light from the bolt in that direction will arrive at the origin in the train before the light arrives from the other direction. Hence, the lightning strikes do not appear simultaneous to the observers in $S'$. Even a classical physicist would agree to this. A relativist using the Lorentz transformations, Eqs. (2–25) and (2–27), would be able to make a precise prediction. The observer in $S'$ would claim that the bolts hit at positions $L'_+ = \gamma L$ and $L'_- = -\gamma L$ at times $t'_+ = -\gamma\left[Lv/c^2\right]$ and $t'_- = \gamma\left[Lv/c^2\right]$, where the plus and minus signs refer to the forward and backward strike, respectively. The two primed times are unequal, so the observer in $S'$ does not see the strikes as simultaneous. Indeed, $t'_+$ is an earlier time than $t'_-$; that is what the minus sign means here. This scenario is plausible, since the train is moving toward the point $L'_+$.

What did Einstein see here that was missed by the classical physicists? Einstein realized that all measurements of time are really the simultaneous measurement of two events: the event in question and another event with which the first is to be correlated on a periodic system that we may call a "clock." In his 1905 paper on relativity, which was written within a few weeks after he had this insight, Einstein wrote, "We have to take into account that all our judgments in which time plays a part are always judgments of *simultaneous events*. If, for instance, I say, 'That train arrives here at 7 o'clock,' I mean something like this: 'The pointing of the small hand of my watch to 7 and the arrival of the train are simultaneous events.'" Since simultaneity is not an absolute concept, time itself will not be an absolute concept. That is why, in relativity, time transforms in a more complicated way than $t' = t$.

## Consequences of the Relativity of Simultaneity

It is not difficult to dream up apparent puzzles that can keep the neophyte relativist up for nights. We can best understand this statement with an example. Suppose a truck of proper length 10 m wants to fit into a garage of proper length 5 m (•Fig. 2–15). Can the truck move fast enough so that a garage attendant can use the Lorentz contraction to fit the truck in? What speed must this be? And

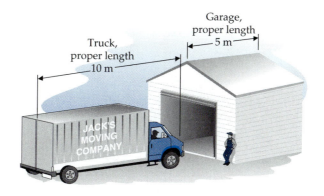

• **Figure 2–15** Can a truck of proper length 10 m long be shown to fit into a garage of proper length 5 m long by the use of the Lorentz contraction? How does this look in the rest frame of the truck?

how does the truck driver view the procedure? The way such problems are resolved is with careful analysis of what we mean by events and how these events are interpreted by different observers; whether two events are simultaneous or not typically plays an important role in the resolution of the problem.

At first sight, the situation is troubling. Let us call frame $S$ the rest frame of the garage and the observer in $S$ the garage attendant, while frame $S'$ is the rest frame of the truck and the observer in $S'$ is the driver. To the attendant, the truck has undergone a Lorentz contraction and could fit nicely if its speed were high enough. But to the driver, things seem to have gotten worse! The garage has undergone a Lorentz contraction and looks even shorter. However, we get into no trouble at all if we use only measurable events that both observers can agree on. Place the garage door at $x = 0$ and its back at $x = L(= 5 \text{ m})$. In the primed (truck) frame, place the front of the truck at $x' = 0$ and its back at $x' = -D(= -10 \text{ m})$. Now one "event" that driver and attendant agree on is the point at which the truck's front arrives at the door—we can always start both clocks at this event, which we place, accordingly, at $t = t' = 0$. But there is also a second event, the one such that the front of the truck arrives at the back wall of the garage; for this event, $x = L$ and $x' = 0$. If the truck is moving with speed $v$ with respect to the garage, then, according to the garage attendant, this second event occurs at time $t = L/v$. According to the truck driver, the time at which this second event occurs is, by the temporal Lorentz transformation [Eq. (2–27)],

$$t' = \gamma\left[t - \left(\frac{v}{c^2}\right)x\right] = \gamma\left(\frac{L}{v} - \frac{vL}{c^2}\right) = \sqrt{1 - \frac{v^2}{c^2}}\frac{L}{v}. \qquad (2\text{–}36)$$

In order to say that the truck fits into the garage, the attendant needs to be able to say that the rear of the truck is coincident with the door *at the same time* that the front is coincident with the back wall. Accordingly, let us describe another, third event, the one wherein the rear of the truck is coincident with the door. This event is described by $x = 0, x' = -D$ and occurs at a time $t$ described by the spatial Lorentz transformation [Eq. (2–25b)] such that

$$x' = -D = \gamma(x - vt) = -\gamma vt, \text{ or } t = \sqrt{1 - v^2/c^2}\,\frac{D}{v}.$$

Now, the garage attendant can claim that the truck fits into the garage if this time is exactly the time he measures for the occurrence of the second event, the coincidence of the truck's front with the rear wall of the garage, namely, $t = L/v$. By setting these times equal to each other we can determine the speed the truck must have to fit is the garage, according to the attendant. We have

$$\frac{L}{v} = \sqrt{1 - v^2/c^2}\,\frac{D}{v}, \text{ or}$$

$$\frac{v^2}{c^2} = 1 - \frac{L^2}{D^2} = 1 - \frac{5^2}{10^2} = 0.75.$$

But what about the puzzle of the driver's seeing the garage contracted? We have already seen that the driver sees the second event at a time given by Eq. (2–36). He sees the third event, however, at a different time! Equation (2–27) tells us that, for the driver, the time at which this third event occurs is

$$t' = \gamma\big[t - (v/c^2)x\big] = \gamma\big[L/v - (v/c^2) \times 0\big] = \gamma L/v.$$

This result resolves the problem: The driver's clock indicates that the time of the third event is later than that of the second event; that is, from the driver's point of view, the front and back ends of the truck do not arrive at the two ends of the garage simultaneously. He can indeed determine that the garage is Lorentz contracted, but such a measurement would involve a *simultaneous* measurement of the separation between the door and the back wall according to the driver's clock. In this example, we have seen that a simultaneous measurement in $S$ is not a simultaneous measurement in $S'$. Incidentally, we hope that the truck has good brakes!

The truck-and-garage problem of this subsection is one of a number of famous puzzles of this kind,[10] all resolved by similar arguments.

## 2–6  The Relativistic Doppler Shift and Relativistic Velocity Addition

### The Doppler Shift

Let us return briefly to Eqs. (2–34) and (2–35)—the equations for $t_y$ and $t_x$ respectively, that describe the arrival times in $S$ of a light pulse on what $S'$ measures to be a sphere. Suppose that every $T'$ seconds, as measured in $S'$, a source at rest in $S'$ emits a spherically spreading light pulse. This fact can be ascertained by standing at some distance from the source in $S'$, in any direction, and noting that a light pulse passes every $T'$ seconds. But to the observer in $S$, the frequency of passage of the light waves will depend upon where on the spherical surface one is. For example, if that observer is in the $x$-direction from the source of the wave, the period will be $t_x$ of Eq. (2–35) with $t'$ replaced by $T'$, while if the observer is in the $y$-direction, the period will be $t_y$ of Eq. (2–34). If we multiply $T'$ by $c$, we find $\lambda'$, the wavelength as measured in $S'$. If we multiply the corresponding $t_x$ and $t_y$ by $c$, we obtain the wavelength of the same set of pulses as measured in these directions in $S$. In other words, we have derived the relativistic Doppler shift in the two given directions! We forego deriving the exact expressions here, since, in the next chapter, we derive them in a somewhat more conventional way; we'll also show there that, in relativity, in contrast to the Doppler shift for sound, the situation in which the source is at rest and the observer in motion and the situation in which the observer is at rest and the source in motion are *perfectly*, not just approximately, symmetric.

Looking again at Eq. (2–34), we see that there is a Doppler shift in a direction at right angles to the motion. This shift is expressed in terms of the shift in the quantity $t_y$, which is a period like the tick of a clock. *Such a shift in the trans-*

---

[10] See, for example, E. F. Taylor and J. A. Wheeler, *Spacetime Physics*, W. H. Freeman and Co., New York (1992).

*verse direction is totally absent in classical physics:* as the expression shows, it is a pure time-dilation effect. In classical physics, the angular-dependent Doppler shift for the frequency $f$ takes the form, to order $v/c$, $f' = f(1 - \cos(\theta)v/c)$, and at $\theta = \pi/2$ there is no Doppler shift. In special relativity, the formula becomes

$$f' = f\gamma[1 - \cos(\theta)v/c]. \tag{2–37}$$

Here, $\theta = 0$ corresponds to the source and detector moving apart ($f' < f$), while $\theta = \pi$ corresponds to the source and detector approaching one another ($f' > f$). Note that for $\theta = \pi/2$, we find the result we got by using $t_y$. This so-called transverse Doppler shift was first observed in 1938 by H. E. Ives and G. R. Stilwell. The remarkable thing is that these experimenters—even in that year—did not believe in relativity! They were still ether physicists. Perhaps their own experiment helped to change their minds.

**Example 2–6**  At what speed would a motorist in a very fast car have to go so that he or she would see a red traffic light as green? We assume that the light looks red when the motorist is at rest. Take a wavelength of 650 nm for red light and 530 nm for green.

**Solution**  This is a case of the longitudinal Doppler shift with an angle $\theta = \pi$ (source and observer approach one another), for which the observed frequency $f'$ is related to the proper frequency $f$ by Eq. (2–37), namely,

$$\frac{f'}{f} = \frac{1 + v/c}{\sqrt{1 - v^2/c^2}} = \sqrt{\frac{1 + v/c}{1 - v/c}}.$$

Note that $f' > f$, as is appropriate if $f'$ corresponds to green and $f$ to red. Since the frequency is inversely proportional to the wavelength, the left side of this equation, in terms of wavelengths, is $\lambda/\lambda'$.

In this example, we know the ratio $\lambda/\lambda'$ and want to find $v/c$. So we square the equation and solve for $v/c$. We find that

$$\frac{v}{c} = \frac{(\lambda/\lambda')^2 - 1}{(\lambda/\lambda')^2 + 1} = \frac{(650/530)^2 - 1}{(650/530)^2 + 1} = 0.20.$$

In miles per hour (mph), $v = 0.20\ c = 0.20 \times (3 \times 10^5\ \text{km/s}) \times (0.6\ \text{mi/km}) \times (3600\ \text{s/hr}) = 130$ million mph. This motorist may not be arrested for running a red light, but he or she runs the risk of a speeding ticket!

### The Addition of Velocities

As we returned to the Doppler shift, let us now also return to Eq. (2–13), the Galilean addition of velocities. This equation gives the velocity $u'$, as measured in $S'$, of an object that moves with speed $u$, as measured in $S$, where the two frames $S$ and $S'$ move uniformly with velocity $v$ with respect to one another, as in •Fig. 2–6. The Galilean relation is

$$u' = \frac{dx'}{dt'} = \frac{dx'}{dt}\frac{dt}{dt'} = u - v. \tag{2–38}$$

At the same time, we have seen that the speed of light in vacuum has a very special place in Einstein's theory: It is the unique speed that has the same value in every uniformly moving coordinate system. Clearly, the Galilean addition of velocities is not compatible with the special role of the speed of light, because, applied to light, Eq. (2–38) implies that a material object can catch up to, and even pass, a light wave. For example, if $v = -0.51c$ and $u = 0.51c$, then, according Eq. (2–38), the object has a speed of $1.02c$ in $S'$.

In relativity theory, the Galilean transformations given in Eq. (2–12) are replaced by the Lorentz transformations of Eqs. (2–25) and (2–27), or, explicitly,

$x' = \gamma(v)(x - vt)$ and $t' = \gamma(v)\left[t - (v/c^2)x\right]$, respectively. When we apply these equations to the calculation of the rate of change of $x'$ in $S'$, we know that the second equality of Eq. (2–38), that $dx'/dt' = (dx'/dt)(dt'/dt)^{-1}$, is generally true—it is nothing more than the chain rule of calculus—but we now have (recall that $dx/dt = u$)

$$\frac{dx'}{dt} = \gamma(v)(u - v) \tag{2–39}$$

and

$$\frac{dt'}{dt} = \gamma(v)\left[1 - (v/c^2)u\right]. \tag{2–40}$$

Putting all three equations together, we have the *relativistic addition theorem for velocities,*

$$u' = \frac{u - v}{1 - (v/c^2)u}, \tag{2–41}$$

where $u' = dx'/dt'$ is the speed in $S'$.

This formula verifies the internal consistency of special relativity, reducing to the classical case when $v/c$ is small and very cleverly leaving the speed of light to be the maximum speed attainable. In particular, if we take $u = c$, we have

$$c' = \frac{c - v}{1 - (v/c^2)c} = c. \tag{2–42}$$

This equation is equivalent to the proposition that all uniformly moving observers will report the same speed of light in vacuum.

The verification of the logical consistency of special relativity leaves us free to move on to explore the consequences of the theory. This is the topic of the next chapter. We conclude the present chapter with a couple of examples.

**Example 2–7** A space platform is the staging area for two relativistic rockets. The first rocket leaves the platform in a given direction with a speed of $0.90c$. Later, a second rocket leaves the platform at a speed of $0.98c$ in the same direction as the first one. What is the relative velocity of the rockets?

**Solution** We want to be careful about what we mean by the relative velocity; once we have specified that, and properly identified the symbols in the relative-velocity formula, the problem becomes a matter of simple applications of that formula. The relative velocity in this case is the velocity an observer in the first rocket would measure for the second rocket. After solving the problem, you can redo it to verify that this relative velocity is the negative of that measured for the first rocket by an observer in the second rocket, as symmetry would suggest.

We can call the observer in the first rocket the $S$ observer and the observer in the second rocket the $S'$ observer. We are interested in finding the speed $v$ with which these two frames move, given that the speed of the platform as seen by the $S$ observer is $u = -0.90c$, and the speed of the platform as seen by the $S'$ observer is $u' = -0.98c$. In other words, in this use of the velocity addition theorem we are going to find $v$, given that we know $u$ and $u'$, rather than find $u'$, given that we know $u$ and $v$. This means that we want to invert Eq. (2–41) and solve for $v$. This inversion is simple—the equation is linear in $v$—and gives

$$v = \frac{u - u'}{(1 - uu'/c^2)}.$$

Inserting the values of $u$ and $u'$, we find, for the speed of the second rocket as seen from the first,

$$v = \frac{(-0.90c) - (-0.98c)}{1 - (-0.90c)(-0.98c)/c^2} = \frac{0.08c}{1 - (0.90)(0.98)} = 0.68c.$$

Note the positive sign: The second rocket is approaching the first, although not at the value $0.08c$ that the Galilean form would have given.

As a useful check that we have not made an error, imagine that the second ship is, instead, a light ray, so that the speed $u' = -c$. In that case, our formula gives

$$v = \frac{(-0.90c) - (-c)}{1 - (-0.90c)(-c)/c^2} = \frac{0.10c}{1 - (0.90)} = c.$$

This is the correct limit; like the observer on the platform, the observer on the first spaceship sees a light ray moving with speed $c$.

---

**Example 2–8**   Consider the pair of rockets of Example 2–7. Identical clocks are in each rocket and on the platform, all three of which are started at $t = 0$ when the first rocket leaves the platform. The second rocket leaves the platform at $t = 60$ s according to the platform clock. (a) At what time, in the frame of the platform, will the second rocket overtake the first one? (b) When this event takes place, what does the clock in the first rocket read? What does the clock in the second rocket read?

**Solution**   (a) From the platform's point of view, this is a straightforward problem in non-relativistic kinematics. The displacement of a rocket moving with velocity $v$ in a given time is the velocity times the time. Thus, if we label the first and second rockets with subscripts 1 and 2, respectively, the displacement of rocket 1 after time $t$ is $x_1 = v_1 t$, and the displacement of rocket 2 after time $t$ is $x_2 = v_2(t - t_0)$, where $t_0$ is the time delay before the second rocket is launched. In each case, the value of $x$ is 0 at the platform. The "overtake" event corresponds to the time $T_P$ of the platform clock for which $x_1 = x_2$, that is,

$$v_1 T_P = v_2(T_P - t_0),$$

an equation whose solution is

$$T_P = \frac{v_2 t_0}{v_2 - v_1} = \frac{(0.98c)(60 \text{ s})}{0.98c - 0.9c} = 735 \text{ s}.$$

As an alternative to this algebraic method, we can use a graphical method: •Figure 2–16 plots the positions of the two rockets as seen from the platform; the point where the two curves cross is the $x$-position where they meet. This is a simple check on the algebra.

(b) We can find the reading $T_1$ of the first rocket's clock by noting that when the second rocket at overtakes the first, the platform clock reads $T_P = 735$ s. But to the observer on the first rocket, this is a dilated time; to that observer, the platform clock runs slow by the factor $\gamma(v_1)$. Thus,

$$T_1 = \frac{T_P}{\gamma(v_1)} = T_P \sqrt{1 - v_1^2/c^2} = (735 \text{ s})\sqrt{1 - (0.9c)^2/c^2} = (735 \text{ s})(0.44) = 320 \text{ s}.$$

That $T_1$ is less than $T_P$ is reasonable, given the answer to the previous example, in which we saw that the observer on the first rocket sees the second rocket moving at a relative speed of $0.68c$, not the relative speed of $0.08c$ seen from the platform.

We can supplement this simple calculation with a second approach, one that uses the Lorentz transformations and a careful understanding of the events involved. The Lorentz transformations

$$x_1 = \gamma(v_1)(x_P - v_1 T_P) \text{ and } t_1 = \gamma(v_1)\left(T_P - (v_1/c^2)x_P\right)$$

give the relation between positions and times in the frame of rocket 1, labeled here with the subscript 1, and in the frame of the platform, labeled here with subscript $P$. The "event" in question is the overtaking of the two ships. It occurs at $x_1 = 0$ (the origin of the first rocket's coordinate system) and at $T_P(=735 \text{ s})$. These two pieces of information

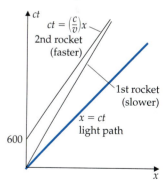

• **Figure 2–16**   The paths of the rockets as seen in the platform frame, plotted on a diagram of $ct$ vs. $x$ (not to scale).

can be used to solve for the remaining unknowns $x_P$ and $T_1$ in the Lorentz transformations. The left side of the first transformation is zero, so $x_P = v_1 T_P$ is the solution to part (a) of the problem. Once we know $x_P$, then, with $\gamma(v_1) = 2.29$, we can simply compute the right side of the second Lorentz transformation:

$$t_1 = \gamma(v_1)\left(T_P - (v_1^2/c^2)T_P\right) = (2.29)\left[(735 \text{ s}) - (0.9)^2(735 \text{ s})\right] = 320 \text{ s}.$$

For the time $T_2$ that the second rocket records, we can use the same reasoning, but must not forget that for the first 60 s the clock on the second rocket and the clock on the platform run in synchrony. Thus, employing the method we used to solve for $T_1$, we obtain

$$T_2 - (60 \text{ s}) = \frac{T - (60 \text{ s})}{\gamma(v_2)} = \left[T - (60 \text{ s})\right]\sqrt{1 - v_2^2/c^2}$$

$$= \left[(735 \text{ s}) - (60 \text{ s})\right]\sqrt{1 - (0.98c)^2/c^2}$$

$$= (675 \text{ s})(0.20) = 134 \text{ s},$$

and it follows that $T_2 = 134 \text{ s} + 60 \text{ s} = 194 \text{ s}.$

---

# QUESTIONS

1. Moving observers see the time of events such as the ticking of a clock differently from one another. The more rapidly the observer moves, the more slowly he or she sees the clock ticking. What kind of clock does this refer to? Is a person a clock? What properties of your body illustrate aspects of a clock?

2. Think about a moving vehicle that emits a light wave into a vacuum. Is there any way an observer in the vehicle could measure any property of that wave, other than the speed of light?

3. A person with a stopwatch stands on the platform of a train station marking the arrival of the head of the train at one end of the platform and the tail of the train at the other end of the platform. This is *not* a good operational way to measure the possible simultaneity of the arrivals of the head and tail in the rest frame of the platform. Why is that? What would be the right way to measure the possible simultaneity?

4. The velocity addition formula shows that one cannot "superpose" two velocities, each of which is greater than $c/2$, such that the result would be an object that one would see moving at a speed greater than that of light. But what about superposing *three* velocities, each of which is greater than $c/3$? Would that work?

5. We have noted that Maxwell suggested a method for measuring the speed of light on the moving Earth, namely, setting up parallel mirrors at a distance $L$ from each other and measuring the time it takes for light to make a round-trip from one mirror to the other and back. Suppose you did this experiment first on Earth and then on a spaceship moving uniformly with respect to Earth. Would you find a difference? Explain.

6. You have been given a superwatch accurate to a nanosecond. You want to set the watch so that it tells the "correct" time as measured by a friend who lives a thousand miles away and has an identical watch. You want to do this remotely—say, by telephone. How would you go about completing your task, and what physical assumptions would be involved?

7. Consider a wire carrying a current. Physically, this means that there are positively charged ions at rest in the laboratory, while electrons move with what is known as the "drift speed." The wire is electrically neutral, which means

that the charge density of the ions is the same as that of the electrons. Will the wire still look electrically neutral to an observer that moves along it, at rest with respect to the electrons?

**8.** Imagine a person rotating with angular velocity $\omega$ while holding a flashlight directed outward. A sequence of light detectors on a circle of radius $R$ is designed to flash upon the arrival of the light. If $R\omega$ were larger than $c$, something that could be easily arranged by making $R$ large enough, the sequence of flashes would propagate with a speed exceeding that of light. Does this scenario violate special relativity?

**9.** A rapidly moving muon (see Example 2–2), or indeed any unstable particle, travels farther than one would expect because to an observer at rest in the laboratory, the unstable particle's "clock" is time dilated. How would an observer moving with the particle explain the same effect?

**10.** A skeptic might argue that, by moving a clock, you have somehow altered its workings, which is why it reads slow. How could you convince such a person by producing the same effect while the clock is safely anchored to, say, Earth?

## PROBLEMS

**1.** ▌▌ Consider light propagating a distance $L$ from point $A$ to a mirror $M$ and back. What would the travel time for the round-trip be if there were an ether wind of velocity $v$ making an angle $\theta$ with the line $AM$? Show that your results reduce to the longitudinal and transverse results of the Michelson–Morley test when $\theta = 0°$ and $90°$, respectively.

**2.** ▌ At what speed would you have to move past a 10-cm ruler so that you would observe its length to be 5 cm?

**3.** ▌ A meter stick moves parallel to its axis with speed $0.96c$ relative to you. What would you measure for the length of the stick? How long does it take for the stick to pass you?

**4.** ▌ Find the speed relative to Earth of a uniformly moving spaceship whose clock runs 5 s slow per hour compared with an Earth-based clock.

**5.** ▌ A supernova at a distance of 175,000 light years from Earth emits particles that travel in a straight line to Earth. If these particles travel with a speed $v$ such that $v/c = 1 - 10^{-6}$, how long will the trip last, as measured by an observer traveling with the particle?

**6.** ▌ If a particle decays with a lifetime of a microsecond as measured in its rest frame, what lifetime would an observer moving past it with half the speed of light observe the particle to have? (For this problem, the lifetime of a particle can be considered to be the tick of a clock, as in the discussion preceding Example 2–2.)

**7.** ▌▌ A beam of muons—see Example 2–2—is injected into a storage ring, a device that uses electromagnetic fields to maintain the muons in uniform circular motion. The ring's radius is 60 m. Find the speed of the muons, as a multiple of $c$, that is needed so that $10^6$ revolutions are possible before the muons decay.

**8.** ▌▌ The radius of our galaxy is approximately $3 \times 10^{20}$ m. A spaceship sets out to cross the galaxy in 25 years, as measured on board the ship. With what uniform speed does the spaceship need to travel? How long would the trip take, as measured by a timepiece stationed on Earth?

**9.** ▌▌ Suppose that you wanted to test time dilation by taking a clock around the world on a commercial airliner. Assuming some reasonable speed for the airplane—1,000 km/hr, say—and some reasonable route, how accurate does your clock have to be to check the dilation formula to an accuracy of 5 percent? (In actuality, the effects of gravity are important here, as we shall learn in Chapter 17.)

**10.** ▌ Sally and Shelly are given identical meter sticks for their birthdays. Shelly gets on a spaceship that leaves Sally behind, moving at a speed of 0.5c relative to Sally. Shelly can ride her jet motorcycle at a speed of 0.1c relative to the ship, and does so, in a direction away from Sally. Shelly carries his meter stick with him, aligned with the motion. What, according to Sally, is the length of Shelly's meter stick?

**11.** ▌▌ Consider an apparatus for performing a Michelson–Morley experiment to measure the speed of sound in the laboratory. A sound wave of frequency 3,600 Hz replaces light. The speed of sound in air is 330 m/s. The arms of the interferometer are 2 m long, and the apparatus is placed in front of a large fan, which blows air along one of the arms at 8 m/s. Estimate the frequency of the beats that occur because of the interference of the waves reflected along the two arms of the interferometer. (*Hint*: Be careful! You are measuring the Doppler shift for sound for the case of a moving medium, and both the speed and the wavelength change.)

**12.** ▌▌▌ It is possible to derive the "transverse" Doppler shift in relativity theory by using nothing more complicated than the Pythagorean theorem, along with some of the basic principles of relativity. Consider a source separated from a receiver by a distance $d$. The time required for a signal to reach the receiver is $d/c$. Suppose now that the receiver is in motion at speed $v$ at right angles to the line between source and receiver. How much time does it take for a signal to reach you this time, and how does the motion of the receiver change the frequency it receives? The answer to this question will give you the transverse Doppler shift. What principles of relativity did you use?

**13.** ▌▌ Start with the expression $x'^2 + y'^2 + z'^2 - c^2t'^2$ and show, with the aid of the Lorentz transformations, that this quantity is equal to $x^2 + y^2 + z^2 - c^2t^2$. This result establishes the invariance of $s^2$ defined by Eq. (2–32).

**14.** ▌▌ Invert Eqs. (2–25) and (2–27) directly to find $x$ and $t$ in terms of $x'$ and $t'$.

**15.** ▌▌ Two relativistic rockets move toward each other. As seen by an observer on Earth, rocket $A$, of proper length 500 m, travels with a speed of 0.8c, while rocket $B$, of proper length 1,000 m, travels with a speed of 0.6c. **(a)** What is the speed of the rockets relative to each other? **(b)** The earthbound observer sets her clock to $t = 0$ when the two noses of the rockets just pass each other. What will the observer's clock read when the tails of the rockets just pass each other?

**16.** ▌▌ Consider the two relativistic rockets described in the previous problem. If the captain of rocket $A$, sitting near the nose of his rocket, sets his clock to $t = 0$ when the two noses pass each other, what will his clock read when he passes the tail of rocket $B$?

**17.** ▌▌ Consider the situation described in the previous problem. If the captain of rocket $B$, sitting near the nose of her rocket, sets her clock to $t = 0$ when the two noses pass each other, what will her clock read when the tail of rocket $A$ passes the tail of rocket $B$?

**18.** ▌▌▌ A relativistic pole-vaulter holds a pole that is 16 ft long in his rest frame. He runs with the pole aligned in the direction of his motion with a speed such that $\sqrt{1 - v^2/c^2} = 1/2$. He approaches a shed that, in its rest frame, is 8 ft long. An observer at rest relative to the shed sees the pole as being only 8 ft long and arranges for gates at the two ends of the shed to slam shut as soon as the front of the pole reaches the far interior end of the shed. This observer sees the entire pole within the shed. On the other hand, the runner sees the shed as having a length of 4 ft and is worried that 12 ft of his pole will be amputated when the gates shut. Should he be worried? Answer this question by thinking carefully about just when each of the pole's ends arrives at the front and rear end of the shed in each of the two frames. Don't worry about what happens to the unfortunate pole-vaulter immediately after the gates shut; remember, this is just a thought experiment!

**19.** ▌ If you move toward an emitter of yellow light ($\lambda = 580$ nm) at half the speed of light, what wavelength would you observe? What would be the answer if the emitter moved toward you?

**20.** ▌ An observer on Earth sends light with frequency $1.2 \times 10^{15}$ Hz to a spaceship traveling with speed $0.8c$ away from Earth. What will be the frequency of the light observed on the spaceship?

**21.** ▌ The spaceship in the previous problem transmits the light received from Earth, at the frequency that is observed, to a spaceship traveling ahead of it, away from Earth, with speed $0.6c$ relative to it. What is the frequency of the light as seen at the second spaceship?

**22.** ▌▌ If the second spaceship in the previous problem were unaware of the existence of the transmitting "middleman" spaceship, its crew would interpret the frequency of the emitted standard frequency of $1.2 \times 10^{15}$ Hz as Doppler-shifted with a shift determined by the speed of the ship relative to Earth. How large would this speed have to be in order for it to agree with the observed frequency, as calculated in the previous problem?

**23.** ▌▌▌ Generalize the result of the previous set of problems. That is, consider the emission of radiation with frequency $f$ from Earth. Rocket $A$ traveling with speed $u$ observes a frequency $f'$ and transmits light with that frequency to rocket $B$, which is traveling with speed $v$ relative to rocket $A$. The frequency observed by rocket $B$ is $f''$. **(a)** What is $f'$? **(b)** What is $f''$? **(c)** Use the relation between $f$ and $f''$ to calculate the velocity of $B$ relative to Earth, and confirm that your result agrees with the formula for the addition of velocities.

**24.** ▌ Two spaceships approach each other. They are each viewed from Earth as having a speed half that of light. What is their speed relative to each other?

**25.** ▌▌ In Section 2–6, we found an expression for $dx'/dt'$ [Eq. (2–41)]. Show that, to an observer traveling with speed $v$, the transverse velocity $dy/dt$ will also be seen to be altered. (*Hint*: Calculate $dy'/dt' = (dy'/dt)/(dt'/dt)$, and use the Lorentz transformation law.)

## APPENDIX

This appendix contains a more conventional (and more formal) derivation of the Lorentz transformations than that given in the text, which assumed the Lorentz contraction. When Einstein wrote his paper on special relativity in 1905, he had not seen Lorentz's paper of 1904 in which the transformations first appeared. Lorentz's derivation was done in the specific context of Maxwell's equations, while Einstein's derivation was of a very general character. We shall reproduce here not the 1905 derivation, but a simpler one that Einstein gave in his more popular writing. As we shall see, the algebra is very simple, but the assumptions must be stated with care.

We start with the usual situation: a system $S$ and a second system $S'$ that is moving uniformly with a speed $v$ in the positive $x$-direction with respect to $S$.

We concern ourselves only with transformations of $x$ and $t$, since, as we mention in the text, the $y$- and $z$-coordinates are unchanged. For the sake of the discussion, we take $y = z = 0$. An "event" is described by the coordinates $(x, t)$ in $S$; our job is to find the coordinates $(x', t')$ in $S'$ of the same event.

**Assumption I**   The transformation between $(x, t)$ and $(x', t')$ is linear and homogeneous. In other words,

$$x' = \alpha x + \beta t \quad \text{and} \quad t' = \delta t + \varepsilon x,$$

where $\alpha$, $\beta$, $\delta$, and $\varepsilon$ are functions of $v$, but not of $x$ or $t$.

**Discussion**   By "linear," we simply mean that $\alpha$, $\beta$, $\delta$, and $\varepsilon$ are not functions of $x$ and $t$. For example, suppose that $\alpha$ were proportional to $x$. Then $x'$ would involve $x^2$, and simple uniform motion such as $x = ut$ would transform into something quite different from uniform motion, even though the two systems move uniformly with respect to one another. This is contrary to our experience—hence the assumption. By "homogeneous," we mean that there is no additive constant in the transformations. For example, we do not have $x' = \alpha x + \beta t + X_0$. The $S$ space–time origin $(x, t) = (0, 0)$ transforms into the $S'$ origin $(x', t') = (0, 0)$. We can always arrange our starting points to satisfy this condition.

**Assumption II**   As viewed from $S$, the origin of $S'$ (the point $x' = 0$) moves according to $x = vt$. Thus, we must have

$$x' = \gamma(v)(x - vt).$$

**Discussion**   The quantity $\gamma(v)$ is not determined by any of the assumptions made so far; that comes shortly.

**Assumption III**   The inverse Lorentz transformation—the transformation that takes you from the primed to the unprimed variables—for $x$ takes the form

$$x = \gamma(-v)(x' + vt').$$

**Discussion**   This assumption follows from the second assumption, together with the symmetry of the situation. The observers in $S$ believe themselves to be stationary, with the observers in $S'$ moving in the positive $x$-direction, while the observers in $S'$ believe *themselves* to be stationary, with the observers in $S$ moving in the negative $x$-direction.

**Assumption IV**

$$\gamma(-v) = \gamma(v).$$

**Discussion**   This assumption can be replaced by others from which the transformations can be derived. For example, an equivalent assumption is that a second coordinate system identical to the first, but with the positive and negative $x$-axes interchanged, leads to the same Lorentz transformation as in the first coordinate system. This assumption, which logically entails assumption IV, is a reasonable one; it merely states that physical results do not depend on the orientation of the coordinate axes. We prefer to simplify the discussion by directly assuming that $\gamma$ depends only on the magnitude of $v$.

**Assumption V**   If a light pulse obeying the equation $x^2 + y^2 + z^2 = c^2t^2$ (that is, the pulse is spherical) is generated at the common origin of $S$ and $S'$ at $t = t' = 0$, then an observer in $S'$ will also see a spherical light pulse, described by $x'^2 + y'^2 + z'^2 = c^2t'^2$.

**Discussion**   This assumption embodies the principle of constancy in making explicit the special character of light propagation in relativity. In the special case in which $y = z = y' = z' = 0$, the assumption reduces to the condition that $x = \pm ct$ implies $x' = \pm ct'$.

Let us proceed to use the foregoing assumptions. First, we derive the function $\gamma$. (We drop the explicit $v$-dependence here.) If $x$ and $x'$ are spatial points on the light pulses, we have

$$ct' = x' = \gamma(x - vt) = \gamma t(c - v) \qquad \text{(A–1)}$$

while

$$ct = \gamma(x' + ct') = \gamma t'(c + v). \qquad \text{(A–2)}$$

We solve Eq. (A–2) for $t'$:

$$t' = \frac{ct}{\gamma(c + v)}.$$

When we substitute this expression back into Eq. (A–1), we find immediately that

$$\gamma^2 = \left(1 - v^2/c^2\right)^{-1}.$$

We choose the positive sign of the square root to find $\gamma$ itself; the sign is chosen so that the Lorentz transformations reduce to the Galilean transformations in the limit of small speeds. Thus, we have derived the spatial Lorentz transformation, Eq. (2–25), viz.,

$$x' = \gamma(v)(x - vt),$$

for points on the light pulse. Since $\gamma$ does not depend on space or time, this is the spatial transformation law for any point.

The temporal transformation follows directly from Eq. (A–1). Solving that equation for $t'$ we find that

$$t' = \gamma t(c - v)/c = \gamma t(1 - v/c) = \gamma(t - vct/c^2)$$

for points on the light pulse. Since for these points the quantity $ct$ can be replaced by $x$, we have

$$t' = \gamma(t - vx/c^2).$$

This equation is indeed linear in $x$ and $t$, as assumption I requires; the equation represents the temporal transformation, Eq. (2–27), under all conditions, not just for points on the light pulse.

# Consequences
# of Relativity

In the previous chapter, we developed the general principles of the special theory of relativity. We noted that because the propagation of light from point to point always takes a finite amount of time, events that one observer claims to have taken place simultaneously will not appear to have occurred simultaneously to another observer moving uniformly with respect to the first. We also noted that lengths of moving objects will appear contracted and that moving clocks will appear to move slower to an observer at rest. Length and time in the "rest" system are called "proper" length and time, and any observer who brings a meter stick or clock to rest will report the same proper length and time as identical meter sticks and clocks which have remained at rest.

Now that we have some feeling for how relativity changes our worldview, we shall present a number of situations that show how the theory operates in the real world. Because we currently have measuring instruments of extraordinary precision and machines that accelerate particles to speeds comparable to that of light, we can no longer make sense out of what these instruments reveal without using Einstein's theory of relativity. Indeed, the theory has become part of the working knowledge of every physical scientist.

We shall describe two categories of physical application: kinematics and dynamics. In the case of kinematics, we do not worry about forces; we use general properties of space and time and the propagation of light. The relativistic Doppler shift is an excellent kinematical example. We also bring in the relations between energy and momentum. In the case of dynamics, we bring in the causes of motion and, in particular, explore what replaces Newton's law, $\vec{\mathbf{F}} = m\vec{\mathbf{a}}$.

## 3–1 Time Dilation and the Decay of Unstable Particles

As we shall discuss in more detail later in the book, there are unstable particles in nature—particles that "fall apart," or decay, into two or more other particles. Radioactive nuclei form one well-known set of examples. Such decays occur at a certain rate, according to the type of particle, following well-known rules. In particular, a set of unstable particles of a given type that are at rest will decay[1] according to the exponential-decay law

$$N(t) = N(0)e^{-t/\tau}, \tag{3-1}$$

where $N(t)$ is the number of particles present at time $t$, given that $N(0)$ is the number present at the initial time $t = 0$. (Note that we cannot predict when a *given* particle in the set will decay. Some will decay immediately and some only after a long time interval; Eq. (3–1) is a *statistical* statement.) The quantity $\tau$ in this equation is characteristic of the particular species of unstable particle. This quantity, which has the dimensions of time, determines how rapidly the population of particles as a whole dies away and it is what we call the particle's lifetime.

From the vantage point of an observer in $S$, the system in which the set of particles is at rest, a fraction of the sample, $N(t)/N(0)$, given by $\exp(-t/\tau)$ will remain at time $t$. Now suppose that at this very time a second observer moving with a velocity $v$ in the $x$-direction with respect to $S$ appears at the position of the sample and measures how much of it remains. To this observer, the time will be, not $t$, but $t' = \gamma(v)t$, where the $\gamma$-factor $\gamma(v) = 1/\sqrt{1 - v^2/c^2}$. But the fractional sample should be the same. If one observer says, for example, that half the sample has decayed, the other observer should agree. For this to be true, the quantity $\tau$ must also be time dilated. In other words, we must have

$$\tau' = \gamma\tau. \tag{3-2}$$

This condition will leave the exponential, and hence the fraction of particles remaining, invariant—the same as viewed by the two observers.

The preceding result presents us with the possibility of seeing the internal consistency of relativity from still another angle. Let us ask how time dilation manifests itself in a typical experiment involving these unstable particles. Take as an example the so-called **muon**, symbolized as $\mu$. This particle is in many respects a "heavy" electron. The electron has a mass of $9.109 \times 10^{-31}$ kg, while the muon is 206.77 times more massive. The muon is unstable and decays with a proper lifetime $\tau = 2.197 \times 10^{-6}$ s. The muon was first detected in cosmic rays—radiation that comes to Earth from outside the solar system. These rays are very energetic, and typically, the muons in them have a speed such that $v/c \approx 0.99$, or $\gamma \approx 7.1$.

We may first ask how far a muon with a $v/c$ of 0.99 will travel over its lifetime. We should point out that the muons that arrive here from outer space are born close to Earth in the decays of other particles that are part of the cosmic radiation (•Fig. 3–1). If relativity were not a factor, such a muon would travel a distance $\tau v$, on the average, before decaying, where $\tau$ is the lifetime and $v$ is the speed of the muon relative to Earth. Ignoring relativity, with $v/c = 0.99$, we have an average path length of

$$d = vt = (0.99 \times 3 \times 10^8 \,\text{m/s})(2.197 \times 10^{-6} \,\text{s}) = 6.53 \times 10^2 \,\text{m}.$$

But if we take time dilation into account, this equation must be multiplied by $\gamma$, giving

$$d = \gamma vt = (7.1)(0.99 \times 3 \times 10^8 \,\text{m/s})(2.197 \times 10^{-6} \,\text{s}) = 4.66 \times 10^3 \,\text{m}.$$

Thus, a muon detector could register muons at a position much farther from the point of their creation than a nonrelativistic treatment suggests—a distance $d = \gamma v\tau$, rather than the distance $v\tau$ that would be expected if there were no

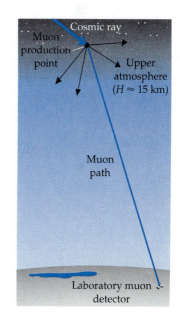

• **Figure 3–1**   High-energy muons are decay products of particles produced in high-energy collisions of cosmic rays with molecules of Earth's upper atmosphere. These muons travel distances on the order of 15 km to detectors on Earth's surface, which means that they live very much longer than expected from a calculation of their proper lifetime.

---

[1] We discuss decays in more detail in Chapter 13.

time dilation. It is the extension of the tracks left by unstable particles before they disintegrate that is one of the best, and indeed was one of the earliest, experimental confirmations of relativistic time dilation.

How does an observer in a coordinate system that moves with the muon account for this situation? To that observer, there is no time dilation. An observer at rest with respect to a decaying particle will report a proper lifetime of $\tau$ and will see the $S$ observer moving backwards at a speed $v$. The distance that the $S$ observer has reported to be $d$, the $S'$ observer will report to be the Lorentz-contracted distance $d\sqrt{1 - v^2/c^2}$, which is just $v\tau$ and is what one would expect to find, since the $S'$ observer moves a distance $v\tau$ with respect to the rest frame $S$ in the time $\tau$. The two descriptions are consistent if we use time dilation in one frame and length contraction in the other.

---

**Example 3–1**    A beam of muons is injected into a storage ring, a device that uses electromagnetic fields to maintain the muons in uniform circular motion. The ring's radius is 60 m, and the muons are injected with a velocity such that $\gamma = 15$. How many revolutions of the ring will an "average" muon make before it decays? The proper lifetime of a muon is $2.2 \times 10^{-6}$ s.

**Solution**    The time-dilation factor is expressed directly in terms of $\gamma$. In particular, we expect that in the frame of the laboratory the muons will last a mean time $\gamma\tau$, where $\tau$ is the proper lifetime. To find out how far the muons travel in this time, we also want their speed, which we can find from $\gamma$ itself. We have

$$\frac{v}{c} = \sqrt{\frac{\gamma^2 - 1}{\gamma^2}} = \sqrt{\frac{15^2 - 1}{15^2}} = 0.998.$$

The distance a muon travels in its time-dilated lifetime is then

$$x = \gamma v \tau = 15 \times (0.998)(3 \times 10^8 \text{ m/s})(2.2 \times 10^{-6} \text{ s}) = 9.88 \times 10^3 \text{ m}.$$

If we divide this distance by the circumference $2\pi R$ of our accelerator, we find the number of revolutions:

$$\text{number of revolutions} = \frac{x}{2\pi R} = \frac{9.88 \times 10^3 \, m}{2\pi \times 60 \, m} = 26.$$

---

In the preceding example, we used the time dilation for accelerated motion. Although generally this is not a valid use, for the special case of uniform circular motion the formula does apply.

## 3–2  The Relativistic Doppler Shift

Let us return now to the relativistic Doppler shift. We present here a conventional derivation, one we can use to understand some important features of the relativistic world. In particular, we show the symmetry between moving source and receiver.

The Doppler shift involves a wave source of definite frequency in the source's rest frame—the **proper frequency**—and an observer. The source and observer can move with respect to one another, and this relative motion leads to the observer's measuring a shifted frequency. We assume that the resting source emits a light wave with period $T$ and wavelength $\lambda$. We know that $\lambda$ and $T$ are connected by the equation $\lambda = cT$, where $c$ is the speed of light. Let us start with the Doppler shift for an observer and a source in relative motion along the $x$-direction. Relativity would suggest there is no way to distinguish whether the source or the receiver is moving, and we shall verify that this is the case by looking at both possibilities. If the observer and the source are at rest with respect to each other, and the wavelength of the light is $\lambda$, then it takes a time $T = \lambda/c$ for each wave crest to pass. But if the observer is moving away from the source with a speed $v$, there are two effects to be considered in computing the time $T'$ between crests. First, the observer will have moved away a distance $vT'$, and second, the original distance $\lambda$ will have Lorentz contracted to $\lambda\sqrt{1 - v^2/c^2}$. Putting these effects together, we see that the total distance the light will travel between pulses is given by

$$cT' = vT' + \lambda\sqrt{1 - v^2/c^2}.$$

Replacing $\lambda$ by $cT$ and then solving for $T'$ in terms of $T$, we find that

$$T' = T\sqrt{\frac{1 + v/c}{1 - v/c}}. \tag{3-3a}$$

Note that since the frequency $f$ is the inverse of the period $T$, Eq. (3-3a) also gives us an expression for the shifted frequency, namely,

$$f' = f\sqrt{\frac{1 - v/c}{1 + v/c}}. \tag{3-3b}$$

Now suppose that the source rather than the observer is in motion; again, the two are moving apart. Classically, we would argue that since a crest is emitted every $T$ seconds, the light would travel a distance $cT' = cT + vT = (c + v)T$ to arrive at the observer. But according to relativity theory, the $T$ seconds will be time dilated to $\gamma T$ seconds. Thus, the distance the light travels, $cT'$, is given by

$$cT' = \gamma T \times (c + v),$$

*which once again gives Eq. (3-3).* We can see that this is the case by direct calculation:

$$cT' = \gamma T(c + v) = cT\frac{1 + v/c}{\sqrt{1 - v^2/c^2}} = cT\sqrt{\frac{1 + v/c}{1 - v/c}}.$$

All this should be contrasted with the classical Doppler shift for sound. There, as we have pointed out, a definite preferred frame exists: that of the frame at rest with respect to the medium that carries the wave. The classical Doppler shift for sound gives different results for the observer versus the source in motion. In relativity, the two situations are perfectly symmetrical.

Here we can mention a practical application: *Doppler radar.* This term is used for radar systems that not only detect the time delay of reflected radiation in order to measure distance, but, in addition, detect any frequency shift in the reflected radar beam. Such a shift describes, by the Doppler mechanism, the movement of the object that is being detected.

**Example 3-2**    The most distant objects known in the universe are the so-called quasars. Their light is highly redshifted, a fact that is usually taken to mean that quasars are receding from us rapidly. A measure of this redshift is given by a parameter $Z$, defined in terms of the wavelength $\lambda$ of a known spectral line emitted by the quasar and the wavelength $\lambda'$ that we observe this line to have. Specifically, $Z = \lambda'/\lambda - 1$. Quasars have been observed with $Z = 4.9$. Apply the relativistic Doppler shift to find $v/c$ for such a quasar.

**Solution**    This is a case of the Doppler shift wherein source and receiver recede from one another. Then $T = \lambda/c$ is the period of the light wave, and we can find $\lambda/\lambda' = T/T'$ from $Z$, where upon we can invert the Doppler shift formula, Eq. (3-3), to find $v/c$. We have, from that equation,

$$\left(\frac{T'}{T}\right)^2 = \left(\frac{\lambda'}{\lambda}\right)^2 = \frac{1 + v/c}{1 - v/c},$$

which is a linear equation that is easily solved for $v/c$:

$$\frac{v}{c} = \frac{(\lambda'/\lambda)^2 - 1}{(\lambda'/\lambda)^2 + 1} = \frac{(Z + 1)^2 - 1}{(Z + 1)^2 + 1} = \frac{5.9^2 - 1}{5.9^2 + 1} = 0.94.$$

These quasars are highly relativistic.

### The Traveling Twins

The relativistic Doppler shift allows us to study and understand a famous co-nundrum, the so-called traveling-twins problem. This problem has been the subject of an immense amount of commentary ever since Einstein first invent-ed it. We suppose we have twins, one of whom sets out on a voyage. Which twin ages more, the one that travels or the one that doesn't? For our purposes, we can think of the twins as identical harmonic oscillators that can continually send out and receive pulses of radiation. Their "ages" are simply a measure of the number of pulses they have emitted. We know by now that in specifying the properties of these oscillators, we must be careful to state whether they are in motion with respect to an observer or whether they are at rest. We shall assume that each oscillator has the same proper frequency $f_0$ when it is measured at rest in its own frame of reference. Now we send one of the oscillators—the travel-ing twin, say, twin $B$—on its trip. The trip is specified in terms of the rest sys-tem of the other oscillator—the stay-at-home twin, or twin $A$—by the following five sequential steps:

1. The oscillator $B$ accelerates to a speed $v$ in a negligible time.
2. It moves away at this constant speed for a long time.
3. It reverses its direction in a negligible time.
4. It returns to the starting point at the same constant speed $v$.
5. It stops in a negligible time.

The stay-at-home twin sees the traveling twin's clock running slow, by time dilation, and would claim that the traveling twin returns having aged less. But we have emphasized the symmetry of relativistic effects. Why can't the travel-ing twin claim that in fact *he* is at rest, while the stay-at-home twin is moving? From this point of view, it is the stay-at-home twin who will have aged less. Which one ages less? This is a question that must have a definite answer, be-cause when the twins get together at the end, they can literally compare their ages! What is wrong with this superficial analysis is that the situations of the twins simply are *not* symmetric.

We assume that throughout the trip each oscillator is continually emitting radiation at its natural frequency $f_0$. We first analyze the problem from the standpoint of the stay-at-home twin $A$. From this point of view, neglecting the small distances traveled during the accelerations, the total distance of the trip is given by $2L_0$, where $L_0$ is the distance, as measured by twin $A$, from the start-ing point to the turnaround point. Thus, twin $A$ will claim that the trip lasts—again neglecting the small acceleration terms—a time $T = 2L_0/v$. During this time, twin $A$ sends out a total number of oscillator beats given by

$$N_{\text{sent}} = f_0(2L_0/v). \tag{3–4}$$

If we want to be fanciful about the situation, we can say that $N_{\text{sent}}$ represents the number of "heartbeats" of the stay-at-home twin. This is the "aging" the stay-at-home twin experiences while his traveling twin is gallivanting about space.

Twin $A$ is also constantly receiving pulses from the traveling twin $B$. The number of *these* pulses represents the aging factor twin $A$ would assign to the traveling twin. In computing this number, two effects will be important: the Doppler shift and the finite speed of light.

The reason that the Doppler shift enters is that part of the time twin $B$ is moving away, so that frequency of the beats he or she emits is Doppler shifted down, and part of the time twin $B$ is coming home, so that the Doppler shift raises the frequency. The finite speed of light comes in when we figure out how long each of the periods of shifted frequencies lasts. This works as follows: Radiation from twin $B$ is received by twin $A$ with the downshifted frequency $f'_- = f_0 \sqrt{\dfrac{(1 - v/c)}{(1 + v/c)}}$ until twin $B$ reaches the endpoint of the trip at $x = L_0$. But, because of the finite speed of light, it will take a time $t = L_0/c$ for this last bit of radiation to return to twin $A$. During this time, twin $B$ will have moved *backwards* a distance $v \times (L_0/c)$. It is only after twin $B$ has traveled this distance that twin $A$ will start to receive signals at the upshifted frequency $f'_+ = f_0 \sqrt{\dfrac{(1 + v/c)}{(1 - v/c)}}$. Therefore, during a time $t_- = L_0/v + L_0/c$, twin $A$ will receive beats with a frequency $f'_-$. Thus, the number of beats received during this time will be given by

$$n_- = f'_- t_- = \frac{L_0}{v} f_0 \sqrt{1 - v^2/c^2}. \tag{3–5}$$

The remaining number of beats will be received during a time $t_+ = L_0/v - L_0/c$ with a frequency $f'_+$. The number of these beats will be given by

$$n_+ = f'_+ t_+ = \frac{L_0}{v} f_0 \sqrt{1 - v^2/c^2}. \tag{3–6}$$

This number is the same as $n_-$, the shorter time having been compensated for by the greater frequency. The total number of beats received by twin $A$ is then given by

$$N_{\text{rec}} = n_+ + n_- = \frac{2L_0}{v} f_0 \sqrt{1 - v^2/c^2}. \tag{3–7}$$

This number is less than $N_{\text{sent}}$, form Eq. (3–4), which means that according to twin $A$, twin $B$ has aged *less*, a result that is in accord with time dilation. It is what we would get if we allowed twin $B$ to travel a time $2L_0/v$ as measured by twin $A$, but with an oscillator whose period has been time dilated by the usual factor $\gamma$.

Now that we have seen how twin $A$ analyzes the situation, let us consider the same problem from the point of view of the traveling twin $B$. As long as twin $B$ concludes from a study of the number of pulses he receives and sends that each twin has aged by the *same* amounts that twin $A$ has deduced, there will be no problem. How does the analysis go? We use primes to indicate quantities measured by twin $B$. From that twin's point of view, the Lorentz–Fitzgerald contraction has shortened the distance of the trip; thus, the duration of the voyage, neglecting the accelerations, is given by $T' = 2(L_0 \sqrt{1 - v^2/c^2})/v$. Thus, the total number of beats sent by twin $B$ is given by

$$N'_{\text{sent}} = \frac{2L_0}{v} f_0 \sqrt{1 - v^2/c^2}. \tag{3–8}$$

This is the age that twin $B$ assigns to himself at the end of the trip, and it is exactly the same as the age that twin $A$ assigned to him from Eq. (3–7). So far, so good—the twins agree.

Finally, we must ask how many beats twin $B$ receives from twin $A$. This number is the age twin $B$ will assign to twin $A$. For that analysis, we note that as soon as twin $B$ turns around, the pulses he receives from twin $A$ will change their frequency due to the Doppler shift. There is no time delay. Thus, the trip will be divided into two parts of equal length characterized by different received frequencies. The number of beats received by twin $B$ from twin $A$ on the way out is

$$n'_+ = \frac{L_0}{v}\sqrt{1 - v^2/c^2} \times f_0\sqrt{\frac{1 - v/c}{1 + v/c}}, \tag{3–9}$$

while the number of beats received by twin $B$ from twin $A$ on the way home is

$$n'_- = \frac{L_0}{v}\sqrt{1 - v^2/c^2} \times f_0\sqrt{\frac{1 + v/c}{1 - v/c}}. \tag{3–10}$$

Hence, the total number of beats received—the age twin $B$ assigns to twin $A$—is

$$N'_{rec} = \frac{L_0}{v}f_0\sqrt{1 - v^2/c^2}\left\{\sqrt{\frac{1 + v/c}{1 - v/c}} + \sqrt{\frac{1 - v/c}{1 + v/c}}\right\}$$

$$= \frac{2L_0}{v}f_0. \tag{3–11}$$

This result is identical to the age the stay-at-home twin assigns to himself from Eq. (3–4). The twins have arrived at the same conclusion.

The more superficial reasoning at the beginning of the discussion assumed that the two twins were in symmetrical situations. But they are not: One undergoes acceleration and the other does not; there is no symmetry in the situation of the two twins.

While this result has never been tested by putting an actual twin in a spaceship, we see that it is actually a matter of comparing any pair of identical clocks. In the purest test, we use as a clock a set of unstable subnuclear particles, whose lifetime can be regarded as the tick of a clock. Pairs of identical unstable particles can be produced, one of which is constrained to remain at rest while the other is speeded up as it travels around a particle accelerator. The lifetimes of two such sets of particles are compared, and the results confirm Einstein's analysis: The traveling particle lasts longer before decaying.

**Example 3–3**   Scientists have two clocks that keep time accurately enough so that a difference of $10^{-9}$ s is detectable. They wish to perform an experiment in which one of the clocks is put on a train that can go 200 km/hr on a long straight track, while the other clock sits at the station. The train goes out $x$ kilometers, then quickly stops, and comes back at the same speed along the same track. Ignore the (short) periods of acceleration, and find the minimum value of $x$ for which there will be an observable difference in the time elapsed on the clocks when they are compared at the end.

**Solution**   As we have seen in our discussion of the twins problem, the clock on the train will arrive back from its round-trip having run slowly by an amount of time that corresponds to a standard time-dilation factor. Thus, if the time for the train's round-trip is $t$ (approximately the same time for both clocks), the standard clock will have run for time $t$, but the clock on the train will have run for a time $\gamma t$. The time *difference* between the clocks will then be

$$\Delta t = t[\gamma - 1].$$

The factor $v^2/c^2$ is small, so we can use the small $\delta$ approximation $1/\sqrt{1-\delta} \approx 1 + \delta/2$ and write $\gamma \approx 1 + v^2/2c^2$. This gives us

$$\Delta t = t(v^2/2c^2).$$

Finally, if $x$ is the distance out, the time $t$ is $2x/v$, so that the relation between $\Delta t$ and $x$ is

$$\Delta t = xv/c^2, \text{ or } x = (\Delta t)c^2/v.$$

The last relation allows us to find $x$ given a minimum observable time difference of $10^{-9}$ s. With a train speed of $(200 \text{ km/hr})(1/3600 \text{ hr/s}) = 0.055$ km/s, we have

$$x = [(10^{-9}\text{ s})(3 \times 10^5 \text{ km/s})^2]/(0.055 \text{ km/s}) = 1.6 \times 10^3 \text{ km}.$$

The scientists will have to use the American Great Plains to perform the experiment!

## 3–3 Mass, Momentum, and Energy

Several times, we have indicated that the speed of light is the maximum speed that is attainable in the universe. The appearance of quantities like $\sqrt{1 - v^2/c^2}$ reinforces this idea, because Eq. (3–3) for the Doppler shift, for example, leads to an *imaginary* number if $v > c$, to say nothing of the fact that when $v = c$ the equation becomes infinite! But, you may ask, can't we simply apply a force and accelerate an object so that its speed exceeds that of light? In fact, it is not possible to accelerate a massive object to (and beyond) the speed of light, according to special relativity. This is because the effect of a force is to change the momentum of the object, and as we shall see, the relativistic momentum contains a factor $\gamma$. As $v \to c$, $\gamma \to \infty$, and a result, it becomes harder and harder to accelerate a massive object as it approaches the speed of light, so that such an object can never quite achieve that speed. An important exception to this rule is that it is in fact *necessary* that a massless object move *with* the speed of light. Indeed, light itself consists of massless objects that move with the speed of light: **photons**. There may be other massless particles as well, and if so, they, too, must always move with the speed of light—*they cannot be slowed down.*

### The Relativistic Momentum

Momentum $\vec{p}$ is a central quantity in mechanics. Momentum is conserved in collisions or, indeed, in any isolated system, and it is $d\vec{p}/dt$ that enters into the most general form of Newton's second law. (See Chapter 1 for a review of Newtonian mechanics.) Here, we shall give a definition of momentum that can be justified by thinking about the relativistic version of mass times velocity. This definition must be such that it reduces to $m\vec{v}$ when the velocity of the body in question moves at nonrelativistic speeds. Dimensional arguments then would suggest that relativistic momentum have the form $\vec{p} = m\vec{v}f(v/c)$, with $f(0) = 1$. Note that we have divided $v$ by $c$ in order to have a dimensionless argument: $c$ is the natural quantity with the dimensions of speed that occurs in relativity. We also expect that the sign of $v$ does not enter into $f$, so that $f = f(v^2/c^2)$. As to the exact form of $f$, Einstein's approach was to look to the transformation of the force, which is $d\vec{p}/dt$ and deduce $\vec{p}$ from that. In other words, knowing the transformation of the force, one can read off, from the equation $\vec{F} = d\vec{p}/dt$ the form the momentum must have. We work this out in the appendix to the chapter. The result is

$$\vec{p} = \gamma m\vec{v}. \tag{3–12}$$

There are various other ways of obtaining this result; in particular, this is the quantity that is conserved in collisions, as has been confirmed in countless experiments with subatomic particles.

## Work and Energy in Relativity

We can use our expression for the momentum and a generalization of Newton's second law in special relativity to develop the notion of relativistic energy. This is an immensely useful tool for relativistic problems, as we shall see later. We derive the relativistic energy by defining the work as in Newtonian mechanics and then defining the kinetic energy through the usual relation between work done and kinetic energy gained.

Recall that in Newtonian mechanics, if an object moves some distance $dx$ in one dimension under the sole influence of a force $F$, then the infinitesimal work done is

$$dW = F\, dx = \frac{dp}{dt}\, dx. \tag{3–13}$$

From this expression, the rate of doing work is

$$\frac{dW}{dt} = \frac{dp}{dt}\frac{dx}{dt} = \frac{dp}{dt}\, v, \tag{3–14}$$

where, to save writing, we call $dx/dt$, $v$. It is then convenient to change variables from $t$ to $v$. We can write Eq. (3–14) as

$$\frac{dW}{dv}\frac{dv}{dt} = \frac{dp}{dv}\frac{dv}{dt}\, v, \tag{3–15}$$

or

$$\frac{dW}{dv} = \frac{dp}{dv}\, v. \tag{3–16}$$

We now integrate both sides of this equation for a motion that, for simplicity, begins at rest and proceeds to a final speed that we will call $v_f$:

$$\int_0^{v_f} \frac{dW}{dv}\, dv = W(v_f) - W(0) = \int_0^{v_f} v\frac{dp}{dv}\, dv$$

$$= -\int_0^{v_f} p\, dv + pv\Big|_0^{v_f}. \tag{3–17}$$

In the last line, we have integrated by parts. At this point, we can use the *relativistic* expression for momentum, Eq. (3–12), so that

$$-\int_0^{v_f} p\, dv = -\int_0^{v_f} \gamma(v) mv\, dv = -mc^2(1 - \sqrt{1 - v_f^2/c^2})$$

and

$$pv\Big|_0^{v_f} = \gamma(v_f) mv_f^2 = \gamma(v_f)\frac{v_f^2}{c^2} mc^2.$$

In the last expression, we have multiplied and divided by $c^2$. With this result, we find the net work, namely, the sum of the two terms in Eq. (3–17):

$$W(v_f) - W(0) = mc^2\left(-1 + \sqrt{1 - v_f^2/c^2} + \frac{v_f^2/c^2}{\sqrt{1 - v_f^2/c^2}}\right)$$

$$= \frac{mc^2}{\sqrt{1 - v_f^2/c^2}}\left(-\sqrt{1 - v_f^2/c^2} + 1 - \frac{v_f^2}{c^2} + \frac{v_f^2}{c^2}\right)$$

$$= mc^2(\gamma - 1).$$

Now, by the Newtonian work–energy theorem, the net work done is the kinetic energy gained. We take this statement as the *definition* of the **relativistic kinetic energy**, which we call (•Fig. 3–2)

$$T = mc^2(\gamma - 1). \tag{3–18}$$

We see that $T$ vanishes when $v_f$ goes to zero, so that $T$ is really a *kinetic* energy.

If this definition of kinetic energy makes sense, it should reduce to the nonrelativistic form for values of $v_f$ that are small compared with $c$. To check, we use the small-$x$ approximation $1/\sqrt{1 - x} \approx 1/(1 - x/2) \approx 1 + x/2$, so that

$$T \approx mc^2\left\{1 + \frac{v_f^2}{2c^2} - 1\right\} = \frac{1}{2}mv_f^2, \tag{3–19}$$

exactly the nonrelativistic kinetic energy.

The relativistic kinetic energy is the difference of two terms, one of which we call $E(v)$ and the other $E(0)$; that is, $T = E(v) - E(0)$. It is natural to interpret the function $E(v)$ as the **total energy** a particle of mass $m$ would have if it were moving with speed $v$. The function $E$ evaluated at $v = 0$ is $E(0) = mc^2$, and it is natural to call this quantity the **rest energy**. The kinetic energy is then the change in energy of the particle as it is accelerated from rest to speed $v$:

$$E(v) = T + E(0) = T + mc^2. \tag{3–20}$$

The term $mc^2$ cancels on the right when we use Eq. (3–18). For simplicity, we drop the argument, so that

$$E = \gamma mc^2. \tag{3–21}$$

This form of $E$ allows us to connect the energy to the momentum. Using the definition of $p$, we have

$$E^2 = \frac{m^2c^4}{1 - v^2/c^2} = \frac{m^2v^2c^2}{1 - v^2/c^2} + m^2c^4 = p^2c^2 + m^2c^4. \tag{3–22}$$

(You can verify the third step by rationalizing and adding the two terms.) Thus, as •Fig. 3–3 shows,

$$E = \sqrt{p^2c^2 + m^2c^4}. \tag{3–23}$$

This expression allows us to consider a new possibility that does not exist in classical physics: We can have a particle with mass zero that nonetheless has a finite energy $E$. For these particles, Eq. (3–23) becomes

$$E = pc.$$

The results we have derived finally allow us to give an expression for the speed of a moving object in terms of momentum and energy. By taking the ratio of Eq. (3–12) to Eq. (3–21), we see immediately that

$$\frac{\vec{v}}{c} = \frac{\vec{p}c}{E}. \tag{3–24}$$

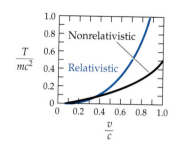

• **Figure 3–2**  The kinetic energy in units of $mc^2$, plotted as a function of $v/c$. The nonrelativistic curve corresponds to the Newtonian form $mv^2/2$.

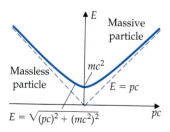

• **Figure 3–3**  The relativistic energy as a function of momentum for massive and massless particles.

This expression is often useful, representing a practical way of measuring the speed of a highly relativistic particle. The momentum and energy can be measured by a variety of means—for example, the curvature of the trajectory of a charged particle in a magnetic field is proportional to the particle's momentum—and their ratio provides a much more sensitive way to tell the difference between a speed of $.999c$ and $.9999c$ than does timed flight.

We notice from our expression for the speed that, for massless particles, $v/c$ is always unity. We can divide the world up into two kinds of objects: those with mass and those that are massless. The particles with zero mass are characterized by an energy $E$, and, as we stated earlier, they always move with a speed $v = c$.

---

**Example 3–4**    An electron is accelerated by a potential of 100 kV, starting from rest. Find the electron's speed after the acceleration, and compare your result with the speed the electron would have without the inclusion of relativistic effects. The mass of the electron is $m_e = 9.1 \times 10^{-31}$ kg, and the charge on the electron is $e = -1.6 \times 10^{-19}$ C.

**Solution**    The details of the acceleration do not interest us; we need only know that, when an electron passes through a potential difference $V$, it gains energy $eV$. We must include the mass term in the electron's energy, however. Thus, the final energy of the electron is

$$E = m_e c^2 + eV.$$

To find the electron's speed according to the relativistic equation (3–24), we also need the electron's momentum, which can be found from the momentum–energy relation of Eq. (3–22), which gives

$$p^2 c^2 = E^2 - m_e^2 c^4.$$

We now have, from Eq. (3–24),

$$v^2 = \frac{p^2 c^4}{E^2} = \frac{(E^2 - m_e^2 c^4)c^2}{E^2} = c^2\left(1 - \frac{m_e^2 c^4}{E^2}\right) = c^2\left(1 - \frac{m_e^2 c^4}{(m_e c^2 + eV)^2}\right).$$

As a last step, we divide both numerator and denominator of the second term by $m_e^2 c^4$ to obtain

$$\frac{v^2}{c^2} = 1 - \frac{1}{\left(1 + \dfrac{eV}{m_e c^2}\right)^2}.$$

When numbers are put into this expression, one finds that $v = 1.65 \times 10^8$ m/s, 55% of the speed of light. In comparison, the nonrelativistic expression $m_e v^2/2 = eV$ for the energy would give $v = 0.62c$. The relativistic effect acts to keep the speed down, as we expect.

It is worthwhile to give some thought to the size of the relativistic effect in this example. If we define $x \equiv eV/m_e c^2$, then our expression is of the form

$$v^2/c^2 = 1 - (1 + x)^{-2}.$$

The ratio $v/c$ depends, in other words, only on the size of $eV$ relative to that of $m_e c^2$.

When $eV \ll m_e c^2$, we expect to be in the nonrelativistic regime. We can verify this by noticing that, in this limit, $x$ is small. We can use a small-$x$ expansion as described in Appendix B–2, namely, $(1 + x)^{-2} \approx 1 - 2x$, which gives us, in this limit,

$$\frac{v^2}{c^2} \approx 1 - (1 - 2x) = 2x = \frac{2eV}{m_e c^2}.$$

When we cancel the factor $c^2$ on both sides, the result is

$$v^2 = 2eV/m_e.$$

But this is precisely the nonrelativistic expression that follows from $m_e v^2/2 = eV$.

The extreme relativistic limit, in contrast, would correspond to $eV \gg m_e c^2$. In this case, $x \to \infty$, and our expression for $v^2/c^2$ approaches 1.

Once we see that it is the ratio $eV/m_e c^2$ that determines the size of the relativistic effect, we can check the value of that ratio for the particular case presented in the example. In that case, $eV = (10^5 \text{ electron volts})(1.6 \times 10^{-19} \text{ J/electron volt}) = 1.6 \times 10^{-14} \text{ J}$, whereas $m_e c^2 = (9.1 \times 10^{-31} \text{ kg})(9.0 \times 10^{16} \text{ m}^2/\text{s}^2) = 8.2 \times 10^{-14} \text{ J}$. Thus, we expect the relativistic effect to be significant, but not dominant. This is in accord with the numerical results we obtained.

Incidentally, the foregoing discussion may suggest to you that the use of the Joule as a unit of energy may not be the most practical one. The next subsection gives units for a better set.

## The Equivalence of Mass and Energy

There is a beautiful and simple way, often attributed to Einstein, to see the equivalence of mass and energy. This derivation makes graphic just what the equivalence really implies. Einstein imagined a box that sits initially at rest on a frictionless surface (•Fig. 3–4). At the left end of the box is an emitter of electromagnetic radiation—a laser, for example. This radiation shoots across the box and is absorbed at the other end. Classical electromagnetism tells us that light carries both momentum and energy. Because of this momentum, light exerts a pressure. We can think of the radiation as being composed of massless particles—any object that moves with the speed of light must have no mass—carrying momentum to the right. These particles are, of course, the photons we mentioned earlier, and the relation between energy and momentum for them (or, for that matter, for *any* massless particle) is $E = pc$. Let us suppose that the photons are emitted in a burst at $t = 0$ and carry momentum $p$. To conserve momentum, the box, whose mass is $M$, must recoil. If $M$ is large enough, we can treat the momentum of the recoil nonrelativistically; that is, this momentum is $-Mv$, where $v$ is the recoil speed and the minus sign expresses the fact that the box recoils to the left. Because the total initial momentum is zero, the law of conservation of momentum gives us

$$|-Mv| = p = \frac{E}{c}.$$

Once the radiation reaches the right side of the box, it is absorbed, and the same calculation shows that the box comes to a halt.

Let us now ask what has happened to the center of mass during this procedure. There is no external force on the box—the burst of radiation involves an *internal* force only—so we expect the center of mass of the box with its contents to stay put. There is no reason to think that relativity will change this fact. On the other hand, we have seen that, because of the conservation of momentum, the box transports its mass to the left. If nothing compensates for this motion, the center of mass will move—a very unhappy situation. But what could

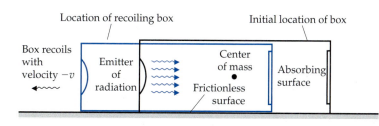

Location of recoiling box        Initial location of box

Box recoils with velocity $-v$ — Emitter of radiation — Center of mass — Frictionless surface — Absorbing surface

• **Figure 3–4** A box on a frictionless surface has a radiation emitter on one end and an absorbing surface on the other. When a burst of radiation is emitted, the box recoils, stopping only when the radiation is absorbed at the other end. The center of mass will not move if there is mass–energy equivalence.

compensate for the motion? The only thing one can imagine is that the radiation transports mass to the right! But, you say, the radiation consists of massless components. How can it transport mass? The radiation contains an energy $E$, and—you guessed it!—this energy must be associated with a mass $E/c^2$ to save the situation.

To demonstrate this argument in detail, we first ask how far the box moves before the radiation is reabsorbed at the other end. From the law of conservation of momentum, the speed of the box is $E/Mc$. The time it takes for the radiation to move from one side to the other is

$$t = L/c,$$

where $L$ is the length of the box.[2] Hence, the distance the box moves is

$$d = vt = v(L/c) = (E/Mc^2) \times L.$$

But the overall center of mass of the box plus the radiation cannot move, since all the forces are internal. This is possible only if the radiation acted like a mass $m$, traveling to the right the length $L$ of the box. The condition that the overall center of mass does not move is then

$$M \times \begin{pmatrix} \text{displacement of box} \\ \text{to the left} \end{pmatrix} = m \times \begin{pmatrix} \text{displacement of light} \\ \text{to the right} \end{pmatrix},$$

or

$$Md = M\left(\frac{E}{Mc^2}\right) \times L = mL.$$

When this equation is solved for $m$, we find that $m = E/c^2$. The energy of the radiation is, as promised, equivalent to a mass according to the Einstein formula $E = mc^2$.

### Observing the Rest Energy

As with all definitions in physics, the usefulness of our definition of energy can be decided only by its relevance to experiment. In other words, what are the consequences of saying that a particle at rest has a residual energy? The reason this energy was not noticed before Einstein had to do with the type of processes that 19th-century physicists studied. Typically, these were "billiard ball" collisions—collisions among atoms—in which the identity of the particles was essentially preserved. That meant that the rest energy was always balanced from one side of the equation to another. It was only at the end of the century that radioactivity, in which a particle spontaneously decays into other particles, was discovered. In *those* processes, the fact that the masses are different before and after decay makes the rest energy apparent.

Let us write the conservation-of-energy equation for a billiard-ball process of the type $A + B \rightarrow A + B$. We use the notation $T_X$ for the kinetic energy of a particle of type $X$ and the notation $m_X$ for its mass. Thus, the conservation-of-energy equation reads

$$T_A + m_A c^2 + T_B + m_B c^2 = T_A' + m_A c^2 + T_B' + m_B c^2. \qquad (3\text{--}25)$$

The important feature of this result is that the rest energies *cancel out* on the right- and left-hand sides of the equations. If we are concerned only with bil-

---

[2] If we wanted to be extremely accurate, we would take into account the fact that the box is moving towards the radiation, so that the distance it travels before hitting the right side is a little less than $L$. But the radiation moves so quickly that this little distance is tiny compared with $L$.

liard-ball collisions, the rest energies never show up and have no consequences. Contrast this with a decay of the form—to take a typical example—$A \rightarrow B + C + D$. Let us suppose that $A$ is at rest when it decays, so that $T_A = 0$. Now the energy-conservation equation reads

$$m_A c^2 = m_B c^2 + m_C c^2 + m_D c^2 + T_B + T_C + T_D. \qquad (3\text{–}26)$$

This means that the kinetic energy shared by the decay products is determined by the difference in rest energy, $(m_A - m_B - m_C - m_D)c^2$. Einstein understood this condition and proposed that it be used as a test of his then-new theory. Indeed, the theory is tested *and confirmed* each time a particle is observed to decay. As a matter of fact, until Einstein made this proposal, the source of energy in radioactive decays was a total mystery. It took some years before it was acknowledged that these decays were processes involving individual nuclei undergoing a spontaneous loss of mass.

## A Word about Units

While mass is measured in kilograms, these are not practical units for subatomic particles, whose masses may range from around $10^{-30}$ kg (the electron) to around $10^{-27}$ kg (the proton). It has become usual in the language that deals with such particles to give values of $mc^2$—the rest energy—rather than the masses themselves. The practical unit for these energies is the *electron volt*, abbreviated eV. One eV is the energy gained by an electron when it moves through a potential difference of one volt, so that $1 \text{ eV} = 1.6022 \times 10^{-19}$ J. In these units, the particles of subatomic physics have "masses"—rest energies—in millions of eV (MeV) or billions of eV (GeV). For example, an electron has a mass of roughly $9.1 \times 10^{-31}$ kg, so that its rest energy is

$$mc^2 = (9.1 \times 10^{-31} \text{ kg})(3.0 \times 10^8 \text{ m/s})^2 \left[ \frac{1 \text{ eV}}{(1.6 \times 10^{-19} \text{ J})} \right] = 0.51 \text{ MeV}.$$

**Example 3–5**   One process observed among subatomic particles is the decay of the so-called $\pi$-meson, a particle whose rest energy is 139.57 MeV, into a muon and a neutrino. The process is symbolized as $\pi \rightarrow \mu + \nu$ (•Fig. 3–5). We have already discussed the muon, which has a rest energy 206.77 times that of the electron, namely, 105.45 MeV, and which decays with a lifetime of $\tau_\mu = 2.197 \times 10^{-6}$ s. The neutrino is a particle that has a mass much smaller than the other masses entering into the process, and we can treat it as massless.

**(a)** Assuming that the $\pi$-meson decays at rest, what is the momentum of the muon?

**(b)** What is the average track length made by the muon in a detector?

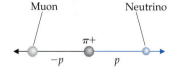

• **Figure 3–5**   A charged pion decays into a muon and a neutrino. In the rest frame of the pion, the muon and the neutrino go off back-to-back, with equal and opposite momenta.

**Solution**   What must we know in order to solve this problem? We have already seen, in the discussion following Eq. (3–2), how far a muon with a given velocity can travel: It lasts longer than it would at rest, due to time dilation, and thus travels farther than its mean lifetime alone would indicate. In particular, we found that if the muon has a speed $v$, it travels a distance $d = \gamma v \tau_\mu$. We can express this relationship in terms of the muon's momentum $p_\mu$—a quantity that we can find from the information on the muon's decay—using $p_\mu = \gamma m_\mu v$. Thus, the mean travel distance is

$$d = \frac{p_\mu c}{m_\mu c^2} c \tau_\mu.$$

(Here we have multiplied and divided by $c^2$; we shall see that this maneuver simplifies the calculation.)

**(a)** Let us now turn to the problem of finding $p_\mu$. Since the $\pi$-meson decays at rest, conservation of momentum reads

$$0 = \vec{\mathbf{p}}_\mu + \vec{\mathbf{p}}_\nu.$$

This equation tells us that the neutrino and the muon emerge from the decay with equal and opposite momenta—momenta whose magnitude we call $p$. (There is no information about the direction of the momenta, but that does not matter.) To find $p$, we use the energy-conservation equation, which, from Eq. (3–23) and the observation that $m_\nu = 0$, reads

$$m_\pi c^2 = \sqrt{p^2 c^2 + m_\mu^2 c^4} + pc.$$

By isolating the square root on one side and squaring, we can solve immediately for $pc$:

$$pc = \frac{(m_\pi^2 - m_\mu^2)c^2}{(2m_\pi)} = \frac{(m_\pi^2 c^4 - m_\mu^2 c^4)}{(2m_\pi c^2)}.$$

Numerical substitution gives

$$pc = \left[(139.57)^2 - (105.45)^2\right]/(2 \times 139.57)\ \text{MeV} = 29.95\ \text{MeV}.$$

**(b)** We can now use the expression for a muon's average travel distance in terms of its momentum:

$$d = \frac{p_\mu c}{m_\mu c^2}\, c\tau_\mu = \frac{29.95\ \text{MeV}}{105.45\ \text{MeV}} (3.00 \times 10^8\ \text{m/s})(2.197 \times 10^{-6}\ \text{s}) = 187\ \text{m}.$$

If you had left out the time-dilation factor and just used the form $d = v\tau_\mu$, you would have found this same answer multiplied by a factor $(2m_\mu m_\pi)/(m_\pi^2 + m_\mu^2) = 0.96$.

## Binding Energy and Mass

Most of us are familiar with the idea that energy can be interchanged among its different forms: Mechanical energy can be converted to electrical energy, which can be converted to chemical energy, and so forth. One type of conversion is potential energy into kinetic energy and vice versa. We have now seen that mass is still another manifestation of energy, which means that *mass by itself is not conserved*, just as, for example, electrical energy by itself is not conserved. It is the *total* energy that is conserved.

To illustrate how this aspect of energy conservation shows up in bound systems, consider a planetary system consisting of a sun and a planet of masses $M$ and $m$, respectively. Because of the gravitational attraction between them, the two bodies form a bound system. In the study of gravitation in introductory physics courses, you learned that such a system is said to have negative energy. What is meant by this statement is that the energy of the system is less than the energy of the two bodies separated by a large distance, with both bodies at rest. Once we take the rest masses into account, the energies of the two separated bodies are $Mc^2$ and $mc^2$, respectively, so that the statement that the two bodies form a bound system translates into the statement that

$$E < Mc^2 + mc^2,$$

where $E$ is the energy of the bound system. We may write this relation in the form

$$E = Mc^2 + mc^2 - B,$$

where $B$ is a positive quantity that we call the *binding energy*—the minimum energy that must be added to the system to separate the bodies. The quantity $-B$ contains the effect of a negative potential energy plus a positive contribution of the kinetic energies of the two bodies moving in their orbits, but the former must outweigh the latter, since otherwise the two bodies would not be bound and would separate.

For many purposes, we may consider the bound system to be a single body. If the center of mass of our two-body system is at rest, that "single body" is at rest, and its energy may be written as $M_b c^2 = Mc^2 + mc^2 - B$. Dividing by $c^2$, we find, for the mass of the bound system,

$$M_b = M + m - B/c^2. \tag{3-27}$$

The right-hand side of this equation is less than $M + m$.

Although we have introduced the notion of the bound system in planetary terms, and using two objects bound together, it applies equally well to any collection of bodies bound together by attractive forces. The conclusion still applies: *The mass of a bound system is less than the sum of the masses of its constitutents.* The question then arises: Why do chemists refer to the conservation of mass? The answer is that, in everyday cases and, in particular, in atomic and molecular physics, the effects are very tiny, because the binding energy is very small compared with the rest energies of the particles involved. For example, to break up a mole of hydrogen molecules into two moles of hydrogen atoms takes an energy of $4.58 \times 10^5$ J. This is the binding energy per mole of hydrogen molecules, and the binding energy per molecule is then $B/N_A$, where $N_A$ is Avogadro's number. With $N_A = 6.02 \times 10^{23}$, we have

$$B = 7.6 \times 10^{-19} \text{ J.}$$

On the other hand, the mass of a hydrogen molecule is, to a good approximation, twice the mass of a proton, that is, $2 \times 1.7 \times 10^{-27}$ kg, and this multiplied by $c^2$ gives

$$\left(2 \times 1.7 \times 10^{-27} \text{ kg}\right)\left(3.0 \times 10^8 \text{ m/s}\right)^2 = 3.1 \times 10^{-10} \text{ J.}$$

Thus, in this case, the ratio $B/M_b c^2 = 2.5 \times 10^{-9}$, an effect too small to be measured directly.

---

**Example 3–6**    As we shall see in later chapters, the binding energy of a hydrogen atom, which is a bound system of a proton and an electron, is approximately $2 \times 10^{-18}$ J. By how much does the mass $m_H$ of such an atom differ from the sum of its constituent masses?

**Solution**    Equation (3–27) directly gives us

$$m_H - m_p - m_e = \frac{B}{c^2} = \frac{2 \times 10^{-18} \text{ J}}{\left(3 \times 10^8 \text{ m/s}\right)^2} \approx 2 \times 10^{-35} \text{ kg.}$$

This number is roughly $2 \times 10^{-5}$ of the electron mass and $10^{-8}$ of the proton mass. The effect is very small in atomic physics.

---

The preceding example shows that the relativistic mass deficit in atoms is too small to measure directly. The binding energy due to the Coulomb attraction is much too small, in comparison to the masses themselves (or, more precisely, to the terms $mc^2$ in the expression for the energy). Gravitational binding energies are even less significant. Do we know of a bound system in which the attractive forces are large enough to make the mass deficit measurable? The atomic nucleus, which is a bound state of protons and neutrons, is such a system. (See Chapter 15.) The forces that bind the protons and neutrons to form the nucleus are so large that the mass deficit in these systems is on the order of one percent. In discussing these masses, it is unnecessary to distinguish between atoms containing the nucleus and the nucleus itself, because the nucleus makes up almost all the mass of the atom.

To take an example, one variety of carbon nucleus is made up of six protons and six neutrons. The mass of a proton and the mass of a neutron are very nearly the same—roughly 1.01, in so-called atomic mass units. The mass of the carbon atom containing the nucleus we are interested in is 12.00 atomic mass units. (Indeed, this particular nucleus is used to define these units, which is a unit specialized to the study of the masses of atoms.) Thus, 12 times the mass of a proton or neutron is greater by one percent than the mass of the carbon nucleus in question.

## 3–4 The Uses of Relativistic Momentum and Energy

The expressions we developed in the previous section turn out to have important applications in nuclear and particle physics. For example, the technology of large accelerators involves relativistic quantities throughout. Because these accelerators are designed to take relativity into account, the fact that they actually work is one of the best general confirmations of relativity! In this section, we shall first take a detour to look at how momentum and energy appear in different inertial frames. Then we shall discuss relativistic particle production, an idea that has meaning only because of the equivalence between mass and energy.

### Transformations of Momentum and Energy

It is very useful to be able to use momentum and energy in different frames of reference—in other words, to be able to transform these quantities. Consider our usual frames $S$ and $S'$ moving with speed $v$ with respect to one another along their respective $x$-axes. The transformation rules for Newtonian physics are simple: The momentum of an object as seen in frame $S$ is $m\vec{u}$, where $\vec{u}$ is the velocity of the object as measured in frame $S$; the object's kinetic energy in $S$ is $mu^2/2$; and we have the simple velocity addition rule $\vec{u}' = \vec{u} - \vec{v}$, where $\vec{u}'$ is the velocity of the object as measured in frame $S'$. The momentum of the object in frame $S'$ is then $m\vec{u}'$, while its kinetic energy is $mu'^2/2$.

In special relativity, things are considerably more complicated, but we will be in for a pleasant surprise. We have seen how the quantities $x$ and $t$ transform under Lorentz transformations. We now show that the relativistic momentum $\vec{p} = \gamma(u)m\vec{u}$ and the relativistic energy $E = \gamma(u)mc^2$ transform like $\vec{x}$ and $t$, respectively. This relationship will simplify our work immensely. We henceforth concentrate on the longitudinal variables—the components along the direction of motion—and drop the boldface notation. We begin by reminding the reader of Eq. (2–41), the relativistic expression for $dx'/dt' = u'$:

$$u' = \frac{u - v}{1 - uv/c^2}. \tag{3–28}$$

Note the difference between $v$ and $u$: The former is the relative speed of the two frames, while the latter is the speed—without the prime, in frame $S$; with the prime, in frame $S'$—of the object whose energy we are interested in. We now write down a remarkable algebraic identity that the reader can verify (see Problem 17) by using Eq. (3–28) and multiplying out:

$$1 - \frac{u'^2}{c^2} = \frac{(1 - u^2/c^2)(1 - v^2/c^2)}{\left(1 - \dfrac{uv}{c^2}\right)^2}. \tag{3–29}$$

This relation makes it simple to derive the transformations of $E$ and $p$. Using it to replace the square-root factor in the $\gamma$-factor in the expression for $E'$, we find that

$$E' = \frac{mc^2}{\sqrt{1 - u'^2/c^2}} = \frac{mc^2(1 - uv/c^2)}{\sqrt{1 - v^2/c^2}\sqrt{1 - u^2/c^2}} = \gamma(v)\frac{mc^2(1 - uv/c^2)}{\sqrt{1 - u^2/c^2}}$$

$$= \gamma(v)\left(E - \left(\frac{v}{c}\right)(pc)\right). \tag{3-30}$$

Here we have been able to recognize the definitions of $E$ and $p$. Similarly, we can start with the expression for $p'$ and express it in terms of $E$ and $p$:

$$p' = \frac{mu'}{\sqrt{1 - u'^2/c^2}} = \frac{m(1 - uv/c^2)}{\sqrt{1 - v^2/c^2}\sqrt{1 - u^2/c^2}}\frac{(u - v)}{1 - uv/c^2}$$

$$= \frac{m(u - v)}{\sqrt{1 - v^2/c^2}\sqrt{1 - u^2/c^2}}$$

$$= \frac{p - (v/c^2)E}{\sqrt{1 - v^2/c^2}} = \gamma(v)[p - (v/c^2)E]. \tag{3-31}$$

This time we have used Eq. (3–28) for $u'$, as well as Eq. (3–29).

Now compare Eqs. (3–30) and (3–31) with Eqs. (2–25) and (2–27), the Lorentz transformation laws for time and space. We see that $E$ transforms like $ct$, while $pc$ transforms like $x$, where, here, $p$ corresponds to the momentum component along the $x$-direction. Also, the transverse components of the momentum transform like $y$ and $z$; that is, $p'_y = p_y$ and $p'_z = p_z$. The four quantities represented by $(E, c\vec{p})$ thus transform like the four quantities represented by $(ct, \vec{r})$. We can therefore conclude that the combination $E^2 - p^2c^2$, like the combination $c^2t^2 - r^2$, is invariant. But recall that $E^2 - p^2c^2 = m^2c^4$, a quantity that is indeed invariant. Everything holds together!

## A Useful Invariant

If we have two different particles with energies and momenta $(E_1, p_1)$ and $(E_2, p_2)$, respectively, then each of the energy and momenta transform as do the quantities $E$ and $p$. That means that their sums $E = E_1 + E_2$ and $p = p_1 + p_2$ also transform that way, by simple addition of the transformation equations. In turn, the quantity $s$ defined by

$$\left(E_1 + E_2\right)^2 - c^2\left(p_1 + p_2\right)^2 \equiv s^2 \tag{3-32a}$$

*is an invariant.* The physical meaning of $s$ is seen clearly in the frame in which $p_1 + p_2 = 0$. In that frame (which is the frame of the center of mass),

$$s = E_1 + E_2. \tag{3-32b}$$

The quantity $s$ is therefore the rest energy of the object formed when particles 1 and 2 collide and stick together. This result is easily generalized to more particles.

## Particle Production

We now make use of our results on momentum and energy. One of the important discoveries of modern physics, something for which relativity is absolutely essential, is the fact that new particles can be produced in collisions, subject to the satisfaction of conservation laws. We can have a reaction of the form $A + B \rightarrow C + D + E + \ldots$, where $A$, $B$, and so on stand for the particles in question. The conserved quantities include dynamical variables such as momentum and energy, as well as some other quantities, like electric charge. In general, if there is more rest energy in the final particles than in the initial particles, then the initial particles must have enough kinetic energy to allow the law of conservation of energy to be satisfied. People who design accelerators need to know how much kinetic energy is required to produce a given particle or particles. We shall not try to examine this issue in general, but we shall analyze it in detail in a special case that represents one of the very significant discoveries in recent physics: the discovery of the **antiproton**.

Let us start with a brief discussion of the antiproton itself and of the kind of reaction in which it can be produced. In 1931, the English theoretical physicist P. A. M. Dirac suggested that electrons have antiparticles, now dubbed positrons; these were discovered in 1932 by the American experimental physicist Carl D. Anderson. It was predicted that for *every* type of elementary particle, there would exist a type of "mirror" particle, its **antiparticle**. The proton would also have an antiparticle, called the antiproton, which we denote by the symbol $\bar{p}$. Dirac's suggestion grew out of his attempt to make quantum mechanics and relativity consistent. Much to his surprise—and indeed, everyone else's—this required a world in which every particle was accompanied by its antiparticle. For our purposes, we can think of an antiparticle as being like its particle, except that the electric charge is reversed in sign. Thus, the antiproton would have the same mass as, and be stable like, the proton. However, the antiproton would have charge $-e$ rather than $+e$, where $e$ is the magnitude of the charge on the electron (Why the proton and electron have charges of the same magnitude remains a deep mystery.)

Until 1954, when an experiment to detect antiprotons was begun at the University of California at Berkeley (•Fig. 3–6), no antiproton had been observed, although the antielectron—Anderson's **positron**—had been studied. It was of crucial importance to verify this idea of particle–antiparticle duality by finding an antiproton. How could one be made? It was known how to construct machines to accelerate protons to high kinetic energies, and these "projectile" protons could crash into hydrogen targets. In the interaction of the projectile protons with the stationary protons that composed the nuclei of the hydrogen, antiprotons could be produced. The predicted reaction with an initial state of two protons and the fewest number of particles in the final state, at least one of which was an antiproton, it turns out, is the reaction (•Fig. 3–7)

$$p + p \rightarrow p + p + p + \bar{p}. \tag{3–33}$$

It is easy to see that this is the minimum reaction involving protons and antiprotons that will still conserve electric charge—indeed, antiparticles must be produced in pairs with their associated particles to conserve charge and other similar quantities. You might be tempted to put a positron in place of one of the

• **Figure 3–6**   The Berkeley Bevatron, the accelerator at which the first laboratory-produced antiprotons were observed.

• **Figure 3–7**   A proton–antiproton pair is produced in the collision of two protons. (a) A schematic representation of the collision and its end products in the laboratory frame, in which the target proton is at rest. (b) A schematic representation of the collision in the center-of-mass frame, in which the protons in the initial state have equal and opposite momenta. The common magnitude of these momenta is such that the proton–antiproton pair, as well as the remaining two protons, all end up at rest in this frame. The protons are denoted by black circles, the antiproton by an open circle.

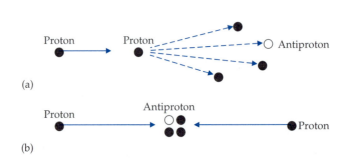

final protons, but that reaction is forbidden by an additional conservation law. The Berkeley scientists wanted the reaction involving the fewest number of particles produced, because that required the least energy and, therefore, the cheapest accelerator. We can now ask, Just what is the minimum energy of the projectile proton that will allow the reaction of Eq. (3–33) to proceed?

To answer this question, we employ a very useful technique: We first study the kinematics of the reaction in a frame of reference in which the answer is simple, namely, the center-of-mass frame. (In this frame, the two protons have equal and opposite momenta.) It is particularly easy to calculate the value of the invariant $s$ described in Eq. (3–32a) in the center-of-mass frame. By its invariant nature, the value of $s$ is the same in the laboratory frame as in the center-of-mass frame, and we can use this value to reverse the process and find the threshold energy in the laboratory frame. In what follows, we illustrate the process explicitly.

In the center-of-mass frame, the question of the minimum value of $s$ necessary to make the reaction given in Eq. (3–32a) possible corresponds to the situation in which all four final particles are produced *at rest*, a situation made possible by the fact that the total momentum in the center-of-mass frame is zero. In other words, all the incoming energy goes into mass energy in the final state. This means that the total energy is just $4mc^2$; that is, if we call the minimum, or "threshold," value $s_0$, then, from the generalization of Eq. (3–32b), we have

$$s_0 = 4mc^2.$$

This is the value of $s$ in the laboratory frame as well. In that frame, there is a target proton with the label 2 at rest, so that $E_2 = mc^2$ and $p_2 = 0$, and a projectile proton with the label 1 with energy $E_1$ and incoming momentum $p_1$ such that $p_1 c = \sqrt{E_1^2 - m^2 c^4}$, the general relation between momentum and energy in special relativity. We want to find $E_1$, and to do so, we use the invariance of $s$. We have

$$s^2 = \left(E_1 + E_2\right)^2 - c^2\left(p_1 + p_2\right)^2$$
$$= \left(E_1 + mc^2\right)^2 - p_1^2 c^2 = \left(E_1 + mc^2\right)^2 - \left(E_1^2 - m^2 c^4\right)$$
$$= 2mc^2 E_1 + 2m^2 c^4.$$

But when the left side is the square of the threshold value $s_0 = 4mc^2$, we have

$$\left(4mc^2\right)^2 = 2mc^2 E_1 + 2m^2 c^4,$$

or, solving for $E_1$,

$$E_1 = 7mc^2.$$

In turn, the kinetic energy of the projectile proton is

$$T_1 = E_1 - mc^2 = 6mc^2.$$

The accelerator, the so-called Berkeley Bevatron, was built to accelerate protons above this threshold. It produced antiprotons with exactly the expected properties.

Incidentally, today there exists a class of accelerators in which beams of particles with equal and opposite momenta are allowed to collide with each other—"colliding beam" accelerators. For such accelerators, the center-of-mass frame and the laboratory frame are one and the same. In 1954, these accelerators did not exist.

**Example 3–7**    Find the threshold energy for the process $\gamma + p \rightarrow \pi^0 + p$ in which a single $\pi$ meson, or pion (see Example 3–5), is produced when an energetic photon—a so-called gamma ray—strikes a proton at rest. The threshold energy is the minimum energy of the incoming photon. The rest energy of a neutral pion is 135 MeV, while the rest energy of a proton is 938 MeV.

**Solution**    By definition, the threshold energy is the energy at which the pion and proton are both at rest in the center-of-mass frame. In this frame, the quantity $s$ defined by Eq. (3–32a) has the threshold value

$$s^2 = \left(m_p c^2 + m_\pi c^2\right)^2.$$

But $s$ is an invariant, and we can use this fact by calculating $s$ in the initial state in the laboratory frame, the frame in which the proton target is at rest. Thus, if in this frame the momentum of the initial photon is $p_\gamma$, then its energy is $p_\gamma c$, and $s$ is given by

$$s^2 = \left(p_\gamma c + m_p c^2\right)^2 - c^2 p_\gamma^2 = 2m_p p_\gamma c^3 + m_p^2 c^4.$$

By equating our two expressions for $s$, we have a relation that can be solved to find $p_\gamma$:

$$p_\gamma c = \frac{2m_p m_\pi c^4 + m_\pi^2 c^4}{2m_p c^2} = \frac{2(938 \text{ MeV})(135 \text{ MeV}) + (135 \text{ MeV})^2}{2(938 \text{ MeV})} = 144.7 \text{ MeV}.$$

It is interesting that the first accelerators to produce pions did not quite need this amount of energy. The reason was that the protons were in atomic nuclei, and protons within nuclei move, supplying some energy.

---

In our discussion so far, we have concentrated on particle production. The conservation of energy and momentum, whether relativistic or not, puts constraints on processes that go beyond particle production, as the next example shows.

---

**Example 3–8**    From the point of view of quantum mechanics, light is carried by photons, symbolized $\gamma$. According to the quantum mechanical rules of electricity and magnetism, a photon can be absorbed by a charged particle such as an electron. Show, by applying the laws of conservation of energy and momentum, that a truly free electron cannot absorb a photon; that is, show that the process $\gamma + e \rightarrow e$ is not possible.

**Solution**    The freedom to work in a convenient frame of reference is a great help in a problem such as this, where there is an obvious frame to choose: the rest frame of the final electron. In this frame, the momentum of the final electron is zero. As for the initial state, if the electron approaches from the right with momentum $-p_e$, and if the photon has momentum $p_\gamma$, then conservation of momentum gives

$$p_\gamma - p_e = 0,$$

or

$$p_e = p_\gamma.$$

Now we attempt to apply conservation of energy. The energy of the initial photon is $E_\gamma = p_\gamma c$, while the energy of the initial electron is $\left(p_e^2 c^2 + m_e^2 c^4\right)^{1/2} = \left(p_\gamma^2 c^2 + m_e^2 c^4\right)^{1/2} = \left(E_\gamma^2 + m_e^2 c^4\right)^{1/2}$. Since the energy of the final electron in this frame is $m_e c^2$, the energy-conservation equation reads

$$E_\gamma + E_e = m_e c^2,$$

or

$$E_\gamma + \left(E_\gamma^2 + m_e^2 c^4\right)^{1/2} = m_e c^2.$$

But $E_\gamma$ is a positive quantity, so the left side of this equation is larger than $m_e c^2$, and the equation has no solution for $E_\gamma$. Only if the electron is bound or can otherwise give up either energy or momentum to another system can both energy and momentum be conserved in this type of process.

# *3–5  Forces in Relativity

In the domain in which it is applicable, Newton's second law,

$$\vec{F} = m\vec{a}, \tag{3–34}$$

or, more generally,

$$\vec{F} = \frac{d\vec{p}}{dt}, \tag{3–35}$$

governs motion. Here, $\vec{p}$ is the momentum, given in Newtonian mechanics by $\vec{p} = m\vec{v}$. Equations (3–34) and (3–35) constitute one of the most successful scientific laws ever formulated. It accounts for motions of objects ranging from air molecules to nearby celestial bodies. When speeds become comparable to $c$, we no longer expect the law to hold as stated, if only because it implies that a massive object can be accelerated to any speed at all. The relativistic modification must somehow prohibit accelerations of massive objects to the speed of light and beyond.

As we mentioned earlier, Einstein's 1905 theory of relativity applies only to non-accelerated motions. That is why it is called the special theory of relativity. (The general theory of relativity, which Einstein published in 1916, deals with arbitrary motions, including accelerations—see Chapter 17.) Thus, when we want to deal with accelerations in the special theory of relativity, we have to resort to ingenuity and approximations. For example, when we treated the "traveling twins," we made the approximation that the accelerated motion took place in a negligible time. In order to find the relativistic equivalent of Newton's second law, we use a different device, one that enables us to take advantage of what we know about Newtonian mechanics: We consider the motion at some time $t'$ as measured in a frame of reference $S'$ with respect to which the moving mass is *instantaneously* at rest. At any instant $t'$, we can always find such a frame.

At the instant $t'$, the particle is at rest in $S'$, and there is no question of anything moving at speeds approaching $c$ or, indeed, at any speed at all. Thus, we assume that Newton's law holds at time $t'$ in the system $S'$:

$$\vec{F}' = m\vec{a}'. \tag{3–36}$$

Here, $\vec{F}'$ and $\vec{a}'$ are the force and the acceleration, respectively, as measured at time $t'$ in $S'$.

The transformation of the left side of Eq. (3–36) to a system $S$ with respect to which $S'$ is moving with a speed $v$ gives the force as seen in $S$; the transformation of the right side of Eq. (3–36) is like the transformation of velocities, but somewhat more involved. For that reason, we present it in the appendix to the chapter and give only the result here. We restrict ourselves to motions along the $x$-axis only. As is shown in the appendix,

$$F = \frac{dp}{dt}, \tag{3–37}$$

where

$$p \equiv \gamma m v \tag{3–38}$$

and $F$ is the force in the transformed frame of reference. We see the presence of a factor $\gamma$ that grows as $v$ approaches $c$, making a given force less "effective" at accelerating the particle as the particle's speed approaches $c$. As with the non-relativistic form of Newton's law, the relativistic generalization we have derived here is useful only if we have independent information on how the force itself is given in terms of the appropriate coordinates.

Note that the momentum is identical to the momentum that we had written on a heuristic basis, Eq. (3–12). The appendix puts this expression for the momentum on a more solid footing.

### Constant Force

A special and very instructive case is that of a *constant force*. Suppose that a force of constant magnitude $F_0$ acts in the $x$-direction on an object whose mass is $m$, starting from $t = 0$, when the object is at rest. We can use Eq. (3–37) to find the speed of the object as seen from its original rest frame as a function of time; in particular, we shall see that the speed of the object cannot exceed $c$.

We want to solve the expression

$$F_0 = \frac{d}{dt}(\gamma(u)mu) \tag{3–39}$$

for the object's speed $u$, where we have used the notation $dx/dt = u$, with $F_0$ constant. We can integrate both sides of this equation with respect to $t$. Since $F_0$ is a constant, we obtain

$$(F_0 t)/m = \gamma(u)u + C. \tag{3–40}$$

Here, $C$ is a constant of integration, and it must be determined by the initial conditions; if we take $u = 0$ at $t = 0$ (i.e., the object on which the force acts starts from rest), then $C = 0$. We can now solve for $u$. By squaring both sides Eq. (3–40), we get

$$u = \frac{(F_0 t/m)}{\sqrt{1 + \dfrac{F_0^2 t^2}{m^2 c^2}}}. \tag{3–41}$$

We have plotted this expression in •Fig. 3–8. It has the property that as $t \to \infty$, $u \to c$. Thus, the velocity of the particle will never exceed that of light. For short times, Eq. (3–41) reduces to the familiar Newtonian expression.

We note that in Newtonian mechanics a constant force leads to a motion with a speed that increases linearly with time $t$, a speed that would eventually surpass the speed of light and would indeed approach infinity as $t$ ap-

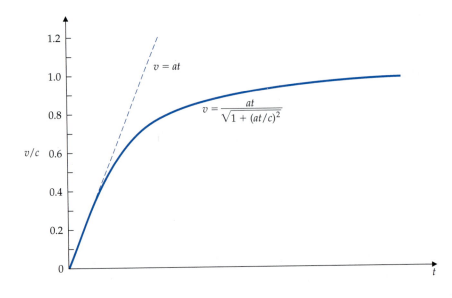

• **Figure 3–8** The speed of a particle subject to a constant force, as a function of time. The parameter $a = F_0/m$. Note that as $t \to \infty$, the curve approaches, but does not reach, $c$.

proaches infinity. The next example finds the position of the object as a function of time. The answer is an interesting one.

---

**Example 3–9** Using Eq. (3–41), which relates the velocity and the time for a constant force with magnitude $F_0$, find the position of an object of mass $m$ on which the constant force acts as a function of time. Examine your answer in the extreme relativistic and non-relativistic limits.

**Solution** If we integrate Eq. (3–41) with respect to time, the left side is the position. The integral of the right side over time can be done exactly. Assuming that $x = 0$ at $t = 0$, a table of integrals gives us

$$x(t) = \frac{mc^2}{F_0} \left( \sqrt{1 + \frac{F_0^2 t^2}{m^2 c^2}} - 1 \right).$$

The position of the object is plotted as a function of time in •Fig. 3–9. There are two interesting limits. For short times, we have, simply,

$$x(t) = \frac{1}{2} \left( \frac{F_0}{m} \right) t^2,$$

which is the familiar Newtonian answer. But for long times, the $t^2$ term in the square root dominates, and in this limit $x(t) = ct$: The particle, whatever its mass, propagates at the speed of light—but no faster.

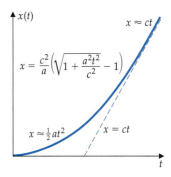

• **Figure 3–9** The displacement of a particle subject to a constant force, as a function of time. In the limit as $t \to \infty$, the curve approaches, but never reaches, the straight-line expression $x = ct$.

---

# SUMMARY

The theory of relativity has a number of consequences that can be tested in the laboratory. When the speeds of the objects involved become comparable to that of light, the effects are dramatic. These effects can be divided into two categories: kinematic and dynamic. Kinematic effects involve general principles such as the conservation of energy and momentum. Dynamic effects involve specific forces.

Among the most important kinematic effects are the following:

- A collection of unstable systems, such as radioactive nuclei or certain elementary particles, forms a clock whose time unit is the time it takes for half the population of the system to decay. When these particles move, this time unit is dilated in the fashion described in Chapter 2. Time dilation is also illustrated with the example of twins, one of whom remains at rest while the other takes a round-trip: The traveling twin ages less.

- Lorentz transformations relate space and time, and they similarly relate energy and momentum. Both energy and momentum accordingly take new forms, which appear in the conservation laws. These laws are powerful tools for analyzing collisions and reactions at speeds which are large enough that special relativity plays a role.

- In special relativity, there is a term in the expression for energy that does not exist in classical physics. A particle at rest has an energy that is proportional to its mass and to the square of the speed of light. The laws of conservation of energy and momentum must be modified to include this new term. The fact that mass represents a new form of energy means that particles can be created from the excess energy that is present in any reaction. For example, antiparticles can be created from the collisions between particles. Indeed, this is how the antiproton was discovered.

As for dynamics, we note that the classical Newton's law allows a particle to be accelerated to any speed at all—even to speeds greater than light. In special relativity, this is not allowed, and Newton's laws are modified accordingly. In relativistic dynamics, a particle becomes more difficult to accelerate as it moves faster. The particle behaves as though its mass had increased, approaching infinity as its speed approaches that of light. In this way, the speed of the particle can never exceed that of light. The example of a constant force is a simple laboratory in which to illustrate just how this principle works.

## QUESTIONS

1. We resolved the absence of symmetry in the situation of the "traveling twins" by pointing out that one of the twins undergoes an abrupt (and large) acceleration, thereby distinguishing that particular twin. Could we avoid the distinction if one of the twins undergoes an arbitrarily small acceleration during the entire round-trip?

2. A newspaper account of the photon stated that the zero mass of the photon would be the mass it had when it was brought to rest. Is this account correct? How would you explain the correct version to the reporter?

3. Classical momentum and energy are separately conserved. Will this still be true when the conditions of special relativity apply? How about the conservation of mass, a useful principle in classical mechanics?

4. One way to check the validity of the proposition that the speed of radiation emitted by a source is independent of the motion of the source is to study radiation from double stars. These stars rotate in orbit about their center of mass. The speed of one of the stars can be obtained from the Doppler shift of spectral lines in the emitted radiation. How would a curve of the Doppler shift as a function of time (that is, as a function of the position of the star in it's orbit, assumed here to be circular) look if the preceding proposition were correct? How would the curve look if the light emitted when the star approached us moved more quickly and the light emitted when the star moved away from us moved more slowly?

5. Light of wavelength $\lambda$ is reflected from a mirror that moves away from the source with speed $v$. What is the wavelength of the light seen at the source after reflection? Suppose the light is in the 400-nm range and the reflector is on the order $4 \times 10^8$ m away, and apply what you learned about the accuracy of the Michelson–Morley apparatus in Chapter 2 to estimate how slow a motion could be detected this way.

6. Any truly massless object must move with the speed of light. But suppose that instead of being massive, a particular particle had only one-millionth the mass of the lightest previously known object. Does this object always move with nearly the speed of light? Or could it be brought to rest?

7. The minimum energy for the production of an additional $p\bar{p}$ pair in a proton–proton collision in the center-of-mass frame (so that each of the colliding protons has the same energy) is $2m_p c^2$ per proton. In the text, the calculated value in the laboratory frame, in which a moving proton incident on a proton that is at rest creates the pair, is $7m_p c^2$. Why, physically, should the second value be larger than the first?

8. Sally and Sal have been given electrons for their birthdays. But Sally lives permanently in a frame of reference that moves uniformly with half the speed of light relative to Sal. Will Sally measure the same mass for her electron as Sal

measures for his? When Sally measures the mass of Sal's electron, will it be the same as the mass he measures for hers?

**9.** A spring is placed between two identical masses, and the system of these objects is compressed by means of an elastic band enclosing it. Both the spring and the elastic band are of negligible mass. When the elastic band is cut with a scissors, the two masses fly apart. Is the mass of the original system less than or greater than the sum of the masses of the two objects? Is the mass of the system just after the elastic band is cut, but before the masses have started to move apart, less than or greater than the sum of the masses of the two objects?

## PROBLEMS

**1.** ▌ An unstable particle has a lifetime of $0.80 \times 10^{-10}$ s in its rest frame. An experiment with such a particle requires it to travel 30 m in the laboratory before decaying. What velocity must the particle have in the laboratory to make the experiment feasible?

**2.** ▌ A quasi-stellar object exhibits a Doppler shift such that $(\lambda_{obs} - \lambda)/\lambda = 1.95$, where $\lambda$ is the wavelength that would be measured by an observer at rest relative to the object. Is the object moving towards us or away from us? Assuming that the object moves either directly away from or directly towards us, what is its speed relative to us?

**3.** ▌ A source flashes with a frequency of $10^{15}$ Hz. The signal is reflected from a mirror moving away from the source at 10 km/s. What is the frequency of the reflected radiation, as received at the source?

**4.** ▌ A spaceship approaches an asteroid and sends out a radio signal with proper frequency $6.5 \times 10^9$ Hz. The signal bounces off the asteroid's surface and returns shifted by $5 \times 10^4$ Hz. What is the relative velocity of the spaceship and the asteroid?

*The next three problems lead to a derivation of the Doppler shift different from that presented in Section 3–2. Assume that all motion is aligned with the x-axis.*

**5.** ▌▌ Using the Lorentz transformations, show that the quantity $px - Et$ is invariant, where $p$ and $E$ are the momentum and energy, respectively, of an object at position $x$ at time $t$.

**6.** ▌▌ A photon is the "quantum of light." Like a classical light wave, it has a particular frequency $f$, yet at the same time it behaves in many respects as a chargeless, massless particle whose energy $E$ is related to the frequency by $E = hf$. What is the momentum of the photon in terms of $h$, $f$, and $c$?

**7.** ▌▌ Using the results of the previous two problems, find the rule for the transformation of the frequency $f$. This is an alternative derivation of the Doppler shift.

**8.** ▌▌ Rocket $A$ moves with a speed of $0.75c$ in a northerly direction relative to an origin. Rocket $B$ moves to the west relative to the same origin with speed $0.4c$. As it moves, rocket $B$ emits radiation with a wavelength 100 nm normal to its line of motion, in the northerly direction. What is the wavelength of the radiation, as seen by rocket $A$?

**9.** ▌ Twins Abby and Bill separate. Abby travels at a uniform speed (except for the brief turn-around period) to a star 4.2 light years away and, upon returning, finds that she is 5.2 years younger than Bill. How fast was Abby traveling? By how much has Bill aged during the period Abby was traveling?

**10.** ▌ Find the kinetic energy of a particle with a mass of one gram moving with half the speed of light. Compare your answer with the one you would find using the nonrelativistic formula.

**11.** ▍ Find the momentum of a particle with a mass of one gram moving with half the speed of light. Compare your answer with the result you would find using the nonrelativistic formula for the momentum.

**12.** ▍ A certain particle is at rest with respect to you. Assuming that you measure its mass to be one gram, what is its energy?

**13.** ▍ What is the total energy of a particle with a rest mass of one gram moving with half the speed of light?

**14.** ▍ If the rest mass of a particle is one gram, what mass would an observer moving past it at half the speed of light measure the particle to have?

**15.** ▍ A 20-megaton hydrogen bomb explodes. Given the equivalence that 1 ton of TNT equals $4.18 \times 10^9$ J, how much "mass" was used up in the explosion?

**16.** ▍ A radio station broadcasts with a power of 500 kW. How much "mass" does the station use up in a year?

**17.** ▍▍▍ Prove the identity of Eq. (3–29).

**18.** ▍ Suppose that we could use the energy released when 1 g of antimatter annihilates 1 g of matter to lift a mass 1 km from the Earth's surface. How much mass could we lift?

**19.** ▍▍ The nuclear reaction $p + d \rightarrow {}^3\text{He} + \gamma$ can occur even when the initial particles have zero kinetic energy—we say the reaction is exothermic. If the gamma ray has energy 5.5 MeV when the reaction occurs with the initial particles at rest, what is the mass of the ${}^3\text{He}$ nucleus? Observe that $m_p = 1.6724 \times 10^{-27}$ kg and $m_d = 3.3432 \times 10^{-27}$ kg.

**20.** ▍▍ A beam of pions ($\pi$-mesons, with rest energy $m_\pi c^2 = 140$ MeV) consisting of $10^{10}$ pions/cm$^2 \cdot$ s moves with speed such that the total energy of each pion is $10^5$ MeV. What is the pressure the beam exerts on a surface that absorbs it and brings it to rest?

**21.** ▍▍ A gamma ray is a quantum of light energy, also known as a photon. For purposes of energy–momentum conservation, it can be regarded as a chargeless, massless particle. Show that energy–momentum conservation forbids the reaction $\gamma \rightarrow e^+ + e^-$, the creation of an electron and a positron from a single quantum of radiation.

**22.** ▍ The $\psi'$ particle (with a mass of 3700 MeV) decays into the $J/\psi$ particle (whose mass is 3100 MeV) through the reaction $\psi' \rightarrow J/\psi + \gamma$. Supposing that the decay occurs with the $\psi'$ at rest, find the energy of the gamma ray.

**23.** ▍▍ Find the minimum energy a gamma ray must have to initiate the reaction $\gamma + p \rightarrow \pi^0 + p$ if the target proton is at rest. Note that $m_p c^2 = 940$ MeV and $m_\pi c^2 = 140$ MeV.

**24.** ▍▍ An atom of mass $M$ decays from an excited state to the ground state—a process described by quantum mechanics—with a change in mass of $\Delta M \ll M$. In the decay process, the atom emits a photon, a quantum of light. Use the laws of conservation of energy and momentum to determine the energy of the photon, assuming that the atom decays from rest.

**25.** ▍ A particle with a rest energy of 2400 MeV has an energy of 15 GeV ($15 \times 10^9$ eV). Find the time (in Earth's frame of reference) necessary for this particle to travel from Earth to a star four light-years distant.

**26.** ▍▍▍ Suppose that a bomb was exploded in outer space, one million km from Earth, and that as a result gamma rays (photons) with energies from 1 eV to 1 MeV arrive at a detector on Earth. All of the rays arrive at the detector at the same time to within the detector's time resolution, which is $10^{-9}$ s. Use this information to find the largest possible value for the mass of a photon. [*Hint*: See the previous problem.]

**27.** ▮▮ Consider a two-body collision in which both momentum and energy are conserved in a particular reference frame $S$. Show that an observer in a frame moving with velocity $v$ relative to $S$ also sees energy and momentum to be conserved only if mass also is conserved in the process.

**28.** ▮▮ The $K^0$ and the $\Lambda^0$ are elementary particles with masses $m_K$ and $m_\Lambda$, respectively. Find the threshold energy for the reaction $\pi^- + p \rightarrow K^0 + \Lambda^0$ in the frame of reference in which the proton is at rest. Express your answer in terms of the masses of the particles.

**29.** ▮▮ If the electron $e^-$ and the positron $e^+$ both have a rest energy of 0.51 MeV, find the minimum energy a particle of light (a photon), $\gamma$, can have in the reaction $e^+ + e^- \rightarrow \gamma + \gamma$. Assume the center-of-mass frame.

**30.** ▮▮ Electrons and positrons (see the previous problem) can be sent at each other as colliding beams so that, effectively, the collision takes place in the center-of-mass frame. Find the minimum energy each incident particle must have in this frame in order that the reaction $e^+ + e^- \rightarrow p + \bar{p}$ can take place.

**31.** ▮▮ A particle known as the $\pi^0$ has a lifetime of $8.4 \times 10^{-17}$ s in its rest frame. In a high-energy collision, a $\pi^0$ is produced with an energy such that the relativistic gamma factor for the particle is $\gamma = 340$. How far will the $\pi^0$ travel before it decays?

**32.** ▮▮ Suppose the $\pi^0$ (previous problem) decays into two massless particles. If one of the particles moves in the same direction that the $\pi^0$ was moving in, what is the direction of motion of the second one? What are the energies of each of the massless particles in terms of the momentum of the original $\pi^0$?

**33.** ▮ A particle whose rest energy is $Mc^2 = 765$ MeV decays into a $\pi^0$ with a rest energy of 135 MeV, and another particle, a $\pi^+$, with rest energy of 140 MeV. What will be the momenta of the decay products?

**34.** ▮▮▮ A particle of mass $M$ decays at rest into two particles of mass $m_1$ and $m_2$. Show that energy and momentum conservation imply that the energy of particle 1 is given by

$$E_1 = \frac{\left(M^2 + m_1^2 - m_2^2\right)c^2}{2M}.$$

What are the momenta of the particles?

**35.** ▮▮ In a system of two particles with energies and momenta $(E_1, p_1)$ and $(E_2, p_2)$, respectively, the quantity

$$s^2 = \left(E_1 + E_2\right)^2 - c^2\left(p_1 + p_2\right)^2$$

is invariant; that is, it has the same numerical value in all inertial frames. **(a)** Consider a center-of-mass collision of a proton and an antiproton $\left(mc^2 = 938.3 \text{ MeV}\right)$. What is the minimum momentum required to produce a particle with mass $Mc^2 = 91.2 \times 10^3$ MeV? **(b)** In a fixed-target accelerator, an antiproton projectile collides with a proton target at rest. What is the minimum energy that the antiproton must have to create the new particle of part (a)?

**36.** ▮▮ Consider a collision of two particles $A$ and $B$. In the laboratory, the energies and momenta of the particles are respectively given by

$$E_A = m_A c^2, p_A = 0;$$

and

$$E_B = \left(p^2 c^2 + m_B^2 c^4\right)^{1/2}, p_B = p.$$

Show that the velocity of an observer who sees the collision in the center-of-momentum frame is

$$v = \frac{pc^2}{E_B + m_A c^2}.$$

Evaluate this equation in the ultrarelativistic frame in which $E_B = pc$ is a very good approximation $(m_B c/p \ll 1)$.

**37.** ▌▌▌ Consider a collision of the form $A + B \rightarrow C + D + \dots$. In the center-of-momentum frame, in which $A$ and $B$ have equal and opposite momenta along the $x$-axis, one of the final particles emerges with energy $E^*$ and momentum $\vec{q}^*$. The momentum makes an angle $\theta$ with the axis of collision, so that

$$q_x^* = q^* \cos\theta \quad \text{and} \quad q_y^* = q^* \sin\theta.$$

In the laboratory frame, which moves with speed $v$ (calculated in the previous problem) relative to the center-of-momentum frame, the components of the momentum are given by $q_x$ and $q_y$. Show that the angle the particle makes with the $x$-axis in the laboratory frame is given by

$$\tan\theta_L = \frac{q_y}{q_x} = \sqrt{1 - v^2/c^2} \, \frac{q^* \sin\theta}{q^* \cos\theta + vE^*/c^2}.$$

Evaluate this angle in the ultrarelativistic limit, and show that the angle is squeezed into the forward direction by a factor $(1 - v^2/c^2)^{1/2}$. This squeezing of the particle production into the forward direction is known as the "searchlight effect."

**38.** ▌▌ An observer at rest in a laboratory sees particles emitted from a source. The first bunch of particles is emitted with speed $v = 0.92c$; the second bunch is emitted $10^{-3}$ s later, with $v = 0.99c$. How long, according to the observer, will it be until the second bunch catches up with the first?

**39.** ▌▌▌ An observer at rest sees two particles. One, with speed $0.9c$, moves along the $x$-axis; the other, with speed $0.5c$, moves along the $y$-axis. What is the relative speed of the particles?

## APPENDIX

In this appendix, we derive the relativistic form of momentum, as well as the relativistic version of Newton's law. The technique will be to start with Newton's law, Eq. (3–36), in a frame $S'$ in which the object on which the force acts is instantaneously at rest. Then we transform our result to a system $S$ with respect to which $S'$ is moving with a speed $v$. It is from $S$ that we observe the motion. To simplify the algebra, we consider the very special situation in which the forces, the accelerations, and the velocity $\vec{v}$ are all in one direction, which we can call the $x$-direction. We use the notation in which an object in motion has velocity $dx/dt = u$ in frame $S$ and velocity $dx'/dt' = u'$ in frame $S'$.

We already know the transformation rule for the velocity in this situation; it is given by Eq. (2–41). To treat forces, we also require the rule for the acceleration. To find the rule for transforming the accelerations, we proceed as we did for the velocity. Thus, we have, generally,

$$a' = \frac{du'}{dt'} = \left(\frac{du'}{dt}\right)\left(\frac{dt'}{dt}\right)^{-1}.$$

Note that we have not yet assumed that the object in question is at rest; that is, $u'$ is not yet zero. Now, the first term is the derivative of $u'$, given by Eq. (2–41); hence, we have

$$\frac{du'}{dt} = \frac{d}{dt}\left(\frac{u - v}{1 - uv/c^2}\right) = \frac{(du/dt)}{1 - uv/c^2} + \frac{v}{c^2}\frac{du}{dt}\frac{u - v}{\left(1 - uv/c^2\right)^2},$$

and $dt'/dt$ is evaluated in Eq. (2–40) as $dt'/dt = \gamma\left[1 - (v/c^2)u\right]$. When we put these equations together, we obtain

$$a' = \left[\frac{(du/dt)}{1 - (v/c^2)u} - \frac{u - v}{\left(1 - (v/c^2)u\right)^2}\left(-\frac{v}{c^2}\right)\frac{du}{dt}\right]\frac{\sqrt{1 - v^2/c^2}}{1 - (v/c^2)u}.$$

Finally, we specialize to the case in which the object in question is at rest in frame $S'$, which implies that $u = v$; this equality greatly simplifies the result, leaving us with

$$a' = \frac{(du/dt)}{\left(1 - v^2/c^2\right)^{3/2}} = \frac{a}{\left(1 - v^2/c^2\right)^{3/2}}. \tag{A–1}$$

To cast things in a more familiar form, we first replace the left side of Eq. (A–1) by $F'/m$, as follows from Eq. (3–36). As for the right side, note the following equality:

$$\frac{d}{dt}\left(\frac{u}{\sqrt{1 - u^2/c^2}}\right) = \frac{du}{dt}\frac{1}{\left(1 - u^2/c^2\right)^{3/2}}.$$

You can check that this equation holds by simply carrying out the differentiation. If we replace $u$ by $v$, as is appropriate here, then the right-hand side is just the right-hand side of Eq. (A–1). As for the left side, it is, by definition, the force in the $S$-frame, $F$, divided by the mass $m$. Thus,

$$F = m\frac{d}{dt}\left(\frac{v}{\sqrt{1 - v^2/c^2}}\right).$$

This results suggest that we should define the relativistic momentum by

$$p \equiv \gamma m v.$$

If we do, then the relativistic generalization of Newton's law takes the form

$$F = dp/dt.$$

Our derivation of the relativistic version of Newton's law was restricted to the special situation in which the force and the acceleration are in the same direction. This corresponds to the situation of a constant force that we treated in the body of the text. In general, the force law is more complicated, and its study goes beyond the ambitions of this text.

# Quantum Mechanics

From the mid-19th through the early 20th centuries, scientists studied a set of new and puzzling phenomena concerning the nature of matter and, indeed, of energy in all its forms. The program that brought these questions to the point where we are today has provided some of the most remarkable success stories in all of science. This is the history of quantum mechanics, which began in mystery and confusion, yet at the end of the century has come to dominate the economies of modern nations.

## Historical Introduction

In many parts of this book, a historical method of presentation is quite appropriate. Such an approach ties physics and its history together in a single package. With quantum mechanics, such an approach is less suitable. So many false starts and wrong paths were pursued in the creation of quantum mechanics that a person just learning the subject can easily get lost in this labyrinth. Thus, we will begin with a purely historical overview that will take us from the 1850s to the 1920s. In this introduction, we will not go into details about the physics, although we will introduce many of the main ideas. Afterwards, we will present the physics with less concern for historical order. That way, you will get an understanding both for these remarkable concepts and for the effort it took to create them.

### Blackbody Radiation and Classical Physics

The path to quantum mechanics begins in 1859 with the work of the German physicist Gustav Kirchhoff, who was studying what happened when the so-called dark D-lines from the light emitted by the Sun—individual frequencies in the continuous spectrum that appear dark compared with the rest of the spectrum—were passed through a sodium flame produced in a Bunsen burner (•Fig. 4–1). He discovered that the lines became darker still, which he correctly interpreted as showing that the D-lines came from the absorption of light from the interior of the Sun by sodium atoms at its surface. This reasoning became a powerful tool for studying the chemical composition of stars.

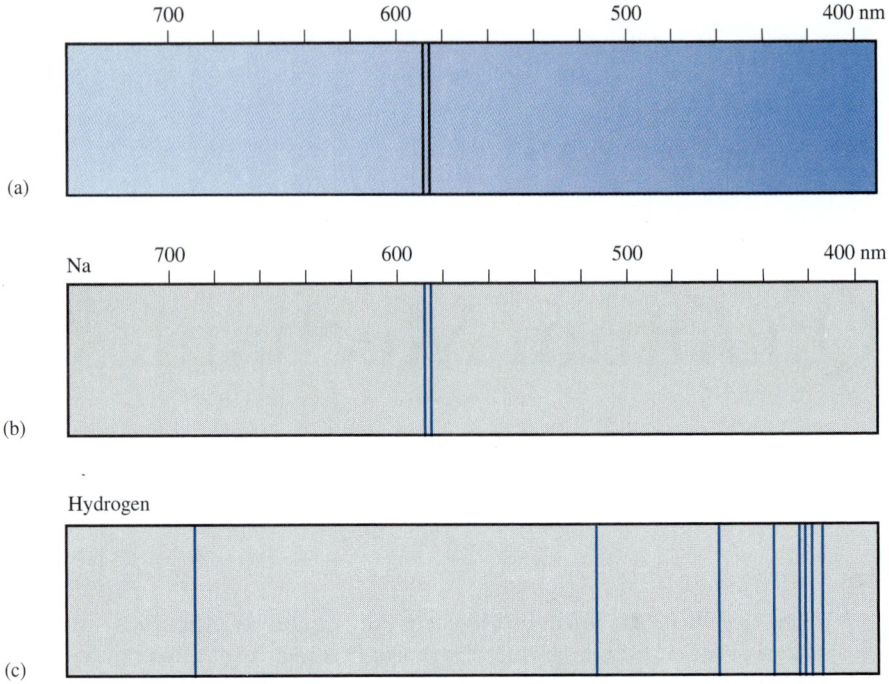

700    600    500    400 nm

(a)

Na    700    600    500    400 nm

(b)

Hydrogen

(c)

• **Figure 4–1**  Light emitted from atoms comes in a series of discrete wavelengths called *lines*. Similarly, light is absorbed by atoms at discrete wavelengths. Here we show a sampling of emission and absorption lines. (a) Absorption lines in sodium; (b) emission lines in sodium; (c) emission lines in hydrogen.

Kirchhoff could not explain such selective absorption—that would have to wait for the quantum theory—but he began to consider the emission and absorption of radiation by heated materials in general. In appreciating what Kirchhoff did, one should recall how little he knew by our standards. Maxwell had not even begun to formulate his electromagnetic equations. Statistical mechanics did not exist and thermodynamics was in its infancy. Nonetheless, by using the principles of thermodynamics, Kirchhoff was able to pose the problem that eventually led to the quantum theory.

Kirchhoff imagined a container—a cavity—whose walls were heated up so that they emitted radiation that was trapped in the container. Within the cavity, there is a distribution of radiation of all wavelengths, which we can describe by the intensity, a quantity that measures the rate at which energy falls on a unit area of surface. In this regard, we consider the intensity distribution $K(\lambda)$, where the intensity between the wavelengths $\lambda$ and $\lambda + d\lambda$ is $K(\lambda)\,d\lambda$. Now we bring the walls into the picture. Suppose that the rate at which radiation energy is emitted by a unit area of the wall in the interval from $\lambda$ to $\lambda + d\lambda$ is $e_\lambda\,d\lambda$. The quantity $e_\lambda$ is termed the *emissivity*. The walls can also absorb energy. We can call $a_\lambda$ the coefficient of absorption. It is a dimensionless measure of how efficiently radiant energy in the interval from $\lambda$ to $\lambda + d\lambda$ can be absorbed by the walls of the cavity. This means that the rate at which energy can be absorbed per unit area of a wall in the interval $\lambda$ to $\lambda + d\lambda$ is $a_\lambda K(\lambda)\,d\lambda$. Kirchhoff defined equilibrium as the situation in which the rate of total emission from a wall is just balanced by the rate of absorption by the wall. Mathematically,

$$\int e_\lambda\,d\lambda = \int a_\lambda K(\lambda)\,d\lambda \qquad \text{(H–1)}$$

where the integrations are over all values of the wavelength. Kirchhoff then assumed that these functions were such that this relation holds separately for each wavelength, so that one can replace equality of integrals with the equation

$$e_\lambda/a_\lambda = K(\lambda). \qquad \text{(H–2)}$$

It is at this point that Kirchhoff's remarkable physical insight enters. All our quantities could depend on the temperature and on other properties, such as the material of the container. But even if $e_\lambda$ and $a_\lambda$ depend, in general, on properties of the material of the container, Kirchhoff argued, their ratio in the equilibrium state does not. It does not even depend on where in the container one measures the intensity. In fact, $K(\lambda)$ depends only on the wavelength and the temperature $T$. Here is how Kirchhoff reasoned: Suppose we have two containers of different shapes and made out of different materials. Both containers are in equilibrium with the radiation in their interiors at the temperature $T$. Suppose now that the radiation had different intensity distributions. Then we could connect the containers and allow the radiation in the container with the larger intensity in some wavelength interval to flow into the other container. But this would constitute a spontaneous transfer of energies between the two containers at a common temperature, which is forbidden by the second law of thermodynamics. (See Chapter 1.) Likewise, if the distribution varied from place to place within one container, we could envision a similar setup within that container. Thus, in equilibrium, $K$ is a function only of wavelength and temperature; it is universal—it does not depend on the material or the shape of the container.

Kirchhoff then made a second observation. Suppose we can produce a system that is a perfect absorber; that is, for this system, $a_\lambda = 1$. In that case,

$$e_\lambda = K(\lambda, T). \tag{H-3}$$

This perfectly absorbing body became known as a **blackbody** and the radiation inside it as **blackbody radiation**. There are many systems in nature—including, as we shall see later in the book, the universe as a whole—that are good approximations to a blackbody. If, for example, we make a small hole in a container and allow radiation to enter it, the radiation will get trapped, and the container will effectively absorb all the radiation incident on it. If we sample the radiation inside, we can, in principle, measure Kirchhoff's function $K(\lambda, T)$, although actually doing so, something that Kirchhoff challenged people to do, took several decades, and the first theory that really accounted for the form of $K$ from first principles was not created until the mid-1920s!

The next step was a theoretical one taken by the great Austrian theoretical physicist Ludwig Boltzmann in the mid-1880s. By this time, Maxwell had formulated his equations, and it was known that electromagnetic radiation, the same radiation found in a blackbody, produces a pressure. Therefore, Boltzmann knew that if the radiation in the blackbody had an energy density given by $u(\lambda, T)$, so that there was a total energy density $u_{\text{tot}}(T)$ gotten by integrating over all wavelengths, then this radiation would exert a force on any unit area—a pressure $p$—of, say, a cylinder containing this radiation. The radiation pressure is proportional to $u_{\text{tot}}$. Boltzmann also showed that in the blackbody $K(\lambda, T)$ and $u(\lambda, T)$ were simply proportional to each other; therefore, like $K$, $u$ was also a universal function. Armed with these results, Boltzmann could consider the expansion of a cylinder with a piston on one end that reflects the radiation. Using the laws of thermodynamics for such an expansion, Boltzmann was able to show that

$$u_{\text{tot}} = \sigma T^4, \tag{H-4}$$

where $\sigma$ is a universal constant known as the *Stefan–Boltzmann constant*. (The Austrian Josef Stefan had guessed at this result prior to Boltzmann's proof.)

The next important steps forward were taken a decade later by the German Wilhelm Wien, who made two contributions towards finding Kirchhoff's

❚ The role of blackbody radiation in our understanding of the universe as a whole will be discussed in Chapter 18.

• **Figure 4–2**  Max Planck. Although he was born in Kiel in Germany in 1858 to a very distinguished—read conservative—family of academics and jurists, he was an integral part of the quantum revolution, a revolution that changed the face of the world.

function $K(\lambda, T)$, one specific and one general. The specific contribution was based on an analogy between the Boltzmann energy distribution for a classical gas consisting of particles in equilibrium and the radiation in the cavity. (This distribution is described in detail in Chapter 12.) The Boltzmann energy distribution describes the relative probability that a molecule in a gas at temperature $T$ has a given energy $E$. This probability is proportional to $\exp(-E/kT)$, where $k$ is Boltzmann's constant, so that higher energies are less likely, and the average energy rises with temperature. Wien's analogy suggested that it was also less likely to have radiation of high frequency (small wavelength) and that an exponential involving temperature would play a role. In fact Wien's analogy is not very good. It represents, however, the first attempt to "derive" Kirchhoff's function from classical physics, something that, as we shall see, is impossible. Nonetheless, the form Wien found fits the small-wavelength (or, equivalently, the high-frequency) part of the blackbody spectrum that experiments were beginning to reveal. Wien's distribution is given by

$$K_{\text{Wien}}(\lambda, T) = b\lambda^{-5}\exp(-a/\lambda T), \tag{H–5}$$

where $a$ and $b$ are constants to be determined experimentally. Similar distributions had been tried before Wien's, but he had the Stefan–Boltzmann equation to guide him. Hence, he was able to fix on the power of $\lambda$ in his result by insisting that the total energy—the integral of $u(\lambda, T)$ over all $\lambda$—behave as the fourth power of the temperature. You can verify that this works by integrating over all wavelengths of $K_{\text{Wien}}$. You can also verify by a change of variables that *any* function of $\lambda T$ multiplying a factor $\lambda^{-5}$ will give the same result. This was Wien's general observation; that is, on the basis of thermodynamics alone, one can show that Kirchhoff's function, or equivalently, the energy density function $u(\lambda, T)$, is of the form $\lambda^{-5}\varphi(\lambda T)$. But this is as far as thermodynamics can go; it cannot determine the function $\varphi$. At this point, enter Max Planck (•Fig. 4–2).

## Planck and the Blackbody Spectrum

Planck was certainly a "reluctant revolutionary." He never intended to invent the quantum theory, and it took him many years before he began to admit that classical physics was wrong. In fact, before he got into physics he was advised *not* to study the subject, because all the problems had been solved! Planck studied under Kirchhoff at the University of Berlin, and after Kirchhoff's death in 1887, Planck succeeded him as professor of physics there. Planck had a great interest in laws of physics that appeared to be universal. Hence, it is not surprising that he became fascinated by the subject of blackbody radiation and the determination of Kirchhoff's function, on which he started working seriously in the mid-1890s. His original goal was to derive Wien's law from Maxwell's electromagnetic theory and thermodynamics. The derivation kept eluding him. In hindsight, this is natural, because such a derivation is impossible!

In early 1900, Planck's colleagues Otto Lummer and Ernst Pringsheim at Berlin did precise and beautiful measurements of the infrared—long-wavelength—region of the blackbody spectrum (•Fig. 4–3). These observations of a previously unexplored regime showed that the Wien law broke down there. A second Berlin group consisting of Heinrich Rubens and Ferdinand Kurlbaum pushed the results even further into the infrared, a finding that they published in October of 1900. Again the Wien form, which is good at short wavelengths, failed. Planck knew about both groups' findings prior to their publication, and he set about trying to find a Kirchhoff function that would fit the data. How he succeeded is something of a mystery that historians of science have tried to reconstruct. But in the end, it was inspired guesswork, coupled with years of

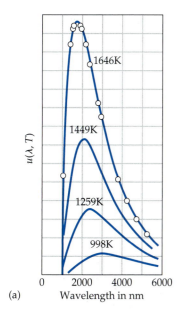

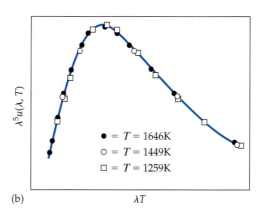

• **Figure 4–3**  (a) Data of Lummer and Pringsheim on the blackbody energy density $u(\lambda, T)$ as a function of wavelength. (b) Data of the same experimenters showing that $\lambda^5 u(\lambda, T)$ is a function of $\lambda T$ only.

thinking about the problem and the inspiration of new experimental data. The result that Planck came up with can be written as

$$K(\lambda, T) = \frac{b}{\lambda^5} \frac{1}{\exp(a/\lambda T) - 1}, \qquad \text{(H–6)}$$

with $a$ and $b$ constants to determined. We see at once that when $a/\lambda T \gg 1$ —that is, for small wavelengths—we can neglect the 1 in the denominator and recover the Wien law. The equation also is in the general form suggested by Wien. Finally, it fits the new data perfectly and has done so ever since. Thus, Kirchhoff's problem was resolved, at least experimentally. But where does this formula come from? That problem was to occupy Planck for the next decade.

From the beginning of his effort to derive the Wien form of the Kirchhoff function, Planck had in mind a model of a blackbody that contained what he called "resonators," which were charges that could oscillate harmonically. From our present point of view, we can think of these as the electrons belonging to the atoms that make up the walls of our container. As Planck knew from Maxwell's equations, oscillating charges can emit and absorb radiation. In thermal equilibrium, the emission and absorption from the oscillators balance. Planck showed that in equilibrium, the energy density of the oscillators was proportional to the energy density of the radiation, $u(\lambda, T)$. Hence, one could study the theoretical energy density of these oscillators in equilibrium—that is, having an average temperature equal to that of the radiation—to determine the Kirchhoff function.

By applying methods of statistical physics introduced by Boltzmann in the 1870s, Planck obtained the Kirchoff formula. He had to make a drastic, quite unjustified assumption: that the oscillators could only emit and absorb energy of frequency $f$ in units of $hf$, where $h$ is a new universal constant with dimensions of energy multiplied by time. Planck called these energy units **quanta**. He was very troubled by his assumptions and spent the next decade trying to derive his distribution from classical physics. He had, of course, no success.

Before going on, we shall take a very brief detour to discuss the contribution of the Englishman John Strutt, better known as Lord Rayleigh. Rayleigh's contribution was to pin down precisely what classical physics predicted for the

Kirchhoff function. His first brief paper on the subject was in 1900 and was published some months before Planck's own paper. Rayleigh's idea was to focus on the radiation and not on Planck's material oscillators. He knew that if he considered this radiation as being made up of standing electromagnetic waves, he could calculate that the energy density of these waves would be equivalent to the energy density of a collection of harmonic oscillators. These are not material oscillators; rather, they are a representation of the energy of the waves when the waves are broken down into their basic "modes"—a decomposition known as a Fourier analysis. What is relevant is that there is no difference between these oscillators and the material ones in the amount of thermal energy that they can share in equilibrium at a temperature $T$, which energy is $kT$.[1] Hence, the average energy per oscillator is $kT$. That is that! But Rayleigh realized that if he integrated this answer over all possible frequencies, he would get infinity for the energy density of the radiation in the cavity—something that came to be known as the *ultraviolet catastrophe*. The name is appropriate because, when the classical integrations were performed to determine the total energy from the energy distribution, the integrals diverged at high frequencies—the "ultraviolet" limit. In 1905, Rayleigh presented the full derivation of his classical answer. Soon after, the British astronomer James Jeans noticed a numerical mistake in Rayleigh's formula, so that the correct classical answer is known today as the *Rayleigh–Jeans law*. And now, enter Albert Einstein.

### Einstein and Quanta

Great physicists know both what problems to pursue and what problems not to pursue. Of one thing Einstein was sure in 1905, when he did the work we discuss here, and that was that it was impossible to derive Planck's formula—which he took as correct—from classical physics. It would be something like trying to derive the theory of relativity from Newtonian mechanics. It simply can't be done.

Let us look again at Planck's formula, Eq. (H–6). But now we are going to re-write it (as Planck did shortly after he discovered the original version) in terms of the frequency $f$, and we are going to put in the constants that Planck determined. Using the same symbol $u$ for the energy density, the result, which will be derived in detail in Chapter 12, is

$$u(f, T) = \frac{8\pi h f^3}{c^3} \frac{1}{\exp(hf/kT) - 1}, \qquad \text{(H–7)}$$

where $h$ is the constant introduced by Planck and $c$ is the speed of light. Now suppose we consider a low frequency limit—opposite to the Wien limit—in which $hf/kT \ll 1$. This can happen if we take $f$ small or $T$ large, or if we imagine a world in which $h$ tends to zero. But a world in which $h \to 0$ is the classical world, so we should get the classical answer if we use the appropriate expansion, namely, $\exp(x) \approx 1 + x$ for small $x$. Thus, in this limit,

$$u(f, T) \approx \frac{8\pi h f^3}{c^3} \frac{1}{1 + (hf/kT) - 1} = \frac{8\pi f^2 kT}{c^3}. \qquad \text{(H–8)}$$

This is exactly Rayleigh's classical answer. Planck's constant has dropped out. We also see the ultraviolet catastrophe looming: Integrating this result over the

---

[1] In fact, as we show in Chapter 12, the temperature is $kT/2$ per "degree of freedom." In the case we are examining, there is one degree of freedom associated with the kinetic energy and one associated with the potential energy.

frequencies gives infinity, due to the large-frequency behavior of the integrand. Hence, Einstein understood with absolute clarity that the correctness of the full Planck formula meant the end of classical physics.

So what ought Einstein to do? He decided to explore the consequences of the Wien end of the Planck formula—the large-frequency end—assuming its truth and using general principles of statistical mechanics and thermodynamics which he felt must hold in this new, nonclassical world. What Einstein discovered was that the statistical mechanics of the radiation at this end of the spectrum was analogous to the statistical mechanics of a "gas" of independent *quanta* of light whose energy $E$ was proportional to their frequency (i.e, $E = hf$). This may seem like a rehash of what Planck had been saying, but in fact, the difference was profound. We can understand the difference with the use of a homely analogy. Planck was saying—in terms of beer—that his oscillators absorbed "beer" only in "pints." But Einstein was saying that the radiation at the Wien end of the spectrum *was to be found* only in "pints." The "barrel of beer" was divided up into these discrete units and no further! This was a terribly radical proposal, because it meant that at the Rayleigh–Jeans, or low-frequency, end of the spectrum, the usual Maxwell description in terms of waves worked—indeed, Einstein had used it in his special theory of relativity, published that same year—while at the Wien, or high-frequency, end of the spectrum, one had to think of radiation as a "gas" of quanta. One is tempted to say that radiation sometimes acts like particles and sometimes like waves, but that is getting ahead of the story.

Einstein proposed a test of these new ideas. If light beams really consisted of quanta of energy satisfying the relation $E = hf$ for a monochromatic beam, then if such a beam were to fall on a metallic surface, it should be able to liberate electrons with an energy that depends only on the frequency—the color— of the light. This phenomenon is referred to as the *photoelectric effect*, and we'll describe it in more detail later. Actually, the effect had already been seen when Einstein proposed it, but it was not until about 1915 that the American experimental physicist Robert Millikan—one of the most important figures in American science in the first quarter of the 20th century—demonstrated with finality that Einstein's idea was right. The photoelectric effect involves only the conservation of energy. But the quanta also carry a momentum—they must, since light carries momentum as well as energy. This momentum has the magnitude $p = hf/c$ and must be conserved in any radiation process.

It was not until 1922 that the American Arthur Compton considered the implications of both the conservation of energy and the conservation of momentum in an experiment on the scattering process $\gamma + e \rightarrow \gamma + e$, where we now introduce the notation "$\gamma$" for the quantum of light. Compton knew about earlier results which suggested that scattered X rays underwent shifts in wavelength. Compton thought of looking at these results by using monochromatic X rays. He indeed found wavelength shifts in the scattered X rays and tried, unsuccessfully, to explain those shifts by classical arguments. But sometime later, a perfect explanation came to him. It depended on treating the X rays as particles with a definite momentum and energy, with the wavelength characterizing the momentum. In line with the fact that momentum is exchanged in a classical collision of particles, the X rays would change their momentum, and hence their wavelengths, in collisions with target electrons. Compton had introduced the *photon*, the quantum of light, as a full-fledged elementary particle, putting the finishing touch on a process of discovery that had begun many years earlier. It is at this point that we take up our physics account, working back and forth without special regard to the historical events.

# C H A P T E R

# 4

# Waves As Particles and Particles As Waves

In introductory physics, we learned that visible light is just a part of the spectrum of electromagnetic radiation that the human eye responds to. This radiation is described classically as a wavelike solution to Maxwell's equations. The clear evidence for the wave nature of radiation is manifested through experiments involving interference and diffraction. As we saw in our historical introduction, however, there is also evidence that *radiation comes in quanta*. And there is more: *Matter has wave characteristics associated with it.* Somehow, both light and matter share wave and particle properties. Just what is the evidence for these remarkable facts, and on what path does it lead us?

## 4–1 The Nature of Photons

As we have mentioned, the "particles" that make up radiation are called **photons**. In this section, we summarize some of their properties.

First, why isn't it immediately obvious to us that light consists of photons? The answer to this rests with our biology. We cannot tell with our eyes that radiation consists of particles, because our eyes are not quite sensitive to the effect of single photons. They are actually pretty close: An increase in the sensitivity of the eye by a factor of 10 would allow the night-adapted eye to respond to single photons. Then we might say, "Of course light comes in little energy packets." Modern electronic technology has made possible the construction of "superhuman eyes" called *photomultipliers*, that do clearly detect single photons (•Fig. 4–4).

Being quanta of light, photons must travel with the speed of light, $c \approx 3 \times 10^8$ m/s. Recall from our discussion of special relativity in Chapter 2 that a particle which travels with the speed of light is massless. For such particles, the energy $E$ and the momentum $\vec{\mathbf{p}}$ are related by

$$E = |\vec{\mathbf{p}}|c. \tag{4–1}$$

The work of Planck and of Einstein established the fact that the energy of a photon is linearly dependent on the frequency $f$ of the light with which it is associated, namely,

$$E = hf, \tag{4–2}$$

where $h$, the constant introduced by Max Planck, has dimensions $[ML^2T^{-1}]$ and magnitude

$$h \approx 6.63 \times 10^{-34} \, \text{J} \cdot \text{s}. \qquad (4\text{–}3)$$

Combining Eqs. (4–1) and (4–2) and using the wave relation $\lambda f = c$, where $\lambda$ is the wavelength of the light associated with the photon we are interested in, we find that the momentum $p$ of a single photon is inversely proportional to the wavelength:

$$p = \frac{E}{c} = \frac{hf}{c} = \frac{h}{\lambda}. \qquad (4\text{–}4)$$

We note that the energy of a single photon, given by Eq. (4–2), can alternatively be expressed with the angular frequency $\omega = 2\pi f$ as

$$E = \hbar\omega \qquad (4\text{–}5)$$

where

$$\hbar = h/2\pi \approx 1.05 \times 10^{-34} \, \text{J} \cdot \text{s} \qquad (4\text{–}6)$$

is what is usually called[2] **Planck's constant**.

This picture suggests that the intensity of a beam of radiation—the rate at which the radiation delivers energy per unit area—of a given frequency is a question only of the *number* of photons in the radiation. The more intense the radiation, the larger is the number of photons. The question of our ability to see single photons is a question of our ability to see light of very low intensity.

**Example 4–1**    Suppose that a 60 W lightbulb radiates primarily at a wavelength $\lambda \approx 1000$ nm, a number just above the optical range. Find the number of photons emitted per second.

**Solution**    If we divide the total energy per second by the energy per photon, we will have the number of photons per second. We know the total energy per second is 60 W. The frequency of the light is $f = c/\lambda \approx 3 \times 10^{14}$ Hz, and the energy per photon is $E = hf$. Then the number of photons emitted per second is

$$n = \frac{60 \, \text{W}}{hf} = \frac{60 \, \text{W}}{(6.63 \times 10^{-34} \, \text{J} \cdot \text{s})(3 \times 10^{14} \, \text{s}^{-1})} = 3 \times 10^{20} \, \text{photons/s}.$$

**Example 4–2**    A star emits approximately $2 \times 10^{45}$ photons/s, 80% of which are in the visible range. The star is 1200 light-years away, and the night-adapted eye, whose pupil makes an opening 4 mm in diameter, can barely see the star. How many photons per second are necessary to trigger the eye as a detector?

**Solution**    The photons are emitted equally in all directions. Thus, the number of photons per second incident on an area $A$—in this case, the area presented by the eye, a circle of diameter 4 mm—a distance $R$ from the star is

$$N = (0.8 \times 2 \times 10^{45} \, \text{photons/s})[A/(4\pi R^2)],$$

where $4\pi R^2$ is the area of a sphere of radius $R$. We have $A = \pi(2 \times 10^{-3})^2 \, \text{m}^2$. In units of meters,

$$R = (1.2 \times 10^3 \, \text{light-years})(3.15 \times 10^7 \, \text{light-seconds/light-year})(3 \times 10^8 \, \text{m/s})$$

$$= 1.14 \times 10^{19} \, \text{m}.$$

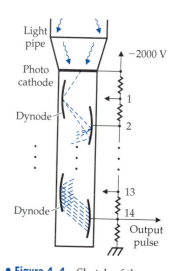

**Figure 4–4**    Sketch of the functioning of a photomultiplier. A photon enters through a "light pipe" and strikes a surface known as a photocathode. An electron is then released from this surface by the photoelectric effect. (See the historical introduction and the more detailed discussion of this phenomenon in Section 4–2.) The electron is accelerated by a voltage to a surface (dynode) where, now much more energetic, it knocks out two or more lower energy electrons. These in turn are again accelerated by a voltage, striking another dynode, producing more and more electrons until there are enough of them to give a large pulse of current. This pulse might be viewed as the analogue of a nerve impulse generated by stimulating the optic nerve. The original single electron produced by a single photon ends up as a member of a cascade with as many as $10^9$ electrons. (From *Subatomic Physics*, 2nd ed., by Hans Frauenfelder and Ernest Henley. Copyright 1991 by Prentice-Hall, Inc., Upper Saddle River, NJ.)

[2] In the older literature, that name was reserved for $h$, the quantity mentioned in the historical introduction, without the division by $2\pi$. Planck's constant $\hbar$ is pronounced "h-bar."

Thus, $A/(4\pi R^2) = 7.7 \times 10^{-45}$, and $N = 12$ photons/s. Since the star is barely visible, we may conclude that 12 photons per second is the order of the minimum number of photons necessary to trigger a signal to the brain.

---

**Example 4–3**    An experimentalist wishes to remove an electron from the atom to which it is bound by sending in a single photon to collide with it. The binding energy of the electron in question is 6.05 eV. (Recall that the energy unit 1 electron volt = 1 eV = $1.6 \times 10^{-19}$ J; this unit was described in Chapter 3.) What is the maximum wavelength of light necessary to remove the electron?

**Solution**    The electron is ejected if it takes the energy of a photon of energy $E_{min} = 6.05$ eV or higher. We have seen that the energy of a photon is *inversely* proportional to its wavelength, according to $E = hf = hc/\lambda$, where we have used the wave relation $f = c/\lambda$. Inverting, we have $\lambda = hc/E$, so that there is a maximum wavelength that will allow the electron to be ejected, namely,

$$\lambda_{max} = \frac{hc}{E_{min}} = \frac{(6.63 \times 10^{-34}\,\text{J}\cdot\text{s})(3.00 \times 10^8\,\text{m/s})}{6.05\,\text{eV}} = \frac{1.99 \times 10^{-25}\,\text{J}\cdot\text{m}}{6.05\,\text{eV}}$$

$$= \frac{1.99 \times 10^{-25}\,\text{J}\cdot\text{m}}{6.05\,\text{eV}}\,\frac{1\,\text{eV}}{1.6 \times 10^{-19}\,\text{J}} = 2.05 \times 10^{-7}\,\text{m}.$$

This wavelength, 205 nm, is in the ultraviolet range.

---

The preceding example suggests that it is useful to have an idea of the energies of photons corresponding to various wavelengths. In Table 4–1 we give a representative sampling.

**Table 4–1    Orders of Magnitudes of Single-Photon Energies Corresponding to Different Parts of the Electromagnetic Spectrum.**

| | Frequency $(\text{s}^{-1})$ | Wavelength (m) | Photon Energy (J) | Photon Energy (eV) |
|---|---|---|---|---|
| AM radio | $10^6$ | 300 | $7 \times 10^{-28}$ | $5 \times 10^{-9}$ |
| FM radio, TV | $10^8$ | 3 | $7 \times 10^{-26}$ | $5 \times 10^{-7}$ |
| microwaves | $10^{10}$ | $3 \times 10^{-2}$ | $7 \times 10^{-24}$ | $5 \times 10^{-6}$ |
| visible light | $6 \times 10^{14}$ | $5 \times 10^{-7}$ | $4 \times 10^{-19}$ | 2.5 |
| X rays | $10^{18}$ | $3 \times 10^{-10}$ | $7 \times 10^{-16}$ | $5 \times 10^3$ |
| gamma rays from nuclear decay | $10^{21}$ | $3 \times 10^{-13}$ | $7 \times 10^{-13}$ | $5 \times 10^6$ |

At this point you may well ask, What about Maxwell's equations, and what about interference and diffraction? What happened to all the experimental evidence that light was a wave? Were these experimenters wrong? How was Einstein able to reconcile the particle nature of light with its wave nature, something that is exhibited even in the basic equation $E = hf$, whose left side apparently refers to the energy of a particle and whose right side refers to a wave property? The answers to such questions are provided by quantum mechanics, and we shall be thinking about them in chapters to come. First, however, let us discuss some results that follow from the particle nature of radiation.

## 4-2 The Photoelectric Effect

Metals contain a large number of free electrons (we denote their mass as $m_e$ and their electric charge as $-e$), about one or two per atom. These electrons are called free because they are not bound to the atoms in the same way that plan-

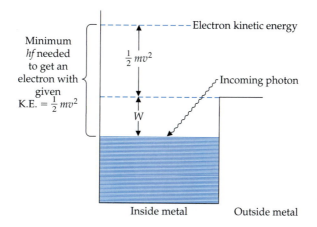

• **Figure 4–5**   The mechanism of the photoelectric effect as envisaged by Einstein.

ets are bound to the sun. However, they are not free to leave the metal: Metals do not "leak" electrons. It takes a certain amount of energy to get an electron out of a metal. You can visualize the situation as a collection of marbles rolling around in a frying pan. The marbles are free to roll around inside, but they do not have enough energy to scale the sides of the frying pan.

When a photon strikes one of these free electrons, the photon may, under the right circumstances, be absorbed by the electron, transferring enough energy to the electron to allow it to escape the metal. Electrons emitted by a metal subject to radiation are called **photoelectrons**. Following Einstein's 1905 paper, we now make the simple assumption that a single photon can be absorbed only by a single electron. A truly free isolated electron cannot absorb a photon and remain an electron, since this would violate the conservation of energy or of momentum. (See Example 3–8.) But this is not a problem here, because the struck electron can transfer momentum to the metal as a whole.

On the basis of this picture (•Fig. 4–5) we can say something about the current of photoelectrons:

- If the photon's energy is too small to boost the electron out of the metal, there will be no current at all. The energy required to do this job—analogous to that required to boost a marble the height of the sides of the frying pan— is known as the **work function** $W$ of the metal; it varies from metal to metal and can also depend on the condition of the surface. Typical values of the work function range from 2 to 8 eV. If the frequency $f$ of the light shining on the metal is such that $hf < W$, there will be no photoelectric current. In other words, for some metals, a weak beam of blue light produces a photocurrent, while a very intense red light produces none. If $hf$ is larger than $W$, then the electrons will emerge with a speed $v$ such that

$$\frac{1}{2} m_e v^2 = hf - W. \tag{4–7}$$

Thus, the energy of the photoelectrons from a particular metal depends only on the frequency of the radiation, and once the threshold frequency is exceeded, the dependence of the electron's kinetic energy on the frequency of the photoelectron is linear.[3] The kinetic energy of the electron is *independent*

---

[3] In accordance with the particular metal involved, the electrons within have a variety of energies, and accordingly, the electrons that come out when light of a given frequency hits the material also have a variety of energies. Thus, the linear dependence on the frequency is revealed only by referring to the *maximum* photoelectron energy. For simplicity, we shall continue to refer in the text to the kinetic energy, rather than the maximum kinetic energy.

of the intensity of the radiation beam—that is, of the number of photons. This independence follows from the fact that an electron will be liberated by a single photon when the photon hits the electron. How is the kinetic energy of the photoelectron measured? This can be done by sending the photoelectron into a region in which there is a potential difference. If that potential difference $V$ is adjusted so that no current flows, then $mv^2/2 = eV$. Once one knows the photoelectron energy, one can plot it against the frequency of the incident light. This plot should be—and is—a straight line.

Contrast this picture with the classical one, in which the energy carried by light depends on the square of the amplitude of the fields. No matter how small the frequency of the light, no matter how small the intensity, if one waits long enough, electrons will accumulate enough electromagnetic radiation to overcome the work function and escape from the metal.

- The foregoing reasoning also indicates that the *number* of photoelectrons emitted is proportional to the intensity of the radiation beam—that is, to the number of photons that shine on the metal. Again, this is not at all characteristic of the classical picture.

- There should be no time interval between the impact of the photon beam on the metal and the beginning of the emission of photoelectrons. This notion is in contrast to the classical picture, in which radiant energy arrives continuously, accumulating until there is enough to liberate an electron.

The German Heinrich Hertz, the same man who was able to verify Maxwell's classical prediction of electromagnetic waves, discovered the photoelectric effect in 1887. It was one of those accidental discoveries, made in this case while Hertz sought to improve his measurements of Maxwell's waves. Hertz's observations were of currents that were produced when light shone on metals. But not only did Hertz not know about photons; he didn't even know about the electron! Hertz's photoelectric effect attracted wide attention and further experiments, the sequence of which led eventually to J. J. Thomson's discovery of the electron in 1897. Many of the phenomena associated with the photoelectric effect were seen by Hertz. Probably, he and others were not much bothered by the difficulties of understanding the data in terms of Maxwell's classical picture of radiation without an understanding of the electron.

The final, definitive experiments on the photoelectric effect (•Fig. 4–6) were carried out in Chicago in the years 1904–1913 by Robert Millikan (•Fig. 4–7). Until these measurements, Einstein's explanation was not very widely accepted, because the experiments themselves were rather imprecise. In particular, systematic tests of the dependence of the photoelectric effect on frequency were

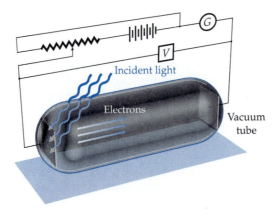

**• Figure 4–6** Robert's Millikan's experiments established the photoelectric effect in detail. The voltage $V$ is adjusted to stop the current, so that the kinetic energy of the fastest electrons is just $eV$. The most important aspect of Millikan's work was his recognition of the necessity of avoiding oxidation of the surface by working in vacuum.

very difficult, and for that matter, the identification of the constant $h$ in Eq. (4–7) with the constant introduced by Planck in connection with blackbody radiation was tenuous at best. In performing the measurements, there were problems of surface contamination. Worse, the range of light frequencies typically available made it important to use alkali metals such as sodium or potassium—ordinary metals require ultraviolet frequencies—and the alkalis are very dangerously flammable. To solve these problems, Millikan constructed what he called a "machine shop *in vacuo.*" In a vacuum, he could cut shavings of these metals to expose fresh surfaces; he manipulated both the knife and the shavings by means of electromagnets. His experiments of the variation of the photoelectron energy with frequency were so precise that he could verify the linearity of this energy and measure its slope well enough to provide what was then the best measurement of Planck's constant that was available (•Fig. 4–8). Initially, Millikan did not care very much for Einstein's ideas on photons—in fact, he even commented that he thought, erroneously, that Einstein himself had abandoned those ideas—but despite himself, he succeeded admirably in verifying the ideas.

• **Figure 4–7**   Robert Millikan, in company with Albert Einstein and Marie Curie.

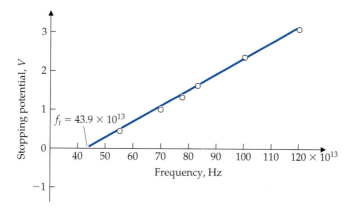

• **Figure 4–8**   Millikan's data on the photoelectric effect. Note the linearity in the plot of the energy of the photoelectron versus the frequency of the incident light.

**Example 4–4**   An experiment shows that when electromagnetic radiation of wavelength 270 nm falls on an aluminum surface, photoelectrons are emitted. The most energetic of these are stopped by a potential difference of 0.406 volts. Use this information to calculate the work function of aluminum in electron volts.

**Solution**   The kinetic energy of the most energetic photoelectrons is given by the electron charge times the potential that stops the photoelectrons:

$$K = eV = (1.6 \times 10^{-19}\,\text{C})(0.406\,\text{V}) = 0.65 \times 10^{-19}\,\text{J}.$$

The photon energy is

$$E = hf = \frac{hc}{\lambda} = (6.63 \times 10^{-34}\,\text{J·s})(3.00 \times 10^{8}\,\text{m/s})/(270 \times 10^{-9}\,\text{m}) = 7.37 \times 10^{-19}\,\text{J}.$$

The difference is the work function:

$$W = E - K = 6.72 \times 10^{-19}\,\text{J} = (6.72 \times 10^{-19}\,\text{J})/(1.6 \times 10^{-19}\,\text{J/eV}) = 4.2\,\text{eV}.$$

The photoelectric effect has some important applications, inasmuch as it lies behind many devices that react to light signals. The camera exposure meter and light-activated keys for automobiles, distant television controls, or garage-door openers are some examples that come to mind. The photomultiplier described in •Fig. 4–4 depends on the photoelectric effect to initiate the cascade of electrons within it.

## 4–3 The Compton Effect

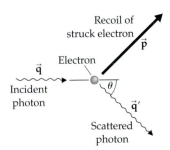

Recoil of struck electron $\vec{p}$

Electron $\vec{q}$

Incident photon

$\theta$

Scattered photon $\vec{q}'$

• **Figure 4–9**   The kinematics of the collision of a photon with an electron that is initially at rest.

If light consists of photons, collisions between photons and particles of matter should be possible. Let us think about the possible collisions between photons and the free electrons of a metal—that is, $\gamma + e \rightarrow \gamma + e$. The kinematics of such collisions[4] are as follows (•Fig. 4–9): Let the initial electron be at rest, with zero momentum and (relativistic) energy $m_e c^2$. Initially, the photon has energy $hf$, and associated with it is a momentum $\vec{q}$ whose magnitude is $hf/c$. After the collision, the photon has energy $hf'$ and momentum $\vec{q}'$ whose magnitude is $hf'/c$. The final electron momentum is $\vec{p}$ and its energy is $\sqrt{p^2 c^2 + m_e^2 c^4}$. Energy and momentum conservation gives

$$\vec{q} = \vec{q}' + \vec{p} \tag{4–8}$$

and

$$hf + m_e c^2 = hf' + \sqrt{p^2 c^2 + m_e^2 c^4}. \tag{4–9}$$

The question we want to answer is, What is the wavelength $\lambda'$ of the scattered photon? First, Eq. (4–8) shows that

$$p^2 = (\vec{q}' - \vec{q})^2 = q'^2 + q^2 - 2\vec{q}' \cdot \vec{q}$$

$$= \left(\frac{hf'}{c}\right)^2 + \left(\frac{hf}{c}\right)^2 - 2\left(\frac{hf'}{c}\right)\left(\frac{hf}{c}\right)\cos\theta, \tag{4–10}$$

where $\theta$ is the angle between the initial and final photon momentum vectors. Next, after isolating the square root, squaring, and dividing by $c^2$, we find that Eq. (4–9) gives

$$p^2 + m_e^2 c^2 = \left(\frac{hf}{c} - \frac{hf'}{c} + m_e c\right)^2$$

$$= \left(\frac{hf}{c}\right)^2 + \left(\frac{hf'}{c}\right)^2 - 2\left(\frac{hf}{c}\right)\left(\frac{hf'}{c}\right) + 2m_e h(f - f') + m_e^2 c^2. \tag{4–11}$$

We now substitute into Eq. (4–11) the value of $p^2$ found in Eq. (4–10), to give

$$2\left(\frac{hf}{c}\right)\left(\frac{hf'}{c}\right)(1 - \cos\theta) = 2m_e hc\left(\frac{f}{c} - \frac{f'}{c}\right). \tag{4–12}$$

By using the general relation $f = c/\lambda$, we easily find that

$$\lambda' - \lambda = \frac{h}{m_e c}(1 - \cos\theta). \tag{4–13}$$

This equation expresses the *change* $\Delta\lambda$ in the wavelength of light that scatters from free electrons. The parameter $h/(m_e c)$ that appears on the right-hand side of Eq. (4–13) has the dimensions of a length. It is called the **Compton wavelength of the electron**, and its magnitude is about $2.4 \times 10^{-12}$ m.

Arthur H. Compton made the foregoing calculation in 1922 in response to his inability to explain his experimental results on the subject with the use of

---

[4] See Chapter 3 for more on relativistic collisions.

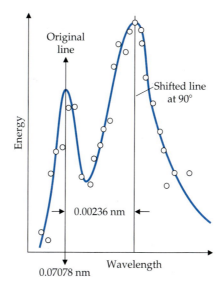

**• Figure 4–10**   In the Compton effect, photons scattered from metal foils exhibit two wavelengths, one associated with scattering from electrons and the other associated with scattering by the heavy ions. The latter is essentially equal to the wavelength of the incoming photon.

classical ideas. Compton sent X rays[5] (high-frequency photons) through thin metallic foils and looked for radiation scattered at 90°. He found that the scattered photons had two wavelengths (•Fig. 4–10). One set of photons had a wavelength shift exactly as predicted for scattering from electrons. A second set had an unshifted wavelength. This set was due to scattering from the positively charged ions. The mechanism for both sets was the same, except that for the unshifted set, the electron mass in Eq. (4–13) was replaced by the ion mass. Because the ion mass is many thousands of times larger than the electron mass, the shift is tiny. This experiment was of great historical importance, confirming photons as real particles with momentum as well as energy.

**Example 4–5**   In a Compton scattering experiment, an incoming X ray of wavelength $\lambda = 5.53 \times 10^{-2}$ nm is scattered and detected at an angle of 35°. Find the fractional shift in the wavelength of the scattered X ray.

**Solution**   If $\lambda$ is the incoming wavelength and $\lambda'$ is the wavelength of the scattered X ray, then according to Eq. (4–13) the fractional change in wavelength is given by

$$\frac{\lambda' - \lambda}{\lambda} = \frac{h}{m_e c \lambda} (1 - \cos\theta)$$

$$= \frac{(6.63 \times 10^{-34}\,\text{J}\cdot\text{s})(1 - \cos(35°))}{(0.91 \times 10^{-30}\,\text{kg})(3.00 \times 10^{8}\,\text{m/s})(5.53 \times 10^{-11}\,\text{m})} = 7.9 \times 10^{-3},$$

or about a 1% shift.

The Compton effect can be a nuisance. It is only because X-ray films are thin that the recoil electrons from the Compton effect do not ruin the resolution of the image. On the positive side, the Compton effect does play an important

---

[5] Why X rays and not, say, visible light? The photons must carry enough energy to be able to treat the scattering as scattering from free, and not bound, electrons. Since, as we know, the energy of a photon is proportional to the frequency, photons at X ray frequencies are much more energetic than visible-light photons. And there is another reason: The *difference* in wavelength in Eq. (4–13) is visible experimentally only if it is sizable compared with $\lambda$ itself, and this demands that $\lambda$ be small, as is the case for X rays.

role in cancer therapy. X-ray photons penetrate to a tumor, where they produce showers of electrons through Compton scattering. In this way, and through further scatterings of these electrons, energy can be deposited in the core of the tumor.

## 4–4 Blackbody Radiation

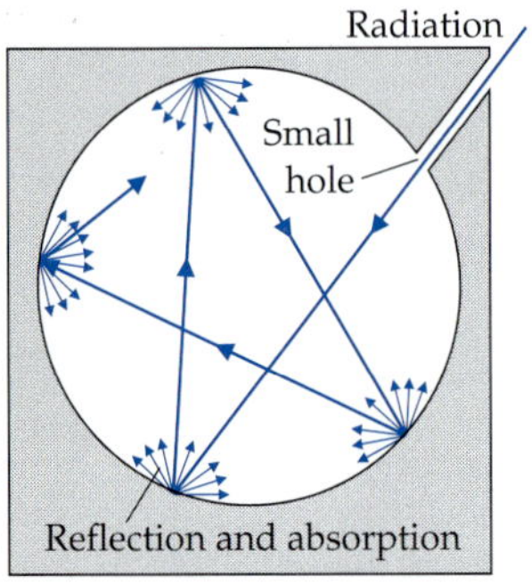

• **Figure 4–11**   A photon entering a cavity through a small hole is effectively absorbed, so that the cavity represents a blackbody.

In the historical introduction to Part II, we described the important role played by blackbody radiation in the eventual understanding of quantum mechanics. Here, we can look at this problem in more detail, armed with the new knowledge that electromagnetic radiation consists of photons.

To create a blackbody, we can cut a small hole in a cavity maintained at a temperature $T$. Any radiation that enters the cavity through the hole is almost guaranteed to be absorbed, entering into thermal equilibrium within (•Fig. 4–11). The small amount of radiation that escapes from within through the hole is very close to perfect blackbody radiation. Kirchhoff focused our attention away from the cavity walls to the radiation in the cavity. He concentrated on the **energy density** $u(f, T)$ of the radiation inside the cavity. The quantity $u(f, T)\, df$ is the energy per unit volume in the frequency range from $f$ to $f + df$. The radiation in the cavity had to be *isotropic*—flowing in no special direction—and *homogeneous*—the same at all points inside the cavity. Futhermore, the radiation had to be "universal,": the same in all cavities for a given $T$ and for each frequency $f$, no matter how each cavity was constructed.

The Rayleigh–Jeans treatment of the energy density showed that classical ideas lead inevitably to a serious problem in understanding blackbody radiation. However, where classical ideas fail, the idea of radiation as photons succeeds. According to this idea, we think of radiation as a gas of photons of varying frequencies, with the energy of each photon given by $hf$.

Consider only photons of a certain fixed frequency $f$. The cavity may contain 0, 1, 2, 3 ... photons of this frequency, and in fact, the energy density depends on the function that describes[6] the probability $P_n$ that there are $n$ photons of frequency $f$ at temperature $T$ in the cavity. This probability function follows from the law first discovered by Boltzmann (one of the subjects of Chapter 12) which states that

$$P(E) = A(T)\exp\left(-\frac{E}{kT}\right),$$

where $k$ is Boltzmann's constant and $A(T)$ is such that

$$\sum_E P(E) = 1.$$

Here $P(E)$ is the probability that the constituents of a thermal system have energy $E$. This last equation says that the system has to have *some* energy.

Applied to the system we are examining, the equation for $P(E)$ yields

$$P_n = A(T)\exp\left(-\frac{hfn}{kT}\right). \tag{4–14}$$

---

[6] If you are unfamiliar with probability, see Appendix B–3 at the end of the text.

We have

$$\sum_{n=0}^{\infty} P_n = 1 = A(T) \sum_{n=0}^{\infty} \left[ \exp\left(-\frac{hf}{kT}\right) \right]^n$$

$$= A(T) \frac{1}{1 - \exp(-hf/kT)};$$

hence,

$$A(T) = 1 - \exp\left(-\frac{hf}{kT}\right),$$

and

$$P_n = \left[1 - \exp\left(-\frac{hf}{kT}\right)\right] \exp\left(-\frac{hfn}{kT}\right). \tag{4-15}$$

Now, given the probability function $P_n$, we can calculate the average number of photons $\langle n \rangle$:

$$\langle n \rangle = (0 \times P_0) + (1 \times P_1) + (2 \times P_2) + (3 \times P_3) +$$

$$= \sum_{n=0}^{\infty} n P_n = \left(1 - e^{-hf/kT}\right) \times \sum_{n=0}^{\infty} n e^{-hfn/kT}.$$

To compute this quantity, we simplify the notation by letting $x = \exp(-hf/kT)$. Then

$$\langle n \rangle = (1 - x) \sum_{n=0}^{\infty} n x^n = (1 - x) \sum_{n=0}^{\infty} x \frac{d}{dx} x^n = (1 - x) x \frac{d}{dx} \sum_{n=0}^{\infty} x^n$$

$$= (1 - x) x \frac{d}{dx} \frac{1}{1 - x} = \frac{x}{1 - x} = \frac{1}{\dfrac{1}{x} - 1}$$

$$= \frac{1}{\exp(hf/kT) - 1}. \tag{4-16}$$

Since we know the energy associated with these photons, we can find the average energy contained in photons of frequency $f$, as well as the corresponding energy density. The average energy is

$$\langle E(f) \rangle = \langle n(f) \rangle hf = \frac{hf}{\exp(hf/kT) - 1}, \tag{4-17}$$

and the energy probability distribution $u(f, T)\, df$ is found by multiplying the average energy per frequency by the number of different *modes* contained in the interval $df$. The modes refer to the possible classical standing waves that can be established in the cavity, and counting these is a classical problem. We are going to put off the problem of just how to do this counting for later (see Chapter 12) and just quote the result here. The number of independent normal modes in the frequency range $(f, f + df)$ is

$$2 \times \frac{4\pi f^2}{c^3}\, df. \tag{4-18}$$

Therefore

$$u(f, T) = \frac{8\pi f^2}{c^3} \langle E(f) \rangle = \frac{8\pi f^2}{c^3} \frac{hf}{\exp(hf/kT) - 1}. \tag{4-19}$$

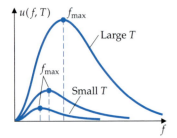

**• Figure 4–12**  The blackbody radiation spectrum as a function of frequency for several values of the temperature $T$.

This is Planck's formula, Eq. (H–7), derived here from the photon point of view.

The **blackbody spectrum** given by Eq. (4–19) has some interesting features. It is plotted as a function of frequency for various temperatures in •Fig. 4–12. We see that it has a maximum at a frequency $f_{max}$ where $du/df = 0$:

$$\frac{d}{df}u(f,T) = 0 = \frac{d}{df}\frac{8\pi h f^3}{c^3}\frac{1}{\exp(hf/kT)-1}$$

$$= \frac{24\pi h f^2}{c^3}\frac{1}{\exp(hf/kT)-1} - \frac{8\pi h f^3}{c^3}\frac{h}{kT}\frac{\exp(hf/kT)}{(\exp(hf/kT)-1)^2}.$$

$$(4\text{–}20)$$

By multiplying by the factor $[\exp(hf/kT)-1]^2$, we can see that this equation takes the transcendental form $y = 3[1 - \exp(-y)]$, where $y \equiv hf/kT$. The solution lies close to $y = 3$, and to a good approximation, we have $y = 3[1 - \exp(-3)] = 2.82$. Thus, the blackbody spectrum peaks at roughly

$$f_{max} = (2.8)\frac{kT}{h} = (5.9 \times 10^{10}/\text{K}\cdot\text{s})T,$$

a result known as the *Wien displacement law*. Since measurements with gratings involve wavelengths, one should convert the distribution in frequency to one in wavelength. This distribution also has a maximum, of course, at a maximum wavelength, which, as you can work out in Problems 16 through 20 at the end of the chapter, occurs at (Note that $\lambda_{max} \neq c/f_{max}$!)

$$\lambda_{max}T = 2.9 \times 10^{-3}\,\text{m}\cdot\text{K}.$$

This presence of such a maximum is what gives the predominant color to the radiation of a blackbody. For example, the Sun's surface is at some 6000 K. The maximum wavelength is then about 480 nm, in the middle of the visible range for the human eye, 400 to 700 nm. This is not a coincidence, since evolutionary forces would make the eye most sensitive to the dominant frequencies in the environment.

**Example 4–6**    Find the energy density of blackbody radiation at $T = 6000$ K in the range from 450 to 460 nm. Assume that this range is so narrow that the energy density function $u(f,T)$ does not vary much over it.

**Solution**    The fact that $u$ does not vary much over a range of frequencies $\Delta f$ means that the energy density is given simply by $u(f,T)\Delta f$ rather than the corresponding integral over a range of frequencies. We first find the range $\Delta f$. Using $f = c/\lambda$, we have

$$f_{max} = (3 \times 10^8\,\text{m/s})/(450 \times 10^{-9}\,\text{m}) = 6.67 \times 10^{14}\,\text{Hz},$$

while

$$f_{min} = (3 \times 10^8\,\text{m/s})/(460 \times 10^{-9}\,\text{m}) = 6.52 \times 10^{14}\,\text{Hz}.$$

Thus,

$$\Delta f = f_{max} - f_{min} = 1.45 \times 10^{13}\,\text{Hz}.$$

As for $u$, we can simply evaluate it at any frequency in the range, and we do so at the midpoint $f_{av} = 6.60 \times 10^{14}$ Hz. We have

$$\frac{hf_{av}}{kT} = \frac{(6.63 \times 10^{-34}\,\text{J}\cdot\text{s})(6.60 \times 10^{14}\,\text{Hz})}{(1.38 \times 10^{-23}\,\text{J/K})(6000\,\text{K})} = 5.28,$$

and the exponential of this is $\exp(5.28) = 197$. Therefore, from Eq. (4–19),

$$u(f_{av}, T) = \frac{8\pi(6.63 \times 10^{-34}\, \text{J·s})(6.60 \times 10^{14}\, \text{Hz})^3}{(3 \times 10^8\, \text{m/s})^3(197 - 1)} = 5.7 \times 10^{-20}\, \text{J·s/m}^3.$$

In turn, the energy density in the range we are interested in is

$$u(f_{av}, T)\Delta f = 8.2 \times 10^{-7}\, \text{J/m}^3.$$

The **total energy density**—the energy density integrated over all frequencies—for blackbody radiation is a function of temperature alone:

$$U(T) = \int_0^\infty u(f, T)\, df = \int_0^\infty \frac{8\pi h f^3}{c^3} \frac{1}{\exp(hf/kT) - 1}\, df. \qquad (4\text{–}21)$$

It is often easiest to handle an integral such as this by making the argument of the exponential a new variable. Therefore, we substitute $x = hf/kT$, from which it follows that $df = dx\, kT/h$, and

$$U(T) = \frac{8\pi h}{c^3} \frac{(kT)^4}{h^4} \int_0^\infty \frac{x^3\, dx}{\exp(x) - 1} = (7.52 \times 10^{-16}\, \text{J/m}^3 \cdot \text{K}^4)T^4, \qquad (4\text{–}22)$$

where we have used the fact that the integral over $x$ is $\pi^4/15$. (See Appendix B–2.) This dependence on the fourth power of temperature had already been foreseen in the form of the *Stefan–Boltzmann law*, before the blackbody spectrum was known. Of course, it was not possible to calculate the constant multiplying the $T^4$ factor until Planck's work, because that constant depends on $h$.

---

**Example 4–7**    Find the average energy contained in photons of a given frequency at temperature $T$ in the two limits $hf \gg kT$ and $hf \ll kT$.

**Solution**    Solving this problem is a matter of taking limits of the more general result for the average energy for a given frequency, namely Eq. (4–17). For the limit $hf \gg kT$, the exponential factor in the denominator swamps the factor of 1, and the average energy is $hf \exp(-hf/kT)$. For the classical limit $hf \ll kT$, the exponential factor is approximately $\exp(hf/kT) \approx 1 + hf/kT$, and

$$\langle E(f) \rangle \approx \frac{hf}{1 + hf/kT - 1} = kT.$$

We see that in this limit we recover the average energy $kT$ that was used in the development of the Rayleigh–Jeans expression for the low-frequency part of the spectrum.

---

## 4–5 Conceptual Consequences of Light As Particles

The fact that $h$ is small explains why we do not see individual photons under ordinary circumstances. As we showed in Example 4–1, a 60-W lightbulb emits more than $10^{19}$ photons per second. Puzzles—or at least conflicts with our intuition—come when we think about individual photons and about how the properties of individual photons mesh with phenomena that are described well by Maxwell's equations. Consider two examples:

1. A beam of light is directed at a barrier with two slits in it. On a distant screen, we expect to see—and we do see—an interference pattern that is described by wave theory. (See Chapter 1.) The pattern does not depend at all on the intensity of the beam; rather, it depends only on the difference in path lengths

between a point on the screen and the two slits. But we have just finished explaining that light is composed of photons. Suppose the intensity of the beam is reduced so that only one photon per second passes through the slits. Then individual photons will hit the second screen in what at first will appear to be a random pattern. After a sufficiently long time—it could be millennia—the "correct" interference pattern will emerge. A careful experiment carried out by the British physicist I. G. Taylor in 1909 showed that this is indeed the case. The question is, *How do photons that arrive at such different times "know" how to build up the pattern?* However hard it is to comprehend, the only real answer to this question is that each photon "knows" that there are two slits; so, in effect, each photon interferes with itself! The reason this is so hard to swallow is that a "normal" particle goes through one slit or the other and is indifferent to the slit it doesn't go through.

2. Light passes through a sheet of Polaroid™ and then through a second one that makes an angle $\alpha$ with the first. As we remarked in Chapter 1, the intensity of the polarized light entering the second sheet, $I_0$, is reduced to $I = I_0 \cos^2 \alpha$ by that sheet. If the intensity is now reduced until photons arrive just one at a time at the sheets, then the $\cos^2 \alpha$ factor still obtains. This is very peculiar. After all, if a photon acted like the sort of particles we are familiar with—a billiard ball, say—it could only be transmitted or blocked; it is not possible that only part of it would get through both sheets of Polaroid™.

One cannot get out of these difficulties by thinking that photons can split into smaller pieces. All of the evidence, starting with the work of Compton and continuing to present-day experiments with radiation of all possible wavelengths, clearly shows that photons cannot be split. They are particles that can be created and absorbed, but not divided. Once we have recognized the impossibility of subdividing photons, neither of the phenomena described here—and there are many others like them—is comprehensible in terms of classical physics. By contrast, we shall see that quantum mechanics does describe them.

## 4–6 Matter Waves and Their Detection

In 1923, in a 16-page doctoral thesis, the French nobleman and physicist Louis de Broglie (•Fig. 4–13) proposed the possibility that matter has wavelike properties. In search of the possible relationship, he took as his starting point the particle properties of radiation. Recall from Eq. (4–4) that photons obey a relation between their momentum $\vec{p}$ and their wavelength $\lambda$ that involves the constant $h$. De Broglie suggested that *this relation is a perfectly general one, applying to radiation and matter alike.* We repeat the relation here:

$$\lambda = \frac{h}{|\vec{p}|} \tag{4–23}$$

The wavelength in this equation, when applied to matter, is known as the **de Broglie wavelength**. Observe that $p = mv$ if the particle in question is nonrelativistic and $p = mv\left(1 - v^2/c^2\right)^{-1/2}$ if the particle is relativistic. For photons, this relation is not new: It follows from $\lambda f = c$, the relativistic relation between energy and momentum, $E = |\vec{p}|c$, appropriate for massless particles, and the Planck formula $E = hf$.

De Broglie's thesis, which his advisor sent to Einstein, soon attracted much attention, and suggestions were made for verifying the existence of the de Broglie waves through the observation of electron diffraction. Waves are diffracted when they move around obstacles, or "turn corners." What de Broglie

• **Figure 4–13**   Louis de Broglie.

was proposing is that electrons or other particles also can pass around obstacles, which was utterly counter to everything we knew about particles at the time! The actual experiments, which were carried out in 1927 by C. J. Davisson and L. H. Germer[7] in the United States and by G. P. Thomson in Great Britain, test that aspect of diffraction which predicts interference maxima and minima when waves pass through gratings. (See Chapter 1.) Recall that the effects of diffraction are most evident when the wavelength of the wave is comparable to the spacing on the grating. A quick estimate of the electron wavelength for electrons that might be used for this test (see Example 4–8) shows that the wavelength is hundreds of times shorter than the wavelengths of visible light. Thus, a very different grating had to be used, and that was supplied by the regular array of atoms making up a crystal. The experiments consist in looking for preferential scattering in certain directions—diffraction maxima—when electrons are incident on the surface of a crystal. The conditions for these interference effects are those of classical optics, as we see in the next subsection.

### Conditions for Interference in Crystals

The derivation of the interference conditions for waves involves the difference in phase between waves reflected from adjacent scattering planes (•Fig. 4–14). In the figure, this difference is given by

$$\Delta\varphi = \left(\frac{2\pi}{\lambda}\right) \times (\text{difference in path length}) = \left(\frac{2\pi}{\lambda}\right)(ABC - AD). \quad (4\text{–}24)$$

Now,

$$ABC - AD = AB + BC - (AX + XD),$$

and since $AX = BC$, the path difference is

$$AB - XD = \frac{a}{\sin\theta} - \frac{a}{\sin\theta}\cos 2\theta = \frac{a}{\sin\theta}(1 - \cos 2\theta) = 2a\sin\theta.$$

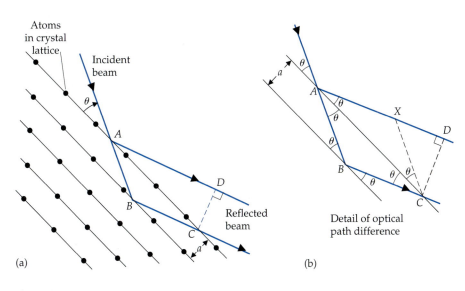

(a)

(b)

Atoms in crystal lattice

Incident beam

Reflected beam

Detail of optical path difference

• **Figure 4–14**  (a) The schematics of a diffraction experiment for electrons reflected by a crystal. (b) Details of the interference conditions for waves, involving the phase difference, and hence the path difference, between waves reflected from adjacent planes.

---

[7] In fact, anomalous results on electron scattering from crystals was first reported some years before the de Broglie hypothesis, and even after de Broglie's work the idea that the hypothesis could be tested by electron scattering from crystals took some time to develop. Thus the Davisson–Germer results actually preceded the theory, and it took some time to understand what the interference effects they saw were due to. Once the connection between theory and experiment was finally made, the theory received strong support.

When we insert this expression into Eq. (4–24), we find that

$$\Delta\varphi = \left(\frac{2\pi}{\lambda}\right)(2a\sin\theta) = \left(\frac{4\pi a}{\lambda}\right)\sin\theta. \tag{4–25}$$

As is true of all types of waves—see Chapter 1—there will be constructive interference whenever this phase difference is equal to $2\pi n$, where $n$ is an integer. Thus the condition for constructive interference—one refers to *diffraction maxima*—is

$$\left(\frac{4\pi a}{\lambda}\right)\sin\theta = 2\pi n, \tag{4–26}$$

or

$$\lambda = \frac{2a}{n}\sin\theta. \tag{4–27}$$

We call the condition for diffraction maxima the **Bragg condition**, after the British father–son team of W. H. and W. L. Bragg, who shared the 1915 Nobel prize in physics for their work on X-ray diffraction by crystals. For us, a crucial feature of the Bragg condition is that it is a general condition for waves scattering from a lattice. It does not matter whether the waves are classical electromagnetic radiation or the "waves" formed by electrons.

---

**Example 4–8**   In a diffraction experiment in which electrons of kinetic energy 110 eV are scattered from a crystal, a first maximum in the intensity of the scattered electrons occurs at an angle $\theta = 10.7°$, where $\theta$ is as defined in •Fig. 4–14. **(a)** How many peaks will there be in the interference pattern? **(b)** What is the spacing between the crystal planes?

**Solution**   **(a)** We begin with the expression that tells us where maxima in the scattering occur, namely, Eq. (4–27). In this case, knowing the location of the first peak, we want to count the total number of peaks. Equation (4–27) can be rewritten in the form

$$\frac{n\lambda}{2a} = \sin\theta_n,$$

where we have put a subscript on the angle to help us count the number of maxima.
   With $\theta_1 = 10.7°$, we have $\sin\theta_1 = 0.186$. The peaks are equally spaced in $\sin\theta_n$ according to our expression, and since the largest possible value of $\sin\theta_n$ is 1, we find the number of maxima by finding the number of times 0.186 fits into 1; the answer is $1/0.186 = 5.4$, so the largest possible value of $n$ is $n = 5$.
**(b)** In this part, knowing that $\sin\theta_n = 0.186$ for $n = 1$, we want to find the value of $a$; we can again rewrite Eq. (4–27), in this case in the form

$$a = \frac{\lambda}{(2\sin\theta_1)} = \frac{\lambda}{(2\times 0.186)}.$$

Thus to find $a$ we need only find the electron wavelength $\lambda$.
   The wavelength of the electron is found by using the de Broglie expression given in Eq. (4–23), namely

$$K = \frac{p^2}{2m} = \frac{h^2}{2m\lambda^2},$$

where $K$ is the kinetic energy of the electron. Hence, we have

$$\lambda^2 = \frac{h^2}{2mK} = \frac{(6.63\times 10^{-34}\,\text{J·s})^2}{2(0.91\times 10^{-30}\,\text{kg})(110\,\text{eV})(1.6\times 10^{-19}\,\text{J/eV})} = 1.37\times 10^{-20}\,\text{m}^2,$$

or

$$\lambda = 1.17 \times 10^{-10} \text{ m}.$$

In turn, this gives

$$a = \frac{1.17 \times 10^{-10} \text{ m}}{2 \times 0.186} = 3.15 \times 10^{-10} \text{ m}.$$

### Testing the Wave Character of Electrons

In the Davisson–Germer experiment (•Fig. 4–15), the spacing between adjacent planes of a crystal—the Bragg planes—was first measured to be $0.91 \times 10^{-10}$ m by X-ray diffraction techniques. In other words, the spacing of the crystal planes was measured, as in Example 4–8, by looking for maxima corresponding to the Bragg condition for X rays, which constitute a particular region of the electromagnetic spectrum. Davisson and Germer then looked for diffraction maxima in the scattering of electrons from the same crystal. A first diffraction maximum was indeed observed in the electron scattering at 65°. With $n = 1$, the Bragg condition predicts that

$$\lambda = 2a \sin \theta = 2(0.91 \times 10^{-10} \text{ m})(\sin 65°) = 1.65 \times 10^{-10} \text{ m}.$$

From the de Broglie condition, this equation implies that the electron momentum is

$$p = \frac{h}{\lambda} = \frac{6.63 \times 10^{-34} \text{ J} \cdot \text{s}}{1.65 \times 10^{-10} \text{ m}} = 4.02 \times 10^{-24} \text{ kg} \cdot \text{m/s}.$$

In turn, this corresponds to an electron kinetic energy of

$$K = \frac{p^2}{2m_e} = \frac{(4.02 \times 10^{-24} \text{ kg} \cdot \text{m/s})^2}{2(0.91 \times 10^{-30} \text{ kg})} \frac{1 \text{ J}}{1.6 \times 10^{-19} \text{ eV}} = 56 \text{ eV},$$

a result that was in good agreement with the independently measured incident electron energy value of 54 eV.

Matter-diffraction experiments have been carried out with more massive particles, such as neutrons and helium atoms (•Fig. 4–16). In each case, the results agree with de Broglie's idea.

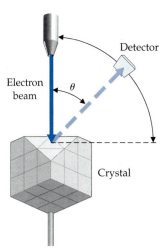

• **Figure 4–15**   Schematic of the Davisson–Germer experiment designed to measure interference patterns in the scattering of electrons from crystals.

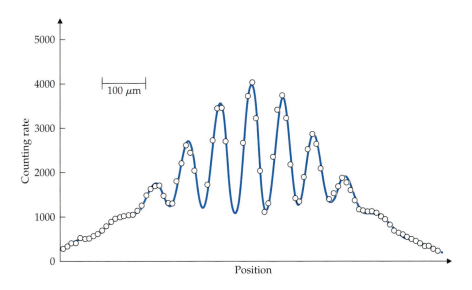

• **Figure 4–16**   Neutrons produce a diffraction pattern just as classical waves do. (From A. Zeilinger et al., *Rev. Mod. Phys*, **60**, 1067 (1988) by permission.)

## 4–7  Conceptual Consequences of Particles As Waves

Just as the small size of $h$ hides the fact that photons exist, so it hides the wave properties of matter from our everyday experience. A dust particle of mass $10^{-4}$ g traveling at 1 m/s has a momentum of $10^{-6}$ kg·m/s and a wavelength of

$$\lambda = \frac{h}{p} = \frac{(6.6 \times 10^{-34} \text{ J·s})}{(10^{-6} \text{ kg·m/s})} = 6.6 \times 10^{-28} \text{ m}.$$

This number is so small—the diameter of an atom is about $10^{-10}$ m—that it is impossible to detect even with the finest instruments, let alone with our human senses.

Nevertheless, on first hearing, the idea that particles can behave like waves is a truly bizarre one. Worse, as was the case for the photon, some serious difficulties arise when we think carefully about de Broglie's discovery. For example, when J. J. Thomson discovered the electron, he established the fact that it was a particle—an object whose position could be described by a coordinate $\vec{r}(t)$. Moreover, the position had to satisfy an equation of the general form $m \, d^2\vec{r}/dt^2 = \vec{F}$, where $\vec{F}$ might be given by, for example, the Lorentz force law. The particular form of the equation is not as important as the fact that there is a solution to it and, therefore, a well-defined path $\vec{r}(t)$. To see why this leads to an apparently paradoxical result, think about the transmission of electrons through a pair of slits in a screen. A classical interference pattern on a screen (see •Fig. 4–17a) is produced when one part of a wave passes through one slit and another part through the other slit. *But a particle with a*

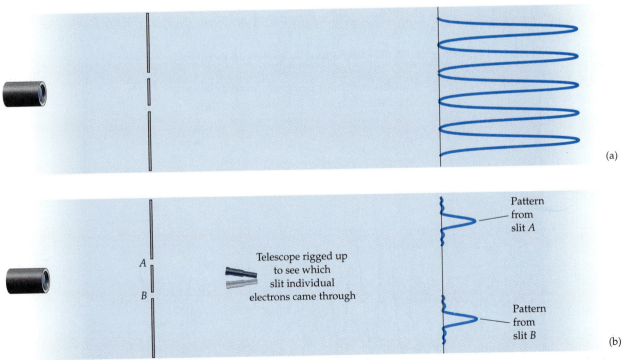

• **Figure 4–17**    (a) Interference pattern on a screen due to electrons passing through two slits. (b) When one slit is open, or, equivalently, when one identifies which slit the electron passes through, the interference pattern disappears. If a monitor identifies which slit any individual electron passes through, the pattern is a sum of the patterns due to electrons passing through single slits.

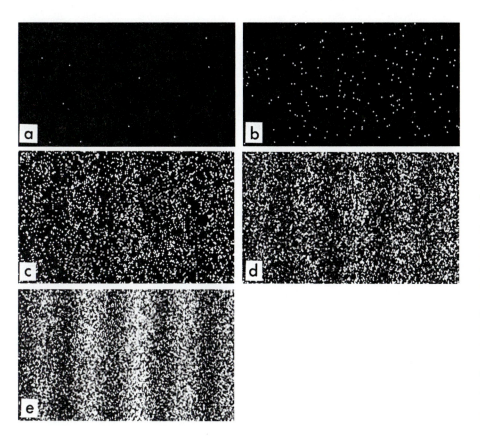

• **Figure 4–18** Individual electrons coming at a rate of 1000 per second (which, at the energies the electrons possess, corresponds to a spatial separation of 150 km) nevertheless create an interference pattern. The photos, which illustrate how the pattern is built up as the number of electrons increases, correspond to 10, 100, 3000, 20,000, and 70,000 electrons respectively having passed the slits. (Courtesy of A.Tonomura, Hitachi Advanced Research Laboratory).

*well-defined path passes through only one slit or the other*. The observation of two-slit interference with electrons is not, as one might at first think, connected with interference between *different* electrons. We can reduce the number of electrons that pass through the slits per unit time to the point where it is clear that the effect is due to the passage of one electron after another (•Fig. 4–18).

A further troubling aspect enters when we employ a detection apparatus that can tell when an electron passes through one slit or the other. We could even tell, by measuring the electron's precise direction of travel, which slit it was going to pass through and, in anticipation, shut the other slit. But when we do this, *the interference pattern disappears* (•Fig. 4–17b). Somehow, the measurement itself has decided whether the electron has behaved like a particle or like a wave. We are forced to the conclusion that the problem is deeper than the classical description ever imagined. The resolution of the problem had to wait for the quantum mechanical revolution.

## SUMMARY

The laws of classical radiation and thermodynamics are in conflict with the way radiation is actually distributed in the interior of a heated cavity at a given temperature—a blackbody. To explain this distribution, we introduce the idea that this radiation comes in quanta—discrete packets of energy. The energy contained in such a packet is proportional to the frequency of the radiation, and the constant of proportionality is a fundamental constant of nature called Planck's constant. Not only does this idea explain the distribution of radiation in a blackbody, but it also leads to the following developments:

- Quanta of light carry both energy and momentum, and these are proportional to each other. If we think of these quanta as particles, then special relativity implies that they are massless, always moving at the speed of light.
- The quantum nature of light can be tested by the photoelectric effect: When one shines light on a metal surface, electrons are liberated from the metal. The quantum hypothesis suggests that the kinetic energy of the liberated electrons is proportional to the frequency of the light, but does not depend on its intensity.
- The hypothesis is also successfully tested by means of the Compton effect, in which collisions between energetic quanta of radiation and electrons obey relativistic kinematic energy and momentum conservation laws.
- When Planck's constant is set equal to zero in the expressions for blackbody radiation, those expressions reduce to classical results. It is the very small magnitude of $hf$ compared with macroscopic energies that signals the apparent classical nature of the world up till the discoveries described in this chapter.
- As light shows particle characteristics, so matter shows wave characteristics. The wavelength of a particle in motion is equal to Planck's constant divided by the momentum of the particle. For objects like baseballs, this wavelength and any associated wave properties are so small as to be unobservable, but for electrons in atoms the wave effects are quite visible. Electrons impinging on suitable diffraction gratings show the diffraction patterns characteristic of waves.

## QUESTIONS

**1.** What is there about the quantity $\hbar \approx 1.05 \times 10^{-34}$ J·s which suggests that it has something to do with angular momentum? In later chapters, we shall see that $\hbar$ is indeed closely associated with angular momentum.

**2.** If you peer through a small hole in a very hot oven, you see a glow that clearly suggests the presence of light waves. This is the blackbody radiation in the oven. If you peer through a hole in a freezer, will there be blackbody radiation inside? What, if anything, distinguishes this radiation from the radiation you see in the oven?

**3.** How would you estimate the temperature of a red-hot piece of steel, given that the wavelength of red light is in the neighborhood of 700 nm? What assumption would you have to make for your estimate?

**4.** If Earth were only half its actual distance from the Sun, would the average temperature of Earth change? If so, by about what factor? State the assumptions you must make to be able to give a rough estimate of the factor.

**5.** Suppose you have a source that emits a beam of light at some frequency $f$ that impinges on a metal plate. What happens to the energy of the photoelectrons that are emitted when you are moving the source closer to the plate?

**6.** In our treatment of the photoelectric effect, we neglected the conservation of momentum. Why is this approach justified? Suppose you take conservation of momentum into account. What qualitative change would it make to the energy of the electron that is emitted?

**7.** Without working out the algebra in detail, why is the process $\gamma + e \rightarrow e$ forbidden by the energy–momentum conservation laws, while the process $e + p \rightarrow \gamma + H$ is allowed? Here, $p$ stands for the proton and H for hydrogen. The allowed process is known as the *radiative capture of electrons*.

**8.** Consider •Fig. 4–17, which shows the interference effects that occur when a beam of electrons passes through two slits. Is there enough information in the figure and the text to figure out how far the slits are from the screen on which the interference pattern is found?

## PROBLEMS

**1.** ❚ The night-adapted eye can detect as few as several photons per second—say five to be definite. Assuming that the pupil of the eye is 0.6 cm in diameter, from what distance would it be possible for the naked eye to detect the 60-W lightbulb of Example 4–1?

**2.** ❚ Chemical processes typically involve energies on the order of 1 eV. What, then, is the typical wavelength of electromagnetic radiation emitted in the course of chemical reactions? Nuclear processes involve energies on the order of 1 MeV. Where in the spectrum of electromagnetic radiation are the photons that may be emitted in a nuclear reaction?

**3.** ❚ It is often convenient to express Planck's constant $\hbar$ in units of MeV·s rather than J·s. Find $\hbar$ in MeV·s.

**4.** ❚ The quantity $\hbar c$ occurs frequently in calculations in quantum mechanics. Express this quantity in MeV·fm. (1 fermi = 1 fm is $10^{-15}$ m. It is a happy coincidence that "fm" can stand both for "femto-," the Greek-based prefix that indicates $10^{-15}$, and Fermi.)

**5.** ❚ Solar radiation falls on Earth's surface at a rate of 1400 W/m². Assuming that the radiation has an average wavelength of 550 nm, how many photons per square meter per second fall on the surface?

**6.** ❚❚ A black wall absorbs all the photons that strike it. How many photons per second would have to be emitted by a laser that fires a beam of cross-sectional area 1 mm² perpendicularly at the wall in order to exert a pressure of 1 atm on the wall? The laser emits light of wavelength 600 nm. What would be the power (energy per unit time) in this beam?

**7.** ❚ By doubling the intensity of his monochromatic light source, an experimenter hopes to increase the speed of the photoelectrons emitted by a given sample of metal. Will the experimenter succeed? What, if anything, will change in the resulting emission of photoelectrons?

**8.** ❚ Light of frequency $8.5 \times 10^{15}$ Hz falls on a metal surface. If the energy of the resulting photoelectrons is 1.7 eV, what is the work function of the metal?

**9.** ❚ A metal has work function 4.7 eV. What is the kinetic energy of a photoelectron if radiation of wavelength 200 nm falls on the surface of the metal?

**10.** ❚ Radiation of wavelength $\lambda = 250$ nm falls on a metal surface. Obtain a numerical expression for $W + K$, where $W$ is the work function of the metal and $K$ is the kinetic energy of a photoelectron, both in units of eV.

**11.** ❚❚ A sodium surface emits $6.25 \times 10^7$ photoelectrons per square centimeter per second. Assume that sodium atoms are regularly spaced (sodium, with atomic weight 23, has a mass density of 0.97 g/cm³) and that the photoelectrons are uniformly supplied by the top 10 layers of atoms. Find how many atoms are needed to produce one photoelectron per second.

**12.** ❚❚ **(a)** A 210-MeV photon collides with an electron at rest. What is the maximum energy loss of the photon? **(b)** Repeat part **(a)**, but with a proton target rather than an electron. Is the difference between your results reasonable?

**13.** ▮ In a Compton scattering experiment, the wavelength of the incident X rays is $7.078 \times 10^{-2}$ nm while the wavelength of the outgoing X rays is $7.314 \times 10^{-2}$ nm. At what angle was the scattered radiation measured?

**14.** ▮▮ In a Compton scattering experiment, a detector is set at an angle of 57°. What must the frequency of the incoming X rays be in order to produce a final X ray with a frequency 1% less than the initial frequency?

**15.** ▮▮ In blackbody radiation, the only quantities that enter are $\hbar$, the radiation angular frequency $\omega$, and the temperature in the combination $kT$. Use the fact that $\hbar\omega$ has dimensions of energy to estimate the temperature of blackbody radiation that is predominantly in the range $f = 10^{16}$ Hz.

**16.** ▮▮▮ The function $u(f, T)$ is the distribution of blackbody radiation in terms of frequency; $u(f, T)\, df$ is the energy contained in the frequency interval from $f$ to $f + df$. Use the relation between frequency and wavelength to find the function $Y(\lambda, T)$ that describes the distribution in wavelength; $Y(\lambda, T)\, d\lambda$ is the energy contained in a wavelength interval from $\lambda$ to $\lambda + d\lambda$.

**17.** ▮▮▮ In the previous problem you will have found the distribution in wavelength of blackbody radiation,

$$Y(\lambda, T) = \frac{8\pi hc}{\lambda^5} \frac{1}{\exp\left(\dfrac{\hbar c}{\lambda kT}\right) - 1}.$$

Find, by graphical or other numerical means, the wavelength $\lambda_{max}$ for which $Y(\lambda, T)$ is a maximum. (*Hint*: You should find that the maximum occurs when $(5 - y) = 5e^{-y}$, where $y = \hbar c/(\lambda kT)$. A solution to this transcendental equation lies close to $y = 5$.)

**18.** ▮▮ Upon exploding, a hydrogen bomb develops a temperature on the order of $10^8$ K. Assuming that the fireball behaves as a blackbody in some time interval, what is the value $\lambda_{max}$ at which the distribution has a maximum? (See Problem 17.) What is the energy of the corresponding photon?

**19.** ▮ The cosmic background radiation is that of a blackbody at 2.7 K. What is the value $\lambda_{max}$ at which the distribution has a maximum? (See Problem 17.) What is the energy of the corresponding photon?

**20.** ▮▮ The Sun radiates approximately as a blackbody at a temperature of about 6000 K. How much energy is emitted per square centimeter per second in the (narrow) range of wavelengths from 579 nm to 581 nm?

**21.** ▮▮ The Sun radiates approximately as a blackbody at a temperature of about 6000 K. What is the total power radiated by the Sun? Assuming that this energy is produced by chemical processes and that the mass of the Sun is $2 \times 10^{30}$ kg, how long would you estimate the Sun could burn? Are you assuming that it keeps radiating at 6000 K? Is this assumption reasonable?

**22.** ▮▮▮ The density of the energy of radiation in an enclosure at temperature $T$ has the form $u(T) = aT^4$. Suppose the enclosure is a sphere whose radius grows at a rate $dr/dt = v_0$. Assuming that no energy enters or leaves the enclosure, will the temperature grow or decrease, and at what rate?

**23.** ▮▮ The night-adapted eye can detect the glow of a burning cigarette at some 500 m. Assuming that the pupil of the eye is 0.6 cm in diameter and that the cigarette tip is a hemisphere 1 cm in diameter glowing as a blackbody at a temperature of 1000 K, estimate the rate at which the eye receives photons from the cigarette.

**24.** ▮▮ The Sun radiates approximately as a blackbody at a temperature of about 6000 K and subtends an angle of 1° at the distance of Earth. A satellite forming a sphere 1 m in diameter orbits the Sun at the distance of Earth. The satellite receives radiation from the Sun and at the same time radiates this energy as a blackbody. Assuming that there is no internal generation of energy in the satellite and no mechanism for energy loss other than via blackbody radiation, what is the temperature of the satellite? (*Hint*: Find the power absorbed by the satellite from the Sun, as well as the power radiated if the satellite is at a temperature $T$. The two powers are equal at equilibrium.)

**25.** ▮▮ **(a)** Use the technique of Problem 24 to estimate the surface temperature of the Moon. **(b)** Use the same technique to estimate Earth's average temperature, with the assumption that only 70% Sun's radiant energy is absorbed by Earth and its atmosphere. (The rest is reflected.) Satellite measurements give an average temperature of 255 K, colder than what we experience on the surface, because the satellite measurements really measure the temperature in the middle of the atmosphere.

**26.** ▮ An electron with energy $10^3$ MeV collides with a photon whose wavelength corresponds to the maximum wavelength of the 2.7-K cosmic background radiation. What is the maximum energy loss the electron can suffer as a result of the collision? What is the wavelength of the photon after this maximal collision?

**27.** ▮▮ A photon of frequency $f$ collides with a photon whose wavelength corresponds to the wavelength maximum of the 2.7-K cosmic background radiation. As a result, an electron–positron pair is produced. (A positron, denoted $e^+$, is identical to an electron except for its charge, which is $+e$ rather than $-e$.) The process referred to here is symbolized as $\gamma + \gamma \rightarrow e^- + e^+$.) The frequency $f$ is such that when the collision is head-on, there is just enough energy to produce the electron–positron pair at rest in the center-of-mass frame. What is $f$?

**28.** ▮ Find the de Broglie wavelength of **(a)** an electron with kinetic energy of 1 eV; **(b)** an electron with kinetic energy of 1 keV; **(c)** an electron with kinetic energy of 10 MeV; **(d)** a neutron with kinetic energy of $kT$, where $T = 300$ K; **(e)** a neutron with kinetic energy of 10 MeV.

**29.** ▮ For what kinetic energy will a particle's de Broglie wavelength equal its Compton wavelength?

**30.** ▮ In order to study structures of size $a$, it is necessary to have a probe with wavelength $\leq a$. Find the minimum kinetic energy of an electron that can be used as a probe to resolve such a structure.

**31.** ▮▮ Even a grapefruit has a de Broglie wavelength. Estimate its value if the grapefruit has been thrown with a speed of 10 m/s. For the mass of the grapefruit, use 500 g. If the grapefruit is thrown toward a wall with two holes separated by 50 cm, find the angular separation between successive maxima of the resulting interference pattern. Treat the grapefruit as pointlike!

**32.** ▮▮ Neutrons $(m = 1.67 \times 10^{-27}$ kg$)$ pass through a crystal and exhibit an interference pattern. If the neutrons have a kinetic energy of 1.7 eV, and the separation between successive maxima in the interference pattern is $6.4 \times 10^{-2}$ rad, what is the separation of the crystal planes that produce the interference pattern?

**33.** ▮ A certain crystal has a planar spacing of 0.25 nm. What energies are necessary to observe up to three interference maxima for **(a)** electrons and **(b)** neutrons?

**34.** ▮▮ The spacing between scattering planes in nickel is $2.15 \times 10^{-10}$ m. What is the scattering angle at which 80 eV electrons have a diffraction maximum?

# Atoms and the Bohr Model

While philosophers going all the way back to the ancient Greeks conjectured that matter consisted ultimately of indivisible units—atoms—it was only in the second half of the 19th century that the study of matter became quantitative. It produced a number of very puzzling results that were crucial to the 20th-century development of quantum mechanics. We begin with these puzzles, and then we shall see how Niels Bohr explained them in terms of quantum ideas. The clues provided by atomic behavior were the ones that led most directly to the theory that explained the strange effects to be described here and those discussed in Chapter 4. Only quantum mechanics can correctly explain the features of atoms.

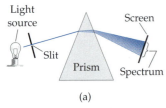

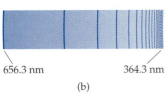

656.3 nm        364.3 nm

(b)

• **Figure 5–1** (a) The principle of a spectrometer: Here, a prism is used to spread the spectrum of light from a source composed of a variety of different wavelengths onto a screen. In practice, reflection gratings are used. (b) The series of spectral lines known as the Balmer lines in the spectrum of hydrogen.

## 5–1 The Behavior and Structure of Atoms

In the mid-19th century, major advances in the study of atoms came from experiments on the light spectra emitted by hot gases. These spectra were observed using diffraction gratings in instruments known as spectroscopes (•Fig. 5–1). We have already described the distribution of blackbody radiation. When a blackbody was observed through a flame in which a particular substance was burned, black lines—gaps in the otherwise continuous distribution of frequencies—appeared in the spectrum of the blackbody. (The word "line" refers to the appearance of the spectrum when viewed through a diffraction grating that spreads the radiation. A particular frequency that was emitted would show up as a bright line parallel to the direction of the diffraction slits; a particular frequency that was blocked from an otherwise continuous spectrum would show up as a black, or absent, line.) The black lines were due to the absorption of light of *particular* frequencies by the atoms in the flame. It was found that every absorption spectrum was characteristic of a given element. Each element had its own spectrum, and this finding made a powerful tool for the identification of the chemical composition of different substances. Characteristic frequencies again made their appearance when substances were excited by an electric discharge. There, the so-called spark spectra showed isolated spectral lines. When solar light was observed, it was found that the Sun, too, produces individual lines, and this work led to the discovery in the solar spectrum of terrestrial elements, the first of which, as we mentioned earlier, was sodium, found by Gustav Kirchhoff in the Sun's spectrum in 1859. Perhaps more remarkably, the English astronomer Joseph Lockyer discovered helium in 1869 by observations of lines in the solar spectrum unlike any that had been observed on Earth. At the time, helium was not a known terrestrial element.

For our purposes, the most interesting experimental results were those relating to the spectrum of hydrogen. In 1853, the Swede Anders Ångstrom first determined that a set of discrete frequencies was present in the radiation emitted by hydrogen. The Swiss high school teacher Johann Balmer analyzed the data on hydrogen in 1885 and showed that these frequencies—or, what is equivalent, their corresponding wavelengths—formed a definite pattern, which soon came to be known as the *Balmer series*. He found by an empirical fit that the particular wavelengths of the spectral lines satisfied a relation that is a special case of what we call today the **Rydberg–Ritz formula**,

$$\frac{1}{\lambda} = \mathrm{Ry}\left(\frac{1}{n^2} - \frac{1}{m^2}\right). \tag{5–1}$$

Here, $n$ and $m$ are positive integers, with $n < m$ to keep the result positive, and the constant $\mathrm{Ry} = 1.09737 \times 10^7\,\mathrm{m}^{-1}$. Balmer's special case was for $n = 2$. The constant is named after the Swedish physicist Johannes Rydberg, who verified and generalized this result during the 1890s. We call Ry the **Rydberg constant** or, simply, the **Rydberg**. The Swiss physicist Walter Ritz emphasized that the Balmer result gives the frequency of any spectral line as the *difference* between two terms, with each term characteristic of the hydrogen atom (•Fig. 5–2).

What is true for hydrogen is in fact true for all radiation emitted or absorbed by individual atoms or molecules. This radiation comes in discrete frequencies that are so thoroughly characteristic of the atoms or molecules in question that they in effect form a fingerprint. Accordingly, there is an important application in which the measurement of the frequencies emitted or absorbed by a material allows us to identify components that are present in the material, even in very small amounts.

Let us turn now from questions of discrete frequencies to questions of atomic structure. In Section 1–6, we described Rutherford's classical experiments. These experiments established the existence of an atom that resembles a miniature solar system. But just how close is this resemblance? Several observations reveal that the atom behaves very differently from an electric equivalent of a mechanical solar system and that Newtonian physics is very far from providing an explanation of atomic structure. We expand on two such observations that we briefly mentioned in Section 1–6.

- When classical electric charges accelerate, they radiate, losing energy in the process. A classical electron in orbit about a nucleus undergoes acceleration, and as it radiates and loses energy, it will spiral into the nucleus. A classical calculation shows that the electron should be absorbed into the nucleus in only $10^{-10}$ s!

- A classical picture cannot explain why all atoms of an element are *the same*, since, in the classical planetary picture, orbital energies, for example, depend on the initial conditions and can vary by arbitrarily small amounts. This classical variability is contradicted by empirical fits like those described by Eq. (5–1), which were interpreted by Ritz as being due to differences in what he called *terms*, but which we shall see are associated with energies that can have only distinct, discrete values.

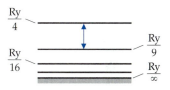

• **Figure 5–2** The horizontal lines describe the Ritz terms. The vertical line shows one of many wavelengths in the hydrogen spectrum, here corresponding to the difference between the terms Ry/4 and Ry/9.

• **Figure 5–3** Niels Bohr early in his career. After the triumph of his atomic model, Bohr returned to Denmark and was given his own institute for physics in Copenhagen. Physicists from all over the world came there to study with Bohr, and in the 1930s it became a haven for refugees from Hitler.

## 5–2 The Bohr Atom

In 1913, the Danish physicist Niels Bohr (•Fig. 5–3) proposed a model that, although it retained classical elements such as visualizable orbits, broke radically with the classical picture. We shall see that Bohr's model of the atom was entirely successful at accounting for the types of data described in Section 5–1. As a result, the Bohr theory attracted a great deal of interest. Nevertheless, some very distinguished physicists balked at the superimposition of new ideas onto well-known classical physics—for example, while only certain orbits are allowed, the ones that are allowed are classical. Other physicists were held back by the fact that the Bohr theory could not answer some very obvious questions: How does an electron "know" that it should behave in the fashion that Bohr's model would require? How does a photon emitted by an atom "know" which direction to go in when it is emitted? These questions, and others like them, would not be answered until a complete version of quantum mechanics was discovered in the late 1920s. But such doubts did not deter a small determined band from working with Bohr on extending his ideas. It took them about a decade to formulate quantum mechanics in its first recognizable form.

Bohr was one of the towering figures of 20th-century physics. His work on the atom was done in England, just after he received his doctor's degree. He was fortunate to have had the opportunity to work with Ernest Rutherford. (See Chapter 1.) Rutherford understood that the shy, almost inarticulate young man he first met had the stuff of genius and gave him the enouragement and support he needed. Bohr reported that his reaction to the Balmer formula, which he had just learned about, led him to his model. He offered the following postulates:

1. Atoms can exist only in certain allowed "states." A state is characterized by, among other things, having a definite (discrete) energy, and any change in the energy of a system, including the emission and absorption of radiation, must take place as transitions between states.

2. The radiation absorbed or emitted during the transition between two allowed states whose energies are $E'$ and $E''$ has a frequency $f$ given by

$$hf = E' - E'', \qquad (5\text{--}2)$$

where $h = 2\pi\hbar$, in which $\hbar$ is Planck's constant, the same constant that appears in the treatment of blackbody radiation.

3. Some of the allowed states—the ones that, we shall see, correspond to classical circular orbits—have energies determined by the condition that *their angular momentum is quantized as an integral multiple of Planck's constant $\hbar$*; that is,

$$L = n\hbar, n = 1, 2, \ldots. \qquad (5\text{--}3)$$

The integer that appears here will be reflected in all atomic properties. We call this integer a **quantum number**.

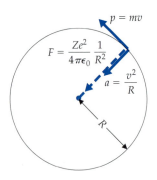

• **Figure 5–4** Force and acceleration for a circular orbit of radius $R$ in a hydrogenlike atom.

These postulates determine the possible values for the radii $R$ of the *allowed circular orbits* and their associated energies. Here is how: Let $m$ be the mass of the electron, and apply Newton's second law to the electron, with a centripetal acceleration $a = v^2/R$. The corresponding inward force is the Coulomb force, the force on an electron, with charge $-e$, due to a nucleus of charge[1] $+Ze$ (•Fig. 5–4). This force has magnitude $F = Ze^2/(4\pi\varepsilon_0 R^2)$. Thus,

---

[1] Real hydrogen has $Z = 1$. Why, then, make $Z$ general? The answer is that this gives us an additional handle on things and, moreover, allows us to treat atoms of general $Z$ that have been ionized to the point of having only one electron present; we can call these "hydrogenlike" atoms.

$$\frac{mv^2}{R} = \frac{Ze^2}{4\pi\varepsilon_0 R^2}. \tag{5–4}$$

If we recall that the angular momentum of our electron is $L = mvR$, then we can substitute for $R$ using $R = L/mv$. This converts Eq. (5–4) into an equation for $v$ in terms of $L$ that is easily solved for $v$:

$$v = \frac{Ze^2}{4\pi\varepsilon_0 L}.$$

Finally, substitution of the quantized values of $L$ from Eq. (5–3) gives the speed $v_n$ of the electron in the orbit labeled by the quantum number $n$:

$$v_n = \frac{Ze^2}{4\pi\varepsilon_0 n\hbar}. \tag{5–5}$$

The presence of the subscript $n$ on the speed emphasizes that the speed has been quantized; that is, that only certain discrete values of the speed appear. This is a restriction that is not at all classical—a classical orbit can always be adjusted as finely as one likes, and there will be a continuum of speeds corresponding to the continuum of orbits. The appearance of integers and the **quantization** of physical quantities is a feature repeated throughout the quantum theory.

It is often useful to express results in terms of the dimensionless quantity known as the **fine-structure constant**, defined by

$$\alpha \equiv \frac{e^2}{4\pi\varepsilon_0 \hbar c}. \tag{5–6}$$

Using the best values of the physical constants that appear in this equation, we find that the numerical value of $\alpha$ is $1/137.035982$. It is useful to remember an approximate form for $\alpha$, namely,

$$\alpha \approx \frac{1}{137}.$$

In terms of $\alpha$, the quantized values of the speed of the electron in its orbit are

$$v_n = \frac{Z\alpha c}{n}, \quad n = 1, 2, \ldots. \tag{5–7}$$

Equation (5–7) for $n = Z = 1$ gives us a characteristic electronic orbital speed: $\alpha c$, or about one percent of the speed of light.

### The Atomic Radius

Once we have the allowed speeds $v_n$, we can find the allowed radii $R_n$ by using the relation $mvR = L$. Substituting the allowed values of $L$, namely $n\hbar$, we find that $R_n = n\hbar/mv_n$. Thus, the radius of the $n$th atomic orbit is

$$R_n = \frac{n\hbar}{mv_n} = \frac{n\hbar}{m(Z\alpha c/n)} = \frac{n^2}{Z}\left(\frac{\hbar}{mc\alpha}\right). \tag{5–8}$$

Once again, we see the quantization phenomenon: The allowed orbits are characterized by integers and take on only certain discrete values. A "basic" orbit is characterized by $n = 1$ and $Z = 1$ (as in hydrogen). This orbit's radius, known as the **Bohr radius**, is given by

$$a_0 = \frac{\hbar}{mc\alpha} = \frac{(1.05 \times 10^{-34}\,\text{J}\cdot\text{s})}{(0.91 \times 10^{-30}\,\text{kg})(3.0 \times 10^{-8}\,\text{m/s})(1/137)}$$

$$= 0.53 \times 10^{-10}\,\text{m}. \tag{5–9}$$

We thus have a remarkable result: *The typical atomic size is correctly given in the Bohr model.* (This value had been estimated well over 100 years earlier; see Section 1–6.)

The Bohr radius gives us a scale for atomic radii. In terms of it, the atomic radius of the $n$th orbit is, according to Eq. (5–8),

$$R_n = \frac{n^2}{Z} a_0. \tag{5–10}$$

## The Atomic Energy

Using Eqs. (5–7) and (5–10), we can find the energy of the electron in the $n$th orbit. This energy is generally given by $E = K + V$, where $K$ is the kinetic energy and $V$ is the potential energy. The potential energy of a particle of mass $m$ and charge $-e$ a distance $r$ from a heavy nucleus of charge $+Ze$ is, as usual,

$$V = -\frac{Ze^2}{4\pi\varepsilon_0 r},$$

where we have set the zero of the potential energy at an infinite distance.

In the Bohr approach, both the orbital kinetic and potential energies are quantized, because the speed, which enters into the kinetic energy, and the orbital radius, which enters into the potential energy, are quantized. Thus, the total energy is also quantized, carrying a subscript $n$:

$$E_n = \frac{1}{2}mv_n^2 - \frac{Ze^2}{4\pi\varepsilon_0 R_n} = \frac{1}{2}m\left(\frac{Z\alpha c}{n}\right)^2 - \frac{Ze^2}{4\pi\varepsilon_0}\frac{Zmc\alpha}{\hbar n^2}$$

$$= \frac{1}{2}m\left(\frac{Z\alpha c}{n}\right)^2 - m\left(\frac{Z\alpha c}{n}\right)^2$$

$$= -\frac{1}{2}m\left(\frac{Z\alpha c}{n}\right)^2. \tag{5–11}$$

As in the Kepler problem of planetary orbits, the energies of the electron bound to the nucleus are negative. That is because we defined the potential energy to be zero when the electron is infinitely far from the nucleus. This makes it evident that positive energy must be supplied to remove the electron from the atom.

The lowest (most negative) allowed energy corresponds to $n = 1$. This lowest energy is known as the **ground state**. As $n$ increases, and the states become more and more **excited**, the allowed energies, sometimes referred to as the **energy levels**, crowd more and more closely together towards an energy value of zero, which corresponds to a just-bound electron. •Figure 5–5 illustrates these levels. Once the energy of the electron–nucleus system becomes zero or positive, the atom is said to be *ionized*: The electron moves independently of the nucleus.

We can consider our basic energy unit to be the ground-state energy for hydrogen ($n = Z = 1$), and this energy is

$$-\frac{1}{2}mc^2\alpha^2 = -\frac{1}{2}(0.51\,\text{MeV})\frac{1}{137^2} = -13.6\,\text{eV}. \tag{5–12}$$

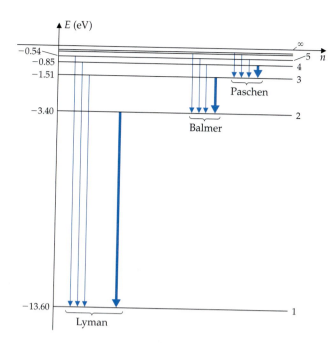

• **Figure 5–5** The energy levels of atomic hydrogen according to the Bohr model. The vertical arrows show the transitions ending with $n = 1$ (Lyman series), $n = 2$ (Balmer series), and $n = 3$ (Paschen series). Note that the energies of these *bound states* are negative. The heavy lines in which the change in $n$ is one are the most intense ones, as discussed in Section 5–4.

This result tells us that the electron volt is the typical energy scale of atomic physics. In terms of the basic energy, the energy of the $n$th orbit for "hydrogen" with a nucleus of charge $+Ze$ is

$$E_n = (-13.6 \text{ eV})Z^2/n^2. \tag{5–13}$$

It is sometimes convenient to express the energies in terms of the Bohr radius $a_0$. With the use of Eq. (5–9), we find that

$$E_n = -\frac{\hbar^2}{2ma_0^2}\frac{Z^2}{n^2}. \tag{5–14}$$

You might note that with the use of the fine-structure constant $\alpha$, the dimensional character of various quantities is clear. Speeds [Eq. (5–7)] are proportional to $c$, radii [Eq. (5–9)] are proportional to $\hbar/mc$, and energies [Eq. (5–11)] are proportional to $mc^2$. In addition to the dimensional quantities, factors of the dimensionless parameters $Z$ and $\alpha = e^2/(4\pi\hbar c) \approx 1/137$ appear.[2]

De Broglie made an interesting observation when, a decade after the Bohr theory was published, he introduced the wave properties of electrons. He noted that if we take a circular orbit of radius $r$ and put a "wave"—actually an electron—of wavelength $\lambda$ on this orbit, then, in general, the wave will interfere destructively with itself as it goes around the orbit, *unless an integral number of wavelengths fits neatly into the orbit*. This condition reads (see • Fig. 5–6)

$$2\pi r = n\lambda, \tag{5–15}$$

where $n$ is an integer. When Eq. (5–15) is combined with the de Broglie condition $\lambda = h/p = 2\pi\hbar/p$, we get

$$pr = n\hbar. \tag{5–16}$$

• **Figure 5–6**    A standing de Broglie wave in a circular orbit. Bohr's states correspond to the situation in which an integral number of waves fit along the orbit.

---

[2] We stress that the appearance of $c$ in the various equations is due only to its introduction in the definition of $\alpha$. There is nothing relativistic about our problem; see, for example, Eq. (5–5).

Since $pr$ is the angular momentum in a circular orbit, this is exactly the Bohr angular momentum condition given in Eq. (5–3). We have seen how the level structure follows from it.

---

**Example 5–1**   Free electrons are most easily captured into an atomic orbit when their kinetic energy matches the kinetic energy associated with that orbit. What is the kinetic energy, in electron volts, that a free electron should have to facilitate its capture into the Bohr orbit for $n = 2$ in hydrogen?

**Solution**   The orbital kinetic energy is determined by the orbital speed, which is given by Eq. (5–7) with $Z = 1$. We then have, for the kinetic energy in the given orbit,

$$K_{n=2} = \frac{1}{2}mv_{n=2}^2 = \frac{1}{2}m\left(\frac{\alpha c}{2}\right)^2 = \frac{\alpha^2 mc^2}{8}$$

$$= \frac{0.51 \text{ MeV}}{8(137)^2} = 3.4 \text{ eV}.$$

Note that by using the combination $mc^2$ we immediately find the result in electron volts.

---

## Atomic Transitions in the Bohr Model

Equation (5–2) allows us to calculate the frequency of the radiation emitted when an electron makes a transition (•Fig. 5–5) from an orbit of higher energy with quantum number $n_2$ to one of lower energy with quantum number $n_1$. We take the difference in energies of the levels and use the fact that $E = hf$, so that the frequency $f_{n_2 \to n_1} = \dfrac{E_{n_2} - E_{n_1}}{h}$:

$$f_{n_2 \to n_1} = \frac{m(Z\alpha c)^2}{2h}\left(\frac{1}{n_1^2} - \frac{1}{n_2^2}\right). \tag{5–17}$$

We can find the wavelengths corresponding to this result by using the relation $f/c = 1/\lambda$. We therefore have a formula of the form

$$\frac{1}{\lambda_{n_2 \to n_1}} = \frac{m(Z\alpha c)^2}{2hc}\left(\frac{1}{n_1^2} - \frac{1}{n_2^2}\right). \tag{5–18}$$

*This equation has the same form as the Rydberg–Ritz formula* given in Eq. (5–1). In particular, we can set $Z = 1$ and thereby identify the Rydberg constant Ry in Eq. (5–1), namely,

$$\text{Ry} = \frac{m(\alpha c)^2}{2hc} = \frac{(mc^2)\alpha^2}{4\pi\hbar c}. \tag{5–19}$$

This formula is easily evaluated, using $mc^2 = 0.511$ MeV for the electron. We also note the very useful value

$$\hbar = 6.58 \times 10^{-22} \text{ MeV} \cdot \text{s}, \tag{5–20}$$

as well as the equally useful product

$$\hbar c = \left(6.58 \times 10^{-22} \text{ MeV} \cdot \text{s}\right)\left(3.00 \times 10^8 \text{ m/s}\right) = 1.97 \times 10^{-13} \text{ MeV} \cdot \text{m}. \tag{5–21}$$

Equation (5–21) allows us to find

$$\text{Ry} = \frac{(0.511 \text{ MeV})(1/137)^2}{4\pi(1.97 \times 10^{-13} \text{ MeV} \cdot \text{m})} = 1.10 \times 10^7 \text{ m}^{-1}, \tag{5–22}$$

which is in very good agreement with experiment.

**Example 5–2** The so-called alpha-Lyman line of hydrogen is radiation emitted in the transition from the $n = 2$ level to the $n = 1$ level. What is the wavelength corresponding to that line?

**Solution** This is a simple application of Eq. (5–18) with $Z = 1$, $n_1 = 1$, and $n_2 = 2$. The coefficient of the factor $(1/n_1^2 - 1/n_2^2)$ is just the Rydberg. Thus,

$$\frac{1}{\lambda_{n_2=2 \to n_1=1}} = \text{Ry}\left(\frac{1}{1^2} - \frac{1}{2^2}\right) = \frac{3}{4}\text{Ry} = 0.825 \times 10^7 \text{ m}^{-1}.$$

Inverting, we find that the sought-after wavelength $1.21 \times 10^{-7}$ m $= 121$ nm, in the ultraviolet range of the spectrum.

**Example 5–3** In the spectrum of radiation emitted by hydrogen atoms, there is a series of transitions to the level with $n = 3$ from levels with $n > 3$. This series of transitions is known as the *Paschen series*. What is the longest wavelength observed in that series?

**Solution** The longest wavelength corresponds to the smallest frequency, because the wavelength is inversely proportional to the frequency. Thus, we want the shortest frequency, and because the frequency of the emitted radiation is proportional to the difference in energy of the levels, we want the transition with the smallest energy difference. That transition is from $n = 4$ to $n = 3$. We thus want to use Eq. (5–18) with $n_2 = 4$ and $n_1 = 3$:

$$\frac{1}{\lambda_{n_2=4 \to n_1=3}} = \text{Ry}\left(\frac{1}{3^2} - \frac{1}{4^2}\right) = \frac{7}{144}\text{Ry} = 5.35 \times 10^5 \text{ m}^{-1}.$$

Inverting, we find a value of $1.87 \times 10^{-6}$ m $= 1870$ nm for the wavelength, in the infrared region.

**Example 5–4** Consider an atom of lead, with $Z = 82$. By bombarding the atom with a beam of electrons, a collision occurs that removes an electron from its ground-state $(n = 1)$ position. An electron from the $n = 2$ level subsequently makes a spontaneous transition to the vacant state. What is the frequency of the photon that is emitted?

**Solution** This is the same kind of transition as described in Example 5–2, except that the value of $Z$ is 82 instead of 1. The frequency $f$ is proportional to $Z^2$ and so is $(82)^2$ times larger than the corresponding result in the case of hydrogen. From Example 5–2 we found the hydrogen wavelength to be 121 nm; the frequency for hydrogen is then

$$f = c/\lambda = (3.00 \times 10^8 \text{ m/s})/(1.21 \times 10^{-7} \text{ m}) = 2.48 \times 10^{15} \text{ Hz}.$$

The frequency for the transition in lead (Pb) is then

$$f_{\text{Pb}} = (82)^2 f_{\text{H}} = 6724 \times (2.48 \times 10^{15} \text{ Hz}) = 1.67 \times 10^{19} \text{ Hz}.$$

Radiation of this frequency lies in the X-ray region of the spectrum.

Problems involving atoms or ions with more than one electron are generally not well described by the Bohr theory, because it does not take into account the fact that electrons repel each other. Thus, this example does not give an answer that is completely trustworthy, but it does give a correct order of magnitude.

In the preceding examples and text, we referred to the historical names of certain series of frequencies corresponding to certain transitions—Balmer, Lyman, and Paschen; in the problems, we treat other examples.

*The Effects of Reduced Mass* A refinement cemented the special place of the Bohr approach. The hydrogen atom is a two-body problem, and as Newtonian mechanics implies, it is equivalent, as far as its internal behavior is concerned, to a single body under the influence of a true central force if the mass that appears in the single-body problem is the reduced mass $\mu$. The reduced mass is determined by the relation

$$\frac{1}{\mu} = \frac{1}{m} + \frac{1}{m_{nuc}},$$

where, as usual in this chapter, $m$ is the mass of the electron and $m_{nuc}$ is the nuclear mass—in the case of most hydrogen atoms, the mass of a single proton. The preceding equation can be rearranged to give

$$\mu = \frac{m m_{nuc}}{m + m_{nuc}}. \tag{5–23}$$

The mass of the proton is some 2000 times larger than that of the electron, so that the denominator in this expression is very nearly $m_{nuc}$, and $\mu$ is very close to the electron mass $m$. (In other words, we have a single-body problem in which the body has mass $m$ if $m_{nuc}$ is infinitely large.) That is why the use of the electron mass in what we have done so far gives good numerical agreement with experiment.

We can make an approximation that takes into account the large nuclear mass as compared to the mass of the electron by the following steps: First, we write the denominator factor in Eq. (5–23) as

$$m_{nuc} + m = m_{nuc}(1 + m/m_{nuc})$$

We can then recognize that the ratio $m/m_{nuc} \ll 1$ and use the approximation

$$\frac{1}{1 + x} \approx 1 - x,$$

which is good for $x \ll 1$. In this way, we find that

$$\mu = \frac{m m_{nuc}}{m_{nuc}\left(1 + \dfrac{m}{m_{nuc}}\right)} = \frac{m}{1 + \dfrac{m}{m_{nuc}}} \approx m\left(1 - \frac{m}{m_{nuc}}\right). \tag{5–24}$$

We can see the utility of this way of writing the reduced mass by turning to Rydberg's constant. Let us enlarge the labeling of Eq. (5–19) to include the mass of the nucleus in the form $m_{nuc} = A m_p$, where $m_p$ is the mass of the proton (and neutron), and the $Z$-value of the particular "hydrogen" atom. This labeling takes the form Ry$(A, Z)$. In addition, we alternatively label Ry with the name of the atom. For example, we write Ry$_H$ = Ry(1, 1) for ordinary hydrogen, Ry$_{He}$ = Ry(4, 2) for helium with a single electron, and so forth. We also write Ry$_\infty$ for the value of the Rydberg with an infinitely heavy nucleus with $Z = 1$.

Using our results to this point, we can now read off the value of Ry$_\infty$ from Eq. (5–19); it is just what we labeled Ry in the text to this point:

$$\mathrm{Ry}_\infty = \mathrm{Ry} = \frac{(mc^2)\alpha^2}{4\pi\hbar c}. \tag{5–25}$$

As for Ry$(A, Z)$, we can read these off by replacing $m$ by the reduced mass—we shall use the approximate form given in Eq. (5–24) and by including a factor of $Z^2$:

$$\mathrm{Ry}(A, Z) = \frac{(mc^2)\alpha^2}{4\pi\hbar c} Z^2\left(1 - \frac{m}{m_{nuc}}\right) = \frac{(mc^2)\alpha^2}{4\pi\hbar c} Z^2\left(1 - \frac{m}{A m_p}\right)$$

$$= Z^2\left(1 - \frac{m}{A m_p}\right)\mathrm{Ry}_\infty. \tag{5–26}$$

Since this is the value of the Rydberg constant that controls the frequencies of emitted and absorbed radiation, Eq. (5–26) offers new ways to test Bohr's model. The reduced-mass correction is small, but experimentally significant.

**Example 5–5**   An experimentalist measures, in a precise manner, the frequency of radiation emitted in transitions for singly ionized helium and for atomic hydrogen. What does the Bohr model predict for the ratio of the Rydberg constants, measured in these two sets of transitions?

**Solution**   The fact that we have found that in the Bohr model both Rydberg constants are multiples of $Ry_\infty$ simplifies this problem considerably. The factor $Ry_\infty$ will cancel in the ratio. Thus, using Eq. (5–26), we have

$$\frac{Ry_{He}}{Ry_H} = \frac{Z_{He}^2\left(1 - \dfrac{m}{A_{He}m_p}\right)}{Z_H^2\left(1 - \dfrac{m}{A_H m_p}\right)} = \frac{2^2\left(1 - \dfrac{m}{4m_p}\right)}{1^2\left(1 - \dfrac{m}{m_p}\right)}.$$

The ratio $m/m_p$ is approximately $5.46 \times 10^{-4}$. Thus, if the Bohr model is correct, the experimentalist should find that

$$\frac{Ry_{He}}{Ry_H} = 4\frac{\left(1 - \dfrac{5.46 \times 10^{-4}}{4}\right)}{(1 - 5.46 \times 10^{-4})} = 4.0016.$$

The experiment described in Example 5–5 actually was carried out in Bohr's time. Balmer's original measured value gave the product $c \times Ry_H$ to be $3.29163 \times 10^{15}$ Hz. (The present best value is $3.28984184 \times 10^{15}$ Hz.) Using what he knew about the electron mass and the other constants in his model, Bohr obtained $3.1 \times 10^{15}$ Hz for this same quantity, which was inside the uncertainty due to experimental error. We see that with the knowledge of the constants available to him, Bohr found it hard to predict a precise value for the Rydberg itself. However, Bohr recognized that the reduced-mass effect could give a precise and measurable prediction for $Ry_{He}/Ry_H$. Bohr correctly attributed the so-called Pickering lines to the spectrum of singly ionized helium. He noted that the Rydberg constant for helium should be four times that for hydrogen ($Z^2 = 4$ for helium). When it was pointed out to Bohr that the experimental ratio was actually 4.0016, he was able to explain it using the reduced-mass effect. This was powerful evidence that Bohr was on an interesting path.

Finally, we remark that the Bohr rules described in this section apply only to the quantization of circular orbits. But as with the planets, in classical physics the $1/r$ Coulomb potential implies the possibility of elliptical orbits. With considerable effort, it is also possible to develop rules for the quantization of these orbits, and these rules describe, correctly, spectral lines other than the ones we have discussed. It is not worth our while to follow this development in detail, because the work of Bohr and his collaborators led within a decade to an even more radical breakthrough: the creation of quantum mechanics, an approach that no longer needed the crutch of classical orbits.

## The Franck–Hertz Experiment

In 1914, just after the Bohr theory was proposed, an experiment carried out by the German physicists James Franck and Gustav Hertz provided direct evidence for the existence of energy levels in atoms. In this experiment (•Fig. 5–7a), a voltage was set up across a tube containing mercury vapor, and electrons

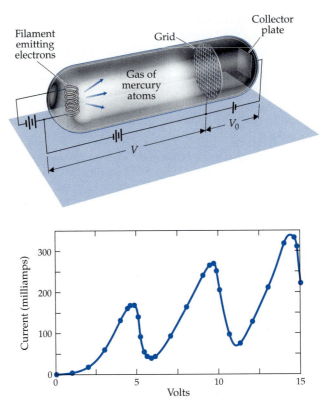

• **Figure 5–7** (a) Schematic diagram of the Franck–Hertz experiment. (b) Data showing the drops in the current passing through a tube of mercury gas. The sudden changes in the current are associated with specific allowed energies.

were accelerated through the tube. As the accelerating voltage was increased, the current rose, but at certain well-defined values of the voltage the current dropped very sharply before resuming its upward trend (•Fig. 5–7b). The rise in the current with increased voltage is certainly expected; it has to do with an increase in the velocity of the electrons. But why the sudden drops? When the electron energy reaches a certain threshold that exactly matches the difference in energy between atomic levels, the electron is able to give up that amount of energy to excite the atom—to induce a transition between the levels. The electron then proceeds to accelerate again through the tube. In mercury, the first threshold occurred at 4.9 eV. Franck and Hertz found that when the electron energy was below that value no lines were produced in the spectrum of mercury. But above that energy, an emission line appeared in the mercury spectrum, with wavelength 253.6 nm. This wavelength corresponds exactly to an energy of 4.9 eV, according to the photon energy relation $E = hf$. In some ways, the Franck–Hertz experiment does no more than ordinary spectroscopy; it does, however, show explicitly and quantitatively where the excitation energy in the atoms comes from.

## 5–3 Application of Bohr's Ideas to Other Systems

The Bohr postulates for the description of atomic spectra rest on two notions: first, the idea of allowed states with discrete energy values; and second, the heuristic rules that employ classical physics to construct an expression for the angular momentum of a system and then quantize that angular momentum in terms of what turns out to be the *same* Planck's constant that appears in the seemingly totally unrelated problem of blackbody radiation. From this perspective, Bohr's work can be seen as something more than a set of rules exclu-

■ In Chapter 9, the quantization of angular momentum in the form $L = n\hbar$ will be shown to be a very general consequence of quantum mechanics.

sively for use in atomic physics: The ideas can be applied to systems other than electrons orbiting about a nucleus. Indeed, Bohr himself recognized that *molecular* spectra had to have the same origin as the more easily treated Balmer series for hydrogen.

In this section, we apply Bohr's approach to two systems quite different from hydrogen: rotating diatomic molecules and a particle moving in circular motion under the influence of a spring force—the two-dimensional harmonic oscillator. Even though the Bohr model was totally superseded by quantum mechanics, the hybrid rules we have outlined have a practical utility in a first try at estimating the $n$-dependence of energies, and hence the range of frequencies, of emitted radiation.

## Rotations of Diatomic Molecules

We model a diatomic molecule such as $H_2$ as a kind of dumbbell (•Fig. 5–8): two pointlike atoms—actually, nuclei, because the nuclei carry most of the atomic mass—each of mass $M$ at opposite ends of a rigid rod of length $R$. (From the known size of molecules, we expect that $R$ is about twice the Bohr radius.) The rotational inertia of this system about an axis perpendicular to the rod and bisecting it is

$$I = M\left(\frac{R}{2}\right)^2 + M\left(\frac{R}{2}\right)^2 = \frac{MR^2}{2}$$

Thus, if our dumbbell rotates with an angular speed of $\omega$ about this same axis, its angular momentum is

$$L = I\omega = \left(\frac{MR^2}{2}\right)\omega. \tag{5–27}$$

The energy is $(1/2)I\omega^2$, which can be written in the form

$$E = \frac{L^2}{2I}. \tag{5–28}$$

If we now set $L = n\hbar$, we see that the energy values are quantized, taking the form (•Fig. 5–9)

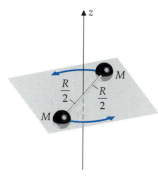

• **Figure 5–8** "Dumbbell" model of a diatomic molecule, with two atoms of mass $M$ each rotating about an axis through the midpoint of a massless rigid rod of length $R$ that holds the atoms together.

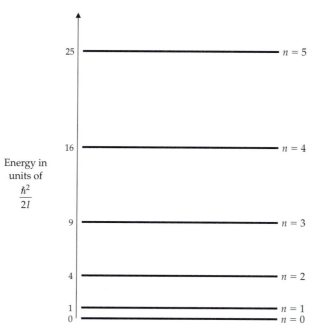

• **Figure 5–9** Energy levels of a diatomic molecule.

$$E_n = \frac{(n\hbar)^2}{(MR^2)} \tag{5-29}$$

As for the numerical values of these levels, we can substitute $a_0$ for $R$ and see by comparison with Eq. (5–14) that the energy levels are separated by gaps that are on the order of $m/M$ smaller than the gaps in the energy levels in, for example, hydrogen. This factor of $1/2000$ makes the effects of these *molecular rotational levels* quite different from those of atomic levels. We shall discuss this point more fully in later chapters, but it can also be demonstrated with the next example.

**Example 5–6**    A particular molecule of chlorine ($Cl_2$) can be treated as a dumbbell-shaped object with an average separation of $1.99 \times 10^{-10}$ m between the chlorine atoms.[3] The atoms have nuclei that consist of 17 protons and 20 neutrons; for the purposes of this example, you may suppose that protons and neutrons have the same mass, $1.67 \times 10^{-27}$ kg. What is the separation, in eV, between the energies of the rotational ground state and the first excited rotational state of this molecule?

**Solution**    This situation is exactly the one we described in the discussion just before this example. That discussion gives, for the difference in energies,

$$E_1 - E_0 = E_1 = \frac{\hbar^2}{MR^2}$$

$$= \frac{\left(1.05 \times 10^{-34} \text{ J} \cdot \text{s}\right)^2}{\left(37 \times 1.67 \times 10^{-27} \text{ kg}\right)\left(1.99 \times 10^{-10} \text{ m}\right)^2}$$

$$= \frac{4.19 \times 10^{-24} \text{ J}}{1.6 \times 10^{-19} \text{ J/eV}} = 2.62 \times 10^{-5} \text{ eV}.$$

This energy-level spacing is more than 10,000 times smaller than the typical 1-eV spacing of electronic levels in hydrogen.

In the Bohr approach, once we have supplied the energy levels, transitions between these levels occur with the emission of radiation, with the Bohr rule relating the frequency to the difference in energy between the two states participating in the transition. Thus, we expect that when a diatomic molecule makes a transition from a rotational state characterized by the integer $n + k$ to the state labeled by $n$, the frequency will be given by

$$f_{n+k \to n} = \frac{E_{n+k} - E_n}{2\pi\hbar} = \frac{\left[(n+k)^2 - n^2\right]\hbar^2}{MR^2(2\pi\hbar)}$$

$$= \frac{(2nk + k^2)\hbar}{2\pi MR^2}. \tag{5-30}$$

Bohr's insights into molecular spectra went well beyond the quantization of classical systems. After all, the two objects that form the "dumbbell" are electrically neutral, and classically there should be no radiation at all. Quantum mechanics shows that the electronic structure of the atoms that form a molecule lies behind the radiation.

*A Simple Demonstration*    The rotational levels of molecules are much more closely spaced than the atomic levels. This fact can easily be demonstrated in

---

[3] Numbers like this one are easily accessible in references such as the *American Institute of Physics Handbook*. These references are available in any engineering or science library and are indispensable tools.

the laboratory or lecture hall by what may be viewed as an inverse Franck–Hertz experiment. Two metal plates are placed about 1 cm apart, and a certain voltage, say 3500 V, is placed across them. This voltage is too low to cause a breakdown, and there are no sparks. Now helium is injected between the plates, and suddenly sparks appear. But the ionization potential of helium is higher than that of oxygen or nitrogen, so that the sparks cannot be directly associated with the ionization of helium. What, then, is happening? The explanation of the effect is the following: There are always some free electrons in the air, and they are accelerated by the electric field. The electrons collide with the molecules of oxygen or nitrogen, and because the energy differences associated with the *rotational* quantum states are so low, the electrons can lose their energy rather quickly and steadily after many successive collisions. Helium does not, however, form diatomic molecules. Thus, the electrons can lose energy only through the excitation of the atomic levels of the helium atoms. But that requires a great deal of energy; in helium, the electrons do not steadily lose energy through collisions. Instead, they can accumulate enough energy in the accelerating potential to ionize the oxygen and nitrogen atoms and give rise to an electrical discharge.

## The Harmonic Oscillator

We can apply Bohr's reasoning to *any* central three-dimensional attractive force, because any such force will support circular orbits. Recall that if we have a central force $F(r)$, we can find the constant speed $v(r)$ of a particle acted upon by that force and rotating in a circular orbit at radius $r$ by using Newton's second law in the form $F(r) = mv^2/r$. These orbits lie in a plane because of the conservation of angular momentum. Then the quantization of angular momentum will allow us to find the energy levels. A simple, fundamental, and very instructive example is provided by a mass tethered by a spring (•Fig. 5–10). In this case, the attractive central force is a Hooke's-law force—a linear restoring force of magnitude $kr$, where $k$ is the spring constant. For a circular orbit of radius $r$, Newton's law takes the form

$$\frac{mv^2}{r} = kr. \tag{5–31}$$

This expression, which is the analogue of Eq. (5–4), tells us that $v$ is proportional to $r$. It is conventional to write $k$ in the form $k = m\omega^2$, in which case Eq. (5–31) yields $v = \omega r$. We then have, for the angular momentum of the mass, $L = mvr = m\omega r^2$. When this equation is combined with the angular-momentum condition $L = n\hbar$, we find a condition for $r$ that quantizes the orbital radii, namely

$$r_n^2 = \frac{n\hbar}{m\omega}. \tag{5–32}$$

Here we have added the subscript on $r$ to emphasize its quantization. Finally, by recognizing that the potential energy of the linear restoring force is $kr^2/2 = m\omega^2 r^2/2$, we find that the allowed energies are

$$E_n = \frac{1}{2}mv_n^2 + \frac{1}{2}kr_n^2 = \frac{1}{2}m(\omega r_n)^2 + \frac{1}{2}(m\omega^2)r_n^2 = m\omega^2\frac{n\hbar}{m\omega}$$

$$= n\hbar\omega. \tag{5–33}$$

Because circular motion can be resolved into two superposed one-dimensional oscillatory motions in the plane of the circle, it is perhaps not too surprising that this result also holds for the one-dimensional oscillator. A more careful

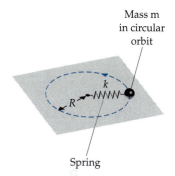

Mass m in circular orbit

Spring

• **Figure 5–10**   Circular orbit of a mass *m* attached to a spring of spring constant $k = m\omega^2$. The spring is tethered to a fixed point.

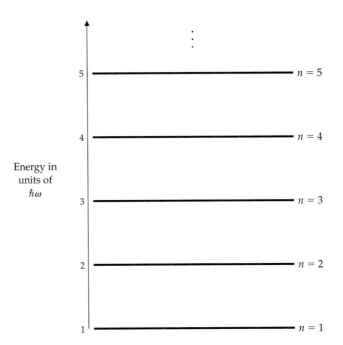

• **Figure 5–11** Energy levels of a harmonic oscillator.

treatment based on the Schrödinger equation (see Chapter 6) corrects Eq. (5–33) only with the addition of a constant term:[4]

$$E_n = \hbar\omega(n + 1). \tag{5-34}$$

This result (see •Fig. 5–11) will turn out to be quite useful—see, for example, the discussion of molecules in Chapter 11. The case of $n = 0$ is especially interesting. Equation (5–34) says that the energy of the ground state of the oscillator is not zero. This is a purely quantum mechanical effect and has important experimental consequences. It means that, in some sense, oscillators can never be quite at rest. They always have some residual oscillation energy.

## 5–4  The Correspondence Principle

Bohr was aware that his rules formed an ad hoc patchwork of classical and quantum ideas. He formulated the **correspondence principle** to guide him in the further development of what is now called the "old" quantum theory. The principle states that whatever the form of the quantum rules, *the theory should agree with classical physics in the limit in which quantum effects become unimportant*. We might be tempted to say that in this limit one can safely set Planck's constant to zero. In fact, however, one needs to be a little more careful than this; if, for example, we simply set $\hbar = 0$, the electronic speed, given by Eq. (5–5), would become infinite! What is more properly meant by a classical limit is a regime in which quantized variables are much larger than their minimum quantum size. In the case of the the hydrogen atom, the classical limit means that the angular momentum $L$ is much larger than $\hbar$. Since the atomic orbits with large angular momentum are those for which $n$ is large, the correspondence princi-

---

[4] The particular constant added to $n$ depends on the number of dimensions of the harmonic oscillator. For a one-dimensional spring, the constant is $1/2$; for a two-dimensional system such as is needed for describing circular orbits, the constant is 1, as in Eq. (5–34); for a three-dimensional oscillator, the constant is $3/2$.

ple tells us to expect to recover the classical behavior of orbits if $n$ is large. This reasoning is consistent with the fact that, for large values of $n$, the energies of the Bohr orbits—recall that these energies are proportional to $1/n^2$—are so close together that for all practical purposes they become indistinguishable from the continuum allowed for classical orbits.

With the use of the correspondence principle, it is possible to derive constraints on the types of transitions that the atom can make. In that way, one establishes **selection rules**. To see how this works, consider the transition from the level labeled with $n + k$ to the one labeled with $n$, with $n$ large. For $n \gg k$, according to the quantum rule given in Eq. (5–17), the frequency of the emitted light is

$$f = \frac{m(Z\alpha c)^2}{2h} \left( \frac{1}{n^2} - \frac{1}{(n + k)^2} \right).$$

We can expand this expression if $n \gg k$. In particular, we write $(n + k)^2 = n^2(1 + k/n)^2$. We then use the small-$x$ expansion

$$\frac{1}{(1 + x)^2} \approx 1 - 2x$$

with $x = k/n$. This gives us

$$\frac{1}{(n + k)^2} \approx (1/n^2)(1 - 2k/n) = 1/n^2 - 2k/n^3,$$

and the $1/n^2$ terms cancel:

$$f = \frac{m(Z\alpha c)^2}{2h} \left[ \frac{1}{n^2} - \left( \frac{1}{n^2} - \frac{2k}{n^3} \right) \right] = \frac{m(Z\alpha c)^2}{2h} \frac{2k}{n^3}.$$

Finally, using $L = n\hbar$, we can substitute the angular momentum for $n$ in the denominator, yielding

$$f \approx \left( \frac{e^2}{4\pi\varepsilon_0} \right)^2 \frac{mZ^2}{2\pi L^3} k. \tag{5–35}$$

Now, according to the classical theory of electricity and magnetism, a charged particle rotating with frequency $f$ emits radiation with that frequency; that is, $f_{cl} = 1/T_{cl}$, where $T_{cl}$ is the classical period. We can find the period by dividing the speed, given by Eq. (5–7), by the circumference $2\pi R_n$, of the orbit, where $R_n$ is as given in Eq. (5–8). The result is

$$T = \left( \frac{4\pi\varepsilon_0}{e^2} \right)^2 \frac{2\pi L^3}{mZ^2}. \tag{5–36}$$

Therefore, we get $f = f_{cl} = 1/T$, provided that $k = 1$. In other words, for large $n$, transitions occur only between adjacent orbits ($\Delta n = 1$). Bohr used this result as a guide to what happens even for low values of $n$, postulating that *all* the transitions are limited, as in •Fig. 5–5, to $\Delta n = 1$.

Now the Balmer series of transitions, Eq. (5–1), does not restrict the change in $n$ to 1, nor do any of the other series of radiated frequencies. However, if one looks at the intensity of the radiation in the entire series, one sees that the $\Delta n = 1$ transitions are the most intense. Thus the selection rule whereby $\Delta n$ should equal 1 is approximately in agreement with experiment.

The information provided by the correspondence principle was certainly an important addition to the tools that would be needed to develop quantum mechanics. The principle proved to be a powerful guidepost in the development of what was to come. But by itself it hardly provides the solid foundation of a real theory. That foundation is the subject of the next chapter.

---

**Example 5–7**    It follows from classical electricity and magnetism that a charge rotating in a circle with angular velocity $\omega$ emits radiation with frequency $f = \omega/2\pi$. Show that, for the harmonic oscillator, the correspondence principle holds for all values of $n$.

**Solution**    We treated the harmonic oscillator in Section 5–3, arguing in particular that the energies allowed had the form

$$E_n = \hbar\omega(n + const).$$

If a photon of frequency $f$ is emitted in an atomic transition, then its energy is $hf$. Therefore, using energy conservation, we find that the frequency $f$ of a photon emitted when the harmonic oscillator makes a transition from the state $n + k$ to the state $n$ is given by

$$hf = 2\pi\hbar f = \left[\hbar\omega(n + k + const)\right] - \left[\hbar\omega(n + const)\right] = \hbar\omega k.$$

Thus, the frequency of the photon is $f = \omega k/2\pi$.

Now, we know that for large value of $n$, where the correspondence principle applies, we must have $f = \omega/2\pi$, so that, by comparison with the quantum mechanical value, $\omega k/2\pi$, we have $k = 1$. But the quantum mechanical value is independent of $n$, so the correspondence result holds for all values of $n$.

---

## * Experiments on Nearly Classical Atoms

In so-called **Rydberg atoms**, one or sometimes two electrons are excited to states corresponding to $n = 10, 20$, or even several hundred. These atoms, which are in states that should connect closely with classical atoms, have been the object of experimental attention since the 1930s and have been observed in outer space.

Although Rydberg atoms have some interesting properties, they are not easily studied. The radius of the Rydberg atom exceeds the radius of the atom in its ground state by a factor of $n^2$. A Rydberg atom with $n = 25$ is some 500 times larger than the ground-state atom. At ordinary densities, the orbits of adjacent Rydberg atoms overlap. Collisions allow the electrons to jump into other orbits without the emission of radiation. If one wants to study radiative transitions, one has to work with gases of very low density. Rydberg atoms can only be studied in a high vacuum. In fact, Bohr himself noted very early that atoms with large values of $n$ were more likely to be found in extrastellar atmospheres, where atomic densities were much lower than could be obtained at the time in the laboratory.

And there is another complication: The spacing between the levels of Rydberg atoms is proportional to $1/n^3$ [see Eq. (5–35)], so that for values of $n$ on the order of 20, the spacings are some thousand times less than those of ordinary atoms. This finding has some interesting effects. Experimenters were puzzled when seemingly random transitions played havoc with their Rydberg atoms. These effects turned out to be due to the presence of blackbody radiation at 300 K, room temperature. Such radiation is always present at room temperature and is strong at the frequencies that correspond to the differences in energy levels of Rydberg atoms. Ordinary atoms are not affected by this blackbody radiation, because its frequencies are concentrated at too small values for that. Ordinary atoms *are* affected at temperatures of thousands of degrees kelvin. But room-temperature blackbody radiation has frequencies that are just right for influencing Rydberg atoms. Thus, experiments on these atoms must be done not only in a high vacuum, but at only a few degrees kelvin.

**Example 5–8**  Consider a very dilute gas of hydrogen atoms that are excited to a state with $n = 25$. **(a)** Calculate the radii of these atoms. **(b)** Calculate the difference in energy between adjacent levels in that regime.

**Solution**  **(a)** The radius is, according to Eq. (5–10), $n^2 = 625$ times larger than the ground-state (Bohr) radius. So

$$r_n = a_0 n^2 = (0.53 \times 10^{-10}\,\text{m})(25)^2 = 3.3 \times 10^{-8}\,\text{m}.$$

**(b)** Given that the energies take the form $(-13.6\,\text{eV})/n^2$, the magnitude of the energy difference between, say, the level $n = 25$ and the level $n = 24$ is

$$\Delta E = (-13.6\,\text{eV})\left[(25)^{-2} - (24)^{-2}\right] = 1.85 \times 10^{-3}\,\text{eV}.$$

(Of course, the $n = 25$ level lies higher than the $n = 24$ level.) This value is a very small difference compared with the value of the energy itself (i.e., $E_{25} = -2.18 \times 10^{-2}\,\text{eV}$).

## SUMMARY

When gases of different elements radiate, the frequencies emitted are discrete and characteristic of each element. In classical physics, with a nuclear atom, not only does the radiation emerging from electrons spiraling into the nucleus have a continuous wavelength, but the end process of the spiraling is the catastrophic collapse of the atom. Bohr explained the experimental observations by constraining the classical theory of an electron orbiting a positively charged nucleus with certain ad hoc rules. His theory had the following important elements:

- Electrons in the atom can occupy only certain allowed "orbits"; that is, they can have only certain allowed energies.
- The orbit corresponding to the lowest energy is stable in that any electrons that occupy it cannot move to other orbits without having energy supplied to them from an outside source.
- Electrons in the orbits of higher energy may make spontanous transitions to orbits with lower energy, and when they do, radiation is emitted whose frequency is proportional to the difference in energy between the two orbits.
- Bohr's idea of energy quantization applies to all bound systems, with the allowed energies characteristic of the binding force. An example is the simple harmonic oscillator.
- There are limits in the orbits of bound systems in which quantum effects become insignificant and the system behaves classically. The statement asserting the existence of such limits, known as the *correspondence principle*, provides a guide to the connection between the quantum and the classical worlds.

## QUESTIONS

**1.** In studying the Bohr model in this chapter, we ignored special relativity. Why, then, does the speed of light appear in our expression for the Bohr radius, Eq. (5–9)? (*Hint*: If the speed of light were 10 times what it is, by how much would the numerical value of the Bohr radius change?)

**2.** In addition to ordinary hydrogen, there is heavy hydrogen—deuterium—with a nucleus composed of one neutron and one proton, and superheavy hydrogen—tritium—with a nucleus composed of two neutrons and one proton. Let us suppose that the neutron and proton masses are identical. If an experiment that is accurate to one part in two thousand is accurate enough to measure the reduced-mass corrections to the hydrogen spectrum, is the same experiment accurate enough to measure the reduced-mass corrections to the spectra of heavy hydrogen and superheavy hydrogen?

**3.** Given that the atom's mass is roughly the mass of the nucleus, that the atom's size—about the Bohr radius—is determined by the electronic structure of the atom, and that the density of matter is in the range $10^3$ kg/m$^3$ to $10^4$ kg/m$^3$, what would you estimate the nuclear mass to be compared with the electron mass? Outline your assumptions and approach before you actually make any calculations.

**4.** According to the second Bohr hypothesis, the minimum value of angular momentum that an atomic state can have is $\hbar$. Later, when we treat quantum mechanics more precisely, we shall see that atomic states can also have an angular momentum of zero. What kind of orbital would that represent in classical terms? And what is there about the orbitals we have considered in this chapter that would rule out zero angular momentum?

**5.** Suppose that instead of the Coulomb potential, electrons were attracted to protons with a potential of the form $g^2/4\pi\varepsilon_0 r^2$. If you constructed a fine-structure constant $\alpha'$ using this $g$, Planck's constant, and the speed of light in the same combination as you do for $\alpha$, would $\alpha'$ be dimensionless?

## PROBLEMS

**1. ▮** Assume that the orbit of Earth around the sun is circular with a radius of $1.5 \times 10^{11}$ m. What is the angular momentum of Earth (mass $= 5.9 \times 10^{15}$ kg) in units of $\hbar$?

**2. ▮** The attractive force between an electron and proton due to gravity is $Gm_p m_e/r^2$, where $m_e = 0.9 \times 10^{-30}$ kg, $m_p = 1.67 \times 10^{-27}$ kg, and $G = 6.67 \times 10^{-11}$ m$^3$/kg·s$^2$. What is the lowest gravitational Bohr radius?

**3. ▮** Suppose that a hydrogen atom in its ground state absorbs a photon whose wavelength is 180 nm. Will the electron be excited to another level, or will it be set free? [*Hint*: The ionization energy for hydrogen in the ground state is given in Eq. (5–12).]

**4. ▮** Suppose that a hydrogen atom in the ground state absorbs a photon of wavelength 15 nm. Will the atom be ionized? If so, what will be the kinetic energy of the electron when it gets far away from its atom of origin?

**5. ▮** If, in Eq. (5–1), you set $n = 1$ and take $m$ greater than 1, you generate what is known as the *Lyman series*. **(a)** Find the wavelengths of the first four members of this series. **(b)** You can observe from your answer to part (a) that the wavelengths of successive members of the Lyman series approach a common limit as $m \to \infty$. What is this limit?

**6. ▮** The spectral lines due to transitions from $n > 4$ to $n = 4$ form the *Brackett series*. What is the lowest frequency in that series of transitions? The highest?

**7. ▮** Suppose that the value of $n$ in a radiative transition in hydrogen changes only by unity, as is suggested by the correspondence principle. Estimate for what range of values of $n$ the photon that is emitted will have wavelengths in the visible range of 400 to 650 nm?

**8. ▮** The muon, with mass $m_\mu = 209m_e$, acts as a heavy electron. The muon can bind to a proton to form a muonic atom. Calculate the ionization energy of this atom, and calculate the radius of the muonic atom in its ground state. Ignore reduced-mass effects.

**9. ▮▮** A negative muon (see the previous problem) essentially at rest (i.e., its kinetic energy can be taken to be zero) enters a gas of hydrogen atoms. It is captured by one of the protons and ends up in the ground state, with the photons that were emitted while the muon descended by steps of $\Delta n = 1$ having escaped freely. The electron that was originally bound to the proton suddenly is in the situation of having had the proton "neutralized," because the proton and the muon form a neutral system much smaller than the original atom consisting of an electron and a proton. What will be the kinetic energy of the suddenly freed electron?

**10.** ❚ A positively charged muon (charge $+e$) enters a gas of hydrogen atoms. The muon can pick up an electron and form an "atom," $\mu^+ e^-$; this bound state is called *muonium*. What is the binding energy of the electron in muonium? (Neglect reduced-mass effects.)

**11.** ❚ We have treated the motion in the Bohr atom nonrelativistically. For what value of $Z$ will this treatment break down? (Calculate $Z$ for which the nonrelativistic value of $v$ would be $c$ for $n = 1$).

**12.** ❚ The one-electron Bohr theory also applies to heavier atoms from which all but one electron have been stripped. We have referred to this kind of atom as a generalized hydrogen atom. Consider carbon, for which $Z = 6$, with only one electron present. What is the wavelength of the photon emitted in the transition from $n = 3$ to $n = 2$? Compare this value with the corresponding one for ordinary hydrogen.

**13.** ❚❚ What is the ratio of the Rydberg value for doubly ionized lithium, a one-electron atom with a nucleus of charge $3e$, to that of hydrogen? Include reduced-mass effects in your calculations.

**14.** ❚❚ The reduced-mass effect is much more important in muonium (Problem 5–10) than it is in ordinary hydrogen. Redo Problem 5–10, and *include* reduced-mass effects. Will the electron be more strongly or more weakly bound in muonium than in ordinary hydrogen?

**15.** ❚❚ The electron and its antiparticle, the positron, can form a bound system called positronium. The system is actually unstable, because the electron and positron can annihilate each other, producing two very energetic photons via the reaction $e^+ + e^- \rightarrow \gamma + \gamma$. Before the decay occurs, positronium exhibits a hydrogenlike spectrum. **(a)** What is the difference in energy between the ground state and the first excited state? **(b)** What is the radius of the ground state of positronium? Note that reduced-mass effects cannot be neglected here!

**16.** ❚❚ Hydrogen, which consists of a single electron bound to a nucleus of charge $Z = 1$, comes in several varieties. One of the varieties is ordinary hydrogen, in which the nucleus is a proton. Two other varieties exist: deuterium, for which the nucleus has $Z = 1$ and a nuclear mass $M \approx 2m_p$, and tritium, for which $Z = 1$ and $M \approx 3m_p$. Calculate the wavelength for the transition $n = 2$ to $n = 1$ in **(a)** deuterium and **(b)** tritium. **(c)** Compare your results with the value for the wavelength for the same transition in hydrogen. A measurement of a shifted spectrum in this transition led to the discovery of deuterium by the American chemist Harold Urey.

**17.** ❚❚ The general quantization of motion in circular orbits is obtained by combining the equation of motion $mv^2/r = |dU(r)/dr|$ with the angular-momentum quantization condition $mvr = n\hbar$. Here, $U(r)$ is the potential energy. Use this procedure to calculate the spectrum for circular motion in the potential $U = F_0 r$.

**18.** ❚❚❚ Consider the potential $U(r) = V_0(r/a)^N$. **(a)** Sketch the potential as a function of $r$, and show that it approaches a sharp box of side $a$ for large $N$. **(b)** Use the general quantization procedure outlined in the previous problem to calculate the energy spectrum for the potential $U(r)$. In particular, give the limit as $N \rightarrow \infty$.

**19.** ■ The lowest electron voltage at which neon emits radiation in a Franck–Hertz experiment is 18.672 volts. What is the wavelength of the light that is emitted?

**20.** ■ The first spectral line of atomic lithium observed in a Franck–Hertz experiment has a wavelength of 670.8 nm. What is the voltage of the electron beam leading to this emission?

**21.** ■■■ The correspondence principle allows us to calculate how long it takes for an electron to make a transition from $n + 1$ to $n$ in hydrogen, as follows: **(a)** Given $v = \alpha c/n$, and $r = \hbar n^2/mc\alpha$, calculate the acceleration of the electron in a circular orbit. **(b)** Use the expression from classical electromagnetism for the power radiated by an accelerating electron, viz.,

$$P = \left(\frac{2}{3}\right)\left(\frac{e^2}{4\pi\varepsilon_0}\right)\left(\frac{a^2}{c^3}\right),$$

to calculate the power radiated in terms of $\alpha, m, c, n,$ and $\hbar$. Here, the use of the correspondence principle suggests that $n$ is large. **(c)** Calculate the energy of the photon emitted in the $n + 1 \rightarrow n$ transition for large $n$. **(d)** Calculate the length of time the electron emits energy from the relation

$$\text{time} = \frac{\text{energy}}{\text{power}}.$$

**(e)** Replace $n$ by unity in your expression to find an approximate value of the time that an $n = 2$ state lasts before it decays to the $n = 1$ state.

**22.** ■■ The energy of a rotator (a dumbbell-shaped object) is given by $L^2/2I$, where $L$ is the angular momentum and $I$ is the rotational inertia (moment of inertia) of the dumbbell. Consider a hydrochloric acid (HCl) molecule, in which the atomic weights of hydrogen and chlorine are 1 and 35, respectively, and the internuclear separation is 0.127 nm. Calculate the energy required to excite the molecule from the ground state to the first rotational excited state.

**23.** ■■ Suppose you are given a molecular spectrum. The wavelengths make up a series of the form $\lambda_n = \lambda_0/(2n + 1)$, for $n = 1, 2, \ldots$. What kind of transitions does this spectrum represent? Give an expression for $\lambda_0$ in terms of the properties of the molecule.

**24.** ■ Consider a hydrogen ($H_2$) molecule, in which the internuclear separation is 0.074 nm. What are the energies of the first and second rotational excited states?

**25.** ■■ The value of the angular-frequency variable characteristic of the vibration of an oxygen ($O_2$) molecule is $3.98 \times 10^{14}$ rad/s. What is the difference in energy between the ground state for vibrations and the first excited vibrational state, in electron volts? Compare this value with that obtained if the energy is split between the ground state and the first excited rotational state.

**26.** ■■ The relationship between the spring constant $k$ and the characteristic angular frequency $\omega$ for the vibrational motion of a diatomic molecule is given by $k = m_{red}\omega^2$, where $m_{red}$ is the reduced mass of the two nuclei at the ends of the "spring." Use the data of the previous problem to calculate the spring constant for the $O_2$ molecule.

**27.** ■■ Consider a container full of a gas of hydrogen atoms, each of which is in a state with $n = 30$. If the interatomic distance is to remain the same as in hydrogen in the $n = 1$ state at atmospheric pressure, how low should the pressure in the container be?

**28.** ■■ In a gas of hydrogen under normal conditions, the interatomic spacing is $1.6 \times 10^{-8}$ m. Assume the gas is made of atomic, not molecular, hydrogen. For what $n$-value of the hydrogen atoms is the size of the atom comparable to the interatomic spacing?

**29.** ∎∎ Consider a gas of hydrogen atoms in a container at atmospheric pressure $(1.01 \times 10^5 \text{ N/m}^2)$ and temperature 300K. Calculate the average volume occupied per atom, using the ideal gas law $pV = N_A kT$, where $N_A = 6.02 \times 10^{23}$ atoms/mole and $k$ is Boltzmann's constant. What is the ratio of the radius of the hydrogen atom to the interparticle spacing?

**30.** ∎∎ Consider a gas of hydrogen atoms, all excited to states with $n = 10$. Use the results of the previous problem to calculate the pressure at which the ratio of the radius of the excited hydrogen atoms to the interparticle spacing is on the order of $1/100$.

**31.** ∎ The process in which a hydrogen atom absorbs a photon and liberates the bound electron is called the photodisintegration of hydrogen. What is the maximum wavelength of a photon that can eject an electron from the ground state of hydrogen? If you supply more energy than that associated with this maximum wavelength, what happens to this extra energy?

**32.** ∎ A reaction in which energy is given off is called *exothermic*. Is the reaction in which an electron is captured at rest by a proton to yield hydrogen in its ground state plus a light quantum exothermic? If so, what is the wavelength of the photon?

**33.** ∎∎∎ The generalization of the Bohr rule to periodic motion more general than a circular orbit states that

$$\oint \vec{p} \cdot d\vec{r} = nh = 2\pi n\hbar,$$

where $\vec{p} \cdot d\vec{r}$ is the dot product of the momentum $\vec{p}$ and the vector interval $d\vec{r}$ along a path that follows the orbit for one circuit. **(a)** Show that, for circular orbits, the generalized rule is identical to the Bohr condition discussed in this chapter. **(b)** Using the generalized rule, show that the spectrum for the one-dimensional harmonic oscillator, for which $E = p^2/2m + (1/2)m\omega^2 x^2$, is $E = n\hbar\omega$. **(c)** Using this rule, show for a particle that moves along the $x$-axis and bounces freely between two walls at $x = 0$ and $x = L$ respectively, with an energy between the walls given by $E = mv^2$, that $E = n^2\pi^2\hbar^2/(2mL^2)$.

# CHAPTER
# 6
# The Schrödinger Equation

The descriptions we have thus far provided for quantum phenomena have all been rather *ad hoc*; they have not been based on any coherent dynamical scheme, as, for example, classical phenomena are based on diverse applications of Newton's second law and Maxwell's equations. It was an extraordinary group of physicists working mainly in Europe in the 1920s that created the desired foundation for quantum physics. In this chapter, we shall outline that foundation and how it is applied.

## 6–1  Wave Functions and Probabilities

In Chapter 4, we described some conceptual difficulties that arise from the idea that radiation has particlelike properties while matter has wavelike properties. In particular, we noted that when photons or electrons are sent through a two-slit apparatus at such a slow rate that their arrivals are separated by very long times, they nevertheless gradually give rise to a characteristic wavelike interference pattern. (See •Figs. 4–16 and 4–17.) Any particular "hit" on the screen by a photon or electron appears to be random, but the result of the accumulation of "hits" produces the pattern.

The situation is analogous to the tossing of an unbiased coin. Each toss is random, but after the coin has been thrown many times, a pattern becomes evident: To a better and better approximation, half the tosses will produce heads and half will produce tails. We describe this by saying that each coin toss is a random event, but that the coin has a *probability* of 50% of coming up heads.

In the case of photons or electrons passing through a two-slit apparatus, the particles are so far apart that any one photon or electron cannot "know" what the others will do. One is led to the inevitable conclusion that each particle in some sense *interferes with itself*. The questions we have before us are twofold: How do we describe a particle interfering with itself, and how do we describe the notion of a probability that a photon or electron ends up on a screen in a certain pattern?

Let us start with the behavior of photons. In the classical limit, there are no photons; instead, we have the classical electromagnetic wave that follows from Maxwell's equations. This wave can be fully described by specifying the electric field it contains. In other words, the collective effect of many photons is described by an electric field $\vec{\mathbf{E}}(\vec{\mathbf{r}}, t)$. The interference pattern appears on the screen because there is an electric field $\vec{\mathbf{E}}_1$ associated with the part of the wave coming through slit 1 and an electric field $\vec{\mathbf{E}}_2$ associated with the part of the wave coming through slit 2, and the net classical field on the far side of the slits will have the form suggested by the principle of superposition, namely,

$$\vec{\mathbf{E}}(\vec{\mathbf{r}}, t) = \vec{\mathbf{E}}_1(\vec{\mathbf{r}}, t) + \vec{\mathbf{E}}_2(\vec{\mathbf{r}}, t).$$

Recall from Chapter 1 that the pattern on the screen is an *intensity* pattern, proportional to $\vec{\mathbf{E}}(\vec{\mathbf{r}}, t)^2 = \left(\vec{\mathbf{E}}_1(\vec{\mathbf{r}}, t) + \vec{\mathbf{E}}_2(\vec{\mathbf{r}}, t)\right)^2$. This is where the interference pattern comes from: Since the fields $\vec{\mathbf{E}}_1$ and $\vec{\mathbf{E}}_2$ have a sign indicating their direction, they sometimes reinforce and sometimes cancel each other.

Now let us translate this idea to the language of photons. The intensity of the wave is proportional to the square of the field $\vec{\mathbf{E}}(\vec{\mathbf{r}}, t)$, which is in turn proportional to the the number of photons, $N$, at point $\vec{\mathbf{r}}$ at time $t$. As we turn down the intensity of the light shining on the slits, the number of photons decreases until we are dealing with one photon at a time. But we have argued that even single photons carry with them the interference properties characteristic of $N$ photons. This notion suggests that we try to describe a single photon by a kind of one-photon electric field $\vec{\mathbf{e}}(\vec{\mathbf{r}}, t)$ that, like the ordinary electric field, obeys the rules of superposition. Thus, the electric field at the screen due to a single photon in a two-slit experiment will have the form

$$\vec{\mathbf{e}}(\vec{\mathbf{r}}, t) = \vec{\mathbf{e}}_1(\vec{\mathbf{r}}, t) + \vec{\mathbf{e}}_2(\vec{\mathbf{r}}, t), \tag{6–1}$$

where the two terms respectively describe the field due to a photon that came through slit 1 with slit 2 closed and the field that came through slit 2 with slit 1 closed. The quantum analog to the classical intensity is then

$$I \propto \left[\vec{\mathbf{e}}(\vec{\mathbf{r}}, t)\right]^2 = \left[\vec{\mathbf{e}}_1(\vec{\mathbf{r}}, t) + \vec{\mathbf{e}}_2(\vec{\mathbf{r}}, t)\right]^2. \tag{6–2}$$

An interference term is immediately apparent here: the cross term $2\vec{\mathbf{e}}_1(\vec{\mathbf{r}}, t) \cdot \vec{\mathbf{e}}_2(\vec{\mathbf{r}}, t)$. Note that if one of the slits—slit 2, say—is closed, then $\vec{\mathbf{e}}_2$ vanishes, and there is no interference pattern.

In classical physics, the electric fields in an electromagnetic wave can be reduced as fine as we like, and we can divide the incident intensity so that part of the fields pass through slit 1 and part through slit 2. The quantum case is quite different: When we speak of a single photon, we cannot speak of some fraction of a photon being at $\vec{\mathbf{r}}$ at a time $t$, since *photons are indivisible*. The way out is to describe the quantity $I$ in Eq. (6–2) as *the probability[1] that the photon is to be found* at the point $\vec{\mathbf{r}}$ at time $t$; more precisely, $I$ will represent a probability distribution function, meaning in this case that

$$\left[\vec{\mathbf{e}}(\vec{\mathbf{r}}, t)\right]^2 d^3\vec{\mathbf{r}}$$

is proportional to the probability that a photon is in a box of size $d^3\vec{\mathbf{r}}$ around the point $\vec{\mathbf{r}}$.

What must we expect of the "fields" $\vec{\mathbf{e}}(\vec{\mathbf{r}}, t)$? We must expect that when many photons are involved, the individual fields somehow combine to create the classical electric field $\vec{\mathbf{E}}$, which obeys Maxwell's equations. And there is a second property that is crucial for interference effects to appear: The "field" $\vec{\mathbf{e}}(\vec{\mathbf{r}}, t)$ must obey a *linear* equation, so that the sums of such fields also obey the same equation and the rules of superposition are satisfied. In this way, the net

---

[1] If you are not familiar with elementary notions of probability, you may want to see Appendix B–3 for a review. We will, in any case, emphasize the basic notions as they apply to quantum mechanics here.

effect of two slits is to produce a sum of two terms, and an interference pattern can result from the square of this sum.

The foregoing comments, which apply to photons, suggest how one might deal with the interference effects observed with electrons (and other particles). There is no classical field connected with electrons in the same way that photons are associated with an electric field, so one must invent something that plays the same role. Accordingly, we assume that associated with each electron is a **wave function** $\psi(\vec{r}, t)$ that must obey some linear equation. This equation, which will be discussed later, is the **Schrödinger equation**. The linearity is crucial: It implies that if $\psi_1(\vec{r}, t)$ and $\psi_2(\vec{r}, t)$ are solutions of the Schrödinger equation, then so is

$$\psi(\vec{r}, t) = A\psi_1(\vec{r}, t) + B\psi_2(\vec{r}, t), \tag{6-3}$$

where $A$ and $B$ are arbitrary complex constants. Thus if $A\psi_1(\vec{r}, t)$ is the wave function for an electron that came through slit 1 and $B\psi_2(\vec{r}, t)$ the wave function for an electron that came through slit 2, then the wave function for the electron at the screen on the far side of the slits is the sum of these two wave functions. The square of the sum, which, we shall argue, is associated with a probability for finding an electron, includes interference terms, and as we have discussed in Chapter 4, interference terms are indeed needed for a proper description of the passage of an individual electron through a two-slit screen.

There is something about wave functions that is different from the behavior of electric fields. A wave function $\psi(\vec{r}, t)$ can be a complex function. If you are not comfortable with complex numbers, you may want to review them; a summary is presented in Appendix (B–4). Equation (B.4–3) shows that we may write any complex function $\psi$ in the form

$$\psi(\vec{r}, t) = R(\vec{r}, t)\exp\left[iS(\vec{r}, t)\right], \tag{6-4}$$

where $R$ and $S$ are both real functions (•Fig. 6–1). The function $R(\vec{r}, t)$ is the magnitude of the wave function and $S$ is its phase. From Eq. (B.4–6), we have

$$R(\vec{r}, t)^2 = \psi^*\psi = \left|\psi(\vec{r}, t)\right|^2. \tag{6-5}$$

Here $\psi^*$ represents the complex conjugate of $\psi$, which is a standard shorthand for $\psi$ with $i$ replaced everywhere by $-i$.

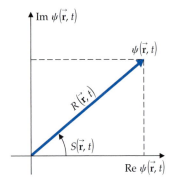

• **Figure 6–1** A complex number (or function) can be written in terms of its real and imaginary parts, or equivalently, in terms of a magnitude and a phase relative to some axis. This is analogous to describing a point in a plane in Cartesian coordinates or in cylindrical coordinates.

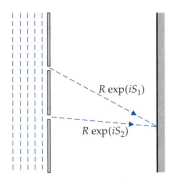

• **Figure 6–2** An electron beam incident on a wall with two slits gives rise to a wave function whose properties are revealed by the probability of arrival of an electron on the screen. The wave function is the sum of two individual wave functions, of the forms $R\exp\left[iS_1(x)\right]$ and $R\exp\left[iS_2(x)\right]$, associated with the passage of the electron through each individual slit.

---

**Example 6–1** The wave function at a point $x$ on a screen of a particle that passed through the slit labeled 1 in a two-slit apparatus is $R(x)\exp\left(iS_1(x)\right)$. The wave function at the same point of a particle that passed through the slit labeled 2 is $R(x)\exp\left(iS_2(x)\right)$. The three functions $R(x)$, $S_1(x)$, and $S_2(x)$ are all real-valued. Show that if both slits are open, the absolute square of the wave function at the screen exhibits an interference pattern.

**Solution** With both slits open, the wave function at the screen is (•Fig. 6–2)

$$\psi(x) = R(x)\exp\left(iS_1(x)\right) + R(x)\exp\left(iS_2(x)\right).$$

Therefore, the square of the absolute value is

$$\left|\psi(x)\right|^2 = \left|R(x)\right|^2\left[\exp\left(iS_1\right) + \exp\left(iS_2\right)\right]\left[\exp\left(-iS_1\right) + \exp\left(-iS_2\right)\right]$$

$$= 2\left|R(x)\right|^2\left[1 + \cos\left(S_1 - S_2\right)\right].$$

As the two phases $S_1$ and $S_2$ vary with position, a systematic cosine dependence will run $\left|\psi(x)\right|^2$ between 0 and $2\left|R(x)\right|^2$. This is a standard interference pattern, with regions of destructive and constructive interference.

## The Probabilistic Interpretation

Electrons, like photons, are indivisible, and if we are to be guided by our tentative discussion of the intensity of photons, we need to introduce an analog to the probabilistic interpretation of the square of the photon "field," $\vec{e}(\vec{r}, t)^2$. Since we have already noted that wave functions have to be complex—the reasons for this will become clear in Section 6–2—we must modify our proposition a little: Given a wave function $\psi(\vec{r}, t)$, *the probability of finding the electron at time* t *in a box the size $d^3\vec{r}$ around the point $\vec{r}$ is*

$$\left|\psi(\vec{r}, t)\right|^2 d^3\vec{r}. \tag{6–6}$$

The reason that the absolute value appears is that probabilities must be positive numbers, and the quantity $\psi^2$, with $\psi$ complex, not only is not positive, but is itself a complex quantity. From Eq. (6–4), we see that only the magnitude $R(\vec{r}, t)$ of $\psi$ enters into the probability:

$$\left|\psi(\vec{r}, t)\right|^2 = R(\vec{r}, t)^2. \tag{6–7}$$

At this point, the significance of Example 6–1 becomes clear: Since the square of the absolute value of the wave function is a probability, and since Example 6–1 shows that in the two-slit experiment that quantity exhibits interference, the two-slit experiment reveals itself in an enhanced (constructive interference) or suppressed (destructive interference) *probability* of the arrival of electrons on the screen.

The important idea that the square of the absolute value of the wave function, Eq. (6–6), measures probability is due to the German theoretical physicist Max Born. In 1926, Erwin Schrödinger (•Fig. 6–3), extending de Broglie's 1924–25 suggestions about symmetry between radiation and matter, and in particular about the wavelike character of matter, wrote an equation for the wave function. Born's suggestion, made in the summer of 1926, was an *interpretation* of the wave function that Schrödinger had introduced. The Schrödinger equation itself followed by some six months another approach, due to Werner Heisenberg, to the problems we have described. The Heisenberg and the Schrödinger approaches, which appeared to be completely different, were soon afterwards shown to be equivalent mathematical representations of the same theory.

Since, at any given time, the probability of finding our electron *somewhere* in space is unity, it follows from our interpretation of $\left|\psi(\vec{r}, t)\right|^2 d^3\vec{r}$ that we must have

$$\int_{\text{all space}} \left|\psi(\vec{r}, t)\right|^2 d^3\vec{r} = 1. \tag{6–8}$$

But what if we should find a wave function that does not satisfy this condition? The linearity of the equation obeyed by $\psi$ guarantees that a complex constant C times $\psi$ is also a solution of the equation. This means that we can always multiply the wave function we found originally by a constant chosen to ensure that the function satisfies Eq. (6–8). (See •Fig. 6–4.) This **normalization** process—making $\psi(\vec{r}, t)$ obey Eq. (6–8)—ensures that Eq. (6–6) describes a probability and not a constant times the probability. We shall study the normalization process in more detail in Section 6–3, together with an example.

The use of wave functions to describe "particles" like the electron shows how such particles can display wavelike properties, although it is important to understand that the waves involved are waves of probability. In places where their amplitudes are small, the probability of locating the particle is also small. Because they have phases, these probability waves can interfere with each other

• **Figure 6–3**  In 1926, Erwin Schrödinger discovered the equation that now bears his name. Schrödinger succeeded Max Planck in Berlin and remained there until he was forced out by Hitler. He became a professor at the Institute for Advanced Studies in Dublin.

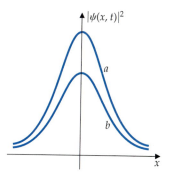

• **Figure 6–4**  The normalization condition—that the integral over all space of $\left|\psi(x, t)\right|^2$ is unity—implies that the area under curve (a) is unity. If it is not, $\left|\psi(x, t)\right|^2$ can be multiplied by a constant (scaled) to form curve (b) such that the area under it *is* unity.

▌ This question is treated in detail in the next chapter.

like any other kind of wave. But how is such a wave picture consistent with, for example, the fact that when we do find an electron, it appears to be a localized object with particlelike properties? Let us postpone a real discussion of this point until we get more deeply into quantum mechanics. At this stage, it is enough to say that we should be careful about just what it is we mean when we say that we have a particle. Presumably, we mean not only that we have something which is pretty well localized in space, but also that the position about which the localization occurs moves at least approximately according to Newton's second law. We also require a particle to have a momentum and an energy that it can exchange with another particle, so that we can produce the kind of billiard-ball collisions that explained the Compton effect. On the basis of our experience with the classical world of waves, there is some hope that this is not impossible with wave functions: We see single-wave pulses travel on a rope, we bounce radar off airplanes, and we can talk about rays in geometrical optics even though waves are behind all of these phenomena. So if, somehow, a particle could be thought of as a wave pulse with a finite extension in space, we can see how a particle description might arise.

## Towards an Equation for the Wave Function

The correct equation for $\psi(\vec{r}, t)$ was discovered by the Austrian physicist Erwin Schrödinger in 1926. Schrödinger based his equation on an analogy with the classical relation between physical optics (waves) and geometrical optics (rays). For Schrödinger, the "waves" were those of de Broglie (Chapter 4), and the "rays" played the role of particles. We shall make use of the Born probability interpretation to suggest how the Schrödinger equation might have arisen if, somehow, the Born proposal had been known to Schrödinger before his work.

Let us return to the probability interpretation of $\psi$ and the normalization condition of Eq. (6–8). We expect that equation to be true at all times, so its derivative with respect to time is zero:

$$\int_{\text{all space}} \left( \frac{\partial \psi^*(\vec{r}, t)}{\partial t} \psi(\vec{r}, t) + \psi^*(\vec{r}, t) \frac{\partial \psi(\vec{r}, t)}{\partial t} \right) d^3\vec{r} = 0. \qquad (6\text{–}9)$$

Note the use of the partial derivative. This is merely a reminder that the wave function depends both on space and time and that here we are interested only in the time aspect.

Equation (6–9) contains important information. As we recalled in Chapter 1, in classical physics we are more used to quantities that obey second-order wave equations. For example, in a discussion of waves on a string aligned in the $x$-direction, the quantity obeying the wave equation was the lateral displacement $y$ of the string. This quantity becomes a function of $x$ and $t$ that is determined by the wave equation

$$\frac{\partial^2 y(x, t)}{\partial t^2} - v^2 \frac{\partial^2 y(x, t)}{\partial x^2} = 0.$$

In such an equation, $\partial y(x, t)/\partial t$ is the transverse velocity of a mass point on the string at position $x$ (*not* the wave velocity—the velocity at which a disturbance propagates along the string). The transverse velocity can be given at a certain time as part of the initial conditions. For example, it can be determined at $t = 0$ by how we start the string vibrating at that time: We may hit the string in a particular location with a hammer of a certain mass going at a certain speed. In contrast, if we look at Eq. (6–9) at time $t = 0$, we see that we would *not* be allowed to set $\partial \psi(x, t)/\partial t$ at $t = 0$ to anything we wish, because then the integral need not come out to be zero. Instead, we can satisfy Eq. (6–9) at all times by insisting that the wave function obeys an equation of the type

$$\frac{\partial \psi(x, t)}{\partial t} = [\text{something acting on}]\psi(x, t) \qquad (6\text{–}10)$$

and that there is a suitable cancellation of the two terms under the integral in Eq. (6–9). The "something" that we have put in brackets in Eq. (6–10) must keep that equation linear, so the "something" cannot involve $\psi$ itself. It can, however, contain constant terms, functions of $x$ and $t$, derivatives that act on the $\psi$ function to the right, and so on. Simplicity and reliance on the de Broglie relation will lead us to this "something" and hence to the Schrödinger equation.

## 6–2 The Form of the Schrödinger Equation

Our aim in this section is to find the "something" in Eq. (6–10) such that Eq. (6–9) is satisfied. To do this, we need some help regarding the form of the wave function itself. Accordingly, we go back to a simplified discussion of the two-slit interference experiment. In an *optical* experiment, we would describe the waves that are involved in harmonic terms—cosines, say—and the combination of the two waves would have the form

$$\cos(kr_A - \omega t) + \cos(kr_B - \omega t), \qquad (6\text{–}11)$$

where $r_A$ and $r_B$ are the distances from the slits $A$ and $B$ to the point on the screen. We choose the screen to be far enough away so that the beams are parallel. Here, $k = 2\pi/\lambda$ and $\omega = 2\pi f$, where $\lambda$ and $f$ are, respectively, the wavelength and frequency of the wave. For light, we would have the relation $c = \lambda f = \omega/k$.

If the optical experiment is to serve as a model for the interference of massive particles—electrons, say—then we could hope to find wave functions that, in one spatial dimension, would have the form

$$\psi(x, t) = A \cos(kx - \omega t), \qquad (6\text{–}12)$$

with $k$ and $\omega$ positive. However, this attempt will not work, for several reasons. With this wave function, a single time derivative is

$$\frac{\partial \psi(x, t)}{\partial t} = A\omega \sin(kx - \omega t).$$

The problem is that there is no simple way to reproduce $A\omega \sin(kx - \omega t)$ by something acting on $\cos(kx - \omega t)$. We could try to use the fact that $\sin x = \sqrt{1 - \cos^2 x}$, but $(A^2 - \psi^2)^{1/2}$ destroys the linearity of the equation. We could, then, try to reproduce $A\omega \sin(kx - \omega t)$ with $-(\omega/k)(\partial\psi(x, t)/\partial x)$; this operation does indeed reproduce $A\omega \sin(kx - \omega t)$, as you can verify for yourself by taking the derivative with respect to $x$, but it has the problem that the equation would have to be changed for waves traveling in the $-x$-direction. Moreover, the ratio $\omega/k$, which is generally the speed of a wave with angular frequency $\omega$, is not generally the same for all frequencies, as it is for light. This would mean, in essence, that we would have a different equation for each frequency.

We would have the same set of difficulties if we were to try a wave function of the form $\psi(x, t) = A \sin(kx - \omega t)$. So where does this leave us? It is here that we are saved by the possibility of *complex* wave functions. For the complex combination

$$\psi(x, t) = A[\cos(kx - \omega t) + i \sin(kx - \omega t)], \qquad (6\text{–}13)$$

where $i = \sqrt{-1}$, has properties that enable us to make a consistent equation. It is simple to check that one time derivative of Eq. (6–13) gives us back something proportional to the original trial wave function:

$$\frac{\partial \psi}{\partial t} = A[\omega \sin(kx - \omega t) - i\omega \cos(kx - \omega t)] = -i\omega\psi(x, t). \quad (6\text{–}14)$$

We have used the property that $i^2 = -1$. Moreover, the fundamental complex-number equation (B.4–2) in Appendix B reveals that $\psi$ in Eq. (6–13) takes on the particularly simple form

$$\psi(x, t) = A \exp\left[i(kx - \omega t)\right], \quad (6\text{–}15)$$

▌ We shall study the plane wave in more detail in Chapter 7.

and in this form Eq. (6–14) is even easier to derive. We refer to the functional form in Eq. (6–15) as a *plane wave*.

The plane-wave form of Eq. (6–15) can be written in another form if we use the de Broglie formula and a second, related, one. We impose the condition

$$k = \frac{2\pi}{\lambda} = 2\pi\left(\frac{p}{h}\right) = p/\hbar, \quad (6\text{–}16)$$

where $p$ is the momentum associated with the particle. Then, in analogy with the Einstein rule for photon energy, $E = hf = \hbar\omega$, we impose the general association

$$\omega = \frac{E}{\hbar} \quad (6\text{–}17)$$

Thus the plane-wave form of the electron wave function is

$$\psi(x, t) = A \exp\left[\frac{i(px - Et)}{\hbar}\right]. \quad (6\text{–}18)$$

Finally, we note that, for an electron moving with no force acting on it,

$$E = \frac{p^2}{2m}. \quad (6\text{–}19)$$

Now let us look again at our first time derivative, Eq. (6–14). We have

$$\frac{\partial \psi}{\partial t} = -i\omega\psi(x, t) = -\frac{i}{\hbar}\frac{p^2}{2m}\psi(x, t). \quad (6\text{–}20)$$

We can now pull down the factor $p^2$ from space derivatives of our plane wave—this is what is wanted for the right-hand side—by taking *two* such derivatives. More precisely,

$$\frac{\partial^2 \psi}{\partial x^2} = -\hbar^2 p^2 \psi(x, t).$$

With this result, we can put Eq. (6–20) in the form

$$\frac{\partial \psi(x, t)}{\partial t} = -\frac{i}{\hbar}\left(-\frac{\hbar^2}{2m}\frac{\partial^2}{\partial x^2}\right)\psi(x, t),$$

or, equivalently,[2]

$$i\hbar\frac{\partial \psi(x, t)}{\partial t} = -\frac{\hbar^2}{2m}\frac{\partial^2}{\partial x^2}\psi(x, t). \quad (6\text{–}21)$$

---

[2] We note here that the appearance of the second derivative with respect to $x$ means that the form of Eq. (6–21) does not change when we replace $x$ with $-x$. Earlier, we mentioned that this would have been a problem if there were only a single derivative with respect to $x$.

This is the **Schrödinger equation**—sometimes called the *time-dependent* Schrödinger equation—for a particle of mass $m$ moving in the absence of any force. We emphasize two crucial features: the presence of complex numbers and the matching of a single time derivative on the left with a double space derivative on the right. The second feature is necessary if we want to make the relation between energy and momentum come out right.

If the particle moves under the influence of a potential $V(x)$, then a generalization in which we replace $E = p^2/2m$ by $E = p^2/2m + V(x)$ is in order:

$$i\hbar \frac{\partial \psi(x, t)}{\partial t} = -\frac{\hbar^2}{2m}\frac{\partial^2}{\partial x^2}\psi(x, t) + V(x)\psi(x, t). \tag{6–22}$$

This is the general form of the Schrödinger equation. Note that what we called "something acting on $\psi$" in Eq. (6–10) is now very specific: What acts on $\psi$ is the combination

$$-\frac{\hbar^2}{2m}\frac{\partial^2}{\partial x^2} + V(x).$$

Equation (6–22), together with the interpretation that $|\psi(x, t)|^2\, dx$ represents the probability of finding the electron between $x$ and $x + dx$ at time $t$ [Eq. (6–6)], are to be taken as *postulates*. Note that what we have presented is not a derivation; it only suggests how Schrödinger might have gone about finding his equation. Our upcoming task will be to see what we can learn about physical systems from the Schrödinger equation and the probabilistic interpretation of the wave function.

# 6–3  Expectation Values

If the wave function is to have an interpretation in terms of probabilities, and if wave functions are going to describe physical systems, then the results of the measurement of physical quantities must involve probabilities. Just how does this happen? In order to answer this question, let us first look a little more carefully at the probabilistic aspect of the wave function itself.

### Normalization

If $|\psi(x, t)|^2\, dx$ represents the probability of finding a particle, then we expect that quantity to be normalized so that when it is integrated over all space to find the total probability, the total probability is unity, as given by Eq. (6–8). This would mean that the probability of finding the particle *someplace* is 1. Is the normalization automatic? Not necessarily, but because of the linearity of the Schrödinger equation, $C\psi(x, t)$ is a solution of that equation if $\psi(x, t)$ is, where $C$ is an arbitrary constant. This means that if we find a $\psi(x, t)$ that does not satisfy Eq. (6–8), we can always multiply $\psi$ by a constant, adjusted so that the normalization to unity follows (•Fig. 6–4). Thus if our solution obeys the equation

$$\int_{-\infty}^{\infty} |\psi(x, t)|^2\, dx = N,$$

then the new wave function $\psi' = \left(\sqrt{N}\right)^{-1}\psi$ will be normalized. Note that $N$ must be independent of time, since we have argued that the whole basis of the equation was the time independence of the integral. For a detailed demonstration, see the Appendix to this chapter.

Henceforth, we shall generally assume that $\psi(x, t)$ is normalized.

### Expectation Values

In this subsection, we are going to think about the consequences of the fact that the wave function is associated with a probability. (Again, see Appendix B.3 for a discussion of probabilities.) For this purpose, let us suppose that we have a normalized probability distribution function $P(x)$—in our case, $P(x) = |\psi(x, t)|^2$, although we shall couch our discussion in more general terms. Given any probability distribution function $P(x)$, the probability of finding the particle described by this function through some measurement of the position $x$ in a range from $x$ to $x + dx$ is $P(x)\, dx$. We can use $P(x)$ to find the *expectation value*, or *average*, of a set of measurements made on the position of the particle. For example, the average value of $x$—the place where the particle is to be found, on average—is

$$\langle x \rangle = \int x P(x)\, dx, \tag{6–23}$$

and more generally,

$$\langle f(x) \rangle = \int f(x) P(x)\, dx. \tag{6–24}$$

The limits of integration would be over all values of $x$ for which the probability is not zero; equivalently, one could just integrate over all space and let the values of position for which the probability is zero be eliminated because $P(x)$ is zero for those values. For quantum mechanics, the probability distribution is determined by the wave function, and Eq. (6–24) takes the form

$$\langle f(x) \rangle = \int_{-\infty}^{\infty} \psi^*(x, t) f(x) \psi(x, t)\, dx, \tag{6–25}$$

subject to the normalization condition of Eq. (6–8), here in the form

$$\int_{-\infty}^{\infty} \psi^*(x, t) \psi(x, t)\, dx = 1. \tag{6–26}$$

We shall see in a moment why we broke up the square of the absolute value so that in Eq. (6–25) the wave function and its complex conjugate appear on opposite sides of $f(x)$. Note that $\langle f(x) \rangle$ depends on time, since $\psi(x, t)$ does. Of course, that is a desirable feature: If the wave function were that of a free electron, and we were looking at $\langle x \rangle$, the expectation value of the electron's position, we would like to be able to describe the fact that the electron moves.

How can we find expectation values for variables other than the position—momentum, for example? We can learn what to do by going back to our plane-wave solution, Eq. (6–18), $\psi(x, t) = A \exp[i(px - Et)/\hbar]$. We can extract the momentum $p$ from this wave function by taking a space derivative,

$$-i\hbar \frac{\partial}{\partial x} \psi(x, t) = p \psi(x, t). \tag{6–27}$$

In fact, if we adopt the rule that

$$p = -i\hbar \frac{\partial}{\partial x}, \tag{6–28}$$

then it would seem reasonable to define the expectation value of the momentum as

$$\langle p \rangle = \int_{-\infty}^{\infty} \psi^*(x, t) p \psi(x, t)\, dx = -i\hbar \int_{-\infty}^{\infty} \psi^*(x, t) \frac{\partial}{\partial x} \psi(x, t)\, dx. \tag{6–29}$$

(At this point it becomes clear why we broke up the square of the absolute value in the probability distribution: In Eq. (6–29), the derivative acts on $\psi$ but not on $\psi^*$.) This equation will be seen to be a correct definition—it will satisfy the correspondence principle in the sense that we expect the average values of quantities like position and momentum to obey classical laws—if we can show that

$$\frac{d\langle x \rangle}{dt} = \frac{\langle p \rangle}{m}.$$
(6–30)

We do so in the appendix to the chapter.

We can now generalize to the expectation value of any power of momentum:

$$\langle p^n \rangle = \int_{-\infty}^{\infty} \psi^*(x, t)\left(-i\hbar \frac{\partial}{\partial x}\right)^n \psi(x, t)\, dx.$$
(6–31)

For example,

$$\langle p^2 \rangle = \int_{-\infty}^{\infty} \psi^*(x, t)\left(-i\hbar \frac{\partial}{\partial x}\right)^2 \psi(x, t)\, dx = -\hbar^2 \int_{-\infty}^{\infty} \psi^*(x, t) \frac{\partial^2}{\partial x^2} \psi(x, t)\, dx.$$
(6–32)

---

**Example 6–2**  Consider the pulselike wave function $\psi(x) = A\exp(-x^2/2a^2)$, where $a$ is a constant with the dimensions of length. What value of $A$ is needed to normalize this wave function?

**Solution**  The constant $A$ is determined by the requirement that

$$1 = \int_{-\infty}^{+\infty} |\psi(x)|^2\, dx = A^2 \int_{-\infty}^{+\infty} \exp\left(\frac{-x^2}{a^2}\right) dx.$$

The integral is a standard integral found in Appendix B.2:

$$\int_{-\infty}^{+\infty} \exp\left(\frac{-x^2}{a^2}\right) dx = a\sqrt{\pi}.$$

With this result, our first equation reads

$$1 = A^2 a(\pi)^{1/2},$$

an equation easily solved for $A$:

$$A = (a)^{-1/2}(\pi)^{-1/4}.$$

---

**Example 6–3**  Calculate $\langle p \rangle$, $\langle p^2 \rangle$, and $\langle p^3 \rangle$ for the normalized wave function of Example 6–2. (Such quantities are the legitimate aim of experimental measurement, because a full set of expectation values for a physical system contains all the information about the system.)

**Solution**  Let us start with $\langle p \rangle$, which for this wave function is given by

$$\langle p \rangle = A^2 \int_{-\infty}^{+\infty} \exp\left(\frac{-x^2}{2a^2}\right)\left(-i\hbar \frac{\partial}{\partial x}\right)\exp\left(\frac{-x^2}{2a^2}\right) dx$$

$$= \left(\frac{i\hbar A^2}{a^2}\right)\int_{-\infty}^{\infty} \exp\left(\frac{-x^2}{2a^2}\right) x \exp\left(\frac{-x^2}{2a^2}\right) dx$$

$$= \left(\frac{i\hbar A^2}{a^2}\right)\int_{-\infty}^{\infty} x \exp\left(\frac{-x^2}{a^2}\right) dx.$$

This formula looks like it might give a complex number. In general, that would be very troublesome, since we expect the average value of the momentum to be real. The momentum is an observable quantity, and its expectation value should be a real number.

But we are saved from this embarrassment by the nature of the integrand. Taken as a whole, the integrand, say $f(x)$, has the property that it is odd about the origin—that is, that $f(-x) = -f(x)$. This means that, as we are integrating from minus infinity to infinity, the total integral will be zero: The integration from minus infinity to zero will cancel against the integration from zero to infinity, and $\langle p \rangle = 0$. Exactly the same argument will work for $\langle p^3 \rangle$.

For $\langle p^2 \rangle$, we must take two powers of the quantity in Eq. (6–28), and the result is real. Using

$$\frac{\partial^2}{\partial x^2} \exp\left(\frac{-x^2}{2a^2}\right) = \left(\frac{x^2}{a^4} - \frac{1}{a^2}\right)\exp\left(\frac{-x^2}{2a^2}\right),$$

we see that we have an integrand, which we may call $g(x)$, that is even; that is, $g(x) = +g(-x)$. Thus we can integrate from zero to infinity and simply double the answer:

$$\langle p^2 \rangle = 2(-i\hbar)^2 A^2 \int_0^\infty \left(\frac{x^2}{a^4} - \frac{1}{a^2}\right)\exp\left(\frac{-x^2}{a^2}\right) dx.$$

We reduce the integral in this equation to a standard integral by the substitution $y = x/a$, so that

$$\langle p^2 \rangle = 2\hbar^2 A^2 \frac{1}{a} \int_0^\infty (1 - y^2)\exp(-y^2)\, dy = \frac{\hbar^2}{2a^2}.$$

In the last step, we used Appendix B.2 for the standard integrals and also substituted for $A^2$, the quantity we calculated in Example 6–2.

Note that since $a$ is the only length in the problem, we would expect, on purely dimensional grounds, that the magnitude of $p^2$ is proportional to $(\hbar/a)^2$.

---

We can also consider the expectation values of mixed quantities. The energy of our electron (or of whatever particle is being discussed) is $E = p^2/2m + V(x)$, which involves both $x$ and $p$. The average value of $E$ is given by

$$\langle E \rangle = \int_{-\infty}^\infty \psi^*(x, t)\left(-\frac{\hbar^2}{2m}\frac{\partial^2}{\partial x^2} + V(x)\right)\psi(x, t)\, dx. \tag{6–33}$$

The pattern is clear: For any quantity $Q = Q(x, t)$ that depends on position and time, the expectation value $\langle Q \rangle$ is the integral over all space of Q centered between $\psi^*$ and $\psi$.

It is customary to call the combination $-(\hbar^2/2m)(\partial^2/\partial x^2) + V(x)$ that acts on the wave function to the right of it in the Schrödinger equation the **Hamiltonian**. This quantity is usually denoted by the letter $H$:

$$H = -\frac{\hbar^2}{2m}\frac{\partial^2}{\partial x^2} + V(x). \tag{6–34}$$

We can think of the Hamiltonian in a new way: it is an **operator**. In classical mechanics, the quantity $p^2/2m + V(x)$ is also called the Hamiltonian, and the classical equations of motion can be formulated in terms of it. But the quantum mechanical Hamiltonian contains derivatives, as well as ordinary functions, of $x$. This is the way the quantum mechanical Hamiltonian differs from the classical one. The concept of an operator generalizes to other quantities that *act* (or *operate*) on the wave function. Their physically measurable counterparts are represented by the expectation values of these operators in ways that we shall describe shortly.

To find expectation values, it is necessary to know the wave function. This is determined by Eq. (6–22), the time-dependent Schrödinger equation, which, in terms of the Hamiltonian operator takes the form

$$i\hbar \frac{\partial \psi(x, t)}{\partial t} = H\psi(x, t). \tag{6–35}$$

Whether we can solve this equation depends on the form of the potential $V(x)$. We shall look at some examples in the next chapter; first, however, consider a significant simplification.

## 6–4 The Time-Independent Schrödinger Equation

The time-dependent Schrödinger equation (6–35) is a partial differential equation involving both time and position. While many partial differential equations can be difficult to solve, in this case we can solve the equation by using the method of **separation of variables**. This method consists of supposing that the wave function can be written as a function of time only, $T(t)$, times a function of position only, $u(x)$:

$$\psi(x, t) = u(x)T(t). \tag{6–36}$$

Then the Schrödinger equation takes the form

$$i\hbar u(x) \frac{dT(t)}{dt} = T(t)\left(-\frac{\hbar^2}{2m}\frac{d^2}{dx^2}u(x) + V(x)u(x)\right).$$

Note that we are no longer using partial derivatives, since the functions in question now depend on only one variable. Dividing by $T(t)u(x)$, we get

$$i\hbar \frac{1}{T(t)}\frac{dT(t)}{dt} = \frac{1}{u(x)}\left(-\frac{\hbar^2}{2m}\frac{d^2}{dx^2}u(x) + V(x)u(x)\right).$$

Now we see—and this is the crucial step in the separation-of-variables technique—that the two sides of this equation depend on entirely independent variables, the left side on time and the right side on position. (If the potential were dependent on time, then the technique would no longer work.) The only way to make the equation hold for all time and all positions is for each side to equal a time- and space-independent constant, the same for both sides. Later, we shall see why we label that constant $E$. Now we concentrate on the fact that we have two equations, which can be put into the form

$$i\hbar \frac{dT(t)}{dt} = ET(t) \tag{6–37}$$

and

$$-\frac{\hbar^2}{2m}\frac{d^2}{dx^2}u(x) + V(x)u(x) = Eu(x). \tag{6–38}$$

The first of these equations isolates the time dependence of the Schrödinger equation, and the second of these is the so-called **time-independent Schrödinger equation**. In fact the time-dependent equation (6–37) has a direct solution:[3]

$$T(t) = \exp\left(\frac{-iEt}{\hbar}\right). \tag{6–39}$$

---

[3] We could have included a multiplicative constant in the function $T(t)$, but that would affect the normalization of the spatial part $u(x)$. It is conventional to put the normalization condition into $u(x)$ by choosing $T(t)$ as in Eq. (6–39).

This last step is important because it means that *we need only solve the time-independent Schrödinger equation*. Once we have done so, the solution to the time-dependent equation is

$$\psi(x, t) = u_E(x)\exp(-iEt/\hbar). \tag{6-40}$$

Note that we have put a subscript on $u(x)$. The reason for that is that $u$ depends on $E$, just as $T$ does.

We next introduce a useful piece of terminology. We have already defined the Hamiltonian operator $H$ and know that Eq. (6–38) takes the form

$$Hu_E(x) = Eu_E(x). \tag{6-41}$$

The solution $u_E(x)$ of this equation has a special property: When the operator $H$ acts on it, it reappears, multiplied by the constant $E$. We say that the function $u_E(x)$ is an **eigenfunction** of the operator $H$, and $E$ is the corresponding **eigenvalue** of that operator. Note many eigenfunctions and eigenvalues may correspond to a given $H$. The problem of solving the Schrödinger equation now comes down to finding eigenfunctions and eigenvalues of the operator $H$. When this is done in the context of, for example, the Coulomb potential, one finds Bohr's atomic energy levels.

## 6–5 An Example: The Infinite Well

In subsequent chapters, we will look at detailed solutions of the time-independent Schrödinger equation and the properties and implications of those solutions. In this section, we shall briefly take up a particularly simplified example. That way, we can visualize a little road map of what is involved in a more extensive study.

Consider an electron confined to a kind of one-dimensional "jail cell." Infinite walls restrict the motion of the electron to the region $0 < x < L$; in that region, the electron is free. The infinite walls at $x = 0$ and $x = L$ may be described by saying that $V(x) = \infty$ to the left of $x = 0$ and to the right of $x = L$ (•Fig. 6–5). We shall generally refer to this potential as the *infinite well*. Outside the free region, the fact that $V$ is infinite means that, according to the Schrödinger equation, $u$ must be zero. In turn, this means that the probability density vanishes outside the infinite well. In the interior region, $V = 0$. This example turns out to be more than fanciful: With modern techniques involving semiconductors, it is possible to produce potentials very close to the one described.

In the region $0 < x < L$, Eq. (6–38), the time-independent Schrödinger equation, is

$$-\frac{\hbar^2}{2m}\frac{d^2}{dx^2}u(x) = Eu(x), \tag{6-42}$$

and we must solve this equation subject to the boundary conditions that

$$u(0) = u(L) = 0. \tag{6-43}$$

Rather than trying to solve Eq. (6–42) in a general way, let us just guess one class of solutions, namely, that *u is sinusoidal*. That is a reasonable guess, be-

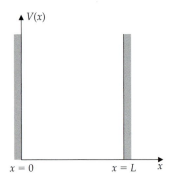

**• Figure 6–5** The infinite-well potential has infinitely high potential-energy "walls" that confine a particle under its influence to the region $0 < x < L$. It is convenient to choose the bottom of the well to correspond to zero potential energy.

cause two derivatives of a sine function do give back the sine function with a minus sign, as Eq. (6–42) requires. Thus we can try

$$u(x) = A \sin(kx),$$

where $A$ is a constant that we can eventually use for normalization. When we plug this into Eq. (6–42), we satisfy that equation provided that

$$E = \frac{\hbar^2 k^2}{2m}. \qquad (6\text{–}44)$$

At this point, we try to satisfy the boundary conditions. The boundary condition at $x = 0$ is automatic; but the one at $x = L$ is not. We can make $\sin kL = 0$ only if $k$ takes on certain values, namely

$$kL = n\pi, \text{ with } n = 1, 2, \dots.$$

In other words, $E$ takes on only certain discrete values (•Fig. 6–6):

$$E = \frac{\hbar^2 \pi^2 n^2}{2mL^2}, \qquad \text{where } n = 1, 2, \dots. \qquad (6\text{–}45)$$

These solutions are drawn in •Fig. 6–7a. We see a strong resemblance to the pictures one finds of standing waves on a string (•Fig. 6–7b). Mathematically, the solutions are certainly standing waves: The function with the time dependence multiplies the function describing the space dependence. The presence of an integer in standing-wave solutions is a standard feature of classical wave motion. Let us not, however, lose sight of the fact that electron motion is not

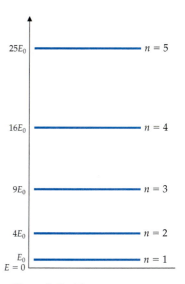

• **Figure 6–6**  The energy spectrum (i.e., the allowed, or possible, energy values) for a particle of mass $m$ in the infinite well of •Fig. 6–5. The ground-state energy is $E_0 = \dfrac{\hbar^2 \pi^2}{2mL^2}$.

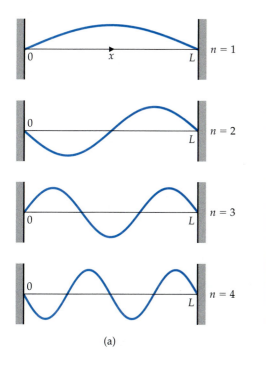

(a)

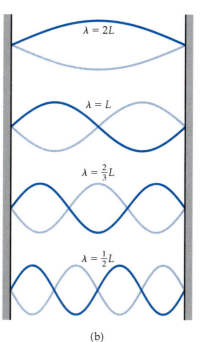

(b)

• **Figure 6–7**  (a) Sketch of the eigenfunctions $u_n(x)$ for the infinite well of •Fig. 6–5. (b) Allowed modes of oscillation of a classical string with its ends fixed.

wave motion in classical physics: The Schrödinger equation and the interpretation of the wave function have introduced wavelike behavior, but of waves of *probability*. That, after all, was our original motivation for introducing the concept of the wave function.

Equation (6–45) shows us a lowest energy, the *ground state*, and a set of higher energies, corresponding to *excited states*. This is just what Bohr postulated for a different system: the atom. While the details are quite different—in particular, the discreteness (the presence of integers) emerges from the Schrödinger equation, without any *ad hoc* quantization rules—the results are qualitatively quite similar. Just as de Broglie proposed for the electrons in the atom, we could have obtained the same result by insisting that a finite number of half-wavelengths fit into the box. (See Problem 6–21.)

In what follows, we shall limit our example further: We work with the ground state, $n = 1$, so that $u(x) = A\sin(\pi x/L)$. Our solution contains what is apparently an unknown constant, $A$. However, we have forgotten about normalization, which requires that

$$\int_0^L \left[ A\sin\left(\frac{\pi x}{L}\right) \right]^* \left[ A\sin\left(\frac{\pi x}{L}\right) \right] dx = |A|^2 \int_0^L \left[ \sin\left(\frac{\pi x}{L}\right) \right]^2 dx = 1.$$

The rightmost integral gives $L/2$, so that $A$ can be chosen to be $\sqrt{2/L}$. We have made the choice for $A$ real. The normalization condition would still be satisfied if we had chosen $A = \sqrt{2/L}\exp(i\varphi)$, but since, as we have noted in Eq. (6–7), an overall factor of the type $\exp(i\varphi)$ does not appear in physical probabilities, it is unnecessary to include such a factor in $A$. The properly normalized ground-state wave function is therefore

$$u_1(x) = \sqrt{\frac{2}{L}} \sin\frac{\pi x}{L}. \tag{6–46}$$

Note the subscript, which is the value of $n$.

---

**Example 6–4**   Suppose a particle of mass $m$ is free within the region $0 < x < L$, but cannot go beyond that region because of high potential walls. Suppose, moreover, that the particle is in the ground state of this one-dimensional box, so that its wave function is given by Eq. (6–46). Calculate $\langle x \rangle$, $\langle p \rangle$, and $\langle p^2 \rangle$ for the particle.

**Solution**   The ground-state wave function is known, so we can start right in on expectation values. Let us start with $\langle x \rangle$. We have

$$\langle x \rangle = \frac{2}{L} \int_0^L x \sin^2(\pi x/L)\, dx = \frac{2}{L}\left(\frac{L}{\pi}\right)^2 \int_0^\pi z \sin^2(z)\, dz$$

$$= \frac{2}{L}\left(\frac{L}{\pi}\right)^2 \left[ \frac{z^2}{4} - \frac{z\sin 2z}{4} - \cos 2z \right]_{z=0}^{z=\pi} = \frac{2}{L}\left(\frac{L}{\pi}\right)^2 \frac{\pi^2}{4} = \frac{L}{2}.$$

We have made the transformation $z = \pi x/L$ in order to simplify the integral. This result is certainly sensible: The particle spends as much time to the left of center as to the right, and on average, is to be found in a position in the middle. We could very well have guessed this expectation value from the drawing of the wave function. We see from •Fig. 6–8 that the wave function, and therefore its square, peaks at $L/2$, and therefore the *most probable* position of the particle is there. Moreover, the wave function is symmetrical about $L/2$, so the *average* position is there as well. All this is not very classical; in a classical situation, we could say *exactly* where the particle was if we knew some initial conditions!

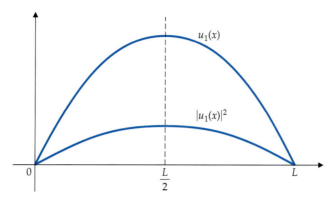

• **Figure 6–8**   The ground-state wave function $u_1(x)$, as well as the associated probability distribution $|u_1(x)|^2$, for the infinite well is symmetric about the midpoint $x = L/2$.

As for the average momentum, it would be much harder to guess its value from a casual look at the shape of the wave function. We have

$$\langle p \rangle = \frac{2}{L} \int_0^L \sin(\pi x/L)\left(-i\hbar \frac{d}{dx}\right)\sin(\pi x/L)\, dx$$

$$= \frac{2}{L}\left(-i\hbar \frac{\pi}{L}\right)\int_0^L \sin(\pi x/L)\cos(\pi x/L)\, dx = -i\frac{\hbar}{L}\int_0^\pi \sin(2z)\, dz = 0.$$

Fortunately, the integral vanishes, since otherwise we would get an imaginary value for the average momentum! The fact that the expectation value of $p$ vanishes agrees with our physical intuition. The particle bounces back and forth, spending half its time going to the right and half its time going to the left. Thus, it is reasonable for the average momentum to be zero. But, unlike a particle that simply obeys Newton's laws, the quantum mechanical particle discussed here does not have an exact momentum. (This idea will be discussed in much more detail shortly.) The particle does, however, obey an "average" Newton's law in that $d\langle x \rangle/dt = \langle p \rangle/m = 0$. [Recall Eq. (6–30).]

The last quantity we want is

$$\langle p^2 \rangle = \frac{2}{L}\int_0^L \sin(\pi x/L)\left(-i\hbar \frac{d}{dx}\right)^2 \sin(\pi x/L)\, dx$$

$$= \frac{2\hbar^2}{L}\left(\frac{\pi}{L}\right)^2 \int_0^L \sin^2(\pi x/L)\, dx = \frac{2\hbar^2}{L}\left(\frac{\pi}{L}\right)^2 \frac{L}{2} = \frac{\pi^2 \hbar^2}{L^2}.$$

This quantity is positive and real, again as we would expect, and because the wave function is the $n = 1$ eigenstate of $H = p^2/2m$, it is just $2mE_1$.

## The Physical Meaning of Eigenfunctions and Eigenvalues

We can see what is significant about having an eigenfunction of a particular operator by considering expectation values. If we ask for the expectation value of the energy in a series of energy measurements when the wave function is given by an eigenfunction of the energy, what will we find? We know how to calculate this expectation value; it is

$$\langle \text{energy} \rangle = \int \psi_E^* H \psi_E\, dx,$$

where $\psi_E(x)$ is the eigenfunction of energy with eigenvalue $E$ and where the operator $H$ that appears in the integral is the Hamiltonian operator,

$$H = -\frac{\hbar^2}{2m}\frac{d^2}{dx^2} + V(x).$$

But because $\psi_E$ is an eigenfunction of the energy with eigenvalue $E$, the action of the Hamiltonian operator on $\psi_E$ is easy to figure out—by definition, it is just $E$ times $\psi_E$, *where $E$ is just an ordinary algebraic quantity contained in the wave function.* Thus

$$\langle \text{energy} \rangle = E \int \psi_E^* \psi_E \, dx = E,$$

where we have assumed, as usual, that the wave function is normalized, so that the integral of $\psi_E^* \psi_E$ is unity. By the same token,

$$\langle \text{energy}^n \rangle = \int \psi_E^* H^n \psi_E \, dx = E^n \int \psi_E^* \psi_E \, dx = E^n.$$

To find this result, we simply apply $H$ to $\psi_E$ $n$ times in succession; each time, a factor of the ordinary algebraic constant $E$ is produced.

Why is this result useful? When dealing with probabilities, one has the important concept of the *standard deviation* $\sigma$. As we illustrate in detail in Appendix B.3, the standard deviation describes the amount of "spread" in the distribution of a measured quantity such as energy: The larger the value of $\sigma$, the broader is the distribution, while the smaller the value of $\sigma$, the more closely the distribution is concentrated around the average value. If the measured statistical quantity is $h$, then $\sigma$ is defined by

$$\sigma^2 \equiv \langle h^2 \rangle - \langle h \rangle^2. \tag{6–47}$$

Given our calculation of $\langle \text{energy}^n \rangle$, we can easily find the standard deviation in energy when the wave function is an energy eigenfunction $\psi_E$. The result is immediate:

$$\langle \text{energy}^2 \rangle - \langle \text{energy} \rangle^2 = E^2 - E^2 = 0. \tag{6–48}$$

In other words, there is *no* spread away from the average value of energy: $E$ is the only value ever measured. The probability of finding the algebraic value $E$ is unity, and the probability of finding any other value is zero. *In other words, the particle described by an eigenfunction $\psi_E(\vec{r}, t)$ has a definite energy $E$.*

This discussion makes it clear why we spoke in the last chapter of wave functions of definite energy: When a wave function satisfies the Schrödinger equation, it is an eigenfunction of the Hamiltonian operator with eigenvalue given by $E$. Such waves represent particles with *definite energy*.

It is *generally* true that, for an eigenfunction of an operator representing some physical variable (energy in the preceding discussion), a measurement of that variable will always yield the eigenvalue. In other words, what was shown for the energy is true for eigenfunctions of other operators. For example, suppose that we have an eigenfunction $u_p(x)$ of momentum with eigenvalue $p$. This means that $u_p(x)$ satisfies the eigenvalue equation

$$-i\hbar \frac{d}{dx} u_p(x) = p u_p(x).$$

We can now imagine calculating the average value of powers of the momentum, given that our system is described by this wave function. The calculation proceeds just as that for the energy:

$$\langle \text{momentum}^n \rangle = \int_{-\infty}^{+\infty} u_p^*(x) \left( -i\hbar \frac{d}{dx} \right)^n u_p(x) \, dx = p^n \int_{-\infty}^{+\infty} u_p^*(x) u_p(x) \, dx = p^n.$$

We have assumed that the eigenfunction $u_p(x)$ of momentum is normalized. When we calculate the standard deviation for a measurement of the momentum, we therefore find that

$$\langle\text{momentum}^2\rangle - \langle\text{momentum}\rangle^2 = \langle p^2\rangle - \langle p\rangle^2 = 0.$$

We come to a conclusion similar to the one we came to for energy: When the system is described by the wave function $u_p(x)$, only one value for the momentum is ever measured, and that is $p$. When a quantum mechanical system is described by an eigenfunction of momentum, it means that the system has a definite value of momentum, and that value is the eigenvalue.

Finally, we can remark that if a particular wave function is *not* an eigenfunction of an operator, *then the physical observable corresponding to that operator does not have a definite value—instead we must rely on expectation values to describe the physical observable of the system.* For example, the infinite-well wave function of Example 6–4 is an eigenfunction of energy, not momentum. That is why we did not find in that example that $\langle p^2\rangle = \langle p\rangle^2$; that would be true only if the wave function were an eigenfunction of momentum.

# 6–6  The Schrödinger Equation in Three Dimensions

The change from one dimension to three dimensions in classical mechanics is simple: It corresponds to a change from $E = p^2/2m + V(x)$ to

$$E = \frac{\vec{\mathbf{p}}^2}{2m} + V(\vec{\mathbf{r}}) = \frac{p_x^2}{2m} + \frac{p_y^2}{2m} + \frac{p_z^2}{2m} + V(\vec{\mathbf{r}}).$$

To extend this idea to quantum mechanics, we postulate a wave function in three dimensions with the interpretation that $\left|\psi(\vec{\mathbf{r}}, t)\right|^2 d^3\vec{\mathbf{r}}$ is the probability of finding the particle at $\vec{\mathbf{r}}$ in a small box of volume $d^3\vec{\mathbf{r}} = dx\,dy\,dz$ near $\vec{\mathbf{r}}$. This wave function satisfies a Schrödinger equation in which we have generalized our notion of momentum according to the rule

$$p_x \to -i\hbar\frac{\partial}{\partial x}, \quad p_y \to -i\hbar\frac{\partial}{\partial y}, \quad p_z \to -i\hbar\frac{\partial}{\partial z}. \tag{6–49}$$

Then the **three-dimensional Schrödinger equation** is

$$i\hbar\frac{\partial\psi(x, y, z, t)}{\partial t} = -\frac{\hbar^2}{2m}\left(\frac{\partial^2}{\partial x^2} + \frac{\partial^2}{\partial y^2} + \frac{\partial^2}{\partial z^2}\right)\psi(x, y, z, t) + V(x, y, z)\psi(x, y, z, t).$$
$$\tag{6–50}$$

This equation is often written in the more compact form

$$i\hbar\frac{\partial\psi(\vec{\mathbf{r}}, t)}{\partial t} = -\frac{\hbar^2}{2m}\nabla^2\psi(\vec{\mathbf{r}}, t) + V(\vec{\mathbf{r}})\psi(\vec{\mathbf{r}}, t), \tag{6–51}$$

where $\nabla^2 = \dfrac{\partial^2}{\partial x^2} + \dfrac{\partial^2}{\partial y^2} + \dfrac{\partial^2}{\partial z^2}$ is standard notation for the sum of the three second-order partial derivatives, an operator that mathematicians call the *Laplacian.*

We examine the three-dimensional situation in later chapters. Of course, we live in a three-dimensional world; nevertheless, many physical ideas and consequences follow from mathematically simpler one-dimensional systems.

## SUMMARY

We have seen that particles can exhibit wavelike properties. Classical waves, such as electromagnetic and sound waves, obey equations that determine how the waves propagate in space with the passage of time. The Schrödinger equation is the counterpart of these equations for quantum mechanical particles. It is a partial differential equation involving second-order spatial derivatives and first-order temporal derivatives. The following are the important ideas associated with this equation:

- The wave function of a particle under the influence of various forces satisfies the Schrödinger equation. The wave function is a complex-valued function of space and time. Its physical significance is that the square of the absolute value of the function describes the probability that the particle is at a given location in space at a given time.
- Knowledge of the wave function allows us to find the probabilities that the particle has certain values of momentum, energy, or other physically measurable quantities.
- The time dependence of the Schrödinger equation can often be factored out of the equation, leaving the so-called time-independent Schrödinger equation, which determines the allowed values of the energy, or energy eigenvalues. In situations in which the particle is bound or confined, these energies are restricted to certain discrete, or quantized, values. For each allowed energy, there is a specific time-independent wave function, forming the set of energy eigenfunctions.

One example of how the Schrödinger equation can be used is that of a deep square potential-energy well in one spatial dimension. For this example, one can find all the wave functions and their associated energies explicitly.

## QUESTIONS

**1.** Suppose a particle in an infinitely deep well were in the ground state of energy. Would this energy change with time? Can you think of any expectation value that also would change with time in this situation? How about the wave function itself? Are your answers to the last two questions consistent?

**2.** The wave function of an electron in a one-dimensional infinite well forms standing harmonic waves. Sketch this wave function for large values of $n$, and use your sketch to indicate the probability of finding the electron in different parts of the well. What would you expect the classical distribution of electron positions to be in this situation, in which the electron bounces back and forth between the walls? Are the classical and large-$n$ quantum probabilities related? Would you expect this result?

**3.** In the Hamiltonians we have considered, do any two energy eigenfunctions that are proportional to each other have the same energy eigenvalues?

**4.** In a two-slit interference experiment, a beam of electrons produces an interference pattern on a distant screen when it passes through a pair of slits in front of the screen. The pattern is analogous to the pattern produced by light when it passes through a double slit. But light also produces an interference pattern when it passes through a single slit. Will this be true for electrons?

**5.** From the discussion in this chapter, can you find two wave functions that are not proportional to each other, yet still have the same energy eigenvalues?

What distinguishes the functions? (*Hint*: Think about right- and left-moving plane waves.)

**6.** Do all linear combinations of the wave functions you found in the answer to the last question still have the same energy eigenvalues?

**7.** If the ground-state energy of an electron in a box were of the same magnitude as that of hydrogen in the ground state, how would the width of the box compare with the Bohr radius?

**8.** In Section 6–5, we described the quantized energies that result for an infinite well with edges at $x = 0$ and $x = L$. We also gave the wave function for the ground state. What would change if the well instead had its edges at $x = -L/2$ and $+L/2$?

**9.** Suppose you have a box a few centimeters wide and a electron with the energy of a few electron volts (i.e., of the same order of a magnitude as an electron in a Bohr orbit). In what range would the $n$-values for the electron in the box lie?

## PROBLEMS

**1.** ∎∎ A time-independent wave function has the form $\psi(x) = C \exp(-x^2/2a^2)$. Find $C$ such that $\int |\psi(x)|^2 \, dx = 1$.

**2.** ∎∎ Consider the normalized wave function of Problem 1. Calculate the expectation values **(a)** $\langle x \rangle$, **(b)** $\langle x^2 \rangle$, **(c)** $\langle p \rangle$, and **(d)** $\langle p^2 \rangle$ for this wave function. Give a physical justification of your results for (a) and (c).

**3.** ∎∎ Show that, for the wave function of Problem 1, the product $\langle x^2 \rangle \langle p^2 \rangle$ is independent of $a$ and is proportional to $\hbar^2$. This is an illustration of a very general quantum mechanical result, which is that (a) $\langle x^2 \rangle$ and $\langle p^2 \rangle$ are correlated in that if one is large, the other is small, and (b) the correlation disappears when $\hbar \to 0$.

**4.** ∎∎ Consider a wave function written in the form

$$\psi(x, t) = R(x, t) \exp[iS(x, t)],$$

which effectively breaks the wave function into a magnitude and a phase. **(a)** What form does the normalization condition, Eq. (6–8), take? **(b)** What is the expectation value of the momentum, $\langle p \rangle$, for this wave function?

**5.** ∎ Consider an electron, mass $m = 0.9 \times 10^{-30}$ kg, in an infinite well of width 0.05 nm. **(a)** What is the ground-state energy of the electron, in electron volts? **(b)** What is the energy of the first excited state?

**6.** ∎ What is the wavelength of a photon emitted when the electron in the infinite well of Problem 5 makes a transition from the first excited state to the ground state?

**7.** ∎ Consider an electron, mass $m = 0.9 \times 10^{-30}$ kg, in an infinite well that is 2 cm wide. For what value of $n$ will the electron have an energy of 1 eV?

**8.** ∎ Obtain an expression for the wavelength of a photon emitted when an electron in an infinite well of width $L$ makes a transition from a state of quantum number $n$ to the ground state.

**9.** ∎ A beam of photons in a range of wavelengths $\lambda = 9.0 \pm 1.0$ nm impinges on an electron in an infinite well of width 1.0 nm. The electron is in the ground state. To what excited states can it be boosted (i.e., to what range of values of $n$ can the electron get excited)?

**10.** ▮▮ Consider the situation described in the previous problem. What is the energy spacing between adjacent levels for $n$ corresponding to an electron energy of 1 eV? Are we justified in treating an electron in a macroscopic box as if it had a continuum of energy values?

**11.** ▮▮ Consider the eigenfunction for the infinite well of width $L$ corresponding to the $n = 2$ (first excited) state. What is the average value $\langle x \rangle$ of $x$? Sketch the form of the probability density $|u_1(x)|^2$ to explain your result on physical grounds.

**12.** ▮▮ Consider an electron in an infinite well of width $L$ and in an energy eigenstate characterized by the quantum number $n$. Calculate $\langle x^2 \rangle$ and $\langle p^2 \rangle$. Show that the product $\langle x^2 \rangle \langle p^2 \rangle$ is independent of $L$. How does that product depend on $n$?

**13.** ▮▮▮ Prove that the eigenfunctions for the infinite-well potential

$$u_n(x) = \sqrt{2/L} \sin(n\pi x/L)$$

have the property that

$$\int_0^L u_n(x)u_m(x)\, dx = 0 \text{ for } n \neq m.$$

This property is called the *orthogonality* property of the eigenfunctions for different eigenvalues.

**14.** ▮▮▮ Consider a wave function of the form

$$\psi(x) = Au_1(x) + Bu_2(x),$$

where $u_1$ and $u_2$ are eigenfunctions for the infinite-well potential. **(a)** Use the requirement that $\psi$ is properly normalized to show that $|A|^2 + |B|^2 = 1$. (*Hint:* Use the results of the previous problem.) **(b)** Calculate the expectation value of the energy $p^2/2m$ for this wave function, and show that it is equal to

$$\langle E \rangle = |A|^2 E_1 + |B|^2 E_2,$$

where $E_1$ and $E_2$ are the energy values corresponding to $n = 1$ and $n = 2$, respectively. (*Hint:* Use the results of Problem 13.)

**15.** ▮▮▮ The results of Problem 14 suggest that $|A|^2$ is the probability that one finds the energy $E_1$ when making an energy measurement for this superposition of two different eigenfunctions, and that $|B|^2$ is the probability of finding the energy $E_2$. Show that this condition is in fact true by comparing it with the probabilistic equations

$$P(E_1) + P(E_2) = 1$$

and

$$E_1 P(E_1) + E_2 P(E_2) = \langle E \rangle.$$

Show that this interpretation will also work for a wave function of the form

$$\psi(x) = Au_1(x) + Bu_2(x) + Cu_3(x).$$

(*Hint:* Use the results of Problem 13.)

**16.** ▮▮ Consider a wave function of the form $\psi(x) = C(\sin \pi x/L)^3$. Calculate $C$ such that $\psi(x)$ is properly normalized. (*Hint:* Use the binomial expansion of $(1/2i)[\exp(iy) - \exp(-iy)]^6$ to obtain the identity $(\sin y)^6 = (1/64)$ $[20 - 15\cos(2y) + 6\cos(4y) - 2\cos(6y)]$.) Assume an infinite well from $x = 0$ to $x = L$.

**17.** ▮▮▮ Use the identity $\sin^3 y = (1/4)[3\sin y - \sin(3y)]$ and the interpretation suggested in Problem 15 to calculate the probability that a measurement of the energy for the wave function of Problem 16 yields $E_1$.

**18.** ▮▮ Consider the wave function $(1/\sqrt{2})[u_1(x) + u_2(x)]$, where the $u_i$ are the eigenfunctions of the infinite-well potential. This function does not depend on time. Assuming that the wave function represents $\psi(x, 0)$ at time $t = 0$, what is the form of the function at time $t$? (*Hint*: Look back at Eq. (6–40).)

**19.** ▮▮ What is the probability that an electron in the infinite well in the state $u_2(x)$ is found in the region between $x = 0$ and $x = L/2$?

**20.** ▮▮ Consider the wave function $\psi(x, t) = (1/\sqrt{2})[u_2(x)\exp(-iE_2t/\hbar) + u_3(x)\exp(-iE_3t/\hbar)]$. Calculate the probability that the electron is in the range $(0, L/2)$ as a function of time. What is the period of oscillation of the probability?

**21.** ▮▮ Show that the energy eigenvalues of the one-dimensional infinite well are correctly quantized by insisting that a finite number of half-wavelengths fit into the box. (*Hint*: Convert the condition into one for momentum and, hence, on $E$.)

**22.** ▮▮▮ Show that $d\langle p \rangle/dt = -\langle dV(x)/dx \rangle$, which is Newton's law in terms of quantum mechanical expectation values. (*Hint*: Start with $\langle p \rangle$, and take the time derivatives under the integral. When you use Schrödinger's equation, notice that the second spatial derivatives neatly collect themselves into a total spatial derivative that you can integrate away. A judicious integration by parts of what remains will produce the answer.) This is a very important result, since it shows us that quantum mechanics does not require us to throw out 300 years of physics. The old physics works, so long as we are not concerned about deviations from the average behavior. The infinite-well potential illustrates this result. In the interior, $V$ is constant and there is no force. However, at the cell walls, the potential jumps up, and the derivative becomes huge, indicating that at the walls the particle is sharply accelerated backwards.

**23.** ▮▮ In the text, we worked out the infinite-well problem in one spatial dimension; that is, we found the eigenfunctions and eigenvalues of energy. Follow this discussion, and work out the same problem in two dimensions. In this case, the potential is zero except on a square of sides of length $L$, where it is infinite. The key here is that we can regard the Hamiltonian as the sum of two independent Hamiltonians belonging to particles moving in the $x$- and $y$-directions respectively. Thus, the system wave function is a product of wave functions of $x$ and $y$.

**24.** ▮▮ Consider an electron in a two-dimensional infinite well that is a square of side $L$. Show that the energy eigenvalues are of the form

$$(\text{constant})(n_1^2 + n_2^2),$$

where $n_1$ and $n_2$ are integers. What is the value of the constant?

**25.** ▮▮ Consider an electron in a square infinite well whose sides have length 0.05 nm. (See the previous problem.) What are the values of the lowest eight eigenvalues? Notice that some are duplicated. Use a plot of your result for the previous problem to explain the duplication.

The following three problems introduce new material.

**26.** ▮▮▮ The classical expression for the electric current carried by a single electron is $-ev$, where $-e$ is the electric charge, and $v$ is the speed, of the electron. Now consider quantum mechanics, in which the single electron is described by the wave function $\psi(x, t)$. **(a)** Obtain a quantum mechanical expression for the electric current in terms of $\psi(x, t)$, given that $\int_{-\infty}^{+\infty} \psi^*(x, t)\psi(x, t)\, dx = 1$. (*Hint*: Instead of the speed, use what can be calculated in quantum mechanics, namely, the expectation value of the speed.) Verify that your result is a real number. **(b)** One can also use the form $-e(\langle v \rangle + \langle v \rangle^*)/2$ for the current. Show that this quantity has the form

$$-\frac{e\hbar}{2im} \int_{-\infty}^{+\infty} \left[ \frac{d\psi^*}{dx} \psi - \psi^* \frac{d\psi}{dx} \right] dx.$$

**27.** ▮▮▮ The quantity in the square brackets in the last expression of the previous problem is called the **current density**, denoted by $j(x, t)$. Calculate $dj(x, t)/dx$ and show that if $\psi(x, t)$ satisfies the Schrödinger equation with a real potential energy and does not depend on time, then $j(x)$ is a constant, independent of $x$. This means that no charge is lost along the $x$-axis.

**28.** ▮▮▮ Calculate the current density for an electron described by the wave function $\psi(x, t) = N\exp(-ax^2/2)\exp(ikx - i\omega t)$. Take into account the fact that $N$ needs to be calculated using the normalization condition $\int_{-\infty}^{+\infty} \psi^*(x, t)\psi(x, t)\, dx = 1$.

## APPENDIX

### Derivation of Eq. (6–30)

We have

$$\frac{d\langle x \rangle}{dt} = \frac{d}{dt} \int \psi^* x \psi\, dx = \int \left[ \frac{\partial \psi^*}{\partial t} x\psi + \psi^* x \frac{\partial \psi}{\partial t} \right] dx.$$

Note that the only time dependence is in $\psi(x, t)$; the $x$ under the integral sign is just a dummy variable that does not depend on time, in contrast to the expectation value of $x$. We now use the Schrödinger equation in the form

$$\frac{\partial \psi(x, t)}{\partial t} = i\frac{\hbar}{2m} \frac{\partial^2}{\partial x^2} \psi(x, t) - \frac{i}{\hbar} V(x)\psi(x, t),$$

as well as the complex conjugate of this relation (we assume that the potential $V(x)$ is real valued), viz.,

$$\frac{\partial \psi^*(x, t)}{\partial t} = -i\frac{\hbar}{2m} \frac{\partial^2}{\partial x^2} \psi^*(x, t) + \frac{i}{\hbar} V(x)\psi^*(x, t),$$

to find

$$\frac{\partial \psi^*}{\partial t} x\psi + \psi^* x \frac{\partial \psi}{\partial t} = -\frac{i\hbar}{2m} \left( \frac{\partial^2 \psi^*}{\partial x^2} x\psi - \psi^* x \frac{\partial^2 \psi}{\partial x^2} \right)$$

$$= -\frac{i\hbar}{2m} \frac{\partial}{\partial x} \left[ \frac{\partial \psi^*}{\partial x} x\psi - \psi^* x \frac{\partial \psi}{\partial x} - \psi^*\psi \right] - \frac{i\hbar}{m} \psi^* \frac{\partial}{\partial x} \psi.$$

We can now integrate over $x$. In this integration, the first term on the right-hand side of the foregoing equation vanishes; it is just the quantity in square brackets evaluated at $\pm\infty$, and $\psi$ must vanish there if the normalization integral is to be finite. We are then left with

$$\int_{-\infty}^{\infty} \left( \frac{\partial \psi^*}{\partial t} x\psi + \psi^* x \frac{\partial \psi}{\partial t} \right) dx = -\frac{i\hbar}{m} \int_{-\infty}^{\infty} \psi^* \frac{\partial}{\partial x} \psi \, dx.$$

The left-hand side of this equation is the derivative of $\langle x \rangle$ with respect to time, while the right-hand side is $\langle p \rangle$ divided by $m$. This is what we set out to prove.

Incidentally, the same manipulations would allow us to show that

$$\frac{d}{dt} \int \psi^* \psi \, dx = \frac{i\hbar}{2m} \int \left[ \frac{\partial}{\partial x} \left\{ \psi^* \frac{\partial \psi}{\partial x} - \frac{\partial \psi^*}{\partial x} \psi \right\} \right] dx.$$

The right-hand side of this expression is just the quantity in curly brackets evaluated at the limits of integration, and this is zero because $\psi$ and its derivatives vanish as $x \to \pm\infty$. In this way, we show that the normalization factor $N$ is unchanging with time. (See Section 6–3.)

# CHAPTER
# 7

# Wave Packets and the Uncertainty Principle

W e now have the Schrödinger equation and the interpretation of the wave function in hand. The consequences for a particle in an infinite well were worked out in Chapter 6, where we found that the particle displayed some very nonclassical features—in particular, the quantization of its energy. In this chapter we turn to the description of a free particle— a particle that is not subject to any forces. Such particles should, at least in some way, exhibit all the classical properties: They carry momentum and energy, and they appear to be localized, in the sense that if they carry charge—think of the particle as an electron—they leave well-defined tracks in a suitably designed detector, such as a hydrogen bubble chamber (•Fig. 7–1). But how can a solution of the Schrödinger *wave* equation look like a particle? The answer to this ques-

• **Figure 7–1** The trajectory of an electron in a hydrogen bubble chamber. The electron does not move in a straight line, because of the presence of a strong magnetic field. The electron ionizes the hydrogen atoms in its path, and these start the growth of bubbles. The bubbles are illuminated and photographed. The gradual energy loss of the electrons leads to their spiral motion, which in turn allows us to determine the direction of motion of the electrons and, given the direction of the magnetic field, the sign of the charge on the electron. In the photo, two spirals start at the same vertex and go in opposite directions. The vertex is where an electron–positron pair is produced. The small spirals show low–energy electrons that have been knocked out of the hydrogen atoms.

tion ultimately leads us to a set of fundamental relations of quantum mechanics: the **Heisenberg uncertainty relations**. These relations place limits on how well we can apply our classical intuitions about position and momentum to quantum phenomena. We shall concentrate mainly on one-dimensional motion, which contains most of the necessary information without too much mathematical complication.

## 7–1  A Free Electron in One Dimension

Let us recall the Schrödinger equation for the free particle and solve it. (When we motivated the discussion of the equation in the first place, we actually started with the solution, but it will be helpful here to repeat the process in reverse.) The free electron obeys the time-dependent Schrödinger equation [Eq. (6–22)],

$$i\hbar \frac{\partial}{\partial t} \psi(x, t) = -\frac{\hbar^2}{2m} \frac{\partial^2}{\partial x^2} \psi(x, t). \tag{7–1}$$

This equation can be solved by first writing

$$\psi(x, t) = u(x) \exp\left(-\frac{iEt}{\hbar}\right). \tag{7–2}$$

Substitution of this form into Eq. (7–1), with subsequent cancellation of the exponential factor $\exp(-iEt/\hbar)$, shows that $u(x)$ satisfies the time-*independent* Schrödinger equation [Eq. (6–38)],

$$Eu(x) = -\frac{\hbar^2}{2m} \frac{d^2}{dx^2} u(x). \tag{7–3}$$

This equation takes the simple form

$$\frac{d^2}{dx^2} u(x) + k^2 u(x) = 0 \tag{7–4}$$

if we introduce the parameter $k$ defined by

$$k^2 = \frac{2mE}{\hbar^2}. \tag{7–5}$$

Equation (7–5) implies that $\hbar^2 k^2 = 2mE$, and if we recall that the momentum $p$ of a particle is related to the particle's energy by $p^2 = 2mE$, then we may conclude that $p = \hbar k$. This is the de Broglie relation, and $k$ is the wave number characterizing that relation.

Equation (7–4) is an equation we are familiar with from the study of the simple harmonic oscillator. Newton's second law for this classical system takes the form

$$\frac{d^2}{dt^2} x(t) + \omega^2 x(t) = 0.$$

By comparing this equation for the motion of a mass on the end of a spring with Eq. (7–4), we see immediately that we need only replace some symbols to find solutions of the Schrödinger equation in terms of the solutions of the equations of motion of a simple harmonic oscillator. The solutions for the mass on the spring are well known: sines, cosines, and linear combinations thereof. Because we have a second-order differential equation, we know there are two independent solutions, which we can take as a variety of combinations of the harmonic functions.

By studying the significance of the parameter $k$, we find that the two solutions of most interest to us are the complex combinations of sine and cosine that form pure exponentials, namely

$$u(x) = \exp(ikx) \tag{7-6}$$

and

$$u(x) = \exp(-ikx). \tag{7-7}$$

To see why, we note that, from its definition [Eq. (7–5)], the parameter $k$ has the following meaning:

$$E = \frac{\hbar^2 k^2}{2m}.$$

But for a free particle, $E = p^2/2m$, so, by comparison,

$$p = \pm \hbar k. \tag{7-8}$$

Now recall that in Chapter 6 we found that in quantum mechanics the momentum $p$ is represented by the differential *operator* $-i\hbar(d/dx)$. If we apply this operator to the harmonic function $\sin kx$, for example, we find that

$$-i\hbar(d/dx)\sin kx = -i\hbar k \cos kx.$$

In other words, we get the cosine, a different function than the one we started with and on which the momentum operator acted. Therefore we can say that $\sin kx$ is *not* an eigenfunction of the momentum operator. Recalling the discussion in Chapter 6 about the significance of eigenfunctions and eigenvalues, this result shows that if we describe the electron with a pure sine or pure cosine, or indeed—as we shall immediately show—any combination of sine and cosine other than $\exp(\pm ikx)$, the momentum of the electron will not have a definite value.

In contrast, the particular combinations of sine and cosine that form $\exp(\pm ikx)$, namely, $\cos kx \pm i \sin kx$, *are* eigenfunctions of the momentum, because

$$-i\hbar(d/dx)\exp(+ikx) = +\hbar k \exp(+ikx) \tag{7-9a}$$

and

$$-i\hbar(d/dx)\exp(-ikx) = -\hbar k \exp(-ikx). \tag{7-9b}$$

The eigenvalues of momentum associated with these eigenfunctions are $\pm\hbar k$. Thus an electron described by the wave functions $\exp(\pm ikx)$ is an electron with a definite value of momentum, namely $p = \pm\hbar k$. The sign in this one-dimensional treatment indicates whether the electron is moving to the right ($+$) or the left ($-$). These solutions also correspond to definite values of the energy, namely $E = \hbar^2 k^2/(2m)$; that is because $u(x)$ is an eigenfunction of energy as well, as Eq. (7–3) shows.

We can now pull the time-dependent and space-dependent pieces together using Eq. (7–2): The solution of the Schrödinger equation for a free electron moving in one dimension with momentum $p$ and energy $E = p^2/2m$—the electron's wave function—is

$$\psi(x, t) = A \exp\left[\frac{i(px - Et)}{\hbar}\right]. \tag{7-10}$$

Here, the linearity of the Schrödinger equation allows us to include the arbitrary constant $A$. [In Chapter 6, we postulated this result; see Eq. (6–18).]

We refer to this wave function as a **plane wave**. According to the sign of $p$, the function represents a wave traveling in the positive or negative $x$-direction.[1] Note that in the form $\exp(ikx)$, the solution repeats when we change $x$ to $x + 2\pi/k$. For a wave, such a periodicity in space describes the wavelength $\lambda$. We thus have

$$\lambda = \frac{2\pi}{k} = \frac{2\pi}{p/\hbar} = \frac{2\pi\hbar}{p}$$

$$= \frac{h}{p}.$$

(7–11)

This is just the de Broglie relation described in Chapter 4.

There are two important problems to resolve in connection with the solution written in Eq. (7–10). First, the square of the absolute value of $\psi(x, t)$ is supposed to give the probability density for finding the electron at position $x$. But for any $a$,

$$|\exp(ia)|^2 = \exp(-ia)\exp(+ia) = \exp(-ia + ia) = \exp(0) = 1.$$

That means, in turn, that

$$|\psi(x, t)|^2 = |A|^2,$$

*independent* of position: The electron has equal probability of being *anywhere* between minus and plus infinity. Thus our solution cannot describe a localized particle, even if it does describe something with a constant, fixed momentum. The second problem is closely related to the first, and it has to do with the normalization of—or, more precisely, the difficulty of normalizing—the solution. The normalization condition requires that the integral over all space of $|\psi(x, t)|^2$ be unity. But this means that

$$\int_{-\infty}^{+\infty} |\psi(x, t)|^2\, dx = |A|^2 \int_{-\infty}^{+\infty} dx = 1.$$

Since the integral is infinitely large, this equation can be satisfied only if $|A|^2$ is infinitely small!

The second problem is an immediate consequence of the first. We begin by solving the first problem—that is, by finding out how to describe an electron that is *localized*. We shall see that solving this problem also solves the second one.

## 7–2  Wave Packets

The goal of this section is to understand how we can make a wave function in space that describes a localized free particle. In the previous section, we found a way to make a wave function correspond to something with definite momentum. But we also found out that this wave function, the plane wave, describes something that is not localized at all: The probability of finding a particle described by the plane wave is the same at all points in space. This result seems to take us farther from our original goal of seeing particlelike behavior in quantum mechanics. So just how is it that a wave function can look like a particle? To answer this question, we note that the linearity of the Schrödinger equation implies that a superposition of plane waves of the form of Eq. (7–10) is still a

---

[1] Strictly speaking, this solution is a plane wave in three dimensions, since the wave function is the same in the entire $yz$-plane, and hence, the peaks and troughs lie on planes advancing in the $x$-direction. We use the same nomenclature, however, even in one dimension.

solution of the equation, provided only that $E$ and $p$ continue to be related by $E = p^2/2m$. Such a sum, which, in principle, could have a coefficient $A$ that is different for each value of $p$, and so is written as $A(p)$, can take the general form

$$\psi(x, t) = \int_{-\infty}^{+\infty} A(p)\exp\left[\frac{i(px - Et)}{\hbar}\right] dp. \qquad (7\text{--}12)$$

(We shall continue to call the wave function $\psi$—of course, this is not the same $\psi$ as that of the plane wave!) The reader can easily check by differentiating inside the integral sign that this wave function still satisfies the Schrödinger equation for a free particle, provided that $E = p^2/2m$.

The wave function described by the preceding superposition is *not* an eigenfunction of momentum—it does not describe a particle with a well-defined momentum. Physically, the reason for this is evident: We have constructed the wave function as a sum of plane waves with *different* momenta. The information about "how much" of each momentum goes into the superposition is contained in the way that the coefficients $A(p)$ vary with momentum. For this reason, we can refer to the coefficients $A(p)$ as *momentum weights*. Mathematically, we can verify this by operating on $\psi(x, t)$ with the momentum operator $-i\hbar(d/dx)$:

$$-i\hbar \frac{\partial}{\partial x}\psi(x, t) = -i\hbar \frac{\partial}{\partial x}\int A(p)\exp\left(i\frac{px - Et}{\hbar}\right) dp$$

$$= -i\hbar \int \frac{ip}{\hbar} A(p)\exp\left(i\frac{px - Et}{\hbar}\right) dp$$

$$= \int p A(p)\exp\left(i\frac{px - Et}{\hbar}\right) dp.$$

The integral on the right is *not* a number times the original wave function, Eq. (7–12); rather, it is a different mixture of plane waves than the one we started with. Thus, the wave function is not an eigenfunction of momentum.

Without going into the mathematical details, we can give a more precise and physical interpretation of the weights. In quantum mechanics, when we say that a wave function has a mixture of momenta, we mean that there are different probabilities of measuring different values of the momentum in a series of measurements. Just the fact that we have not specified the sign of $A(p)$, or even whether it contains the complex number $i$, tells us that $A(p)$ alone cannot be proportional to a probability of finding a given momentum in a distribution of momenta. But we know that probabilities for measurements of position are associated with $|\psi(x)|^2$, and only a small extension of this idea suggests that, since the square of the absolute value of the wave function is where the probabilities come in, the quantity which describes the probability that a measurement of momentum picks out that momentum is $|A(p)|^2$. More precisely, one can show that in one dimension

$|A(p)|^2 dp$ is proportional to the probability that the momentum
will be found in a window of width $dp$ around the value $p$. $\qquad$ (7–13)

The words "proportional to" can be transformed to "equal to" if we normalize $A(p)$ properly. In other words, $D^2|A(p)|^2$ is the probability distribution for momentum in the wave function we are considering, where the constant $D^2$ is determined by the condition

$$D^2 \int_{\text{all momenta}} |A(p)|^2 dp = 1. \qquad (7\text{--}14)$$

We next state a simple and, it turns out, general result[2] regarding $D^2$: $D^2 = 2\pi\hbar$, so that

$2\pi\hbar|A(p)|^2\,dp$ is equal to the probability that the momentum
will be found in a window of width $dp$ around the value $p$.     (7–15)

This statement is a mathematical consequence of the relation between $\psi(x)$ and $A(p)$ given in Eq. (7–12) and the normalization condition on $\psi(x)$. While the proof of the statement is beyond the scope of this book, we can at least verify that it is true. (See Example 7–1.)

Finally, one very important type of momentum distribution will interest us in this chapter. We can imagine that $A(p)$ is centered about some particular value $p_0$ of the momentum, with $A(p)$ falling off as we depart from $p_0$. We say that $A(p)$ is *localized* about a central value of momentum. *How* localized the weights are depends on what we can refer to as a **width** $\Delta p$ of momenta about the central value $p_0$. By this statement, we mean that there is little probability of finding a momentum value larger than $p_0 + \Delta p/2$ or smaller than $p_0 - \Delta p/2$, as in •Fig. 7–2. The plane wave that we studied in Section 7–1 can be thought of as a limiting case with a width that is infinitely small. If we insist that the particle have a *perfectly* definite momentum, then we are stuck with the plane wave, which, as we have seen, is highly unlocalized in space. But if we can be content with a particle with a *nearly* definite momentum, the way to do this is to use weights centered about a particular value. The narrower the width described by the weights $A(p)$, the more precisely the momentum is constrained.

## Making a Pulse

If the wave function of Eq. (7–12) is not an eigenfunction of momentum, why is it interesting? We can now show that the superposed function describes something that is in some sense "localized" in space, at least if the momentum weights $A(p)$ are localized in momentum. That this should be possible is known from classical wave theory: A clap of thunder, which is sharply localized in time, can be analyzed into a superposition of pure tones, each of which is of the form $\sin\omega t$, with the sum over the $\omega$-values running over a very large range of frequencies. This particular superposition of sound waves is a **pulse**. Here, we want to make plausible the pulselike behavior of the wave function without getting involved in wave theory (which may not be part of everybody's background). To do so, we start off with a simple example.

Consider a superposition of waves whose momenta lie between $p_0 + \Delta p/2$ and $p_0 - \Delta p/2$, with the weight factor $A(p)$ independent of momentum in this allowed range. We thus have (•Fig. 7–3)

$$A(p) = \begin{cases} 0 & \text{for} \quad p < p_0 - \Delta p/2 \\ C & \text{for} \quad p_0 - \Delta p/2 < p < p_0 + \Delta p/2. \\ 0 & \text{for} \quad p > p_0 + \Delta p/2 \end{cases} \quad (7\text{–}16)$$

According to our earlier discussion, for this $A(p)$ there is an equal probability of finding the particle with any momentum[3] in the range $p_0 - \Delta p/2$ to $p_0 + \Delta p/2$. This result follows from our assumption that the waves being added have equal amplitude.

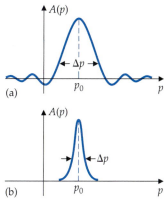

• **Figure 7–2**   Two examples of a weight function for the momentum $A(p)$, here chosen to be real, that reaches a peak near a central value $p_0$ with most of the weight concentrated between $p_0 - \Delta p/2$ and $p_0 + \Delta p/2$. (a) A broad $A(p)$, with a large spread about the average value $p_0$; (b) a narrow $A(p)$.

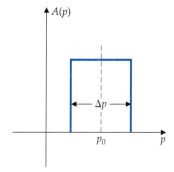

• **Figure 7–3**   The weight function $A(p)$ for Eq. (7–16).

---

[2] This result is a statement that applies to one-dimensional motion only. For three-dimensional motion, a factor $(2\pi\hbar)^3$ and a factor $dp_x\,dp_y\,dp_z$ enter into the situation.

[3] Strictly speaking, since we are dealing with a continuum, we mean by this statement that there is an equal probability of finding the particle in a tiny slice $dp$ anywhere in the finite range $-\Delta p/2$ to $+\Delta p/2$.

Let us turn now to the wave function, Eq. (7–12), for the choice of weights. To simplify matters, we consider the wave function at $t = 0$, so that the factor $\exp(iEt/\hbar)$ is replaced by unity. The quantity under the integral sign in Eq. (7–12) is easily integrated. We have

$$\psi(x) = \int_{-\infty}^{\infty} A(p)\exp\left(\frac{ipx}{\hbar}\right) dp = C\int_{p_0-\Delta p/2}^{p_0+\Delta p/2} \exp\left(\frac{ipx}{\hbar}\right) dp.$$

By changing the variable of integration from $p$ to $q = p - p_0$, we get

$$\psi(x) = C\exp\left(\frac{ip_0x}{\hbar}\right)\int_{-\Delta p/2}^{+\Delta p/2} \exp\left(\frac{iqx}{\hbar}\right) dq$$

$$= C\exp\left(\frac{ip_0x}{\hbar}\right)\left(\frac{\hbar}{ix}\right)\left[\exp\left(i\frac{x\Delta p}{2\hbar}\right) - \exp\left(-i\frac{x\Delta p}{2\hbar}\right)\right] \qquad (7\text{–}17)$$

$$= 2C\exp\left(\frac{ip_0x}{\hbar}\right)\left(\frac{\hbar}{x}\right)\sin\left(\frac{x\Delta p}{2\hbar}\right).$$

(Note that we have used the mathematical identity $\sin\theta = \left(e^{i\theta} - e^{-i\theta}\right)/2i$.)

We now need to analyze the content of our wave function. First, let us consider the probability distribution in $x$, which is described by $|\psi(x)|^2$. We have

$$|\psi(x)|^2 = \frac{4C^2\hbar^2}{x^2}\sin^2\left(\frac{x\Delta p}{2\hbar}\right). \qquad (7\text{–}18)$$

*This quantity is not a constant in space*, in sharp contrast to the square of the absolute value of $\psi$ for the plane wave. Before discussing the form of Eq. (7–18), we note that the normalization requirement easily determines $C$, again in sharp contrast to the case of the plane wave, about which we remarked that normalization was a problem. Normalization requires that

$$\int_{-\infty}^{+\infty} |\psi(x)|^2 dx = 1,$$

and this in turn implies that (see Problem 7–6)

$$C^2 = \frac{1}{(2\pi\Delta p\hbar)}. \qquad (7\text{–}19)$$

---

**Example 7–1**   Calculate the expectation value of the momentum for the wave function described by Eq. (7–16).

**Solution**   There are two possible approaches to this problem. We could, given $A(p)$, find the wave function $\psi(x)$ and then find the expectation value of $p$ by taking the integral over all space of the product $\psi^*(x)(-i\hbar\partial/\partial x)\psi(x)$, as in Eq. (6–29). This approach requires us to use the quantum mechanical momentum operator $-i\hbar\partial/\partial x$. Or we could use the fact that the probability distribution of the momentum is directly given by $|A(p)|^2$, according to Eq. (7–15). This second way is much simpler, so we shall use it here.

First, if $|A(p)|^2$ is a probability distribution, it must be properly normalized, meaning that it must be adjusted by a constant so that its integral over all momenta equals unity. We therefore include a factor $D^2$, as in Eq. (7–14). We then have

$$D^2\underset{\text{all momenta}}{\int} |A(p)|^2 dp = D^2\int_{-\Delta p/2}^{+\Delta p/2} C^2 dp = D^2\frac{1}{2\pi\hbar\,\Delta p}\int_{-\Delta p/2}^{+\Delta p/2} dp$$

$$= D^2\frac{1}{2\pi\hbar\,\Delta p}\Delta p = D^2\frac{1}{2\pi\hbar}.$$

When we set the last quantity equal to unity, we verify that $D^2 = 2\pi\hbar$, in agreement with the general statement of Eq. (7–15).

Now that we are sure of the momentum distribution function, we can evaluate the average momentum, $\langle p \rangle$. We have

$$\langle p \rangle = 2\pi\hbar \int p|A(p)|^2 dp = 2\pi\hbar C^2 \int_{p_0 - \Delta p/2}^{p_0 + \Delta p/2} p\, dp$$

$$= 2\pi\hbar C^2 \left[ \frac{(p_0 + \Delta p/2)^2}{2} - \frac{(p_0 - \Delta p/2)^2}{2} \right]$$

$$= 2\pi\hbar C^2 [p_0 \Delta p] = 2\pi\hbar \frac{1}{2\pi\hbar\Delta p} p_0 \Delta p$$

$$= p_0.$$

This constant momentum is expected: It is the mean value of the momentum distribution given by Eq. (7–16).

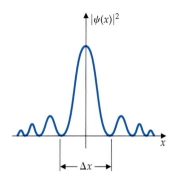

• **Figure 7–4** The probability density $|\psi(x)|^2$ obtained in Eq. (7–18) that results from the superposition of plane waves with the weight function $A(p)$ of Eq. (7–16). Our choice of "width"—the distance between the first two minima of the distribution—is shown in the figure.

Let us now turn to the form of the spatial probability distribution associated with our wave function, Eq. (7–18). We plot this distribution in •Fig. 7–4. It peaks at $x = 0$, falls off to zero when $x\Delta p/2\hbar = \pm\pi$, and then oscillates with a rapidly diminishing maximum height. Roughly speaking, the bulk of the probability density is located within the central peak, between the first two zeros. The width in $x$ of this region is given by

$$\Delta x = 4\pi\hbar/\Delta p. \tag{7–20}$$

*A particle described by such a probability density is localized.* While the point $x = 0$ is the place the particle is most likely to be found, there is a substantial probability that it will be found in a region of width $\Delta x$ about $x = 0$. The spread in momentum values is $\Delta p$, and we can say from Eq. (7–20) that the more precise the momentum—that is, the smaller the value of $\Delta p$—the more spread out the pulse is in space. In fact, as $\Delta p$ approaches zero, the spread $\Delta x$ approaches infinity, and we get arbitrarily close to a plane wave. This is perfectly reasonable; we have already noted that a plane wave has no spread in momentum whatsoever.

A superposition of plane waves made up of weights that cover a finite range of momenta $\Delta p$ is known as a **wave packet**. With such a wave packet, we can calculate the probability density for finding the particle, and this distribution will have a spread $\Delta x$ in space. The inverse relationship between $\Delta x$ and $\Delta p$, as in Eq. (7–20), is a general feature of wave packets. We shall explore it further a bit later.

We finish this subsection by noting that the normalization of the plane wave no longer poses a difficult problem. That is because, in any real physical situation, there is always some uncertainty in momentum. When we produce free electrons, for example, by stripping them from atoms and accelerating them, there is always some "smearing" of the momentum distribution. And no matter how many slits and shutters we use to narrow the velocity distribution, we can never get a state that has a precise momentum. So we can always imagine forming a pulse in space, even if it is a very wide one, and treating the physics of plane waves through a limiting process in which the momentum distribution in the pulse becomes very narrow.

### The Free Particle Moves

There is one hole to be filled in before we have constructed a complete picture with our example: We want to see how time dependence enters. Up to now we have considered the special case of $t = 0$. If we include the full time dependence of Eq. (7–12), we have a time-dependent pulse. In that equation we would

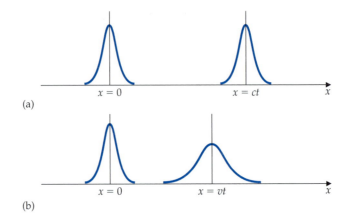

• **Figure 7–5** The propagation of a wave packet along the $x$-axis. (a) A packet of light waves $(E = pc)$ at time $t = 0$ and at a later time $t$. This packet retains its shape. (b) A packet of matter waves describing a nonrelativistic electron $(E = p^2/2m)$ spreads with the passage of time.

substitute $E = pc$ for light or any particle moving ultrarelativistically and $E = p^2/2m$ for electrons or other particles of mass $m$ moving with nonrelativistic velocities.

Let us first consider the form of a pulse of light waves. With $E = pc$, we have

$$\psi(x, t) = \int A(p)\exp\left[\frac{ip(x - ct)}{\hbar}\right]dp. \qquad (7\text{–}21)$$

But we see immediately that

$$\psi(x, t) = \psi(x - ct, t = 0). \qquad (7\text{–}22)$$

In other words, if, for example, at $t = 0$ $\psi(x, 0)$ peaks at $x = 0$, then at a later time the same peak will be at $x - ct = 0$ (i.e., at $x = ct$). This pulse moves at the speed of light without changing its shape (•Fig. 7–5a).

When $E = p^2/2m$, the situation is more complicated: The integral

$$\psi(x, t) = \int A(p)\exp\left[(ipx - ip^2t/2m)/\hbar\right]dp \qquad (7\text{–}23)$$

cannot be evaluated in any simple function. In general, one can say that the pulse will peak at a different place when $t$ is not equal to 0, and it will change shape as well, but one cannot say much more. We can, however, state a general result whose proof lies beyond the scope of this treatment: When $A(p)$ strongly peaks about some value $p_0$, then the pulse peak travels with a speed $p_0/m$, and the peak spreads out as time grows (•Fig. 7–5b). The spreading has to do with the fact that each of the momentum components in the pulse moves with a different speed. One should not interpret the spreading as meaning that the electron itself spreads out with time in quantum mechanics! Rather, the probability of finding an electron away from a moving center grows with time. This is a perfectly acceptable notion.

## 7–3  Uncertainty Relations

The example that we treated in the previous section allows us to make an important numerical observation. From Eq. (7–20), the product of the spread in $x$ and the spread in $p$ is given by

$$(\Delta x)(\Delta p) = 4\pi\hbar. \qquad (7\text{–}24)$$

What is significant about this result is that it is *independent* of $\Delta p$. In other words, if the wave function in our example contains a spread $\Delta p$ in momentum, then the spread in space of the probability density is inversely proportional to $\Delta p$, with the proportionality constant containing a single power of Planck's constant.

We can generalize the result of this example only if we standardize what we mean by $\Delta p$ and $\Delta x$. This is done in the next subsection.

### Evaluation of Widths in Position and Momentum

In our discussion of widths to this point, we have relied on sensible estimates. In the special case described by Eq. (7–16), there is a natural measure of $\Delta p$, and we chose $\Delta x$ to be given by the width of the first peak in the oscillating function obtained in Eq. (7–18). There is a conventional way of defining these widths in more general cases. We make use of what is known as the **standard deviation** $\sigma$ defined in Eq. (6–47) and in Appendix B.3. The standard deviation does just what we want: It measures the spread in a distribution. Accordingly, we define the width in $x$ to be the square root of the standard deviation in the space distribution:

$$(\Delta x)^2 = \sigma(x) = \langle x^2 \rangle - \langle x \rangle^2. \tag{7–25}$$

Similarly, the width in momentum is the square root of the standard deviation in the momentum distribution:

$$(\Delta p)^2 = \sigma(p) = \langle p^2 \rangle - \langle p \rangle^2. \tag{7–26}$$

We illustrate how to use these more precise definitions of widths, together with a better understanding of the relation between $\Delta p$ and $\Delta x$, with several examples.

---

**Example 7–2**   Consider a particle of mass $m$ in an infinite well that runs from $x = 0$ to $x = L$. In Chapter 6, we saw that the normalized wave function corresponding to the energy $E = \hbar^2 n^2 \pi^2 / (2mL)^2$ is given by

$$\psi_n(x) = \sqrt{\frac{2}{L}} \sin\left(\frac{n\pi x}{L}\right).$$

**(a)** Calculate $\langle p^2 \rangle - \langle p \rangle^2 = (\Delta p)^2$ for this wave function.
**(b)** Calculate $\langle x^2 \rangle - \langle x \rangle^2 = (\Delta x)^2$ for this packet.
**(c)** Use your results from parts (a) and (b) to evaluate the product $\Delta x \Delta p$ for this wave function. For what value of $n$ is $\Delta x \Delta p$ smallest?
(*Hint*: You may find it useful to use the relations

$$\sin^2 x = (1 - \cos 2x)/2 \quad \text{and} \quad \int y^2 \cos y \, dy = 2y \cos y + (y^2 - 2)\sin y.$$

In addition, $\langle p \rangle$ and $\langle x \rangle$ do not really need to be calculated. Why is that?)

**Solution**   **(a)** To find the expectation value of $p^n$ given the wave function $\psi(x)$, we take the integral over all space of the product $\psi_n^*(x)(-i\hbar\partial/\partial x)^n \psi_n(x)$, as in Eq. (6–29). In this case, neither $\langle p \rangle$ nor $\langle p^2 \rangle$ requires a complicated calculation. First, we observe that, for any real wave function, such as in the one under consideration, $\langle p \rangle = 0$, since otherwise $\langle p \rangle$ would be purely imaginary, given the factor of $i$ in the momentum operator $-i\hbar\partial/\partial x$. This statement can be verified by an exact calculation of the integral.

The expectation value $\langle p^2 \rangle$ is also easily calculated, because $E = p^2/2m$; that is, $p^2 = 2mE$. But the wave function is an eigenfunction of the energy operator—it has a precise energy—and hence, our expectation value is

$$\langle p^2 \rangle = \int \psi_n^*(x)(2mE)\psi_n(x)\,dx = 2m\left[\frac{\hbar^2 n^2 \pi^2}{2mL^2}\right] \int \psi_n^*(x)\psi_n(x)\,dx$$

$$= 2m\left[\frac{\hbar^2 n^2 \pi^2}{2mL^2}\right] = \frac{\hbar^2 n^2 \pi^2}{L^2}.$$

Note that we have used the fact that the wave function is normalized.

As a result of our two "calculations," we find that

$$(\Delta p)^2 = \frac{\hbar^2 n^2 \pi^2}{L^2}.$$

**(b)** We have, generally, $\langle x^n \rangle = \int_{-\infty}^{+\infty} x^n |\psi_n(x)|^2 \, dx$. Thus, to do this part of the problem, we need only perform integration. In this case, the limits on the integration run from 0 to $L$. But without doing any integration, we can immediately see that $\langle x \rangle = L/2$, since the wave function is symmetric about the midpoint of the well. The reader can easily check this in a direct calculation.

As for $\langle x^2 \rangle$, here we must actually perform the calculation. We have

$$\langle x^2 \rangle = \int_{-\infty}^{+\infty} x^2 |\psi_n(x)|^2 \, dx = \left(\sqrt{\frac{2}{L}}\right)^2 \int_0^L x^2 \sin^2\left(\frac{n\pi x}{L}\right) dx.$$

We use the relation $\sin^2 x = (1 - \cos 2x)/2$ under the integral. There are then two terms to integrate. The first term is $(2/L)$ times $(1/2)$ times the integral from 0 to $L$ of $x^2$, which equals $L^3/3$. This term gives $(2/L)(1/2)(L^3/3) = L^2/3$. For the second term, we change the variable of integration from $x$ to $u = 2n\pi x/L$. Then $dx = (L/2n\pi)\, du$ and the integral over $u$ runs from 0 to $2n\pi$. This term then gives

$$\frac{2}{L}\frac{1}{2}\int_0^{2n\pi}\left(\frac{uL}{2n\pi}\right)^2 (\cos u)\left(\frac{L}{2n\pi}\, du\right) = \frac{2}{L}\frac{1}{2}\left(\frac{L}{2n\pi}\right)^3 \int_0^{2n\pi} u^2 (\cos u)\, du$$

$$= \frac{L^2}{(2n\pi)^3}\int_0^{2n\pi} u^2 (\cos u)\, du.$$

We then evaluate

$$\int_0^{2n\pi} u^2 (\cos u)\, du = \left(2u \cos u + (u^2 - 2)\sin u\right)\Big|_0^{2n\pi} = 4n\pi.$$

Pulling all our terms together, we find that

$$\langle x^2 \rangle = \frac{L^2}{3} - \frac{L^2}{(2n\pi)^3} 4n\pi = L^2\left(\frac{1}{3} - \frac{1}{2n^2\pi^2}\right).$$

Using our results for $\langle x^2 \rangle$ and with $\langle x \rangle^2 = L^2/4$, we have, finally,

$$(\Delta x)^2 = L^2\left(\frac{1}{3} - \frac{1}{2n^2\pi^2} - \frac{1}{4}\right) = L^2\left(\frac{1}{12} - \frac{1}{2n^2\pi^2}\right).$$

**(c)** For this part, we simply take the product

$$\Delta x \Delta p = L\sqrt{\left(\frac{1}{12} - \frac{1}{2n^2\pi^2}\right)}\frac{\hbar n\pi}{L} = \hbar n\pi\sqrt{\left(\frac{1}{12} - \frac{1}{2n^2\pi^2}\right)}.$$

Note that the product is independent of the size of the well, which provides the scale for $\Delta x$. Or equivalently, the product is independent of the width $\Delta p$. It is, however, linearly dependent on $\hbar$. The argument of the square root is safely positive, and the answer takes on the smallest value for $n = 1$, in which case

$$\Delta x \Delta p = \hbar\pi\sqrt{\left(\frac{1}{12} - \frac{1}{2\pi^2}\right)} \cong 0.57\hbar.$$

---

### The Heisenberg Uncertainty Relation

The relationship between the "widths" of $|A(p)|^2$ and $|\psi(x)|^2$, found in both Example 7–1 and Example 7–2, can be summarized as

$$\Delta p \Delta x \geq \hbar/2. \tag{7–27}$$

It is possible to show—although the derivation would be more formal than we would like in this text—that this relation, the **Heisenberg position-momentum uncertainty relation**, is very general. It was discovered in the framework of quantum mechanics by Werner Heisenberg in 1927 and played a critical role in the interpretation of quantum mechanics and in showing that there could be no conflict between quantum and classical physics in their respective domains of applicability.

There is one particular wave packet for which the relatonship in Eq. (7–27) is realized as an equality: the **Gaussian wave packet**. For this packet, we have the *maximum simultaneous localization* in position and momentum, in the sense that the product $\Delta x \Delta p$ is as small as it can be. This is a special feature that makes the Gaussian function worth exploring further. We shall do so in Examples 7–3 through 7–5 with interspersed comments.

The Gaussian packet is defined by specifying the weights

$$A(p) = C \exp\left[-\frac{a^2(p - p_0)^2}{4\hbar^2}\right], \tag{7–28}$$

where $C$ is a constant that can ultimately be determined by the normalization of the wave function. •Figure 7–6 shows $A(p)$, which peaks at $p_0$ and has a spread, or width—marked on the figure—that is proportional to $\hbar/a$. This means that as $a$ is made larger, the width of the peak shrinks.

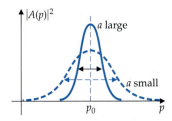

• **Figure 7–6**  The probability density for momentum $|A(p)|^2$ for the Gaussian wave packet of Eq. (7–28), plotted as a function of $p$, with large (narrow distribution) and small (wide distribution) values of the length parameter $a$.

---

**Example 7–3**  Use the **Gaussian wave packet** defined in Eq. (7–28) to calculate the wave function $\psi(x) \equiv \psi(x, t = 0)$ from Eq. (7–12).

**Solution**  The momentum integral for the wave function, Eq. (7–12), can be integrated explicitly if $A(p)$ is a Gaussian function in momentum. As we shall see here, the integration gives back another Gaussian, this time in space.

In this case, Equation (7–12) reads

$$\psi(x) = C \int_{-\infty}^{\infty} \exp\left[-\frac{a^2(p - p_0)^2}{4\hbar^2} + i\frac{px}{\hbar}\right] dp.$$

We approach this integral by first changing the variable of integration from $p$ to $k = (p - p_0)/\hbar$. As $p$ runs from $-\infty$ to $+\infty$, so does $k$, and $dp = \hbar\,dk$. Thus

$$\psi(x) = C\hbar \exp\left(\frac{ip_0 x}{\hbar}\right) \int_{-\infty}^{\infty} \exp\left[-\frac{a^2 k^2}{4} + ikx\right] dk.$$

One more change of variables gets us where we want to go: We change variables from $k$ to $q = (a/2)k - i(x/a)$. We need only do a little algebra at this point:

$$\psi(x) = C\hbar \exp\left(\frac{ip_0 x}{\hbar}\right) \int_{-\infty}^{\infty} \exp\left[-\frac{a^2}{4}k^2 + ikx + \frac{x^2}{a^2} - \frac{x^2}{a^2}\right] dk$$

$$= C\hbar \exp\left(\frac{ip_0 x}{\hbar}\right) \int_{-\infty}^{\infty} \exp\left[-\left(\frac{ak}{2} - i\frac{x}{a}\right)^2 - \frac{x^2}{a^2}\right] dk$$

$$= \frac{2C\hbar}{a} \exp\left(\frac{ip_0 x}{\hbar}\right) \exp\left(-\frac{x^2}{a^2}\right) \int_{-\infty}^{\infty} \exp(-q^2)\,dq.$$

The remaining integral is just a number, $(\pi)^{1/2}$ (see Appendix B.2), so that

$$\psi(x) = \frac{2C\hbar\sqrt{\pi}}{a} \exp(ip_0 x/\hbar) \exp\left(-\frac{x^2}{a^2}\right). \tag{7–29}$$

---

The result of the preceding example merits further discussion. Earlier, we argued that for the Gaussian wave packet the width in momentum—the momentum spread over which the weights $A(p)$ are significant—is proportional

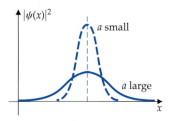

• **Figure 7–7** The probability density $|\psi(x)|^2$ as a function of $x$ for the packet of Eq. (7–29). The distribution is narrow, corresponding to sharp localization, for small $a$, when the momentum distribution is broad. The distribution is wide for large $a$ when the momentum probability distribution is sharply localized.

to $\hbar/a$. The result of the calculation in Example 7–3 allows us to see what our object looks like in space (•Fig. 7–7). First of all, we see immediately that $|\psi(x)|^2$, which represents the probability density of finding our particle at some position $x$, peaks (or is localized) at $x = 0$. The peak falls off more slowly as $a$ increases, as we can see from the plots of $|\psi(x)|^2$ for various values of $a$ in the figure. In fact, as $a$ becomes very large, $|\psi(x)|^2$ becomes spread out over all of space, and $\psi(x)$ itself approaches a plane wave with momentum $+p_0$. Conversely, as $a$ decreases, the width of $\psi(x)$ decreases, meaning that the particle described by this wave packet becomes more and more precisely located.

Thus, we have found, at least qualitatively, that for the Gaussian wave function, $\Delta p$ is proportional to $\hbar/a$, while $\Delta x$ is proportional to $a$. This result confirms the claim that we made at the outset of the section: The product $\Delta x \Delta p$ is proportional to $\hbar$. Later, in Example 7–5, we shall see that when we use our more precise definitions of $\Delta x$ and $\Delta p$, the Gaussian wave packet realizes the smallest possible value for the product $\Delta x \Delta p$, namely, $\hbar/2$. To use those more precise definitions, we want to be able to calculate averages for the Gaussian wave packet. This we do in Example 7–4.

**Example 7–4**    Consider the Gaussian wave function calculated in Example 7–3, namely,

$$\psi(x) = B \exp\left(-\frac{x^2}{a^2}\right) \exp(ikx),$$

where $B$ is a (real) constant to be determined by normalization. **(a)** Find the value of $B$. **(b)** Calculate $\langle p \rangle$ for this wave function.

**Solution**    **(a)** The normalization condition is

$$1 = \int_{-\infty}^{+\infty} |\psi(x)|^2 \, dx = B^2 \int_{-\infty}^{+\infty} \left( \exp\left(-\frac{x^2}{a^2}\right) \exp(-ikx) \right) \left( \exp\left(-\frac{x^2}{a^2}\right) \exp(ikx) \right) dx$$

$$= B^2 \int_{-\infty}^{+\infty} \exp\left(-\frac{2x^2}{a^2}\right) dx.$$

Thus

$$B^2 = \left[ \int_{-\infty}^{+\infty} \exp\left(-\frac{2x^2}{a^2}\right) dx \right]^{-1}.$$

The integral is given in Appendix B.2; the result is

$$B^2 = \left[ \frac{a\sqrt{\pi}}{\sqrt{2}} \right]^{-1} = \sqrt{\frac{2}{\pi}} \frac{1}{a}.$$

For the remainder of this example, we will not need the explicit value of $B$.
**(b)** In contrast to what we did in Example 7–1, here we are given $\psi(x)$ directly, so we work with the momentum operator on the wave function:

$$\langle p \rangle = \int_{-\infty}^{+\infty} \psi^*(x) \left(-i\hbar \frac{d}{dx}\right) \psi(x) \, dx$$

$$= B^2 \int_{-\infty}^{+\infty} \exp\left(-\frac{x^2}{a^2}\right) \exp(-ikx) \left\{ \left(-i\hbar \frac{d}{dx}\right) \exp\left(-\frac{x^2}{a^2}\right) \exp(ikx) \right\} dx$$

$$= -i\hbar B^2 \int_{-\infty}^{+\infty} \exp\left(-\frac{x^2}{a^2}\right) \exp(-ikx) \left(-\frac{2x}{a^2} + ik\right) \exp\left(-\frac{x^2}{a^2}\right) \exp(ikx) \, dx$$

$$= -i\hbar B^2 \int_{-\infty}^{+\infty} \left(-\frac{2x}{a^2} + ik\right) \exp\left(-\frac{2x^2}{a^2}\right) dx.$$

There are two terms to be integrated. We can see right away that the result of integrating the first term is zero, because the integral runs from $-\infty$ to $+\infty$ and the integrand is odd. This term of the integration had better vanish! Otherwise, as we have already remarked in Example 7–2, its contribution to $\langle p \rangle$ would be proportional to $i$—that is, imaginary. Accordingly, we have

$$\langle p \rangle = -i\hbar(ik)B^2 \int_{-\infty}^{+\infty} \exp\left(-\frac{2x^2}{a^2}\right) dx.$$

But $B^2$ times the integral is unity, as we see from part (a). Hence, with $(-i)(+i) = +1$, we conclude that

$$\langle p \rangle = \hbar k.$$

If we now recall that $\exp(ikx)$ is a plane wave with momentum $\hbar k$, we see that we have shown that even when the plane wave is multiplied by a real function of $x$—in this case, $\exp(-x^2/a^2)$, it retains, *on the average*, the same momentum.

---

**Example 7–5**   Consider the Gaussian wave packet with $p_0 = 0$, viz.,

$$A(p) = C \exp\left[-a^2 p^2/4\hbar^2\right].$$

This weight distribution is properly normalized—that is, it has the interpretation of Eq. (7–15)—if $C$ is given by (see Problem 22)

$$C^2 = \frac{a}{(2\pi)^{3/2}\hbar^2}.$$

Find $(\Delta x)^2$ and $(\Delta p)^2$, using the definitions of Eqs. (7–25) and (7–26). You may want to consider the results of Example 7–3 for $\psi(x)$. In turn, calculate the product of the uncertainties, $\Delta x \Delta p$, for this wave packet.

**Solution**   There are really several calculations to be done here. We start with the calculation of $\Delta p$, an easy one because $2\pi\hbar|A(p)|^2$ is directly the momentum distribution. Thus,

$$\langle p^n \rangle = 2\pi\hbar \int_{-\infty}^{+\infty} p^n |A(p)|^2 dp.$$

We know that $|A(p)|^2$ is a function that is even in $p$, so that when it is multiplied by an odd function of $p$ and then integrated from $-\infty$ to $+\infty$, the result is zero. For $\sigma(p)$, we are interested in a single power:

$$\langle p \rangle = 0.$$

The standard deviation also involves $\langle p^2 \rangle$, and this is a little more complicated:

$$\langle p^2 \rangle = 2\pi\hbar \int_{-\infty}^{+\infty} p^2 |A(p)|^2 dp = 2\pi\hbar C^2 \int_{-\infty}^{+\infty} p^2 \exp\left(-\frac{2a^2 p^2}{4\hbar^2}\right) dp.$$

The last integral can be found in Appendix B.2 and gives us

$$\langle p^2 \rangle = \frac{\hbar^2}{a^2}.$$

Finally, then,

$$(\Delta p)^2 = \langle p^2 \rangle - \langle p \rangle^2 = \frac{\hbar^2}{a^2}.$$

We also want the standard deviation in $x$, and this calculation is no more difficult than that of the momentum. Here we want the wave function itself, properly normalized, which we get from Example 7–4. In particular, the square of the wave function is

$$|\psi(x)|^2 = \sqrt{\frac{2}{\pi}}\frac{1}{a} \exp\left(-\frac{2x^2}{a^2}\right).$$

Because this expression is normalized,

$$\langle x^n \rangle = \int_{-\infty}^{+\infty} x^n |\psi(x)|^2 \, dx.$$

The same reasoning as for the momentum tells us that $\langle x \rangle = 0$, and, again as for the momentum,

$$\langle x^2 \rangle = (\Delta x)^2 = \frac{a^2}{4}.$$

The desired product of the uncertainties is then immediate:

$$\Delta x \Delta p = \frac{\hbar}{2}.$$

This result verifies that the Gaussian wave packet "saturates" the Heisenberg uncertainty relation, Eq. (7–27).

---

The Heisenberg uncertainty relation is really a very general property of wave packets and, viewed properly, is not restricted to quantum mechanics. In particular, if, as the de Broglie relation would suggest we should do, we write $\Delta p / \hbar = \Delta k$, where $k$ is a wave number, then the Heisenberg relation becomes

$$\Delta k \Delta x \geq 1/2. \tag{7–30}$$

This inequality applies equally to pulses of sound! It is only because we could interpret $|\psi(x)|^2$ as the probability density for the position of a particle and $2\pi\hbar |A(p)|^2$ as the probability density for the momentum of the particle that we were led to a quantum mechanical interpretation of the relation. As we discuss in more detail in the next section, the Heisenberg uncertainty relation sets limits on the precision with which we can determine momentum and position in physical systems.

## 7–4 The Meaning of the Uncertainty Relations

From the Heisenberg uncertainty relation, we can conclude that *we cannot simultaneously measure the position and the momentum of a particle with arbitrary precision*. This indeterminacy is not related to any indeterminacy in the Schrödinger equation; that equation is well defined, and so are its solutions. Rather, the indeterminacy reflects the limits beyond which we can no longer use the intuitive classical notions of the trajectory $x(t)$ of a particle and its momentum $m \, dx(t)/dt$. The small size of $\hbar$ guarantees that no classical experiment is going to be affected by this uncertainty. We can see this with an example.

**Example 7–6**   Consider a grain of dust of mass $10^{-7}$ kg moving with a velocity of around 10 m/s. Suppose that the measuring instruments available to us leave the velocity uncertain within a range of $10^{-6}$ m/s (i.e., one part in $10^7$). Given this instrumental uncertainty in the velocity, find the intrinsic quantum mechanical uncertainty of a position measurement of the dust grain.

**Solution**   The instrumental uncertainty in the momentum is

$$\Delta p = m \Delta v = (10^{-7} \, \text{kg})(10^{-6} \, \text{m/s}) = 10^{-13} \, \text{kg} \cdot \text{m/s}.$$

Hence, according to the uncertainty relation, the position could at best be measured to within a window

$$\Delta x \cong \frac{\hbar}{\Delta p} = \frac{(1.05 \times 10^{-34} \, \text{J} \cdot \text{s})}{(10^{-13} \, \text{kg} \cdot \text{m/s})} \cong 10^{-21} \, \text{m}.$$

This is such a small number that we would never measure it in any conceivable experiment—it is about $10^{11}$ times smaller than the size of one of the approximately $10^{19}$ atoms that make up the dust particle!

In contrast to its role on the macroscopic scale, the uncertainty relation plays a crucial role on the atomic scale. Consider an electron in a Bohr orbit. The velocity of the electron is given by $\alpha c/n$ (recall that $\alpha = e^2/4\pi\varepsilon_0\hbar c \cong 1/137$), so that an ignorance regarding which orbit the electron is to be found in implies that $\Delta v = \alpha c$ and, therefore, $\Delta p = mc\alpha$. This much uncertainty in the momentum in turn implies, by Heisenberg's relation, an uncertainty in the location of the electron on the order of $\hbar/mc\alpha$. But that distance is comparable to the radius of the Bohr atom and therefore is in the range of relevant physical magnitudes.

The uncertainty relation is not dependent on the construction of wave packets; it emerges naturally from the structure of quantum mechanics. It is forced on us by the need to reconcile the complementary particle and wave properties of radiation and matter. We can illustrate this need by a *thought experiment* devised by Heisenberg himself (•Fig. 7–8), who identified the **uncertainty principle** as the basic principle that underlies the structure of quantum mechanics.

A thought experiment is an imaginary experiment that may be difficult (or even impossible) actually to carry out, but one that is totally consistent with the laws of physics. Imagining how one might measure the speed of a seismic wave on a star is a thought experiment—but imagining optics in a rocket traveling at twice the speed of light is nonsense. The idea is to think of something— a "gimmick"—that, in one way or another, would shed light on some logical question. Thought experiments are a wonderful way to clarify the issues raised by the probability interpretation of quantum mechanics and the uncertainty principle. The thought experiment we describe here is called the **Heisenberg microscope.**[4]

This "experiment" is designed to measure an electron's $x$-position and the $x$-component of the electron's momentum simultaneously (•Fig. 7–9). Suppose

• **Figure 7–8**  Werner Heisenberg wrote his first paper on the old quantum theory at the age of 20. A few years later he became an active member of the Bohr group, and in June of 1925 he wrote the paper that created quantum mechanics. Heisenberg's discovery of the uncertainty relations played a major role in helping Niels Bohr clarify the interpretation of the new theory.

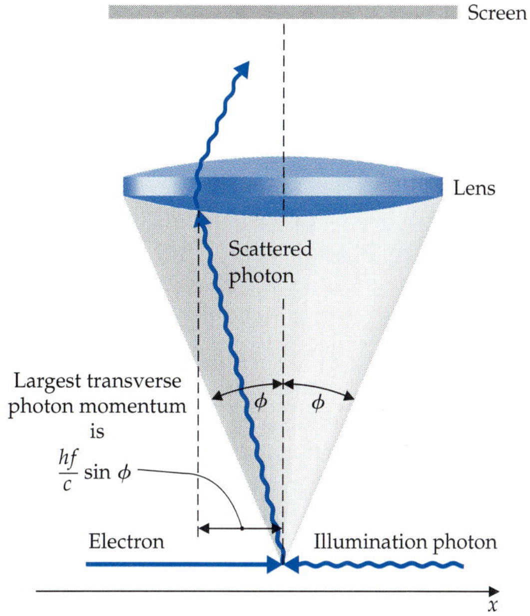

• **Figure 7–9**  The Heisenberg microscope. The position of the electron is to be determined by having it photographed. The illumination consists of photons, one of which scatters off the electron and goes through the lens onto the screen. The most precise localization uses photons of the shortest wavelength, but short-wavelength photons transfer the largest amount of unpredictably directed momentum to the electron.

Screen

Lens

Scattered photon

$\phi$ $\phi$

Largest transverse photon momentum is

$$\frac{hf}{c}\sin\phi$$

Electron

Illumination photon

$x$

[4]On his Ph.D. oral examination, Heisenberg could not explain how an actual microscope worked. Some years later, when he created this example, he still hadn't gotten it quite right, and he had to be straightened out by Bohr.

an electron moves from left to right with an initial momentum $p_x$. We assume that this momentum is known precisely—the electron's initial wave function is composed of a sharp peak in momentum. The electron's position is to be observed by shining light on it. The light comes in the form of a single photon with a precisely known momentum (a precisely known wavelength) coming from the right. The timing of the collision between the electron and the photon is arranged so that it takes place under the lens of a microscope. The observation takes place if the photon scatters off the electron and passes through the lens onto a photographic plate. If, after the collision, the photon has a wavelength $\lambda$ (a frequency $f$), then classical lens optics tells us that the resolution of the lens depends on both this wavelength and the angle that the lens subtends at the object being observed. The resolution—the ability to locate the electron in space—gives us $\Delta x$, the "uncertainty" in the location of the electron. The classical optics formula for the resolution is

$$\Delta x \cong \frac{\lambda}{\sin \varphi}, \tag{7–31}$$

where $\varphi$ is as shown in the figure. Note that the quantity $\Delta x$ can be reduced as much as we like, either by making $\lambda$ smaller or by increasing $\varphi$ (for example, by making the lens larger).

Now in the collision a certain amount of momentum is transferred to the electron. By itself this does not mean that the electron's momentum becomes uncertain. After all, the amount of the momentum transferred to the electron could in principle be obtained from the initial momenta of the electron and photon and a measurement of the final momentum of the photon. Thus the uncertainty in the electron's momentum just after the collision, when its position is measured, is the same as the uncertainty in the photon's momentum. Let us concentrate on the $x$-component of the electron's momentum; we therefore want to know the $x$-component of the *photon's* momentum after the collision. *But this quantity is uncertain*, because all we know is that the photon has passed somewhere through the lens, and according to the figure, there is a range of directions the photon could move in and still pass through the lens. If the total momentum of the photon is $p = hf/c$, then, from the figure, the uncertainty in the $x$-component of the photon's momentum associated with the lens opening is

$$\Delta p_x|_{\text{photon}} = 2p \sin \varphi = 2\left(\frac{hf}{c}\right) \sin \varphi.$$

As we have argued, this is the uncertainty in the $x$-component of the electron's momentum; that is,

$$\Delta p_x \cong 2\frac{hf}{c} \sin \varphi. \tag{7–32}$$

Note that this quantity can also be reduced, either by decreasing the photon's frequency or by making $\varphi$ smaller (for example, by making the lens smaller). *But both of these steps would increase the uncertainty in the electron's position.*

The play between the uncertainty in the $x$-position and the $x$-component of momentum is summarized in the product $\Delta x \Delta p_x$. We have

$$\Delta x \Delta p_x \cong \frac{\lambda}{\sin \varphi} \frac{2hf}{c} \sin \varphi = 2h = 4\pi\hbar. \tag{7–33}$$

Here, we have used the relation between wavelength and frequency for light: $\lambda f = c$. Equation (7–33) is independent of any details of the system and takes the general form of the uncertainty relation. In other words, our thought experiment does not allow us to escape the consequences of Heisenberg's uncertainty relation.

In this analysis of the position and momentum of an electron we needed to use both the wave and particle properties of photons. We could have turned the argument around: The complementary wavelike and particlelike properties of radiation can be reconciled only within limits imposed by the uncertainty principle. Effectively, the uncertainty principle always saves us from contradiction. We can illustrate this notion again in a discussion of the two-slit interference experiment.

## The Two-Slit Experiment

The uncertainty principle plays another role in setting limits on our use of intuitive classical concepts. The uncritical use of these concepts can lead to apparent contradictions. An example is provided by the so-called two-slit experiment. Electrons passing through a pair of slits produce a two-slit interference pattern, *even if they pass with such low intensity that they pass only one electron at a time*. This is a problem in classical reasoning, because if it were possible to tell which slit the electron came through, it would be possible to treat each transmission of an electron as an event in which only one slit is open. That would lead to an accumulation of electrons arriving at the screen as the sum of two experiments, in each of which one slit was open and the other closed. Such experiments do *not* give a two-slit diffraction pattern: It seems that just knowing which slit the electron went through destroys the interference pattern! And, indeed, the uncertainty principle ensures that this is exactly the case. By the uncertainty principle, the presence of a monitor that is good enough to identify the slit that an electron has passed through disturbs the experiment sufficiently to destroy the interference pattern. Let us now demonstrate this.

Our experiment is described by •Fig. 7–10. The condition for constructive interference is that

$$a \sin \theta_n = n\lambda, \qquad n = 0, \pm 1, \pm 2, \ldots . \qquad (7\text{--}34)$$

The separation between adjacent maxima on the detection screen is therefore

$$d \sin \theta_{n+1} - d \sin \theta_n = \frac{\lambda d}{a}. \qquad (7\text{--}35)$$

For small angles, the left-hand side of this equation is, more simply, $d(\theta_{n+1} - \theta_n)$.

Let us now assume that a monitor just behind the slits determines the position of the electron to an accuracy sufficient to tell which slit the electron came

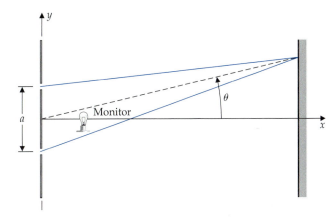

• **Figure 7–10** An unspecified monitor is designed to detect which of two slits a photon passes through. Whatever the mechanism, the monitor must be able to resolve the location of the photon after its passage to an accuracy better than the separation between the slits.

through. This is equivalent to a measurement of the $y$-component of the electron's position wherein the precision is better than the separation between the slits; that is,

$$\Delta y < \frac{a}{2}. \tag{7-36}$$

Such a monitor might, for example, be a source of photons, which, by bouncing off the electron, reveals its position. The necessary precision is ensured by making the photon have a wavelength small enough. Of course, the smaller the wavelength of the photon, the larger is the momentum it carries. Now, any measurement of the position of the electron in this way will give the electron a kick—a momentum transfer—in the $y$-direction and thereby introduces an uncertainty $\Delta p_y$ in the electron's $y$-momentum. We can estimate the minimum size of $\Delta p_y$ by means of the uncertainty principle:

$$\Delta p_y > \frac{(\hbar/2)}{(a/2)} = \hbar/a. \tag{7-37}$$

Having introduced an uncertain transverse component of momentum, we have automatically introduced an uncertainty in the arrival spot on the detection screen. If the electron came through carrying a (longitudinal) momentum $p$, then the electron moves off the two slits at an angle

$$\Delta\theta = \frac{\Delta p_y}{p} = \frac{\hbar}{ap} = \frac{\lambda}{2\pi a}. \tag{7-38}$$

Finally, this angular uncertainty translates into an uncertainty in the arrival point on the detection screen: The transverse arrival position is uncertain by

$$\Delta y = d\Delta\theta = \frac{d\lambda}{2\pi a}. \tag{7-39}$$

Comparing Eqs. (7–35) and (7–39), we see that our monitor has disturbed the electron enough to wipe out the interference pattern. The argument we have presented here could, of course, be turned on its head: Logical consistency demands that $\Delta y \Delta p_y > \hbar$.

## 7–5  The Time–Energy Uncertainty Relation

There is another uncertainty relation that is quite useful—one involving time and energy. We can find it by using the momentum–position relation for free particles, for which $E = p^2/2m$. We have $\Delta E = (2p\Delta p)/2m = p\Delta p/m = v\Delta p$. Then

$$\Delta E = v\Delta p \geq v\frac{\hbar/2}{\Delta x} = \frac{\hbar/2}{(\Delta x/v)} = \frac{\hbar/2}{\Delta t},$$

or

$$\Delta E \Delta t \geq \hbar/2. \tag{7-40}$$

This is the **time–energy uncertainty relation**. It is more general than our plausibility argument might suggest. It asserts that a state of finite duration $\Delta t$ cannot have a precisely defined energy, with the uncertainty in $E$ specified by Eq. (7–40). Thus if an excited atomic state has a lifetime[5] $\tau$, the excited state does not have a precise energy $E_1$; rather, its energy is uncertain by an amount

-------

[5] We described the lifetime of an unstable state at the beginning of Section 3–1, and we shall revisit the subject later. Here, you can just think of the lifetime as the average time that passes before an unstable state decays in a collection of such states.

$\Delta E_1 \cong \hbar/\tau$. This uncertainty manifests itself when the state decays to, say, the ground state, with energy $E_0$; the frequency of the radiation emitted in the decay, $f = (E_1 - E_0)/h$, will be spread by an amount $\Delta f = (\Delta E_1)/h \cong (\hbar/\tau)/h = 1/(2\pi\tau)$. The spectral line is *broadened* (•Fig. 7–11), a quantum mechanical phenomenon. The width of the spectral line is its *natural line width*.

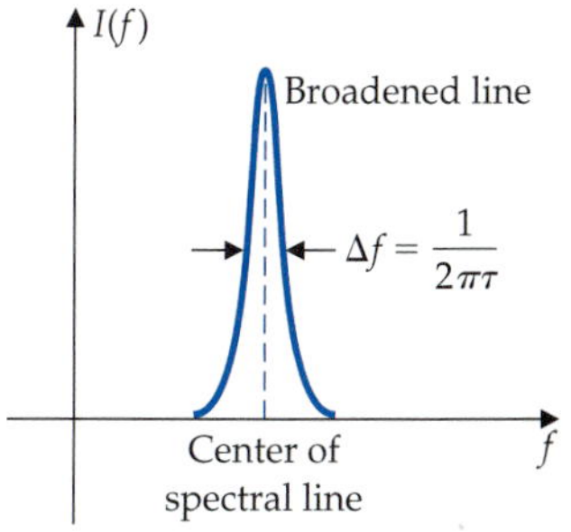

**Example 7–7**   Truly monochromatic light—light of one frequency only—takes the mathematical form of a simple plane wave. Because of this, the wave is infinitely long in extent; that is, it is spread out over the entire line of its motion. One way of making trains of light waves that are of finite extent in time (and in space) is to "chop" the monochromatic light by sending it through shutters that are timed to open and close for a certain time interval $\Delta t$. One can make such a shutter with a rotating toothed wheel, with the light aimed at the circumference of the wheel (•Fig. 7–12). Every time a tooth passes in front of the light beam, the wave train is interrupted. Consider such a wheel, with 360 teeth on the circumference, rotating at 3,600 revolutions per minute. What is the spread in the frequency of monochromatic light of frequency $f$ after it has passed through the wheel?

**Solution**   We want to calculate the time that it takes for one tooth to be followed by another in crossing the light beam—this will be the duration of each wave train. If we take this duration to represent the spread in time of the light pulse, then the uncertainty, or spread, in energy of the light is given by the uncertainty principle. And the spread in energy is directly related to the spread in frequency for light by the relation $E = hf$ for photons.

The time necessary for one tooth to move into the space vacated by a previous tooth is

$$\Delta t = \frac{\text{angle between two teeth}}{\text{angular velocity}}.$$

The angle between two teeth is $2\pi/360$ rad. The angular velocity of the wheel is

$$\omega = (2\pi \times 3{,}600 \text{ rad/min})(1 \text{ min}/60 \text{ s}) = 120\pi \text{ rad/s}.$$

Thus

$$\Delta t = \frac{2\pi/360 \text{ rad}}{120\pi \text{ rad/s}} = 4.6 \times 10^{-5} \text{ s}.$$

Given this value of $\Delta t$, we can estimate the spread in energy that has been introduced into the "chopping" process from the uncertainty principle. That spread is determined by $\Delta E \Delta t \cong \hbar/2$, and with $\Delta f = \Delta E/\hbar$, we estimate the spread in frequency in the "chopped" wave train to be

$$\Delta f \cong \frac{1}{2\Delta t} = 1.1 \times 10^4 \text{ Hz}.$$

**•Figure 7–11**   Spectral lines emitted in the transition of electrons from excited states to other states are broadened, and precision measurements verify this phenomenon. The width of the line is inversely proportional to the lifetime of the excited states.

$I(f)$

Broadened line

$\Delta f = \dfrac{1}{2\pi\tau}$

Center of spectral line

$f$

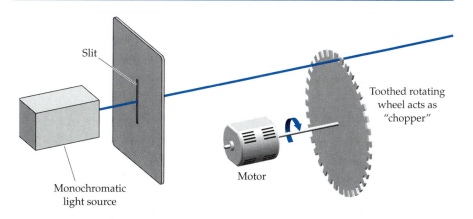

Slit

Toothed rotating wheel acts as "chopper"

Motor

Monochromatic light source

**•Figure 7–12**   A rotating toothed wheel acts as a "chopper" for a train—in principle, infinitely long—of monochromatic waves.

The presence of the quantity $\Delta E$ in the time–energy uncertainty relation might mistakenly be taken to indicate that energy is not being conserved. Actually, energy is *always* conserved, but a physical state that lasts only for a short time does not have a well-defined energy. The extent to which the energy is not well defined is reflected in the spread of the spectral line. We can crudely rephrase this by saying that we can insist on energy conservation only to a precision of $\Delta E$ if we deal with times so short that $\Delta t \cong \hbar/\Delta E$. For example, the only thing that prevents a photon from turning into an electron–positron pair is the principle of conservation of energy: The minimum energy needed to make the pair is $2m_e c^2$. But uncertainty suggests that such a pair *can* be made, as long it does not survive very long—the electron and positron annihilate and recreate the original photon within a time $\Delta t$, and the length of time such a pair can exist is $\Delta t \cong \hbar/(2m_e c^2)$. We shall see later in the text that such effects have measurable consequences.

---

**Example 7–8**   Calculate how long a virtual electron–positron pair can exist. Repeat the calculation for a baseball–antibaseball pair.

**Solution**   The mass of a particle and its antiparticle are the same. The minimum amount of energy that needs to be "borrowed" to make a pair of anything of mass $M$ is $\Delta E = 2Mc^2$. By the uncertainty relation $\Delta E \Delta t \geq \hbar/2$, the maximum time for which such "borrowing" can go on is $\Delta t \cong (\hbar/2)/(2Mc^2)$. For electrons and positrons, the mass is $m_e = 0.9 \times 10^{-30}$ kg, so that

$$\Delta t \cong \frac{1.05 \times 10^{-34} \text{ J} \cdot \text{s}}{4(0.9 \times 10^{-30} \text{ kg})(3 \times 10^8 \text{ m/s})^2} = 3.3 \times 10^{-22} \text{ s}.$$

This is indeed a brief time, but compare the time for baseballs, for which $M = 0.145$ kg:

$$\Delta t \cong \frac{(1.05 \times 10^{-34} \text{ J} \cdot \text{s})}{4(0.145 \text{ kg})(3 \times 10^8 \text{ m/s})^2} = 2 \times 10^{-51} \text{ s}(!)$$

---

# 7–6  Estimating Energies

The uncertainty principle does more than tell us what we cannot do: Among other things, we can use it to estimate the ground-state energies of physical systems. Consider, for example, a harmonic oscillator, a system whose classical energy is

$$E = \frac{p^2}{2m} + \frac{1}{2}m\omega^2 x^2, \tag{7–41}$$

where $\omega$ is the angular frequency of oscillation. (See Chapter 1.) Classically, the minimum energy is zero, which occurs when the kinetic energy is zero ($p = 0$) and the particle is at rest at a position corresponding to the bottom of the potential-energy well. The uncertainty principle, however, forbids a situation in which momentum and position are both precisely known. If the uncertainty in position is $\Delta x = a$, then the momentum will have an uncertainty of $\Delta p = \hbar/2a$. Near the lowest possible energy, where, classically, $p = 0$, the uncertainty in the momentum is the momentum itself. The energy is then given by

$$E = \frac{(\Delta p)^2}{2m} + \frac{1}{2}m\omega^2(\Delta x)^2 = \frac{\hbar^2}{8ma^2} + \frac{1}{2}m\omega^2 a^2. \tag{7–42}$$

This estimate for the lowest energy is a function of $a$ (•Fig. 7–13), and we can imagine varying $a$ so that our estimate is a minimum. This requires that

$$\frac{dE}{da} = -\frac{\hbar^2}{4ma^3} + m\omega^2 a = 0. \tag{7-43}$$

We can solve Eq. (7–43) to find the value of $a$ for which $E$ has a minimum. We obtain $a^2 = \hbar/2m\omega$, and the minimum value of $E$ is then

$$E = \frac{\hbar\omega}{2}. \tag{7-44}$$

We refer to this minimum energy, which is, obviously, different from the classical value of zero, as the **zero-point energy**. The uncertainty principle requires that a little residual motion remain in any physical system—a kind of built-in "nervosity." We have already remarked on this idea in the case of the harmonic oscillator, in Chapter 5. One thing the zero-point energy does is essentially prevent helium from solidifying at low temperatures. Normally, a collection of interacting atoms forms a crystal at low temperatures, because there is a minimum in the energy into which the atoms settle. In helium, the potential-energy minimum is relatively shallow, and because the mass of the helium atom is small, the kinetic energy is relatively large. Classically this does not matter, because the atoms can go all the way to a state of zero motion. However, at the minimum quantum mechanical energy—the zero-point energy—the motion is large enough so that the helium remains liquid as the temperature drops towards zero.

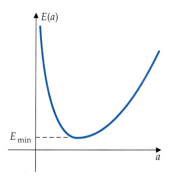

• **Figure 7–13**    Plot of the harmonic oscillator's energy as a function of the uncertainty $\Delta x = a$ in the position of the oscillator.

**Example 7–9**    Consider an electron in a one-dimensional infinite well; that is, the electron is free to move along a length $L$ and is forbidden from moving to regions beyond this length. (See Chapter 6.) Use the uncertainty principle to estimate the ground-state energy of such an electron.

**Solution**    The electron could be anywhere in the bottom of the well, meaning that $\Delta x = L$. This means that the uncertainty principle implies that $\Delta p > \hbar/2L$, and for our estimate, we take this value for the momentum itself. As for the electron's total energy $E$, we just take the expression for the free electron. Thus,

$$E = \frac{p^2}{2m} = \frac{\hbar^2}{8mL^2}.$$

Compare this result with the true ground-state energy [Eq. (6–45) with $n = 1$], namely $E = (\hbar^2\pi^2)/(2mL^2)$. While our estimate is off by a numerical factor, all the qualitative features are correct.

## SUMMARY

The fact that quantum mechanical systems are described by the solutions of the Schrödinger equation and that these solutions are related to probabilities leads to limitations on what experiments can tell us. These limitations are characteristic of all wavelike behavior and are summarized in the Heisenberg uncertainty relations. The following are some important properties of the uncertainty relations:

• The position and the momentum of a particle cannot be simultaneously measured with unlimited accuracy in a given experiment.

- The magnitude of the position–momentum effect is proportional to Planck's constant, and the restriction would vanish entirely if that constant were equal to zero. Thus Planck's constant once again determines the magnitude of quantum mechanical effects.

- One consequence of the uncertainty principle is that physical systems have energies that cannot become smaller than a particular value, different for each system. This result can be used to estimate the ground-state energies of systems, as is illustrated by estimates of minimum energies for the harmonic oscillator and hydrogenlike atoms.

- There is also an uncertainty relation between energy and time: The more precise an experiment that measures the energy of a system, the more time such an experiment will necessarily take.

These properties can be seen in some exact solutions of the Schrödinger equation. For a free particle in one dimension, the simplest wave function is a so-called plane wave, characterized by having a precise value of momentum while being spread over all space—that is, all positions are equally likely, and the position of the particle is completely uncertain. The plane wave thus represents an extreme manifestation of the uncertainty principle. To find solutions that have a simultaneous uncertainty in both position and momentum, we construct combinations of plane waves such that there is a significant probability of finding the particle only in limited regions of space—in other words, the waves represent particles that are localized. These "wave packets" can be analyzed to determine the probability of finding the particle with different values of momentum, and, in accordance with the uncertainty principle, such an analysis yields a limited range of allowed momenta. And, again in accordance with the uncertainty principle, if one reduces the width of a packet so that the position of the particle is known, one finds that all momenta become equally likely and the momentum is completely uncertain.

In addition to its usefulness in estimating ground-state energies of bound systems, the uncertainty relation eliminates many of the "paradoxical" aspects of quantum mechanics by making it impossible to confront possible inconsistencies between the wave and particle pictures of matter. This effect can be illustrated by an "experiment" in which one attempts to observe the position of a particle using a microscope. To observe this position more precisely, photons of shorter wavelength must be used. But these photons carry higher momenta, so they disturb the momentum of the particle to a greater degree—an illustration of the uncertainty principle. This same argument allows one to see how the wavelike behavior of matter is eliminated when one attempts to find which of two slits a particle passes through by observing it closely.

## QUESTIONS

**1.** Does the wave function for a free particle—that is, a particle with a definite, precise momentum—have nodes—in other words, points at which the probability of finding the particle is zero?

**2.** Recall the Heisenberg microscope. On the basis of the observation that the final momentum of the scattered photon was uncertain, since we did not know in which direction it had been scattered, we were led to the uncertainty relation. Suppose we raise the following objection: Identifying the place where the photon exposes the grain of silver on the photographic plate, we are able to determine exactly the trajectory of the photon, and hence, there will be no uncertainty in $\Delta p_x$. Analyze what is wrong with this objection by turning your attention to the photographic plate. What does the uncertainty relation tell you about the plate?

**3.** When the Schrödinger equation was first discovered, it was thought that the waves it described might be classical waves that "guided" the motion of the particle to which they were attached. However, if you take a Gaussian wave packet of the kind described in the text and initially confine it to the dimensions of an atom, you can show that it will spread out over the entire solar system in a few hours, even if its magnitude falls off very rapidly in space. How would we interpret this spread if we accept the fact that the wave function describes probabilities?

**4.** Can the group velocity of a wave packet, a quantity defined as $v_g = dE/dp$, ever exceed the speed of light?

**5.** The uncertainty relation would seem to suggest that the quantum mechanical world is not a precise one. Is there anything imprecise about the rules of quantum mechanics, as you have seen them up to this point? About the predictions?

**6.** Here is a suggestion of how you might try to beat the uncertainty principle: You take a microscope and shorten the wavelength of light it uses until it is comparable to or less than, say, the circumference of a Bohr orbit. Then you simply watch the electron go around. Will this idea work? If not, why not?

**7.** Consider the following proposal for violating the uncertainty relation (•Fig. 7–14): Photons of momentum $p$ are sent down a very long pipe of length $L$ and diameter $D$. The inside of the pipe is coated with a totally absorbing substance, so that only photons that never touched the walls emerge. The geometry shows that the transverse momentum of these photons must be less than $(D/L)p$. Thus the uncertainty in the transverse momentum is $2(D/L)p$, and this can be made as small as you like by making $L$ large. On the other hand, the uncertainty in the transverse location is $D$, so that the uncertainty relation appears to be violated. Look at the assumptions that have been made in setting up this apparatus, and see whether there is a loophole that saves the uncertainty relation, which *must* be satisfied.

**8.** Photographic shutters are getting better and better, and the electronic revolution promises much improvement in the speed at which they open and close the aperture of a lens. At what point will the improvement become counterproductive for ordinary snapshots?

**9.** Consider •Fig. 7–1 and its caption. Two high-energy particles are produced at a vertex. One is an electron, the other a positron—a positively charged electron. Can you use other evidence in the photo to determine which is the electron's track? The magnetic field is perpendicular to the plane of the photograph. Does it point up or down?

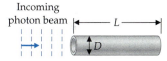

• **Figure 7–14** Plane wave of photons approaches a pipe of length $L$ and diameter $D$.

# PROBLEMS

**1.** ∎∎ Consider the very simple wave packet

$$\psi(x) = A \exp\left[\frac{i(p + \Delta p)x}{\hbar}\right] + A \exp\left[\frac{i(p - \Delta p)x}{\hbar}\right].$$

Show that $\psi(x)$ takes the form of a single plane wave of momentum $p$ multiplied by a function. What is that function?

2. ■■ Repeat the calculation of Problem 1 for the wave packet

$$\psi(x) = A \exp\left[\frac{i(p + \Delta p)x}{\hbar}\right] + 2A \exp\left[\frac{ipx}{\hbar}\right] + A \exp\left[\frac{i(p - \Delta p)x}{\hbar}\right].$$

3. ■■ Consider the wave packet of Problem 1. The time-dependent form of the packet is

$$\psi(x, t) = A \exp\left[\frac{i(p_1 x - E_1 t)}{\hbar}\right] + A \exp\left[\frac{i(p_2 x - E_2 t)}{\hbar}\right],$$

where $p_1 = p + \Delta p$, $p_2 = p - \Delta p$, $E_1 = E + \Delta E$, and $E_2 = E - \Delta E$. **(a)** Show that $\psi$ takes the form of a plane wave times a time-dependent modulating factor. **(b)** Show that the modulating factor has a time dependence that can be interpreted as the propagation of an "envelope" moving with speed $v = \Delta E / \Delta p$. The **group velocity** of a packet of waves is generally given as

$$v_g = dE/dp.$$

4. ■ Consider the definition of the group velocity given in the previous problem. Show that for the propagation of light, the group velocity is the speed of light.

5. ■■ Consider waves for which the momentum–energy relation is $p^2 = 2m(E - V_0)$. Calculate, in terms of $E$, the group velocity of a packet made of these waves. Show that the wave packet speeds up when it passes over a region in which $V_0$ is negative (i.e., over a potential energy "hole").

6. ■ Calculate $C^2$ in Eq. (7–18) from the normalization condition. Use

$$\int_{-\infty}^{+\infty} \frac{\sin^2 u}{u^2} = \pi.$$

7. ■ The resolving power of a microscope depends on the wavelength of the photon used to examine the object that is observed. What is the energy of the photons that must be used to resolve a molecule whose size is 0.8 nm?

8. ■ Suppose one wishes to use an electron beam to resolve a molecule whose size is 0.8 nm. What is the kinetic energy of electrons that have the requisite wavelength?

9. ■ An electron microscope operates with a beam of electrons, each of which has an energy of 60 keV. What is the smallest size that such a device could resolve? What must the energy of each neutron in a beam of neutrons be in order to resolve the same size of object?

10. ■■■ A beam of electrons traveling with speed $0.8 \times 10^8$ m/s passes through a slit of width $10^{-5}$ m. Because of the uncertainty in the lateral position of the beam, there will be an uncertainty in the transverse momentum as well. Estimate this uncertainty, and use it to calculate the spread of the image of the electron beam on a photographic plate placed perpendicular to the beam at a distance of 2.0 m beyond the slit.

11. ■■ A beam of neutrons with a kinetic energy of $2.4 \times 10^{-4}$ eV falls on a slit of width $10^{-4}$ m. What will be the angular spread of the beam after it passes through the slit?

12. ■■■ Use the uncertainty relation to estimate the ground-state energy of a particle in a one-dimensional potential given by $V(x) = \lambda x^4$. Check the dimensions of your answer.

13. ■■■ Consider a particle of mass $m$ in a potential of the form $V(x) = \infty$ for negative $x$ and $V_0 \times (x/a)$ for positive $x$. $V_0$ and $a$ are constants with the dimensions of energy and length, respectively. Use the uncertainty relation to estimate the ground-state (lowest) energy of the particle.

**14.** ▋▋ The binding energy of an electron in a crystal lattice is $1.2 \times 10^{-4}$ eV. Use the uncertainty relation to estimate the size of the spread of the electron wave function. How many lattice sites will that include, assuming that the spread is spherically symmetric and that the spacing between the ions is 0.12 nm?

**15.** ▋ The size of a lead nucleus is approximately $7.8 \times 10^{-15}$ m. Use the uncertainty relation to estimate the energy of an electron that can be emitted from the nucleus, assuming that the electron was localized inside the nucleus before it emerged. Express your answer in MeV.

**16.** ▋ Monochromatic light of wavelength 683 nm passes through a fast shutter that opens for $10^{-10}$ s. The light that emerges will no longer be monochromatic. What will be the spread in wavelengths?

**17.** ▋ An atomic state has a mean life of $2.6 \times 10^{-10}$ s. What is the uncertainty in the energy value of that state?

**18.** ▋ A photon of energy 0.85 MeV is emitted by a nucleus, which ends up in its ground state. The lifetime of the state is measured to be $0.4 \times 10^{-9}$ s. What is the natural width of the energy level from which the decay takes place?

**19.** ▋ A subatomic particle produced in a nuclear collision is found to have a mass such that $Mc^2$ is 1,228 MeV, with an uncertainty of $\pm 56$ MeV. Estimate the lifetime of this state. Assuming that when the particle is produced in the collision it travels with a speed of $10^8$ m/s, how far can it travel before it disintegrates?

**20.** ▋▋▋ Suppose one undertook a study of the orbits in hydrogen. According to the Bohr theory, the circular orbits have radii $r_n = n^2 a_0$, where the Bohr radius $a_0$ is roughly 0.05 nm. To map the orbits with some reasonable precision, it is necessary to localize the electrons with a precision $\Delta x = 0.5a_0$. What amount of kinetic energy (in eV) is likely to be transferred to the electron that is under observation with a probe that can localize the electrons with this precision? Will the atom remain relatively undisturbed by the observation? Your answer to this question illustrates why the idea of an atomic orbit is not a very clean one.

**21.** ▋▋ The time–energy uncertainty principle permits one to create particles in a **virtual** form for brief times. These are particles the production of which violates energy conservation, but only for times short enough that $\Delta E \Delta t \cong \hbar$. During this short time, the virtual particles can propagate and interact, but then they must be annihilated again, so that, for times larger than $\Delta t$, energy is conserved. Suppose that a virtual particle for which $mc^2 = 140$ MeV is created, and during the time of its existence, it travels with a speed close to the speed of light. Roughly how far will the particle travel during its brief existence? (That is, how far will this virtual particle exert its influence?)

**22.** ▋▋ In Example 7–4, we calculated the normalization of the Gaussian wave function. **(a)** Starting with this normalized wave function, show that the constant $C$ appearing in the Gaussian momentum distribution and given in Example 7–5 follows. **(b)** Using this value of $C$, verify that the Gaussian momentum distribution is properly normalized—that is, that $2\pi\hbar/A(p)^2$, integrated over all momenta, is equal to unity.

The next four problems make use of a wave packet for which the momentum weights are given by

$$A(p) = C \exp\left(-\frac{a|p|}{\hbar}\right).$$

**23.** ∎∎ Consider the wave packet for $A(p) = C \exp(-a|p|/\hbar)$. Find the value of the constant $C$ that will make this packet properly normalized—that is, so that

$$2\pi\hbar \int |A(p)|^2 \, dp = 1.$$

(*Hint*: Divide the integral over all values of $p$ into two integrals, one from $-\infty$ to 0 and the second from 0 to $+\infty$.)

**24.** ∎∎ Consider the wave packet for which $A(p) = C \exp(-a|p|/\hbar)$, with the value of $C$ as calculated in Problem 23. **(a)** Calculate the wave function $\psi(x)$ that results according to

$$\psi(x) = \int A(p) \exp\left(\frac{ipx}{\hbar}\right) dp.$$

Show that the quantity $|\psi(x)|$ is correctly normalized to be the probability distribution for position.

**25.** ∎∎ Consider the wave packet for which $A(p) = C \exp(-a|p|/\hbar)$. Using $A(p)$, show that $\langle p \rangle = 0$, and using $\psi(x)$ from Problem 24, show that $\langle x \rangle = 0$. You should not have to evaluate any integrals in detail.

**26.** ∎∎ Consider the wave packet for which $A(p) = C \exp(-a|p|/\hbar)$, with the value of $C$ as calculated in Problem 23. Using $A(p)$, calculate $\langle p^2 \rangle$, and using $\psi(x)$ from Problem 24, calculate $\langle x^2 \rangle$. Then, taking $\langle p \rangle$ and $\langle x \rangle$ from Problem 25, show that

$$\Delta x \Delta p > \frac{\hbar}{2}.$$

(*Hint*: You may want to use $\int z^2 \exp(-\alpha z) \, dz = d^2/d\alpha^2 \int \exp(-\alpha z) \, dz$.)

**27.** ∎∎∎ Consider the wave packet $\psi(x) = \int_{-\infty}^{+\infty} A(p) \exp(ipx/\hbar) \, dp$, with the

Gaussian form $A(p) = C \exp[-\alpha(p - p_0)^2/\hbar^2]$. At a later time, this wave function changes, with $px$ replaced by $px - Et$ in the plane-wave argument, where $E = p^2/2m$. **(a)** Use this information to find $\psi(x, t)$. (*Hint*: We have, immediately,

$$\psi(x, t) = \int_{-\infty}^{+\infty} \exp\left\{\frac{-\alpha(p - p_0)^2 - ip^2\hbar^2 t/2m}{\hbar^2}\right\} \exp\left(\frac{ipx}{\hbar}\right) dp.$$

The procedure for integrating this function by successive changes of variable is already outlined in the text.) **(b)** Calculate $|\psi(x, t)|^2$, and show that the peak of the wave packet travels with speed $p_0/m$, which is the group velocity $v_g$ at the peak of the wave packet (Problem 3). This calculation confirms the particle interpretation of the packet.

C H A P T E R

# 8

# Barriers and Wells

he quantum mechanical nature of matter has some surprising consequences, a number of which are revealed in experiments that involve either the scattering of one particle by another or the use of spectroscopy
to study energy levels. The Schrödinger equation can be employed to predict
the results of such experiments, provided that the potential energy is known,
as is the case in simple atomic systems. The behavior seen in realistic experiments and applications can be illustrated in highly simplified systems which
contain one dimensional potentials with sharp edges.

  We first describe "scattering," which in one dimension means that a
particle of fixed momentum and energy approaches a potential barrier. We shall
see that in quantum mechanics a certain fraction of the incident particles is reflected, while the remaining fraction is transmitted. This is already a deviation
from classical expectations, for which there is total transmission if the energy
of the incoming particle is larger than the maximum height of the potential barrier and total reflection if the energy of the incoming particle is smaller than the
maximum height of the potential barrier. The interesting phenomenon of transmission even for energy less than the barrier height is known as "tunneling"
through the barrier. Tunneling is a common feature of many physical systems
and devices; for example, nuclear fission, the scanning tunneling microscope,
and the tunnel diode all depend on the phenomenon.

  We shall also study the properties of energy levels, or bound states, in
finite potential wells, a generalization of the energy levels of the infinite well
studied in Chapter 6. In doing so, we are able to see more precisely how these
bound states are determined.

**8–1** Particle Motion in the
Presence of a Potential Barrier

**8–2** Wave Functions in the
Presence of a Potential Barrier

**8–3** Tunneling through the
Potential Barrier

**8–4** Applications and Examples
of Tunneling

**8–5** Bound States

## 8–1 Particle Motion in the Presence of a Potential Barrier

A one-dimensional **potential barrier** is formed by a potential-energy function
(•Fig. 8–1) of the form

$$V(x) = \begin{cases} 0, & \text{for} & x < -a & \text{(region I)} \\ V_0, & \text{for} & -a < x < a & \text{(region II)} , \\ 0, & \text{for} & x > +a & \text{(region III)} \end{cases} \qquad (8\text{–}1)$$

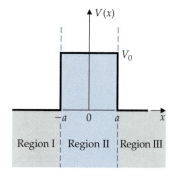

• **Figure 8–1** A square barrier,
formed by a potential-energy
function of width $2a$ and
height $V_0$.

**203**

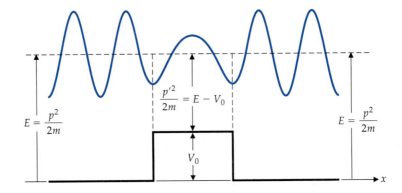

• **Figure 8–2** When a classical particle has enough energy to pass over a barrier, the quantum mechanical description of the particle shows that it has a changed wavelength in the region where $V \neq 0$.

where $V_0$ and $a$ are each positive. We shall refer to this potential as a **square barrier**. $V_0$ is the *height* of the barrier and $2a$ is its *width*. We start out with a particle of total energy $E$ approaching from the left—from $x = -\infty$. According to the laws of classical mechanics, there are two possible motions:

- If $E > V_0$, then the particle will pass the barrier. As it passes over the top of the barrier, the relation $E = p^2/2m + V_0$ shows that its momentum is reduced from $\sqrt{2mE}$ to $\sqrt{2m(E - V_0)}$, but it returns to its original momentum when it reaches $x = a$.
- If $E < V_0$, then the particle hits a wall and is reflected back to the left.

If, in contrast, we think in terms of de Broglie waves, then the particle has a wavelength that depends on its momentum. For $x < -a$ and for $x > a$,

$$\lambda = \frac{2\pi\hbar}{p} = \frac{2\pi\hbar}{\sqrt{2mE}},$$

while for the region $-a < x < a$ we have

$$\lambda = \frac{2\pi\hbar}{\sqrt{2m(E - V_0)}}.$$

The two possibilities that we described for classical motion look quite different in terms of wave mechanics:

- For $E > V_0$, the wavelength $\lambda$ is always real, although it is increased in the region where the barrier sits (•Fig. 8–2). Now for classical waves a change in wavelength is associated with a change in the index of refraction, and when a classical wave impinges on a surface at which the index of refraction changes, there is usually *reflection* in addition to *transmission*. Thus we might expect that in some way the particle will both reflect from, and pass across, the barrier.
- For $E < V_0$, the wavelength in region II becomes imaginary. In classical physics, this is a phenomenon we associate with the *evanescent waves* that occur when light is reflected from a region with a larger index of refraction to a region with a smaller index of refraction at an angle larger than the critical one. If region II were infinitely large, there would be *total internal reflection*. However, region II is finite in extent. In region II, the fact that $\lambda$—and hence $p/\hbar$—is imaginary implies that the wave function falls off *exponentially* there (•Fig. 8–3). The falling exponential remains finite, if small, at the far end of a finite barrier, and the wave then resumes its progress, albeit with an attenuated amplitude. Again, both transmission and reflection occur. In this case we refer to the transmission as **barrier tunneling**.

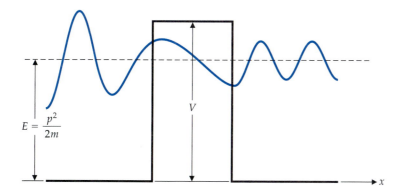

• **Figure 8–3**   When a classical particle has too little energy to pass over a barrier, the quantum mechanical description of the particle shows that it has a wave function that forms a falling exponential shape in the region for which $V > E$. When this "evanescent" wave emerges into a region in which $V = 0$, the wave function returns to the form it had when it entered the barrier, with the same wavelength (corresponding to the same energy), but an attenuated amplitude.

A simple laboratory demonstration illustrates tunneling in classical wave optics. As shown in •Fig. 8–4, a laser beam that strikes a Plexiglas™ rod at a small angle of incidence will undergo total internal reflection and emerge at the far end of the rod. If a second piece of Plexiglas™ is brought close enough to the original optical guide, the beam "tunnels" to the second piece with an intensity that varies exponentially with the thickness of the "barrier"—the thickness of the layer of air or other material between the two pieces of Plexiglas™. The amount of tunneling depends on the index of refraction of the material between the two pieces of Plexiglas™. This classical effect depends on the wave nature of light.

Tunneling is a quantum mechanical phenomenon. It occurs because of the wavelike properties of matter, in a manner analogous to the way classical waves

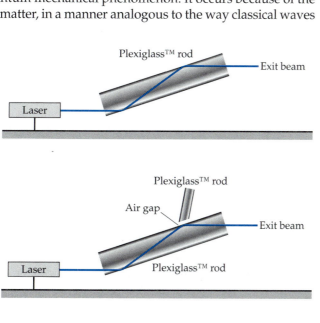

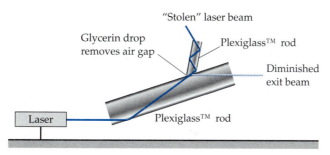

• **Figure 8–4**   A demonstration of tunneling in classical optics. The indices of refraction of air and glycerine correspond to barriers of different heights for the classical waves that travel down the tube shown at the right. Air presents a large barrier and glycerine a small one.

such as light exhibit the tunneling behavior described in the previous two paragraphs.

## 8–2 Wave Functions in the Presence of a Potential Barrier

We can see how what is described qualitatively in Section 8–1 is realized in detail in quantum mechanics by solving the time-independent Schrödinger equation for our potential. We already know the solution for the wave function in regions I and III: It will be in the form of plane waves moving either to the right or to the left.

With the notation $k = p/\hbar$, we can write down general solutions in the three regions. We do so with the boundary condition such that if the potential were not present, the solution would represent an electron moving to the right with momentum $p$—that is, a plane wave $\exp(ikx)$. In region I the Schrödinger equation takes the form

$$\frac{d^2u(x)}{dx^2} + \frac{2m}{\hbar^2} Eu(x) = 0, \qquad (8\text{–}2)$$

where $E = \hbar^2k^2/2m$. We expect that in region I, because of the presence of the potential, we will have a wave function that represents the incoming wave and a possible reflected wave (•Fig. 8–5). This wave function—and, indeed, all of those we shall study—will have the same time-dependent factor $\exp(-iEt/\hbar)$, which we accordingly can drop. Thus we write

Region I:
$$u(x) = \exp(ikx) + R\exp(-ikx) \qquad (8\text{–}3)$$

Now we found in Chapter 7 that functions of the form $\exp(\pm ikx)$ are difficult to normalize. This means that there may be an overall factor in front of the right side of Eq. (8–3) that can be determined only as a limiting process and that we have ignored. But this will not be a problem here, because, as we shall see, the *relative* probabilites of finding an electron with momentum $p = -\hbar k$ and finding it with momentum $+\hbar k$ are determined by the factor $R$, and the physics of the situation determines $R$ uniquely. We can determine $R$ precisely, and at the end of this section we shall give an unambiguous interpretation of its meaning. At this point it suffices to say that $R$ is a measure of how much of the incident wave is reflected.

In region II, the time-independent Schrödinger equation reads

$$\frac{d^2u(x)}{dx^2} + \frac{2m}{\hbar^2} (E - V_0)u(x) = 0. \qquad (8\text{–}4)$$

Now, this looks just like the free-particle Schrödinger equation (8–2) if, instead of $k$, we work with a different quantity $q$ defined by

$$q^2 = \frac{2m}{\hbar^2} (E - V_0). \qquad (8\text{–}5)$$

• **Figure 8–5** An incident wave with amplitude unity will divide into a reflected wave with amplitude $R$ and a transmitted wave with amplitude $T$ when it encounters a potential barrier. The wavy lines are purely symbolic: The wave functions that describe the system are complex.

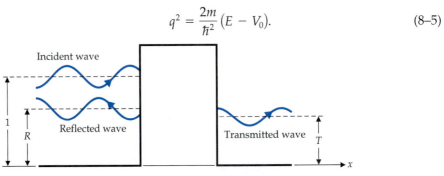

In this language, it is evident that the solution takes the same general form as the solution in region I, but with unknown coefficients for the right- and left-moving parts, corresponding to $\exp(+iqx)$ and $\exp(-iqx)$, respectively. Both terms arise because of multiple reflections at the boundaries. Thus

*Region II*: $\qquad\qquad u(x) = A\exp(iqx) + B\exp(-iqx).$ $\qquad$ (8–6)

We shall need to specify the physical conditions that enable us to find $A$ and $B$.

Finally, for region III we note that in the absence of a potential there would be only a solution like $\exp(ikx)$. The potential *cannot* produce a wave function with behavior described by $\exp(-ikx)$, since such a function would represent a wave with momentum $-p$ coming from the right. Hence, the solution in region III must be of the form (see •Fig. 8–5)

*Region III*: $\qquad\qquad u(x) = T\exp(ikx).$ $\qquad$ (8–7)

Later we shall show how to determine and interpret $T$ precisely; for now, we say only that $T$ is a measure of how much of the incident wave is transmitted. If there were no potential barrier—if $V_0$ were zero—we would expect that $R = 0$ and $T = 1$. Whatever method we use to determine these quantities should give those results in this limit.

## Continuity Conditions

We now have solutions in the three regions. These solutions contain some as yet undetermined constants, so our next task is to find them. We can do that by imposing physical restrictions on the wave functions. We observe that because we want the probability density to be continuous, we must insist that *the wave function be continuous everywhere*. We also insist that *the first derivative of the wave function be continuous everywhere*. This second condition is a little less intuitive. We can understand it by noting that if $du/dx$ is discontinuous somewhere, meaning that it changes abruptly there, then at that point $d^2u/dx^2$ is *infinite* (•Fig. 8–6). But the Schrödinger equation relates the second derivative to the potential, and a physically realizable potential is never infinite.[1]

Let us apply these conditions at the various boundaries. We first equate the wave functions given by Eqs. (8–3) and (8–6), as well as their derivatives on the two sides of the boundary at $x = -a$, to obtain

$$\exp(-ika) + R\exp(ika) = A\exp(-iqa) + B\exp(iqa) \qquad (8–8)$$

and

$$ik\exp(-ika) - ikR\exp(ika) = iqA\exp(-iqa) - iqB\exp(iqa). \qquad (8–9)$$

[The first equation, which matches the functions, is straightforward to implement; for the second, which matches the derivatives of the functions, we need only use $d/dx[\exp(\pm ikx)] = \pm ik\exp(\pm ikx)$.]

Similarly, at $x = +a$ the continuity conditions imply, successively, that

$$A\exp(iqa) + B\exp(-iqa) = T\exp(ika) \qquad (8–10)$$

and

$$iqA\exp(iqa) - iqB\exp(-iqa) = ikT\exp(ika). \qquad (8–11)$$

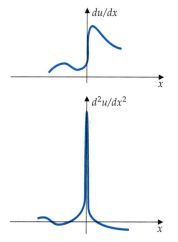

• **Figure 8–6** When a function, here $du/dx$, has a very sharp change such as a sudden rise in its value at a certain point, then its derivative, here $d^2u/dx^2$, has a very large spike at that point. If in the limit as $x \to 0$ the change in $du/dx$ is discontinuous, $d^2u/dx^2$ has an infinitely thin spike that goes to $+\infty$.

---

[1] In examples such as those in Chapter 6, we do have infinite potentials—the infinite well, for instance—and there the wave functions do have discontinuous first derivatives at the walls. This does not matter, since the wave functions vanish at the boundaries and everywhere beyond them. Such examples would be more realistic if we considered just very high, rather than infinite, walls, but few of the conclusions would change.

We thus have four linear equations for the four unknowns $A$, $B$, $R$, and $T$. Of course, we are interested only in $R$ and $T$, since it is these quantities that describe what is observed where we would place detectors, outside the region of the potential. We leave the solution of these equations for the appendix to this chapter; here we just quote the results, viz.,

$$R = \frac{i(q^2 - k^2)\sin(2qa)}{2kq\cos(2qa) - i(k^2 + q^2)\sin(2qa)} \exp(-2ika) \qquad (8\text{--}12)$$

and

$$T = \frac{2kq}{2kq\cos(2qa) - i(k^2 + q^2)\sin(2qa)} \exp(-2ika). \qquad (8\text{--}13)$$

### Properties of the Solution for $E > V_0$

The interesting features of our results reside in the coefficients $R$ and $T$:

1. We have $q^2 = k^2 - 2mV_0/\hbar^2$, so that even for $E > V_0$, R *is not zero*. In other words, at energies for which, classically, the particle would not be reflected, quantum mechanically it *is* reflected.

2. As $V_0 \to 0$—that is, as $q \to k$—the coefficient $R \to 0$. This is as it should be: In the absence of a barrier, there is no reflection. In this same limit, $T \to 1$, again as is reasonable. (See Problem 4.) The same behavior holds for $E \gg V_0$, in which case $q \cong k$ in the denominators, and then $R \cong V_0/E$. Similarly, $|T| \cong 1$ in this limit.

3. The absolute value of the reflection coefficient ($|R|$) can never exceed unity, as Problem 3 shows.

4. Some algebra (see Problem 5) shows that

$$|R|^2 + |T|^2 = 1. \qquad (8\text{--}14)$$

To appreciate the significance of these results, we must be able to interpret $R$ and $T$. Let us start by recalling our treatment of a plane wave $A\exp(ikx)$, with $A$ constant (Chapter 7). With this wave function, the probability density of finding an electron at $x$ is $|A|^2$. Because this probability density is independent of $x$, normalization poses a problem: It involves an integral of the constant $|A|^2$ over all space—here, an integral from $-\infty$ to $+\infty$. One way out of this difficulty is to realize that in a real experiment we deal with a beam of electrons and to adjust the factor $A$ to reflect that situation. Suppose, then, that we have a beam of electrons with $N$ electrons per unit length and each electron is described by the wave function $A\exp(ikx)$. Then the wave function must be normalized so that in a finite interval of length $b$ there will be $Nb$ electrons; that is, we require that the integral of $|\psi|^2$ over an interval of length $b$ give $Nb$. This means that

$$\int_0^b |\psi(x)|^2\, dx = \int_0^b |A|^2\, dx = |A|^2 \int_0^b dx = b|A|^2 = Nb.$$

It then follows that $|A|^2 = N$. In other words, with $N$ electrons per unit length, the properly normalized wave function is $(N)^{-1/2}\exp(ikx)$. We thus note that in the wave function we have used to represent the incoming electrons, we have tacitly assumed a beam with an electron density of 1 per unit length.

Because it is better to refer to a beam with a certain density of moving electrons, it is better to use a language of fluid flow to interpret $R$ and $T$. That is, we want to talk about the **flux** of electrons. In fluid mechanics, the flux describes the mass or volume of a fluid that crosses a unit area per second. In a one-dimensional flow, the flux would be the number of particles that pass a certain point per second. It is more natural to measure the number of particles that pass a certain point per second than to look at a slice of the $x$-axis and count the number of electrons in it. If the particles move with speed $v$ and the number of particles per unit length is $N$, then the number of particles that pass a point per unit time is $vN$. This is easily visualized if you picture a set of beads spaced uniformly along a piece of string that is passing through your fingers.

For a wave function $A \exp(ikx)$, the flux is then $vN = v|A|^2$, and with our choice of $N = 1$ corresponding to the term $\exp(+ikx)$ in the wave function of Eq. (8–3) in region I, the flux of particles traveling to the right is $vN = v = (p/m)$. And we can now interpret $R$: The term $R \exp(-ikx)$ in the wave function represents particles moving to the left with a flux $v|R|^2$. Similarly, the flux of particles moving to the far right in region III is $v|T|^2$. The principle of conservation of flux—conservation of matter in fluid mechanics and conservation of particles in our present context—implies that all the particles coming in from the left be accounted for in the part reflected and the part transmitted, or, mathematically,

$$v = v|R|^2 + v|T|^2.$$

But this is just Eq. (8–14). The argument also shows that it is physically reasonable that $|R|^2$ (and $|T|^2$) is bounded by 1. In other words, you cannot have more flux reflected back (or transmitted through) than came in.

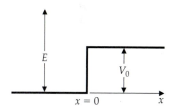

**Example 8–1**   Consider a beam of electrons traveling to the right along the $x$-axis with energy $E$. The potential energy is $V = 0$ for $x < 0$, but at $x = 0$ there is a potential "step," and the potential energy increases to (positive) $V_0$ for $x > 0$ (• Fig. 8–7). Assuming that $E > V_0$, **(a)** calculate the reflection and transmission coefficients, and **(b)** show that flux is conserved.

• **Figure 8–7**   Example 8–1. A potential step of height $V_0$. The energy of the incident particle is $E$.

**Solution**   **(a)** As before, we introduce $k$ and $q$ defined by

$$k^2 \equiv \frac{2mE}{\hbar^2} \text{ and } q^2 \equiv 2m\frac{(E - V_0)}{\hbar^2},$$

where both $k$ and $q$ are real numbers. The momentum on the left side is $\hbar k$, and on the right side it is $\hbar q$.

For $x < 0$, we hypothesize an incoming and reflected probability wave:

$$\psi(x) = \exp(ikx) + R\exp(-ikx), \qquad x < 0.$$

And for $x > 0$, we set a transmitted wave:

$$\psi(x) = T \exp(iqx), \qquad x > 0.$$

We now apply boundary conditions at $x = 0$. For continuity of the wave function at $x = 0$,

$$1 + R = T.$$

For the continuity of the derivative of the wave function at $x = 0$, we note that $d(\exp(ikx))/dx = ik \exp(ikx)$ and so forth, so that when we set $x = 0$, the continuity of the derivatives reads

$$ik - ikR = iqT.$$

The last two equations are linear equations in $R$ and $T$. Their solutions are, as you can easily verify,

$$R = \frac{k - q}{k + q} \quad \text{and} \quad T = \frac{2k}{k + q}.$$

**(b)** The velocity of the electrons is $p/m$, where $p$ is the momentum of the electrons. Thus the speed on the left is $\hbar k/m$, while on the right it is $\hbar q/m$. Conservation of flux then implies that $(\hbar k/m)(1 - |R|^2) = (\hbar q/m)|T|^2$. But from our explicit expressions, we have

$$k(1 - |R|^2) = k\left(1 - \frac{(k - q)^2}{(k + q)^2}\right) = \frac{4k^2 q}{(k + q)^2}$$

and

$$q|T|^2 = q\left(\frac{2k}{(k + q)}\right)^2.$$

These expressions are identical; hence flux is conserved.

Note that it is the conservation of the flux of electrons—that is, the number of electrons passing by any point per unit time—that is the relevant conservation law. It is *not* true here that $|T|^2 + |R|^2 = 1$; that equation holds only if the speeds in the relevant regions are the same, as indeed they are in our discussion of the finite barrier [Eq. (8–1)].

## 8–3 Tunneling through the Potential Barrier

Let us again consider the square barrier of Eq. (8–1), but this time we suppose that the incoming energy $E$ of the electron is less than $V_0$. Classically, the electron will bounce back from such a barrier in perfect reflection. In quantum mechanics we proceed as before, except that now

$$q^2 = \frac{2m}{\hbar^2}(E - V_0) < 0. \tag{8–15}$$

If $q^2 < 0$, $q$ must be imaginary. We thus take over the formulas of the previous section with the replacement $q \rightarrow i\kappa$, where

$$\kappa^2 = \frac{2m}{\hbar^2}(V_0 - E) \tag{8–16}$$

is positive. With this change,

$$\cos 2qa = \frac{\exp(2iqa) + \exp(-2iqa)}{2} \rightarrow \frac{\exp(-2\kappa a) + \exp(2\kappa a)}{2}$$

and

$$\sin 2qa = \frac{\exp(2iqa) - \exp(-2iqa)}{2i} \rightarrow \frac{i(\exp(2\kappa a) - \exp(-2\kappa a))}{2}.$$

The coefficient $T$ then takes the form

$$T = \frac{2k(i\kappa)e^{-2ika}}{2k(i\kappa)(e^{2\kappa a} + e^{-2\kappa a})/2 + (k^2 - \kappa^2)(e^{2\kappa a} - e^{-2\kappa a})/2}$$

$$= \frac{4ik\kappa e^{-2\kappa a}}{2ik\kappa(1 + e^{-4\kappa a}) + (k^2 - \kappa^2)(1 - e^{-4\kappa a})} e^{-2ika}. \tag{8–17}$$

*This coefficient is not zero;* in effect, the electron has succeeded in **tunneling** through the barrier. The full expression is rather complicated and, for our purposes, not particularly interesting. To see through it, let us calculate $|T|^2$ assuming that $4\kappa a$ is large enough that we can neglect $\exp(-4\kappa a)$ compared with unity. Then (see •Fig. 8–8)

$$|T|^2 = \frac{16k^2\kappa^2}{(k^2 + \kappa^2)^2} \exp(-4\kappa a). \tag{8–18}$$

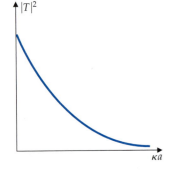

• **Figure 8–8**   A plot of $|T|^2$ for $E < V_0$, when there is tunneling. $|T|^2$ drops *exponentially* in this region.

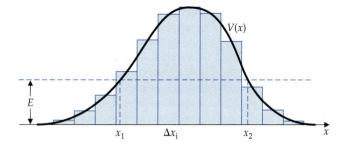

• **Figure 8–9**   Under many circumstances, tunneling through a barrier of arbitrary shape can be analyzed in terms of successive tunnelings through a series of thin square barriers.

The important—and characteristic—factor here is the exponential $\exp(-4\kappa a)$. The prefactor $16k^2\kappa^2/(k^2 + \kappa^2)^2$ is special to our sharp-edged barrier and will change if the barrier changes shape. But an analog to the factor $\exp(-4\kappa a)$ appears for any barrier and sets the scale for the degree of tunneling.

In the more general case in which the barrier height is a function of position, we proceed by dividing the irregular barrier into thin slices, each of which may be viewed as a square barrier (•Fig. 8–9). Then the probability of penetrating the $i$th barrier has the form

$$P_i = F_i \exp(-4\kappa_i \Delta x_i),$$

where $\Delta x_i$ is the width of that barrier, $V_i$ is its height, and

$$\kappa_i = \sqrt{\frac{2m(V_i - E)}{\hbar^2}}.$$

Here we leave $F_i$, the quantity that multiplies the exponential factor, unspecified; moreover, we shall assume that its dependence on $\Delta x_i$ is small compared with the exponential factor. We now argue that the probability of going through a series of barriers is the product of the probabilities of going through the individual barriers.[2] If we use the fact that

$$\exp(a_1) \times \exp(a_2) \times \exp(a_3) \times \ldots \times \exp(a_n) = \exp(a_1 + a_2 + a_3 + \ldots + a_n),$$

then we end up with an overall probability of the form

$$P = F \exp\left(\sum_i -2\Delta x_i \sqrt{\left(\frac{2m}{\hbar^2}\right)(V(x_i) - E)}\right)$$

$$\rightarrow F \exp\left(-2\int dx \sqrt{\left(\frac{2m}{\hbar^2}\right)(V(x) - E)}\right). \qquad (8\text{–}19)$$

In the last step, we have gone to the limit of very thin slices and replaced the sum by an integral.

The limits on the integral are given by the so-called turning points $x_1$ and $x_2$, points at which

$$V(x_1) = V(x_2) = E.$$

The prefactor $F$ cannot be obtained by this crude method, but it is not important for us to know it precisely, because the argument of the exponential, always of the general form

$$-2(\text{a measure of } \kappa)(\text{a measure of width}),$$

---

[2] This will be the case if going through a barrier is improbable, so that we can ignore the possibility that a particle is reflected by the barrier, then is reflected again in the forward direction, and has a second or even third crack at penetrating the barrier.

is largely what determines the dependence of the probability of transmission of the particle on its energy and potential.

• **Figure 8–10** Example 8–2. The truck is moving so slowly that it can go past the barrier only by tunneling.

**Example 8–2** Consider a 2,000-kg truck moving towards a bump in the road whose center is at $x = 0$ and whose height is given by 0 for $x < -0.1$ m, $(0.05$ m$)\cos(5\pi x)$ for $-0.1$ m $< x < +0.1$ m, and 0 for $x > +0.1$ m (•Fig. 8–10). Assuming that the truck moves so slowly that its kinetic energy is insufficient to climb the bump to any height at all, estimate the probability for the truck to tunnel through the bump.

**Solution** The probability of transmission is $|T|^2$, given by Eq. (8–18), with the change made in Eq. (8–19). For purposes of estimation, we shall worry only about the exponential factor. Then, with $V(x) = mgy$ and $y = (0.05$ m$)\cos(5\pi x)$ in the region $-0.1$ m $< x < +0.1$ m and zero elsewhere, we have

$$|T|^2 \cong \exp\left(-\frac{2}{\hbar}\int\sqrt{2mV(x)}\,dx\right)$$

$$= \exp\left(-\frac{2\sqrt{2(2,000\text{ kg})(2,000\text{ kg})(9.8\text{ m/s}^2)(0.05\text{ m})}}{1.05\times10^{-34}\text{J}\cdot\text{s}}\int_{-0.1\text{ m}}^{+0.1\text{ m}}\cos(5\pi x)\,dx\right).$$

The integral is given by

$$\frac{1}{5\pi}\left[\sin(0.5\pi) - \sin(-0.5\pi)\right] = \frac{2}{5\pi} \cong 0.13.$$

Substituting into the expression for $|T|^2$ and numerically evaluating the rest of the terms in the expression shows that the argument of the exponent is $5\times10^{36}$, so that $|T|^2 \cong \exp(-5\times10^{36})$. This is an incomprehensibly small number, which is why we can ignore any factors of 2 or 10 or 100 that might come from the prefactor of the exponential.

What the preceding example illustrates is that Planck's constant, $\hbar$, is so small that the exponential factor is effectively zero in macroscopic situations. Part of you walking through a wall would be, at the very least, disconcerting! But on the atomic scale, the effect can be not only large, but useful; shortly, we shall describe a couple of examples of great importance.

## 8–4 Applications and Examples of Tunneling

Tunneling is ubiquitous in quantum physics. We briefly discuss a few examples here, but note that we will run into others as we proceed in our discussion of various aspects of the structure of matter.

### Nuclear Physics

The earliest application of tunneling actually arose in one of the initial successes of quantum mechanics: its application to nuclear physics, a field in its infancy in the late 1920s. The Russian-born theoretical physicist George Gamow and, independently, Ronald Gurney and Edward U. Condon at Princeton University studied a type of radioactivity called **alpha decay**. In this kind of decay, an unstable parent nucleus converts into a daughter nucleus with the emission of an alpha particle—a helium nucleus, [4]He, denoted by the symbol[3] $\alpha$. For ex-

---

[3] The designation $\alpha$ goes back to the early days of radioactivity, when the nature of these emissions was not known. Likewise, $\beta$ stood for the emission of electrons and $\gamma$ for the emission of energetic electromagnetic quanta—photons.

ample, consider the very heavy element americium, which decays into neptunium according to the process

$$^{241}\text{Am} \rightarrow {}^{237}\text{Np} + \alpha$$

(generically, ${}^{A}Z \rightarrow {}^{A-4}(Z-2) + \alpha$). This process manifests itself experimentally in the emission of a (detectable) $\alpha$-particle with positive kinetic energy. To estimate the rate at which such a decay takes place, let us imagine the *inverse* process, in which an $\alpha$-particle impinges on a ${}^{237}\text{Np}$ nucleus. The coulomb potential—associated with the repulsive coulomb force—between the $\alpha$-particle (charge $Z_1 e = +2e$) and the daughter nucleus [charge $Z_2 e = 93e$ for Np or, more generally, $(Z-2)e$] has the form

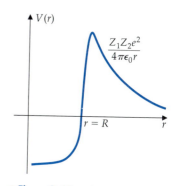

$$V_{\text{coulomb}}(r) = \frac{Z_1 Z_2 e^2}{4\pi\varepsilon_0 r}. \tag{8-20}$$

This potential extends from the nuclear radius $R$ outward and is greatest at $R$. Within the radius the attractive nuclear force is dominant (•Fig. 8–11). Now, measurement of the decay shows that the $\alpha$-particle's kinetic energy is less than the coulomb potential energy at the radius $R$. Thus, classically, the coulomb potential energy would prevent the $\alpha$-particle from penetrating to the inner core of the nucleus and being absorbed in the formation of the parent nucleus, ${}^{241}\text{Am}$ in this case. Thus the parent nucleus can form only if the $\alpha$-particle tunnels through the barrier—called the **coulomb barrier**—formed by the combination of coulomb and nuclear potential energies. The actual decay, as opposed to the inverse process, involves penetration of the same barrier, but this time from within. The probability of transmission of the $\alpha$-particle through the barrier is given by Eq. (8–18) for $|T|^2$, with the modification of Eq. (8–19):

• **Figure 8–11** The potential energy of an $\alpha$-particle in the region of a nucleus. The potential due to a combination of the coulomb force and the nuclear force forms a barrier for $\alpha$-decay.

$$|T|^2 = F \exp\left(-2\int dx \sqrt{\left(\frac{2m}{\hbar^2}\right)(V(x) - E)}\right). \tag{8-21}$$

Since the lower limit on the integral is the nuclear radius $R$, knowledge of $|T|^2$ can give us information about $R$. A detailed examination of the probabilities of alpha decay for a large number of nuclei showed for the first time that the radius of a nucleus of atomic weight $A$ is given by

$$R \cong 1.5 A^{1/3} \text{ fm}. \tag{8-22}$$

This early result revealed that the volume $4\pi R^3/3$ of the nucleus was proportional to its atomic weight, so that the nuclear density was almost constant. The result also demonstrated just how small the nucleus was. (A more detailed treatment of the nuclear decay rate is found in Chapter 15.)

*Nuclear Fusion*   Another context in which tunneling comes up is the study of the fusion of light nuclei. This has a potentially important technological application to the production of clean nuclear power. An important reaction involves the **fusion** of two deuterons (a deuteron is a nucleus made of a proton and a neutron; it has the charge, and roughly double the mass, of a proton) to make a triton (a nucleus made of a proton and two neutrons) and a neutron, with the release of a great deal of energy. Symbolically, we write the reaction as

$$^2\text{H} + {}^2\text{H} \rightarrow {}^3\text{H} + n + 6.4 \times 10^{-13} \text{ J}.$$

This is a process that is inhibited from happening because of the coulomb repulsion between the two deuterons. The reaction rate can become useful only by a combination of two effects: the fact that the coulomb barrier can be tunneled through and the fact that, by raising the gas of (ionized) deuterons to a

• **Figure 8–12** The Tokomak fusion reactor is designed to confine highly energetic charged particles by strong and cleverly arranged magnetic fields. This photograph is of the JET Tokomak and it's auxiliary systems.

higher temperature, the number of deuterons with higher energy grows (See Chapter 12). More energetic deuterons are closer to the top of the barrier, and there the barrier is less wide than for low $E$. Thus it is more likely that deuterons of higher energy will tunnel through the barrier separating them. Detailed calculations show that it is necessary to achieve temperatures on the order of $10^7$ K to have a practical reaction rate. In a hydrogen bomb such temperatures are achieved by means of compression initiated by an atomic bomb. To get to these temperatures in an unexplosive manner and to confine such a hot gas for a time long enough to extract the generated energy are the two great challenges facing this field[4] (•Fig. 8–12). While much progress has been made over the last 40 years of research in this field, no immediate breakthrough is in sight.

### Molecular Physics

Tunneling plays a role in the behavior of the ammonia molecule, $NH_3$, among many other molecules. In ammonia, the three hydrogen nuclei are at the corners of an equilateral triangle, with the nitrogen nucleus at the apex of a pyramid (•Fig. 8–13). The nuclei are enveloped in a cloud of electrons, three from each of the hydrogen atoms and seven from the nitrogen atom. In Chapter 11, we shall discuss the structure of molecules and, in particular, how the electrons can keep the nuclei together, but for the time being we need only note that the specific structure we have described, a pyramid, is energetically the most favorable one. More precisely, the potential energy of the nitrogen nucleus is a function of the distance $x$ along an axis going though the middle of the equilateral triangle, and for large $x$, the potential energy rises significantly. The potential energy also rises when $x$ is very small—when the nitrogen atom would

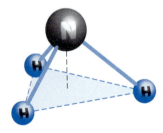

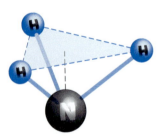

• **Figure 8–13** Two arrangements of the nuclei of the ammonia ($NH_3$) molecule, each classically stable, place the nitrogen nucleus above and below, respectively, the plane of the equilateral triangle formed by the hydrogen nuclei.

---

[4] For example, the confinement must be done with magnetic fields, because no material vessel could stand the high temperatures.

be in the plane of the equilateral triangle—since the coulomb repulsion between all the positively charged nuclei is then largest. Thus the potential energy has minima at $x = \pm a$ (•Fig. 8–14), symmetrically distributed about the plane $x = 0$. The reason for the symmetry is obvious: Things look exactly the same whether the pyramid points up or down.

Suppose now that at some time $t = 0$ the nitrogen nucleus is on the right side, in the potential hole about $x = +a$. It finds a barrier on the left, but that barrier is not nearly as large as the barrier on the right. *Then the nitrogen nucleus will tunnel across to the hole about $x = -a$.* Estimating the characteristic tunneling time would require a calculation more elaborate than we want to carry through here; let us simply label it as $\tau$. At that time the nitrogen nucleus is on the left side, and the tunneling repeats itself. Thus the nitrogen nucleus oscillates back and forth between the two minima with a total period of $2\tau$, corresponding to a frequency $f = 1/(2\tau)$. Classically, an oscillating charge will radiate with frequency $f$, and in quantum mechanics, in accordance with the correspondence principle, one expects ammonia to emit radiation with the same frequency. In quantum mechanical language, this radiation occurs through the emission of photons with energy

$$ hf = \frac{2\pi\hbar}{(2\tau)} = \frac{\pi\hbar}{\tau}. $$

A calculation of the tunneling time shows that $f \cong 2.4 \times 10^{10}$ Hz. This corresponds to a photon energy

$$ hf = (6.6 \times 10^{-34}\,\text{J}\cdot\text{s})(2.4 \times 10^{10}\,\text{Hz}) $$
$$ = (1.6 \times 10^{-23}\,\text{J})(1.6 \times 10^{-19}\,\text{J/eV}) = 9.9 \times 10^{-5}\,\text{eV}. $$

Radiation of this frequency is indeed observed for ammonia. In fact, this radiation can be used as an observational tool: It has been detected in interstellar space, signaling the presence of ammonia molecules between the stars.

In quantum mechanics, radiation is associated not with a moving charge, but instead with transitions between energy levels. This is a crucial element of the Bohr model. Let us see how it is realized in ammonia. We expect that there are two wave functions, $\psi_L$ and $\psi_R$, corresponding to the nitrogen nucleus being, respectively, on the left or on the right of the plane of the hydrogen nuclei. These wave functions are eigenfunctions of the energy, and because the potential is symmetric about $x = 0$, the energy values are the same for both wave functions. However, the fact that there is tunneling changes this picture: Instead of $\psi_L$ and $\psi_R$, the correct energy eigenfunctions turn out to be

$$ \psi_\pm = \psi_L \pm \psi_R. $$

The eigenfunction with the plus sign, $\psi_+$, is symmetric about $x = 0$, and the eigenfunction with the minus sign, $\psi_-$, is antisymmetric about $x = 0$. Both of these, together with $\psi_L$ and $\psi_R$ alone, are sketched in •Fig. 8–15. From this figure, we can see that $\psi_-$ has more curvature in it, and thus, on the average, a bigger slope, than $\psi_+$. This implies (see Problem 29) that the energy associated with $\psi_-$ is larger than the energy associated with $\psi_+$. There is a difference in the energies, and it is transitions between $\psi_+$ and $\psi_-$ which have the frequency that is observed. The energy difference is just the energy of the emitted photons that we estimated earlier.

## Electronics

Tunneling plays an important role in electronics. Conductors allow electrons to flow within them; insulators do not. Nevertheless, tunneling can allow electrons to flow from the surface of metals or, equivalently, between metals that

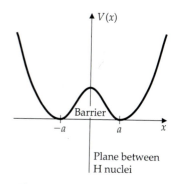

**• Figure 8–14** The potential energy of the nitrogen nucleus in ammonia as a function of the distance from the midpoint of the plane of the three hydrogen nuclei. The double well corresponds to two classically stable positions.

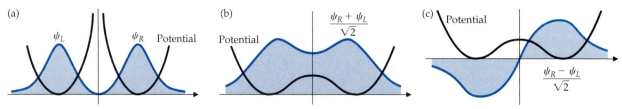

• **Figure 8–15**    (a) The wave functions for the nitrogen nucleus in ammonia in either of the two stable positions (potential-energy minima) if there were no mixing of the two due to tunneling. Tunneling implies that the wave functions with fixed energies are (b) the symmetric (even) superposition of the two wave functions in (a) or (c) the antisymmetric (odd) superposition of the two wave functions in (a).

are separated by insulators. Here we shall describe this process in more detail and look at some examples.

A metal is quite well described as a crystal lattice of ions with free electrons moving about in the background of positive charge. If we consider a one-dimensional model of a metal, we may describe it as a deep and very wide potential well. There will be very many bound states (levels), well described by the wave functions of the infinite well discussed in Chapter 6. For reasons that will be discussed in Chapter 9, electrons are stacked up in the levels, two to a level. No more than two electrons can occupy each level. •Figure 8–16a is a schematic representation of the situation. The wall of the potential extends above the energy level of the electron with the highest energy. The height of this wall is the work function $W$ discussed in Chapter 4 in connection with the photoelectric effect. It takes energy $W$ to free an electron from the metal. If an external potential associated with an electric field of magnitude $\mathcal{E}$ that attracts electrons is applied to the metal, the potential energy is modified as shown in •Figure 8–16b, and tunneling becomes possible. Beyond the edge of the metal, the potential energy takes the form

$$V(x) = W - e\mathcal{E}x. \tag{8–23}$$

If we measure the electron energy from the top of the occupied levels, then, in the application of Eq. (8–21), we can set $E = 0$. The integral in that equation runs from 0 to $a$, where $a$ is the place at which $V(x) = 0$; that is, at $x = a = W/e\mathcal{E}$.

A direct comparison with experiment is difficult, since the theoretical calculation assumes that the surface of the metal is without blemishes. Because $|T|^2$ is very sensitive to the value of the effective width of the barrier $a$, such imperfections make a quantitative comparison problematic. Nevertheless, recent technological advances in fabricating very precise surface layers have allowed a check on the correctness of the ideas presented here.

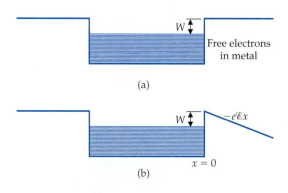

• **Figure 8–16**    (a) Valence electrons in a metal act as though they were in a well, with only two electrons allowed at each level. (b) When an external electrical potential is present that attracts the electrons, a barrier is formed, and the electrons can escape the metal by tunneling.

**Example 8–3** (a) Calculate the argument of the exponential function in [Eq. (8–21)] for an electron energy of $E = 0$ and for the potential given by

$$V(x) = \begin{cases} W - e\mathcal{E}x, & \text{for } 0 < x < W/e\mathcal{E} \\ 0, & \text{elsewhere} \end{cases}.$$

(b) Find the value of the corresponding exponential function if $W = 5$ eV and $\mathcal{E} = 10^9$ V/m. (These are realistic numbers: While this field is enormous, it typically acts over such a short distance that the total potential difference is only 10 to 100 V.) (c) How is your answer changed when $\mathcal{E}$ is instead $1.1 \times 10^9$ V/m?

**Solution** (a) The argument of the exponential function in Eq. (8–21) takes the form

$$-2 \int dx \sqrt{\left(\frac{2m}{\hbar^2}\right)(V(x) - E)} = -\frac{2\sqrt{2m}}{\hbar} \int_0^X \sqrt{W - \beta x}\, dx,$$

where $X = W/e\mathcal{E}$ and $\beta = e\mathcal{E}$. We therefore require the integral

$$\int \sqrt{A + Bx}\, dx = \frac{2}{3B}(A + Bx)^{3/2}.$$

Evaluation at the upper and lower limits quickly gives

$$-\frac{4}{3}\left(\frac{2mW}{\hbar^2}\right)^{1/2}\frac{W}{(e\mathcal{E})}.$$

for the exponent.

(b) We require only numerical evaluation here. The term

$$(2mW/\hbar^2)^{1/2} = \frac{[2(0.9 \times 10^{-30}\,\text{kg})(5\,\text{eV})(1.6 \times 10^{-19}\,\text{J/eV})]^{1/2}}{1.05 \times 10^{-34}\,\text{J}\cdot\text{s}} = 1.1 \times 10^{10}\,\text{m}^{-1}.$$

We have, in addition,

$$\frac{W}{(e\mathcal{E})} = \frac{5\,\text{eV}}{(e)(10^9\,\text{V/m})} = 5 \times 10^{-9}\,\text{m}.$$

Thus the exponent is

$$-\frac{4}{3}(1.1 \times 10^{10}\,\text{m}^{-1})(5 \times 10^{-9}\,\text{m}) = -73.$$

Consequently, the flux of transmitted electrons is

$$|T|^2 \propto \exp(-73) \cong 10^{-32}.$$

(c) The field appears inversely in the exponent. If we increase the field by 10%, then the exponent decreases by roughly 10%, from 73 to 66. But this drop has a large effect on the flux of transmitted electrons: The factor $10^{-32}$ becomes $\exp(-66) \cong 10^{-29}$. A 10% change the width of the barrier changes the probability of transmission by a factor of nearly 1,000.

---

Tunneling across a space between two metals plays an important role in **scanning tunneling microscopy**. A weak positive potential is placed on an extremely fine tungsten needle. When the top of the needle is a very small distance from a metallic surface, a tunneling effect takes place. As indicated in the preceding example, the number of electrons that flow from the surface to the tip per unit time—that is, the current—is very sensitive to the distance between the tip and the surface. This effect allows one to study the surface of metallic samples. •Figure 8–17 shows how that is done. The tungsten needle is mounted on a piezoelectric support. With elastic properties that depend on applied electric fields, these piezoelectric materials are arranged in such a way that the support senses the magnitude of the tunneling current and moves up or down

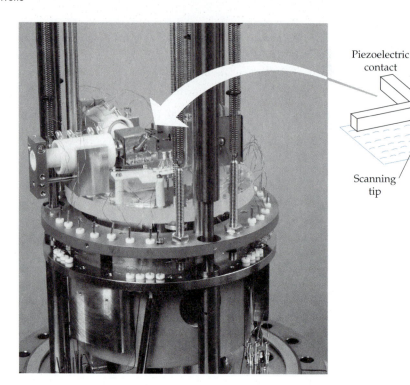

**• Figure 8–17** A picture of a scanning tunneling microscope used in the Electron Physics Group at the National Institute of Standards and Technology. The microscope is suspended for vibration isolation and operates in an ultra-high vacuum. The white screws are piezoelectric motors that can move the sample horizontally. The insert shows the tip and its sensors. The microscope was designed and constructed by Dr. J. A. Stroscio at the NIST.

in a manner that keeps the current constant, thus maintaining a constant distance of the needle from the surface. The voltage supplied to the support to carry out the up or down motion is recorded, thereby providing a record of the surface of the material being scanned. The resolution with which the details of the surface can be mapped depends on how fine the tungsten needle tip is. By heating the needle and applying a strong electric field, one can effectively pull off tungsten atoms from the tip layer by layer, till one is left with a tip that consists of a single atom, of size 0.1 nm.

Another important application of the scanning tunneling microscope depends on the fact that the electric field due to the needle can induce a charge separation of the neutral atoms in the metallic surface, and thus induce a dipole moment. This dipole responds to the nonuniform field, and the needle can, in effect, lift single atoms out of the metallic surface, one at a time. This effect (•Fig. 8–18) promises to be of great value in the construction of ultrasmall circuits and the creation of new, artificial molecules.

## 8–5 Bound States

We know from classical physics that when a potential is *attractive*, as it is in the Kepler problem, objects subject to the potential may be trapped in orbits. In Chapter 6, we looked at the possible trapped states that occur in an infinite well. In this section, we want to go a little further, with potentials that are a little more realistic, forming *finite* potential wells. Our interest is in seeing how the Schrödinger equation determines the energies of the trapped states. In the infinite well, these states are determined by insisting that half-wavelengths "fit" the well. By considering only a slightly more general potential, we can get a much more complete understanding of why and how it is that the Schrödinger equations leads to a *discrete* set of bound-state energies.

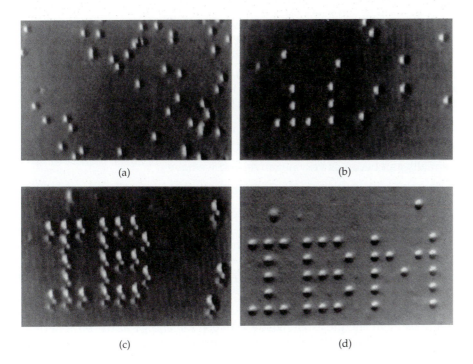

(a)                          (b)

(c)                          (d)

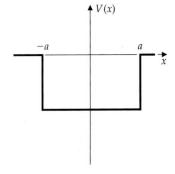

• **Figure 8–18**   The scanning tunneling microscope can be used to lift *single atoms* off a surface and place them elsewhere on the surface. Here xenon atoms placed on a nickel surface are moved around by a scanning tunneling microscope to spell out the name of the company at which a great deal of research on surface physics is done with such microscopes.

We can use the potential of Eq. (8–1) as an instructive example if we simply *reverse the sign of* $V_0$ (•Fig. 8–19). Thus, in our equations, we set $V_0 \rightarrow -V_0$. This finite potential well is called a *square well*. It is very similar to the infinite well, which takes the same form, but with $V_0$ infinitely large.

There is another difference, but it is innocuous: In the infinite well of Chapter 6, we set the bottom of the well at an energy of zero, with the walls infinitely far above zero. Here, we shift the zero of the energy so that the potential energy outside the well is zero and the potential inside the well is $-V_0$. No physics will depend on the choice of zero for the energy.

We refer to the states in which the particle involved is trapped as **bound states**. For the potential drawn as in •Fig. 8–19, with the zero of energy shown, the bound states are those[5] for which $E < 0$. Then, for regions I and III (outside the well), the Schrödinger equation takes the form

$$\frac{d^2u(x)}{dx^2} - \frac{2m}{\hbar^2}|E|u(x) = 0. \tag{8–24}$$

As opposed to Eq. (8–2), the additional minus sign in Eq. (8–24) changes the solutions from harmonic to exponential behavior. If we define $\kappa$ by

$$\kappa^2 = \frac{2m|E|}{\hbar^2}, \tag{8–25}$$

then the equation takes the form

$$\frac{d^2u(x)}{dx^2} - \kappa^2 u(x) = 0. \tag{8–26}$$

The solutions of this equation are linear combinations of $\exp(\kappa x)$ and $\exp(-\kappa x)$. There are no plane waves in this case! But that is fine; we are looking for bound states of the attractive potential.

• **Figure 8–19**   The square-well potential is an example of a finite potential well. Here the depth of the well is $V_0$.

---

[5]Subsequently, we shall comment briefly on $E > 0$ for this potential.

Which of the two exponential solutions apply? Let us begin with region III $(a < x < \infty)$. There the solution must be of the form $A \exp(-\kappa x)$, because a contribution of the form $\exp(\kappa x)$ grows without limit for large $x$ and is impossible to normalize. Aside from that, a solution of the form $\exp(\kappa x)$ would make it more probable for the particle to be farther away from the potential, a physically unacceptable proposition. Thus for region III,

Region III:
$$u(x) = A \exp(-\kappa x). \tag{8–27}$$

For the very same reason, the solution in region I $(-\infty < x < -a)$ *falls* exponentially as $x$ becomes very negative, and we have

Region I:
$$u(x) = B \exp(\kappa x). \tag{8–28}$$

Note that the coefficients $A$ and $B$ of the falling exponentials can be either positive or negative.

Next consider region II, "inside" the well. There we have

$$\frac{d^2 u(x)}{dx^2} + \frac{2m}{\hbar^2}(-|E| + V_0)u(x) = 0. \tag{8–29}$$

Bound-state solutions will have $-|E| + V_0$ positive, meaning that the possible solutions have energies somewhere between the energy at the bottom of the well, $-V_0$, and that at the top of the well, 0. (See Problem 8–23.) Accordingly, we define the real quantity $q$ by

$$q^2 = \frac{2m(-|E| + V_0)}{\hbar^2} = \frac{2mV_0}{\hbar^2} - \kappa^2. \tag{8–30}$$

The resulting equation is

$$\frac{d^2 u(x)}{dx^2} + q^2 u(x) = 0, \tag{8–31}$$

with the general solution

Region II:
$$u(x) = C \cos qx + D \sin qx. \tag{8–32}$$

### Even and Odd Solutions

Having found the general solutions in the three regions, we now match the wave functions and their derivatives at the two boundaries $x = \pm a$. Here we recall a pattern that we observed with the infinite well: As •Fig. 6–7a shows, the energy eigenfunction with the lowest energy in the infinite well is symmetric about the centerline of the well. The eigenfunction corresponding to the first excited state is antisymmetric about that line, the eigenfunction for the second excited state is again symmetric, and so on, with continued alternation between symmetric and antisymmetric eigenfunctions. The *only* eigenfunctions are either symmetric or antisymmetric eigenfunctions. This pattern repeats for the square well and, indeed, for all potentials that have a central axis about which there is symmetry. By defining $x = 0$ to be at the centerline, we need only consider eigenfunctions that are symmetric or antisymmetric in $x$. We shall consider these separately. In Eq. (8–32), they correspond to solutions with $D = 0$ (pure cosine—symmetric, or even) or $C = 0$ (pure sine—antisymmetric, or odd). Note that for the pure cosine solutions, the matching of the wave functions at $x = \pm a$ says that the falling exponentials in the solution outside the well are of the same magnitude and sign $(B = A)$. For the pure sine solutions, $B = -A$.

*Even Solutions* We generally want to match the solutions at $x = \pm a$, but because the solutions are even, what happens on the left is just a mirror image of what happens on the right, and we need only worry about matching at $x = a$. As $x$ approaches $a$ from the inside, we have

$$u(a) = C \cos qa \qquad (8\text{--}33a)$$

and

$$du/dx|_{x = a} = -qC \sin qa. \qquad (8\text{--}33b)$$

As $x$ approaches $a$ from the outside, we have

$$u(a) = A \exp(-\kappa a) \qquad (8\text{--}34a)$$

and

$$du/dx|_{x = a} = -\kappa A \exp(-\kappa a). \qquad (8\text{--}34b)$$

The magnitudes of $C$ and $A$ can ultimately be determined by normalization, which we can ignore in the present context. We can get rid of both $C$ and $A$ by dividing $du/dx$ by $u$ at $x = a$, both inside and outside, and equating the two ratios. This gives us the condition

$$\frac{-qC \sin qa}{C \cos qa} = \frac{-\kappa A \exp(-\kappa a)}{A \exp(-\kappa a)},$$

or

$$q \tan qa = \kappa. \qquad (8\text{--}35)$$

This relation involves the three quantities $|E|$, $V_0$ [see Eq. (8–30)], and $a$. If we regard $V_0$ and $a$ as fixed, then *only certain values of E will satisfy Eq. (8–35)*. This is one way to determine the allowed energies for our well. This approach to the problem is studied further in Problems 20 and 21.

But there is another way to see how these energies are determined. Let us look again at the matching problem. We can always make the wave functions themselves continuous at $x = a$ by choosing the scale $C/A$ appropriately. But what about the derivatives of the wave functions? Suppose we vary $|E|$, always for $V_0$ and $a$ fixed, starting with $|E|$ nearly zero. Because $\kappa$ is proportional to the square root of $|E|$, the exponential outside is almost flat; that is, the derivative is small. On the other hand, $q$ has its largest possible value, $q_{max} = (2mV_0)^{1/2}/\hbar$, so that the cosine function generally has a large derivative, and there is no hope of matching the derivatives. As we increase $|E|$, the exponential drops off more and more, so that its slope increases, while $q$ decreases, so that the slope of the cosine function decreases as well. Eventually, $|E|$ will reach a value such that the derivatives are matched. This value of $|E|$ is the first eigenvalue. If $|E|$ is increased further, the wave function gets flatter on the inside and steeper on the outside, so again, no match is possible ($\bullet$Fig. 8–20).

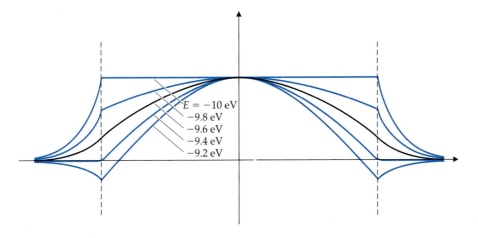

E = −10 eV
−9.8 eV
−9.6 eV
−9.4 eV
−9.2 eV

$\bullet$ **Figure 8–20** Finding the ground-state energy for the square well. As $E$ increases from a minimum, the (even) wave functions show kinks, except for the smooth—and therefore allowed—wave function that corresponds to some value of $E = E_0$ larger than the starting value. This value of $E$ represents the first energy eigenvalue, or ground-state energy. The corresponding wave function has no nodes.

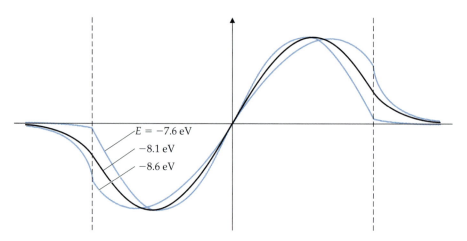

**• Figure 8–21**  The lowest allowed state—one with a smooth wave function—for the odd solutions in the square well is found with an energy $E_1$ higher than that of $E_0$, namely −8.1 eV. This is the second-lowest allowed energy and thus corresponds to the first excited state. Note that the associated wave function has a single node.

We can note two points:

- There will be a bound state for any value of $V_0$, no matter how small.
- The wave functions are real. We took advantage of this by sketching them.

*Odd Solutions*  We can proceed as we did for the even states, except now the inside solution is

$$u(x) = B \sin qx.$$

If we draw wave functions, we see that the odd solutions can be treated in the same way that we treated the even solutions. The only difference is that the inside wave function is a sine function, and thus it starts at zero for $x = 0$. This means that if $q$ is small, the function is rising at $x = a$, and there is no way of matching the slope there with the slope of a falling exponential. In fact, $qa$ must be larger than $\pi/2$ in order that the inside wave function slope downward at $x = a$ and can match a falling exponential (•Fig. 8–21).

We leave further discussion of the bound states to the problems.

### Nodes and Energies

In our discussion of bound states to this point, we have dealt with a single bound state. Because the infinite well has an infinite number of bound states, you should expect that if the finite potential well is deep enough, there will be more than one bound state. Here we want to see how this occurs. For definiteness, let us consider the even wave functions.

We therefore look at the derivative-matching problem for a much larger value of $V_0$, or equivalently for a much larger value of $q_{max}a$. Two things can happen:

1. The wave function inside the well can have a sharper hump, and such a wave function can tie onto a very rapidly falling exponential. This means that for a deep potential the binding energy $|E|$ will be larger.
2. The (harmonic) wave function inside the well can have such a sharp hump that it actually becomes negative at some $x$-value within the well. Every time the wave function crosses the $x$-axis in this way, we describe the crossing point as a **node**. For example, if $q_{max}a$ is increased to $\pi$, the wave function inside will be negative (and flat) at $x = a$. This wave function has a single node. Such a wave function can match onto a flat outside wave function, one for which the binding energy is small. This is a second bound state, whose binding energy is zero. More generally, we get the second bound state whenever $q_{max}a > \pi$. This is illustrated in •Fig. 8–22.

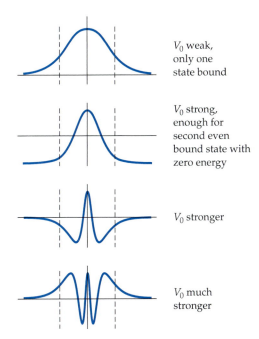

$V_0$ weak, only one state bound

$V_0$ strong, enough for second even bound state with zero energy

$V_0$ stronger

$V_0$ much stronger

• **Figure 8–22** Even bound-state solutions for a square well. At $q_{max}a = \pi$, the inside solution ties onto a flat line on the outside, corresponding to a state with zero binding energy.

With the aid of more sketching of wave functions, one can see that for a two-node wave function—the function changes sign twice—three bound states are possible, and so on. The more nodes, the higher are the energies of the bound states (i.e., the more loosely they are bound). We can understand why higher energy wave functions have more nodes if we note that the momentum is related to $d\psi/dx$, so that, roughly speaking, the more a wave function wiggles, the higher is the average value of the slope, and accordingly the higher is the average kinetic energy. The sign of the slope is not important, since the kinetic energy involves the square of the momentum. (See Problem 29.) But the more the number of nodes, the more the amount of "wiggle" there is in the wave function. We conclude that for a very deep potential, a zero-node wave function will have a large net negative energy, a one-node wave function will have a smaller net negative energy, and so on, until the kinetic energy gets to be larger than the potential energy and the particle is no longer confined to the well.

## SUMMARY

The wavelike aspect of matter produces some surprising results. These effects are evident for potential energies with a steplike structure: wells, walls, and barriers. The calculation of wave functions for these situations is simple, involving basically free-particle behavior together with the application of continuity of the wave function at boundaries between different values of the potential energy. The result is that the effects observed for classical waves in these situations is reproduced for particles as well. One such feature of special interest is the penetration of potential-energy barriers. In classical physics, particles cannot penetrate through barriers; in quantum mechanics, by contrast, the wave character does allow the penetration, even if the probability of this tunneling phenomenon is small. That probability, as well as that of the complementary property of reflection from the barrier, can be calculated in terms of the wave function.

The tunneling phenomenon is important for understanding certain physical processes and because of its applications to technology. Among both of these uses we cite the following:

- In the alpha decay of heavy nuclei, an $\alpha$-particle tunnels through the coulomb barrier formed by the combination of coulomb and nuclear potential energies. The process is perfectly explained by the tunneling phenomenon.
- The fusion of deuterons accompanied by the release of energy can occur only by tunneling. This process may have important technological applications.
- The study of material surfaces by means of scanning tunneling microscopes depends critically on the tunneling of electrons across the potential peak that holds the electrons in the material.
- A wide range of tunneling phenomena is used in deliberate ways in semiconductor devices, a subject to be discussed in Chapter 14.

If the barrier is replaced by a well then solutions with well-defined negative energies appear. These are the allowed bound states. By using and further developing many of the solution techniques employed in the understanding of tunneling phenomena, we can learn a good deal about general properties of the discrete bound states in attractive potentials. In particular, we mention the following:

- When the potential is symmetric about some central point, the eigenfunction of energy is either odd or even about that point.
- The lowest energy (ground) eigenstate is even, with eigenstates corresponding to an increasing succession of energies alternating between odd and even.
- The number of nodes in the successive eigenstates increases, starting with no nodes in the ground state, one node for the first excited state, etc. More generally, the more "wiggles" in a bound state wave function, the higher is the energy.

## QUESTIONS

**1.** In the wave function that describes a particle colliding with a barrier in one dimension we have elements that describe momenta going in opposite directions in the same wave function. Does this mean that the particle "does not know what it is doing"? What is the probability interpretation of this wave function?

**2.** Consider a potential that takes the form $V(x) = Ax^3 + Bx^2$, with $-\infty < x < +\infty$. Can such a potential ever appear in the description of a physical system? Explain your reasoning.

**3.** What is the minimum number of nodes the odd solution of the Schrödinger equation has for the attractive square well? What does this number tell you about how the lowest energy eigenvalues for the even and odd solutions compare?

**4.** In the study of transmission across a square barrier, one finds in Eq. (8–12) that the barrier does not reflect when $\sin 2qa = 0$—that is, when $qa = n\pi/2$ for $n$ a positive integer. Give a physical explanation of how this perfect transmission is achieved. Can you think of a practical application of the phenomenon?

**5.** For a particle in a well, the possible values of energy correspond to wave functions with nodes, and the probability of finding the particle at those nodes is zero. How, then, does the particle get from one side of a node to the other?

**6.** In describing the various transmission and reflection phenomena, we have made frequent comparisons between particles and light. But in this chapter the potentials describe *forces* on *particles*. Is there a force that acts on light when it enters a medium? Discuss the parallels and differences between particles and light for the matters discussed in the chapter.

**7.** For bound states of the attractive square well, the wave function falls off exponentially as the coordinate $x$ goes to infinity. Is this behavior unique to the square well? What property must the potential $V(x)$ have for the behavior to occur? (*Hint*: Look at the one-dimensional Schrödinger equation.)

**8.** We can control barrier heights and thicknesses for the electronic applications described in the chapter. Can you see a way to control any of the tunneling phenomena in nuclear physics?

**9.** Why is it that a square well, no matter how shallow, will always have at least one bound state, while this is not so if there is an infinitely high wall on one side of the well?

## PROBLEMS

**1.** ▮ A beam of electrons is sent along the $x$-axis from $-\infty$ with kinetic energy $E = 4.2$ eV. The beam encounters a potential barrier of height $V_0 = 3.2$ eV and width $2a = 1.2$ nm. What fraction of the incident beam is reflected?

**2.** ▮ Consider a potential of the form shown in •Fig. 8–23. Write down the general solutions in all of the regions, assuming that a beam of right-moving particles is sent in from $-\infty$ with a wave function of the form $\exp(ikx)$. What will be the general form of the wave function in the leftmost region? What will it be in the rightmost region?

**3.** ▮▮ Show that the reflection coefficient, given by Eq. (8–12), can never exceed unity. Is this a reasonable result?

**4.** ▮ Show that in the limit as $V_0 \rightarrow 0$, $T$, as given by Eq. (8–13), becomes unity. Is this reasonable?

**5.** ▮▮ Prove that $|R|^2 + |T|^2 = 1$ for $R$ and $T$ in Eqs. (8–12) and (8–13).

**6.** ▮▮▮ Find an expression for the transmission coefficient of a particle with a fixed, finite energy approaching a barrier in the limiting case as $V_0 \rightarrow \infty$ and $a \rightarrow 0$ such that $aV_0$ is a constant. What is $|R|^2$? (*Hint*: You can simplify your algebra by defining a parameter $\lambda \equiv k^2a^2$.)

**7.** ▮▮ Consider the step potential discussed in Example 8–1. Solve the Schrödinger equation for the case where $V_0 > E$, and discuss the meaning of your answer.

**8.** ▮▮ Consider the step potential discussed in Example 8–1, with the condition that a particle of energy $E = \hbar^2q^2/2m$ comes in from $-\infty$ with amplitude 1. What is the wave function for negative $x$?

**9.** ▮ Consider a potential barrier of the type shown in Eq. (8–1). Assume that a left-moving plane wave of the form $\exp(-ikx)$ is sent in from $+\infty$. Write down a form for the reflected wave in region III and a form for the transmitted wave in region I, where the regions are those defined in Eq. (8–1).

**10.** ▮▮▮ Consider the square barrier, and send in left-moving particles from the right, so that the wave function in region III is $\exp(-ikx) + R'\exp(ikx)$, while that in region I is $T'\exp(-ikx)$. Follow the procedure outlined in the text to calculate $R'$ and $T'$. How are $R'$ and $T'$ related to $R$ and $T$ obtained in the text? Show that $|R'|^2 = |R|^2$.

**11.** ▮▮ Consider the process of tunneling through a general barrier, as shown in •Fig. 8–9. Use the figure to show that the limits on the integral in Eq. (8–19)

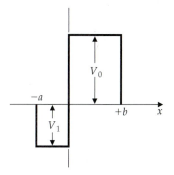

• **Figure 8–23** Problem 2.

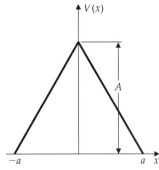

• **Figure 8–24**   Problem 12.

are the places at which $V(x) = E$. These are called *classical turning points*, since it is at those points that classical particles would turn around in encountering the barrier.

**12.** ▮▮ Consider the triangular potential shown in •Fig. 8–24. Write an expression for $V(x)$, noting that in the region between $-a$ and $0$ the potential is of the form $A + Bx$, and in the region between $0$ and $a$ it is of the form $A - Bx$.

**13.** ▮▮ **(a)** Write down an approximate expression for $|T|^2$ for low-energy electrons (with $E \ll A$) in the potential of Problem 12. **(b)** Calculate $|T|^2$ for $E = A/2$. **(c)** Using the electron mass $m_e = 0.9 \times 10^{-30}$ kg, estimate your result for part (b) numerically if $A = 10$ eV and $a = 0.05$ nm.

**14.** ▮▮ Consider the emission of electrons from a metal subject to an electric field, with $V(x) = W - e\mathcal{E}x$ and $E = 0$. Work out the integral to estimate $|T|^2$, the probability of transmission of electrons. Note that the upper limit on the integral is given by $a = W/e\mathcal{E}$.

**15.** ▮▮ Estimate how large an electric field $\mathcal{E}$ is needed to have $|T|^2 \cong 10^{-12}$ in the context of the previous problem.

**16.** ▮▮▮ Consider tunneling through a barrier that has the form

$$V = \begin{cases} 0, & \text{for } r < R \\ \hbar^2 L^2/(2mx^2), & \text{for } r > R. \end{cases}$$

Here $r$ is a radial variable—you can think of it as a one-dimensional variable like $x$, except that $r$ must always be positive. Show that, for energies small enough to make $kR = \sqrt{(2mER^2)/\hbar^2} \ll L$, the exponential suppression—the typical exponential falloff of the tunneling probability with the width of the barrier—becomes a power-law suppression of the form $(kR)^L$. (*Hint:* You may want to use the indefinite integral)

$$\int \sqrt{1 - y^2} \frac{dy}{y} = \sqrt{1 - y^2} - \ln\left(\frac{2 + 2\sqrt{1 - y^2}}{y}\right).$$

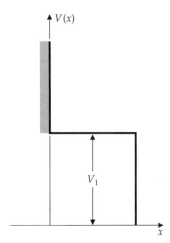

• **Figure 8–25**   Problem 17.

**17.** ▮▮ Consider the potential shown in •Fig. 8–25. Write down the solutions of the Schrödinger equation in the various regions.

**18.** ▮▮ Consider the potential shown in •Fig. 8–26. The wave function for zero binding energy is

$$u(x) = \begin{cases} 0, & \text{for } x < 0 \\ A \sin qx, & \text{for } 0 < x < a, \\ \text{constant}, & \text{for } x > a \end{cases}$$

where $q^2 = 2mV_1/\hbar^2$ for $E = 0$. Work out the matching conditions at $x = a$. If the width of the potential is $2a = 0.2$ nm, what is the depth of the potential in electron volts, given that the particle is an electron (mass $= 0.9 \times 10^{-30}$ kg).

**19.** ▮▮ Consider the bound-state problem for a square well. The even solutions inside the well have the form $A \cos qx$, and those outside the well on the right have the form $B \exp(-\kappa x)$. Write out the matching conditions which require that $u(x)$ and $du(x)/dx$ on the two sides be the same at $x = a$, and show that these lead to the condition $q \tan qa = \kappa$. This transcendental relation is called the *eigenvalue condition*, because it can be satisfied only for certain values of energies, viz., the energies of the bound states.

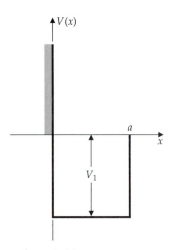

• **Figure 8–26**   Problem 18.

**20.** ▮▮▮ Analyze the eigenvalue condition (see Problem 19) for even solutions, $q \tan qa = \kappa$, as follows: Define $y = qa$ and $\lambda = 2mV_0a^2/\hbar^2$. **(a)** Show that the matching condition can be written in the form

$$\tan y = \frac{\sqrt{\lambda - y^2}}{y}.$$

**(b)** Sketch the two sides of the equation for the range $0 < y < 4\pi$ and for the values $\lambda = 5, 12$, and 45. (Note that the lines plotting the right side terminate at $y^2 = \lambda$.) The intersection points can then be used to find the bound-state energy eigenvalues.

**21.** ▮▮▮ Use the sketch in Problem 20 to show that, for $\lambda$ very large, the values of $y$ approach $y = (n + 1/2)\pi$. Show that these values correspond to the even eigenvalues for the infinite potential well discussed in Chapter 6.

**22.** ▮▮ Consider the bound-state problem for a square well. The odd solutions inside the well have the form $A \sin qx$, and those outside the well on the right have the form $B \exp(-\kappa x)$. Write out the matching conditions which require that $u(x)$ and $du(x)/dx$ on the two sides be the same at $x = a$, and show that these conditions lead to a new eigenvalue condition, $-q \cot qa = q \tan(qa + \pi/2) = \kappa$.

**23.** ▮▮ Consider the square-well potential defined by •Fig. 8–19. Show, by studying the possible solutions and the requirement of matching the wave functions and their derivatives across the boundary, that there are no solutions if $E < -V_0$.

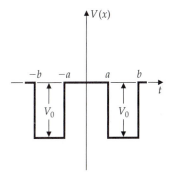

•**Figure 8–27** Problem 24.

**24.** ▮▮▮ Sketch the wave function for the lowest state of an electron in a double square-well potential (•Fig. 8–27). Show that the matching of wave functions and their derivatives allows one to match to a more steeply falling exponential on the outside, because the exponential falloff in the region between the two potentials does not have to be the same as that on the outside. This shows that two wells bind a particle more strongly than a single well.

**25.** ▮▮▮ One example involving a one-dimensional potential energy that we did not explicitly consider is the potential well formed by the reverse of Eq. (8–1), but with the interacting particle having positive energy: a finite square well in which $V_0$ has a negative value in the region between $-a$ and $+a$ and is zero everywhere else, with $E > 0$. Calculate the transmission and reflection coefficients. What would you expect classically? For what energies do you get the classical result? What happens in the limit of an infinitely deep well? (*Hint*: Use the results of the barrier calculation with $V_0 \rightarrow -V_0$!)

**26.** ▮▮▮ Examine the reflection coefficient (one of the quantities you calculated in Problem 25) for a potential well of depth $V_0$ and width $2a$. For certain energies, there will be *no reflection*. What are these energies?

**27.** ▮▮▮ Consider a particular example of the potential well of Problem 26, one for which the depth is 1.1 eV and the width is 0.5 nm. What is the lowest kinetic energy for which an electron will undergo no reflection? For what other energies will there be no reflection?

**28.** ▮▮▮ An electron of kinetic energy 0.1 eV is incident on a potential well of width 1.0 nm. What possible value(s) must the well depth have if the electron undergoes no reflection? (See Problem 26.)

**29.** ▮▮ Show that the average kinetic energy of a particle with a given wave function $\psi(x)$ can be written in the form

$$K = \frac{\hbar^2}{2m} \int |d\psi/dx|^2 \, dx.$$

This equation shows that a wave function with a lot of "wiggles" corresponds to a particle with higher energy. (*Hint*: Write out $\langle p^2 \rangle$ and use integration by parts.)

---

## APPENDIX

### Derivation of Eqs. (8–12) and (8–13)

We are interested in finding $R$ and $T$, given the four linear equations (8–8) through (8–11) for $A$, $B$, $R$, and $T$. We start by defining $X = A \exp(iqa)$ and $Y = B \exp(-iqa)$. Then Eqs. (8–10) and (8–11) become, respectively,

$$X + Y = T \exp(ika) \tag{A–1}$$

and

$$X - Y = \left(\frac{k}{q}\right) T \exp(ika). \tag{A–2}$$

We can solve for $X$ and $Y$ by taking the sum and difference, respectively, of these equations. Once we have $X$ and $Y$, we have, in turn, $A = X \exp(-iqa)$ and $B = Y \exp(+iqa)$. These can be substituted into Eqs. (8–8) and (8–9), respectively giving

$$\exp(-ika) + R\exp(ika) = \frac{q+k}{2q} T \exp[i(k - 2q)a]$$
$$+ \frac{q-k}{2q} T \exp[i(k + 2q)a] \tag{A–3}$$

and

$$\exp(-ika) - R\exp(ika) = \frac{q+k}{2k} T \exp[i(k - 2q)a]$$
$$- \frac{q-k}{2k} T \exp[i(k + 2q)a]. \tag{A–4}$$

If we now take the sum of these two equations, we get a linear equation for $T$ alone, which we can easily solve, giving Eq. (8–13). In turn, we substitute our result for $T$ into either Eq. (A–3) or Eq. (A–4), which then becomes a linear equation for $R$, and this immediately gives Eq. (8–12).

# Angular Momentum and the Hydrogen Atom

To this point, we have concentrated on quantum mechanics in one dimension. Although certain physical systems are well modeled in one dimension, the real world is three dimensional. In three dimensions important new concepts enter, including some that have no classical counterpart. These new concepts are centered around the quantum mechanical treatment of angular momentum, something that we have already encountered in our discussion of the Bohr model. We can see features associated with angular momentum in the hydrogen atom, the simplest two-body problem. In hydrogen, the coulomb force between the electron and the proton is well understood, and the experimental spectrum is very well known. Hydrogen offers us an entry into the general problems and applications of atomic and molecular physics. The atomic model formulated by Niels Bohr in 1913 required an assertion about the quantization of angular momentum. If the circular orbit discussed in Chapter 5 lies in the $xy$-plane, then the Bohr quantization rule states that the $z$-component of the angular momentum must be an integer times the Planck constant $\hbar$. Does this relation emerge from the three-dimensional Schrödinger equation?

To deal with these issues in an orderly matter, we shall proceed as follows: First we set up the Schrödinger equation for a particle in a central potential such as the coulomb potential. Then we sketch out a way to solve this equation and find out the physical meaning of its eigenfunctions and eigenvalues. We shall see that the angular momentum appears naturally. Next, we apply this approach to hydrogen itself. The dependence of the energy of hydrogen on angular momentum can be elucidated by studying the properties of the hydrogen atom in a magnetic field. We shall see how the observed spectrum of hydrogen, together with the behavior of the element in a magnetic field, leads one to the idea that the electron (and the proton) carries angular momentum even when it is at rest, a purely quantum mechanical effect we refer to as spin. We finish with some consequences of the spin and with an important application: magnetic resonance imaging.

# 9–1 The Schrödinger Equation for Central Potentials

■ We wrote the Schrödinger equation for three dimensions in Section 6–6.

The three-dimensional Schrödinger equation in Cartesian coordinates $x$, $y$, and $z$ is ill adapted to handle central forces. A central force acting on an object of mass $\mu$—here we are thinking ahead to the coulomb force acting on the electron in hydrogen—depends only on the magnitude of the radial distance of the object from the origin where the center of the force sits and is directed along the line from the origin to the object.[1] In classical physics, a central force has an important feature: There is no torque relative to the origin on the object under the influence of the force and, therefore, angular momentum does not vary with time; that is, *angular momentum is conserved*. We expect something like this to be true in quantum mechanics as well.

Central forces are best studied in the spherical coordinates $r$, $\theta$, and $\varphi$ (•Fig. 9–1). That is because the potential $V$ for a central force depends *only* on the radial variable $r$. How do we implement this change of variables in the Schrödinger equation? We start with the full time-dependent three-dimensional Schrödinger equation as presented in Eq. (6–51). Because we are going to be interested in eigenstates of energy, we remove the time dependence by writing

$$\psi(\vec{r}, t) = \psi(\vec{r}) \exp\left(-\frac{iEt}{\hbar}\right). \tag{9–1}$$

(See Section 6–4 for a more complete treatment; the separation-of-variables technique we employed there is also useful here.)

When this form is inserted into the time-dependent equation, the time dependence cancels, just as in the one-dimensional case, and we are left with the time-independent Schrödinger equation,

$$-\frac{\hbar^2}{2\mu}\left[\frac{\partial^2}{\partial x^2} + \frac{\partial^2}{\partial y^2} + \frac{\partial^2}{\partial z^2}\right]\psi(\vec{r}) + V(r)\psi(\vec{r}) = E\psi(\vec{r}). \tag{9–2}$$

The vector $\vec{r}$ is the vector from the center of force to the particle of mass $\mu$ under the influence of the stated force. Note that the potential depends only on the magnitude of $r$.

We next make the change of variables from the original Cartesian (rectangular) coordinates $(x, y, z)$ to spherical coordinates $(r, \theta, \varphi)$ (•Fig. 9–1):

$$x = r\sin\theta\cos\varphi, \, y = r\sin\theta\sin\varphi, \, z = r\cos\theta. \tag{9–3}$$

The angle $\theta$ is the polar angle and $\varphi$ is the azimuthal angle.

We now want to write the Schrödinger equation in terms of the spherical coordinates. The radial variable $r$ already appears explicitly in the potential. To make the new variables appear in the derivatives, we use chain-rule relations. This is a technical exercise that can be found in any calculus text; here we just quote the result. In spherical coordinates, the time-independent Schrödinger equation becomes

$$-\frac{\hbar^2}{2\mu}\left[\frac{\partial^2}{\partial r^2} + \frac{2}{r}\frac{\partial}{\partial r} + \frac{1}{r^2}\left(\frac{\partial^2}{\partial\theta^2} + \cot\theta\frac{\partial}{\partial\theta} + \frac{1}{\sin^2\theta}\frac{\partial^2}{\partial\varphi^2}\right)\right]\psi(\vec{r})$$
$$+ V(r)\psi(\vec{r}) = E\psi(\vec{r}). \tag{9–4}$$

---

[1] The physical system is actually a two-body system with the central force on the line between the bodies. After introducing the reduced mass into the Schrödinger equation (see Problems 9–2 to 9–5), one finds a force fixed at the center of mass of the system acting on an object with the reduced mass of the two-body system. For hydrogen, the lone (and massive) proton sits very close to the center of mass of the system, and the reduced mass is very close to the mass of the electron.

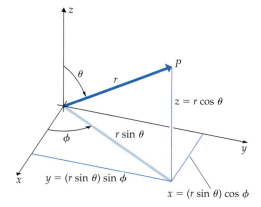

• **Figure 9–1**   Spherical (polar) coordinates, in which a point $P$ is labeled by the radial distance $r$ from the origin, an angle $\theta$ (measured down from the z-axis) and an angle $\varphi$ about the z-axis (measured from the x-axis).

The left side of Eq. (9–4) is the Hamiltonian, in spherical coordinates, operating on the wave function, with the term in square brackets the Laplacian in spherical coordinates; the right side is a number, $E$, multiplying the wave function. Except for some considerable differences in detail, we want to solve the equation, given the potential, just as we did the Schrödinger equation in one dimension. We look for the particular functions (the eigenfunctions) that, when operated on by the energy operator, give back the same functions multiplied by particular numbers (the eigenvalues). The eigenvalues are the allowed energies, and if we want to know the allowed bound-state energies, we are interested in the allowed *negative* energies. Later, we shall confirm that the corresponding eigenfunctions may be used to calculate the probability distribution for finding the particle in space for the given allowed energy.

### Reduction and Partial Solution of the Schrödinger Equation

While Eq. (9–4), which is a partial differential equation, looks complicated, in fact it can be reduced to a set of one-dimensional differential equations in $\theta$ and $\varphi$, and these equations can be solved directly. To proceed, we use the same technique we employed in Section 6–4 to remove the time dependence, namely separation of variables. We first write a trial solution of the form

$$\psi(\vec{\mathbf{r}}) = R(r)Y(\theta, \varphi). \tag{9–5}$$

We refer to the function $R(r)$ as the **radial wave function**, while $Y(\theta, \varphi)$, which, as we shall see, we can find explicity, is known as a **spherical harmonic**.

When Eq. (9–5) is inserted into Eq. (9–4), the derivatives taken with respect to $r$ act only on $R(r)$, and the derivatives taken with respect to $\theta$ and $\varphi$ act only on $Y(\theta, \varphi)$. Aside from these derivatives, the functions $R$ and $Y$ are purely multiplicative. Thus, when we next divide our equation by the product $R(r)Y(\theta, \varphi)$, we obtain

$$-\frac{\hbar^2}{2\mu} \frac{1}{R(r)} \left( \frac{\partial^2}{\partial r^2} + \frac{2}{r} \frac{\partial}{\partial r} \right) R(r)$$

$$-\frac{\hbar^2}{2\mu} \frac{1}{r^2} \frac{1}{Y(\theta, \varphi)} \left( \frac{\partial^2}{\partial \theta^2} + \cot\theta \frac{\partial}{\partial \theta} + \frac{1}{\sin^2\theta} \frac{\partial^2}{\partial \varphi^2} \right) Y(\theta, \varphi) + V(r) = E.$$

If in addition we now multiply by $r^2$, we simplify enough to see easily that the second term on the left is exclusively dependent on $\theta$ and $\varphi$, and the first and third terms on the left, as well as the $E$-term on the right, are exclusively dependent on $r$. We can group all the $r$-dependent terms on the left side and all the angle-dependent terms on the right. But then—and this is the key step in the separation-of-variables technique—*each side must separately be a constant*, since otherwise the equation cannot possibly be satisfied. Call this constant $\lambda\hbar^2/2\mu$. You can easily verify that at this point we are left with two equations, namely

$$\left(\frac{\partial^2}{\partial\theta^2} + \cot\theta\,\frac{\partial}{\partial\theta} + \frac{1}{\sin^2\theta}\frac{\partial^2}{\partial\varphi^2}\right)Y(\theta,\varphi) = \lambda Y(\theta,\varphi) \qquad (9\text{–}6)$$

and

$$-\frac{\hbar^2}{2\mu}\left(\frac{d^2R(r)}{dr^2} + \frac{2}{r}\frac{dR(r)}{dr} + \lambda\frac{R(r)}{r^2}\right) + V(r)R(r) = ER(r). \qquad (9\text{–}7)$$

Notice that in Eq. (9–7) we have recognized the presence of the single variable $r$ and have replaced the partial derivatives by ordinary derivatives. In other words, that equation for $R$ is an ordinary differential equation. Moreover, it is the type of equation we have become used to dealing with in one-dimensional problems: It is an eigenvalue equation that has solutions only for certain values of the energy $E$. But we can do no more with this equation until we specify the potential (Section 9–3).

Equation (9–6) is more technical. We can make progress towards its solution by utilizing still another separation of variables: We try the solution

$$Y(\theta,\varphi) = F(\theta)\Phi(\varphi). \qquad (9\text{–}8)$$

When this solution is substituted into Eq. (9–6), we find, with by-now familiar reasoning, two equations, one for $\Phi$ and one for $F$:

$$\frac{d^2\Phi(\varphi)}{d\varphi^2} = -m^2\Phi(\varphi) \qquad (9\text{–}9)$$

and

$$\frac{d^2F(\theta)}{d\theta^2} + \cot\theta\,\frac{dF(\theta)}{d\theta} - m^2F(\theta) = \lambda F(\theta). \qquad (9\text{–}10)$$

Here $m^2$ is our new "separation constant," for the separation of $Y$ into $F$ and $\Phi$, the quantity analogous to $\lambda$ for the separation of $\psi$ into $R$ and $Y$. In writing $m^2$ rather than $m$, we have anticipated that this constant turns out to be positive. We shall justify this expectation later.

Each of equations (9–9) and (9–10) is an ordinary differential equation, involving the single variables $\varphi$ and $\theta$, respectively. And there is no reason we cannot solve these equations, even without knowledge of the potential, because the potential does not appear in either equation!

### Probabilistic Interpretation of the Wave Function

Recall the probabilistic interpretation of the wave function $\psi(x)$ of a particle in one dimension:

$$|\psi(x)|^2\,dx = \text{probability of finding the particle}$$
$$\text{within a gap of width } dx \text{ at the location } x.$$

The generalization of this result to three dimensions is as follows:

$$|\psi(\vec{\mathbf{r}})|^2\,dx\,dy\,dz = \text{probability of finding the particle}$$
$$\text{within a box of volume } d^3\vec{\mathbf{r}} = dx\,dy\,dz \text{ about the point } \vec{\mathbf{r}} = (x,y,z).$$

The problem now is merely to translate the volume of the box to spherical coordinates. This is a standard exercise in calculus which we indicate in

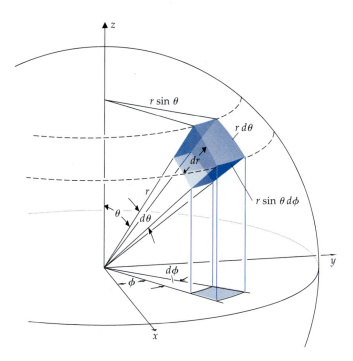

• **Figure 9–2** An infinitesimal volume element in terms of $dr$, $d\theta$, and $d\varphi$.

•Figure 9–2. From this figure, we see that the volume of the box in spherical coordinates is $r^2\, dr \sin\theta\, d\theta\, d\varphi$, and

$$|\psi(\vec{r})|^2 r^2\, dr \sin\theta\, d\theta d\varphi = \text{probability of finding the particle}$$
within a box of volume $r^2\, dr \sin\theta\, d\theta\, d\varphi$ about the point $\vec{r}$.  $\qquad$ (9–11)

As with one dimension, this interpretation leads to a normalization condition: that the sum of $|\psi(\vec{r})|^2 r^2\, dr \sin\theta\, d\theta\, d\varphi$ over every point in space is unity. Using Eq. (9–5), we find that

$$1 = \int_{\text{all space}} |R(r)|^2 |Y(\theta, \varphi)|^2 r^2\, dr \sin\theta\, d\theta\, d\varphi = \int_0^\infty |R(r)|^2 r^2\, dr \times$$

$$\int_{\text{all angles}} |Y(\theta, \varphi)|^2 \sin\theta\, d\theta\, d\varphi.$$

$\qquad$ (9–12)

From this expression, it is reasonable to infer that

$$|R(r)|^2 r^2\, dr = \text{probability of finding the particle within a gap}$$
of width $dr$ about the radial distance $r$.  $\qquad$ (9–13)

and

$$|Y(\theta, \varphi)|^2 \sin\theta\, d\theta\, d\varphi = \text{probability of finding the particle}$$
within an area $d\theta\, d\varphi$ about the angular location $(\theta, \varphi)$.  $\qquad$ (9–14)

There are then two separate normalization conditions to apply: that the probability of finding the particle at some radius among all possible radii is unity and that the probability of finding the particle at some angle among all possible angles is unity. Mathematically,

$$\int_0^\infty |R(r)|^2 r^2\, dr = 1 \qquad (9\text{–}15)$$

and

$$\iint\limits_{\text{all angles}} |Y(\theta, \varphi)|^2 \sin\theta \, d\theta \, d\varphi = 1, \qquad (9\text{--}16)$$

where "all angles" means that the integral over $\theta$ runs from $-\pi$ to $+\pi$ and the integral over $\varphi$ runs from 0 to $2\pi$.

We shall have good reason to refer to these results, because they will eventually give us probabilities for finding the electron at various locations in the atom.

### *Solving for the Spherical Harmonics

We can start with the equation for $\Phi(\varphi)$, which is in fact straightforward to solve. The solution is

$$\Phi(\varphi) = C \exp(im\varphi), \qquad (9\text{--}17)$$

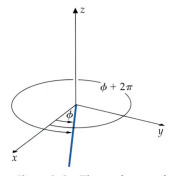

• **Figure 9–3**  The angles $\varphi$ and $\varphi + 2\pi$ represent the same location for a given $r$ and $\theta$.

where $C$ is any constant. But there is a constraint on $m$ which follows from the fact that when we increase the angle $\varphi$ by $2\pi$, we are going around a circle about the origin in the $xy$-plane (•Fig. 9–3). Under these circumstances, we do not want the solution to give us something new. In other words, we require that the wave function $\Phi(\varphi)$ be *single valued*.[2] This means that

$$\Phi(\varphi + 2\pi) = \Phi(\varphi),$$

or

$$C \exp[im(\varphi + 2\pi)] = C \exp(im\varphi),$$

or

$$\exp(2\pi im) = 1.$$

The last equation can be satisfied only if $m$ is an integer—that is,

$$m = 0, \pm 1, \pm 2, \pm 3, \dots.$$

The solution of Eq. (9–10) for $F(\theta)$ is more difficult, and we shall content ourselves here with a very brief discussion without paying too much attention to proving anything. Suffice it to say that the equation is solved using standard techniques for differential equations. Let us start with some conditions on its solution. We see that both $m$ and $\lambda$ appear in the equation. We have already discovered restrictions on $m$ from the equation for $\Phi$, so we can say that only $\lambda$ appears as a parameter in the equation for $F$. But in fact this equation can be satisfied in a physically acceptable manner only if

$$\lambda = -\ell(\ell + 1), \qquad (9\text{--}18)$$

where

$$\ell = 0, 1, 2, 3, \dots \qquad (9\text{--}19)$$

is a positive integer. Moreover, Eq. (9–10) imposes a further restriction on $m$, namely, given $\ell$, rather than being able to take on all integer values, $m$ can run only from $-\ell$ to $+\ell$; that is,

$$m = -\ell, -\ell + 1, -\ell + 2, \dots, -1, 0, +1, \dots, \ell - 2, \ell - 1, \ell. \qquad (9\text{--}20)$$

Thus, if $\ell = 0$, then $m$ must also be 0; if $\ell = 1$, then $m$ can be only $-1, 0$, and $+1$; if $\ell = 2$, then $m$ can take on the values $-2, -1, 0, +1$, and $+2$; and so forth.

---

[2] This is the same reasoning that went into establishing an international date line. If there were no date line, you could, by running rapidly around a very small circle enclosing the North Pole, get yourself back to the days of Julius Caesar in a relatively short time—but the newspaper headlines and the reality would be that of today.

The structure of the Schrödinger equation has forced two parameters, $\ell$ and $m$, to be integers. We therefore refer to them as **quantum numbers**. Later we shall see that they are associated with the quantization of angular momentum.

Given a value of $\ell$ chosen from the range of possible positive integers, the function $F(\theta)$ is an $\ell$th-order polynomial in $\sin\theta$ and $\cos\theta$. These polynomials also depend on what $m$-value has been chosen. Rather than presenting $F(\theta)$ separately, we shall give the combination $Y(\theta, \varphi)$, the product of $\Phi$ and $F$. We have stated that $F$ depends on both $\ell$ and $m$; $\Phi$ depends on $m$ alone. Thus the form of the spherical harmonic $Y(\theta, \varphi)$ depends on both $\ell$ and $m$ and is accordingly labeled $Y_{\ell m}(\theta, \varphi)$. Table 9–1 lists spherical harmonics for $\ell = 0, 1,$ and 2.

We can systematically verify by direct substitution that the $Y$'s are solutions of Eq. (9–6). The simplest case is that in which $Y$ is a constant, namely the case $Y_{00}$. This means that the effect of the derivatives with respect to $\theta$ in Eq. (9–6) is to produce zero. The equation could then be satisfied if both $\ell$ and $m = 0$. Thus we have at once verified the $\ell$-value corresponding to a constant solution and the restriction on $m$. As to just what constant value $Y_{00}$ takes, that is a matter of normalization. The normalization requirement of Eq. (9–16) is in this case

$$\iint\limits_{\text{all angles}} |Y_{00}(\theta, \varphi)|^2 \sin\theta \, d\theta \, d\phi = \left(\frac{1}{\sqrt{4\pi}}\right)^2 \iint\limits_{\text{all angles}} \sin\theta \, d\theta \, d\varphi.$$

But the last integral is just the total solid angle about a sphere, $4\pi$. You can verify this as follows:

$$\iint\limits_{\text{all angles}} \sin\theta \, d\theta \, d\varphi = \iint\limits_{\text{all angles}} d(\cos\theta) \, d\varphi = \int_{-1}^{+1} d(\cos\theta) \int_0^{2\pi} d\varphi = 4\pi.$$

Thus the equation is satisfied—$Y_{00}$ is correctly normalized.

**Table 9–1   Spherical Harmonics**

$$Y_{00} = \frac{1}{\sqrt{4\pi}}$$

$$Y_{11} = \sqrt{\frac{3}{8\pi}} \sin\theta \exp(i\varphi)$$

$$Y_{10} = \sqrt{\frac{3}{4\pi}} \cos\theta$$

$$Y_{22} = \sqrt{\frac{15}{32\pi}} \sin^2\theta \exp(2i\varphi)$$

$$Y_{21} = -\sqrt{\frac{15}{8\pi}} \sin\theta \cos\theta \exp(i\varphi)$$

$$Y_{20} = \sqrt{\frac{5}{16\pi}} (3\cos^2\theta - 1).$$

*Generally, $Y_{\ell-m} = (-1)^\ell Y_{\ell m}$.

---

**Example 9–1**   An electron in an atom has a wave function whose angular part is $Y_{21}$.
**(a)** How much more (or less) likely is it to find the electron in an angular gap $d\theta \, d\varphi$ at $\theta = \pi/8$ and $\varphi = 0$ than in an angular gap $d\theta \, d\varphi$ of the same size at $\theta = \pi/4$ and $\varphi = 0$?
**(b)** Than in the same gap at $\theta = \pi/2$ and $\varphi = 0$?

**Solution**   **(a)** According to our interpretation of the angular part of the wave function, Eq. (9–14), the ratio of the probability of finding the electron at the location $\theta = \pi/8$ and $\varphi = 0$ in a gap $d\theta \, d\varphi$ to the probability of finding the electron at the location $\theta = \pi/4$ and $\varphi = 0$ in a gap $d\theta \, d\varphi$ is

$$\frac{\left|Y_{21}\left(\frac{\pi}{8}, 0\right)\right|^2 \sin\left(\frac{\pi}{8}\right) d\theta \, d\varphi}{\left|Y_{21}\left(\frac{\pi}{4}, 0\right)\right|^2 \sin\left(\frac{\pi}{4}\right) d\theta \, d\varphi}.$$

The gap factors $d\theta \, d\varphi$ cancel. From Table 9–1, we can read off the values of $Y_{21}$. Note that the pure phase factor $\exp(i\varphi)$ cancels when the absolute value is taken. Then the ratio of the probabilities is

$$\frac{\sin^3\left(\frac{\pi}{8}\right)\cos^2\left(\frac{\pi}{8}\right)}{\sin^3\left(\frac{\pi}{4}\right)\cos^2\left(\frac{\pi}{4}\right)} = 0.27.$$

**(b)** The function $Y_{21}$ contains a factor of $\cos\theta$, and $\cos(\pi/2) = 0$. Thus the probability of finding the electron at the equator of our system is zero.

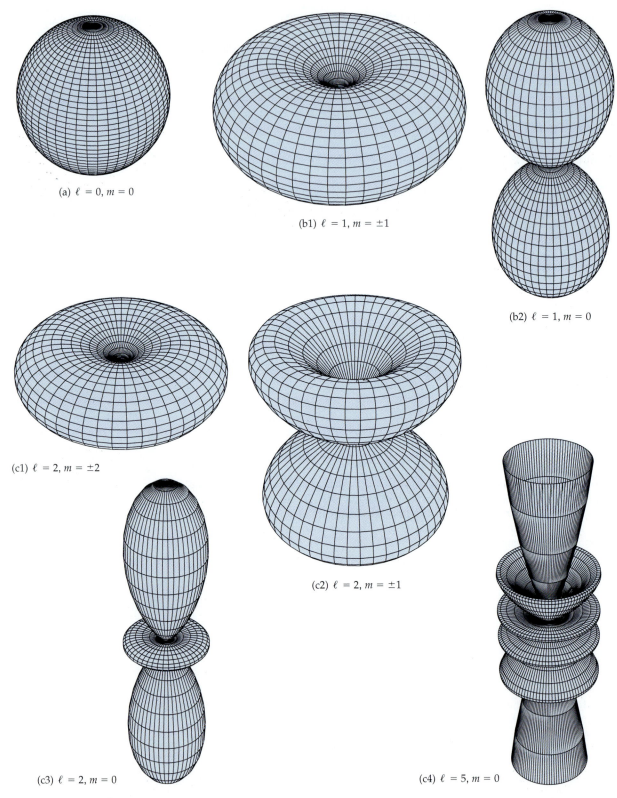

(a) $\ell = 0,\, m = 0$

(b1) $\ell = 1,\, m = \pm 1$

(b2) $\ell = 1,\, m = 0$

(c1) $\ell = 2,\, m = \pm 2$

(c2) $\ell = 2,\, m = \pm 1$

(c3) $\ell = 2,\, m = 0$

(c4) $\ell = 5,\, m = 0$

• **Figure 9–4**   $|Y_{\ell m}|^2$ represents the probability of finding the electron in a solid angle $\sin\theta\, d\theta\, d\varphi$. We plot the distributions for **(a)** $\ell = 0,\, m = 0$; **(b)** $\ell = 1,\, m = \pm 1,\, \ell = 1,\, m = 0$; and **(c)** $\ell = 2,\, m = \pm 2,\, m = \pm 1$, and $m = 0$. We also include $\ell = 5,\, m = 0$, to show the growing complexity of the distribution.

The preceding example shows that the probability distribution function in terms of an angle is straightforward to calculate. We can note generally that if the wave function of a particle under the influence of a central force—and, as usual, we are thinking of the electron in hydrogen—is proportional to $Y_{\ell m}(\theta, \varphi)$, then the probability distribution is independent of the azimuthal angle $\varphi$. That is because $|Y_{\ell m}(\theta, \varphi)|^2$ contains $|\exp(im\varphi)|^2$, and this quantity is unity. •Figure (9–4) is a plot of the probability distribution for the $\ell$ and $m$ values of Table 9–1 as a function of $\theta$.

## 9–2 Angular Momentum

The Schrödinger equation (9–4) is simpler than it appears to be, and the solutions of the angular part—the spherical harmonics—have a special physical significance. Both of these statements have to do with angular momentum. In order to see how, we take a detour to look at the quantum mechanical version of angular momentum. The classical definition of the angular-momentum vector of a particle relative to some point $P$ is

$$\vec{L} = \vec{r} \times \vec{p}, \tag{9–21}$$

where $\vec{p}$ is the momentum of the particle and $\vec{r}$ is the radius vector of the particle from the fixed point $P$ (•Fig. 9–5). The vector product is defined in the usual way, using the right-hand rule. (For a brief review, see Appendix B–1.) Thus the direction of the classical angular-momentum vector is perpendicular to both $\vec{r}$ and $\vec{p}$. The vector product can be written component by component as follows:

$$L_x = yp_z - zp_y,\ L_y = zp_x - xp_z,\ L_z = xp_y - yp_x. \tag{9–22}$$

How would we form the quantum mechanical operator that corresponds to the classical angular momentum? All that is needed is to use the expressions of Eq. (9–22), with the momenta replaced by their representation in terms of derivatives, as given by Eq. (6–28) or, for three dimensions, Eq. (6–49), namely

$$p_x = -i\hbar\frac{\partial}{\partial x},\ p_y = -i\hbar\frac{\partial}{\partial y},\ p_z = -i\hbar\frac{\partial}{\partial z}. \tag{6–49}$$

Then

$$L_x = -i\hbar\left(y\frac{\partial}{\partial z} - z\frac{\partial}{\partial y}\right),\ L_y = -i\hbar\left(z\frac{\partial}{\partial x} - x\frac{\partial}{\partial z}\right),\ L_z = -i\hbar\left(x\frac{\partial}{\partial y} - y\frac{\partial}{\partial x}\right). \tag{9–23}$$

We shall also be interested in the square of the angular momentum:

$$L^2 = L_x^2 + L_y^2 + L_z^2. \tag{9–24}$$

To best study this quantity, one makes a change of variables to spherical coordinates in Eqs. (9–23) and (9–24). This change is the same type of technical exercise that was necessary for the Schrödinger equation, and once again we give only the result here:

$$L^2 = -\hbar^2\left(\frac{\partial^2}{\partial\theta^2} + \cot\theta\frac{\partial}{\partial\theta} + \frac{1}{\sin^2\theta}\frac{\partial^2}{\partial\varphi^2}\right). \tag{9–25}$$

Notice that this operator involves *only* the angles and *not* the radial coordinate.

The $z$-component of the angular momentum takes a particularly simple form in spherical coordinates, one that will turn out to be interesting in what follows. This quantity works out to be

$$L_z = -i\hbar\frac{\partial}{\partial\varphi}. \tag{9–26}$$

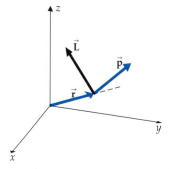

• **Figure 9–5** The angular momentum $\vec{L} = \vec{r} \times \vec{p}$ of a particle at position $\vec{r}$ moving with momentum $\vec{p}$. The vector $\vec{L}$ is perpendicular to both $\vec{r}$ and $\vec{p}$, and its direction is specified by the right-hand rule.

$L_z$ is simple because the $z$-axis plays a special role in spherical coordinates: The polar angle $\theta$ is measured from it.

### Eigenvalue Equations for $L^2$ and $L_z$

We can imagine the existence of wave functions that are eigenfunctions of the quantum mechanical operators $L^2$ and $L_z$. These wave functions would obey the respective equations

$$(L^2)(\text{wave function}) = (\text{eigenvalue of } L^2)(\text{wave function}) \qquad (9\text{--}27)$$

and

$$(L_z)(\text{wave function}) = (\text{eigenvalue of } L_z)(\text{wave function}). \qquad (9\text{--}28)$$

At this point, we know the significance of a wave function that obeys either of these equations: A particle described by that wave function has a definite value for the square of the angular momentum or for the $z$-component of the angular momentum, according to the equation that is obeyed. This means that a measurement of the square of the angular momentum or of the $z$-component of the angular momentum if the system is described by that wave function gives a single, definite value of the respective quantity.

Now, *we already know the eigenfunctions and possible eigenvalues of $L^2$ and $L_z$.* We start by comparing Eq. (9–6) with Eq. (9–27), where the operator $L^2$ is given by Eq. (9–25). We see immediately that $Y_{\ell m}(\theta, \varphi)$ is an eigenfunction of $L^2$. Moreover, from Eq. (9–18), the corresponding eigenvalues of $L^2$ are $+\hbar^2 \ell(\ell + 1)$. [Note that we can already see from Eq. (9–25) that $L^2$ is of the form $\hbar^2 \times$ (a dimensionless quantity), so the eigenvalues must be of the same form.] That is,

$$L^2 Y_{\ell m}(\theta, \varphi) = \hbar^2 \ell(\ell + 1) Y_{\ell m}(\theta, \varphi), \text{ where } \ell = 0, 1, 2, 3, \dots. \qquad (9\text{--}29)$$

Given the significance of the eigenvalues of the square of the angular momentum, we can see why the quantum number $\ell$ is called the **angular-momentum quantum number**.

Now, what about $L_z$? For this, it suffices to square Eq. (9–28), which then becomes

$$L_z^2(\text{eigenfunction of } L_z) = (\text{eigenvalue}^2)(\text{eigenfunction of } L_z). \qquad (9\text{--}30)$$

But from Eq. (9–26),

$$L_z^2 = -\hbar^2 \frac{\partial^2}{\partial \varphi^2}. \qquad (9\text{--}31)$$

We can now compare Eq. (9–30), where $L_z^2$ is given by Eq. (9–31), with Eq. (9–9). All the $\varphi$ dependence in $Y_{\ell m}(\theta, \varphi)$ is contained in $\Phi(\varphi)$. We conclude that $Y_{\ell m}(\theta, \varphi)$ is an eigenfunction of $L_z$ with a corresponding eigenvalue $\hbar m$. That is,

$$L_z Y_{\ell m}(\theta, \varphi) = \hbar m Y_{\ell m}(\theta, \varphi), \text{ where } m \text{ is any integer from } -\ell \text{ to } +\ell. \qquad (9\text{--}32)$$

The physical role of the quantum number $m$ will become clear later in this chapter. For now we can say that our results confirm, and even go beyond, Bohr's inspired guess. The two possible signs for $m$—the fact that $m$ can be either positive or negative—correspond to two possible senses of rotation about the $z$-axis: counterclockwise for positive eigenvalues of $L_z$ and clockwise for negative eigenvalues.

↓ **Example 9–2**   Verify that the spherical harmonic $Y_{11}(\theta, \varphi)$ is an eigenfunction of $L^2$ and of $L_z$ with quantum numbers $\ell = 1$ and $m = 1$ by using the explicit form of $Y_{11}$ (Table 9–1).

**Solution**   From the table, we have

$$Y_{11}(\theta, \varphi) = \sqrt{\frac{3}{8\pi}}\, \sin\theta \exp(i\varphi),$$

and we want to act on this function with the operators $L^2$ [Eq. (9–25)] and $L_z$ [Eq. (9–26)]. The $L^2$ equation is then

$$L^2 Y_{11}(\theta, \varphi) = -\hbar^2\left(\frac{\partial^2}{\partial\theta^2} + \cot\theta\,\frac{\partial}{\partial\theta} + \frac{1}{\sin^2\theta}\frac{\partial^2}{\partial\varphi^2}\right)\sqrt{\frac{3}{8\pi}}\,\sin\theta \exp(i\varphi)$$

$$= -\hbar^2\sqrt{\frac{3}{8\pi}}\left[-\sin\theta + \cot\theta(\cos\theta) + \frac{1}{\sin^2\theta}(\sin\theta)i^2\right]\exp(i\varphi).$$

The product $\cot\theta(\cos\theta) = (\cos\theta/\sin\theta)\cos\theta = \cos^2\theta/\sin\theta$; in addition, $i^2 = -1$. Then the quantity in square brackets is

$$-\sin\theta + \frac{\cos^2\theta}{\sin\theta} - \frac{1}{\sin\theta} = -\sin\theta + \frac{1}{\sin\theta}(\cos^2\theta - 1)$$

$$= -\sin\theta + \frac{1}{\sin\theta}(-\sin^2\theta) = -2\sin\theta.$$

Thus

$$L^2 Y_{11} = 2\hbar^2\sqrt{\frac{3}{8\pi}}\,\sin\theta \exp(i\varphi).$$

We can recognize $Y_{11}$ on the right again, verifying that it is an eigenfunction of $L^2$. The eigenvalue is the factor multiplying $Y_{11}$, namely $2\hbar^2$. If we compare this factor with the form $\hbar^2\ell(\ell + 1)$, we see that $\ell = 1$.

We next want to do the same for $L_z$:

$$L_z Y_{11}(\theta, \varphi) = -i\hbar\,\frac{\partial}{\partial\varphi}\sqrt{\frac{3}{8\pi}}\,\sin\theta \exp(i\varphi) = -i\hbar(i)\sqrt{\frac{3}{8\pi}}\,\sin\theta \exp(i\varphi)$$

$$= \hbar\sqrt{\frac{3}{8\pi}}\,\sin\theta \exp(i\varphi).$$

Again, $Y_{11}$ reappears on the right side, showing that it is an eigenfunction of $L_z$, and the eigenvalue is $+\hbar$. Comparing this with $\hbar m$, we see that $m = 1$.

Our results, which hold for any particle in a central potential, and hence for an electron under the influence of the coulomb force supplied by a nucleus, are of such great importance that we want to emphasize them by repetition. We refer here to an electron. The rules of quantum mechanics imply that if the angular part of the electron wave function is $Y_{\ell m}(\theta, \varphi)$, then

a measurement of the square of the angular momentum of the electron in this state will give the definite value $\hbar^2\ell(\ell + 1)$, where $\ell = 0, 1, 2, 3, \ldots$, while a measurement of the z-component of the angular momentum of the electron in its orbit will give the definite value $\hbar m$, where $m = \ell, (\ell - 1), \ldots - (\ell - 1), -\ell$.

In Section 9–4, we shall discuss how the quantization of angular momentum can be experimentally verified. The procedure makes use of the fact that charged

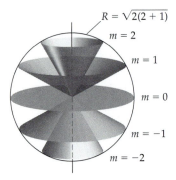

$R = \sqrt{2(2 + 1)}$

$m = 2$

$m = 1$

$m = 0$

$m = -1$

$m = -2$

• **Figure 9–6**    A semiclassical depiction of the angular-momentum vector for angular-momentum quantum number $\ell = 2$ and its possible $z$-components. The vector lies on the surface of one of the cones that terminate on the five circles specified by $m = 2, 1, 0, -1$, and $-2$, respectively. The presence of these cones implies that the $x$- and $y$-components of the angular momentum are not separately specified.

particles with angular momentum, such as electrons in an atom, act as magnetic dipoles and thus can be manipulated with a magnetic field.

The limitation that the absolute value of the quantum number $m$ is limited by the quantum number $\ell$ comes from the fact that the $z$-component of the angular momentum cannot exceed the magnitude of the angular momentum vector itself. In classical mechanics, it is possible for $L_z$ and $|\vec{L}|$ to coincide—this happens when $\vec{L}$ points along the $z$-axis. But since in quantum mechanics the magnitude of the angular-momentum vector is $\hbar[\ell(\ell + 1)]^{1/2}$, a quantity larger than the maximum value of $L_z$, we conclude that in quantum mechanics $L_z$ and $|\vec{L}|$ cannot coincide. In other words, the electron in its orbit cannot have an angular momentum that is perfectly aligned along the $z$-axis. We can picture this limitation by drawing a sphere, the square of whose radius is $\ell(\ell + 1)$—the common factor $\hbar^2$ can be dropped for simplicity. The vector $\vec{L}$ that is maximally aligned with the $z$-axis does not point along that axis. Rather, its tip lies on a <u>cone around</u> the $z$-axis whose height is $\ell$. The radius of the cone is $\sqrt{\ell(\ell + 1) - \ell^2} = \sqrt{\ell}$ (•Fig. 9–6). The angular-momentum vector with $m = \ell - 1$ lies on another cone, as do all the other possible vectors, ending with the $m = -\ell$ cone pointing downwards.

It is clear that for $\ell$ large the classical approximation that an angular momentum is a vector that can point in any direction you like is very good. For angular momenta on the order of $1 \text{ kg} \cdot \text{m}^2/\text{s}$, $\ell = O[10^{34}]$!

Finally, for a given value of $\ell$, $m$ can take on the values $0, \pm 1, \pm 2, \dots, \pm \ell$. Thus for a given value of $\ell$, there are $(2\ell + 1)$ different eigenfunctions $Y_{\ell,\ell}, \dots, Y_{\ell,\ell-1}, Y_{\ell,-\ell}$. In general, whenever there are a number of different solutions that correspond to a fixed eigenvalue, we say that we have a **degeneracy**. In this case, we have a $(2\ell + 1)$-fold degeneracy.

## 9–3    Allowed Energies and Electron Spatial Distribution in the Hydrogen Atom

To this point we have been quite general in our treatment of central forces. Now we want to concentrate on hydrogen, the simplest atom. Hydrogen consists of a single electron, with mass $m_e$ and charge $-e$, bound to a nucleus some 2,000 times more massive than the electron with charge $+Ze$, where $Z = 1$.[3] By studying the dynamics of this comparatively simple system, not only will we be seeing whether we can understand hydrogen, but we will also be laying the groundwork for a description of multielectron atoms.

We have already seen that the Schrödinger equation leads to a wave function whose angular part is described by $Y_{\ell m}(\theta, \varphi)$. The angular momentum of the hydrogen electron is accordingly quantized, with the electron having the two angular-momentum quantum numbers $\ell$ and $m$. This part of the electron's wave function describes how the probability of finding the electron varies with the angle. We now want to concentrate on the radial part of the Schrödinger equation, Eq. (9–7), with $\lambda = -\ell(\ell + 1)$, $\mu = m_e$, and $V(r) = -Ze^2/(4\pi\varepsilon_0 r)$:[4]

$$-\frac{\hbar^2}{2m_e}\left[\frac{d^2}{dr^2} + \frac{2}{r}\frac{d}{dr} - \frac{\ell(\ell + 1)}{r^2}\right]R(r) - \frac{Ze^2}{4\pi\varepsilon_0 r}R(r) = ER(r). \quad (9\text{–}33)$$

As for the one-dimensional Schrödinger equation studied earlier, only certain values of $E$ and corresponding radial wave functions $R(r)$ will satisfy that equa-

---

[3] For hydrogen, $Z = 1$, but the same potential will describe any nucleus to which only a single electron is bound. We thus leave $Z$ arbitrary.

[4] See footnote 1—and don't confuse the electron mass $m_e$ with the $L_z$ quantum number $m$!

tion—it is an eigenvalue equation. Note the term $\ell(\ell + 1)/r^2$, which is the remnant of the angular dependence in the original operator $p^2/2m_e$ in the energy. This term is often referred to as a **centrifugal barrier**. The term merits this name because it appears in the same position as the potential $V(r)$, here the coulomb potential, and it has a sign opposite to that potential. It therefore acts like a potential for a repulsive force that increases with increasing angular momentum (•Fig. 9–7). As the angular momentum increases, the electron is in a sense "encouraged" to stay away from the nucleus by the centrifugal barrier.

We are interested in bound-state solutions of Eq. (9–33)—solutions with $E < 0$. Only certain values of $E$—the eigenvalues—will allow solutions, and these will be the allowed energies of hydrogen. (There are certainly solutions for $E > 0$ as well. These solutions correspond to electrons moving towards the origin from $r = \infty$ and back to $\infty$ under the influence of the coulomb potential, but they are not of interest to us here.) As for the importance of the radial wave function $R(r)$ itself, we can keep in mind that it gives the probability of finding the electron at a given radius according to Eq. (9–13).

Our eigenvalue equation is an ordinary second-order differential equation. ("Ordinary" refers to the fact that the equation is in just one variable, "second order" to the fact that no more than second derivatives enter into the equation.) Such equations have two independent solutions; for example, the equation $d^2u/dx^2 - u(x) = 0$ has the solutions $\exp(x)$ and $\exp(-x)$. The *physical* requirement that the solutions be normalizable constrains the possible solutions and eigenvalues.

We shall be less interested in working out the details of solving Eq. (9–33) than in understanding the important features of these solutions and where these features come from. [It is always possible to verify that our solutions are good ones, and to verify the eigenvalues, by direct substitution into Eq. (9–33).] But we start our discussion with a look at the eigenvalues that are associated with the eigenfunctions.

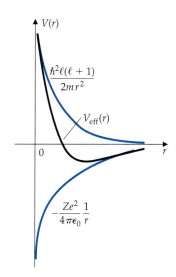

• **Figure 9–7** The effective potential is the sum of the attractive coulomb term $-Ze^2/(4\pi\varepsilon_0 r)$ and the repulsive centrifugal barrier potential $\ell(\ell + 1)\hbar^2/(2m_e r^2)$. The effect of the centrifugal barrier, which is to keep the electron away from the proton, increases with angular momentum.

## Energy Eigenvalues for Hydrogen

The possible values of energy—the eigenvalues—that emerge from the radial Schrödinger equation are given by

$$E = -\frac{1}{2} m_e \left(\frac{Ze^2}{4\pi\varepsilon_0 \hbar}\right)^2 \frac{1}{(n_r + \ell + 1)^2} = -\frac{1}{2} m_e c^2 \left(\frac{Ze^2}{4\pi\varepsilon_0 \hbar c}\right)^2 \frac{1}{(n_r + \ell + 1)^2}. \quad (9\text{–}34)$$

($E$ is actually independent of $c$, since the system is nonrelativistic, but the second form is particularly convenient.) We recognize the presence of the *fine-structure constant* $\alpha = e^2/(4\pi\varepsilon_0 \hbar c)$ first introduced [Eq. (5–6)] in the discussion of the Bohr model. In terms of $\alpha$,

$$E = -\frac{1}{2} m_e c^2 (Z\alpha)^2 \frac{1}{(n_r + \ell + 1)^2}. \quad (9\text{–}35)$$

Here the new **radial quantum number** $n_r = 0, 1, 2, \ldots$, and since $\ell \geq 0$, we must have $n_r + \ell + 1$ always a positive integer. We can then define the **principal quantum number**

$$n \equiv n_r + \ell + 1, \text{ with } n \text{ a positive integer.} \quad (9\text{–}36)$$

In terms of the principal quantum number,

$$E = -\frac{1}{2} m_e c^2 (Z\alpha)^2 \frac{1}{n^2}. \quad (9\text{–}37)$$

This is exactly the form the bound-state energies take in the Bohr model. We illustrate the spectrum in •Fig. 9–8. Note that for a given $n$, the largest allowed value of $\ell$ is $n - 1$, because $n_r + \ell + 1$ is greater than or equal to zero.

We saw that for angular momentum a $(2\ell + 1)$-fold degeneracy was associated with the $2\ell + 1$ wave functions possible for a given $\ell$. The radial equation has this same degeneracy, in that it does not at all depend on the quantum number $m$. In addition, the fact that the energy depends on the combination $n = n_r + \ell + 1$ and not on the $n_r$ and $\ell$ quantum numbers separately, while the wave function depends on both $n$ and $\ell$, implies further degeneracy. To explore this new degeneracy, let us count the number of possible states, one by one. In this count, we shall label the radial eigenfunctions by $n$ and $\ell$, namely as $R_{n\ell}(r)$. The states are as follows:

- $n = 1$: This state is possible only with $n_r = 0$ and $\ell = 0$, which represents just one state. The electron wave function for the $n = 1$ state is $R_{10}(r)Y_{00}(\theta, \varphi)$.
- $n = 2$: Among several possibilities, we have $n_r = 1$ and $\ell = 0$, a single state with wave function $R_{20}(r)Y_{00}(\theta, \varphi)$. In addition, we have $n_r = 0$ and $\ell = 1$. This pair of values corresponds to *three* states, namely those for which $m = +1, 0,$ and $-1$. These states all have the same energy. The three wave functions are $R_{21}(r)Y_{1m}(\theta, \varphi)$ with the three values of $m$. In all, there are four degenerate states, each with energy corresponding to $n = 2$.
- $n = 3$: The possibility $n_r = 2$ and $\ell = 0$ corresponds to the one state $R_{30}(r)Y_{00}(\theta, \varphi)$. We also have $n_r = 1$ and $\ell = 1$, which corresponds to the three states $R_{31}(r)Y_{1m}(\theta, \varphi)$ with $m = 1, 0, -1$. Finally, we have $n_r = 0$ and

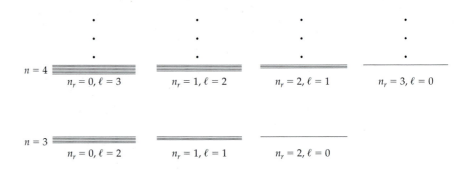

•**Figure 9–8**  Energy levels for hydrogen. The degenerate levels are drawn slightly apart to show their multiplicity, but to the approximation treated here, they are strictly degenerate. The spacing between the nondegenerate levels is not drawn to scale.

$\ell = 2$, which corresponds to the five states $R_{32}(r)Y_{2m}(\theta, \varphi)$ with $m = 2, 1, 0,$ $-1,$ and $-2$. Thus there are a total of $1 + 3 + 5 = 9 = 3^2$ degenerate states, all with the energy corresponding to $n = 3$.

If we had made a similar count for $n = 4$, we would find 16 states, and for any $n$, we would find $n^2$ states. We will have to return to the problem of how we observe such a degeneracy, since, on the face of it, the frequency of radiation emitted in a transition depends only on the energy difference, and not on the number of wave functions that have the same energy.

The $2\ell + 1$ degeneracy in $m$ for a given $\ell$ is common to all central potentials. But the additional degeneracy in $\ell$ is special to the $1/r$ potential. Even a slight modification of this potential will lift the degeneracy. As we shall discuss in Chapter 11, the degeneracy is lifted, or *broken*, in atoms more complex than hydrogen.

We have left out one ingredient that increases the degeneracy further: Electrons have an additional internal property—a form of angular momentum—that makes them come in two states. The energy of the hydrogen atom, at least to the approximation we have employed to this point, is independent of these states. Thus the degeneracy in our hydrogen atom is $2n^2$, not $n^2$. We shall come back to this point when we discuss spin.

## Radial Eigenfunctions for Hydrogen

*Large- and Small-r Behavior*   While a full systematic solution of Eq. (9–33) is more than we want to do, we can see some general features of $R(r)$ without too much difficulty. The next example shows how we can understand the behavior of the wave functions at large $r$.

---

**Example 9–3**   Examine Eq. (9–33) for values of $r$ large enough so that the $1/r$ and the $1/r^2$ terms in the equation can be neglected. Find the general solution under these conditions. What does the normalization condition of Eq. (9–15) have to say about the acceptability of your solutions?

**Solution**   Under the conditions described, Eq. (9–33) takes the form

$$-\frac{\hbar^2}{2m_e}\frac{d^2R(r)}{dr^2} = ER(r).$$

Now for bound states, $E$ is negative; that is, $E = -|E|$. Our equation thus takes the form

$$\frac{d^2R}{dr^2} = \frac{2m_e|E|}{\hbar^2}R.$$

This equation is easy to solve. The function that gives itself back when differentiated is an exponential. If we try

$$R(r) = C\exp(\pm Ar),$$

where $A$ is positive, and if we take two derivatives, we find that

$$\frac{d^2R}{dr^2} = C(A^2)\exp(\pm Ar).$$

Hence the general solution in this case is

$$R(r) = C_1\exp(+Ar) + C_2\exp(-Ar),$$

where

$$A^2 = \frac{2m_e|E|}{\hbar^2}.$$

There are indeed two independent solutions, signaled by the presence of the two arbitrary constants $C_1$ and $C_2$.

The normalization condition plays a crucial role in eliminating one of the two independent solutions. For large $r$, the $C_1$ term is much larger than the $C_2$ term. As a matter of fact, the integral over all space of the $C_1$ term is infinite, due to the growing exponential behavior of the solution at infinity. This term is not renormalizable, and the only way we can have a sensible probabilistic interpretation is to choose $C_1 = 0$.

The correct solution behaves as $\exp(-\lambda r)$ for large $r$.

---

The falling exponential behavior described in the previous example is a very general feature of bound-state wave functions. We can use the result of that example and insert the value of $E$ that is special to the potential energy of the hydrogen atom given by Eq. (9–37). If, in addition, we use the definition of the Bohr radius given in Eq. (5–9), viz.,

$$a_0 = \frac{\hbar}{m_e c \alpha},$$

then with a little algebra we can see that the exponential factor in the hydrogen wave function must take the large-$r$ form, so that

$$R(r) \propto \exp(-Zr/na_0).$$

The treatment of the large-$r$ behavior does not allow us to find polynomial behavior—powers of $r$—that multiplies the falling exponential. But just as we were able to find the large-$r$ behavior, we can also find the small-$r$ behavior. This is done in the next example.

---

**Example 9–4** Examine Eq. (9–33) for $r$-values so small that the constant term $ER(r)$ and the coulomb term can be neglected in comparison with the $1/r^2$ term. What does the solution $R(r)$ look like, and what is the role of the normalization condition? (*Hint:* Try a power solution $r^b$.)

**Solution** Under the conditions described, Eq. (9–33) takes the form

$$\left(\frac{d^2}{dr^2} + \frac{2}{r}\frac{d}{dr} - \frac{\ell(\ell+1)}{r^2}\right)R(r) = 0.$$

The hint suggests that we try $R(r) = Cr^b$. When we insert this attempted solution into the equation, we find that

$$b(b-1)r^{b-2} + 2br^{b-2} - \ell(\ell+1)r^{b-2} = 0$$

or, canceling the factor $r^{b-2}$ and rearranging,

$$b(b+1) - \ell(\ell+1) = 0.$$

This is a quadratic equation for $b$; it tells us what values of $b$ are allowed. You can verify that the two possible values of $b$ that will satisfy the equation are

$$b = \ell \text{ and } b = -\ell - 1.$$

Thus the most general solution of our eigenvalue equation for small $r$ is

$$R(r) = C_1 r^\ell + C_2 r^{-\ell-1}.$$

At this point we want to think about the normalization condition, Eq. (9–15). The integrand in this equation is $r^2 R(r)^2$. Thus for small $r$ (near $r = 0$) the two solutions behave as $C_1 r^{2+2\ell}$ and $C_2 r^{2-2(\ell+1)} = C_2 r^{-2\ell}$. But the integral for the $C_2$ term is infinite at the lower limit, $r = 0$, for $\ell$ a positive integer, so that it is unacceptable, and we must set $C_2 = 0$. Even for $\ell = 0$, a non-zero value of $C_2$ is unacceptable, because with $C_2 \neq 0$ the kinetic energy would have an infinite expectation value.

We conclude that $R$ behaves as $r^\ell$ for very small values of $r$.

The foregoing example, which shows that the radial wave function behaves as $r^\ell$ for small values of $r$, tells us that the probability of finding the electron at the origin is zero for all values of $\ell$ except $\ell = 0$. The larger the value of $\ell$, the less likely it is that the electron is near the origin. This notion is at least consistent with a classical picture, in which a non-zero value of angular momentum implies an orbit about, but not through, the center of force.

*The Structure of the Radial Wave Functions*    Using more advanced methods, one finds that the radial wave function takes the general form

$$R_{n\ell} = C\left(\frac{Z}{na_0}\right)^{3/2}\left(\frac{Zr}{a_0}\right)^\ell$$

$$\times \left(\text{order } (n - \ell - 1) \text{ polynomial in } Zr/a_0\right)\exp(-Zr/na_0) \qquad (9\text{–}38)$$

This function is real; complex factors in the full wave function are associated only with $Y_{\ell m}(\theta, \varphi)$, the angular part of the wave function. Since $n - \ell - 1 = n_r$, the integer $n_r$ may be interpreted as the number of times that the function $R(r)$ goes through zero (aside from $r = 0$); that is, $n_r$ *is the number of nodes in the eigensolution.* This interpretation is in accord with the idea that the higher the number of nodes, the higher is the corresponding energy. (See Chapter 8.) Table 9–2 presents the first few radial functions $R_{n\ell}(r)$, properly normalized.

**Table 9–2    Radial Eigenfunctions (See Also Fig. 9–9)**

$$R_{10} = 2\left(\frac{Z}{a_0}\right)^{3/2}\exp\left(-\frac{Zr}{a_0}\right)$$

$$R_{20} = 2\left(\frac{Z}{2a_0}\right)^{3/2}\left(1 - \frac{Zr}{2a_0}\right)\exp\left(-\frac{Zr}{2a_0}\right)$$

$$R_{21} = \frac{1}{\sqrt{3}}\left(\frac{Z}{2a_0}\right)^{3/2}\left(\frac{Zr}{a_0}\right)\exp\left(-\frac{Zr}{2a_0}\right)$$

*Calculating Averages*    Given the radial wave functions, we can use their probabilistic interpretation to find averages involving the electron radial distance in hydrogen. The next two examples illustrate the process.

**Example 9–5**    Find the average distance $\langle r \rangle$ of the electron from the nucleus for the ground state of hydrogen [i.e., the state with $\hbar^2\ell(\ell + 1) = 0$ and $n = 1$].

**Solution**    Recall that if a state is described by a wave function $\psi(\vec{r})$, then the expectation value of any function of position $f(\vec{r})$ is

$$\langle f(\vec{r}) \rangle = \int \psi^*(\vec{r})f(\vec{r})\psi(\vec{r})\, d^3\vec{r}.$$

In this example, the function of position is just $r$. With no angular dependence in $f$, the angular integration can be done "for free," as follows: The volume element $d^3r$ in spherical coordinates breaks into an angular and a radial part: $d^3r = \sin\theta\, d\theta\, d\varphi\, r^2\, dr$. Then if the angular dependence of the wave function is $Y_{\ell m}(\theta, \varphi)$—here $\ell = 0$ and $m = 0$, but that is not important—the normalization of the wave function guarantees that

$$\iint |Y_{\ell m}(\theta, \varphi)|^2 \sin\theta\, d\theta\, d\varphi = 1,$$

with limits as previously described. That leaves us with an integral found in tables of integrals:

$$\langle r \rangle = \int_0^\infty R_{10} r R_{10} r^2 \, dr = \frac{4Z^3}{a_0^3} \int_0^\infty r^3 \, dr \exp\left(-\frac{2Zr}{a_0}\right) = \frac{3}{2}\frac{a_0}{Z}.$$

The Bohr radius sets the scale for the expectation value of the radial position. The larger the value of $Z$, the smaller $\langle r \rangle$ is: A more highly charged nucleus pulls the electron in.

**Example 9–6**   Calculate $\langle 1/r \rangle$ for an electron in the ground state of hydrogen, and use your result to calculate the average kinetic energy $\langle p^2/2m \rangle$.

**Solution**   We have

$$\left\langle \frac{1}{r} \right\rangle = \int_0^\infty R_{10}(r)\frac{1}{r} R_{10}(r) r^2 \, dr = \frac{4Z^3}{a_0^3} \int_0^\infty \exp\left(-\frac{2Zr}{a_0}\right) r \, dr = \frac{1}{a_0}.$$

Now,

$$\langle V(r) \rangle = \left\langle -\frac{Ze^2}{4\pi\varepsilon_0 r} \right\rangle = -\frac{Ze^2}{4\pi\varepsilon_0}\left\langle \frac{1}{r} \right\rangle = -\frac{Ze^2}{4\pi\varepsilon_0 a_0} = -\frac{Z^2 e^2 mc\alpha}{4\pi\varepsilon_0 \hbar} = -Z^2\alpha^2 mc^2.$$

Hence

$$\left\langle \frac{p^2}{2m} \right\rangle = \langle K \rangle = E - \langle V(r) \rangle = -\frac{\left(Z^2\alpha^2 mc^2\right)}{2} - \left(-Z^2\alpha^2 mc^2\right) = \frac{Z^2\alpha^2 mc^2}{2}.$$

In other words, $\langle K \rangle = -(1/2)\langle V(r) \rangle$. It turns out that this condition obtains not only for the ground state, but for all bound states in the hydrogen atom. It is the quantum mechanical equivalent of what in classical mechanics is known as the "virial theorem," which relates the average over time of the kinetic and potential energies. Our result is an illustration of how quantum theoretical expectation values for operators like position and momentum reproduce the classical results.

*Peaks in the Radial Distribution*   Apart from the exponential drop-off factor characteristic of all bound states, the radial wave function is a polynomial of degree $n_r = n - \ell - 1$, and it thus has $n_r$ zeros, or nodes. In between the nodes there must be bumps, or peaks, in the probability density distribution $r^2[R_{n\ell}(r)]^2$ (•Fig. 9–9). The most interesting case is that where, for a given $n$, $\ell$ takes on its largest value, $n - 1$. For this case, the radial wave function has the form

$$R_{n,\,n-1} = C' r^{n-1} \exp\left(-\frac{Zr}{na_0}\right), \tag{9–39}$$

where the constant $C'$ is determined by normalization. This wave function is zero only at the origin and at infinity, and it has only one maximum. The function corresponds to circular orbits in the Bohr atom, and the probability density $r^2R^2$ will peak when

$$\frac{d}{dr}\left(r^2 r^{2n-2}\exp\left(-\frac{2Zr}{a_0 n}\right)\right) = \frac{d}{dr}\left(r^{2n}\exp\left(-\frac{2Zr}{a_0 n}\right)\right) = 0. \tag{9–40}$$

By taking the derivative, we find that the peak occurs for

$$r = \frac{n^2 a_0}{Z}. \tag{9–41}$$

This is indeed the radius of the orbit in the Bohr atom. [See Eq. (5–10).]

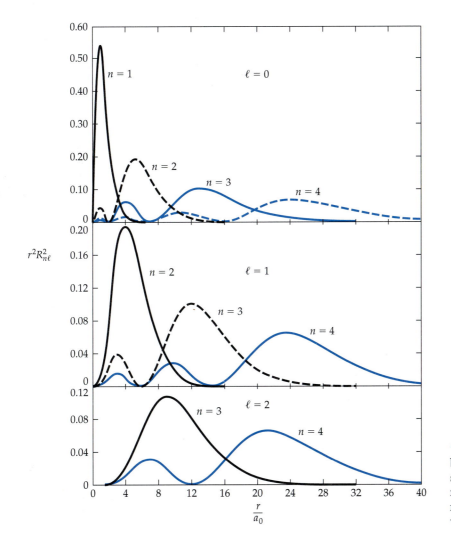

● **Figure 9–9** The radial probability distribution $r^2[R_{n\ell}(r)]^2$ for several values of $n$ and $\ell$. The number of nodes is $n - \ell - 1$, not counting the node at $r = 0$, which is due to the $r^2$ factor.

## 9–4 The Zeeman Effect

We have seen that for any given pair of quantum numbers $(n_r, \ell)$—or equivalently, $(n, \ell)$—there are $(2\ell + 1)$ states with the same energy $E_n$. These states correspond to different values of the $z$-component of the angular momentum—that is, to the degeneracy of the states having the quantum numbers $m = \ell$, $\ell - 1, \ell - 2, \ldots, -\ell$. Is this degeneracy ever detectable? We know how to measure the energy of a state, or, more precisely, the difference between the energies of two states: We observe the frequencies of the radiation emitted when the atom is excited and when it subsequently makes a transition from the higher energy state to the lower energy state. But if the states are degenerate, the frequencies that could otherwise distinguish them are identical.

The tool that is needed to distinguish the states is provided by the fact that when the hydrogen atom is placed in a magnetic field, each of these previously degenerate states has a slightly different energy, depending on the value of $m$. We say that the states are **split**. To see how this works we must understand how an electron in orbit has properties that make it sensitive to a magnetic field.

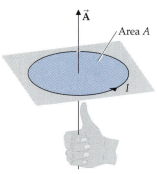

• **Figure 9–10**   If a current makes a small planar loop, the area $A$ that the loop encloses has a direction associated with it. The vector $\vec{A}$ is perpendicular to the loop and points in a direction specifed by a right-hand rule: The fingers of the right hand follow the current, and the right thumb points in the direction of $\vec{A}$.

## The Connection between Magnetic Moments and Angular Momentum

In Chapter 1 we reviewed the magnetic properties of a current loop. A closed circuit carrying a current $I$ and with an area $\vec{A}$ (magnitude $A$ and direction perpendicular to the area, with the right-hand rule associating the positive direction with the flow of the current, as in •Fig. 9–10) acts as a magnetic dipole, with dipole moment[5]

$$\vec{\mu} = I\vec{A}.$$

The magnetic moment [see the discussion immediately after Eq. (1–59)] is a vector pointing in the direction of $\vec{A}$.

Consider an electron, mass $m_e$ and charge $-e$, moving with velocity $v$ in a circle of radius $r$. The magnitude of the current is the rate at which charge flows past a point, and that rate is given by $e/T$, where $T$ is the period of the circulation of the charge. Since $T = 2\pi r/v$, the current $I$ is given by $ev/2\pi r$. The area of the loop is $\pi r^2$, so that the magnetic moment has magnitude

$$\mu = \frac{ev}{2\pi r}\,\pi r^2 = \frac{1}{2}\,evr = \frac{em_e vr}{2m_e} = \frac{e}{2m_e}L, \tag{9–42}$$

where $L$ is the magnitude of the electron's classical angular momentum about the origin. The direction of $L$ is the same as the direction that we ascribe to the area $\vec{A}$ when we use the right-hand rule (•Fig. 9–10). Thus, in Eq. (9–42), we may convert $\mu$ and $L$ to vectors. Since the charge on the electron is negative, $\vec{\mu}$ points in a direction opposite to that of $\vec{L}$. When we are not dealing with a simple point charge in an orbit, the magnetic moment need not have the value that Eq. (9–42) gives. On dimensional grounds it must remain proportional to $(q/2m)\vec{L}$, where $m$ is any generic mass and $q$ is any generic charge, but there may be a numerical factor involved, which we denote by $g$ and which is called the **gyromagnetic ratio**, or $g$-factor. Quite generally, therefore,

$$\vec{\mu} = \frac{gq}{2m}\,\vec{L}. \tag{9–43}$$

We assume that this classical formula remains true in a quantum theory. In quantum mechanical systems such as atoms and molecules the concept of the "orbit" is no longer valid, but, as we have seen, a state does indeed have angular momentum. Thus we can use Eq. (9–43) for a general description of the magnetic dipole moment of a microscopic system.

When the $z$-component of an atomic electron's angular momentum is quantized, a measurement would reveal this quantity to be an integer times $\hbar$, so that a measurement of $\mu_z$ contains the coefficient $e\hbar/2m_e$. This quantity, which is called the **Bohr magneton**, is given by

$$\mu_B \equiv \frac{e\hbar}{2m_e} = 9.274 \times 10^{-24}\ \text{J/T}. \tag{9–44}$$

Note that a system need not have a net charge to have a dipole moment. For example, an atom is electrically neutral, but has a dipole moment because one component of the charge circulates while the other component remains stationary. A neutron, even though it is electrically neutral, also has a magnetic dipole moment whose form is given by Eq. (9–43). The factor $q$, usually taken to be the fundamental charge $e$, is present for dimensional reasons, the mass will

---

[5]From this point on, the Greek letter "mu" ($\mu$) will refer to a magnetic moment, *not* to a reduced mass.

be the neutron mass, the factor $g$ represents the balanced charges circulating in the neutron that are responsible for the dipole moment, and $\vec{\mathbf{L}}$ is the neutron's *internal* angular momentum. Recall from Chapter 1 that the expression for the energy of a current loop (or, equivalently, a magnetic dipole) when the loop is in an external magnetic field $\vec{\mathbf{B}}$ is [Eq. (1–60)]

$$U = -\vec{\boldsymbol{\mu}} \cdot \vec{\mathbf{B}}. \tag{9–45}$$

**Example 9–7**  Consider an atomic electron with angular-momentum quantum number $\ell = 3$, placed in a magnetic field $\vec{\mathbf{B}}$. Show that if only the energy of the dipole interaction is taken into account, then the energy values of the electron are

$$E = \frac{me\hbar B}{2m_e}, \quad \text{where } m = 3, 2, 1, 0, -1, -2, \text{ and } -3.$$

What happens when the field $\vec{\mathbf{B}} \to 0$?

**Solution**  When we say that an electron has angular-momentum quantum number $\ell$, we mean that the eigenvalue of the operator $L^2$ is $\ell(\ell + 1)\hbar^2$. In this case, the eigenvalues of $L_z$ are, in units of $\hbar$, $\ell, \ell - 1, \ell - 2, \ldots, 0, -1, -2, \ldots$, and $-\ell$. We can treat an electron as an orbiting pointlike object, so that $g = 1$ and $q = -e$. We define the z-axis by the direction of the vector $\vec{\mathbf{B}}$, so that the components of $\vec{\mathbf{B}}$ are $(0, 0, B)$. Thus $\vec{\mathbf{L}} \cdot \vec{\mathbf{B}} = L_z B$, and the energy becomes

$$U = \left(\frac{e}{2m_e}\right) B\hbar m.$$

With $\ell = 3$, the $m$-values are $3, 2, 1, 0, -1, -2$, and $-3$.

When $B = 0$, these seven energy levels coalesce into a single one with $U = 0$. The energy levels do not disappear; rather, the energy differences between them drop to zero, and we describe the resulting level as sevenfold degenerate.

### Hydrogen in Magnetic Fields and the Zeeman Effect

At this point we know that an electron in an orbit acts as a magnetic dipole with magnetic dipole moment $\vec{\boldsymbol{\mu}} = e\vec{\mathbf{L}}/2m_e$ and that the energy $U$ of such a dipole in an external magnetic field is given by $-\vec{\boldsymbol{\mu}} \cdot \vec{\mathbf{B}}$. If we define the z-axis as the direction of $\vec{\mathbf{B}}$, then the scalar product between $\vec{\mathbf{L}}$ and $\vec{\mathbf{B}}$ picks out $L_z$, and the energy is

$$U = -\frac{eB}{2m_e} L_z. \tag{9–46}$$

This potential energy must be put into the Schrödinger equation alongside the coulomb potential energy. But that poses no problem: *The angular wave function $Y_{\ell m}(\theta, \varphi)$ that we found for the case $B = 0$ is already an eigenfunction of the operator $L_z$ with eigenvalue $m\hbar$.* This means that the magnetic-energy term in the operator for energy acts as an ordinary algebraic quantity, and the electron's energy in the magnetic field is

$$E = E_n - \frac{eB\hbar}{2m_e} m, \quad \text{where } m = \ell, \ell - 1, \ldots, -\ell. \tag{9–47}$$

We sketch the various values of the energies in •Fig. 9–11. The degenerate energies of each state with a given $\ell$-value have been split into $2\ell + 1$ different energies in the presence of the magnetic field. This splitting is known as the **Zeeman effect**. Each of the levels is split by the same amount from its neighbors, and the amount of the splitting is proportional to the magnetic field $B$.

$n_r = 0, \ell = 2$

$m = -2$
$m = -1$
$m = 0$
$m = 1$
$m = 2$

$m = -1$
$m = 0$
$m = 1$

$n_r = 0, \ell = 1$

• **Figure 9–11** The splittings of otherwise degenerate levels when an atom is placed in a magnetic field. This is the Zeeman effect. All possible transitions are sketched in, although selection rules will make some of the transitions much less probable than the allowed ones. Note that all the level splittings are of the same magnitude, $eB\hbar/2m_e$.

$m = 0$

$n_r = 0, \ell = 0$

For example (•Fig. 9–11), an $\ell = 1$ state is split into three, while an $\ell = 0$ state is not split. This means, in turn, that transitions between the three $\ell = 1$ states to a single $\ell = 0$ state gives light of three distinct frequencies. In other words, there are three spectral lines for such a transition, rather than one. As we shall describe shortly, this effect is observed, and the degeneracy is real.

It is perhaps useful to estimate the size of the splittings compared with typical electronic energies, which are on the order of one to several eV. A reasonable laboratory magnet will produce a field of a couple of tesla. The factor $eB\hbar/2m_e$ then turns out to be roughly $10^{-4}$ eV.

Finally, note that we described the degeneracy in the states of different $m$-values as having something to do with a symmetry, in this case a symmetry in direction. With the inclusion of the external magnetic field, there now is a preferred direction—that of the magnetic field—in the problem, and the energy can depend on how the angular momentum is "aligned" along the preferred direction. The presence of the external magnetic field thus "breaks the symmetry" of the situation.

**Example 9–8**  Consider a gas of hydrogen atoms in the $n = 2$, $\ell = 1$ state. The gas is placed in a constant magnetic field of strength 4.0 T, and the three otherwise degenerate levels corresponding to $m = -1, 0$, and $+1$ are split by the magnetic field. What is the spacing between the three levels? For what value of $B$ would the splitting be comparable to the splitting between the ground state and these $n = 2$, $\ell = 1$ levels in the absence of $\vec{\mathbf{B}}$?

**Solution**  Suppose the $z$-axis is defined by the orientation of $\vec{\mathbf{B}}$. Then the corresponding energy of the electron, which has an orbital angular momentum corresponding to $\ell = 1$, is

$$E_{\text{mag}} = \left(\frac{e}{2m_e}\right)B\hbar m,$$

where the $m$-values are 1, 0, and −1. This equation gives three different energies, according to the value of $m$. The splitting between these energies is then given by the difference in the $m$-values, namely 1:

$$\Delta E_{\text{mag}} = \left(\frac{e}{2m_e}\right) B\hbar = (9.3 \times 10^{-24}\,\text{J/T})(4.0\,\text{T}) = 3.7 \times 10^{-23}\,\text{J} = 2.3 \times 10^{-4}\,\text{eV}.$$

This splitting can be compared with the spacing, in the absence of $\vec{\mathbf{B}}$, between the ground state ($n = 1$, $\ell = 0$), with energy −13.6 eV, and the levels $n = 2$, $\ell = 1$, for which the energy is $(-13.6\,\text{eV})/n^2 = (-13.6\,\text{eV})/4 = -3.4\,\text{eV}$. The spacing between these states is some 10 eV. The strength of the $B$-field would have to be a factor of about $5 \times 10^4$ larger than 4 T to attain this same splitting. Such a $B$-field is not even remotely available—at least not on Earth!

## Experimental Observation of the Zeeman Effect

In 1862, Michael Faraday made an early attempt to observe the effects of a magnetic field on a source of light, but his apparatus was not able to distinguish the small splitting produced by the field. To detect such splitting requires a good diffraction grating, with many slits, to resolve closely spaced spectral lines, and it was not until the 1870s that the American Henry Rowland made that possible with his invention of ruling machines for gratings. It was for the then young Dutch physicist Pieter Zeeman, in 1896, to observe the effect now named after him. In •Fig. 9–12, we show spectral lines split by a magnetic field. H. A. Lorentz, who was working on an electrical theory of matter, convinced Zeeman to look for polarization in these lines; the effect was found, and they shared the Nobel prize in 1902 for their discovery. Zeeman was in no way able to understand his result. But knowledge of the presence of these lines certainly influenced the development of the quantum theory.

### The Stern–Gerlach Experiment

If the magnetic field $\vec{\mathbf{B}}$ is a constant, then so is the potential energy of a dipole within it, and the field exerts no net force on the dipole. The potential energy $U = -\vec{\boldsymbol{\mu}} \cdot \vec{\mathbf{B}}$ also exists when the $B$-field is not homogeneous, but in that case the potential energy is no longer constant. As a result, a net force will act on the dipole, with $F_x = -\partial U/\partial x$, $F_y = -\partial U/\partial y$, and $F_z = -\partial U/\partial z$.

In 1921 Otto Stern proposed an experiment that he subsequently carried out together with Walter Gerlach in 1924. The experiment used the magnetic force we have been describing to show directly the quantization of angular momentum. The magnet depicted in •Fig. 9–13 gives rise to a magnetic field that has as its largest component the $z$-component of the field. Because of the slight bending of the field lines shown in the figure, the $z$-component has a small $z$-dependence. Thus there is a force in the $z$-direction given by

$$F_z = -\frac{\partial U}{\partial z} = \mu_z \frac{\partial B_z}{\partial z}. \tag{9–48}$$

With no magnetic field, a beam of electrically neutral atoms sent through the Stern–Gerlach apparatus would leave a single trace on a photographic plate (•Fig. 9–14a). With the field present, if $\mu_z$ (equivalently $L_z$) were not quantized, each atom in the beam could be pointing in a different direction, and the trace on the plate would become smeared out (•Fig. 9–14b). What Stern and Gerlach found was that the beam was split into *discrete components*, each of which made a separate trace on the plate (•Fig. 9–14c). For example, a beam of oxygen atoms was split into five components, which suggests that the atoms had angular momentum $\ell = 2$, with the five different angles of deflection corresponding to

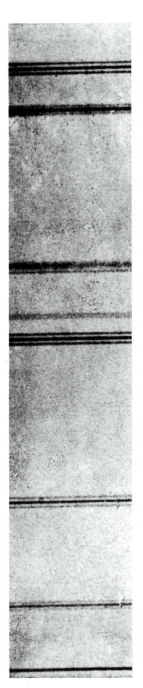

• **Figure 9–12**  Photograph of spectral lines split by a magnetic field (the Zeeman effect).

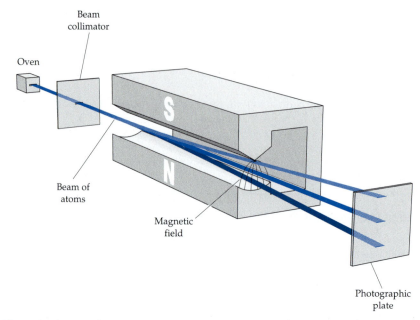

• **Figure 9–13** The Stern–Gerlach apparatus. The magnet's pole pieces are shaped to give a magnetic field that is primarily oriented in the $z$-direction, but varies for different values of $z$.

$m = 2, 1, 0, -1$, and $-2$. When the splitting of the beam into spatially distinct components was first observed, people called the phenomenon *space quantization*, a meaningless name that still appears in some places.

## 9–5 Spin

Several experimental results suggest that we have not yet accounted for all the features of hydrogen. Among them, we can cite the following:

- In atoms with many electrons, the ground state does not consist of all the electrons in the $n = 1$, $\ell = 0$ level; there are at most two electrons in any given atomic state.
- There are terms in the atomic energy levels that we cannot account for with the "ingredients" we have assembled to this point.
- In very strong magnetic fields, there are Zeeman splittings that cannot be accounted for with the $\ell$ quantum number.
- Stern–Gerlach experiments on some atoms reveal beams that split into two (silver is an example) or another even number of pieces.

The last finding suggests that silver atoms carry a magnetic moment with *two* possible values. Since the $z$-component of the magnetic moment is proportional to the $z$-component of the angular momentum, the finding further suggests, in line with the fact that an angular momentum with quantum number $\ell$ has $2\ell + 1$ possible $z$-component values, that silver atoms have an angular momentum quantum number "$\ell$" $= 1/2$. As we shall describe shortly, when we couple this notion with other pieces of evidence, we are led to the conclusion that *single electrons* have a kind of internal angular momentum called **spin**, with a quantum number analogous to $\ell$. We label this quantum number $s$; it has the value $1/2$, and its $z$-component can take on the two values $\pm 1/2$. These two possible states are alternatively labeled "up" and "down," respectively. As we

---

• **Figure 9–14** A beam of atoms in the form of a flat ray, broad in the $x$-direction, but narrow in the $z$-direction, passes through a Stern–Gerlach apparatus. The inhomogeneity in the magnetic field deflects a magnetic dipole in the $z$-direction by an amount proportional to $L_z$. **(a)** In the absence of the magnetic field, the beam passes straight through without deflection. **(b)** With a magnetic field present, the beam would be smeared in the $z$-direction if the angular momentum of the atoms took on classical (i.e., arbitrary) values. **(c)** Since the possible values of $L_z$ are quantized, so are the traces. Here we picture five allowed values of $L_z$.

shall see, the spin of the electron shows up in a variety of ways in atomic energy levels.

The notion that the electron should have spin was first proposed in 1925 by the Dutch physicists Samuel Goudsmit and George Uhlenbeck, while they were still graduate students. The spin has no classical counterpart. For an *orbiting* electron we have a classical angular momentum of the form $\vec{L} = \vec{r} \times \vec{p}$, and even in the correct quantum version, the orbital angular momentum is associated with the spatial motion of the electron. In contrast, *the spin of an electron has nothing to do with its motion.* The spin of an electron is an *intrinsic* quantum mechanical property of the electron that persists even when the electron is brought to rest.

The fact that the spin behaves like an angular momentum permits us to be precise about its properties: The spin of an electron is described by the three components of a vector operator that we denote by $\vec{S}$. When we say that the spin is $1/2$, we mean that the electron is an eigenstate of the operator $S^2$ with eigenvalue

$$S^2 \psi_{el} = s(s + 1)\hbar^2 \psi_{el} = \frac{1}{2}\left(\frac{1}{2} + 1\right)\hbar^2 \psi_{el} = \frac{3}{4}\hbar^2 \psi_{el}. \qquad (9\text{--}49)$$

(This is precisely analogous to an eigenvalue $\ell(\ell + 1)\hbar^2$ of the operator $\vec{L}^2$.) If we were to measure the $z$-component of the operator $\vec{S}$, the result would be one of the two possibilities

$$S_z = \pm\frac{\hbar}{2}. \qquad (9\text{--}50)$$

The two values of $S_z$ describe the "up" and "down" states that we referred to a few paragraphs ago. The value of $S_z$ specifies the projection of the spin onto the $z$-axis.

We might expect—again by analogy with our earlier work—that the electron spin is described by a wave function like $Y_{1/2, \pm 1/2}(\theta, \varphi)$, but in fact that does not work: There are no internal angles that can be marked, and the spin is not describable by a spatial wave function. The way in which we describe the wave function of an electron in, say, a hydrogen atom is simply to tack on a piece that denotes the spin state. Thus

$$\psi_{el}(\mathbf{r}, \pm) = R_{n\ell}(r)Y_{\ell m}(\theta, \varphi)\chi_{\pm}, \qquad (9\text{--}51)$$

where $\chi_{\pm}$ is the piece we have tacked on. $\chi$ is an eigenfunction of $S^2$ with the eigenvalue $(3/4)\hbar^2$ [Eq. (9–49)] and is also an eigenfunction of $S_z$ with the eigenvalues $S_z = \pm\hbar/2$, the last labeled explicitly with the $\pm$ subscript. We have also added the same label $\pm$ on the wave function. We multiplied the original spatial part of the wave function by the spin wave function $\chi_{\pm}$ because joint probabilities are multiplicative. The full wave function $\psi(\vec{r}, \pm)$ determines the probability that the electron is at a given location *and* has one of the two possible spin projections.

### The Magnetic Moment of the Electron and the Anomalous Zeeman Effect

We have already argued that an orbital angular momentum is associated with a magnetic dipole moment in that it forms a current circulating in a loop. We might suppose, then, that any particle with spin will have a magnetic moment that is proportional to the spin. Such an extrapolation suggests that the electron has an intrinsic magnetic dipole moment of the form

$$\vec{\mu}_e = -g_e \frac{e}{2m_e} \vec{S}, \qquad (9\text{--}52)$$

where $\vec{S}$ represents the electron spin. The factor $e/2m_e$ has the right dimensions and is inserted analogously to the factor in Eq. (9–43). The factor $g_e$ in Eq. (9–52) is the gyromagnetic ratio of the electron. It is initially present as a kind of fudge factor whose value could be obtained experimentally. When Dirac first treated quantum mechanics in a relativistic framework, the existence of this dipole moment emerged naturally, with $g_e = 2$.[6] This was an important milestone in the development of relativistic quantum mechanics. The observed departure from $g_e = 2$ (by about 1 part in 1,000) helped in the development of what is known as quantum electrodynamics—the marriage of electromagnetism with quantum mechanics and relativity. In quantum electrodynamics, the value of $g_e$ can be calculated to great precision and is in remarkable agreement with experiment. (See Chapter 16.)

The experiments that first measured the magnetic moment of the electron, together with the g-factor, used the fact that the energy of a magnetic dipole in a magnetic field has energy $U_{\text{mag}} = -\vec{\mu} \cdot \vec{B}$. Thus hydrogen placed in an external magnetic field $B$ that is aligned with the z-axis has a magnetic energy

$$U_{\text{mag}} = \frac{eB}{2m_e}\left(L_z + g_e S_z\right). \tag{9–53}$$

The change in the energy of hydrogen due to the external magnetic field is then

$$\Delta E_{\text{mag}} = \frac{e\hbar B}{2m_e}\left(m + g_e m_s\right), \tag{9–54}$$

where $m$ and $m_s$ are the quantum numbers associated with $L_z$ and $S_z$, respectively. These splittings are described in •Fig. 9–15. The presence of spin makes this pattern an **anomalous Zeeman effect** rather than the normal Zeeman effect. Without the external field, the degeneracy for a given $\ell$-value is $2(2\ell + 1)$, the new factor of two associated with the possibility that the electron spin is up or down. These degenerate levels are split by the external field according to

$$\text{spin "up" } \Delta E_{\text{mag}} = \frac{e\hbar B}{2m_e}\left(m + 1\right), \qquad m = \ell, \ell - 1, \ldots, -\ell \tag{9–55}$$

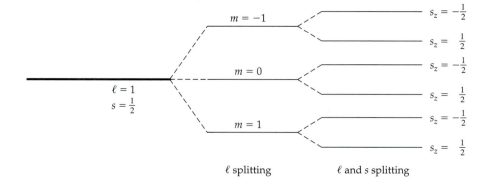

• **Figure 9–15** Atomic energy levels in strong fields have splittings that reveal the presence of an angular momentum with half-integer values (the spin). This is the anomalous Zeeman effect.

---

[6] For protons, which also have spin $1/2$, the gyromagnetic ratio has the value $g_p = 5.5857\ldots$, which shows that $g = 2$ is not a general characteristic of spin-1/2 particles.

and

$$\text{spin "down"} \; \Delta E_{\text{mag}} = \frac{e\hbar B}{2m_e}(m-1), \qquad m = \ell, \ell-1, \ldots, -\ell \quad (9\text{--}56)$$

In these expressions we have used the value $g_e = 2$.

---

**Example 9–9**    Consider the $n = 2$, $\ell = 1$ energy states of hydrogen. A magnetic field of 1 T is applied in a direction that we define to be the z-direction. Find the pattern of splittings. Estimate the ratio of the splitting of levels to the original energy of the states.

**Solution**    We can directly apply Eqs. (9–55) and (9–56). In each case, the values of $m$ are 1, 0, and −1. Thus for spin up, the energy shifts are for these three $m$-values respectively, $e\hbar B/m_e$, $e\hbar B/2m_e$, and 0. For spin down, the energy shifts are respectively 0, $-e\hbar B/2m_e$, and $-e\hbar B/m_e$. These shifts are shown in •Fig. 9–15. Note that there is still a degeneracy for the states $(m = -1, m_s = +1/2)$ and $(m = +1, m_s = -1/2)$ at the value of the original energy level. As for the magnitude of the shifts, the split defined in the figure is given by

$$\Delta E = \frac{e\hbar B}{2m_e} = \frac{(1.6 \times 10^{-19}\,\text{C})(1.05 \times 10^{-34}\,\text{J·s})(1\,\text{T})}{2(9.1 \times 10^{-31}\,\text{kg})} = 9.2 \times 10^{-24}\,\text{J} = 5.8 \times 10^{-5}\,\text{eV}.$$

We can compare this value with the energy $E$ of the unsplit $n = 2$ level, which is $-(13.6\,\text{eV})/n^2 = -3.4\,\text{eV}$. The ratio $\Delta E/E$ is $(5.8 \times 10^{-5}\,\text{eV})/(3.4\,\text{eV}) = 1.7 \times 10^{-5}$.

---

### Modern Measurement of the Electron g-Factor

The best measurements of the electron $g$-factor have been made using the Penning trap. This device has allowed Hans Dehmelt and his collaborators[7] at the University of Washington in Seattle to make measurements of the quantum nature of *individual* ions, as well as of electrons and positrons. The Penning trap (•Fig. 9–16) allows one to contain a single such particle for long periods of time. Powerful amplifiers with resonant circuitry measure the radiation emitted by these particles as they move within the containing fields and make quantum jumps. When inhomogeneity is introduced into the magnetic field, the frequency of an orbital motion for trapped spin-up electrons becomes different from that for trapped spin-down electrons. This difference in frequency, which contains the $g$-factor as the only unknown, is measurable. The result for the $g$-factor is accurate to several parts in $10^{11}$. As we discuss in Chapter 16, this measurement gives us great confidence in the quantum theory of electrons interacting with the electromagnetic field.

### *Addition of Spin and Orbital Angular Momentum

The spin $\vec{S}$ and the orbital angular momentum $\vec{L}$ should be expected to sum to a total angular momentum $\vec{J}$, just as in classical physics; that is,

$$\vec{J} = \vec{L} + \vec{S}. \qquad (9\text{--}57)$$

The quantum mechanical aspect of the quantity $\vec{J}$ would reveal itself in the fact that only certain values would be allowed, rather than a classical continuum. More precisely, because we deal only with the square of the angular momentum and with its z-component in quantum mechanics, the results of adding $\vec{L}$ and $\vec{S}$ must be expressed in terms of such quantities only.

---

[7]See the paper by Dehmelt in the *American Journal of Physics* **58**, p. 1 (1990).

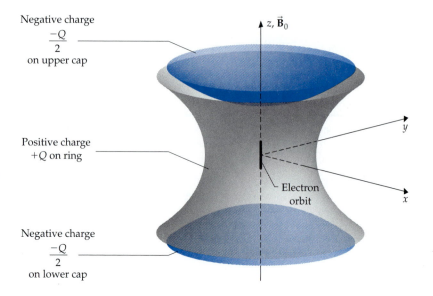

Negative charge
$\dfrac{-Q}{2}$
on upper cap

Positive charge
$+Q$ on ring

Negative charge
$\dfrac{-Q}{2}$
on lower cap

Electron orbit

$z, \vec{B}_0$

$y$

$x$

• **Figure 9–16** Sketch of the principal features of the Penning trap. The simplest motion of an electron in the trap is along the axis of symmetry, in a helical path along a magnetic field line. Each time the electron comes too close to one of the negatively charged caps, it turns around. [after Fig. 1 in *American Journal of Physics*, **58**, 1 (January 1990)]

The addition of $S_z$ and $L_z$ appears to be straightforward. We expect that the operator

$$J_z = L_z + S_z \tag{9–58}$$

will have eigenvalues $\hbar m_j = \hbar m + \hbar(\pm 1/2)$; in other words,

$$m_j = m \pm 1/2. \tag{9–59}$$

Going beyond the calculation of the sum of the $z$-components is more complicated. If $\vec{L}$ and $\vec{S}$ were classical angular momenta with magnitudes $L$ and $S$, respectively, then the magnitude of their sum could take on any value between $J_{max} = L + S$ and $J_{min} = L - S$. These cases correspond respectively to the alignment and antialignment of $\vec{L}$ and $\vec{S}$ with each other. This correspondence is nicely illustrated in the demonstration of a student sitting on a rotating piano stool and holding a rotating bicycle wheel, with the wheel rotating in the same and in the opposite direction as the rotation of the piano stool. For the case of a quantum mechanical $L$ adding with a quantum mechanical $S$, we might also imagine the two possibilities of the spin in alignment and in antialignment.

In the case of alignment, we would have

$$j_+ = \ell + 1/2. \tag{9–60a}$$

The possible values of $m_j$ for this value of $j$ are

$$\ell + 1/2, \ell + 1/2 - 1, \ell + 1/2 - 2, \ldots, -(\ell + 1/2 - 1), -(\ell + 1/2),$$

which add up to $2(\ell + 1/2) + 1 = 2\ell + 2$ different values of $m_j$ for $j = j_+$.

For antialignment, the $j$-value is

$$j_- = \ell - 1/2, \tag{9–60b}$$

with possible $z$-projection values of

$$\ell - 1/2, \ell - 1/2 - 1, \ell - 1/2 - 2, \ldots, -(\ell - 1/2 - 1), -(\ell - 1/2).$$

In this case the number of different values of the $z$-projection quantum number is $2(\ell - 1/2) + 1 = 2\ell$.

There are no other possible $j$-values. We can check this fact by noting that, since $\vec{L}$ and $\vec{S}$ are independent angular momenta, the total number of states of the form $Y_{\ell m}(\theta, \varphi)\chi_\pm$ is $(2\ell + 1) \times 2 = 4\ell + 2$. This number matches precisely the total $2\ell + 2$ and $2\ell$ that we get from adding the $z$-projection quantum numbers from the two possibilities $j_+$ and $j_-$.

In this chapter we are interested in the addition of spin-$1/2$ angular momentum and orbital angular momentum. But we can state here the more general result. The summation of $\vec{\mathbf{L}}$ and spin-$1/2$ angular momenta into two total angular momenta with quantum numbers $j_+ = \ell + 1/2$ and $j_- = \ell - 1/2$ is a special case of the addition of two angular momenta in general. Let us call these two angular momenta $\vec{\mathbf{J}}_1$ and $\vec{\mathbf{J}}_2$, with respective angular-momentum quantum numbers $j_1$ and $j_2$. (These could be orbital angular momenta, spins, or combinations thereof.) Then if the total angular momentum is $\vec{\mathbf{J}}_{\text{tot}} = \vec{\mathbf{J}}_1 + \vec{\mathbf{J}}_2$, $\vec{\mathbf{J}}_{\text{tot}}$ can have the possible angular-momentum quantum numbers

$$j = j_1 + j_2, j_1 + j_2 - 1, \ldots, |j_1 - j_2|. \tag{9–61}$$

Each of these $j$-values will have its full set of possible $z$-component quantum numbers $m_j = j, j - 1, \ldots, -j$.

In the next subsection, we shall see how these results have a direct application to features of the hydrogen spectrum.

### *Spin–Orbit Coupling

In classical physics, a current loop is not only something that responds to a magnetic field; it also *produces* a magnetic field. In this way there is an interaction between two magnetic dipoles—and an energy to go along with that interaction. This interaction energy is proportional to the scalar product of the two magnetic dipole moments. The particular case that interests us here is the interaction of the internal magnetic moment of the electron with the magnetic moment associated with its orbital motion. This kind of interaction comes about because, in the reference frame of the electron, it appears as if the nucleus is moving. This apparently moving charged nucleus generates a magnetic field that can interact with the electron's magnetic moment. Viewed from the reference frame of the electron, the magnetic field is proportional to the orbital angular momentum of the proton. But that is in fact equal to the orbital angular momentum of the electron if we think of the proton as stationary and the electron orbiting. Hence our interaction is an interaction of the electron's spin with its own orbital angular momentum. A careful classical calculation (including relativity) of the interaction of the spin magnetic moment of an electron moving in a circular orbit of radius $r$ through the coulomb field of a proton gives a potential energy of the form

$$U_{\text{spin–orbit}} = \frac{1}{4\pi\varepsilon_0} \frac{1}{2c^2r^3} \vec{\boldsymbol{\mu}}_L \cdot \vec{\boldsymbol{\mu}}_S,$$

where $\vec{\boldsymbol{\mu}}_L$ and $\vec{\boldsymbol{\mu}}_S$ are the magnetic dipole moments associated with the orbital and spin angular momenta, respectively. This potential has the typical $1/r^3$ behavior of a dipole–dipole interaction energy.

When we insert the expressions for the magnetic dipole moments in terms of $\vec{\mathbf{L}}$ and $\vec{\mathbf{S}}$ themselves, we find that

$$U_{\text{spin–orbit}} = \frac{1}{4\pi\varepsilon_0} \left( -\frac{eZ}{2m_e} \vec{\mathbf{L}} \right) \cdot \left( -\frac{g_e e}{2m_e} \vec{\mathbf{S}} \right) \frac{1}{2c^2r^3} = \frac{Ze^2}{4\pi\varepsilon_0} \frac{\vec{\mathbf{L}} \cdot \vec{\mathbf{S}}}{4m_e^2 c^2 r^3} \frac{g_e}{2}. \tag{9–62}$$

As has been the case throughout this chapter, the factor $Z$ emphasizes the role of the nucleus; we can get an expression that applies directly to hydrogen simply by replacing $Z$ by 1.

To proceed to quantum mechanics, we simply interpret this expression [Eq. (9–62)] as an operator. The effect is to add a term to the energy of the hydrogen atom that couples the spin with the orbital angular momentum. That is why we have put the "spin–orbit" subscript on the potential energy. When $U_{\text{spin–orbit}}$ is added to the total energy, that energy depends on both the spin and orbital angular momenta. Spectroscopy can reveal the effect of $U_{\text{spin–orbit}}$ in energy levels, but the measurement must be precise, because with a factor of $1/c^2$ this energy is small compared with the coulomb energy.

The calculation of the actual shift in energy levels, $\Delta E_{\text{spin–orbit}}$, due to the spin–orbit operator is a technical exercise that does not interest us here. We can, however, estimate the energy shift that occurs in hydrogen when the spin–orbit interaction is taken into account. In an atomic state, we expect the effect of the operator $\vec{\mathbf{L}} \cdot \vec{\mathbf{S}}$ to be on the order of $\hbar^2$. In addition, we can replace the operator $1/r^3$ in Eq. (9–62) by the factor $1/(n^2 a_0)^3$ for an electron with principal quantum number $n$, where $a_0$ is the Bohr radius, as suggested by Eq. (5–10). Finally we set $g_e = 2$. These replacements give us the following estimate for the shift in energy levels in hydrogen:

$$\Delta E_{\text{spin–orbit}} \cong -\frac{e^2}{4\pi\varepsilon_0} \frac{\hbar^2}{4m_e^2 c^2} \left(n^2 a_0\right)^{-3}$$

$$= -\frac{e^2}{4\pi\varepsilon_0} \frac{\hbar^2}{4m_e^2 c^2} \frac{1}{n^6} \left(\frac{\hbar}{m_e c\alpha}\right)^{-3} = -\frac{m_e c^2 \alpha^2}{4n^6} \alpha^2.$$

Here we have used the fine-structure constant $\alpha = e^2/(4\pi\varepsilon_0 \hbar c)$ of Eq. (5–6). If we recall that the Bohr energy is $-m_e c^2 \alpha^2/(2n^2)$, then we see that the energy shift due to the inclusion of the spin–orbit coupling is $\alpha^2/2n^4 \cong 10^{-5}$ of the Bohr energy itself.

*Fine Structure*   We have already remarked on the degeneracy associated with a given $n$-value. Since to the leading order the energy depends only on $n$ [see Eq. (9–37)], it is the same for the $n - 1$ different $\ell$-values allowed for a given $n$, for the $2\ell + 1$ different $m$-values associated with a given $\ell$, and for the two different $s_z$-values allowed for each state. *The spin–orbit coupling breaks a large part of this degeneracy.* To see this, we square Eq. (9–57), giving $\vec{\mathbf{J}}^2 = (\vec{\mathbf{L}} + \vec{\mathbf{S}})^2 = \vec{\mathbf{L}}^2 + \vec{\mathbf{S}}^2 + 2\vec{\mathbf{L}} \cdot \vec{\mathbf{S}}$. Thus,

$$\vec{\mathbf{L}} \cdot \vec{\mathbf{S}} = \frac{\vec{\mathbf{J}}^2 - \vec{\mathbf{L}}^2 - \vec{\mathbf{S}}^2}{2},$$

and the eigenvalues of $\vec{\mathbf{L}} \cdot \vec{\mathbf{S}}$ are those of $(\vec{\mathbf{J}}^2 - \vec{\mathbf{L}}^2 - \vec{\mathbf{S}}^2)/2$; that is,

$$\text{eigenvalues of } \vec{\mathbf{L}} \cdot \vec{\mathbf{S}} = \left[ j_\pm(j_\pm + 1) - \ell(\ell + 1) - \frac{3}{4} \right] \frac{\hbar^2}{2}, \qquad (9\text{–}63)$$

where the factor $3/4$ is $s(s + 1)$ and the two $j$-values are those of Eqs. (9–60a and b). Now the eigenvalues of $\vec{\mathbf{L}} \cdot \vec{\mathbf{S}}$ are what enter the expression for the spin–orbit energy shift. Since, from Eq. (9–63), the eigenvalues of $\vec{\mathbf{L}} \cdot \vec{\mathbf{S}}$ depend on the $\ell$-value, the spin–orbit coupling term lifts the $\ell$-degeneracy of the atomic levels.

•Figure 9–17 illustrates the splitting pattern, as well as the typical size of the splitting, associated with spin–orbit coupling in hydrogen. The slight adjustment of the energy levels gives what is known as the **fine structure** of the atomic levels.

Fine structure, together with other details of atomic spectra, with or without the presence of external fields, contains a vast amount of information about the atom. These details even possess information about the structure of the

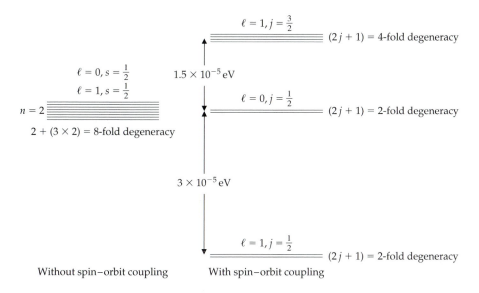

$\ell = 1, j = \frac{3}{2}$

$(2j + 1) = $ 4-fold degeneracy

$\ell = 0, s = \frac{1}{2}$

$\ell = 1, s = \frac{1}{2}$

$1.5 \times 10^{-5}\,\text{eV}$

$\ell = 0, j = \frac{1}{2}$

$(2j + 1) = $ 2-fold degeneracy

$n = 2$

$2 + (3 \times 2) = $ 8-fold degeneracy

$3 \times 10^{-5}\,\text{eV}$

$\ell = 1, j = \frac{1}{2}$

$(2j + 1) = $ 2-fold degeneracy

Without spin–orbit coupling

With spin–orbit coupling

• **Figure 9–17**   The effect of spin–orbit coupling on the $n = 2$ levels in hydrogen. For a given value of orbital angular momentum $\ell$ of the electron, there are two split energy levels, corresponding to $j = \ell + 1/2$ and $j = \ell - 1/2$. These "doublets" are clearly observed. In a magnetic field, much of the remaining degeneracy is removed.

vacuum itself. The study of atomic spectra has turned out to be one of the really fruitful sources for the ideas of 20th-century physics.

## *9–6   Hyperfine Structure and Magnetic Resonance Imaging

There are many contributions to the energy spectrum of hydrogen beyond those we have already discussed; the clarification of some of them has led to the theory of quantum electrodynamics, something we shall discuss in Chapter 16. In this section we concentrate on a particular contribution which arises from the fact that the proton has spin 1/2 and an associated magnetic moment. The latter interacts with the magnetic moment of the electron and gives rise to splittings called **hyperfine structure**. The hyperfine interaction is small, but it has important applications in both radio astronomy and medical technology.

The proton has a magnetic moment given by

$$\vec{\boldsymbol{\mu}}_p = +g_p \frac{e}{2m_p} \vec{\mathbf{I}}, \qquad (9\text{–}64)$$

where the subscript $p$ denotes the proton and $\vec{\mathbf{I}}$ is the proton spin. The constant $g_p$ is the gyromagnetic constant of the proton, and its measured value is 5.56. The reversal in sign compared with Eq. (9–52) comes from the positive charge on the proton. This dipole moment interacts with that of the electron. The potential energy is like that of the fine structure (Section 9–5), in particular containing a $1/r^3$ term, and it is treated in the same fashion. There are, however, some important differences between the two cases. For one, only electrons in the $\ell = 0$ state couple to the nuclear spin to first approximation. The reason is that, as we have seen, for $\ell > 0$ the wave function of the electron vanishes at the origin, while the wave function for the nucleus is concentrated there. With the small probability of the electron being near the proton, there is, to the leading order, no interaction energy. And for another thing, the dipole moment of the proton contains its mass in the denominator, and therefore it is some 2,000 times smaller than the dipole moment of the electron. We therefore expect the splittings to be $O(10^{-3})$ times the fine-structure splittings. (That is why the term "hyperfine" is used.)

Once we have established that only electrons in the $\ell = 0$ state couple to the nuclear spin, the lone remaining part of the electron's total angular momentum is the electron spin $\vec{S}$ itself. Thus the coupling between the proton spin and the electron angular momentum is a coupling between the proton spin and the electron spin. This spin–spin interaction contains the factor $\vec{I} \cdot \vec{S}$. The implications of this fact hark back to our discussion of the addition of angular momenta in the previous section. If we call the total spin of the proton–electron system $\vec{S}_{\text{tot}}$, then Eqs. (9–60a and b) show us that we can find the values of the $\vec{S}_{\text{tot}}$ quantum number $s_{\text{tot}}$ by replacing "$\ell$" by the proton spin quantum number, namely 1/2. This means that $s_{\text{tot}}$ can be either 1 or 0. In other words, two spin 1/2's can add either to a spin 1 or to a spin 0. We refer to these states as triplet and singlet states, respectively, because there are three z-projection quantum numbers for a spin-1 (triplet) state and one z-projection quantum number for a spin-0 (singlet) state.

We now use the technique expressed in Eq. (9–63) for the case where "$\ell$" is 1/2 and "$j_{\pm}$" is either 1 or 0. The possible eigenvalues of $\vec{I} \cdot \vec{S}$ are then $\left[1(1 + 1) - 3/4 - 3/4\right]\hbar^2/2 = [1/2]\hbar^2/2$ for the spin-1 (triplet) state and $\left[0(0 + 1) - 3/4 - 3/4\right]\hbar^2/2 = [-3/2]\hbar^2/2$ for the spin-0 (singlet) state. In other words, the amount of the hyperfine energy splittings depends on the total spin. The result of a complete calculation is that for $\ell = 0$ the total spin-1 (triplet) state lies above the total spin-0 (singlet) state by an amount $\Delta E = hf$, where the frequency $f$ is

$$f = 1,420 \text{ MHz}.$$

The wavelength corresponding to this frequency is 21 cm. Light—actually a radio signal—of this wavelength is emitted when there is a transition between the two levels.

This transition is of great importance in astrophysics. Its importance lies in the fact that such radio waves are not absorbed by interstellar dust, so that we can map hydrogen in space using the presence or absence of the 21-cm radiation. Excitation to the higher total spin-1 state occurs through collisions involving at least one hydrogen atom. The 21-cm radio waves can be detected, and from the intensity, the distribution of atomic hydrogen in the plane of the galaxy can be mapped.

**Nuclear Magnetic Resonance**

In hyperfine structure, the energy comes from a coupling between the nuclear magnetic moment and, in effect, a magnetic field due to the electron's magnetic moment. One can also have couplings between the nuclear spin and an *external* magnetic field $\vec{B}$; the energy is $-\vec{\mu}_p \cdot \vec{B}$. We shall now see how it is possible to use this energy to explore the presence of hydrogen in an environment such as the human body.

We know that a magnetic field exerts a torque on a current loop or, more generally, on a magnetic dipole. [See Chapter 1 and in particular Eq. (1–59)]. The torque is the rate of change of angular momentum. With regard to the internal magnetic moment of the proton, the angular momentum in question is the proton spin $\vec{I}$. The dynamical equation for torque then takes the form

$$\frac{d\vec{I}}{dt} = \vec{\mu}_p \times \vec{B} = +\frac{eg_p}{2m_p}\vec{I} \times \vec{B}. \tag{9–65}$$

If we pick the direction of $\vec{\mathbf{B}}$ to define the $z$-axis and recall that a cross product of two vectors is perpendicular to each, then the right side of Eq. (9–65) has no $z$-component; in other words,

$$\frac{dI_z}{dt} = 0, \tag{9–66}$$

so that the $z$-component of the proton spin is unchanging. Moreover, writing the components of the cross product explicitly, we have

$$\frac{dI_x}{dt} = \frac{eg_p}{2m_p} BI_y \quad \text{and} \quad \frac{dI_y}{dt} = -\frac{eg_p}{2m_p} BI_x. \tag{9–67}$$

These equations describe a two-component vector—the projection of the proton spin on the plane perpendicular to the magnetic field—undergoing *rotation* (•Fig. 9–18). You can verify this statement if you compare the equations for the rotation of the position vector $(x, y, 0)$ about the $z$-axis. The equations for $x$ and $y$ have the same form as those of Eq. (9–67), namely, $dx/dt = \omega y$ and $dy/dt = -\omega x$. Comparing the pairs of equations, you can see that the angular frequency of the rotation in our case is $\omega = eg_p B/2m_p$. This type of motion is known as *precession*; it is followed by a tilted gyroscope under the influence of gravity. To get an idea of the numbers involved, if $B = 1$ T, then, since $g_p \cong 5.6$, $\omega \cong 2.7 \times 10^8$ rad/s.

To understand the relevance of the preceding discussion, we now turn to the question of the energy of the proton in the external magnetic field. We know that the energy is $-\vec{\boldsymbol{\mu}}_p \cdot \vec{\mathbf{B}}$, so that when the spin is pointing up ($I_z = +1/2$), the energy is $-(eg_p\hbar/4m_p)(B) = -(eg_p B/2m_p)(\hbar/2)$, whereas when the spin is pointing down ($I_z = -1/2$), the energy is $+(eg_p B/2m_p)(\hbar/2)$. Thus the spin-up configuration has lower energy than the spin-down configuration. Now the technique we are going to describe next involves changes in energy. But we have seen that as long as $\vec{\mathbf{B}}$ is unchanging, $I_z$ stays the same; that is, *the nuclear spin component $I_z$ cannot change*, and the energy remains fixed. In order to change the energy, we supply an additional weak field $\vec{\mathbf{B}}'$ that lies entirely in the $xy$-plane and that rotates about the $z$-axis with a frequency $\omega_0$. When $\omega_0$ approaches $\omega$, the field $\vec{\mathbf{B}}'$ rotates *along with* the proton spin. But then the spin vector will precess about the direction of the new field $\vec{\mathbf{B}}'$, a precession that involves a change in $I_z$. We say that a **spin flip** occurs. (Note that if the new field is weak enough, it influences the energy only through its role in causing the spin flip.) Of course energy is conserved, and when the spin flips, the proton either absorbs or emits electromagnetic radiation in the form of a photon, depending on whether the transition is to the higher energy or lower energy state. In either case, the magnitude of the photon's energy is

$$\frac{eg_p B}{2m_p}\frac{\hbar}{2} - \left(-\frac{eg_p B}{2m_p}\frac{\hbar}{2}\right) = \hbar\frac{eg_p B}{2m_p} = \hbar\omega.$$

If we recall that a photon has energy $\hbar\omega = hf$, we see that *these photons have the frequency of the spin precession.* The emission or absorption of such photons is detectable and betrays the presence of hydrogen. The entire process is known as **nuclear magnetic resonance**, where "resonance" refers to the fact that the time-varying field must have a frequency that closely matches $\omega$ in order to produce the spin flip.

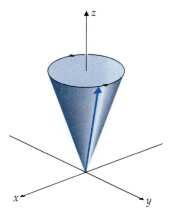

• **Figure 9–18** The equations of motion for the components of the proton spin when the proton is placed in a magnetic field look like the classical equations of a spin vector rotating about the magnetic field and mapping out a cone. This motion is very much like the classical precession of a tilted top under the influence of gravity.

In practice, a sample whose hydrogen content is of interest is placed in a magnetic field $\vec{\mathbf{B}}$. The lower of the two energy levels is more populated, as suggested by the Boltzmann factor. (See Chapter 12.) An alternating current with a frequency that can be controlled is sent through a wire that surrounds the sample, and this current produces an additional time-varying magnetic field. When the frequency of this field matches $\omega$, a spin flip is induced. On the average the spin flip corresponds to transitions from the more populated lower level to the less populated upper level, and photons are absorbed. Energy is thus removed from the electromagnetic field, and the change in energy is sensed in changes in currents.

The process just described is a good way of measuring $g_p$, something of interest in itself. But once we know $g_p$, we can use the intensity of the energy absorption and its spatial dependence to measure the spatial concentration of hydrogen in the sample. In medical technology, where the "sample" is a patient, this process is known as **magnetic resonance imaging** (MRI). The measurement of the distribution of hydrogen in the body translates to an ability to detect abnormalities and to map nonbony tissue. [Bones contain very little hydrogen (•Fig. 9–19).]

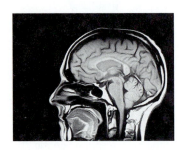

• **Figure 9–19** A nuclear magnetic resonance image.

In order to produce a magnetic resonance image, it is important that the $B$ field not be homogeneous. When that is the case, then, with a fixed photon frequency $\omega_0$, there will be only one place at which resonance will occur, namely, wherever $B$ has the value $2m_p\omega_0/eg_p$. By moving the patient in a known, space-varying $B$ field, the abnormality can be localized.

## SUMMARY

The three-dimensional Schrödinger equation can be solved for the case of a coulomb potential energy—the case of the hydrogen atom, consisting of a single electron orbiting a nucleus—justifying and extending Bohr's treatment of the problem. The significant results of this treatment are as follows:

• Angular momentum plays a special role. For the (central) coulomb potential, angular momentum is quantized in integer multiples of Planck's constant, with a coefficient depending on an angular momentum quantum number $\ell$. Any one component of the angular momentum, usually chosen in the $z$-direction, is quantized as an integer multiple of Planck's constant with a coefficient limited by the angular-momentum quantum number $\ell$.

• The energy eigenvalues, which depend on a principal quantum number, are those of the Bohr model. For a given principal quantum number a limited number of angular-momentum quantum numbers are allowed. The energies are independent of all the possible angular-momentum values, so that there is a significant degree of degeneracy. This degeneracy is special to the coulomb potential.

• The eigenfunctions of hydrogen factor into a piece that depends on the radial coordinate and another piece—a spherical harmonic—that depends on the angular coordinates. These eigenfunctions depend on both the principal and the angular-momentum quantum numbers and express the probabilities of finding the electron in various locations around the atom. In particular, the radial probabilities have a falling exponential behavior that dominates at large distances, a characteristic of all bound systems.

• An electron in a state with nonvanishing angular momentum forms a current loop, or equivalently, a magnetic moment that can be measured by looking at shifts in atomic energy levels in an external magnetic field. The degenerate energy levels are split (the Zeeman effect), and the magnitude of the component of the quantized angular momentum along the direction of the magnetic field can be measured.

- Zeeman spectroscopy shows that even in a state with zero angular momentum the electron has a magnetic moment. This quantity is attributed to an intrinsic angular momentum of the electron, called its spin. The spin has two peculiar properties: First, it is characterized by a quantum number that, in contrast to the integer values of the orbital angular momentum, has the value one-half. Second, the magnetic moment is related in a nonclassical way to the angular momentum. Given these features of the spin, an electron interacts with the magnetic field it experiences because it moves in a coulomb field. This interaction is called spin–orbit coupling. Even without an external field, spin–orbit coupling breaks the degeneracy of the atomic levels, but the effect is small. In addition, the constituents of the nucleus themselves have spin and can also interact with the magnetic field associated with the moving electron. This effect is the basis of magnetic resonance imaging.

# QUESTIONS

**1.** In his semiclassical model of hydrogen, Bohr excluded the zero angular-momentum state. Why? Do these objections apply in the quantum theory?

**2.** In the quantum theory it is generally not possible to specify all three components of the angular momentum simultaneously. Is the $\ell = 0$ angular-momentum state an exception to this rule?

**3.** In classical physics you can imagine balancing an ice pick in a perfectly upright position. Is this possible in the quantum theory? Can you specify an initial condition such that you know the position of the ice pick and that it has exactly zero angular momentum?

**4.** We have seen in this chapter that, in contrast to the Bohr atom (Chapter 5), atomic "orbits" with angular momentum $\ell = 0$ are allowed. Such orbits would correspond classically to the electron passing directly though the center of mass of the atom, which, for all practical purposes, lies on top of the nucleus. Experiments show that when free electrons are incident on nuclei, the resulting collision strongly deflects the electron. Why don't the $\ell = 0$ orbits disrupt the atom?

**5.** Spherical harmonics, which are eigenfunctions of the angular momentum, contain the complex number $i = \sqrt{-1}$. Is it all right for a function that is supposed to be associated with measurable physical quantities to have complex numbers. Why or why not?

**6.** The table of radial eigenfunctions for the hydrogen atom suggests—and this is borne out generally—that, for $\ell \neq 0$, the probability of the electron's being found at the origin is zero. This finding was the basis for the observation that unless $\ell = 0$ there is no hyperfine structure, which involves an interaction between the electron spin moment and the nuclear spin moment. Do you expect this observation to hold exactly? What mechanisms can you think of that would produce a hyperfine structure even for $\ell \neq 0$?

**7.** Consider a penny spinning about an axis through its center at the rate of a few revolutions per second. How would you estimate a value for $\ell$?

**8.** Consider a simplified classical model of an atom, in which an electron, charge $-e$, moves in a circular orbit in a magnetic field perpendicular to the plane of the orbit. In the absence of the magnetic field, the electron can move clockwise or counterclockwise with an angular frequency $\omega_0$. What will be the effect of the magnetic field on the motion of the electron? If you can make an estimate, compare it with the Zeeman effect, as given in Eq. (9–47), for example.

## PROBLEMS

1. ▮▮ We have written the radial Schrödinger equation as an equation for $R(r)$. It takes a somewhat simpler form in terms of $u(r) = rR(r)$. Find the equation obeyed by $u(r)$. Show that that equation looks just like a one-dimensional Schrödinger equation with

$$
V(x) = \begin{cases} \infty, & \text{for } x < 0 \\ -\dfrac{Ze^2}{4\pi\varepsilon_0}\dfrac{1}{x} + \dfrac{\hbar^2\ell(\ell + 1)}{2m_e}\dfrac{1}{x^2}, & \text{for } x > 0 \end{cases}.
$$

Sketch this potential and show that, for $\ell \neq 0$, there is a barrier that keeps the electron away from the origin, so that the only way the electron can be near the origin is through tunneling.

The next four problems introduce you to the quantum mechanical treatment of the motion of the center of mass of a two-body system. In particular we have in mind the hydrogen atom, made of an electron and a proton. We will see that the Schrödinger equation (9–2) is left intact, except for the replacement of the mass of the electron, $m_e$, by the reduced mass $\mu = (m_e m_p)/(m_e + m_p)$, where $m_p$ is the mass of the hydrogen nucleus—the proton. This correction is an important one that is readily detected in spectroscopic measurements.

The starting point of the four problems is the full Schrödinger equation,

$$
\left[-\frac{\hbar^2}{2m_e}\left(\frac{\partial^2}{\partial x_e^2} + \frac{\partial^2}{\partial y_e^2} + \frac{\partial^2}{\partial z_e^2}\right) - \frac{\hbar^2}{2m_p}\left(\frac{\partial^2}{\partial x_p^2} + \frac{\partial^2}{\partial y_p^2} + \frac{\partial^2}{\partial z_p^2}\right) + V(\vec{r}_e - \vec{r}_p)\right]\psi(\vec{r}_e, \vec{r}_p)
$$
$$
= E\psi(\vec{r}_e, \vec{r}_p).
$$

This equation contains the coordinates of the proton, as well as those of the electron. That is, it captures the proton's motion! By taking $V$ to be a function only of the differences in the coordinate's, we simplify everything immensely. Indeed, the equation suggests that we introduce a *relative* coordinate vector $\vec{r} \equiv \vec{r}_e - \vec{r}_p$. In that language, $V$ depends only on $r$; we here note the additional simplification that $V$ depends only on the *magnitude* of $\vec{r}$. In addition, we introduce a vector $\vec{R}$ that describes the position of the center of mass of the electron and proton. Then, in terms of the electron and proton masses and the total mass $M$ that is their sum, $\vec{R}$ has the following components:

$$
X = \frac{(m_e x_e + m_p x_p)}{M}, \quad Y = \frac{(m_e y_e + m_p y_p)}{M}, \quad Z = \frac{(m_e z_e + m_p z_p)}{M}.
$$

Now we pose the problems.

2. ▮▮ Show that the Schrödinger equation, written in terms of the preceding coordinates, takes the form

$$
\left[-\frac{\hbar^2}{2M}\left(\frac{\partial^2}{\partial X^2} + \frac{\partial^2}{\partial Y^2} + \frac{\partial^2}{\partial Z^2}\right) - \frac{\hbar^2}{2\mu}\left(\frac{\partial^2}{\partial x^2} + \frac{\partial^2}{\partial y^2} + \frac{\partial^2}{\partial z^2}\right) + V(r)\right]\psi(\vec{r}, \vec{R})
$$
$$
= E\psi(\vec{r}, \vec{R}),
$$

where the only thus-far undefined quantities are the components of $\vec{r}$, namely, $x, y,$ and $z$.

**3.** ∎∎ The fact that the potential is a function of $r$ and not of $R$ suggests writing the wave function in the form $\psi(\vec{r}, \vec{R}) = u(\vec{r})U(\vec{R})$. Show that a trial separation of this type leads to the two equations

$$\left[ -\frac{\hbar^2}{2\mu} \left( \frac{\partial^2}{\partial x^2} + \frac{\partial^2}{\partial y^2} + \frac{\partial^2}{\partial z^2} \right) + V(r) \right] u(\vec{r}) = Eu(\vec{r})$$

and

$$-\frac{\hbar^2}{2M} \left( \frac{\partial^2}{\partial X^2} + \frac{\partial^2}{\partial Y^2} + \frac{\partial^2}{\partial Z^2} \right) U(\vec{R}) = E'U(\vec{R}).$$

The first of these equations is just Eq. (9–2) with the electron mass replaced by the reduced mass $\mu$. This equation has the very important implication that we can find the correction due to the motion of the proton in quantities like the allowed internal energies of the electron–proton system by making the appropriate replacement. The second equation describes the motion of a free object of mass $M$—that is, the motion of the center of mass of the system.

**4.** ∎∎ In Problem 3, you showed that for an interparticle potential that depends only on the separation between the masses of a two-body system the full Schrödinger equation reduces to two equations, the second of which involves the motion of the center of mass of the system. Solve the second equation for $U(\vec{R})$, the wave function describing the behavior of the center of mass of the system. (*Hint*: You have already seen this equation in one dimension, where it describes the motion of a free particle of mass $M$ along, say, the $x$-axis.)

**5.** ∎∎ In Problem 3, you showed that for an interparticle potential that depends only on the separation between the masses of a two-body system the full Schrödinger equation reduces to two equations, the first of which describes the internal motion of a particle of reduced mass $\mu$ under the influence of a coulomb potential. Show that when the proton mass tends to infinity, this equation reduces to the Schrödinger equation treated in the text, in which a particle of mass $m_e$ (an electron) moves in a fixed coulomb field due to the charge on the proton.

**6.** ∎ A mass of 1 microgram is rotating about an axis at 3,000 revolutions per minute. The radius of the circular orbit is 0.1 mm. What is the value of $m$, the $L_z$ quantum number, for this system?

**7.** ∎ Consider two masses of 0.1 gm each, connected by a rigid rod of length 0.5 cm, rotating about their center of mass with an angular frequency of 800 radians/sec. **(a)** What is the value of $\ell$ corresponding to this situation? **(b)** What is the energy difference between adjacent $\ell$-values for the $\ell$ that you have just calculated?

**8.** ∎ Calculate the frequency of the spectral line corresponding to the transition $n = 1, \ell = 1$ to the state with $n = 1, \ell = 0$ for singly ionized helium (helium with one electron removed). How does this frequency differ from the corresponding frequency in hydrogen?

**9.** ∎∎ Use Eq. (9–37) to evaluate the frequency of the light emitted in a transition between the $n = 2, \ell = 1$ and the $n = 1, \ell = 0$ states. How big is the correction if you replace the electron's rest mass by its reduced mass? (See Problems 2 through 5.) Assume $Z = 1$ (hydrogen).

**10.** ∎∎ Show that $R_{20}$ is a solution of the Schrödinger equation with the coulomb potential, and find the corresponding eigenvalue.

11. ■■ Find $\langle r^2 \rangle$ for the ground state of hydrogen. Compare your result with $\langle r \rangle^2$ for the same state.

12. ■■ Show that, for the state described by $R_{20}$,

$$\langle r^2 \rangle = \frac{21}{8} \left( \frac{4a_0}{Z} \right)^2.$$

How does this answer compare with the one from the previous problem? Why is $\langle r^2 \rangle$ larger in this case than it is for the ground state?

13. ■■ Find the expectation value of the coulomb potential

$$V(r) = -\frac{e^2}{4\pi\varepsilon_0} \frac{1}{r}$$

in the state of hydrogen specified by $n = 2$, $\ell = 1$, and $m = 0$. Use the method outlined in Example 9–6 to find the expectation value of the kinetic energy, $\langle K \rangle = \langle p^2/2m_e \rangle$, in this state. Show that $\langle K \rangle = -(1/2)\langle V(r) \rangle$.

14. ■■ Show that in the state $n = 2$, $\ell = 1$, and $m = 0$, $\langle r \rangle = 5a_0/Z$. Set $Z = 1$, and compare $\langle r \rangle$ with the result found for the ground state of hydrogen in Problem 11. In doing your work, you will notice that the angular integrations are simple, because those parts of the wave functions have been normalized separately.

15. ■■ The radial probability density for an electron is $r^2R^2(r)$. That means that the probability of finding an electron at a radius $r$ within a radial thickness $dr$ is $dr\, r^2R^2(r)$ for an infinitely thin shell and approximately $\Delta r\, r_{av}^2 R^2(r_{av})$ for a shell of finite thickness $\Delta r$. The quantity $r_{av}$ is some average radius within the shell. **(a)** Use the preceding approximation to estimate the probability that an electron in the $n = 1$, $\ell = 0$ state will be found in the region from $r = 0$ to $r = 10^{-15}$ m (about the size of the atomic nucleus and some $10^5$ times smaller than an atom). **(b)** Repeat the calculation of part (a) for the $n = 2$, $\ell = 1$ state. Your answer for part (b) should be much smaller than for part (a). Why?

16. ■ Consider the state $n = 2$, $\ell = 0$ in hydrogen, with $Z = 1$. What is (are) the value(s) of $r$ for which the probability density is zero? Sketch the probability density as a function of $r$ for this state.

17. ■■ As we have seen, the states for which $\ell = n - 1$ are the states in which the radial wave function has no nodes between $r = 0$ and $r = \infty$, meaning that there is a single maximum at some intermediate $r$. Find the value of the radius at which $R_{10}$ and $R_{21}$ are maximum, and compare these values with $\langle r \rangle$ for each state.

18. ■ An electron in the CERN LEP accelerator moves in a circle of radius 5 km with energy 90 GeV (1 GeV = $10^9$ eV; the electron is ultrarelativistic). Find the orbital angular-momentum quantum number $\ell$ of the electron. What is the magnetic moment of the circulating electron in units of the Bohr magneton?

19. ■■ An unknown magnetic field produces a set of lines for a transition from a state with $\ell = 3$ to one with $\ell = 2$, between which there is a maximum energy difference of $6.1 \times 10^{-4}$ eV. How many lines are there? What is the magnitude of the field?

20. ■■ Find an expression, good to $O(B)$, for the *shift* in the frequency of emitted light that occurs when a magnetic field of magnitude $B$ is applied to a set of atoms. Find an expression also good to $O(B)$, for the (approximate) shift in the wavelength of the emitted light.

21. ■■ Consider a hydrogen atom in a magnetic field. Sketch the possible transitions between the $n = 3$, $\ell = 2$ and $n = 2$, $\ell = 1$ levels. Which lines will *not* appear if we require that $\Delta m = 1$, 0, and $-1$ only? How many spectral lines with different frequencies will be seen in the spectrum? What will the frequencies of these lines be if the magnetic field has magnitude $B = 4$ T?

**22.** ∎∎ Consider the $n = 3$, $\ell = 2$ state of singly ionized helium (a single electron bound to a $Z = 2$ nucleus). A magnetic field of magnitude 3.2 T is applied to the system. What is the maximum splitting $\Delta E$ induced between the no-longer-degenerate energy levels, and what is the ratio $\Delta E/E$?

**23.** ∎∎ Consider a hydrogen atom whose orbiting electron forms a magnetic dipole moment of magnitude $\mu_B$. What is the angular-momentum quantum number of the electron? A beam of these atoms passes through a Stern–Gerlach apparatus. Estimate the field gradient $\partial B_z/\partial z$ necessary to make the force on the atom 100 times the weight of the atom. Give a numerical value. What should be the energy of each atom in the beam such that its observed deflection is 5°?

**24.** ∎∎ Rutherford conjectured that in the nucleus there should be a neutral particle with about the same mass as that of the positively charged proton. Not knowing any better, he took this particle to be a bound state of an electron and a proton. Given that the neutron, proton, and electron have spin 1/2, show that Rutherford's scheme cannot possibly describe the neutron. Rutherford did not know about the neutron, but he could have known about the deuteron, which is the nucleus of "heavy hydrogen" and consists of a neutron and a proton bound together. Show that Rutherford's conjecture also could not explain the deuteron, which is known to have a total angular momentum of unity.

**25.** ∎ **(a)** Given that the neutron and proton each have spin 1/2, show that no combination of spins and orbital angular momenta can make the nucleus of $^3$He, which consists of two protons and a neutron, have integer total angular momentum. **(b)** Given that the neutron and proton each have spin 1/2, show that no combination of spins and orbital angular momentum can make the nucleus of $^4$He, which consists of two protons and two neutrons, have a half-integer angular momentum.

**26.** ∎ What are the total angular momentum values that are obtained in the addition of **(a)** $\ell = 2$ and $s = 1/2$? **(b)** $\ell = 1$ and $s = 3/2$? **(c)** $s_1 = 1/2$ and $s_2 = 1/2$?

**27.** ∎ What angular-momentum states result when angular momentum $\ell_1 = 3$ is added to angular momentum $\ell_2 = 2$? Check that the degeneracies $2\ell + 1$ for these states add up to the total number of states $(2\ell_1 + 1)(2\ell_2 + 1)$.

**28.** ∎∎ Suppose we add three angular momenta: $\ell_1 = 1$, $\ell_2 = 2$, and $\ell_3 = 3$. What are the resulting angular momenta? Show that your result is the same whether you first add $\ell_1$ and $\ell_2$ or whether you first add $\ell_2$ and $\ell_3$.

**29.** ∎∎ A moving charge sets up a magnetic field $B = \mu_0 I/2R$ at the center of a circular loop of radius $R$ carrying current $I$. $\mu_0$ is the magnetic permeability of the vacuum. Suppose an atomic electron is in a state with angular-momentum quantum number $\ell$. **(a)** Use the classical expression to find the magnetic field produced by this electron at the nucleus. **(b)** The proton that forms the nucleus has magnetic moment $(g_p e\hbar)/(2m_p)$. The proton's gyromagnetic ratio is $g_p = 2.79$. *Estimate* the energy of the magnetic interaction of the proton with the circulating electron.

**30.** ∎∎∎ The torque exerted by a magnetic field $\vec{\mathbf{B}}$ on a magnetic dipole of moment $\vec{\boldsymbol{\mu}}$ is given by $\vec{\boldsymbol{\tau}} = \vec{\boldsymbol{\mu}} \times \vec{\mathbf{B}}$. Assuming that the classical equation for the rate of change of the angular momentum, namely, $d\vec{\mathbf{L}}/dt = \vec{\boldsymbol{\tau}} = \dfrac{ge}{2m} \vec{\mathbf{L}} \times \vec{\mathbf{B}}$, also holds in quantum mechanics, write out the equations of motion for $L_x$, $L_y$ and $L_z$ for a constant magnetic field $\vec{\mathbf{B}}$ pointing in the z-direction. Show that the angular-momentum vector precesses, and obtain an expression for the frequency of precession. The quantum mechanics equations are exactly the same as the classical ones, except that the angle which $L_z$ makes with the direction of $\vec{B}$ is constrained by the fact that $L_z$ is quantized.

**31.** ▮ The magnetic moment associated with the orbital motion of an atomic electron has magnitude $3.213 \times 10^{-23}$ A·m². What is the angular-momentum quantum number of the electron's orbit?

**32.** ▮▮ Consider positronium, an "atom" formed by the coulomb attraction between an electron and a positron (charge $+e$, mass $m_e$). What is the value of the length scale $a_0$—the quantity analogous to the $a_0$ that appears in the exponent of Eq. (9–38)—for this atom? What are the allowed energies for this atom, in units of the lowest energy of "ideal hydrogen" considered throughout this chapter? The reduced mass is essential for the calculation. (See Problems 2 through 5.)

**33.** ▮▮ Estimate the value of external magnetic field for which the energy splittings due to the Zeeman effect are comparable in magnitude to spin–orbit effects. The relevance of this problem is that for fields much stronger than this, the spin–orbit splitting can be neglected in comparison with the splittings induced by the external field.

**34.** ▮▮ Experimentalists have devised an interesting way to study the motion of the surface of a star. They look at a strong spectral line in potassium ($\lambda = 395$ nm) through a cell containing potassium vapor. Because of the Doppler shift the particular line will not be absorbed. However, by putting the cell in a magnetic field one can get one of the Zeeman-shifted potassium lines to match the frequency of the incoming light. By varying the magnetic field till one gets absorption of the incoming line, one can determine the frequency of the incoming light. Typically, magnetic fields in the laboratory are on the order of 0.01 T. What order of magnitude Doppler shifts can be detected this way?

# Many Particles

We have studied the wave equation obeyed by a single particle and have managed to understand some of the meaning and consequences of the particle's quantum behavior. When many particles are involved, there are two types of effects, one expected and another not at all expected. The expected behavior involves the simple generalization of a one-particle wave function to a many-particle wave function, as well as the generalization of the Schrödinger equation and its solution. The unexpected types of effects concern a new fact: At the microscopic level, there are classes of particles that are, literally, indistinguishable from one another. For example, all electrons are strictly identical. Once we have identical particles we are naturally led to thinking about how we can construct wave functions for them. This leads to new quantum mechanical symmetries whose consequences are direct: These symmetries guide atomic structure and hence all chemistry and biology, as well as the structure of solids, the evolution of stars, and conduction in materials.

## 10–1 The Multiparticle Schrödinger Equation

It is reasonable to expect that an extension of the one-dimensional Schrödinger equation to the description of $N$ particles must involve a wave function of the form $\psi(x_1, x_2, x_3, \ldots, x_N; t)$, where the position variable[1] of particle $i$ is $x_i$. A straightforward extension of the Schrödinger equation has the form

$$i\hbar \frac{\partial \psi(x_1, x_2, \ldots; t)}{\partial t} = H\psi(x_1, x_2, \ldots; t), \qquad (10\text{–}1)$$

where $H$ is the energy operator, which we again refer to as the Hamiltonian. The classical form of the energy is

$$\frac{p_1^2}{2m_1} + \frac{p_2^2}{2m_2} + \cdots + V(x_1, x_2, \ldots, x_N). \qquad (10\text{–}2)$$

---

[1] We have used a notation here that corresponds to each of the particles in one dimension. To generalize to a multiparticle three-dimensional equation, we would have to employ three spatial coordinates for particle 1, three for particle 2, and so forth. This notation, however, would only complicate the presentation at this point.

To go to a quantum mechanical formulation, we take $H$ to have the classical form of the energy, but with the usual transcription $p_1 \to -i\hbar(\partial/\partial x_1)$ and so forth, for each momentum. This leads to the partial differential equation

$$i\hbar \frac{\partial \psi(x_1, x_2, \ldots; t)}{\partial t} = \left( -\frac{\hbar^2}{2m_1} \frac{\partial^2}{\partial x_1^2} - \frac{\hbar^2}{2m_2} \frac{\partial^2}{\partial x_2^2} - \cdots + V(x_1, x_2, \ldots) \right)$$

$$\psi(x_1, x_2, \ldots; t). \qquad (10\text{-}3)$$

The wave function has the interpretation that

$$\left| \psi(x_1, x_2, \ldots; t) \right|^2 dx_1\, dx_2 \ldots dx_N \qquad (10\text{-}4)$$

is the probability of finding particle 1 in the range $(x_1, x_1 + dx_1)$, particle 2 in the range $(x_2, x_2 + dx_2)$, and so forth.

As we have done in Chapter 6 and again in Chapter 9, we reduce the equation to a time-independent one by separation of variables:

$$\psi(x_1, x_2, \ldots; t) = u(x_1, x_2, \ldots, x_N) \exp(-iEt/\hbar). \qquad (10\text{-}5)$$

When we substitute this equation into Eq. (10-3) and cancel the exponential factor, we get an equation that no longer involves $t$:

$$Eu(x_1, x_2, \ldots, x_N) = \left( -\frac{\hbar^2}{2m_1} \frac{\partial^2}{\partial x_1^2} - \frac{\hbar^2}{2m_2} \frac{\partial^2}{\partial x_2^2} - \cdots + V(x_1, x_2, \ldots) \right)$$

$$u(x_1, x_2, \ldots, x_N). \qquad (10\text{-}6)$$

This is an *energy eigenvalue equation*, so-called because it has solutions only for certain values of $E$, where $E$ is the total energy of all the particles. The total energy will take on discrete values if the $N$ particles are bound together into a single, multibody bound system. An atom of gold, consisting of 79 electrons and a single nucleus, is an example of such a system. Or the total energy could be continuous if the system is not bound. A single free electron and a lead atom from which the single electron has been removed is an example of this kind of a system. Equation (10-6) is the multiparticle version of the time-independent one-particle Schrödinger equation (6-38).

## 10–2  Independent Particles

We shall first treat Eq. (10-6) for a very simple case, that of $N$ particles which do not interact with each other, but which each experience an external potential. We say we have independent particles in that case. An example of this might be a collection of one of each type of atom in the periodic table, separated far enough from each other so that any electrical forces between them can be neglected. (Why not just say a collection of hydrogen atoms? We shall see later in this chapter that identical particles—and each hydrogen atom with a single proton as its nucleus is identical to every other hydrogen atom with a single proton as its nucleus—are very special.) For independent particles (i.e., particles that do not interact with each other),

$$V(x_1, x_2, \ldots, x_N) = V_1(x_1) + V_2(x_2) + \ldots + V_N(x_N). \qquad (10\text{-}7)$$

As far as we know at this point, each of our particles could be subject to a *different* potential, and that is why we have put a label relating to the particles on the individual potentials. We can now write

$$Eu(x_1, x_2, \ldots, x_N) = \left[ \left( -\frac{\hbar^2}{2m_1} \frac{\partial^2}{\partial x_1^2} + V_1(x_1) \right) + \left( -\frac{\hbar^2}{2m_2} \frac{\partial^2}{\partial x_2^2} + V_2(x_2) \right) + \cdots \right]$$

$$u(x_1, x_2, \ldots, x_N). \quad (10\text{–}8)$$

The right-hand side consists of $N$ terms, each of which is the kinetic energy plus the potential energy of a particular particle. The particles are independent of each other in that the energy of any one particle does not depend on any other. We thus expect that the probability of finding a particular particle somewhere is independent of the probabilities of finding other particles anywhere else. This in turn suggests that

$$u(x_1, x_2, \ldots, x_N) = u(x_1)u(x_2) \ldots u(x_N), \quad (10\text{–}9)$$

so that the probability of finding particle 1 in the range $(x_1, x_1 + dx_1)$, and simultaneously particle 2 in the range $(x_2, x_2 + dx_2)$, and so forth, is the product of the independent single particle probabilities,

$$\{|u_1(x_1)|^2 \, dx_1\} \{|u_2(x_2)|^2 \, dx_2\} \ldots \{|u_N(x_N)|^2 \, dx_N\}. \quad (10\text{–}10)$$

With this interpretation in hand, our task now is to find the individual functions $u_i$, and to do this we must find the equations they obey. When Eq. (10–9) is substituted into Eq. (10–8), we have, after dividing by $u(x_1)u(x_2) \ldots u(x_N)$,

$$E = \frac{\left( -\frac{\hbar^2}{2m_1} \frac{d^2}{dx_1^2} + V_1(x_1) \right) u_1(x_1)}{u_1(x_1)} + \frac{\left( -\frac{\hbar^2}{2m_2} \frac{d^2}{dx_2^2} + V_2(x_2) \right) u_2(x_2)}{u_2(x_2)} + \cdots$$

$$+ \frac{\left( -\frac{\hbar^2}{2m_N} \frac{d^2}{dx_N^2} + V_N(x_N) \right) u_N(x_N)}{u_N(x_N)}. \quad (10\text{–}11)$$

(Note that we cannot just cancel the functions $u_i(x_i)$ in each of the terms on the right, because in the numerator something operates on the function! Note also that the partial derivative notation is no longer necessary, because only one variable appears in each term.) The left side of Eq. (10–11) does not depend on any of the $x_i$, while the right side is a sum of terms that are functions only of $x_1, x_2, \ldots$. Thus, the equation can be satisfied only if each one of the terms on the right side is a constant. We write these constants as $E_1, E_2, \ldots$ for the respective terms on the right side, and, accordingly, we solve the full equation if we solve each of the $N$ equations

$$\left( -\frac{\hbar^2}{2m_1} \frac{d^2}{dx_1^2} + V_1(x_1) \right) u_1(x_1) = E_1 u_1(x_1),$$

$$\left( -\frac{\hbar^2}{2m_2} \frac{d^2}{dx_2^2} + V_2(x_2) \right) u_2(x_2) = E_2 u_2(x_2),$$

$$\vdots$$

$$\left( -\frac{\hbar^2}{2m_N} \frac{d^2}{dx_N^2} + V_N(x_N) \right) u_N(x_N) = E_N u_N(x_N). \quad (10\text{–}12)$$

The functions $u_i(x_i)$ are therefore solutions of one-particle Schrödinger equations, with potentials $V_i(x_i)$ and one-particle energy eigenvalues $E_i$. This interpretation is cemented because Eq. (10–11) implies that the total energy is the sum of the single particle energies; that is,

$$E = E_1 + E_2 + \ldots + E_N. \quad (10\text{–}13)$$

## 10–3  Identical Particles

One of the most striking features of the microscopic world is the lack of variability of the constituents. On our human scale, no two planets are the same, no two sets of fingerprints match, and no two snowflakes are identical. But this complexity is the result of the many ways of arranging just a few kinds of constituents. At the atomic scale it is quite literally impossible to distinguish among constituents of a specific kind. Spectroscopic evidence tells us that there is only one kind of hydrogen atom[2] and one kind of helium atom. Thus even though each hydrogen atom has an infinite number of excited states, all hydrogen atoms have the same set of excited states. All electrons are identical. Measurements show that they all have exactly the same mass, electric charge, magnetic dipole moment, and so on. Spectroscopy provides us with more evidence: The spectrum of a given element is always the same, and this would not be the case if there were several types of electrons that could lead to different versions of the same element. Similar evidence from nuclear physics suggests that all protons, or all neutrons, are the same. This indistinguishability appears to be a law of nature: There is simply no way to "tag" an electron or a proton—to place, say, a little spot of red paint on one or to follow one of two electrons that have collided with each other with a movie camera (•Fig. 10–1).

What are the implications of this indistinguishability for a system of many identical particles, like a many-electron system? For one thing, we cannot have different electrons having different potential energies. Consider, for simplicity, a two-electron system. The potential energy in Eq. (10–7) might be

$$V(x_1, x_2) = V_1(x_1) + V_2(x_2),$$

where, for example, $V_1$ is an infinite well of width $L_1$ and $V_2$ is an infinite well of width $L_2$, with $L_2 \gg L_1$. But such a potential would immediately allow us to distinguish between the two electrons: One is in a small box, the other is in a large box, and they have different energy spectra. Since the functional forms of the two potentials are different, an interchange of the two identical electrons would lead to the potential energy

$$V(x_2, x_1) = V_1(x_2) + V_2(x_1),$$

which is *different* from $V(x_1, x_2)$. Interchanging the two electrons would allow us to distinguish between that situation and the one in which no exchange has occurred. And this would mean that the electrons are distinguishable. The only way to avoid such a state of affairs is to have

$$V(x_1, x_2) = V(x_2, x_1). \tag{10–14}$$

In turn, we can in general satisfy the resulting relation $V_1(x_1) + V_2(x_2) = V_1(x_2) + V_2(x_1)$ only if the potentials $V_1$ and $V_2$ are *identical*—that is, only if

$$V_1(x) = V_2(x) \equiv V(x). \tag{10–15}$$

In other words, if two particles are truly indistinguishable, they must be subject to the *same* external potential

$$V(x_1, x_2) = V(x_1) + V(x_2). \tag{10–16}$$

---

[2] We are ignoring the possibility that hydrogen can have three different types of nuclei; this is merely a detail for the purposes of our discussion here.

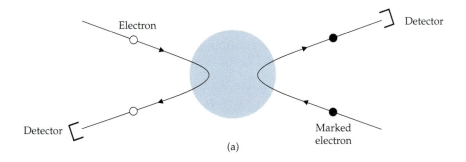

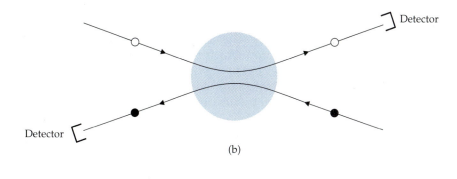

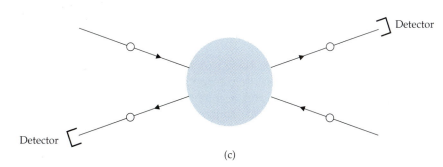

• **Figure 10–1** In a collision involving two classical electrons, two different paths will send electrons into the same detectors. These paths are shown in (a) and (b) for one "marked" and one "unmarked" electron. In one case the marked electron undergoes large angle scattering; in the other the collisions are glancing. Even without the marking, a camera would in principle allow you to follow the electrons and tell which type of collision occurred. In case (c), involving strictly identical electrons, there is no way of marking or following them. The scattering process is described by a single wave function that involves the amplitudes for both the large-angle and the glancing collisions.

We generalize this situation easily to a system of $N$ independent electrons—a situation that we shall encounter later in our discussion of bulk materials. We have

$$V(x_1, x_2, \ldots, x_N) = V(x_1) + V(x_2) + \ldots + V(x_N). \quad (10\text{–}17)$$

We now observe that having the same functional form for all the potentials is not enough. For suppose we (incorrectly) assume that the wave function continues to have the factorized form of Eq. (10–9), viz.,

$$u(x_1, x_2, \ldots, x_N) = u_{n_1}(x_1)u_{n_2}(x_2) \ldots u_{n_N}(x_N). \quad (10\text{–}18)$$

Here the individual pieces $u_i(x)$ of the full wave functions solve the equation

$$\left(-\frac{\hbar^2}{2m}\frac{d^2}{dx^2} + V(x)\right)u_i(x) = E_i u_i(x). \quad (10\text{–}19)$$

The $u_i(x)$ are eigenfunctions of the one-particle Schrödinger equation with potential $V(x)$, distinguished only by their eigenvalues $E_i$. Again, the total energy $E$ is the sum of all the individual energies $E_1, E_2, \ldots, E_N$.

But there is a problem with the solution of the form given in Eq. (10–18): It assigns electron "1" the energy $E_{n_1}$, electron "2" the energy $E_{n_2}$, and so forth, and this assignment would be a way to distinguish the electrons. To understand the problem and its resolution more clearly, let us again concentrate on a two-electron system. Then, if we suppose electron 1 is in the state labeled $m$ and electron 2 in the state labeled $n$, our naive solution takes the form

$$u(x_1, x_2) = u_m(x_1)u_n(x_2). \tag{10-20}$$

With this form, we can say that electron 1 has energy of the $m$th state, $E_m$, and electron 2 has the energy of the $n$th state, $E_n$. On the other hand, if we interchange the two electrons, we have the wave function

$$u(x_1, x_2) = u_m(x_2)u_n(x_1), \tag{10-21}$$

and this also represents a distinguishable situation, since now electron 1 has energy $E_n$ and electron 2 has energy $E_m$. Because electrons are indistinguishable, neither of these wave functions is satisfactory. Here is how we can solve the problem: We take a wave function which is a combination that remains the same when we switch the two electrons or which is a combination that changes in overall sign when we switch the two electrons. The overall sign change is allowed because physical probabilities involve the *square* of the wave function. These two combinations are

$$u_S(x_1, x_2) = \frac{1}{\sqrt{N_S}} \left[ u_m(x_1)u_n(x_2) + u_m(x_2)u_n(x_1) \right] \tag{10-22}$$

and

$$u_A(x_1, x_2) = \frac{1}{\sqrt{N_A}} \left[ u_m(x_1)u_n(x_2) - u_m(x_2)u_n(x_1) \right]. \tag{10-23}$$

(The factors in front are normalization factors; for the problem of electrons in an infinite well, it is easy to show that they are both $1/\sqrt{2}$—see Problem 1 at the end of the chapter.) The subscripts $S$ and $A$ stand for **symmetric** and **antisymmetric**, referring to whether the overall wave function remains the same or changes sign when we switch the two electrons. Explicitly, the switched forms are

$$u_S(x_2, x_1) = +u_S(x_1, x_2), \quad \text{and} \quad u_A(x_2, x_1) = -u_A(x_1, x_2). \tag{10-24}$$

In this context we often say that the symmetric wave function is invariant under the exchange of particles, while the antisymmetric wave function changes sign under such an exchange.

We can extend the notion of symmetric and antisymmetric wave functions to more than two independent particles. For three particles in the states labeled $m$, $n$, and $p$, for example, the antisymmetric form is

$$\begin{aligned} u_A(x_1, x_2, x_3) = \frac{1}{\sqrt{N_A}} \big[ & u_m(x_1)u_n(x_2)u_p(x_3) - u_m(x_2)u_n(x_1)u_p(x_3) \\ & - u_m(x_1)u_n(x_3)u_p(x_2) + u_m(x_3)u_n(x_1)u_p(x_2) \\ & + u_m(x_2)u_n(x_3)u_p(x_1) - u_m(x_3)u_n(x_2)u_p(x_1) \big]. \end{aligned} \tag{10-25}$$

You can verify that this wave function changes its overall sign under particle exchange by exchanging the names of any pairs of variables. The symmetric wave function is of the same form as Eq. (10–25), except that all the signs are positive. When there are even more identical particles, things get yet more complicated, but can be worked out, although we shall not bother with that here.

Nothing in the discussion to now has told us what the correct wave function is for a pair of electrons or a pair of other identical particles. Which do we choose, symmetric or antisymmetric wave functions? Do we have a choice in any given situation? The answer to these questions is provided by a profound law of nature, discovered by Wolfgang Pauli, that has no classical counterpart.

## 10–4 Exchange Symmetries and the Pauli Principle

Wolfgang Pauli (•Fig. 10–2) was an Austrian-born theoretical physicist who was famous for both his genius and his devastating wit. He once said of a piece of research which particularly displeased him that it was "not even wrong." He wrote a masterful review of relativity theory—still worth reading—when he was a teenager. Einstein came to give a lecture with Pauli in attendance; Pauli began some public remarks after the lecture by saying, "What Mr. Einstein has just said is not so stupid!"

In 1924, before the advent of the quantum theory, Pauli, who was then 24, discovered that to re-create the periodic table it was necessary to assume as a principle that no more than two electrons could have a given atomic quantum number. This idea is the **Pauli exclusion principle**, or, more simply, the **exclusion principle**. Pauli also was the first to introduce the notion that an electron has to have an additional quantum number that can have two values, to account for the "two" in the statement that no more than two electrons can be in the same state. A few years later, the Dutch physicists Samuel Goudsmit and George Uhlenbeck proposed the existence of the electron spin. Pauli's two-valued quantum number was identified as the two-valued z-component of the spin—that is, the two possible eigenvalues $\pm \hbar/2$ of the spin operator $S_z$. When this spin quantum number is included in the full collection of quantum numbers to which Pauli's exclusion principle applies, the exclusion principle can be stated in the form

*No more than one electron can have a given set of quantum numbers.*

As Pauli first formulated it, the exclusion principle is a statement about *quantum numbers*. But with the discovery of the Schrödinger equation and wave functions labeled by quantum numbers, the implications of the exclusion principle for *wave functions* of identical particles became clear. We have already noted that identical particles such as electrons must be described by wave functions that are symmetric or antisymmetric under the exchange of the particles. The exclusion principle tells us that the choice is not arbitrary: For electrons, the wave functions must be *antisymmetric*. (We'll see why below.) And there is more: With the discovery of new particles and the understanding of the blackbody radiation spectrum as a consequence of symmetry in the many-photon wave function, the Pauli exclusion principle has a generalization in terms of wave functions that applies to *all* identical particles.

All particles fall into two classes. **Fermions** consist of electrons, protons, neutrons, and, in general, systems whose total angular momentum, including spin, is 1/2, 3/2, 5/2, .... (Fermions were named after Enrico Fermi (•Fig. 10–3), who, with Paul A. M. Dirac (•Fig. 10–4), studied the consequences of the many-body properties of these particles.) **Bosons** consist of photons, helium nuclei, hydrogen atoms, and, in general, systems whose total angular momentum, including spin, is 0, 1, 2, .... (Bosons were named after the Bengali physicist S. N. Bose, who, with Einstein, explored the many-body properties of the most prominent example of this class, the photon.) The generalized Pauli principle, which we may call the **exchange symmetry principle**,[3] states that

• **Figure 10–2** Wolfgang Pauli was born in Vienna in 1900 and died in Zurich in 1958. He was one of the most brilliant theoretical physicists of this century.

• **Figure 10–3** The Italian-born American physicist Enrico Fermi (1901–1954) was both a brilliant theorist and an equally brilliant experimental physicist. Among his many accomplishments was his leadership in the construction of the first nuclear reactor. He is shown here on a hike with Niels Bohr in 1931.

---

[3] The connection between symmetry under the interchange of identical particles and their spin was proven by Pauli in 1940, many years after the development of quantum mechanics. Relativity plays an important role in the proof.

• **Figure 10–4**  Paul A. M. Dirac was one of the pioneers of quantum mechanics. He developed the quantum theory of radiation and did groundbreaking work in the creation of relativistic quantum mechanics. He predicted the existence of anti-matter. The French physicist Leon Brillouin, himself an important contributor, is in the background.

▌ We discuss multiple bosons further in Chapters 12 through 14.

*The wave function of a many-particle system is antisymmetric under the exchange of two identical fermions and symmetric under the exchange of two identical bosons.*

As we shall see, this seemingly innocuous statement—especially the part about fermions—has powerful implications.

An immediate corollary of the exchange symmetry principle when it is applied to fermions is that

*No two identical fermions can be in the same quantum mechanical state.*

In other words, the statement of the exclusion principle in terms of quantum numbers follows immediately from the exchange symmetry principle, which is a statement about wave functions.

It is easy to deduce the corollary above: From Eq. (10–23), the antisymmetric wave function for two electrons in states $a$ and $b$ takes the form

$$u(x_1, x_2) = \mathcal{N}[u_a(x_1)u_b(x_2) - u_a(x_2)u_b(x_1)], \tag{10–26}$$

where $\mathcal{N}$ is a normalization factor. But if our two fermions—electrons, for example—have the same quantum numbers, then they will be described by the same wave function, with the same label. If we make the labels $a$ and $b$ identical, our wave function is zero, verifying that the electrons cannot be in the same state. Note that the labels $a$ and $b$ include the spin labeling. If the two spin states were different, we could not write $a = b$. In the discussion that follows, we symbolize each of the two states by an explicit label according to whether the $S_z$ eigenvalue is $+\hbar/2$ or $-\hbar/2$. We describe these states as "spin up" and "spin down" or with the subscripts "↑" and "↓". The importance of the explicit labeling will be apparent when we discuss the case of two electrons with spin as their only degree of freedom.

The exchange symmetry principle has important implications for bosons as well, but nothing quite as simple as the exclusion principle. With the exception of the next example, we concentrate on fermions in the remainder of this chapter.

**Example 10–1**    Consider two independent spin-zero bosons—there is no interaction between them, and we don't need to worry about a spin label—each in the same one-dimensional infinite well. Recall that the single particle wave function in that case is

$$u_n(x) = \sqrt{\frac{2}{L}} \sin\left(\frac{n\pi x}{L}\right),$$

where $n$ is the quantum number that labels the state. Construct the wave functions of the two-boson system, and show that there is no exclusion principle (i.e., that the wave function is not zero when both bosons are in the same state).

**Solution**    The combination of the product of two single-particle wave functions that is symmetric under the interchange of the two particles has the general form

$$u_{\text{sym}}(x_1, x_2) = \mathcal{N}'[u_m(x_1)u_n(x_2) + u_m(x_2)u_n(x_1)],$$

where $\mathcal{N}'$ is a normalization constant. This is just the form taken by Eq. (10–26), which is the wave function for two fermions, but with a plus sign in place of the minus sign.

When the two bosons are in the same state, $m = n$, and our wave function becomes

$$u_{\text{sym}}(x_1, x_2) = 2\mathcal{N}'u_n(x_1)u_n(x_2) = 2\mathcal{N}'\frac{2}{L}\sin\left(\frac{n\pi x_1}{L}\right)\sin\left(\frac{n\pi x_2}{L}\right).$$

This is not zero.

## The Total Spin of Two Electrons

In the absence of any spatial dependence, the only property that an electron has is its spin state. We can now classify two-electron states by making use of the fact that spins sum as angular momenta. In doing so we shall ignore spatial dependence entirely. According to the general rules developed in the previous chapter, if we add two angular momenta $L_1$ and $L_2$, we obtain a series of total angular momenta ranging from $|L_1 + L_2|$ to $|L_1 - L_2|$. In the case of two spin-1/2 particles, we should be able to form a total spin $S = 1$ and a total spin $S = 0$. There are three $m$-values $[1, 0, \text{and} -1]$ for the $S = 1$ combination and one $m$-value $[0]$ for the $S = 0$ combination—four in all. This is in accord with an approach using the $z$-components of the original spins. If we have two electrons, their $S_z$ values can be $S_{1z} = \pm\hbar/2$ and $S_{2z} = \pm\hbar/2$. These can be combined in $2 \times 2 = 4$ ways to form four states. Two of these are easy to recognize: The state with $S_{1z} = +\hbar/2$ and $S_{2z} = +\hbar/2$ forms a state with a total $z$-component of angular momentum—a *total spin projection*—of $S_z = \hbar(1/2 + 1/2) = +\hbar$, while the state with $S_{1z} = -\hbar/2$ and $S_{2z} = -\hbar/2$ forms a total spin projection $S_z = \hbar(-1/2 - 1/2) = -\hbar$. These must correspond to $S = 1$ and $m = +1$ and $-1$, respectively. The wave functions for these cases can be written as follows:

$$S = 1, m = +1: \qquad \chi_+(1)\chi_+(2); \qquad\qquad (10\text{-}27a)$$

$$S = 1, m = -1: \qquad \chi_-(1)\chi_-(2). \qquad\qquad (10\text{-}27b)$$

Here $\chi_\pm(1)$ or $\chi_\pm(2)$ is the spin part of the wave function for electron 1 and electron 2, respectively, with spin up or spin down according to the subscript $+$ or $-$.

We have now seen how to form the total spin-1 states with spin projection $+1$ and $-1$. The remaining two states are formed with one electron with spin up and the other with spin down, so that both cases have $m = 0$. How, then, is it possible to tell which state corresponds to $S = 1, m = 0$ and which corresponds to $S = 0, m = 0$? A full answer to this question requires more advanced methods than we wish to make use of here. We can, however, observe that the $m = \pm1$ states do not change sign under the interchange of the two electron labels 1 and 2—that is, they are *even* under that interchange. But there is nothing magical about our choice of the $z$-direction, so that if we had projected along any other axis the $S = 1$ states would still be even under the interchange. In other words, all the $S = 1$ projections, including the $m = 0$ part of the $S = 1$ wave function, are even under interchange. Other methods confirm that the $S = 1, m = 0$ state is the *symmetric* combination of one spin up and one spin down, namely,

$$S = 1, m = 0: \qquad \chi_+(1)\chi_-(2) + \chi_-(1)\chi_+(2). \qquad (10\text{-}27c)$$

The symmetry of this wave function under the interchange of the 1 and 2 labels is evident. (The wave function is not properly normalized, but that is of no importance in this context.)

At this point we might guess that the independent combination of spin up and spin down that is *antisymmetric* under interchange—we say that this state is *odd* under interchange—is the $S = 0, m = 0$ state, and that guess would be correct:

$$S = 0, m = 0: \qquad \chi_+(1)\chi_-(2) - \chi_-(1)\chi_+(2). \qquad (10\text{-}28)$$

(Again, we have not worried about normalization.) The three $S = 1$ states of Eqs. (10–27) together are said to form a **triplet** state, while the $S = 0$ state of Eq. (10–28) forms a **singlet** state.

Now if there were no spatial wave function for the electrons, which of these states would be realized? The answer is supplied by the requirement that the overall wave function of two electrons be antisymmetic. This means that with no spatial wave function, the electrons "choose" Eq. (10–28), the singlet state; that is the antisymmetric state.

How can the presence of a spatial wave function change this "selection"? As we have remarked in the previous chapter, the wave function of the electrons is composed of a spatial part multiplied by a spin part, with the degree of freedom of the one independent of the other. Accordingly, let us think about two electrons again, this time including a spatial wave function. In particular, suppose the two electrons are in the ground state of an (ordinary) attractive three-dimensional potential.

We can start by ignoring the exclusion principle and treating the electrons as if they were distinguishable. We can then label one of the electrons as "1" and the other as "2." Let us also denote the single-particle spatial ground-state wave function as $u_0$. In that case, if the electron labeled, say, 1 were independent of the other electron, then its wave function would be $u_0(\vec{r}_1)\chi(1)$. Note that we have not specified whether the spin is up or down. The overall wave function would be the simple product of the wave function of each electron, namely, $u_0(\vec{r}_1)u_0(\vec{r}_2)\chi(1)\chi(2)$.

Now we take the exclusion principle into account. We do this in two steps. First, since two electrons cannot be in the same quantum state, and since both are in the spatial ground state, one must have spin up and the other spin down; that is, we should first write, say, $u_0(\vec{r}_1)u_0(\vec{r}_2)\chi_+(1)\chi_-(2)$. Second, the exclusion principle requires us to make the overall wave function antisymmetric on the exchange of labels—that is, to take our trial wave function and subtract from it the same form, but with the labels reversed. Thus we have

$$\psi = \mathcal{N}\{u_0(\vec{r}_1)u_0(\vec{r}_2)\chi_+(1)\chi_-(2) - u_0(\vec{r}_2)u_0(\vec{r}_1)\chi_+(2)\chi_-(1)\}$$

$$= \mathcal{N}\, u_0(\vec{r}_1)u_0(\vec{r}_2)[\chi_+(1)\chi_-(2) - \chi_-(1)\chi_+(2)].$$

The factor $\mathcal{N}$ is for normalization.

From this expression we see that if the two electrons have a symmetric spatial wave function, then the spins must be antisymmetric. *In this case, the electrons have a total spin of 0.* Is it ever possible, then, for the two electrons to have a total spin of 1? To arrange that, we need only make certain that the spatial part of the wave function is antisymmetric under interchange. For example, we could have one electron in the first excited state, wave function $u_1$, and the other in the ground state, wave function $u_0$. Under those conditions, the following total wave function is indeed antisymmetric, with a total spin of 1 (and a spin projection quantum number $m = +1$):

$$\psi = \mathcal{N}[u_0(\vec{r}_1)u_1(\vec{r}_2) - u_0(\vec{r}_2)u_1(\vec{r}_1)]\chi_+(1)\chi_+(2).$$

One can also construct the equivalent case with total spin 1, but with a spin projection 0.

### Electrons in a Well

Let us look at a simple example, that of one or several electrons in a one-dimensional infinite well of length $L$. We worked out the single-particle wave functions for the case of the infinite well in Chapter 6. The wave functions are

$$u_n(x) = \sqrt{\frac{2}{L}}\sin\frac{n\pi x}{L}, \qquad n = 1, 2, \ldots, \tag{10–29}$$

with corresponding energy levels

$$E_n = \frac{\pi^2 \hbar^2 n^2}{2mL^2}. \qquad (10\text{–}30)$$

The integer $n$ labels the state. The case $n = 1$ corresponds to the lowest energy and therefore represents the ground state for a single electron.

With spin present, *two* electrons can be accommodated for each infinite-well state labeled $n$, one electron with spin up, the other with spin down. A third electron would either have to be spin up or spin down, and by the exclusion principle, it could not be placed in the state labeled $n$ (•Fig. 10–5). The lowest energy state for two electrons in the infinite well is that of each electron being in the state $n = 1$, one with spin up and the other spin down. For three electrons, the lowest state contains two electrons in the $n = 1$ state, but the third electron must be in the $n = 2$ state, with its spin either up or down. All this is made explicit in the next two examples.

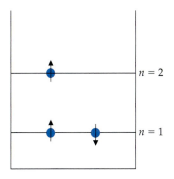

• **Figure 10–5**  When three electrons are in a well, only two can go into the one-particle ground state—one with spin up and one with spin down. To get the lowest energy three-electron state—the ground state—the third electron must be placed into the first excited one-electron state.

**Example 10–2**  **(a)** Write out the wave function for the ground state of a system of two electrons in a one-dimensional potential well. Write your wave function in terms of the single-particle eigenfunctions $u_{n\sigma}(x)$, where $n$ labels the energy (with $n = 1$ labeling the ground state), and $\sigma$ labels the spin. **(b)** What is the total spin of the system?

**Solution**  **(a)** The ground state will have both electrons in their lowest energy state, that is, with $n = 1$. We have to make a combination of products of $u_{1\uparrow}(x_1)$, $u_{1\downarrow}(x_1)$, $u_{1\uparrow}(x_2)$, and $u_{1\downarrow}(x_2)$ that is antisymmetric under the interchange $x_1 \leftrightarrow x_2$. [We have represented the spin label by the arrows ↑ (up) and ↓ (down).] Consider the most general form:

$$A u_{1\uparrow}(x_1)u_{1\uparrow}(x_2) + B u_{1\uparrow}(x_1)u_{1\downarrow}(x_2) + C u_{1\downarrow}(x_1)u_{1\uparrow}(x_2) + D u_{1\downarrow}(x_1)u_{1\downarrow}(x_2).$$

Under the interchange $x_1 \leftrightarrow x_2$, the first term and the last term are unaltered; thus they cannot change sign. This means that $A = 0$ and $D = 0$. However the second term turns into the third and the third into the second under that interchange. This means that we can make an antisymmetric combination with $C = -B$. We see that

$$u_{1\uparrow}(x_1)u_{1\downarrow}(x_2) - u_{1\downarrow}(x_1)u_{1\uparrow}(x_2)$$

is indeed antisymmetric.

This form verifies what you would write merely by using the exclusion principle and placing one electron in the up state and the other in the down state, with a minus sign between the two possible choices for which electron is up and which is down.
**(b)** We can easily see that the total spin of the ground state is zero by changing notation slightly: Instead of combining the space and spin labels by writing the single-particle wave functions as $u_{n\sigma}(x)$, we write $u_n(x)\chi_\pm$, where $u_n(x)$ is the spatial part and $\chi_\pm$ is the spin part, as we did in the case of the total spin of two electrons. In that case, our solution contains an overall factor $u_n^2(x)$, where $n$ corresponds to the ground state, multiplying the combination $\chi_+(1)\chi_-(2) - \chi_-(1)\chi_+(2)$. But we have already shown that this spin combination represents a total spin of zero.

**\*Example 10–3**  Write out the wave function for the lowest energy state of *three* electrons in a potential well. Use the notation $u_{n\sigma}(x)$ for the single-electron wave functions, with $n = 1$ representing the single-electron ground state and $n = 2$ representing the first excited state of a single electron.

**Solution**  By the exclusion principle, two electrons can go into the $n = 1$ state, and the remaining electron must go into the $n = 2$ state. This third electron can be in an up or down state, with the same energy for either. Let us concentrate on the up state. We might naively write the product of the antisymmetric combination $\{u_{1\uparrow}(x_1)u_{1\downarrow}(x_2) - u_{1\downarrow}(x_1)u_{1\uparrow}(x_2)\}$ and $u_{2\uparrow}(x_3)$, namely

$$\{u_{1\uparrow}(x_1)u_{1\downarrow}(x_2) - u_{1\downarrow}(x_1)u_{1\uparrow}(x_2)\}u_{2\uparrow}(x_3).$$

However, this product would not be antisymmetric under the interchange of any pair of electrons. In particular, the $x_3$ term is singled out in the $n = 2$ state. To fix this, we take our product and subtract from it the same expression with $x_3$ and $x_2$ exchanged, as well as a second expression with $x_3$ and $x_1$ exchanged. Explicitly, this combination is

$$\{u_{1\uparrow}(x_1)u_{1\downarrow}(x_2) - u_{1\downarrow}(x_1)u_{1\uparrow}(x_2)\}u_{2\uparrow}(x_3) - \{u_{1\uparrow}(x_1)u_{1\downarrow}(x_3) - u_{1\downarrow}(x_1)u_{1\uparrow}(x_3)\}u_{2\uparrow}(x_2)$$
$$- \{u_{1\uparrow}(x_3)u_{1\downarrow}(x_2) - u_{1\downarrow}(x_3)u_{1\uparrow}(x_2)\}u_{2\uparrow}(x_1).$$

Multiplying out, we obtain

$$u_{1\uparrow}(x_1)u_{1\downarrow}(x_2)u_{2\uparrow}(x_3) - u_{1\downarrow}(x_1)u_{1\uparrow}(x_2)u_{2\uparrow}(x_3)$$
$$- u_{1\uparrow}(x_1)u_{1\downarrow}(x_3)u_{2\uparrow}(x_2) + u_{1\downarrow}(x_1)u_{1\uparrow}(x_3)u_{2\uparrow}(x_2)$$
$$- u_{1\uparrow}(x_3)u_{1\downarrow}(x_2)u_{2\uparrow}(x_1) + u_{1\downarrow}(x_3)u_{1\uparrow}(x_2)u_{2\uparrow}(x_1).$$

It is easy to check that this expression is antisymmetric—it changes sign—under the interchange of any two of the electron labels.

---

The foregoing two examples describe the ground state for two and three electrons in a potential well. What about the first excited state? For simplicity consider the case of two electrons. For the first excited state we have products made of one out of the list $u_{1\uparrow}(x_1)$, $u_{1\downarrow}(x_1)$, $u_{2\uparrow}(x_1)$, and $u_{2\downarrow}(x_1)$ and one out of the list $u_{1\uparrow}(x_2)$, $u_{1\downarrow}(x_2)$, $u_{2\uparrow}(x_2)$, and $u_{2\downarrow}(x_2)$. These products must always involve one term with energy label $n = 1$ (energy $E_1$) and the other with energy label $n = 2$ (energy $E_2$). Accordingly we take the general form

$$Au_{1\uparrow}(x_1)u_{2\uparrow}(x_2) + Bu_{1\uparrow}(x_1)u_{2\downarrow}(x_2) + Cu_{1\downarrow}(x_1)u_{2\uparrow}(x_2) + Du_{1\downarrow}(x_1)u_{2\downarrow}(x_2)$$

and subtract from it the same form with $x_1 \leftrightarrow x_2$; that is, we add

$$-Au_{1\uparrow}(x_2)u_{2\uparrow}(x_1) - Bu_{1\uparrow}(x_2)u_{2\downarrow}(x_1) - Cu_{1\downarrow}(x_2)u_{2\uparrow}(x_1) - Du_{1\downarrow}(x_2)u_{2\downarrow}(x_1).$$

The sum of these terms is automatically antisymmetric under the exchange of $x_2$ and $x_1$. Since $A$, $B$, $C$, and $D$ are not constrained in any way, there are, in fact, four antisymmetric wave functions for this first excited state of the two-electron system, whose energy is $E_1+E_2$. Three of these wave functions will correspond to total spin 1 and will be spatially antisymmetric, and one will correspond to total spin 0 and will be spatially symmetric.

## Exchange Forces

If we look at the preceding examples and discussion, we can see that the exclusion principle leads to an interesting effect: Even if the identical fermions in the potential well do not interact—we have essentially treated them as "chargeless electrons"—there is nevertheless something like a *repulsion* between them. We can similarly show that identical bosons have a type of *attraction* that has only to do with the exchange symmetry principle. We refer in each case to **exchange forces**. These are not forces in the usual Newtonian sense; it is just that the particles behave *as if* there were forces between them. It was Heisenberg who first introduced this idea.

To see in a simple way how these "forces" arise, consider two identical particles in an infinite well that runs from $x = 0$ to $x = L$. One particle is in the state labeled $m$, the other in the state labeled $n$. We now ask the following question: What is the probability that both particles are in the left side of the well—that is, that a measurement shows $0 < x_1 < L/2$ *and* $0 < x_2 < L/2$? If the two-particle wave function is $\psi(x_1, x_2)$, then this probability is

$$P = \int_0^{L/2} \int_0^{L/2} |\psi(x_1, x_2)|^2 \, dx_2 \, dx_1. \tag{10-31}$$

Let us start by examining the question under the assumption that the two particles are not identical. Then the two-particle wave function is

$$\psi(x_1, x_2) = u_m(x_1)u_n(x_2). \tag{10-32}$$

For this case the probability $P$ in Eq. (10–31) immediately factors into the product,[4]

$$P = \int_0^{L/2} (u_m(x_1))^2 \, dx_1 \times \int_0^{L/2} (u_n(x_2))^2 \, dx_2. \tag{10-33}$$

This is just the product of the two indpendent probabilities of finding a single particle on the left of the well. We don't even have to calculate anything: Given the symmetry about the middle of the well, nothing can make it more likely to find a single particle on the left than on the right, so the probability of a single particle being on the left is 1/2. The probability $P$ is then $(1/2) \times (1/2) = 1/4$.

Let us now consider the case of two identical particles. For spinless bosons the wave function is simply symmetric—the symmetric combination of Eq. (10–32). For fermions the spin, which we shall suppose is 1/2, comes in to complicate things, but let us imagine that the spins are parallel. Then the spatial part of the wave function is the antisymmetric combination of Eq. (10–32). Under these conditions,

$$\psi(x_1, x_2) = \left(\frac{1}{\sqrt{2}}\right)\left[u_m(x_1)u_n(x_2) \pm u_m(x_2)u_n(x_1)\right], \tag{10-34}$$

where "plus" and "minus" respectively refer to bosons and to fermions with parallel spins. (To understand the occurrence of the factor $1/\sqrt{2}$, see the remark below Eq. (10–23) on page 274; see also Problem 1 at the end of the chapter.) We now must calculate $P$, as given by Eq. (10–31):

$$P = \frac{1}{2}\int_0^{L/2}\{(u_m(x_1))^2(u_n(x_2))^2 + (u_m(x_2))^2(u_n(x_1))^2$$

$$\pm 2u_m(x_1)u_n(x_2)u_n(x_1)u_m(x_2)\} \, dx_2dx_1.$$

The first of the three terms is just the same as the uncorrelated case, so that the integral is 1/4. The second term is the same as the first—just rename the variables of integration to see this. Thus the first two terms give a contribution of $(1/2)(1/4 + 1/4) = 1/4$ to $P$. The last term is the product of two separate integrations, again with limits of 0 and $L/2$:

$$\pm\frac{1}{2}2\left(\int_0^{L/2}u_m(x_1)u_n(x_1)\,dx_1\right) \times \left(\int_0^{L/2}u_m(x_2)u_n(x_2)\,dx_2\right) =$$

$$\pm\frac{1}{2}2\left(\int_0^{L/2}u_m(x_1)u_n(x_1)\,dx_1\right)^2. \tag{10-35}$$

To get to the last form, we have used the fact that the two integrations on the left are identical—again, just change the name of the variable of integration to see this. Now if the integration ran from 0 to $L$, the integral would vanish (see Problem 1), but for the range from 0 to $L/2$, it does not, in general, vanish. Thus the third term contributes $\pm(1/2)2K = \pm K$ to the probability, where $K$ is the square of the integral in Eq. (10–35) and is therefore *positive*.

We conclude that the probability that we find both particles on the left of the infinite well is

$$P = \frac{1}{4} \pm K. \tag{10-36}$$

---

[4] The $u_i$ are real, so we dispense with absolute values.

For bosons we use the plus sign, and the probability of having the two particles on the same side of the box is *increased* above that of the uncorrelated case. That is, bosons act as if they attract each other. For electrons with parallel spins the probability is *decreased*: These electrons act as if they repel each other. This observation will play an important role in understanding the spectrum of helium (Chapter 11). Ferromagnetism also depends on this effect.

We shall see many applications of the exclusion principle in the second part of this text: The principle plays a critical role in the structure of atoms, molecules, and nuclei, as well as in the technology of semiconductor and laser devices. Metals contain "free" electrons, so that there too the consequences of the exclusion principle come into play. When many identical particles appear together we have *degenerate matter*. We discuss degenerate fermionic matter in the rest of the chapter.

## 10–5  The Fermi Energy

To see why the exclusion principle affects the structure of matter so powerfully, consider electrons in the one-dimensional infinite well. Recall that the single-particle energy levels are given by Eq. (10–30). Suppose that we have $N$ electrons in the well and that we ignore the coulomb repulsion between these electrons. (While this might appear to be a poor approximation, we shall see later that it is not always so bad. In any case, at this stage we are just isolating the impact of the exclusion principle.)

Let us start by supposing that there is no exclusion principle—that is, that our particles are bosons. Then the lowest energy state—the ground state—of the $N$-particle system would be the one in which all $N$ bosons occupied the $n = 1$ level (•Fig. 10–6a). The total ground-state energy would be

$$E_g = NE_1, \tag{10-37}$$

where $E_1$ is given by Eq. (10–30) with $n = 1$. The average energy per particle is then

$$\frac{E_g}{N} = E_1. \tag{10-38}$$

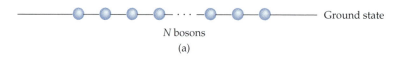

• **Figure 10–6**  (a) The ground state for $N$ bosons has all the particles in the one-particle ground state. (b) The first excited state contains $N - 1$ bosons in the one-particle ground state and a single boson in the first excited one-particle state.

The first excited state would be a state in which $N - 1$ bosons are in the state $n = 1$ and one boson is in the first excited single-particle state, the state with $n = 2$ (•Fig. 10–6b). This $N$-particle excited state has energy $E^* = (N - 1)E_1 + E_2$.

Now we shall see how the exclusion principle changes all of this for fermions. We take our particles to be electrons, say. In constructing the $N$-particle ground state, we cannot throw all $N$ electrons into the single-particle state with $n = 1$. Only two electrons can occupy the $n = 1$ state, two electrons the $n = 2$ state, and so on. With $N$ electrons *all* the levels up to $n = N/2$ are occupied in the ground state (•Fig. 10–7). (We have assumed that $N$ is even for simplicity; for the huge numbers of electrons in real matter, the last electron hardly makes a difference.) Using the fact that the single-particle energies are $E_n = n^2 E_1$, the $N$-particle ground-state energy is

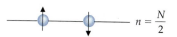

$$E_g = E_1 \left[ 2(1^2) + 2(2^2) + \cdots + 2\left(\frac{N}{2}\right)^2 \right] = 2E_1 \sum_{j=1}^{N/2} j^2. \qquad (10\text{–}39)$$

The sum can be calculated exactly (see Problem 17 for another approach) using the formula $\sum_{j=1}^{n} j^2 = \frac{1}{6}n(n + 1)(2n + 1)$, which is approximately $n^3/3$ for large $n$. In the case at hand the upper limit $n$ is $N/2$, and since $N$ is large the sum is approximately $N^3/24$. Thus

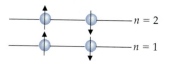

$$E_g \approx \frac{N^3}{12} E_1. \qquad (10\text{–}40)$$

The approximation is valid for the large values of $N$ that are involved in real matter; we have ignored terms of order $N^2$ compared with $N^3$.

From Eq. (10–40), we immediately calculate the average energy per particle in the ground state:

$$\frac{E_g}{N} \approx \frac{N^2}{12} E_1. \qquad (10\text{–}41)$$

• **Figure 10–7**   The lowest energy, or ground, state for $N$ fermions has at most two particles in each single-particle level, one with spin up and one with spin down. All the levels up to $n = N/2$ are filled.

This average energy is quite different from that in the boson case, because, in contrast to that case, in which the average energy is constant, for the fermion case *the average energy grows with $N$*; in particular it grows as $N^2$.

In the construction of the $N$-fermion ground state, the highest energy level to be filled has an energy called the **Fermi energy**, $E_F$. For our one-dimensional well, it is the energy level corresponding to $n = N/2$, so that

$$E_F = \frac{\pi^2 \hbar^2 (N/2)^2}{2mL^2} = \frac{\pi^2 \hbar^2 N^2}{8mL^2}. \qquad (10\text{–}42)$$

The Fermi energy is conveniently expressed in terms of the density of the fermions involved. In our one-dimensional example, the density is $n_e = N/L$, and

$$E_F = \frac{\pi^2 \hbar^2}{8m} n_e^2. \qquad (10\text{–}43)$$

This form is useful because the size of the well doesn't enter into the calculations.

**Example 10–4**   Consider a one-dimensional box 1 cm wide. The Fermi energy is given as 2.0 eV. How many electrons are there in the box?

**Solution**   We use Eq. (10–43) in the form

$$n^2 = \frac{8mE_F}{\hbar^2 \pi^2} = \frac{8(0.91 \times 10^{-30}\, \text{kg})(2.0\, \text{eV})(1.6 \times 10^{-19}\, \text{J/eV})}{(1.05 \times 10^{-34}\, \text{J·s})^2 \pi^2}$$

$$= 2.1 \times 10^{19}\ (\text{electrons/m})^2.$$

Hence the electron density is $n = 4.6 \times 10^9$ electrons/m. Since the box is 1 cm long, the number of electrons is

$$N = nL = \left(4.6 \times 10^9 \text{ electrons/m}\right)\left(10^{-2} \text{ m}\right) = 4.6 \times 10^7 \text{ electrons.}$$

## Three Dimensions

The one-dimensional box demonstrates by example how the effects of the exclusion principle works. But in the real world we are more often concerned with three dimensions; for example, a bulk metal or semiconductor whose conduction properties we wish to understand is a three-dimensional box[5] in which the valence electrons (see Chapter 1) are, for all practical purposes, free. We therefore want to extend our reasoning to this case. The ideas involved are not new, but the situation is technically a bit more complicated.

We start by enumerating the possible single-particle energies in a three-dimensional infinite well formed by a cube whose sides have length $L$. (At the end we shall again express everything in terms of a density, and in this form the precise size and shape of the box will not enter into the calculations.) Now as far as fitting the wave functions and finding the allowed single-particle energy levels is concerned, the three Euclidean dimensions are independent. This means that the single-particle levels are the sum of three allowed one-dimensional energy levels:

$$E_{(n_1, n_2, n_3)} = \frac{\pi^2 \hbar^2}{2mL^2} \left(n_1^2 + n_2^2 + n_3^2\right) = E_1 \times \left(n_1^2 + n_2^2 + n_3^2\right). \quad (10\text{--}44)$$

Here $n_1$, $n_2$, and $n_3$ are independent integers taking on the values $1, 2, \ldots,$ and we have used the one-dimensional single-particle level $E_1$ from Eq. (10–30) as a scale. In other words, there is an allowed energy state corresponding to any triplet of integers $(n_1, n_2, n_3)$.

The lowest single-particle energy state corresponds to the lowest values of the integers—that is, $(1, 1, 1)$. The next single-particle state will have one of the $n_i$ with the value 2. Thus there are actually *three* energy levels with this same energy. The states $(1, 1, 2)$, $(1, 2, 1)$, and $(2, 1, 1)$ are said to be **degenerate**, because they have the same energies, and the **degree of degeneracy** is three in this case. As the energy of the allowed levels increases, so does the degree of degeneracy. In general, the degree of degeneracy grows quickly as we move up the energy scale, because the number of ways in which we can get three integers $n_1$, $n_2$, and $n_3$ such that $n_1^2 + n_2^2 + n_3^2$ adds up to some large integer grows quickly. For example, the states $(1, 2, 6)$, $(1, 6, 2)$, $(2, 6, 1)$, $(2, 1, 6)$, $(6, 1, 2)$, $(6, 2, 1)$, $(3, 4, 4)$, $(4, 3, 4)$, and $(4, 4, 3)$ all have the same energy. The possibility of degeneracy is the major difference between the one- and the three-dimensional cases.

Suppose we take $N$ fermions, where $N$ is large, and fill up the energy levels in the box, two fermions per level, starting at the bottom. The energy of the last electron to be accommodated is the Fermi energy $E_F$. Now we can phrase our problem as *How many states are there with energy less than or equal to some value of E?* Once we know that, we set the number of states equal to $N/2$, and

---

[5] In fact, in modern applications, semiconductor devices often are used in planes so thin that we can consider them to be two-dimensional systems. Systems with one, two, or three dimensions all occur in current technological applications.

that will correspond to $E = E_F$. Using Eq. (10–44), we see that our question is equivalent to another: How many triplets of integers are there such that

$$n_1^2 + n_2^2 + n_3^2 \le \frac{E}{E_1}? \tag{10–45}$$

For bulk materials, we expect $N$ to be large; hence $E_F$ will be large. We can use this information to our advantage in estimating the answer to our question.

For our estimation, we can plot the set of three integers as a cubic lattice (•Fig. 10–8). Each point on the lattice corresponds to some triplet of integers $(n_1, n_2, n_3)$. The condition of Eq. (10–45) may be written as

$$n_1^2 + n_2^2 + n_3^2 \le R^2, \tag{10–46}$$

where

$$R^2 = \frac{E}{E_1}. \tag{10–47}$$

To answer our question, we need to find the number of lattice points that satisfy Eq. (10–46). Since the integer triplets make lattice points with a spacing of one unit in each Cartesian direction, the lattice is formed of cubes of unit volume. Thus we can calculate the number of lattice points in question by calculating the volume of a sphere of radius $R$. Of course, this isn't quite right, because the cubes at the surface do not fit precisely into the sphere; but the error becomes negligibly small as the radius of the sphere gets larger and larger. There is one slight subtlety: We take all the integers $n_i$ as positive. To see why, think back to the wave function for the one-dimensional infinite well, which was proportional to $\sin(n\pi x/L)$. If we make the change $n \to -n$, the wave function just changes sign; it is not a different wave function. Thus if we restrict ourselves to positive $n$ we avoid overcounting. With this restriction, we actually take only one-eighth of the volume (•Fig. 10–9)—that is, $(1/8)(4\pi R^3/3)$. This is the number of states that satisfy Eq. (10–46).

Having taken care of our counting problem, we set the number of states to $N/2$ and $E$ to $E_F$. We then have

$$\frac{N}{2} = \frac{1}{8} \frac{4\pi R^3}{3} = \frac{\pi}{6} \left( \frac{E_F}{E_1} \right)^{3/2}. \tag{10–48}$$

We can solve this equation for $E_F$ to obtain $E_F = E_1(3N/\pi)^{2/3}$. When we insert the explicit expression for $E_1$, we find that

$$E_F = \frac{\hbar^2}{2m} \left( \frac{3\pi^2 N}{L^3} \right)^{2/3}. \tag{10–49}$$

Once again, the fermion density $n_f = N/L^3$ appears in such a way that the size of the box is removed:

$$E_F = \frac{\hbar^2}{2m} \left( 3\pi^2 n_f \right)^{2/3}. \tag{10–50}$$

There is a simple way to think about the Fermi energy. We see from Eq. (10–50) that the magnitude of the electron momentum at that energy is

$$p_F = \sqrt{2mE_F} = \hbar \left( 3\pi^2 n_f \right)^{1/3}. \tag{10–51}$$

This momentum is called the **Fermi momentum**. The de Broglie wavelength corresponding to the Fermi momentum is

$$\lambda_F = \frac{h}{p_F} = \frac{2\pi\hbar}{p_F} = 2 \left( \frac{\pi}{3} \right)^{1/3} n_f^{-1/3}. \tag{10–52}$$

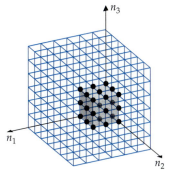

• **Figure 10–8**   The energy states in a three-dimensional well are labeled by a set of three integers. These integers can be depicted as sitting on the intersection points of a three-dimensional cubic lattice.

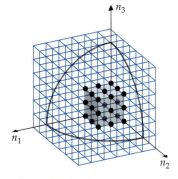

• **Figure 10–9**   Because negative $n$-values for a three-dimensional well do not represent different states from those already labeled by the corresponding positive values, only positive values need be counted. This condition restricts us to one quadrant (one-eighth) of the full lattice. In other words, to find the points that correspond to the energy states and thus allow us to count the number of states with energy up to $E$, we restrict ourselves to points that lie within the octant that contains all positive coordinates of a sphere whose radius is proportional to $E$.

But $n_f^{-1/3}$ is the average interfermion spacing—call it $a$—so that we can state our result in the easily remembered form

$$a \cong \frac{\lambda_F}{2}. \tag{10-53}$$

This result implies that *the closest that two electrons can get to each other is rough-ly half a de Broglie wavelength corresponding to the Fermi energy.*

---

**Example 10–5**   Calculate the Fermi energy for aluminum. For this metal, $A = 27$, the mass density is $\rho = 2.7 \times 10^3 \, \text{kg}/\text{m}^3$, and, on average, three electrons per atom are free. Give your answer in eV.

**Solution**   The expression for $E_F$ in Eq. (10–50) requires us to know the number density of free electrons. Since there are three free electrons per aluminum atom, the number density of free electrons is three times the number density of atoms. We find the number density of atoms by noting that the mass of an atom of atomic number $A$ is $A$ times the mass of one nucleon $= A(1.67 \times 10^{-27} \, \text{kg}) = 4.5 \times 10^{-26} \, \text{kg}$ for aluminum. In turn the number of aluminum atoms per kg is $1/(4.5 \times 10^{-26} \, \text{kg}) = (2.2 \times 10^{25})/\text{kg}$. Thus the number density of aluminum atoms is

$$(\# \text{ of atoms}/\text{m}^3) = (\# \text{ of atoms}/\text{kg})(\text{kg}/\text{m}^3) = (\#\text{atoms}/\text{kg})\rho$$

$$= (2.2 \times 10^{25} \, \text{kg}^{-1})(2.7 \times 10^3 \, \text{kg}/\text{m}^3) = 6.0 \times 10^{28} \, \text{atoms}/\text{m}^3.$$

We can use three times this number for the free-electron (fermion) density $n_f$ in Eq. (10–50), which then gives

$$E_F = \frac{\hbar^2}{2m}(3\pi^2 n_f)^{2/3} = \frac{(1.05 \times 10^{-34} \, \text{J} \cdot \text{s})^2}{2 \times 0.91 \times 10^{-30} \, \text{kg}} \left[3\pi^2(3 \times 6.0 \times 10^{28} \, \text{m}^{-3})\right]^{2/3}$$

$$= 1.9 \times 10^{-18} \, \text{J}.$$

Using $1 \, \text{eV} = 1.6 \times 10^{-19} \, \text{J}$, we find that $E_F = 11.6 \, \text{eV}$. This result is quite close to the experimental value of 11.7 eV.

---

**Example 10–6**   Neutrinos are fermions just like electrons, except that their mass is zero or nearly so, and instead of having a two-valued "spin" label, the spin can only be in one state. Thus there can be only one neutrino in an energy state. The mass of the neutrino, if it has one, is so close to zero that we can take $E = |\vec{p}|c = pc$. Calculate the Fermi energy of a gas of $N$ free neutrinos in a cube with sides of length $L$.

**Solution**   The energy is

$$E = pc = \frac{\hbar c \pi}{L} \sqrt{n_1^2 + n_2^2 + n_3^2}.$$

Thus Eqs. (10–46) and (10–47) together have as their counterpart

$$R^2 = n_1^2 + n_2^2 + n_3^2 = \left(\frac{E_F}{E_1}\right)^2 = \left(\frac{E_F}{\hbar \pi c}\right)^2 L^2.$$

In turn, and with only one neutrino at each positive $(n_1, n_2, n_3)$ lattice point, we find that

$$N = \frac{1}{8}\frac{4\pi}{3} R^3 = \frac{\pi}{6}\left(\frac{E_F L}{\hbar \pi c}\right)^3.$$

Finally, we recognize that $V = L^3$ and that $n = N/V$, so that the solution of this equation is

$$E_F = \hbar \pi c \left(\frac{6}{\pi}\right)^{1/3} n^{1/3}.$$

This result is different from the $E_F$ we found earlier for electrons because neutrinos have a relativistic relation between momentum and energy.

## Examples of Degenerate Matter

On Earth, we see degenerate matter in the form of the electrons in metals. Recall from Chapter 1 that the valence electrons of the atoms in the metal are, for all practical purposes, free. In particular, they do not repel each other very much, because of the positive background charges of the ions remaining after the electrons are removed from the atoms. Thus the valence electrons in a metal behave as a degenerate fermionic system at low temperatures. This behavior explains many experimental features of metals: their heat capacity, their electric conduction properties, the dependence of electron emission on temperature, and so forth. However, more direct experimental verification of an energy distribution that is regulated by the exclusion principle was a long time in coming.

One of the most convincing techniques—one that clearly illustrates the interdependence of the subfields of physics—employs positrons, particles with the mass of electrons, but with a charge $+e$ rather than $-e$. In 1932—the same year that neutrons were found—positrons were discovered in cosmic rays,[6] although it was not until the mid-1970s that they were regularly used as probes for the presence of a Fermi energy. When positrons and electrons collide, they can annihilate each other and produce a pair of photons, and the energy of these photons can be measured. This energy includes the energy available in the mass of the electron and positron, plus whatever kinetic energy the colliding particles have. When positrons enter a metal, interactions with the material bring them very nearly to rest: Since no other positrons are present, they can go to the single-particle ground state. Electrons, on the other hand, have a variety of kinetic energies all the way up to the Fermi energy. The positrons can interact with any of the electrons, so that the photons which are produced upon their mutual annihilation have energies that reflect the distribution of electron energies. Even better, the presence of a maximum electron energy produces a sharp cutoff in the photon energies that are measured. Experiments of this type do indeed verify the idea of degenerate electrons and can reveal many details about electrons in solids.

Another area where degenerate fermions come in is stellar evolution. We'll work out a striking consequence of the exclusion principle in stars in the next section.

# 10–6 Degeneracy Pressure

If the interfermion spacing in degenerate fermionic matter decreases, the wavelength $\lambda_F$ in Eq. (10–53) decreases and the energy of the fermions increases. It takes energy to compress degenerate fermionic matter: Such matter resists being squeezed, and positive work must be done to compress it. We can correctly say that any compression is resisted by the exclusion principle.

## A Back-of-the-Envelope Estimate of Degeneracy Pressure

Just how important is this effect of resistance to compression? We want to see how the total energy varies as we squeeze the box. (The total energy is not to be confused with the Fermi energy; to get the total energy we add the energies of the individual electrons.) To find $E_{\text{tot}}$ we use a special technique: We ensure

---

[6] See Chapter 16 for a detailed discussion of this discovery.

that the $N$ electrons are in separate states by putting them in identical little boxes: cubes of size $a$. If the total volume is $V$, then $a^3 = V/N$. For motion in the $x$-direction, for example, the lowest energy of the electron in its little box is given by Eq. (10–30) and is

$$\frac{p_x^2}{2m} = \frac{\hbar^2 \pi^2}{2ma^2}. \tag{10–54}$$

In the three-dimensional box, we have

$$\frac{p_x^2}{2m} + \frac{p_y^2}{2m} + \frac{p_z^2}{2m} = \frac{3\hbar^2 \pi^2}{2ma^2}. \tag{10–55}$$

If we now assume that each of the electrons in its little box has the same minimum energy, the total energy is $N$ times the result of Eq. (10–55); that is,

$$E_{\text{tot}} = N \frac{3\hbar^2 \pi^2}{2ma^2}. \tag{10–56}$$

We want to see how this energy varies with volume. The volume is not evident in Eq. (10–56), but we can bring it out by noting that $n = 1/a^3$, or $1/a = n^{1/3}$, where $n = N/V$ is the electron density. When we make the substitution for $a$, we obtain

$$E_{\text{tot}} = \frac{3\hbar^2 \pi^2}{2m} N^{5/3} V^{-2/3}. \tag{10–57}$$

Once we know how the total energy depends on the volume for a given number of particles, we can see how an alteration in the volume changes that energy. A measure of this relationship is the ratio of the change in energy to a change in volume, a quantity with dimensions of pressure known as the **degeneracy pressure** and given by

$$p_f = -\frac{\partial E_{\text{tot}}}{\partial V} = \frac{\hbar^2 \pi^2}{m} N^{5/3} V^{-5/3} = \frac{\hbar^2 \pi^2}{m} n^{5/3}, \tag{10–58}$$

where the subscript $f$ stands for "fermion." Note the sign of $p_f$: The pressure is positive. That is because a positive energy is required to *decrease* the volume.

### *A More Accurate Calculation of the Degeneracy Pressure

Let us now find the total energy of $N_{\text{tot}}$ fermions in a box of volume $V$ in the lowest possible energy state by actually summing the energies of each of the fermions. Once we find the total energy, the pressure is a measure of the change in energy for a given change in volume, just as before.

As a preliminary step we find the number[7] of fermions $N$ that have energy less than some intermediate energy $E$. The technique for determining this quantity is exactly the technique that gives $E_F$ of Eq. (10–49). We invert that equation, keeping in mind that $E$, rather than $E_F$, appears:

$$N = \frac{(2mE)^{3/2}}{\hbar^3} \frac{L^3}{3\pi^2} = \frac{\pi}{3} \left( \frac{L}{\pi\hbar} \right)^3 (2mE)^{3/2}. \tag{10–59}$$

We find the number $dN$ of fermions with energy between $E$ and $E + dE$ by taking the difference between the equation evaluated at $E + dE$ and the equation

---

[7] Note that in this optional subsection $N$ is the intermediate number, not the total number, which is $N_{\text{tot}}$.

evaluated at $E$; this procedure is equivalent to differentiation with respect to $E$, and we obtain

$$dN = \frac{\pi}{3}\left(\frac{L}{\pi\hbar}\right)^3 (2m)^{3/2} \frac{3}{2}\sqrt{E}\,dE.$$

Finally we find the *total* energy of the system by integrating the quantity $E\,dN$:

$$E_{\text{tot}} = \int E\,dN = \frac{\pi}{3}\left(\frac{L}{\pi\hbar}\right)^3 (2m)^{3/2}\frac{3}{2}\int_0^{E_F} E\sqrt{E}\,dE$$

$$= \frac{\pi}{5}\left(\frac{L}{\pi\hbar}\right)^3 (2m)^{3/2}E_F^{5/2}.$$

We can now insert our expression for $E_F$, Eq. (10–49), and then replace the factor $L^3$ by the volume $V$ of the material. (We did our counting of states for a cubic box of volume $L^3$, but the number of states is independent of the shape of the container of the fermions if there are very many of them.) Thus we arrive at

$$E_{\text{tot}} = \frac{3^{5/3}\pi^{4/3}}{5}\frac{\hbar^2}{2m}V^{-2/3}N_{\text{tot}}^{5/3}. \tag{10--60}$$

The degeneracy pressure is, accordingly,

$$p_f = -\frac{\partial E_{\text{tot}}}{\partial V} = \frac{6}{5}\left(\frac{\pi^4}{3}\right)^{1/3}\frac{\hbar^2}{2m}N_{\text{tot}}^{5/3}V^{-5/3} = \frac{6}{5}\left(\frac{\pi^4}{3}\right)^{1/3}\frac{\hbar^2}{2m}n^{5/3}, \tag{10--61}$$

where $n$ is the fermion density. Our estimate of Eq. (10–58) is about five times larger than this result, but does have the correct dependence on the variables.

## Astrophysical Applications

The resistance to compression originating in the exclusion principle plays an important role in stellar evolution. Stars start out as hydrogen gas, which "burns" by undergoing a succession of nuclear reactions. (This is only a crude approximation, but it will do for now; the details will be filled in later in the book.) At some point the burning stops. Then the only effects that we have to worry about are the gravitational forces, which work to compress the stellar matter, and the degeneracy pressure of the electrons in the star, which works to hold the electrons, and with them all the stellar matter, apart.

To see how these two effects balance out, we must first estimate a *gravitational pressure*. We make a simplifying assumption here: We assume that the mass density $\rho$ of the star is constant as a function of the radius of the star, which is also assumed to be spherical. We start with a rough estimate.

*A Back-of-the-Envelope Estimate of Gravitational Pressure* To estimate the gravitational pressure, we assume that the mass is distributed at the surface of the sphere. The gravitational energy of a mass $M$ distributed uniformly over a sphere of radius $R$ is

$$U = -\frac{GM^2}{R} = -GM^2\left(\frac{4\pi}{3}\right)^{1/3}V^{-1/3}. \tag{10--62}$$

Now we use this energy to calculate a gravitational pressure. The same reasoning we employed for the degeneracy pressure tells us that the gravitational pressure is

$$p_g = -\frac{\partial U}{\partial V} = -\frac{GM^2}{3}\left(\frac{4\pi}{3}\right)^{1/3}V^{-4/3}. \tag{10--63}$$

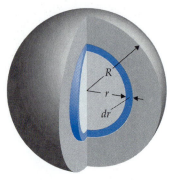

• **Figure 10–10**  To find the gravitational potential energy of a uniform sphere of matter, we add up the separate contributions of thin spherical shells.

*A More Accurate Calculation of the Gravitational Pressure*   It is straightforward to calculate the total gravitational potential energy of a star if the density of the star is uniform. Recall from Chapter 1 that for inverse-square forces such as the gravitational force, and for spherically symmetric mass distributions, all the mass inside a given radius acts as if it were at the geometric center of the object or distribution involved. Thus, because the mass $dm$ of the material in a shell lying between the radius $r$ and $r + dr$ is $\rho(4\pi r^2\, dr)$, and the total mass inside that radius is $m(r) = \rho(4\pi r^3/3)$, the potential energy of the material is (•Fig. 10–10)

$$dU_g = -G\frac{m\,dm}{r} = -G\frac{(4\pi\rho r^3/3)(4\pi\rho r^2\,dr)}{r} = -\frac{16\pi^2 G\rho^2}{3}r^4\,dr. \quad (10\text{–}64)$$

We integrate to find the total gravitational energy:

$$U_g = -\frac{16\pi^2 G\rho^2}{3}\int_0^R r^4\,dr = -\frac{16\pi^2 G\rho^2}{15}R^5, \quad (10\text{–}65)$$

where $R$ is the star's radius. (Note that dimensional analysis—the making up of a quantity with the dimensions of energy using the available physical parameters, here $G$, $\rho$, and $R$—would have given us $-G(\rho R^3)^2/R$. Of course, this analysis does not give us the dimensionless constants in front.)

There is a connection between $\rho$, $R$, and the stellar mass $M$. The mass of the star is almost totally accounted for with $N$ nucleons—neutrons or protons—each of mass approximately $m_p$. Thus $\rho = Nm_p/V$, where $V = 4\pi R^3/3$ is the star's volume. Taking this fact into account, we find that Eq. (10–65) becomes

$$U_g = -\frac{3}{5}\left(\frac{4\pi}{3}\right)^{1/3}G(Nm_p)^2 V^{-1/3}, \quad (10\text{–}66)$$

and the gravitational pressure is

$$p_g = -\frac{\partial U_g}{\partial V} = +\frac{1}{5}\left(\frac{4\pi}{3}\right)^{1/3}G(Nm_p)^2 V^{-4/3}. \quad (10\text{–}67)$$

In this case there is no minus sign, because it takes positive energy to *increase* the volume.

*Balancing the Degeneracy Pressure against the Gravitational Pressure*   At the beginning of the "burned-out" star's collapse, the gravitational pressure is larger than the degeneracy pressure. Indeed, that is why there *is* a collapse. But the degeneracy pressure is of the opposite sign and grows more quickly than the gravitational pressure as the volume decreases. Then at some point in time the gravitational pressure and the degeneracy pressure of the electrons balance. Let us estimate for what value of the star's volume this balance occurs. It is reasonable to assume[8] that the number of protons in the star is approximately $N/2$ and that $N_e$ is therefore equal to $N/2$. Thus the degeneracy pressure of the electrons is, from Eq. (10–61),

$$p_e = \frac{6}{5}\left(\frac{\pi^4}{3}\right)^{1/3}\frac{\hbar^2}{2m_e}\left(\frac{N}{2}\right)^{5/3}V^{-5/3}. \quad (10\text{–}68)$$

When we set the gravitational pressure equal to the degeneracy pressure, we find a condition for the volume or, equivalently, the radius. We spare you the algebraic details and just quote the answer:

$$R = \left(\frac{81\pi^2}{128}\right)^{1/3}\frac{\hbar^2}{Gm_p^2 m_e}N^{-1/3} \cong (1.15\times 10^{23}\ \text{km})N^{-1/3}. \quad (10\text{–}69)$$

---

[8] Actually this is a good assumption only for stars such as the white dwarfs we are treating here. Our sun is to a good approximation neutron free.

Stars that reach the degenerate state described here are familiar to astronomers as **white dwarfs**. For a star the mass of the sun, $M_{sun} \approx 2 \times 10^{30}$ kg, and we have $N = M_{sun}/m_p \approx (2 \times 10^{30} \text{ kg})/(1.7 \times 10^{-27} \text{ kg}) = 1.2 \times 10^{57}$. It then follows that

$$R \approx 1.1 \times 10^4 \text{ km.}$$

Our sun is still in the process of burning its hydrogen and has a radius some 60,000 times larger than $R$! Indeed, a white dwarf with the mass of the sun has a radius only about twice that of Earth.

You may wonder why we used the degeneracy pressure of the electrons in this calculation. After all, protons and neutrons are fermions and are certainly degenerate. But a glance at the expression for the degeneracy pressure tells the story: The mass of the fermion in question appears in the denominator, and since nucleons are some 2,000 times more massive than electrons, their degeneracy pressure is negligible.

Not all stars end their evolution at the white-dwarf stage: The more massive stars have larger numbers of protons or electrons, and the larger the number of electrons, the larger is the electrons' average energy. For large enough numbers, the electrons are, on average, relativistic, and this has an important effect. Instead of being quadratic in the momentum, the electron energy becomes *linear* in the momentum (recall that for highly relativistic particles $E = pc$), and this linearity has the effect of making the degeneracy pressure contain a factor $V^{-4/3}$ rather than $V^{-5/3}$. But $V^{-4/3}$ is the factor contained in the gravitational pressure, and the volume cancels when we try to equate these pressures. In other words, the pressures cannot balance: The gravitational pressure wins out, and the star continues to collapse. This collapse forces all the particles closer together and allows the reaction $e^- + p \rightarrow n + \nu$ to occur. The symbol $\nu$ stands for the neutrino (see Example 10–6), which, as we shall see in Chapter 16, reacts very weakly with other particles and has more than enough energy to escape the star. We are thus left with neutrons. Now neutrons are also fermions, and we again have a degeneracy pressure, the only changes being that $N_e$ is replaced by $N$ (an unimportant modification) and $m_e$ is replaced by $m_n \cong m_p$. Once again equilibrium is possible, and at that point we have a **neutron star**. We find its radius by extending Eq. (10–69) so that $m_e$ is replaced by $m_p$. But $m_e/m_p \cong 1/2,000$, so that the radius of the neutron star is about 2,000 times smaller than the white dwarf's radius; that is, for a neutron star a little more massive than the Sun, the radius is only about 10 km! If the mass is larger than a couple of solar masses, the same thing happens to the neutrons as happened to the electrons: They become relativistic, and gravitational collapse again takes over. This time there is nothing to stop the collapse, and we end up with a **black hole**. (See Chapter 17.)

## SUMMARY

Once we know how to solve the Schrödinger equation for a system involving more than one particle, each under the influence of a common potential, we can confront some very interesting features of nature. In classical physics no two objects are really exactly alike. But in the quantum theory particles can truly be identical; all electrons are identical, for example. This principle has very important consequences:

- The wave function of a system of several identical particles has a symmetry under the exchange of the coordinates corresponding to the individual particles. There are two classes of such symmetries. For fermions—particles with half-integer spin such as electrons—the wave function is antisymmetric, meaning that if you exchange the coordinates of any two identical fermions, the wave function will become the negative of its original value. For bosons—particles with no spin or integer spin such as photons—the wave function is symmetric, meaning that if you exchange the coordinates of any two identical bosons, the wave function is unchanged.

- The antisymmetry of the wave function for identical fermions means that if the coordinates of two such particles are identical, the wave function must vanish, something known as the Pauli exclusion principle.

- In constructing the wave functions for identical particles and applying the symmetry properties under an exchange of labels for the coordinates, both spatial coordinates and spin coordinates must be included. By the spin coordinate we mean the particular value of the spin projection (e.g., whether a spin-1/2 fermion has its spin up or down). Thus a wave function for two electrons is overall antisymmetric under exchange if it is symmetric under exchange of the space coordinates and antisymmetric under exchange of the spin labels or if it is antisymmetric under exchange of the space coordinates and symmetric under exchange of the spin labels.

- Two spin-1/2 fermions in a symmetric spin state have a total spin of unity. Two spin-1/2 fermions in an antisymmetric spin state have a total spin of zero.

Applications of these ideas are of major importance both for the understanding of the physical world and for technological innovation. Later in this text we treat the periodic table, the effects of the stimulated emission of radiation, and the behavior of metals and other solids, subjects all dominated by the multiparticle interactions studied in this chapter. By applying the ideas of the chapter to a collection of electrons in a three-dimensional deep well, we enter into some astronomical issues and begin to make inroads into the question of the solid state. In particular,

- Because of the Pauli principle, only two electrons can be put into each energy eigenstate of a well. The energy of the last state to be filled is the Fermi energy. It is the electrons at the Fermi energy that can respond to the effects of further excitation.

- The exclusion principle implies that energy must be supplied to reduce the size of a box containing a collection of identical fermions. This effect accounts for the compressibility of ordinary matter as well as the properties of white-dwarf stars and neutron stars.

## QUESTIONS

**1.** The photon is a particle with a spin of unity. Some theoretical physicists devised a neutrino theory of light (see Example 10–6 for what you need to know about the neutrino) in which two neutrinos would be bound together in an $\ell = 0$ state to make a photon. Can this be done within the strictures of the Pauli principle? It is likely that to each neutrino there is an associated antineutrino which is distinct from it, although the two share the same magnitude of mass, charge, and spin. Could this combination of neutrinos make a photon if they are bound together in the same state?

**2.** Consider two electrons in a box. What is the Fermi energy for this system?

**3.** Suppose you are sitting at your desk, at a library table, or perhaps at your kitchen table, reading this book. Are there any examples of degenerate matter within your sight?

**4.** According to the exclusion principle, in any system involving more than one electron, the wave function of the system must be antisymmetric under the interchange of the coordinates (and spin labels) of any pair of electrons. At the same time, it does not appear reasonable that in describing the $2p$ state of the electron in New York, it should be necessary to take into account the wave function of an electron in the ground state of hydrogen located on the Moon. Give an argument as to why we can forgo antisymmetry in this instance. When does antisymmetry need to be taken into account? (*Hint*: See Problem 4.)

**5.** How would you estimate how much work must be done to squeeze the hydrogen atom so that its radius is reduced by 5%?

**6.** In this chapter we calculated the Fermi energy on the assumption that the electrons were in a cubical box. Suppose we had to work out the problem assuming that the electrons were in a spherical box. This would involve solving the Schrödinger equation in a spherical box, which would be technically more complicated. Can you give reasons indicating that the calculation that we carried out is accurate even in this case?

## PROBLEMS

**1.** ▮▮ Show that for the case of two particles in an infinite well, $N_A = N_S = 2$ in Eqs. (10–22) and (10–23). In order to do this, you need to show that

$$\frac{2}{L} \int_0^L \sin(n\pi x/L)\sin(m\pi x/L)\, dx = \begin{cases} 0 & \text{for } m \neq n \\ 1 & \text{for } m = n. \end{cases}$$

**2.** ▮▮ Show that for the case of three particles in an infinite well the normalization factor $N_A$ in Eq. (10–25) is 6. Use the same procedure as you did for Problem 1.

**3.** ▮▮ **(a)** Consider the antisymmetic wave function

$$\psi(x_1, x_2) = \frac{1}{(\sqrt{N_A})} \left[ u_1(x_1)u_2(x_2) - u_1(x_2)u_2(x_1) \right]$$

Obtain an expression for $N_A$ from the requirement that

$$\int_{-\infty}^{+\infty} \int_{-\infty}^{+\infty} |\psi(x_1, x_2)|^2\, dx_1\, dx_2 = 1.$$

**(b)** Do the same for the symmetric wave function

$$\psi(x_1, x_2) = \frac{1}{(\sqrt{N_S})} \left[ u_1(x_1)u_2(x_2) + u_1(x_2)u_2(x_1) \right].$$

In both cases assume that the single-particle wave functions $u_i$ are normalized and that

$$K_{12} \equiv \int_{-\infty}^{+\infty} u_1^*(x)u_2(x)\, dx \neq 0.$$

**4.** ▮▮▮ Calculate $K_{12}$ defined in Problem 3 for the case in which electrons 1 and 2 have Gaussian packet wave functions given by

$$u_1(x) = (2\alpha^2/\pi)^{1/4} \exp\left[-\alpha^2(x - L)^2\right] \quad \text{and}$$
$$u_2(x) = (2\alpha^2/\pi)^{1/4} \exp\left[-\alpha^2(x + L)^2\right],$$

where $\alpha$, a constant with dimensions of inverse length, determines the width of the peaks in $u_i$. In particular, show that $K_{12}$ vanishes rapidly as $L$ increases (i.e., as the overlap between the wave functions becomes small). Under those circumstances one can argue that antisymmetrization becomes irrelevant, though in principle still necessary. Thus we do not have to render the wave function of an electron in a hydrogen atom in the laboratory antisymmetric to every other electron in the universe.

**5. ▮▮** Consider two identical free spinless bosons in a one-dimensional infinitely deep well of width $2a$. What is the wave function of the ground state? What is the wave function of the first excited state? Use the result of Problem 1 to calculate the normalization constants for the two wave functions. (*Hint*: Only one of the particles will be in the first excited ($n = 2$) state.)

**6. ▮▮** Given the situation described in Problem 5, and given that the wave function corresponds to the first excited state of the two-particle wave function, find the probability that both particles are in the left side of the well. Compare your result with what you would calculate if the wave function were not symmetric, but instead corresponded to the first excited two-particle state for independent particles. Is your result consistent with the discussion of exchange forces?

**7. ▮▮** Consider the case of two identical fermions, both in the spin-up state in the box of Problem 5. Write the wave function for the lowest energy state. For what value(s) of position does the wave function vanish?

**8. ▮▮** Consider the case of three identical free bosons (e.g., $\alpha$-particles), with no spin label, in a one-dimensional infinitely deep well of length $L$, a case treated in Section 10–4 with regard to fermions. **(a)** Write the wave function for the ground state. **(b)** Write the wave function for the first excited state.

**9. ▮▮** Use the result of Problem 1 to calculate the normalization constants for the two wave functions in Problem 8.

**10. ▮▮** Consider two electrons in a spherically symmetric attractive potential. When the electron–electron repulsion is ignored, one of the electrons is in the 1s state, with spatial wave function $u_1(\vec{r})$, and the other is in the 2p state, with spatial wave function $u_2(\vec{r})$. The spatial wave function will be made up of products of $u_1(\vec{r}_1)u_2(\vec{r}_2)$ and $u_1(\vec{r}_2)u_2(\vec{r}_1)$, each of these multiplied by one of the products $\chi_+(1)\chi_+(2)$, $\chi_+(1)\chi_-(2)$, and so on. There are eight possible products. Write out the combinations of products that are allowed by the exclusion principle.

**11. ▮** In Eq. (10–35) we defined a quantity $K$. Calculate $K$ for the infinite well. [The wave functions are given, among other places, in Eq. (10–29).] Show from your calculation that the probability $P$ calculated in Eq. (10–36) cannot be negative for electrons, a result that to say the least would not be very satisfactory.

**12. ▮** Calculate the Fermi energy for potassium, for which the free-electron density is $1.40 \times 10^{28}$ electrons/m³.

**13. ▮** Calculate the Fermi energy for Cu, for which $A = 63.5$, $\rho = 8.95 \times 10^3$ kg/m³, and approximately one electron per atom is free. Repeat the calculation for lead, for which $A = 207.2$ and $\rho = 11.4 \times 10^3$ kg/m³. Assume in the case of lead that there are two free electrons per atom.

**14. ▮▮** Show that the Fermi energy can be expressed in terms of the density of the metal, $\rho$, its atomic weight $A$, and the number $n_V$ of free electrons per atom by the formula

$$E_F = \text{Const.} \left( \rho n_V / A \right)^{2/3},$$

and calculate the constant. Assume that $\rho$ is given in kg/m³ and that you want $E_F$ in eV.

**15. ▮** The Fermi energy of rubidium ($A = 85.5$, $\rho = 1.53 \times 10^3$ kg/m³) is 1.8 eV. What is the average number of valence (free) electrons?

**16.** ▌ Zinc $(A = 65.4, \rho = 7.1 \text{ g/cm}^3)$ has a valence $n_V = 2$. What is its Fermi energy?

**17.** ▌▌ While the sum in Eq. (10–39) can be calculated exactly, a useful addition to our repertoire of techniques is to recognize that $\sum_{1}^{N/2} j^2 = \sum_{1}^{N/2} j^2 \Delta j$, where $\Delta j = 1$. For $N/2$ large, we can change variables and turn the sum into an integral. If we introduce the new variable $x = j/(N/2)$, then a change of $j$ by unity is a very small change in $x$, so that the sum over $j$ is an integral over $x$. Show that the sum becomes $\sum_{1}^{N/2} j^2 \cong \left(\frac{N}{2}\right)^3 \int_{0}^{1} x^2 \, dx$ and thereby verify the approximate result for the sum given in the text.

**18.** ▌▌ Use dimensional analysis to show that the Fermi energy for nonrelativistic particles must be proportional to $n^{2/3}$. (*Hint*: The energy can only involve $\hbar$, $m$, and the interparticle spacing $a$ and must therefore be of the form $E = \hbar^\alpha m^\beta a^\gamma$.)

**19.** ▌▌ Use dimensional analysis to determine the dependence on volume of the total energy for a fixed number of particles, given that, assuming that the density is fixed, the number of particles is proportional to the box volume. You might wish to read the hint in the previous problem.

**20.** ▌▌ Consider a gas of free massless particles (as in Example 10–6). Use dimensional analysis to show that the Fermi energy must be proportional to $n^{1/3}$. (*Hint*: The energy can involve only $\hbar$, $c$, and the interparticle spacing $a$.)

**21.** ▌▌ Use the procedure of Problem 18 to show that the total energy of a neutrino gas for a fixed number of neutrinos is proportional to $V^{-1/3}$.

**22.** ▌▌ A measure of the compressibility of matter is the parameter $C \equiv \dfrac{\text{fractional change in volume}}{\text{additional pressure exerted}} = (\Delta V/V)/\Delta p$. For the case in which the pressure resisting compression is only the degeneracy pressure $p_f$, $C$ takes the form

$$C = \left(\frac{1}{V}\right)\left(\frac{dp_f}{dV}\right)^{-1}.$$

Show that in this case $C^{-1} = (5/3)p_f$.

**23.** ▌▌ The reciprocal of the quantity $C$ described in Problem 22 is called the **bulk modulus** $B$ of a material. That is, $1/C \equiv B$. Calculate $B$ for copper, assuming that it is all due to degeneracy pressure. Compare the value you obtain with the experimental value of $1.34 \times 10^{13} \text{ N/m}^2$.

**24.** ▌ Calculate the radius of an electron-degenerate star of mass $0.8 \times 10^{30}$ kg.

**25.** ▌▌ Calculate the radius of a neutron star. The calculation is essentially the same as that for a degenerate star, with the following changes: (*i*) Replace $m_e$ by $m_n$; (*ii*) replace the number of electrons $N_e$, which was taken to be $N/2$ (where $N$ is the number of protons plus the number of neutrons in the star), by $N$. Your result should be

$$R = \left(\frac{81\pi^2}{16}\right)^{1/3} \frac{\hbar^2}{Gm_n^3} N^{-1/3}.$$

Calculate the numerical value of the radius for a star of mass $4 \times 10^{30}$ kg.

26. ▌▌▌ Calculate the Fermi energy and the degeneracy pressure for a two-dimensional gas of free electrons. Recall that the possible energy values of a particle in a two-dimensional infinite well are given by

$$E = \frac{\hbar^2 \pi^2}{2mL^2}(n_1^2 + n_2^2), \quad \text{where} \quad n_1 = 1, 2, 3, \ldots, \text{ and } n_2 = 1, 2, 3, \ldots .$$

27. ▌▌▌ Calculate the number of states whose energy lies between $E$ and $E + \Delta E$, where $\Delta E \ll E$, in one, two, and three dimensions. From this result you can calculate the density of states $dn(E)/dE$ for electrons restricted to these numbers of dimensions. This is not at all a purely mathematical exercise, as semiconductor devices make use of electron "gases" of this type in all three types of dimensionality. (*Hint*: For two and three dimensions, construct lattices, as in the text, in the three types of spaces, and look at how many points there will be in the region between the "radius" $R$ and $R + dR$. All that remains is to relate $R$ to the energy $E$.)

# Applications

I n Part 2, we presented an overview of many of the elements of quantum mechanics, a truly revolutionary view of the physical world. As with many revolutions, the consequences of quantum mechanics could hardly have been foreseen by its creators. It has been variously estimated that 20 to 50 percent of the developed economies of the world are based on quantum mechanical applications. In this part of the text, we shall touch on some of these applications, from chemistry to materials science to electronics and beyond.

# Complex Atoms and Molecules

Quantum mechanics was developed in part in response to the challenge posed to classical physics by the spectral lines of atoms and molecules. Even before the discovery of the quantum theory by Heisenberg, Schrödinger, and Dirac, Bohr's ideas about atomic structure indicated that the periodic table could be understood in terms of stationary states of electrons. With the understanding that we have gained from our study of the hydrogen atom, and with our knowledge of the constraints imposed on multielectron systems by the exclusion principle, we are in a position to give a reasonably accurate qualitative account of the atoms of the periodic table. Helium, with one more electron than hydrogen, already provides us with a rich complexity. The addition of more electrons allows us to build up the periodic table of elements in a systematic way. The same notions help us understand the more complex structures of molecules. Modern chemistry is based on the deeper understanding of atoms and molecules that quantum mechanics provides.

## 11–1 Energy in the Helium Atom

The next chemical element after hydrogen is helium, which has two electrons circulating outside a nucleus with two positively charged protons. Even if we consider the nucleus to be so massive as not to play anything but a passive role in the helium atom, this still involves the interactions of two electrons with each other and the nucleus. It took Heisenberg a few years after the invention of quantum mechanics to describe helium in terms of wave functions in a way that is consistent with observation. We will now introduce you to some of the basic ideas.

The potential energy for helium consists of three terms. Two represent the interaction of the two electrons with the nucleus of charge $Z = 2$, one for each electron, and the third term represents the electron–electron repulsion. We begin with a very crude approximation: We ignore the electron-electron repulsion. This puts each electron in a hydrogenlike atom with $Z = 2$. The binding energy is $13.6\,Z^2$ eV for each electron, so that the total binding energy—the energy that would have to be supplied to free both electrons—is $2 \times 13.6 \times 4$ eV $= 108.8$ eV. Each of the electrons is in an orbit of radius $a_0/Z$, where $a_0$ is the Bohr radius.

You might think that in this approximation the wave function is the product of two hydrogenlike ground-state wave functions. Each of these wave functions is proportional to the hydrogen radial wave function for $n = 1$ and $\ell = 0$, namely $R_{10}(r)$; in particular there is no angular dependence, because $\ell = 0$. However, it is not enough to write $\psi(r_1, r_2) = R_{10}(r_1)R_{10}(r_2)$: First, the total wave function must be antisymmetric under electron exchange, and secondly, we have forgotten the spins! We fix this by noting that if the space state of both electrons is $R_{10}$, then the only way to make the overall wave function antisymmetric is to put the two spins into the (antisymmetric) $S = 0$ spin state. Thus we combine the simple product of the $R$'s with the $S = 0$ spin wave function, Eq. (10–23):

$$\psi(r_1, r_2) \propto R_{10}(r_1)R_{10}(r_2)[\chi_+(1)\chi_-(2) - \chi_-(1)\chi_+(2)]. \qquad (11–1)$$

(We write "proportional to" because we haven't been very careful about normalization here.) Since, to the level of approximation of our work here, the spins do not enter into the energy, this state has energy $-108.8$ eV.

Now we consider the first excited state of helium in this approximation—that is, we continue to ignore the electron–electron repulsion. Here, we place one electron in the $n = 1$, $\ell = 0$ state and the other in an $n = 2$, $\ell = 0$ state. The energy is then $(-13.6\,Z^2 - 13.6\,Z^2/2^2)$ eV $= -68.0$ eV. The total spin of the two electrons can be either $S = 1$ or $S = 0$, according to whether we arrange the spatial wave function to be antisymmetric—in which case the spin wave function is $S = 1$—or symmetric—in which case the spin wave function is $S = 0$. Now we come to the point of our argument: When the electron–electron repulsion is taken into account, these states are no longer degenerate. The reason is basically that when the spatial wave function is antisymmetric, the probability that each electron is at the same place is zero. More generally, for an antisymmetric wave function it is more probable that the electrons stay away from each other and the effect of repulsion is decreased. In other words, *the choice of spin state has an important effect on the energy, even though the spin does not enter directly into the expression for the energy.* We described this so-called *exchange force* in Chapter 10, where we showed that the exclusion principle altered the probability of finding particles close together, *even in the absence of interactions.*

As we argued qualitatively in that chapter, when the spins of the two electrons are parallel, the electrons are kept apart by a distance $\lambda/2$, where $\lambda$ is the de Broglie wavelength. This exclusion-principle effect is absent when the two electrons are in a spin singlet state. Thus, roughly speaking, the energy difference between the singlet and triplet states due to coulomb repulsion is the electrostatic energy of a sphere of charge $e$ and radius $\lambda/2$; that is,

$$\Delta E \cong \frac{e^2}{4\pi\varepsilon_0(\lambda/2)} = \frac{e^2}{4\pi\varepsilon_0} \times \frac{p}{\pi\hbar} = \frac{e^2}{4\pi\varepsilon_0\hbar c} \times \frac{1}{\pi} \times \sqrt{p^2c^2}$$

$$= \frac{\alpha}{\pi}\sqrt{2mc^2(p^2/2m)} \cong \frac{\alpha}{\pi}\sqrt{2mc^2\left(\frac{mc^2\alpha^2}{2}\right)} \cong \frac{1}{\pi}mc^2\alpha^2,$$

which is on the order of electron volts. In effect, the exclusion principle creates a spin-dependent interaction of the size of the coulomb potential.

The exchange force is also a factor when we have a crystal consisting of an array of atoms. In that case, too, changing the spin at one site costs energy on the order of electron volts—such a change turns a pair in a singlet state into a pair in a triplet state, and vice versa. This fact plays an important role in understanding the "permanence" of magnets. A magnet of the kind that you might stick to your refrigerator is "permanent" because the vast majority of the atoms, each of which is a tiny magnetic dipole, have their spins aligned together to form a large magnetic dipole.[1] The spins stay parallel to each other because it costs energy to disrupt this collective effect. In such a material there must exist interatomic forces which lead to a potential energy favoring the alignment of spins in the same direction. These forces are the exchange forces that we have argued are important in helium.

We have seen that if we neglect the electron–electron repulsion, the ground-state energy of helium is $-108.8$ eV; the coulomb repulsion between the two electrons will increase this value. Without repulsion, each electron is most likely to be found at a distance $a_0/Z$ from the nucleus. Therefore, we estimate the repulsion effect by assuming that the typical separation between the two electrons lies between $a_0/Z$ and $2a_0/Z$. With $(e^2/4\pi\varepsilon_0)/2a_0 = 13.6$ eV and with $Z = 2$, this corresponds to a repulsion energy between 27.2 eV and 54.4 eV. The binding energy for the ground state then lies between $108.8\text{ eV} - 27.2\text{ eV} = 81.6\text{ eV}$ and $108.8\text{ eV} - 54.4\text{ eV} = 54.4\text{ eV}$. In fact, the experimental value is 79 eV. Another way of putting this is to express the ground-state energy in terms of an effective $Z$, denoted by $Z^*$, which represents the average value of charge that either electron "sees." This will not be equal to $Z = 2$, because of the screening of the nuclear charge by the other electron. With $E(Z^*) = 2 \times (-13.6\text{ eV})Z^* = 79$ eV, we find that $Z^* = 1.7$, which is indeed less than the unscreened nuclear charge of $Z = 2$.

When one of the electrons is removed, the remaining electron is in a $1s$ state of a $Z = 2$ hydrogenlike atom. The binding energy of this remaining electron is $-54.4$ eV, so that the difference of $79 - 54.4 = 24.6$ eV is the single-electron ionization energy—the energy required to remove a single electron from helium.

The first excited state of helium will be either a spatially symmetric (i.e., spin antisymmetric $S = 0$) or spatially antisymmetric (spin symmetric $S = 1$) combination of $(1s)(2s)$ and $(2s)(1s)$. The second of these has the lower energy, since in a spatially antisymmetric state the electrons tend to be away from each other, which has the double effect of reducing the energy-raising repulsion between them and reducing the shielding of the $Z = 2$ nuclear charge. A very rough estimate of the energy of that state can be made by ignoring the $e$–$e$ repulsion and ignoring the screening. Then the energy of a $(1s)(2s)$ configuration is

$$-(13.6\text{ eV})\left[\frac{Z^2}{1^2} + \frac{Z^2}{n^2}\right] = -68\text{ eV for } Z = 2.$$

Thus, it takes approximately $79\text{ eV} - 68\text{ eV} = 11\text{ eV}$ to excite an electron to the first excited state. On the chemical scale this is a great deal of energy: Ordinary chemical reactions proceed by adding or taking away on the order of 1 eV per atom. This is one of the main reasons that helium is so chemically inactive.

## 11–2 Building Up the Periodic Table

An atom of atomic number $Z$ and mass number $A$ has, at its core, a nucleus containing $Z$ protons and $A - Z$ neutrons. As described in Section 11–1, we treat the nucleus, whose charge is $+Ze$, as fixed in space. The $Z$ electrons move rapidly about the nucleus, and from this classical description of the atom we create a picture in which each electron moves in a potential that consists of the effect of the nuclear point charge $+Ze$ and a smeared-out negative charge $-(Z - 1)e$. Since in this picture the $Z - 1$ background electrons are not on top

---

[1]Such substances are **ferromagnetic**. **Antiferromagnetic** substances also exist, for which the potential energy favors alternating antiparallel spins.

of the nucleus, we do *not* have a hydrogenlike potential energy. Each electron "sees" a *net* charge of $+e$, but this charge is distributed in space. The distribution is at least approximately spherical about the nucleus, so that we may write the potential energy of any one electron as $U(r)$, with known small-[2] and large-radius limits, namely

$$U(r) \rightarrow \frac{Ze^2}{4\pi\varepsilon_0 r} \quad \text{as } r \rightarrow 0$$

and

$$U(r) \rightarrow \frac{e^2}{4\pi\varepsilon_0 r} \quad \text{as } r \rightarrow \infty. \tag{11-2}$$

In other words, we may write the potential energy *as if* $Z$ were a function of $r$:

$$U(r) = \frac{Z(r)e^2}{4\pi\varepsilon_0 r}. \tag{11-3}$$

Because the potential energy depends only on $r$, the force is a central force—it points in the radial direction—and therefore angular momentum is conserved. We may thus describe the wave function of the electron under consideration in terms of its angular-momentum quantum numbers: the orbital angular momentum $\ell$, the z-component of the orbital angular momentum $m_\ell$, and the spin, which for $s = 1/2$ is just the value of $s_z$, or equivalently, the choice between an "up" and a "down" spin label. There will also be an analog of the principal quantum number in hydrogen, which we will again label with $n$. Thus the wave function will have the generic form

$$\psi(\mathbf{r}) = R_{n\ell}(r)Y_{\ell m}(\theta, \varphi)\chi_{\pm}. \tag{11-4}$$

This approximation is crude. As was pointed out in Chapter 9, there is a coupling between the spin and orbital angular momenta, called the spin–orbit coupling. This coupling plays an important role in explaining the fine details of atomic structure, but it is not necessary for the purposes of this section, which is to get only a rough picture of the structure of atoms in their ground state.

The potential energy is no longer a pure $1/r$ form. Thus it is no longer true that, for a given value of $n$, all the states with $\ell = 0, 1, 2, \ldots (n - 1)$ have the same energy. In fact, we expect, *in general*, that for a given $n$, the energy eigenvalues will increase with $\ell$. We can argue that this is true because of the so-called centrifugal-barrier term, the term in the radial wave equation (9–17) that has the form

$$U_{\text{centr}}(r) = \frac{\hbar^2\ell(\ell + 1)}{2mr^2}. \tag{11-5}$$

This term represents a positive contribution to the potential energy, a repulsion that acts to keep the electron away from the nucleus. It reduces the magnitude of the binding energy, and it grows with $\ell$. For hydrogen, the pure $1/r$ potential is an attraction that compensates for the repulsion due to the centrifugal barrier in just such a way as to make the energies independent of the value of $\ell$. In more complex atoms, the electrical attraction is smeared out [as indicated in

---

[2] Actually, the nucleus is not infinitely small: It is some $10^5$ times smaller than the typical atomic size. When, in Eq. (11–2), we say $r \rightarrow 0$, we nevertheless assume that $r$ is greater than the typical nuclear radius of about $10^{-15}$ m.

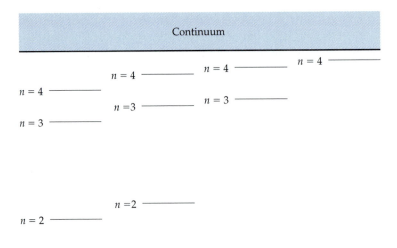

**• Figure 11–1** Schematic sketch of the spectrum in which the $n^2$ degeneracy of the $1/r$ potential is broken. The $2\ell + 1$ levels corresponding to a given $\ell$ are still degenerate.

Eq. (11–3)] and is *weaker* than pure $1/r$ for larger values of $r$. This allows the centrifugal barrier to have an $\ell$-dependence of the energy for a given value of $n$ such that electrons with larger $\ell$-values are less tightly bound (•Fig. 11–1).

### How to Build Up the Periodic Table

In this subsection we systematically examine the structure of elements with Z-values up to 10, starting with hydrogen ($Z = 1$). We use the approximations described earlier, together with the exclusion principle. We introduce a standard nomenclature that employs the labeling $n = 1, 2, \ldots$, but replaces the labeling $\ell = 0, 1, 2, 3, 4, \ldots$ by the letters $s, p, d, f, g, \ldots$, as in Table 11–1. This labeling is a historical artifact but continues in use today. It derives from the early days of atomic spectroscopy, when the series of spectral lines had names like "sharp" and "diffuse." The labeling system is accordingly called **spectroscopic notation**.

| Table 11–1 | Spectroscopic Notation | | | | |
|---|---|---|---|---|---|
| $\ell$-value | 0 | 1 | 2 | 3 | 4 |
| letter label | $s$ | $p$ | $d$ | $f$ | $g$ |

As we go along, we shall want to keep in mind that according to the Pauli principle the number of electrons that can be accommodated in a state with angular momentum $\ell$ is $2 \times (2\ell + 1)$, where the factor $2\ell + 1$ counts the number of orbital angular-momentum projections and the exterior factor of 2 counts the number of spin projections—spin "up" or spin "down"—for the electron. For a schematic picture of the filling of levels as we build up the elements we shall describe, see •Fig. 11–2.

*Hydrogen* $(Z = 1)$   This atom was discussed in some detail in Chapter 9. It has only one electron, and the ground state, with $n = 1$, $\ell = 0$, is a state denoted in spectroscopic notation as 1s. The ionization energy—the energy required to remove the electron entirely from the atom, starting with the ground state—is 13.6 eV, while the energy needed to excite the electron from the ground state to the first excited state is 10.2 eV.

*Helium* $(Z = 2)$   The ground state of helium is one in which both electrons are in 1s states. The spectroscopic notation for this is $1s^2$. To satisfy the exclusion principle, one electron must have spin up while the other has spin down. In

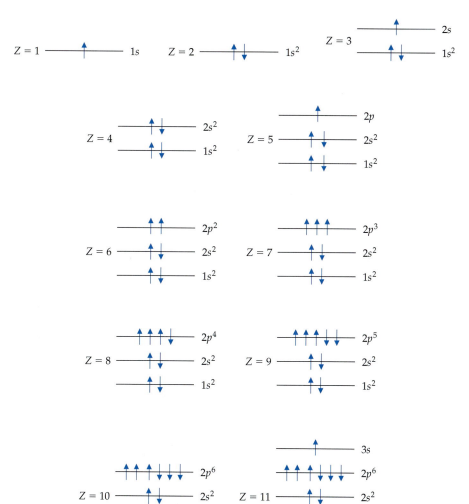

• **Figure 11–2**  How the electron shells are filled for atoms with $Z$ ranging from 1 to 11. The Pauli principle dictates that no more than two electrons can occupy a given atomic state.

fact, as mentioned in Chapter 10, since the wave function for $1s^2$ is symmetric in its spatial coordinates, the two electrons will be in an antisymmetric, or singlet, spin state—the state with total spin $S = 0$. We say that *the electrons have their spins paired up*. We shall see later that atoms in which the outer electrons have their spins paired up into singlet states are less reactive than atoms in which this is not the case. The spin does not enter the energy structure at our level of approximation. Other features of helium were described in Section 11–1.

*Lithium* $(Z = 3)$   The exclusion principle forbids the presence of three electrons in the $1s$ state, a state that can have at most two electrons. The lowest electron configuration will contain two electrons in the $n = 1$, $\ell = 0$ state and one electron in the $n = 2$, $\ell = 0$ state, with labeling $1s^2 2s$. Thus, two electrons fully occupy the $n = 1$ level; for this $n$, the only allowed value of $\ell$ is $\ell = 0$. We say that when all the levels of a given $n$ and $\ell$ are filled, we have a **closed shell.** Hence, for lithium, a single electron is added to a closed shell made up of the two $n = 1$ electrons. The electron outside the shell is called a **valence** electron, which, in lithium, has $n = 2$. Keep in mind that lower $n$-values are associated with electrons that are a smaller average distance from the nucleus. Thus the valence electron would be screened from "seeing" the full charge of the nucleus. If the screening due to the two inner electrons were perfect, then the valence electron would "see" a charge of $+e$, not $+3e$, and its binding energy would be $(13.6\,\text{eV})/n^2 = (13.6\,\text{eV})/4 = 3.4\,\text{eV}$. However, the screening is not perfect, because the valence electron is in an $\ell = 0$ state, which does not undergo centrifugal repulsion. Therefore the wave function is not pushed away from the nucleus by the angular-momentum barrier. This in turn means that there is some reasonable probability that the electron can be close to the nucleus and "see" its full $Z = 3$ charge. The experimental value of the ionization energy for the valence electron is $5.4\,\text{eV}$, which corresponds to its "seeing" an effective charge of 1.26, in agreement with our remarks.

Whenever there is a single electron outside a closed shell—and lithium is the first element in the periodic table that is an example of this—the element can be chemically very active. The reason for this will emerge in our discussion of ionic bonding in Section 11–4.

*Beryllium* $(Z = 4)$   The natural place for the fourth electron to go is the empty slot in the $2s$ state, so that the overall configuration is $1s^2 2s^2$. Thus we have two closed shells. As far as the energy is concerned, the situation is very much like that of helium: If the screening due to the two $n = 1$ electrons were perfect, the outer electrons would "see" a net charge with $Z = 2$, and because the outer electrons have $n = 2$, we would expect the ionization energy for each of them to be about one-fourth that of helium. The value is actually larger, because the screening is not perfect. A 50% increase, similar to the one encountered in lithium, yields approximately 9 eV for this ionization energy. The experimental value is 9.3 eV. The two electrons in the $2s$ state have a spatially symmetric wave function, so that they are in a total spin $S = 0$ state, with antiparallel spins. As with helium, the outer electrons have their spins paired up, so beryllium is not very reactive.

*Boron* $(Z = 5)$   With the available places in the $1s$ and $2s$ configuration all taken, the fifth electron can go either into the $3s$ state or into one of the $2p$ states. Although the $2p$ state lies higher than the $2s$ state, it lies lower than the $3s$ state, so that the fifth electron goes into the $2p$ state. The ionization energy for the removal of the last electron from boron is smaller than that of beryllium, since, as just mentioned, the 2p state lies higher than the 2s state; experiment gives 8.3 eV, as opposed to 9.3 eV for beryllium. With a single electron outside a closed shell, boron is highly reactive.

*Carbon* $(Z = 6)$   The configuration of carbon is $1s^2 2s^2 2p^2$. The structure of the angular wave function allows the second 2p electron literally to stay away from the first 2p electron, minimizing the repulsive energy. Thus the increase in Z from 5 to 6 is accompanied by an *increase* in the ionization energy! Experiment confirms this perhaps counterintuitive state of affairs, with a value of 11.3 eV for carbon.

How does this mechanism work? The three possible $\ell = 1$ wave functions $Y_{11}$, $Y_{10}$, and $Y_{1,-1}$ are proportional to $\sin\theta\, e^{i\varphi}$, $\cos\theta$, and $\sin\theta\, e^{-i\varphi}$, respectively. These wave functions can be recombined into wave functions proportional to $\cos\theta$, $\sin\theta\cos\varphi$, and $\sin\theta\sin\varphi$, the squares of which are shown in •Fig. 11–3. The figure shows a probability distribution that contains three orthogonal "arms," and when two electrons go into different arms, the repulsion is lowered. Note that the electrons do not necessarily form a spin singlet combination, because their spatial wave functions can be completely different. The electrons are in this sense not *paired*. We might therefore expect carbon to have *two* active electrons—in chemical language, we might expect carbon to be *divalent*.

This argument has to be refined because of the subtlety of close-lying levels. It costs relatively little energy to promote one of the 2s electrons to one of the unoccupied 2p levels—into the third orthogonal arm. The new configuration, $1s^2 2s 2p^3$, has *four* unpaired electrons. We will see later in the chapter that electrons are the material that bind atoms together to make molecules and that they can do so only when they are not paired in their original atoms— that is, when they are not in the same spatial state as another electron in an opposite spin state. With four unpaired electrons potentially available for binding with other atoms, it is advantageous for carbon to combine with other atoms, since the energy cost of promoting a 2s electron to the 2p orbital is comparatively small. For this very reason, carbon plays a remarkable role in the formation of large organic molecules—the molecules on which life as we know it is based.

*Nitrogen* $(Z = 7)$   In nitrogen, the configuration is $1s^2 2s^2 2p^3$. Since the chemistry depends very little on the closed shells, this configuration is often abbreviated as $2p^3$. The three valence electrons can have spatial wave functions that do not overlap very much, so the increase in binding energy is expected to be

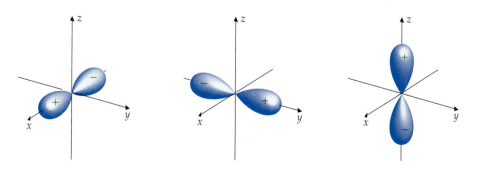

• **Figure 11–3**   The two outer (2p) electrons in carbon distribute themselves among three mutually orthogonal arms. In this figure, the lobes represent the regions of maximum probability of finding an electron. The signs indicate that the wave functions have opposite signs in the paired lobes.

the same as the increase in passing from boron to carbon. This is in agreement with the measured ionization energy value of 14.5 eV.

*Oxygen* (Z = 8)  In our abbreviated notation, the configuration for oxygen is $2p^4$. The fourth electron must go into one of the already occupied p-states and so form a spin singlet state with the electron already in it. With two electrons in the same p-state, the effect of the electron–electron repulsion increases, and even though Z increases by one, we would expect the ionization energy not to increase as fast as in the previous examples. In fact, the ionization energy is 13.6 eV, nearly the same as for nitrogen, showing that the additional positive electron–electron repulsion energy pretty well matches the new (negative) nucleus–electron energy. Because the two electrons in the same 2p state are paired up, only the remaining two electrons participate strongly in chemical reactions, and we expect oxygen to be divalent.

*Fluorine* (Z = 9)  The configuration of fluorine is $2p^5$. The increase in the ionization energy seen with the earlier elements resumes, with the experimental value of 17.4 eV. Fluorine is chemically very active, because it can "accept" an electron from another atom to form the closed shell $2p^6$, which is very stable. One sometimes says that fluorine has a "hole" in an otherwise closed shell.

*Neon* (Z = 10)  With Z = 10, the configuration is $2p^6$, and all electrons are paired off. The 2p shell is thus closed. The ionization energy is 21.6 eV, continuing the upward trend. As with helium, in neon the first available state that an electron can be excited into has a higher n-value and a much higher energy value. Accordingly, it takes quite a lot of energy to excite the atom. Neon shares with helium the property of being an inert gas.

**Example 11–1**   Consider the atom with Z = 9 (fluorine). If we treat the outer electronic configuration as a closed shell with a hole in it, what would you expect the total orbital angular momentum of fluorine to be? What are the possible values of the total angular momentum, given that a closed shell has both a total angular momentum and a total orbital angular momentum of zero?

**Solution**   Since the addition of a single 2p electron would close the shell, leading to an $\ell_{tot}$ value of 0, the $\ell$ quantum number of fluorine must be such that when $\ell$ is added (quantum mechanically) to $\ell = 1$, we get $\ell_{tot} = 0$. Now the allowed values of $\ell_{tot}$ are $\ell + 1$, $\ell$, and $|\ell - 1|$, unless $\ell = 0$, in which case $\ell_{tot} = 1$. This means that $\ell = 1$. Similarly, since the addition of spin 1/2 (of the last 2p electron) would give a total spin of zero, the shell with a single hole in it must have a total spin of 1/2. This is to be expected if we take into account the fact that in each of the fully occupied states, the electron spins must point in opposite directions; that is, the electrons pair up to form an S = 0 state. Only one of the three 2p levels has a single electron, so that the total spin is the spin of this single electron, S = 1/2.

## 11–3  Beyond Z = 10 and General Comments

The discussion of the previous section shows how the exclusion principle brings variety to the structure and chemical properties of the elements. As Z increases beyond 10, the pattern tends to repeat. The next *period* or *series*—the group of atoms that are formed as electrons are added, starting from a set of closed shells, until the next set of shells is filled—again has eight elements in it. First the 3s shell is filled, to give sodium (Z = 11) and magnesium (Z = 12). Then the 3p shell is filled, giving the elements aluminum (Z = 13), silicon (Z = 14), phosphorus ($Z = 15$), sulfur ($Z = 16$), chlorine ($Z = 17$), and, closing the shell, argon ($Z = 18$). The properties of the elements of the $3s$–$3p$ series are similar to the properties of those of the $2s$–$2p$ series. For example, like lithium, sodium is highly reactive, and similarly, like fluorine, chlorine, which contains a closed shell with a hole in it, is highly reactive. For the $Z = 11$ to $Z = 18$ series, the ionization energies are somewhat smaller than those of the $n = 2$ series, as befits the less tightly bound $n = 3$ states, and the breaks in their values match those in the $2s$–$2p$ series.

We generally refer to the elements consisting of a closed shell plus one electron—lithium and sodium are the first two—as **alkali metals**, and those elements consisting of a closed shell with a hole—fluorine and chlorine are the first two—are called **halogens**.

The $n = 3$ levels also should allow a series with $\ell = 2$: the $3d$ shell. The number of electrons that can be accommodated in a state with angular momentum $\ell = 2$ is $2 \times (2\ell + 1) = 10$. However, it is at this stage that the departures from the $1/r$ shape of the electrostatic potential begin to make a qualitative difference. The $3d$ states have (slightly) higher energy than the $4s$ states. Together, the two sets of states form a new period consisting of successive elements with configurations $4s$, $4s^2$, $4s^23d$, $4s^23d^2$, and $4s^23d^3$. Because of the closeness of the levels, the next element has the exceptional configuration $4s3d^5$; then the series resumes with $4s^23d^5$, $4s^23d^6$, ... , $4s^23d^8$; there follows another exceptional configuration $4s3d^{10}$ and, finally, $4s^23d^{10}$. After that, the $4p$ shell is successively occupied, making the configurations $4s^23d^{10}4p^k$, with $k = 1, 2, \ldots, 6$. This period ends with krypton ($Z = 36$). The chemical properties of the elements at the beginning and end of the period are similar to those of elements at the beginning and end of other periods. Potassium ($Z = 19$), with its single $4s$ electron, is an alkali metal. Bromine, coming just before the end of the period, with a configuration $4s^23d^{10}4p^5$, is, like fluorine and chlorine, a halogen, with a single hole in the outer $p$ shell. The intermediate elements for which the $3d$ shell is successively occupied do not differ very much in their chemical properties. The reason is that the probability distribution for the $4s$ electrons is weighted at larger values of $r$ than that of the $3d$ electrons, so that the filled $4s^2$ shell shields the inner $3d$ electrons from influences outside the atom. The same thing happens when the $4f$ shell is successively occupied just after the $6s$ shell is filled. The elements in this latter series are called the *rare earth* elements.

If we look at a periodic table of the elements, we see that the number of stable or nearly stable elements goes up to $Z = 102$. Are there more elements, with higher $Z$? There is no reason in nonrelativistic quantum mechanics that this should not be possible. The main reason that no stable higher $Z$ elements exist is that the *nuclei* with large $Z$ are unstable. We shall discuss the reasons for this instability in our survey of nuclear physics in Chapter 15. Suffice it to say here that the protons in the nucleus repel each other through coulomb forces. This electrostatic repulsion is countered by the attractive nuclear force, but electrostatic repulsion also ensures that when $Z$ gets large enough, the nucleus undergoes spontaneous fission. Nuclear theory does seem to predict that some superheavy nuclei can be more nearly stable. If these are discovered, then their atoms can be treated just as we have done here for lower $Z$.

Most elements have ionization energies for the removal of a single electron in the 5- to 15-eV range. That is because the electrons in the highest shells are the ones most easily ionized, and these are shielded by the inner electrons, so they do not "feel" the full attraction of the nucleus. Thus, the effective $Z$-value is small—in the range from 1 to 3.

As one final observation, note that since the effective potential is not a pure $1/r$ form, parameters such as the energy and radius do not have a simple $n^2$ dependence. In particular, the radii of *all* atoms are on the order of 0.1 nm, even though the values of $n$ increase from 1 to 7!

Table 11–2 lists the name of the element, its Z-value, the electron configuration (e.g., $1s^2 2p^4 \ldots$), the ionization energy in eV, and the radius of atoms with larger Z-values.

**Table 11–2   Electron Configurations for Atoms with $Z = 1$ to $Z = 42$**

| Atom | $Z$ | Electron Configuration | Ionization Energy (eV) | Radius, * $r/a_0$ |
|------|-----|------------------------|------------------------|-------------------|
| H  | 1  | $1s$                     | 13.6  | 1.00 |
| He | 2  | $1s^2$                   | 24.6  | 0.55 |
| Li | 3  | He $2s$                  | 5.4   | 3.00 |
| Be | 4  | He $2s^2$                | 9.3   | 1.96 |
| B  | 5  | He $2s^2 2p$             | 8.3   | 1.47 |
| C  | 6  | He $2s^2 2p^2$           | 11.3  | 1.17 |
| N  | 7  | He $2s^2 2p^3$           | 14.5  | 0.98 |
| O  | 8  | He $2s^2 2p^4$           | 13.6  | 0.85 |
| F  | 9  | He $2s^2 2p^5$           | 17.4  | 0.75 |
| Ne | 10 | He $2s^2 2p^6$           | 21.6  | 0.66 |
| Na | 11 | Ne $3s$                  | 5.1   | 3.23 |
| Mg | 12 | Ne $3s^2$                | 7.6   | 2.42 |
| Al | 13 | Ne $3s^2 3p$             | 6.0   | 2.47 |
| Si | 14 | Ne $3s^2 3p^2$           | 8.2   | 2.02 |
| P  | 15 | Ne $3s^2 3p^3$           | 11.0  | 1.74 |
| S  | 16 | Ne $3s^2 3p^4$           | 10.4  | 1.53 |
| Cl | 17 | Ne $3s^2 3p^5$           | 13.0  | 1.38 |
| Ar | 18 | Ne $3s^2 3p^6$           | 15.8  | 1.25 |
| K  | 19 | Ar $4s$                  | 4.3   | 4.10 |
| Ca | 20 | Ar $4s^2$                | 6.1   | 3.19 |
| Sc | 21 | Ar $4s^2 3d$             | 6.6   | 2.96 |
| Ti | 22 | Ar $4s^2 3d^2$           | 6.8   | 2.79 |
| V  | 23 | Ar $4s^2 3d^3$           | 6.7   | 2.64 |
| Cr | 24 | Ar $4s\, 3d^5$           | 6.76  | 2.73 |
| Mn | 25 | Ar $4s^2 3d^5$           | 7.43  | 2.41 |
| Fe | 26 | Ar $4s^2 3d^6$           | 7.90  | 2.32 |
| Co | 27 | Ar $4s^2 3d^7$           | 7.86  | 2.23 |
| Ni | 28 | Ar $4s^2 3d^8$           | 7.63  | 2.15 |
| Cu | 29 | Ar $4s\, 3d^{10}$        | 7.72  | 2.24 |
| Zn | 30 | Ar $4s^2 3d^{10}$        | 9.39  | 2.02 |
| Ga | 31 | Ar $4s^2 3d^{10} 4p$     | 6.00  | 2.37 |
| Ge | 32 | Ar $4s^2 3d^{10} 4p^2$   | 8.13  | 2.06 |
| As | 33 | Ar $4s^2 3d^{10} 4p^3$   | 10.00 | 1.89 |
| Se | 34 | Ar $4s^2 3d^{10} 4p^4$   | 9.75  | 1.74 |
| Br | 35 | Ar $4s^2 3d^{10} 4p^5$   | 11.84 | 1.60 |
| Kr | 36 | Ar $4s^2 3d^{10} 4p^6$   | 14.00 | 1.51 |
| Rb | 37 | Kr $5s$                  | 4.18  | 4.32 |
| Sr | 38 | Kr $5s^2$                | 5.69  | 3.47 |
| Y  | 39 | Kr $5s^2 4d$             | 6.60  | 3.19 |
| Zr | 40 | Kr $5s^2 4d^2$           | 6.95  | 3.00 |
| Nb | 41 | Kr $5s\, 4d^4$           | 6.77  | 3.00 |
| Mo | 42 | Kr $5s\, 4d^5$           | 7.18  | 2.87 |

*The radii are defined by the calculated peak in the probability distribution of the outermost electrons.

**Example 11–2**    Use the information contained in Table 11–2 to calculate $Z_{eff}$ as defined by the equation

$$\text{ionization energy} = (13.6 \text{ eV}) \left( \frac{Z_{eff}^2}{n^2} \right)$$

for He, Li, F, Ne, Na, and Ca.

**Solution**    Let us denote the ionization energy by I.E. Then the aforesaid definition yields

$$Z_{eff} = n\sqrt{\frac{\text{I.E.}}{13.6 \text{ eV}}}.$$

The $n$ value to be used is the largest in the configuration. The value of I.E. (in eV) and $n$ values for the elements listed are He(24.6; 1), Li(5.4; 2), F(17.4; 2), Ne(21.6; 2), Na(5.14; 3), and Ca(6.11; 3). Thus,

$$Z_{eff}(\text{He}) = \sqrt{\frac{24.6}{13.6}} = 1.34, \qquad Z_{eff}(\text{Li}) = 2\sqrt{\frac{5.4}{13.6}} = 1.26,$$

$$Z_{eff}(\text{F}) = 2\sqrt{\frac{17.4}{13.6}} = 2.26, \qquad Z_{eff}(\text{Ne}) = 2\sqrt{\frac{21.6}{13.6}} = 2.52,$$

$$Z_{eff}(\text{Na}) = 3\sqrt{\frac{5.14}{13.6}} = 1.83, \qquad Z_{eff}(\text{Ca}) = 3\sqrt{\frac{6.11}{13.6}} = 2.01.$$

This pattern shows that the inner electrons are quite effective in screening the full nuclear charge.

## Moseley's Law and the Auger Effect

An early experimental study reveals how some of the issues we have been discussing emerged. After Bohr had published his three early papers on atomic structure in 1913, the English physicist Henry J.G. Moseley, who read these papers with a great deal of interest, began experiments on atomic transitions. At the time, the Z and A values for elements in the periodic table were not known; today we understand that these numbers count, respectively, the number of protons and the number of neutrons plus protons in the nucleus. The $A$-value could be measured by experiments on atomic masses, but no one knew how to determine the Z-values other than by a very uncertain inference based on the chemistry of the element. To give you an idea of the confusion, the suggestion that $Z = A/2$ for all elements was current. Note that the Bohr model did not—indeed, could not—settle this issue, since, as we have seen throughout the chapter, concepts that go far beyond the Bohr model are necessary to understand atoms beyond hydrogen in the periodic table.

Moseley found a way to measure the Z-values systematically, using what are known as secondary X rays. When a beam of X rays passes through matter, atomic electrons can absorb the photons in the beam and be ejected from the atom. Electrons from higher shells can then make transitions to the newly created hole. The frequencies of the radiation emitted (the secondary radiation) when these transitions take place covers a range that corresponds to the difference in energy between the levels in question; Moseley measured[3] the so-called $K_\alpha$-series X rays, which, in retrospect, correspond to the ejected atomic

---

[3] He also looked at the higher frequency $K_\beta$ series, which correspond to the ejected atomic electrons having come from the $n = 1$ shell and the electrons replacing them having come from the $n = 3$ shell, as well as $L$-series X rays, which correspond to the electrons having been ejected from the $n = 2$ shell. Here, we concentrate on the $K_\alpha$ series.

electrons having come from the $n = 1$ (i.e., 1s) shell and the electrons replacing them having come from the $n = 2$ shell. For these transitions, the radiation is in the X-ray range.

Moseley found a simple empirical result for the $K$-series X-ray frequencies. Let the Z-value that marked the place of an element in the periodic series be $Z_{inf}$, where the subscript "inf" abbreviates "inferred." Then Moseley found empirically (•Fig. 11–4) that the frequency of the $K_\alpha$ series radiation was proportional to $(Z_{inf} - 1)^2$. Since Moseley's measurement had nothing to do with chemistry, which is the origin of the periodic table, he argued, on the basis of the Bohr model, that one could explain at least part of this result if $Z_{inf}$ was actually the charge on the nucleus—Z itself. In other words, he confirmed that the assignment of Z by the chemical properties of the periodic table was correct. In addition, there were gaps in his plot at $Z = 43, 61$, and 75, suggesting the existence of as-yet undetected elements with these Z-values. These elements were all subsequently discovered. Neither Moseley nor anyone else at the time understood why $Z - 1$ appeared, rather than Z.

How do the ideas of the Bohr model "explain" Moseley's data? In a hydrogenlike atom of charge Z, the difference in energy between the $n = 2$ level and the $n = 1$ level is given by

$$(13.6\ \text{eV})\left[\left(-\frac{Z^2}{2^2}\right) - \left(-\frac{Z^2}{1^2}\right)\right] = \frac{3}{4}(13.6\ \text{eV})Z^2.$$

This is the $K_\alpha$ photon's energy, equal to $hf_{K\alpha}$ where the frequency is $f_{K\alpha}$. Thus, the frequency of the $K_\alpha$-series photon is, in this picture,

$$f_{K\alpha} = \frac{3}{4}(13.6\ \text{eV})\frac{Z^2}{h} = \frac{3(13.6\ \text{eV})Z^2}{4(4.14 \times 10^{-15}\ \text{eV}\cdot\text{s})} = (2.46 \times 10^{15}\ \text{Hz})Z^2.$$

According to a naive argument, the second electron in the 1s level screens the nuclear charge by one, so that Z in the expression for the photon frequency is changed to $Z - 1$, and this is Moseley's empirical effect.

But it is not so easy to employ this naive argument knowing what we now know. In particular, it is surprising that the single remaining $n = 1$ electron should reduce the $n = 1$ energy term in the expression for the Moseley frequency from Z to $Z - 1$. In fact, that the Bohr form with Z reduced to $Z - 1$ works is really only an accident! The origin of the screening effect is more com-

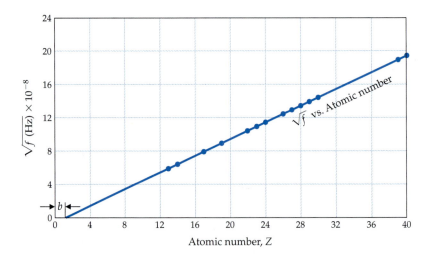

• **Figure 11–4**  Some of Moseley's data showing the linear relation between the square root of the frequency and Z. The intercept $b$ shows that it is approximately $Z - 1$ that enters into $f^{1/2}$ for the $K_\alpha$ series.

plicated than that of the simplest arguments. The electrons that jump from the $n = 2$ level must have $\ell = 1$, because of selection rules for radiative transitions. (See Chapter 13.) But these electrons lie on the average further from the nucleus than the $n = 2$, $\ell = 0$ electrons. Those electrons have a fair probability of penetrating close to the nucleus, adding to the screening both for the energy of the $n = 2$, $\ell = 1$ state from which the electrons jump and for the energy of the $n = 1$, $\ell = 0$ state to which they jump. In retrospect, if Moseley had known he had to worry about all this, he might not so easily have come to his conclusion!

---

**Example 11–3**   The $L_\beta$-series corresponds to transitions in complex atoms from the $n = 3$ levels to the $n = 2$ levels. Write the general Bohr-inspired structure of the frequencies in this series, given that the amount of screening could be different in the two levels. Do you expect the frequencies to be quadratic in $Z$, as is the case for the $K_\alpha$-series?

**Solution**   A simple generalization of the formula based on the Bohr model takes the form

$$hf_{L\beta} = (13.6 \text{ eV}) \left[ \left( -\frac{(Z - s_3)^2}{3^2} \right) - \left( -\frac{(Z - s_2)^2}{2^2} \right) \right],$$

where $s_3$ is the amount of screening of the $n = 3$ electrons and $s_2$ is the amount of screening of the $n = 2$ electrons. Note that this formula continues to be quadratic in $Z$, even if the location of the associated parabola shifts.

---

Moseley's work, together with the Franck–Hertz experiment described in Chapter 5, not only cemented the understanding of the role of the atomic charge $Z$ in chemistry, but also provided a new confirmation of Bohr's work and went far in encouraging Bohr to continue along the path he had forged. Once a correct treatment of screening is worked out, based on the Schrödinger equation and the more sophisticated ideas described earlier in the chapter, it is readily seen that the Moseley data contain useful information on atomic structure.

An effect that is closely related to what Moseley saw is the **Auger effect**, studied by the French physicist Pierre Auger in 1923. When an electron is ejected from an inner shell, creating a vacancy, that vacancy will be filled by an electron from a higher level. The energy given up by the electron that moves to the lower level may appear as radiation, generally an X ray—that is what Moseley studied. Auger discovered that there also exist radiationless transitions,[4] in which energy is given to an electron in one of the higher orbits. This is a quantum mechanical process that was not envisaged by Bohr. These electrons, which are generally given enough energy to be ejected, are known as *Auger electrons*. By studying the intensity of the Auger electrons as a function of energy, spectroscopists see the analogue of spectral lines (•Fig. 11–5). Auger spectra tell us about inner-shell energy levels to a great degree of accuracy. But these energy levels are affected by the outer electrons—those associated with molecular bonds—because the outer electrons have a small probability of penetrating to the inner shells and thus shift these levels a bit. In this way, secondary information about the outer levels is revealed, and many subtle effects of molecular bonds can be studied. For small $Z$ values, the Auger effect is actually much more probable than the emission of Moseley's secondary X rays.

---

[4] We shall see many examples of nonradiative transitions when we study nuclear physics.

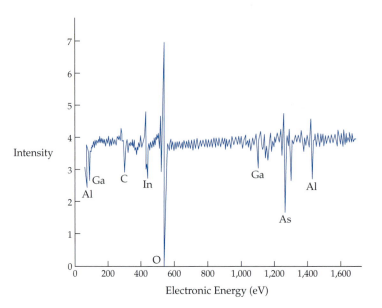

**• Figure 11–5** Spectrum of emitted Auger electrons. The electron energies characterize different elements, and the specific location provides information about the effects of nearby atoms. The sensitivity allows one to use the Auger effect to detect tiny amounts of impurities.

## 11–4 Molecules

A molecule is a stable arrangement of electrons and more than one nucleus. Some simple and familiar examples include HCl, $O_2$, $N_2$, $CO_2$, and $NH_3$. This notation suggests that molecules are bound states of atoms, but generally speaking that is an oversimplification; once a molecule is formed, the electrons cannot be said to belong to any particular nucleus. Here we limit ourselves to diatomic molecules: molecules with two nuclei. Even in this simple case the system is more complicated than two separate atoms. The $H_2$ molecule consists of two protons and two electrons. Once the center of mass is fixed, the electrons can move relative to the center of mass and to each other, and so can the two nuclei. Given this degree of complication, our discussion will be semi-quantitative at best.

As a way to approximate things, we can again use the fact that electrons are at least 2,000 times less massive than nuclei. Thus the nuclear motion is much slower than that of the electrons, and we can use the following procedure:

1. First, we assume that the nuclei are fixed in space. This assumption allows us to treat the motion of the electrons in the electric field of the fixed nuclear charges.
2. The moving electrons create a smeared-out charge distribution. In the next approximation, the nuclei are allowed to move in this charge distribution.

### The $H_2^+$ Molecule

We begin with a discussion of the simplest molecule, $H_2^+$, singly ionized $H_2$, consisting of two protons and one electron. A plot of the energy $E$ of this molecule as a function of the distance $R$ between its protons is revealing. If the protons are very close together—in other words, if $R \rightarrow 0$—the molecule looks exactly like a singly ionized helium atom. The electronic energy of this system is

$$E_{\text{elec}}(0) = -13.6 \, Z^2 \text{ eV} = -54.4 \text{ eV}. \tag{11–6}$$

If the two protons are far apart, then the electron will be bound to one proton or the other as in ordinary hydrogen, so that

$$E_{\text{elec}}(\infty) = -13.6 \text{ eV}. \tag{11–7}$$

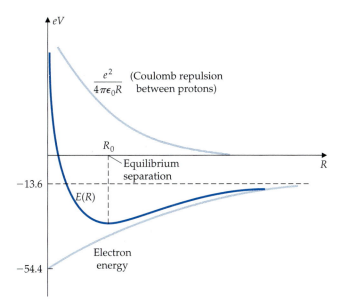

• **Figure 11–6** Contributions of electronic and coulomb energies to the internuclear potential $U(r)$.

The full electronic expression $E_{elec}(R)$ will interpolate between these two values, with $E_{elec}(R)$ approaching $E_{elec}(\infty)$ at a distance of at most a few hydrogen atom radii $a_0$. We have drawn a curve that describes this behavior in •Fig. 11–6.

Now, the electronic energy is not the whole story: There is also a contribution to the energy that comes from the proton–proton repulsion, and that has the form

$$E_{nucl} = \frac{e^2}{4\pi\varepsilon_0 R}. \tag{11–8}$$

We also plot this contribution in the figure. When we add the electronic and nuclear contributions, we get the third curve shown. *There is a minimum at $R = R_0$.* This is a stable equilibrium point, and it characterizes the size of the molecule. The experimental value is $R_0 = 0.106$ nm, and the energy at $R = R_0$ is 2.8 eV below the $-13.6$ eV that is the energy for $R = \infty$.

Let us see how what we have just described is reflected in the wave function of the molecule. The wave function is the solution of a Schrödinger equation of the form

$$\left( \frac{p_e^2}{2m_e} - \frac{e^2}{4\pi\varepsilon_0 r_1} - \frac{e^2}{4\pi\varepsilon_0 r_2} + \frac{e^2}{4\pi\varepsilon_0 R} \right) \psi(\vec{r}, R) = E\psi(\vec{r}, R). \tag{11–9}$$

Here, the first term is the kinetic energy of the electron—recall that $p_e$ is a derivative operator—and there is no term for the kinetic energy of the protons, consistent with our assumption that the nuclei are stationary. The wave function depends on $\vec{r}$ and $R$. The variable $\vec{r}$ is the position of the electron; note, however, that what enters into the energy are the distances $r_1$ and $r_2$ between the electron and the first and second proton, respectively. The variable $R$ is the separation between the nuclei—all that we need, given that their center of mass is assumed to be stationary.

What, qualitatively, do we expect for the wave function? If the electron were localized in some outer region surrounding the two protons, the repulsion between the protons would raise the energy. But if the electron had a high probability of lying between the two protons, the energy would be lowered, because both protons are attracted to the electron. Or, equivalently, the energy would decrease because the protons are screened from one another by the electron between them, and hence their mutual repulsion is less. Thus the situation in which the electron lies between the protons is favored, and the electron is tied not to one proton or the other, but to both of them.

### The H₂ Molecule and Valence Bonds

The $H_2$ molecule involves two electrons. Spin and the exclusion principle enter the structure of this molecule in an important way. The $H_2$ molecule is typical of diatomic molecules whose atoms each have unfilled shells. (The $1s$ shell of the hydrogen atom has two places, so the shell is unfilled.)

Where will the electrons "be?" In other words, what is their wave function like? The overall wave function must be antisymmetric under the interchange of the electrons. If the spatial wave function of the two electrons is symmetric under their interchange, then the spin wave function must be antisymmetric; that is, it must be a spin singlet. If the spatial wave function is antisymmetric, then the spin wave function will be a (symmetric) spin triplet. In fact, the ground-state wave function is the spatially symmetric $S = 0$ state. A spatial wave function that is antisymmetric about the plane that intersects the proton–proton axis as in •Fig. 11–7 has to vanish on the plane. A spatially symmetric wave function is flat at the plane, as shown in the figure. The antisymmetric wave function has more curvature in it, and since the momentum is proportional to the *derivative* of the wave function, the more wiggly a wave function is, the larger is the kinetic energy. For a spatially symmetric wave function, the electrons are more likely to find themselves between the protons, and even though this positioning enhances the effect of the electron–electron repulsion, *each* electron is attracted to *each* of the nuclei there, and this effect wins out. In turn, if the wave function is symmetric in space, it must be antisymmetric in spin, which accounts for the $S = 0$ quantum number.

The German-born British theoreticians Fritz London and Walter Heitler first applied quantum mechanics to molecular bonding in 1927. They were able to carry the preceding arguments considerably farther with the use of what is known as the **valence bond**, which occurs when an electron in an atom is not already paired up with another electron in the same atom to form an antisymmetric spin state ($S = 0$). Then that electron can pair up to form an antisymmetric spin state with an electron from a second atom, whereupon the two atoms form a bond.

Why can't an electron already paired to $S = 0$ with another electron in the same atom take part in the formation of bonds? The reason is that when a third electron is introduced into the system, its spin must be parallel to that of *one* of the paired electrons, and therefore the wave function is antisymmetric in the spatial coordinates of these two electrons. A spatially antisymmetric wave function always implies a higher energy—an effect strong enough to ensure that there are no $H_3$ molecules.

• **Figure 11–7**    The spatially symmetric and antisymmetric wave functions. The latter has more curvature and thus represents a state with a larger expectation value for the kinetic energy.

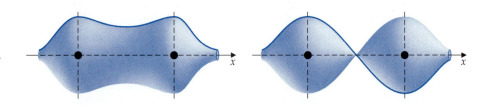

### Ionic Bonding

Another mechanism that determines molecular structure is apparent in the binding of atoms one of which has a single electron outside a shell (as in alkali metals) and the other of which has a single hole in a closed shell (as in halogens). Sodium fluoride is one example of this kind of binding. Under these circumstances, the electron outside the shell in the sodium fills the hole in the shell of the fluorine atom. We can compare the energy required—the ionization energy—to remove the electron of the alkali metal from its orbit outside the closed shell with the energy released—the *electron affinity*—when a single electron falls into the hole in the outer shell of the halogen. In general, the electron affinity is similar in magnitude to the ionization energy. For example, the energy required to remove the electron from sodium is 5.1 eV, while the energy released when an electron falls into the hole in the outer shell of fluorine is 3.5 eV. Thus, it costs only 1.6 eV for a sodium atom to transfer its valence electron to a fluorine atom.

If there is a *positive* energy cost for the electron transfer, why will it happen at all? After the rearrangement, we have a sodium atom with an electron missing (a sodium *ion*, $Na^+$) and a fluorine atom with one electron too many (a fluoride ion, $F^-$). There is now a coulomb attraction between these oppositely charged objects, and since the dimensions of the molecule are roughly of atomic size, we have an additional negative contribution to the energy that is on the order of the ionization potential itself. The *net* effect is that the energy of the bound NaF system is lower than that of the separated Na and F atoms.

But wouldn't the two atoms get closer and closer, thereby decreasing the energy more and more? The answer to this question is no, and the reason has to do with electronic states. To see how it works, note that the electron configuration for sodium ($Z = 11$) is $1s^2 2s^2 2p^6 3s$, while that of fluorine ($Z = 9$) is $1s^2 2s^2 2p^5$. If the nuclei somehow were to coalesce, we would have an atom with $Z = 20$, and the lowest electronic configuration of that atom is $1s^2 2s^2 2p^6 3s^2 3p^6 4s^2$. This would mean that *nine* electrons would have to be promoted to states with higher energies, and such a promotion would cost a lot of energy. We sketch the net potential for NaF as a function of nuclear separation in •Fig. 11–8. As our arguments suggest, it contains a minimum, at a separation of roughly 0.2 nm.

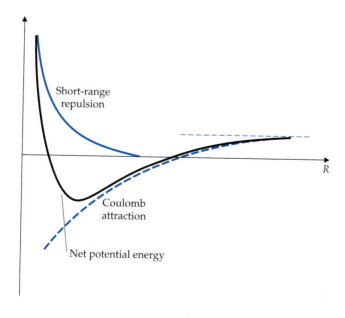

Short-range repulsion

Coulomb attraction

Net potential energy

$R$

• **Figure 11–8**   Potential energy between $Na^+$ and $F^-$ ions. At large distances the coulomb energy vanishes, and what is left is just the positive rearrangement energy; the zero is set for uncharged atoms. At short separation distances there is a repulsion that competes with, and wins out over, the coulomb attraction.

# 11–5 Nuclear Motion and Its Consequences

As we noted at the beginning of the previous section, the huge mass ratio of nuclei to electrons allows for a two-step approach to molecular motion. In the first step, we treat the nuclei as if they were infinitely massive. The electrons' wave function may, in principle, be calculated for any internuclear separation, and the value of that separation which leads to the lowest energy state will be the correct one. In the second step, we take the electron wave function as a function of the separation $R$ as providing a potential energy for the nuclei. If we now take into account the fact that the nuclei are not infinitely massive, we see that the nuclei can move in this potential. The movement gives rise to new terms in the energy and new quantum mechanical energy levels.

Although a correct description calls for a Schrödinger equation, a classical discussion of the motion gives us a physical picture of what happens. Because the nuclei are so massive, their motion is very slow compared with that of the electrons; during this slow motion, the electrons have plenty of time to adapt to the slowly varying internuclear separations. Thus the nuclear motion is described in terms of a potential $U_e(R)$, where the subscript $e$ reminds us that this energy is due to the nuclei being embedded in a cloud of negative charge density associated with the electron distribution. Given a potential, we must treat the system quantum mechanically, but for the low-lying excitations, such a treatment will actually turn out to be very simple. In what follows, we will limit ourselves to a discussion of diatomic molecules, so that $R$ refers to the separation between two point masses.

## Vibrations in Molecules

The correct potential energy $U(R)$ for the nuclei consists of two parts. First, there is a term $U_e(R)$ that describes the interaction of the nuclei with the outer valence electrons, those that are not so strongly bound to their original nucleus and that can move around both nuclei in some way. Second, there is an *effective* coulomb repulsion between the two nuclei, one that is in standard coulomb form, but with screened nuclear charges. The screening comes from the presence of the low-lying electrons, those that do *not* move easily from nucleus to nucleus, but that nevertheless act to screen the nuclear charges from one another. Thus

$$U(R) = U_e(R) + \frac{Z_1 Z_2 e^2}{4\pi\varepsilon_0 R}. \qquad (11–10)$$

The charges $Z_1$ and $Z_2$ of the nuclei are the *screened* charges, usually numbers closer to unity than to the actual atomic charges. •Figure 11–9 illustrates this screening in a schematic way. The figure shows two centers of coulomb potential energy, and the total potential energy presents a barrier to the passage of electrons from one well to the other. The high-lying electrons can *tunnel* back and forth through the barrier and, in effect, cannot be said to be in one well or another. By contrast, the barrier is all but impenetrable to the low-lying electrons.

As we saw in our discussion of $H_2^+$, this potential has a minimum at some $R = R_0$. Now, any potential-energy minimum offers the possibility of oscillations about that minimum, and in this case the possibility will be realized in our equations when we take into account the finite mass of the nuclei. Our next

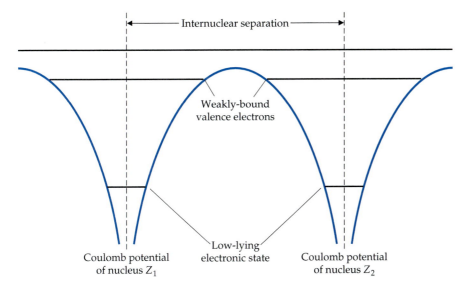

Weakly-bound
valence electrons

Low-lying
electronic state

Coulomb potential
of nucleus $Z_1$

Coulomb potential
of nucleus $Z_2$

• **Figure 11–9**   An effective coulomb barrier inhibits the "sharing" of inner-shell electrons between nuclei.

task is to find the frequency of those oscillations. First, we expand the potential energy about $R = R_0$, using the Taylor expansion (Appendix B–2),

$$U(R) = U(R_0) + \left(\frac{dU}{dR}\right)_{R=R_0}(R - R_0) + \frac{1}{2}\left(\frac{d^2U}{dR^2}\right)_{R=R_0}(R - R_0)^2 + \cdots$$

$$\cong U(R_0) + \frac{1}{2}\left(\frac{d^2U}{dR^2}\right)_{R=R_0}(R - R_0)^2. \tag{11–11}$$

In the second line we have used the fact that $R = R_0$ is a minimum, so that the first derivative is zero at $R_0$. The potential energy is quadratic in this approximation and so corresponds to the potential energy of a harmonic oscillator with equilibrium at $R = R_0$. We can find the frequency of our "oscillator" from the second derivative in Eq. (11–11), which is the "spring constant" $k$. To find this frequency, we require a knowledge of $U_e(R)$. There is a simple way to make an estimate for the spring constant: We use just the coulomb term, setting $Z_1$ and $Z_2$ equal to unity, roughly their effective value when only one electron per nucleus is a valence electron. Thus

$$k = \left(\frac{d^2U}{dR^2}\right)\Bigg|_{R=R_0} = \frac{2e^2}{4\pi\varepsilon_0 R_0^3}. \tag{11–12}$$

This [see Eq. (1–6)] is equal to $\omega^2 M_{\text{red}}$, where $M_{\text{red}}$ is the reduced mass of the two nuclei and $\omega$ is the frequency associated with the harmonic potential. In other words,

$$\omega^2 = \frac{e^2}{2\pi\varepsilon_0 R_0^3 M_{\text{red}}}. \tag{11–13}$$

The vibrations of the two nuclei in our diatomic molecule occur in one dimension, along the line joining the nuclei. The quantized energies of our "spring" are then given by Eq. (5–34),[5] namely

$$E_{n_v} = \left(n_v + \frac{1}{2}\right)\hbar\omega, \qquad n_v = 0, 1, 2, \ldots. \tag{11–14}$$

[5] See also the footnote of Chapter 5.

We now turn our attention to an estimate of the size of this effect. Since the internuclear separation at equilibrium is roughly the diameter of a hydrogen atom, we estimate that

$$R_0 \cong 2a_0,$$

where $a_0$ is, as usual, the Bohr radius. We recall [see Eq. (5–11)] that electronic energy differences are typically given by

$$\Delta E_{\text{elec}} \cong \frac{e^2}{4\pi\varepsilon_0} \frac{1}{2a_0} = m_e c^2 \alpha^2,$$

where the fine-structure constant $\alpha \cong 1/137$. This equation gives us energy separations for vibrational states that are on the order of

$$\Delta E_{\text{vib}} \cong \hbar\omega = \frac{1}{2} m_e c^2 \alpha^2 \sqrt{\frac{m_e}{M_p}} = \frac{1}{2} \Delta E_{\text{elec}} \sqrt{\frac{m_e}{M_p}}. \tag{11–15}$$

The square-root factor is typically on the order of $10^{-2}$, while the coefficient of the square root is 13.6 eV, a typical value for the electronic energy. The vibrational levels are separated by amounts that are roughly one percent of the separation of the atomic levels. This means that the wavelengths of the light emitted when transitions occur between vibrational levels in molecules are some 100 times larger than the wavelengths in the ultraviolet and optical region that characterizes atomic transitions. The vibrational wavelengths lie in the far infrared region. As with all the spectroscopies we have met to this point, the infrared spectrum—detected via infrared spectroscopy—is a rich source of information, in this case about molecular structure.

---

**Example 11–4**   Data on energy differences in vibrational spectra are often presented by giving the value of $1/\lambda$, where $\lambda$ is the wavelength of a photon that would be emitted when a molecule makes a transition from higher to lower energy. For example, $1/\lambda = 2{,}885.9$ cm$^{-1}$ for the transition between the $n_v = 1$ and the $n_v = 0$ states in HCl. Assume that the mass of the chlorine nucleus is approximately 35 times the mass of the proton in the following
**(a)** Use these numbers to calculate the energy difference $E_1 - E_0$, in electron volts, between the two states. **(b)** Calculate the internuclear separation, assuming that $Z_{\text{eff}} = 1$ for both nuclei.

**Solution**   **(a)** For a harmonic oscillator, whose spectrum is given by

$$E = \hbar\omega\left(n + \frac{1}{2}\right) = hf\left(n + \frac{1}{2}\right),$$

the energy differences are

$$E_1 - E_0 = hf = \frac{hc}{\lambda} = (6.63 \times 10^{-34} \text{ J·s})(3 \times 10^8 \text{ m/s})(2{,}886 \times 10^2 \text{ m}^{-1})$$

$$= 5.74 \times 10^{-20} \text{ J} = \frac{5.74 \times 10^{-20} \text{ J}}{1.6 \times 10^{-19} \text{ J/eV}} = 0.36 \text{ eV}.$$

**(b)** Equation (11–13) gives us the necessary connection between the frequency and the internuclear separation $R_0$. It already assumes that the effective charge on each nucleus is 1. We have

$$\omega = 2\pi f = 2\pi \times \frac{c}{\lambda} = \frac{2\pi(3 \times 10^8 \text{ m/s})}{2.886 \times 10^5 \text{ m}^{-1}} = 5.45 \times 10^{14} \text{ rad/s}.$$

The reduced mass in this case is given by

$$\frac{1}{M_{red}} = \frac{1}{M_H} + \frac{1}{M_{Cl}} = \left(\frac{1}{M_H}\right)\left\{1 + \frac{1}{35}\right\} = \left(\frac{1}{M_H}\right)\{1.03\}.$$

By inverting Eq. (11–13), we then get

$$R_0^3 = \left(\frac{e^2}{2\pi\varepsilon_0}\right)\left(\frac{1}{M_{red}\omega^2}\right)$$

$$= (1.6 \times 10^{-19}\,C)^2(1.8 \times 10^9\,kg\cdot m^3\cdot s^{-2}\cdot C^{-2})(1.03)(1.67 \times 10^{-27}\,kg)^{-1}(5.45 \times 10^{14}\,rad/s)^{-2}$$

$$= 0.96 \times 10^{-30}\,m^3,$$

so that $R_0 = 0.98 \times 10^{-10}$ m. This is in the expected range of around 0.1 nm.

### Rotations of Molecules

Even if the vibrational degrees of freedom are not excited, so that the "spring" is in its ground state, there are still lower lying states of excitation for a molecule: The molecule can act as a rigid extended structure and rotate. We can approximate a diatomic molecule as a dumbbell-like object with a moment of inertia, $I$, about an axis (•Fig. 11–10). The energy is $E_{rot} = L^2/2I$, and we quantize the angular momentum in the usual fashion, which in turn gives us quantized rotational energies:

$$E_{rot,\ell} = \ell(\ell + 1)\hbar^2/2I. \qquad (11\text{–}16)$$

In this case, almost all the mass is in the nuclei. If, as before, $R_0$ is the equilibrium separation between the nuclei and $M_{red}$ is their reduced mass, then

$$I = M_{red}R_0^2 \cong 4M_{red}a_0^2. \qquad (11\text{–}17)$$

We may then say that the separation between rotational levels is of the approximate magnitude

$$\Delta E_{rot} \approx \frac{\hbar^2}{2I} = \frac{\hbar^2}{8M_{red}a_0^2}. \qquad (11\text{–}18)$$

The Bohr radius $a_0 = \hbar/(mc\alpha)$, and the typical electronic energy is $O(mc^2\alpha^2)$. Thus,

$$\Delta E_{rot} \approx \frac{1}{8}\frac{m}{M_{red}}(mc^2\alpha^2) = \frac{1}{8}\frac{m}{M_{red}}\Delta E_{elec} = O(10^{-4})\Delta E_{elec}. \qquad (11\text{–}19)$$

In other words, these splittings are another factor of about 100 down from the vibrational splittings, which are themselves about 100 times smaller than the electronic splittings.

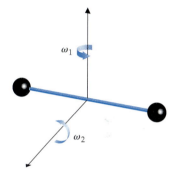

• **Figure 11–10**   To find the moment of inertia of a diatomic molecule, we assume that all the mass is concentrated at the two nuclei, so that the molecule resembles a dumbbell.

### Example 11–5   Find the ratio of the rotational splitting in the hydrogen molecule $H_2$ to that of Hbr. The bromine nucleus has a mass roughly 80 times that of the hydrogen nucleus (the proton).

**Solution**   As Eq. (11–18) shows, the difference between rotational levels in these two molecules is simply that of different reduced masses. In particular,

$$\frac{\Delta E_{rot}(H_2)}{\Delta E_{rot}(HBr)} = \frac{1/M_{red}(H_2)}{1/M_{red}(HBr)}.$$

We have

$$\frac{1}{M_{red}(H_2)} = \frac{1}{M(H)} + \frac{1}{M(H)} = \frac{2}{M(H)}$$

and

$$\frac{1}{M_{\text{red}}(\text{Hbr})} = \frac{1}{M(\text{H})} + \frac{1}{M(\text{Br})} = \left(\frac{1}{M(\text{H})}\right)\left(1 + \frac{1}{80}\right) = \frac{1.01}{M(\text{H})}.$$

Thus,

$$\frac{\Delta E_{\text{rot}}(\text{H}_2)}{\Delta E_{\text{rot}}(\text{HBr})} = \frac{2/M(\text{H})}{1.01/M(\text{H})} = \frac{2}{1.01} = 1.98.$$

In the limit in which the hydrogen atom's partner were infinitely heavy, the splitting in that molecule would be half as large as in the hydrogen molecule itself.

---

Energy splittings are detectable when there are transitions with the emission of a photon. The photon wavelengths will be about 100 times larger than those of photons in vibrational transitions and some 10,000 times larger than the wavelengths of photons in electronic transitions. Since electronic energies are in the electron-volt region, which produces visible light in the transitions, the wavelengths in rotational transitions are in the millimeter range. Radiation in this range provides us with still another spectroscopic tool for understanding molecules.

•Figure 11–11 shows the effects of both vibrations and rotations on the energies of diatomic molecules.

*Effects of the Exclusion Principle* Throughout this chapter, we have concentrated on how the exclusion principle affects the disposition of electrons. But the principle applies to nuclei as well if identical nuclei are present in a given molecule. For diatomic molecules with identical nuclei, the consequence is that, in a given molecule, *the values of rotational quantum numbers ℓ are either all even or all odd*. This is simple to see if we note that the interchange of the two nuclei is equivalent to rotating the molecule about an axis perpendicular to the line connecting them and passing through their center of mass

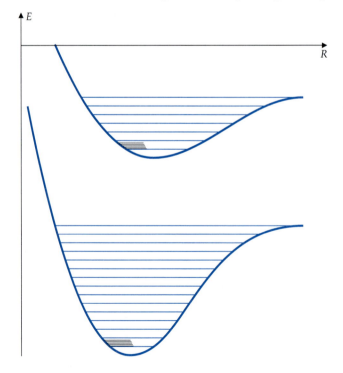

• **Figure 11–11** Schematic picture of vibrational and rotational spectra of a diatomic molecule for the lowest and first excited electronic states. The colored lines are the vibrational levels, and the black lines are the rotational levels. The successive rotational levels are more closely spaced than the successive vibrational levels.

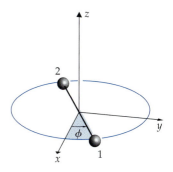

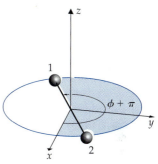

(•Fig. 11–12). Consider an eigenstate of $\vec{L}^2$ maximally aligned along the z-axis, with $L_z = \ell\hbar$. For this state, the dependence on the azimuthal angle is $\exp(i\ell\varphi)$. The interchange of the two nuclei is equivalent to the change $\varphi \rightarrow \varphi + \pi$. Thus, $\exp(i\ell\varphi) \rightarrow \exp(i\ell\pi)\exp(i\ell\varphi)$. This new phase factor must be minus one times the original phase factor for nuclei that are identical fermions—this will ensure that the wave function changes sign under interchange. That is, $\exp(i\ell\pi) = -1$ for identical fermions; in other words, $\ell$ must be odd. For nuclei that are identical bosons, the phase factor must be plus one times the original phase factor, and $\ell$ must be even. As a consequence, *transitions with $\Delta\ell = 1$ are ruled out in diatomic molecules with identical nuclei.* But electromagnetic transitions with $\Delta\ell = 1$ are much more likely than those with higher values of $\Delta\ell$, for reasons having to do with the conservation of angular momentum. (See Chapter 13.) Consequently, rotational transitions are strongly suppressed in diatomic molecules with identical nuclei.

• **Figure 11–12**   The interchange of two identical nuclei is equivalent to a 180° rotation about the z-axis shown in the figure.

---

**Example 11–6**   Consider the NaCl molecule, for which the equilibrium separation between the Na ($A = 23$) and Cl ($A = 35$) nuclei is $2.4 \times 10^{-10}$ m.
**(a)** Calculate, in eV, the energy necessary to excite the first rotational state above the ground state. **(b)** What is the wavelength of the photon emitted in the transition from $\ell = 1$ to $\ell = 0$?

**Solution**   **(a)** In this part, we want to find the energy difference between the ground state and the first rotational state. That difference is given by Eq. (11–18), and we use it to find

$$\Delta E = E_{\text{rot, 1}} - E_{\text{rot, 0}} = \frac{(2 - 0)\hbar^2}{(2I)} = \frac{\hbar^2}{I}.$$

The moment of inertia of the NaCl molecules is given by $I = M_{\text{red}}R^2$, where $M_{\text{red}} = M_{\text{Na}}M_{\text{Cl}}/(M_{\text{Na}} + M_{\text{Cl}}) = M_p[(23)(35.5)/(23 + 35.5)] = 14.0M_p$, in which $M_p$ is the mass of the proton (or neutron). Thus we have

$$I = (14.0)(1.67 \times 10^{-27}\,\text{kg})(2.4 \times 10^{-10}\,\text{m})^2 = 1.34 \times 10^{-45}\,\text{kg}\cdot\text{m}^2.$$

It follows that

$$\Delta E = \hbar^2/I = \frac{(1.05 \times 10^{-34}\,\text{J}\cdot\text{s})^2}{1.34 \times 10^{-45}\,\text{kg}\cdot\text{m}^2}$$

$$= \frac{8.2 \times 10^{-23}\,\text{J}}{1.6 \times 10^{-19}\,\text{J/eV}} = 5.1 \times 10^{-5}\,\text{eV}.$$

**(b)** The relation between the photon's energy and wavelength is $E = hc/\lambda$, so that

$$\lambda = \frac{2\pi Ic}{\hbar} = 2.4 \times 10^{-2}\,\text{m}.$$

---

# SUMMARY

Complex atoms—atoms containing more than one electron—are understood using the following set of principles:

- Electrons are placed in successive states, with two electrons per state ("paired electrons"), according to the Pauli exclusion principle.
- The states with a given principal quantum number are not degenerate, largely because of two effects that are not present in hydrogen. First, electrons in states that leave them closer to the nucleus screen the nuclear charge from electrons lying further out, on the average. Second, the exclusion principle leads to an important spin-dependent effect: When the overall wave function

for the two paired electrons is antisymmetric in space and symmetric in spin, the atomic energy is lower because the electrons remain far from each other and there is less effect from their mutual repulsion. Nevertheless, there is some remnant of the degeneracy, and the set of states associated with a given principal quantum number—the states composing a shell—are closer to one another than to states from different shells, at least until the number of electrons is larger than 25 or so.

- Chemistry is associated with unpaired electrons and with shells that are incomplete. The most chemically inactive elements are those with shells that are filled.

Molecules are formed when there is more than one fixed center of force (the nuclei of the atoms that form the molecule), with electrons in eigenstates of the resulting potential. In other words, electrons cannot be said to belong to any particular nucleus in a molecule. Several features of molecules are worth singling out:

- Molecules can form when there is a minimum in the potential energy of the electrons. By estimating the potential energies, we can estimate electronic energies in molecules.
- The exclusion principle again plays an important role, particularly in valence bonding, in which electrons from different atoms pair up. In the other major type of binding, ionic binding, closed shells form around each nucleus by exchanging electrons between the atoms under favorable energy conditions.
- The complex shapes of molecules allow energy levels other than the electronic levels characteristic of atoms. The two new types of energy levels are vibrational levels, associated with restoring forces around the minimum in the energy of the nuclei, and rotational levels, associated with the rotation as a whole of the nuclei in the molecule. The electronic, vibrational, and rotational energy-level spacing is in the ratio $1:10^2:10^4$.
- The effects of identical particles—in this case, the nuclei—plays an important role in the rotational levels. In particular, the angular-momentum quantum number is restricted to even or odd values, according to whether the nuclei are bosons or fermions, respectively. This property is revealed, as are all the features described in this chapter, in electromagnetic transitions with the emission of radiation.

## QUESTIONS

**1.** One of the great puzzles in heredity before the quantum theory of molecules was invented was how the gene could retain its integrity generation after generation when it was kept at the human body temperature. How does the quantum theory of molecules resolve this problem? (Note that, according to very general ideas about statistical systems, all objects that are part of a thermal system move with energies on the order of $kT$, where $k$ is Boltzmann's constant.)

**2.** In the text we discussed the bonding of hydrogen and oxygen to produce water. From what you know, is there any reason to think you could not bind "heavy hydrogen," in which the proton constituting the nucleus of ordinary hydrogen is replaced by a proton and neutron, to produce "heavy water?" Consider two kinds of heavy water, $D_2O$ and $HDO$, where D symbolizes hydrogen with the deuteron as its nucleus—a nucleus that consists of a neutron and a proton and that has a mass close to twice the proton mass.

**3.** If you read in the newspapers that someone had discovered a helium fluoride compound, would you believe the story? Give the reasons for your answer.

**4.** Experimentalists have been able to construct exotic molecules like $H_2^+$ in which two protons are bound together not by an electron but rather by a particle called a *muon*, which is essentially identical to the electron, except that it has a mass some 200 times that of the electron. How large would such a molecule be, and how much energy would it take to dissociate it?

**5.** At the end of this chapter, we argued that the symmetry or antisymmetry of the nuclear wave function says something about the possible $L$ values for the molecular rotation. Suppose we consider the $H_2$ molecule, in which there are two protons. They can be in either a total spin $S = 1$ state or a total spin $S = 0$ state. What is the "parity" (i.e., evenness or oddness) of the $\ell$ values for the two possibilities?

**6.** In 1928, C. V. Raman discovered that when light of frequency $f$ was scattered by molecules, the outgoing light also could have frequency $f$, but some of the time it had frequency $f \pm f_0$, where $f_0$ was independent of $f$. Bearing in mind that light is absorbed by exciting the scatterer to some excited state and is then emitted by the de-excitation of the scatterer, can you explain this phenomenon? In searching for the answer, consider the different possible excitations of a molecule and the possibility that the molecule in the end state is not in the same state that it was initially.

**7.** The molecule $H_2$ is made up of two hydrogen atoms. Are these atoms identical? If so, what statistics do they obey? What is the consequence, if any, of your answer for the $H_2$ molecule?

**8.** In this chapter, we described building up the periodic table, although we did not go into much detail regarding the larger atoms. A striking fact is that as $Z$ increases, many atoms start to become unstable, and there are no stable atoms beyond $Z = 83$ (although some elements beyond this exist because they have very long lifetimes or they are replenished by the instability of other atoms). Is there anything about the physics described in the chapter that might explain this fact?

**9.** In this chapter, we described the vibrations of molecules. Can single atoms vibrate? Can single atoms rotate? What sorts of experiments could you perform to try to detect the rotations of single atoms?

**10.** How might Moseley have measured the wavelengths of the X rays in his experiments? (*Hint*: The name Bragg would have entered into Moseley's thinking.)

## PROBLEMS

**1.** ∎ Write the configuration for the ground state of the magnesium atom ($Z = 12$). Given that the closed shells always have $L_{tot} = S_{tot} = 0$, what is the total angular momentum of magnesium.

**2.** ∎∎ Consider the ground state of the silicon atom ($Z = 14$). What is the electronic configuration for this state?

**3.** ∎∎∎ What are the orbital angular momentum and spin of the ground state of scandium ($Z = 21$)? What are the possible values of the total angular momentum $J$? Given the fact that there is a spin–orbit interaction, predict which of the $J$ values has a lower energy. (*Hint*: Use what you learned about the spin–orbit coupling in hydrogen in Chapter 9, Section 9–5.)

**4.** ∎∎ The last (most energetic) electron in sodium is in a $3s$ state. What do you expect the order of the energies to be for the last electron in the three possible states $3s$, $3p$, and $3d$?

5. ▌ X rays bombarding heavy atoms can be used to eject electrons from the $1s$ shell in atoms; indeed, this is the starting point for Moseley's experiments. Estimate the maximum wavelength of photons required to eject an electron from the $1s$ shell of copper, for which $Z = 29$.

6. ▌▌ In Moseley's law, the wavelengths for the transition from the $n = 2$ to the $n = 1$ state are labeled as $\lambda_\alpha$. You can assume that the Bohr formula holds with an effective charge $Z_{eff} = Z - s$ to take screening into account. The value of $\lambda_\alpha$ for Ca is measured to be 0.357 nm. Given that Z is known to be 20 for Ca, what is the value of $s$?

7. ▌▌ Assuming that in Moseley's law the value of the screening parameter $s$ (see Problem 6) is always the same for the $n = 2$ to $n = 1$ transition, calculate $\lambda_\alpha$ for Mn with $Z = 25$. You can compare your result with the measured value of 0.2117 nm.

8. ▌▌ Moseley also measured a different set of lines: the $\lambda_\beta$ lines. These are K-series lines as well. Their values for a few elements are Ca $(Z = 20)$: 0.3085 nm; Mn $(Z = 25)$: 0.1923 nm; and Cu $(Z = 29)$: 0.1403 nm. What transitions do these absorption lines correspond to, and what is the associated screening value $s$?

9. ▌▌ In the $O_2$ molecule, the internuclear separation is 0.121 nm, and the splitting between vibrational levels is, in the standard nomenclature, $1/\lambda = 1,580$ cm$^{-1}$. Are these values compatible with Eq. (11–13)? If not, and if this incompatibility is attributed to a modification of Eq. (11–12) with an incomplete screening of the nuclear charge—that is,

$$k = \frac{2q^2}{4\pi\varepsilon_0 R_0^3}, \text{ with } q = Z_{eff}e,$$

then what is $Z_{eff}$?

10. ▌▌ Chlorine comes with nuclei ("isotopes") of two dominant types: $^{35}$Cl, for which $A = 35$, and $^{37}$Cl, for which $A = 37$. The vibrational spectrum of H$^{35}$Cl is characterized by $(1/\lambda) = 2,990$ cm$^{-1}$. What is the value of that parameter for the molecule H$^{37}$Cl? The fact that these numbers are different is often called the *isotope effect*.

11. ▌▌ Consider the NaCl molecule, in which the atomic weights of Na and Cl are 23 and 37, respectively. The internuclear separation is 0.236 nm, and the "spring constant" for vibrations is $k = M_{red}\omega^2 = 10^9$ J/m$^2$. Calculate the energy difference between adjacent vibrational levels.

12. ▌▌ The absorption line for the transition $\ell = 0 \rightarrow \ell = 1$ in CO has a wavelength of 2.603 mm. Find the rotational inertia, as well as the equilibrium separation, of this molecule.

13. ▌▌ Consider the molecule NO, and find the wavelengths of the radiation emitted when there is a transition between rotational levels.

14. ▌▌▌ The CN molecule has a visible absorption line at 387.4 nm, which corresponds to the energy difference between the ground state and the first electronic excitation state. Both of these states have rotational states associated with them, so that the line is split. The $\ell = 0 \rightarrow \ell = 1$ line has wavelength $\lambda = 387.4608$ nm, the $\ell = 1 \rightarrow 2$ line has $\lambda = 387.3998$ nm, and the $\ell = 2 \rightarrow 3$ line has $\lambda = 387.3369$ nm. All of these lines correspond to absorption—that is, transition from a lower $\ell$ *up* to a higher $\ell$. There is also the emission line $\ell = 1 \rightarrow 0$ at $\lambda = 387.5763$ nm. Sketch the energy diagram and the lines.

15. ∎∎∎ The data provided for the CN molecule in Problem 14 allow us to calculate the rotational inertia for the nuclei in the lowest electronic state and for those in the first excited electronic state, since the energies are respectively given by

$$E = E_0 + \frac{\hbar^2\ell(\ell + 1)}{2I_0}$$

and

$$E = E_1 + \frac{\hbar^2\ell(\ell + 1)}{2I_1}.$$

Use the data of Problem 14 to calculate $I_0$ and $I_1$, and use these to find the internuclear separation in the two electronic "potentials."

16. ∎∎∎ The rotational motion of molecules has an effect on the equilibrium separation of the nuclei. If $R_0$ is the separation for $\ell = 0$, calculate the new separation by minimizing

$$E(R) = \frac{1}{2}M_{red}\omega^2(R - R_0)^2 + \frac{\hbar^2\ell(\ell + 1)}{2M_{red}R^2},$$

assuming that $R - R_0$ is small. How will the spectrum of the molecule be affected? [*Hint*: The moment of inertia changes; calculate how.]

17. ∎∎ There exists in nature a particle known as a *muon*. It appears to be nothing more than a heavy electron, of mass $m_\mu = 207m_e$. Consider a molecule that is an analogue of $H_2^+$ (with two protons plus a muon). (a) Estimate the size of such a molecule. (b) If a rotational state could be excited, what would be the wavelength of the radiation emitted in the transition to the ground state?

18. ∎∎ A molecule has a first vibrational state that is 0.022 eV above the energy of the ground state. The first rotational state of the molecule is $9.2 \times 10^{-4}$ eV above the energy of the ground state. How many rotational levels will there be in the gap between the two states?

19. ∎∎ The wavelength of the $n_v = 1 \rightarrow n_v = 0$ transition in carbon monoxide (CO) is $2.93 \times 10^3$ cm, where $n_v$ is the vibrational quantum number. At what temperature would you expect the $n_v = 1$ state in CO to be 1% occupied? You can anticipate a result of the next chapter, viz., that the relative probability of finding states of energy $E$ is proportional to $\exp(-E/kT)$, where $k$ is Boltzmann's constant.

# CHAPTER

# 12

# Statistical Physics

We have by now learned a lot about atoms and molecules. In principle, the Schrödinger equation could be used to find properties of macroscopic systems of many such particles—a mole contains some $6 \times 10^{23}$ of them—interacting with one another. But the idea of solving a Schrödinger equation for that many particles is practically meaningless, and the same statement can be made about using Newton's laws to solve a classical system of that many particles.

In fact, the kind of questions we ask are less ambitious than the detailed description of the wave function, or the motion, of each of the constituent objects. We begin with a system at a temperature $T$, held there by contact with a thermal reservoir. We suppose that the system has been in this condition for enough time that it has "settled down"—we say that the system is in *thermal equilibrium*. We then might be interested only in measuring an *average* molecular speed; we might be interested in seeing whether *on the average* the velocities of two different molecules at different points in space are correlated with each other or are independent. For a system in equilibrium, averages such as these do not change with time. The study of these and similar issues is the domain of **statistical physics**, which together with its offshoot statistical thermodynamics are vast subjects, with large books devoted to them. We can give only a glimpse of the subject, and we do so mainly by concentrating on those aspects that are of use to us in our discussion of semiconductors and other technological applications.

We concentrate on systems that consist of "almost" noninteracting particles. A dilute gas is a good example of such a system. The gas molecules move about freely, except for the very brief moments during which they undergo impulsive collisions with other gas molecules or with the walls of the container. We also deal with both classical and quantum mechanical statistical physics. We shall see that at low temperatures the effects of the Pauli principle—the consequences of the particles being fermions or bosons—are of great importance. At higher temperatures these effects cease to be important. And as we might expect, classical statistical physics and quantum statistical physics merge into each other as the temperature of the system rises. Therefore it is not a hollow exercise to study classical statistical mechanics, a very useful subject when it applies. Nor is it merely theoretical gymnastics to study the quantum situation:

Quantum effects dominate large systems of direct practical interest, including most solids at room temperature.

It will turn out that there is one key idea to keep in mind: The physically measurable quantities that characterize a system of many particles can all be obtained if one knows the probability that for a temperature $T$ the system has a particular energy[1] $E$. The expression for the probability is due to Boltzmann and is a result that is central to much of physics, chemistry, biology, and technology.

## 12–1 The Description of a Classical Gas

As we saw in Chapter 1, some quantities—the pressure and temperature of a gas, for example—can be understood in terms of simple averages. In particular, we noted that there was a simple relation between the pressure and the average kinetic energy of the (pointlike) molecules of a gas. Combined with the ideal-gas equation of state $pV = NkT$, where $k$ is Boltzmann's constant, this relation allowed us to interpret the temperature in terms of the average molecular kinetic energy $\langle K \rangle$:

$$\langle K \rangle = \left( \frac{3}{2} \right) kT.$$

Let us repeat the discussion of pressure of Chapter 1, but this time taking into account the possibility that the molecules may have a variety of velocities and energies.[2] Since the velocities of our molecules can take on a continuum of values, we have to be careful in describing our velocity distribution.[3] In a box of unit volume full of molecules with a variety of velocities, there is *no* molecule with a particular chosen velocity. Instead, there is only a set of molecules with a velocity within a little window *about* the chosen velocity. More precisely, we write the number density—the number per unit volume—of molecules with a velocity between $\vec{v}$ and $\vec{v} + d^3\vec{v}$ as $n(\vec{v})d^3\vec{v}$. We refer to $n(\vec{v})$ as the **number density distribution**. Note that if the total molecular density (the number of molecules per unit volume) is $n$, we have

$$n = \int n(\vec{v}) \, d^3\vec{v}.$$

Alternatively, we can describe probabilities rather than number densities. The probability of finding a molecule with velocity in the range from $\vec{v}$ to $\vec{v} + d^3\vec{v}$ is

$$f(\vec{v}) \, d^3\vec{v} = n(\vec{v}) \frac{d^3\vec{v}}{n}.$$

More specifically,

$f(\vec{v}) \, d^3\vec{v} = f(\vec{v}) \, dv_x \, dv_y \, dv_z = $ the probability of finding a molecule with velocity in the range from $(v_x, v_y, v_z)$ to $(v_x + dv_x, v_y + dv_y, v_z + dv_z)$.

---

[1] We are used to thinking of energy as being conserved, so perhaps this statement, as it stands, is puzzling. The point is that when we consider a system in equilibrium as having a particular temperature $T$, we mean that it is in contact with a thermal reservoir—it is effectively surrounded by a larger system at temperature $T$. Because there can be energy transfer to and from the reservoir, the energy of our system can fluctuate about some average value that depends on $T$.

[2] As we stated in the introduction, we are going to assume that any collisions are brief, so that we can forget about any potential-energy terms, and the "energy" is the molecule's kinetic energy.

[3] In Appendix B–3 we present some of the basic probability concepts.

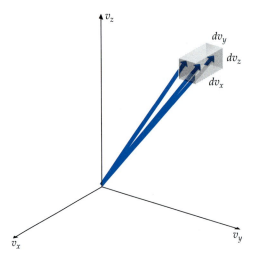

• **Figure 12–1**   Velocity vectors with their tips in the volume $d^3\vec{\mathbf{v}}$. Since $d^3\vec{\mathbf{v}}$ is small, all these vectors are very nearly the same vector $\vec{\mathbf{v}}$.

We call $f(\vec{\mathbf{v}})$ the *probability density function*. We see that we have shown the direct relation

$$f(\vec{\mathbf{v}}) = \frac{n(\vec{\mathbf{v}})}{n}.$$

Finally, we point out that averages can be simply written in terms of $f(\vec{\mathbf{v}})$; for example,

$$\langle \vec{\mathbf{v}}^2 \rangle = \int \vec{\mathbf{v}}^2 f(\vec{\mathbf{v}}) \, d^3\vec{\mathbf{v}}.$$

Suppose then that we have a set of molecules with a velocity in the range from $\vec{\mathbf{v}} = (v_x, v_y, v_z)$ to $\vec{\mathbf{v}} + d^3\vec{\mathbf{v}}$. We can view this collection of molecules as represented by a set of velocity vectors with their tips in a small "volume" $d^3\vec{\mathbf{v}} = dv_x \, dv_y \, dv_z$ (•Fig. 12–1). The number density of such molecules will be given by the volume $d^3\vec{\mathbf{v}}$ multiplied by the number density distribution $n(\vec{\mathbf{v}})$.

To find the pressure exerted on the wall, let us consider a small area $dA$ of the wall. The force exerted on that area is the momentum transferred to the area by molecules that strike it within a time $dt$, divided by $dt$. If the wall is oriented parallel to the $xy$-plane, then a molecule with mass $m$ and velocity $\vec{\mathbf{v}}$ undergoing an elastic collision with the wall will transfer momentum $2mv_z$ to the wall. This momentum can be transferred by slow or fast molecules: The slower ones may have to hit the wall head on, and the faster ones will transfer the momentum in glancing collisions. The important point is that the number of molecules that hit the area $dA$ in a time $dt$ is the number of molecules of any velocity that are contained in a cylindrical volume whose base area is $dA$ and whose height is $v_z \, dt$ (•Fig. 12–2). Thus the momentum transfer due to the molecules with velocity in the range from $\vec{\mathbf{v}}$ to $\vec{\mathbf{v}} + d^3\vec{\mathbf{v}}$ is

$$(2mv_z)(n(\vec{\mathbf{v}}) \, d^3\vec{\mathbf{v}})(dA \, v_z \, dt) = 2m \, dA \, dt \, v_z^2 n(\vec{\mathbf{v}}) \, d^3\vec{\mathbf{v}}.$$

To find the total momentum $\Delta P$ transferred to the wall in time $dt$ we need only sum over the molecules with all the different velocities that are headed toward the wall (molecules with $v_z$ positive):

$$\Delta P = 2m \, dA \, dt \int_{v_z \text{ positive}} d^3\vec{\mathbf{v}} \, v_z^2 n(\mathbf{v}).$$

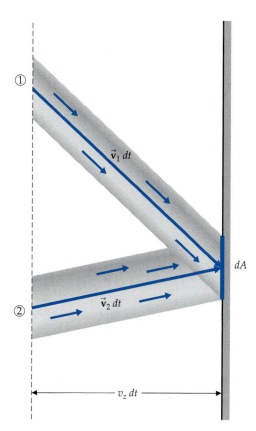

**• Figure 12–2**   The number of molecules that impinge on an area $dA$ of a wall and bounce elastically off the wall can come from a variety of cylinders, each characterized by the velocity vector $\vec{v}$ and of height $v_z \, dt$.

To go from this expression to that for the pressure on the wall, we first find the force (transfer of momentum per unit time) by dividing by the time interval $dt$ and then divide by the area $dA$ to get the force per unit area—that is, the pressure $p$. Thus

$$p = 2m \int_{v_z \text{ positive}} d^3\vec{v} \, v_z^2 n(\vec{v}). \tag{12–1}$$

To go further we would need to know something about $n(\vec{v})$. But very plausible reasoning allows us to say enough to simplify Eq. (12–1). We can assume that there is no preferred direction—a molecule with speed $v$ is equally likely to be moving in any direction—so that $n(\vec{v})$ is independent of the direction of $\vec{v}$. In other words, $n(\vec{v}) = n(v)$, where $v$ is the speed. Two immediate simplifications follow: First, because a molecule is equally likely to be moving left or right, we can remove the restriction that $v_z$ be positive if we divide by 2; second, because there is no special direction, the integral in Eq. (12–1) is unchanged if we write $v_x^2$ or $v_y^2$ instead of $v_z^2$, so that

$$\int d^3\vec{v} \, v_z^2 n(v) = \frac{1}{3} \int d^3\mathbf{v}(v_x^2 + v_y^2 + v_z^2) n(v) = \frac{1}{3} \int d^3\mathbf{v} \, v^2 n(v).$$

Hence

$$p = \frac{1}{3} m \int d^3\vec{v} \, v^2 n(v) = \frac{1}{3} mn \int d^3\vec{v} \, v^2 f(v). \tag{12–2}$$

In the last step, we have replaced $n(v)$ by $nf(v)$. In doing so we recognize that the integral in Eq. (12–2) is the average of the square of the speed, $\langle v^2 \rangle$. This brings us back to the development of Chapter 1. The resulting expression for the pressure is that of Eq. (1–38). More precisely, Eq. (12–2) can be written as

$$p = \frac{2}{3} n \int d^3\mathbf{v} \left( \frac{1}{2} mv^2 \right) f(v) = \frac{2}{3} n \langle K \rangle, \tag{12-3}$$

where $K$ is the kinetic energy of a single molecule. As in Chapter 1 [see Eq. (1–48)], Eq. (12–3) is consistent with the ideal-gas law if we interpret the temperature $T$ according to

$$\langle K \rangle = \left( \frac{3}{2} \right) kT, \tag{12-4a}$$

so that

$$\frac{1}{2} m \langle v^2 \rangle = \frac{3}{2} kT. \tag{12-4b}$$

Here $k$ is Boltzmann's constant. The ideal-gas law is easily seen to follow from this result together with Eq. (12–3): According to that equation, $\langle K \rangle = (3/2)(p/n)$ $= (3/2)(pV/N)$, while according to Eq. (12–4a), $\langle K \rangle = (3/2)kT$. Comparing, we have $pV/N = kT$, just the ideal-gas law.

---

**Example 12–1**   Consider a gas composed of nitrogen molecules ($N_2$) at room temperature $T = 300K$. Given that the mass of a nitrogen molecule is $2 \times 14 \times 1.67 \times 10^{-27}$ kg, estimate the **root-mean-square speed** $v_{rms} \equiv \sqrt{\langle v^2 \rangle}$ of a nitrogen molecule.

**Solution**   We have

$$\langle v^2 \rangle = \frac{3kT}{m} = \frac{3 \times (1.38 \times 10^{-23} \text{ J/K}) \times (300K)}{28 \times 1.67 \times 10^{-27} \text{ kg}} = 26.6 \times 10^4 \text{ (m/s)}^2.$$

The square root of this quantity gives us $v_{rms} = 5.2 \times 10^2$ m/s.

---

**Example 12–2**   Spherical droplets of water of diameter 1.2 μm are distributed in an atmosphere of nitrogen molecules at 300K. Assuming that the water droplets are at this same temperature—that is, that they are in thermal equilibrium with the nitrogen molecules—estimate $v_{rms}$ (see Example 12–1) for the droplets.

**Solution**   We can use exactly the same formula, Eq. (12–4b), for the droplets as we used for the nitrogen in Example 12–1. This formula tells us that, for a given temperature, $v_{rms}$ is inversely proportional to the mass:

$$\frac{v_{rms}(\text{drop})}{v_{rms}(N_2)} = \sqrt{\frac{m(N_2)}{m(\text{drop})}}.$$

To find the mass of a droplet, we recall that the density of water is $\rho = 1.0 \times 10^3$ kg/m$^3$, so that

$$m(\text{drop}) = \left( \frac{4}{3} \right) \pi r^3 \rho = \left( \frac{4}{3} \right) \pi (0.6 \times 10^{-6} \text{ m})^3 \times (10^3 \text{ kg/m}^3) = 0.9 \times 10^{-15} \text{ kg}.$$

We also use the results of the previous example for $v_{rms}$ for nitrogen at 300K, namely 520 m/s. Thus

$$v_{rms}(\text{drop}) = (520 \text{ m/s}) \sqrt{\frac{28 \times 1.67 \times 10^{-27} \text{ kg}}{0.9 \times 10^{-15} \text{ kg}}} = 3.7 \times 10^{-3} \text{ m/s}.$$

This is a small number compared to the rms speed of the nitrogen molecules—a direct consequence of the large mass of the drop compared to the mass of the nitrogen molecules.

## 12–2 The Maxwell Distribution

To calculate quantities such as $\langle v^2 \rangle$, as opposed to merely recognizing their presence, we need to know $f(v)$. In 1860 James Clerk Maxwell found the distribution function $f(v)$, a quantity that is today known as the Maxwell distribution. Maxwell created statistical mechanics with this work. His derivation remains a simple one, and we will use it here. We have already pointed out that $f$ depends only on the speed:

$$f(v_x, v_y, v_z) = f(v_x^2 + v_y^2 + v_z^2). \qquad (12\text{–}5)$$

In addition, we note that without a preferred direction, the motions in the three directions are independent, meaning that the probability of finding a molecule with an $x$-component of velocity in the range from $v_x$ to $v_x + dv_x$ is independent of the probability of finding the molecule's $y$-component of velocity in the range from $v_y$ to $v_y + dv_y$, and so on. When we know the probability of individual events and we want the joint probability of an event in which the individual events are independent of each other, we must multiply the individual probabilities together. For example, the probability of two coins both coming up tails is $1/2 \times 1/2 = 1/4$. Thus Maxwell wrote

$$f(v_x, v_y, v_z) = h(v_x) \times h(v_y) \times h(v_z), \qquad (12\text{–}6)$$

where $h$ is a distribution in the one-dimensional variable that is its argument. In turn,

$$h(v_x) \times h(v_y) \times h(v_z) = f(v_x^2 + v_y^2 + v_z^2). \qquad (12\text{–}7)$$

Note that the distribution function $h$ is the *same* for each of the three Cartesian directions. This follows naturally from symmetry: We could easily relabel the $x$-, $y$-, and $z$-axes, and that should not have any effect on the physical distribution of molecules.

Equation (12–7) is a very restrictive one, not satisfied by just any function.[4] If we take the natural logarithm of both sides we find that

$$\ln[h(v_x)] + \ln[h(v_y)] + \ln[h(v_z)] = \ln[f(v_x^2 + v_y^2 + v_z^2)].$$

This form allows us to see that

$$h(v_x) = \text{constant} \times \exp[-Bv_x^2]. \qquad (12\text{–}8a)$$

In turn,

$$f(v) = C \exp[-B(v_x^2 + v_y^2 + v_z^2)] = C \exp(-Bv^2). \qquad (12\text{–}8b)$$

The constants $C$ and $B$ are still to be determined. We can find $C$ from the normalization: The sum of the probabilities must add to unity, or

$$1 = \int d^3\mathbf{v}\, C \exp(-Bv^2)$$

$$= C \int_{-\infty}^{+\infty} dv_x \exp(-Bv_x^2) \int_{-\infty}^{+\infty} dv_y \exp(-Bv_y^2) \int_{-\infty}^{+\infty} dv_z \exp(-Bv_z^2).$$

Using $\displaystyle\int_{-\infty}^{+\infty} dv_x \exp(-Bv_x^2) = \sqrt{\pi/B}$, we find that

$$C = \left(\frac{B}{\pi}\right)^{3/2}. \qquad (12\text{–}9)$$

---

[4] For example, if you try $g(v_x) = (v_x)^n$, you will not be able to satisfy Eq. (12–7). The form given in Eq. (12–8a), to follow, can be shown by rigorous mathematical means to be unique.

To determine the constant $B$, we can repeat the calculation of $\langle v^2 \rangle$, which, according to Eq. (12–4), is given by $\langle v^2 \rangle = 3kT/m$. We have

$$\langle v^2 \rangle = \left(\frac{B}{\pi}\right)^{3/2} \int d^3\vec{\mathbf{v}}\, v^2 \exp(-Bv^2) = -\left(\frac{B}{\pi}\right)^{3/2} \frac{d}{dB} \int d^3\vec{\mathbf{v}} \exp(-Bv^2)$$

$$= -\left(\frac{B}{\pi}\right)^{3/2} \frac{d}{dB}\left(\frac{\pi}{B}\right)^{3/2} = \frac{3}{2B}.$$

Therefore

$$B = \frac{2}{3}\langle v^2 \rangle = \frac{m}{2kT}. \tag{12–10}$$

In sum, we have found the *Maxwell distribution of molecular velocities*:

$$f(v) = \left(\frac{m}{2\pi kT}\right)^{3/2} \exp\left(-\frac{mv^2}{2kT}\right). \tag{12–11}$$

We note that

- The Maxwell distribution depends only on the magnitude of the velocity, not on its direction.
- For a fixed temperature, the Maxwell distribution drops exponentially with the kinetic energy $K = mv^2/2$ of the molecule, falling by a factor of $e(= 2.72\ldots)$ every time $K$ increases by a factor $kT$.
- As with any probability distribution, the Maxwell distribution can be used to find averages; for example,

$$\langle v^n \rangle = \int v^n f(v)\, d^3\vec{\mathbf{v}}.$$

- The Maxwell distribution is, properly speaking, a velocity distribution, but it can be used to lead to a speed distribution. (See Problem 11 at the end of the chapter.) We label the speed distribution $g(v)$, and its meaning is that $g(v)\, dv$ describes the probability that a molecule will have a speed between $v$ and $v + dv$. It is given by

$$g(v) = 4\pi\left(\frac{m}{2\pi kT}\right)^{3/2} v^2 \exp\left(-\frac{mv^2}{2kT}\right). \tag{12–12}$$

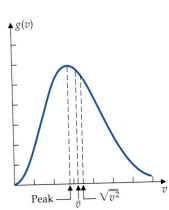

• **Figure 12–3** Plot of the speed distribution for molecules in a gas. There is a maximum at the most probable speed. Also indicated are the values of the average speed and the root-mean-square speed. Note that the probability of finding a molecule at rest is zero and that the probability of finding a very fast molecule falls off exponentially.

•Figure 12–3 shows this distribution plotted as a function of the speed for some choice of $m$ and $T$. It has the property that there is no probability of finding a molecule nearly at rest, nor of finding one moving at an arbitrarily high speed. There is a speed for which the distribution is a maximum that is not far from the average. This most likely speed is calculated in Problem 12 at the end of the chapter. We can use the speed distribution to calculate averages involving the speed alone, as in the next example.

**Example 12–3**  (a) Find the average speed of a gas molecule of mass $m$ in equilibrium at temperature $T$, and compare it with $v_{\text{rms}}$. (b) Apply your result to a gas of helium atoms ($m \approx 4$ proton masses) at room temperature $T = 293\text{K}$.

**Solution**  (a) The average speed is calculated by starting from the speed distribution in the usual way, namely

$$\langle v \rangle = \int_0^\infty v g(v)\, dv = 4\pi\left(\frac{m}{2\pi kT}\right)^{3/2} \int_0^\infty v^3 \exp\left(-\frac{mv^2}{2kT}\right) dv.$$

To perform the integration, we first change variables from $v$ to $x$, using the identity $x \equiv mv^2/2kT$, or $v^2 = 2kTx/m$. Then $2v\,dv = (2kT/m)\,dx$. The integral then becomes

$$I \equiv \frac{1}{2} \int_0^\infty \exp\left(-\frac{mv^2}{2kT}\right) v^2 2v\,dv = \frac{1}{2}\left(\frac{2kT}{m}\right)^2 \int_0^\infty \exp(-x)x\,dx.$$

The remaining integral over $x$ on the right is unity, so that $I = (1/2)(2kT/m)^2$, and it follows that

$$\langle v \rangle = \sqrt{\frac{8}{\pi}\frac{kT}{m}}.$$

This equation can be compared with $v_{rms} = \sqrt{3kT/m}$. We see that $\langle v \rangle$ differs from $v_{rms}$ by a factor of $(8/3\pi)^{1/2} \approx 0.92$. **(b)** Numerical calculations of this type are greatly simplified with a useful number to remember: At room temperature, $kT \approx 1/40$ eV. (We leave it to you to verify this number.) Given this value for $kT$, it pays to insert factors of $c^2$ to go with the masses and use the relativistic mass–energy equivalence. For example, we have

$$v_{rms} = \sqrt{3\frac{kT}{m}} = \sqrt{3\frac{kTc^2}{mc^2}} = c\sqrt{3\frac{kT}{mc^2}}.$$

In these units, the mass of helium is replaced by $m_{He}c^2 \cong 4m_p c^2 \cong 4(938 \text{ MeV}) = 3.75 \times 10^9$ eV. Then

$$v_{rms} = c\sqrt{3\frac{(1/40 \text{ eV})}{3.75 \times 10^9 \text{ eV}}} = (4.5 \times 10^{-6})c = 1{,}340 \text{ m/s}.$$

As for the average speed, we have from part (a) that

$$\langle v \rangle = (8/3\pi)^{1/2}v_{rms} = 1{,}235 \text{ m/s}.$$

## Experimental Verification of the Maxwell Distribution

The Maxwell distribution is subject to direct experimental test in a variety of ways, all of them variants of the one we shall describe here. The technique consists of measuring the distribution of speeds in a beam of gas molecules escaping through a hole into a vacuum. Since, as we shall see, the distribution in the beam is a bit different from that of the gas as a whole, our first job is to find the distribution in the beam. We do so as follows: Suppose that the hole, of area $dA$, is cut in a wall parallel to the $xy$-plane. Then the number $dN$ of molecules that strike the hole in a time interval $dt$ with $z$-component of velocity between $v_z$ and $v_z + dv_z$ is the number obtained in our earlier[5] calculation of the pressure, *integrated* over all possible values of $v_x$ and $v_y$ (but *not* over $v_z$). This number is

$$dN = \int_{-\infty}^{+\infty} \left(dA\,n(\vec{v})\,dv_z\,v_z\,dt\right)dv_x\,dv_y = dA\,dv_z\,dt\,n_0\int_{-\infty}^{+\infty}(v_z f(\vec{v}))\,dv_x\,dv_y,$$

where $n_0$ is the overall number density of the gas. Thus, using the explicit form of the velocity distribution, we find that the number of molecules emerging per unit area per unit time with $z$-component of velocity between $v_z$ and $v_z + dv_z$ is

$$\frac{dN}{dt\,dA} = n_0\left(\frac{m}{2\pi kT}\right)^{3/2}\exp\left(-\frac{mv_z^2}{2kT}\right)v_z\,dv_z\int_{-\infty}^{+\infty}dv_x\,dv_y\exp\left(-\frac{m(v_x^2 + v_y^2)}{2kT}\right).$$

---

[5] See the discussion leading up to Eq. (12–1).

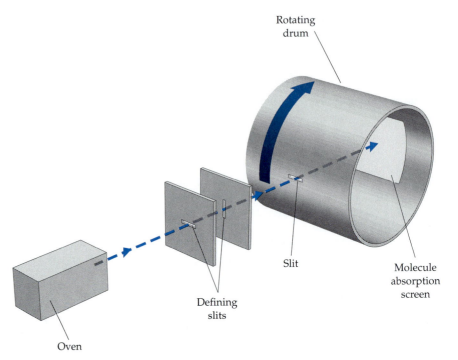

• **Figure 12–4** A schematic diagram of an experimental test of the Maxwell velocity distribution. Molecules of various speeds exit the oven through the slits. The speed of the molecules that can enter the drum depends on the drum's rotation speed. By varying the rotation speed the distribution can be measured.

Rotating drum

Slit

Molecule absorption screen

Defining slits

Oven

Using our standard integral $\int_{-\infty}^{+\infty} dv_x \exp(-Bv_x^2) = \sqrt{\pi/B}$ (see Appendix B–2) twice, we obtain

$$\frac{dN}{dt\,dA} = n_0 \left(\frac{m}{2\pi kT}\right)^{1/2} \exp\left(-\frac{mv_z^2}{2kT}\right) v_z\,dv_z. \qquad (12\text{–}13)$$

Equation (12–13), which follows directly from the Maxwell distribution, represents the distribution of speeds along the direction of a collimated beam of molecules. The measurement of that distribution by means of this equation is a direct test of the Maxwell distribution. It was measured by, among others, the German Otto Stern in 1920. As in •Fig. 12–4, molecules from a small hole in an oven were emitted and were formed into a beam by a pair of slits. The beam fell upon a rotating drum with a slit in it. The passage of the rotating drum in effect opens up a rapid shutter, which lets in a bunch of molecules. The fast ones cross the drum interior rapidly, while the slower ones lag, hitting a different portion of the inside wall of the rapidly rotating drum. The variation in the thickness of the layer of molecules sticking to the wall is thereby translated into a velocity distribution. The result is in excellent agreement with Eq. (12–13).

## 12–3 The Boltzmann Distribution

A glance at Eq. (12–11) shows that the Maxwell distribution can actually be written in terms of energy. The exponential factor is the one containing the dynamical variables, and it takes the form

$$\exp\left(-\frac{E}{kT}\right) \qquad (12\text{–}14)$$

for the case that the molecules are pointlike and noninteracting, so that their energy $E$ is $mv^2/2$.

The Maxwell distribution is, as we shall see, a special case of a more general result: *The probability that a single member of a large number of molecules in equilibrium at temperature T has an energy E is proportional to* $\exp(-E/kT)$. The energy can be a function of the velocity of the molecules (in the pointlike case, $E = mv^2/2$), but it can also contain rotational energy for nonpointlike molecules (e.g., $E = I\omega^2/2$, where $I$ is the rotational inertia and $\omega$ is the angular velocity of a molecule), vibrational energy for nonpointlike molecules (e.g., $E = \kappa x^2/2 + \mu(dx/dt)^2/2$, where $x$ is a "stretch" variable, $\kappa$ is a spring constant, and $\mu$ is an appropriate reduced mass), or even a spatially dependent term (e.g., $mgh$, where $h$ is the height of the molecule above sea level). The factor $\exp(-E/kT)$ is the **Boltzmann factor**, and the distribution proportional to it is known as the **Boltzmann distribution**. To make sure that it represents a probability distribution we must normalize properly. We deal with this case by case.

## An Elementary Derivation of the Boltzmann Distribution

By looking at how equilibrium is established in a gas we can make the Boltzmann distribution plausible. For a gas in equilibrium, we can speak of a number density distribution $n(E)$ in energy: The number of molecules with energies from $E$ to $E + dE$ is $n(E)dE$. Collisions between molecules are relevant to finding $n(E)$, because molecules exchange energy in these collisions.

Consider then a typical collision, in which a molecule with energy $E_1$ collides with a molecule with energy $E_2$. After the collision, the molecules have energies $E_3$ and $E_4$, respectively (•Fig. 12–5a). As we saw in our earlier discussion of collisions of molecules with a wall, the rate of collisions is proportional to the number density of molecules present. If the target is not a wall, but other molecules, the rate of collision must also be proportional to the density of the target molecules. Thus the overall rate for such collisions must be proportional to the product of $n(E_1)$ and $n(E_2)$:

$$R(1 + 2 \rightarrow 3 + 4) = Cn(E_1)n(E_2).$$

The factor $C$ describes the dynamics of the collision. It depends generally on the initial and final velocities of the colliding molecules. Similarly, there will be a rate for the inverse collision (•Fig. 12–5b), one in which molecules with velocities $-\vec{v}_3$ and $-\vec{v}_4$ collide to produce molecules with velocities $-\vec{v}_1$ and $-\vec{v}_2$. This rate will be given by

$$R(3 + 4 \rightarrow 1 + 2) = C'n(E_3)n(E_4).$$

The coefficient $C'$ again depends on the dynamics of the collision.

In the preceding two equations, the factors $n(E_1), \ldots, n(E_4)$ specify that the rates at which collisions occur depend on the number of projectiles and the

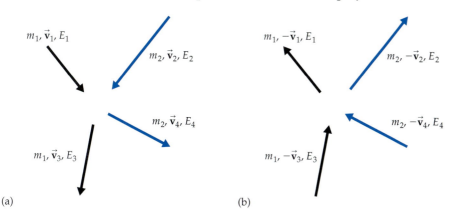

$m_1, \vec{v}_1, E_1$

$m_2, \vec{v}_2, E_2$

$m_2, \vec{v}_4, E_4$

$m_1, \vec{v}_3, E_3$

$m_1, -\vec{v}_1, E_1$

$m_2, -\vec{v}_2, E_2$

$m_2, -\vec{v}_4, E_4$

$m_1, -\vec{v}_3, E_3$

(a)                    (b)

• **Figure 12–5**   (a) A collision between molecules of masses $m_1$ and $m_2$ with initial velocities $\vec{v}_1$ and $\vec{v}_2$ and final velocities $\vec{v}_3$ and $\vec{v}_4$, respectively. (b) The same collision as in (a), but with all molecular motion reversed. At the atomic level the two collisions are equally likely.

number of targets. The factors $C$ and $C'$ represent the rate for a single particle hitting a single target. Thus we expect $C$ and $C'$ each to be functions of the energies $E_1, \ldots, E_4$. Now there is a law in physics known as the **principle of microscopic reversibility**, which effectively states that

$$C' = C. \tag{12–15}$$

This result may be visualized by stating that a film of a collision between two molecules would look exactly the same run in reverse as it would run forward. This statement is true only on the microscopic level, where no energy is dissipated in friction; it is not true in a collision between two cars and is only approximately true in a collision between two billiard balls. (Even the "click" accompanying the collision of the billiard balls sends out energy in the form of an acoustical wave, and a reverse collision would have to have an acoustical wave sent in at just the right moment to make the final energy equal to the initial energy.)

Now for a collection of molecules in equilibrium, as many collisions that go one way will go the other; that is,

$$R(1 + 2 \rightarrow 3 + 4) = R(3 + 4 \rightarrow 1 + 2).$$

Equation (12–15) implies that the $C$-factors cancel, and we are left with

$$n(E_1)n(E_2) = n(E_3)n(E_4). \tag{12–16}$$

This condition, together with the conservation of energy, is enough to lead us to the desired result. We take the logarithm of Eq. (12–16) to obtain

$$\ln n(E_1) + \ln n(E_2) = \ln n(E_3) + \ln n(E_4). \tag{12–17}$$

To this ingredient we add the conservation of energy,

$$E_1 + E_2 = E_3 + E_4. \tag{12–18}$$

When we compare Eq. (12–17) with Eq. (12–18), we see that the only way we can satisfy both is to have $\ln n(E)$ linear in $E$; that is,

$$\ln n(E) = A - \beta E,$$

where $\beta$ is a constant that does not depend on $E$. (The minus sign is just a matter of convenience; in any case we need to find $\beta$.) In turn, this equation gives us

$$n(E) = n(0)\exp(-\beta E). \tag{12–19}$$

Finally, in order to find $\beta$, we note that the distribution $n(E)$ must reduce to the Maxwell distribution when $E$ is the energy of a free particle. This gives us

$$\beta = 1/kT. \tag{12–20}$$

The factor $n(0)$ is a normalization constant that will be discussed in detail when we have to deal with it. At this point, suffice it to say that the total density of molecules is

$$n = \sum n(E),$$

where the sum is over all allowed energies. In atomic and molecular systems we often have degeneracies—for example, a rotator whose energy is $L^2/2I$ has $(2\ell + 1)$ energy levels, each with energy $\hbar^2\ell(\ell + 1)/2I$, and in the absence of an external magnetic field these levels are degenerate. This degeneracy has to be taken into account in the sum. Thus if the degeneracy for a given value of $E$

is $g(E)$ (in the example cited, $g$ is $2\ell + 1$), then the probability of measuring the energy $E$ is

$$P(E) = \frac{n(E)}{n} = \frac{g(E)\exp\left(-\dfrac{E}{kT}\right)}{\sum g(E)\exp\left(-\dfrac{E}{kT}\right)},$$

where, again, the sum in the denominator is over all allowed values of $E$.

## A System of Molecules with Discrete Energies

At this point a brief and simplified illustration is helpful. Consider a "gas" composed of a collection of microscopic compass needles (•Fig. 12–6). Each needle can point either along or opposite an external magnetic field, and these two possible configurations have the two energies $E_1$ and $E_2$, respectively with $E_1 < E_2$. (The energies are those of a magnetic moment in the magnetic field.) We say that there are two **states** possible for the needles. This system is a good learning tool—and there is another reason for studying it: In the quantum mechanical world energies do indeed come in discrete values.

Suppose we put a large collection of these compass needles in a heat bath at a temperature $T$. More concretely we could imagine the needles immersed in a superlarge container of some gas at temperature $T$. The needles are subject to random bombardment by the gas molecules. Some of the collisions will leave a given needle in its original state, while some will change the state—that is, flip it over. In a short time the system will be in an equilibrium characterized by the temperature $T$.

According to the Boltzmann distribution, the fractions of needles in the two states are given by

$$\frac{\langle n_1 \rangle}{N} = \frac{\exp(-E_1/kT)}{\exp(-E_1/kT) + \exp(-E_2/kT)} \tag{12–21a}$$

and

$$\frac{\langle n_2 \rangle}{N} = \frac{\exp(-E_2/kT)}{\exp(-E_1/kT) + \exp(-E_2/kT)}. \tag{12–21b}$$

We have normalized these equations by performing the sum over probabilities explicitly—you can see immediately that the two fractions add to unity.

•Figure 12–7 shows the variation with temperature of these averages. The quantity $kT$ has the dimensions of energy, which suggests some limits that clarify the meaning of the distribution. For very high temperatures, $1/kT \to 0$, we have

$$\frac{\langle n_1 \rangle}{N} = \frac{\langle n_2 \rangle}{N} = \frac{1}{2}. \tag{12–22}$$

This is the result we would get if the energies $E_1$ and $E_2$ were the same, or, since $E_1 \neq E_2$, if the energy difference between them were irrelevant, as would be the case at high temperatures.

The opposite limit, $T \to 0$, is a bit trickier, but no less helpful. Both $E_1/kT$ and $E_2/kT$ become large in this limit, so that the exponentials drop off to zero. But $\exp(-E_2/kT)$ vanishes faster than $\exp(-E_1/kT)$. Therefore, from Eqs. (12–21),

$$\frac{\langle n_1 \rangle}{N} \xrightarrow[T \to 0]{} 1$$

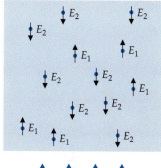

• **Figure 12–6** A collection of compass needles that can take only two positions, "up" and "down," with respect to the direction of an external magnetic field. Each needle has energy $E_1$ and $E_2$, respectively, for these positions. The magnetic moments are assumed to be such that $E_2 > E_1$.

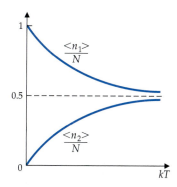

• **Figure 12–7** A plot of $\langle n_1 \rangle/N$ and $\langle n_2 \rangle/N$ as a function of $kT$.

and

$$\frac{\langle n_2 \rangle}{N} \xrightarrow[T \to 0]{} 0.$$

In other words, all the needles are in the lower energy state, which corresponds nicely with our idea of the effects of low temperature: At low temperatures, collisions with the heat bath molecules are insufficiently energetic to lift the needles into the higher energy state.

**Example 12–4**    Consider a collection of identical disks confined to the $xy$-plane that can rotate freely about their own axes, parallel to the $z$-axis. If the rotational inertia of a disk about its axis is $I$, then the energy of a disk rotating with angular velocity $\omega$ is $I\omega^2/2$. Suppose that the disks collide with each other and that in a collision the angular velocities can be exchanged. Suppose further that these angular velocities can range from $-\infty$ to $+\infty$. If the collection of disks is in equilibrium at temperature $T$, find the average magnitude of the angular momentum $I|\omega|$.

**Solution**    To calculate the average angular momentum, we need to know the distribution of angular velocities of the disks. The Boltzmann distribution gives us this; it takes the form

$$f(\omega)\,d\omega = C \exp\left(-I\omega^2/2kT\right)d\omega,$$

where the constant $C$ is determined by the normalization condition $\displaystyle\int_{-\infty}^{+\infty} f(\omega)\,d\omega = 1$. Thus (see Appendix B–2)

$$C^{-1} = \int_{-\infty}^{+\infty} \exp\left(-\frac{I\omega^2}{2kT}\right)d\omega = \sqrt{\frac{2\pi kT}{I}}.$$

Now that we have the distribution of angular velocities, we can find the average of the magnitude of the angular momentum $I\omega$. We want

$$\langle I|\omega|\rangle = C\int_{-\infty}^{+\infty} I|\omega|\exp\left(-\frac{I\omega^2}{2kT}\right)d\omega = 2CI\int_{0}^{+\infty} \omega \exp\left(-\frac{I\omega^2}{2kT}\right)d\omega$$

$$= CI\int_{0}^{+\infty} \exp\left(-\frac{Ix}{2kT}\right)dx = \sqrt{\frac{2IkT}{\pi}}.$$

Note that even though the Boltzmann distribution contains the energy, it can be used to find the averages of other quantities through the appearance of the relevant variables, in this case angular speed.

**Example 12–5**    Consider a gas of hydrogen atoms in equilibrium at a temperature $T$. (You may think we are ignoring chemistry in so doing, because at temperatures we normally deal with hydrogen occurs in nature as the diatomic molecule $H_2$. However, see our comments at the end of the example.) At what temperature would one find that the atoms with the energy of the first excited atomic state number 10 percent of the atoms with the energy of the ground state? Ignore spin-orbit and other fine-structure effects, so that there are two states that have the ground-state energy (spin up or spin down) and eight that have the first excited-state energy (spin up and down, and for each spin the four possibilities $\ell = 0$ and $\ell = 1$, $m = 0$ and $\pm 1$.)

**Solution**    The energies of the two states in question are $E_1 = mv^2/2 - B_1$ and $E_2 = mv^2/2 - B_2$, where $B_1 = 13.6\,\text{eV}$ is the binding energy for the ground state of hydrogen and $B_2 = B_1/4 = 3.4\,\text{eV}$ is the binding energy for the first excited state. The probability of finding a hydrogen atom in one state at energy $E_i$ is

$$\text{const} \times \exp\left(-\frac{E_i}{kT}\right).$$

If $n_i$ is the number of states that have energy $E_i$, then the probability of finding a hydrogen atom in any state at energy $E_i$ is the sum over the probabilities for one state,[6] namely

$$\text{const} \times n_i \times \exp\left(-\frac{E_i}{kT}\right).$$

The constant is for normalization, and we do not need to know it, because we are interested in the *relative* probabilities for $E_2$ and $E_1$. The ratio of the two probabilities is

$$\frac{n_2 \exp(-E_2/kT)}{n_1 \exp(-E_1/kT)} = \frac{8 \exp(-E_2/kT)}{2 \exp(-E_1/kT)} = 4 \exp\left[-\frac{(E_2 - E_1)}{kT}\right].$$

The kinetic-energy terms cancel, leaving the ratio of the two probabilities as

$$4 \exp\left[\frac{B_2 - B_1}{kT}\right].$$

The statement of the problem sets this ratio at 0.10, and we want to solve for the temperature. We do so by taking the natural logarithm,

$$\ln\{4 \exp[(B_2 - B_1)/kT]\} = \ln(0.10),$$

or

$$\ln(4) + (B_2 - B_1)/kT = \ln(0.10).$$

We can solve this equation for the temperature:

$$T = \frac{B_2 - B_1}{k} \frac{1}{\ln(0.10) - \ln(4)} = \frac{B_2 - B_1}{k} \frac{1}{\ln(0.10/4)} = \frac{(3.4 - 13.6)\ \text{eV}}{8.62 \times 10^{-5}\ \text{eV/K}} (-0.271)$$

$$= 3.21 \times 10^4\ \text{K}.$$

This is a very high temperature. It is perhaps interesting to quote also the equivalent energy $kT$, which you can calculate to be 2.8 eV. It is reasonable that this value be on the order of electron volts to be capable of putting a finite fraction of the atoms in the first excited state. The 2.8-eV figure also suggests that molecular hydrogen, whose binding energy into two hydrogen atoms is 4.5 eV, is to a very large extent dissociated into its constituent atoms at temperatures of the order of $10^4$K.

## 12–4  Equipartition and Heat Capacity

In Section 12–1 we argued that the average energy per molecule for a gas of noninteracting point molecules is

$$\langle E \rangle = \frac{3}{2} kT. \tag{12–23}$$

The total energy $U$ of a mole of gas is then $N_A$ times this value, so that, as we described in Chapter 1, the *heat capacity at constant volume*—for one mole this is called the *molar heat capacity at constant volume*—is

$$c_V' \equiv \left.\frac{\partial U}{\partial T}\right|_V = \frac{3}{2} N_A k = \frac{3}{2} R, \tag{12–24}$$

where $R$ is the universal gas constant. This result follows from the fact that when the volume is held constant, no work is done, so that the energy change of the gas is given by the amount of heat energy added, in accordance with the first law of thermodynamics [Eq. (1–41)].

[6] See also the passage from Eq. (12–38) to (12–39), the discussion leading up to Eq. (12–43), and the discussion following Eq. (12–61).

Had we considered motion in one dimension, we would have found that $\langle E \rangle = kT/2$. (See Problem 9 at the end of the chapter.) The result $3kT/2$ is the sum of three contributions of $kT/2$ each from the three independent degrees of freedom—the three Cartesian directions $x$, $y$, and $z$—of a point particle.

Consider now a gas of diatomic molecules such as $O_2$, $N_2$, or $H_2$. Since most of the mass of an atom is at its nucleus, these molecules behave to a good approximation as two point masses connected by a massless and rigid bond. In addition to having a motion in three-dimensional space which is that of a point mass at the molecule's center of mass, a diatomic molecule can also have an *internal motion*. The energy of such a molecule consists of two contributions. One is the kinetic energy of the molecule as a whole, given by $Mv^2/2$, where $M$ is the total mass of the molecule and $v$ is the speed of its center of mass. The second is the energy stored in the internal motion of the molecule—the energy it would have even if the center of mass were at rest. For the kind of diatomic molecule we are interested in here, the only internal motion is rotation about the two axes perpendicular to the rigid axis connecting the atoms. To the extent that the atomic nuclei, which carry the bulk of the mass, are pointlike—a very good approximation to reality—rotations about the axis joining the atoms do not contribute to the energy; the rotational inertia about that axis is effectively zero. The two allowed rotations are independent. Also, the two rotational inertias $I$ are the same for the allowed rotations. Let us label the rotation axes 1 and 2 (•Fig. 12–8) and calculate how these rotations affect the average energy. The rotational contribution to the energy of our diatomic molecule is given by

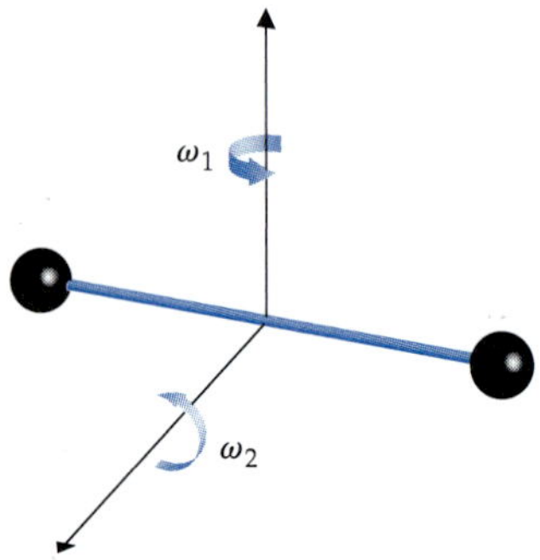

$$E_{\text{rot}} = \frac{1}{2}\left(I\omega_1^2 + I\omega_2^2\right), \tag{12-25}$$

**• Figure 12–8** Rotational degrees of freedom of a diatomic molecule. Only rotations about the two axes shown contribute to the energy, because the rotational inertia about the axis joining the atoms is very small.

where $\omega_1$ and $\omega_2$ are the respective rotational velocity components about the two axes. The fact that these components occur quadratically in the energy, just as do the linear velocity components in the point-particle kinetic energy, is the crucial feature that determines their contribution to the average energy. Whenever we have an independent *quadratic* contribution to the energy, we say we have a **degree of freedom**. The pointlike molecule has three degrees of freedom, while our rigid diatomic molecule has five degrees of freedom—three for the motion of the center of mass and two for the rotational motion.

With our new expression for the energy, the Boltzmann distribution for the gas now gives us a probability distribution for a given value of $\vec{v}$, $\omega_1$, and $\omega_2$:

$$C_{\text{trans}}C_{\text{rot}}\exp\left(-\frac{mv^2}{2kT}\right)\exp\left(-\frac{I\left(\omega_1^2 + \omega_2^2\right)}{2kT}\right). \tag{12-26}$$

Here $C_{\text{trans}}$ and $C_{\text{rot}}$ are normalization constants set so that the integrals over all $\vec{v}$ and over all $\omega_1$ and $\omega_2$ are equal to unity. In particular,

$$
\begin{aligned}
(C_{\text{rot}})^{-1} &= \int_{-\infty}^{+\infty} d\omega_1 \int_{-\infty}^{+\infty} d\omega_2 \exp\left(-\frac{I\left(\omega_1^2 + \omega_2^2\right)}{2kT}\right) \\
&= \left[\int_{-\infty}^{+\infty} d\omega \exp\left(-\frac{I\omega^2}{2kT}\right)\right]^2 = \left(\sqrt{\frac{2\pi kT}{I}}\right)^2.
\end{aligned} \tag{12-27}
$$

(Although a truly infinite $\omega$ is not possible because the total energy of the entire gas is finite, the rapid falloff of the integrand justifies extending the limits of integration to infinity.)

To calculate the average rotational energy by means of Eqs. (12–26) and (12–27), it is convenient to use $\beta \equiv 1/kT$. Then the average rotational energy is

$$\langle E_{\text{rot}} \rangle = C_{\text{rot}} \int_{-\infty}^{+\infty} d\omega_1 \int_{-\infty}^{+\infty} d\omega_2 \left(\frac{I\left(\omega_1^2 + \omega_2^2\right)}{2}\right)\exp\left(-\beta\frac{I\left(\omega_1^2 + \omega_2^2\right)}{2}\right). \tag{12-28}$$

The fact that the exponentials factor into a product of exponentials allows us to write this equation as the sum of two terms, one for each degree of freedom. If we denote each term as $\langle E \rangle_{\text{dof}}$, then we have

$$\langle E \rangle_{\text{dof}} = C_{\text{rot}} \int_{-\infty}^{+\infty} \left( \frac{I\omega^2}{2} \right) \exp\left( -\beta \frac{I\omega^2}{2} \right) d\omega. \qquad (12\text{–}29)$$

The integral can be evaluated by a simple technique:

$$\int_{-\infty}^{+\infty} \left( \frac{I\omega^2}{2} \right) \exp\left( -\beta \frac{I\omega^2}{2} \right) d\omega = -\frac{d}{d\beta} \int_{-\infty}^{+\infty} \exp\left( -\beta \frac{I\omega^2}{2} \right) d\omega$$

$$= -\frac{d}{d\beta} \sqrt{\frac{2\pi}{I\beta}} = \frac{1}{2\beta} \sqrt{\frac{2\pi}{I\beta}}.$$

Hence

$$\langle E \rangle_{\text{rot}} = C_{\text{rot}} \frac{1}{2\beta} \sqrt{\frac{2\pi}{I\beta}} = \frac{1}{2\beta} = \frac{1}{2} kT. \qquad (12\text{–}30)$$

The average rotational energy is thus $\langle E_{\text{rot}} \rangle = 2\langle E \rangle_{\text{dof}} = kT$. For a mole of gas of rigid diatomic molecules, the total energy associated with rotation is then

$$U_{\text{rot}} = RT. \qquad (12\text{–}31)$$

Therefore the total (translational plus rotational) energy of our gas is $U_{\text{tot}} = (5/2)RT$, and the molar heat capacity changes from $3R/2$ to $5R/2$. The gas stores energy not only in translational motion but also in rotational motion. We shall see shortly that our calculations match experiment only at high temperatures. The classical ideas we applied to find our results cannot accommodate this feature of the data.

Note that the amount of energy stored does not depend on details of the individual motion: masses, rotational inertias, and so forth. Indeed, the way we derived the average energy for one degree of freedom makes it clear that the result depends very little on the fact that that energy came from rotational or translational motion: *Any* degree of freedom will contribute an amount $kT/2$ to the average energy per molecule and an amount $R/2$ to the molar heat capacity.

A further example is provided by still other degrees of freedom contained in diatomic molecules. A diatomic molecule is not truly rigid. The bond between the atoms of such a molecule acts like a spring, and *vibrational* motion is possible. If we write the separation between the atomic nuclei as $x_0 + x$, where $x_0$ is the equilibrium separation, the vibrational energy takes the form

$$E_{\text{vib}} = \frac{p^2}{2\mu} + \frac{1}{2} \kappa x^2, \qquad (12\text{–}32)$$

where $p$ is the relative momentum of the nuclei, $\kappa$ is a spring constant, and $\mu$ is a reduced mass. (As in the previous cases, neither $\kappa$ nor $\mu$ will occur in the average energy, so we need not concern ourselves with their precise values.) The quantities $p$ and $x$ represent two new independent quadratic variables, and the general Boltzmann distribution covers their values. This distribution is given by an extension of Eq. (12–26), and such forms automatically contribute a term $kT/2$ to the average energy per molecule:

$$\langle E_{\text{vib}} \rangle = 2 \times \frac{kT}{2}. \qquad (12\text{–}33)$$

Taking all of the degrees of freedom into account, we expect that, for diatomic molecules,

$$U = \frac{7}{2} RT \qquad (12\text{–}34)$$

and

$$c'_V = \frac{7}{2} R. \tag{12–35}$$

To summarize, whenever the energy of a system can be written as a sum of independent terms, each of which is quadratic in a variable that represents an associated degree of freedom, then, when the system is in equilibrium at temperature $T$, each of the terms—each degree of freedom—contributes $kT/2$ to the average energy per molecule. This is known as the law of **equipartition of energy**.

**Example 12–6** A gas of CO molecules is in equilibrium at $T = 300K$. The atomic numbers $A$ of C and O are 12 and 16, respectively, and the separation between the C and O nuclei is 0.113 nm. **(a)** Find the root-mean-square rotational speed. **(b)** Use your result from Part (a) to estimate the speed of the atoms rotating about their common axis. Compare this speed with the root-mean-square speed of the molecules.

**Solution** **(a)** The rotational inertia of two masses $m_1$ and $m_2$ separated by a distance $d$ is given by

$$I = \frac{1}{2} m_{reduced} d^2 = \frac{1}{2} \frac{m_1 m_2}{m_1 + m_2} d^2.$$

The mass of a nucleus is $m = A \times 1.67 \times 10^{-27}$ kg, so that

$$I = \left(\frac{1}{2}\right)\left(\frac{12 \times 16}{12 + 16}\right)(1.67 \times 10^{-27} \text{ kg})(1.13 \times 10^{-10} \text{ m})^2 = 7.31 \times 10^{-47} \text{ kg} \cdot \text{m}^2.$$

Now there are two degrees of freedom associated with the rotation, so the corresponding average energy is $kT = (1/2) I\langle\omega^2\rangle$. Thus

$$\omega_{rms} = \sqrt{\langle\omega^2\rangle} = \sqrt{\frac{2kT}{I}} = \sqrt{\frac{2(1.38 \times 10^{-23} \text{ J/K})(300K)}{7.31 \times 10^{-47} \text{ kg} \cdot \text{m}^2}}$$

$$= 0.75 \times 10^{13} \text{ rad/s}.$$

**(b)** Let us change our notation a bit: Let the subscripts O and C refer to oxygen and carbon, respectively. We are dealing here with the rotation of an asymmetric dumbbell, as in •Fig. 12–9. The rotation is about the center of mass, which, if we take the molecule to be momentarily oriented along the $x$-axis, is at $X = (m_O x_O + m_C x_C)/(m_O + m_C)$. In the figure, this point is located at a distance $r_O = m_C d/(m_C + m_O)$ from the oxygen nucleus and $r_C = m_O d/(m_C + m_O) = (m_O/m_C)r_O$ from the carbon nucleus. In turn, given the angular speed $\omega$, the speeds of the nuclei due to the rotational motion are $v_O = r_O \omega$ and $v_C = r_C \omega = (m_O/m_C)r_O \omega = (m_O/m_C)v_O$ for oxygen and carbon, respectively. Substituting the appropriate numbers, we obtain

$$v_O = \frac{12}{12 + 16} (0.113 \times 10^{-9} \text{ m})(0.75 \times 10^{13} \text{ rad/s}) = 363 \text{ m/s}$$

and

$$v_C = \frac{16}{12} (363 \text{ m/s}) = 484 \text{ m/s}.$$

By comparison, the rms speed of the entire molecule is

$$v_{rms} = \sqrt{\frac{3kT}{m_{mol}}} = \sqrt{\frac{3(1.38 \times 10^{-23} \text{ J/K})(300K)}{(12 + 16)(1.67 \times 10^{-27} \text{ kg})}} = 515 \text{ m/s}.$$

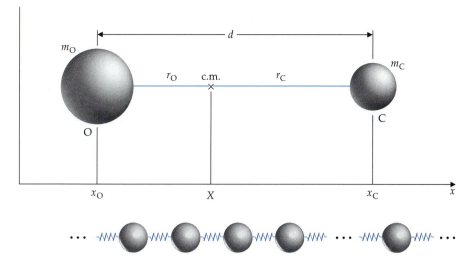

• **Figure 12–10**   A chain of atoms joined by bonds that act as springs.

One may similarly apply the equipartition theorem to the calculation of the specific heat of a crystalline solid. Consider, for simplicity, a *one-dimensional crystal*—a chain of $N$ atoms forming a line—held together by springlike forces with spring constant $\kappa$ that act between neighboring atoms (•Fig. 12–10). The energy of the crystal has the form

$$E = \sum_{i=1}^{N} \left( \frac{p_i^2}{2m} + \frac{1}{2} \kappa \left(x_i - x_{i+1}\right)^2 \right),$$

where the index $i$ refers to the atoms in the chain. The solution of the motion for this system is the subject of classical mechanics and is something we won't take the time for here. The important result is that the motion can be decomposed into that of a set of $N$ independent harmonic oscillators called *normal modes*. If that were the whole story, however, each oscillator would act independently of all the others, and the notion of a crystal in equilibrium at a temperature $T$ would have no meaning. Instead, the expression for the energy ought to contain small additional terms. These terms allow the various oscillators to interact with each other so that they can reach equilibrium at some temperature $T$, just as rare collisions between "free" gas molecules are necessary for the thermal equilibrium of an ideal gas. Once the atoms of the chain reach equilibrium, we can state that the average energy per oscillator is $kT$ and the total energy is $U = NkT$.

This discussion is easily extended to a *three*-dimensional crystal of $N$ atoms. The oscillators are now three rather than one dimensional, so the number of degrees of freedom per atom is $3 \times 2 = 6$ rather than 2. Thus the total energy is $U = 6 \times NkT/2 = 3NkT$. The specific heat of a crystal made of a mole of atoms is then

$$C = \frac{\partial U}{\partial T} = 3N_A k = 3R,$$

a result known as the **law of Dulong and Petit**. We shall see next how well this law—and equipartition in general—is experimentally satisfied.

## Experiments on Equipartition

The primary techniques for tests of equipartition involve calorimetry. The methods of calorimetry are old ones, and central to the subject is the heat capacity defined in Eq. (12–24). The prediction of equipartition for gases is that the heat

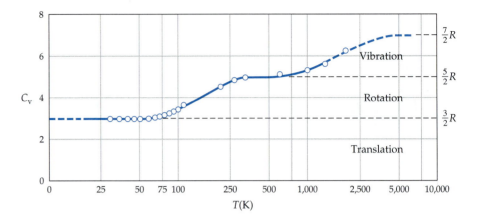

• **Figure 12–11** Measured values of the specific heat of a gas of $H_2$ molecules as a function of temperature. The plateaus correspond to the activation of different degrees of freedom. The units of specific heat are cal/mol · K.

capacity is unvarying with temperature, at a value that depends on the molecular structure. This prediction is very far from being the case, however (•Fig. 12–11). For diatomic molecules, only the amount of heat capacity associated with translational motion, $3R/2$, is present at low temperatures; the amount $2R/2$ associated with rotations enters only at higher temperatures, with an additional $2R/2$ for vibrations coming in at still higher temperatures. Despite what your classical intuition might tell you, at low temperatures collisions between molecules do not appear to get those molecules to rotate—or, for that matter, to vibrate. This phenomenon is very puzzling. Why don't these classical collisions transfer enough energy to induce rotations or vibrations? The reason is that rotational and vibrational energies are quantized, as we saw in Section 11–5. Translational energy—the kinetic energy of the molecules as a whole—is not,[7] and thus an arbitrarily small amount of energy added to the system can be stored in the form of thermal motion of the molecules. However, in order to excite rotational energy states of molecules, the thermal energy $kT$ must be larger than some threshold value, which we saw in Eq. (11–19) is on the order of $10^{-3}$ eV per molecule. In order to excite vibrational levels in molecules, $kT$ must be large enough to exceed the threshold value of 0.1 eV per molecule.

How about the heat capacity of crystalline solids? In 1819 the two young Frenchmen Pierre Louis Dulong and Alexis Thérèse Petit made the unexpected discovery that at room temperature a dozen metals, as well as sulfur, had about the same specific heat, approximately $3R$, some 25 J/mol · K. However by the 1870s investigators noted that carbon, in the form of diamond powder, gave values that differed from $3R$ and, among several measurements, from each other. Heinrich Friedrich Weber, who became one of Einstein's teachers at the Swiss Polytechnique in Zurich, suggested that the values differed within carbon because of the temperature dependence of the specific heat. Weber noted that the two experiments that appeared to disagree with each other had been done at different temperatures. He went on to confirm his suggestion by measuring a systematic temperature dependence of specific heat over a wide range of temperatures. As •Fig. 12–12 shows, all substances have the same type of temperature dependence, so that the law of Dulong and Petit is good only at temperatures that are sufficiently high. Again, as was first pointed out by Einstein in 1906, the (almost) independent oscillators that make up the normal modes of the vibrations of a crystal lattice are also quantized, with energies given by $(n + 1/2)\hbar\omega$. To get beyond the zero-point energy (the ground-state

---

[7] Or at least it is not *noticeably* quantized if the container is large.

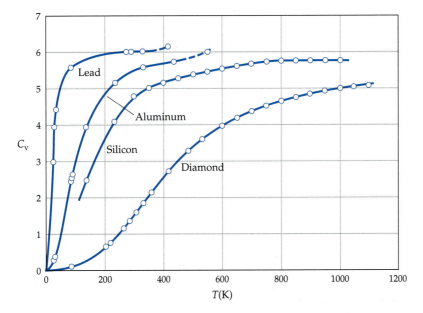

• **Figure 12–12** The specific heats of crystalline solids rise to the value $3R$ predicted by equipartition as the temperature increases. The asymptotic value $3R$ is reached by lead at quite low temperatures and by diamond at high temperatures. Here specific heats are plotted in units of cal/mol · K.

energy of the oscillator, an energy that the oscillator would possess at a temperature arbitrarily close to $T = 0$), the thermal energy $kT$ must exceed $\hbar\omega$ per lattice point. The frequency $\omega$ can be determined from the elastic properties of the lattice, and Einstein's rough, but conceptually correct, theory agrees with the experiments.

## 12–5 The Fermi–Dirac Distribution

In quantum mechanics we are forced to give up the idea of particles to which we can attach individual labels undergoing collisions that we could follow in detail if we chose to. Rather we have, say, two electrons in some initial state that is characterized by a set of quantum numbers $\{q_1\}$ and $\{q_2\}$ making a quantum mechanical transition to a final state characterized by the quantum numbers $\{q_3\}$ and $\{q_4\}$. The set of quantum numbers $\{q\}$ are *all* the quantum numbers that characterize a particle. For example, a free particle inside a three-dimensional box is characterized by the set of integers $(n_1, n_2, n_3)$ that correspond to the allowed wave functions in the box and quantum numbers that characterize the spin state. In due course we will label the state by the energy, but we must keep in mind that a given value of the energy usually describes more than one set of $(q)$'s.

As we saw in Chapter 10, the question of whether these are the same or different electrons from the ones we started with has no meaning in quantum mechanics. There simply is no way of telling. In a deep sense all electrons are the same, and we can't keep track of them in the same way we can keep track of billiard balls. This sameness is what leads to the Pauli exclusion principle for electrons. Similarly, if the particles making up the statistical system are identical bosons, there are consequences associated with the symmetry. In this section we discuss the implications of the exclusion principle for a gas of noninteracting fermions.

A **Fermi–Dirac** system—a gas of electrons in a background of neutralizing positive charge is an example—is composed of identical fermions, particles that obey the Pauli exclusion principle. We have seen that the Pauli principle forbids two electrons from occupying exactly the same quantum state in an atom. We cannot have, for instance, two electrons with the same energy in a state of zero orbital angular momentum with their spins pointing in the same direction. In the electron gas we still have the Pauli principle; this time, however, the electrons do not have the quantum numbers of electrons bound in atoms, but rather those of electrons restricted to a three-dimensional box. These quantum numbers include a spin quantum number, as well as the integers $(n_1, n_2, n_3)$ that correspond to the allowed wave functions in the box—this last set would also characterize the electron energy, at least if we ignore the interactions between the electrons.

To learn the properties of the gas, we do not need to know the set of quantum numbers in detail. Thus suppose that *on average* the number of fermions with the set of quantum numbers we label as $k$ is $n_k$. (If we do our job of finding $n_k$ correctly, we would expect $n_k$ to lie between 0 and 1. The exclusion principle would not allow it to exceed 1!) We now proceed to apply equilibrium by comparing the transition rate for the process $1 + 2 \rightarrow 3 + 4$ with that for the process $3 + 4 \rightarrow 1 + 2$, where the numbers 1, 2, ... are just some particular choice of $k$. The rate for the first process is, as in the classical case, proportional to the probabilities that fermions have the quantum numbers 1 and 2. However—and this is a new feature—this transition would be impossible if the states 3 and 4 already contained fermions. The transition would then be "blocked." Hence we must also have factors for the probability that there are no fermions in states 3 and 4, and by the ordinary laws of probability these factors are respectively proportional to $1 - n_3$ and $1 - n_4$. Therefore the rate for the process $1 + 2 \rightarrow 3 + 4$ is of the form $C n_1 n_2 (1 - n_3)(1 - n_4)$.

Now in equilibrium the rate for the process $1 + 2 \rightarrow 3 + 4$ is equal to that for the process $3 + 4 \rightarrow 1 + 2$. (See •Fig. 12–5.) Since, as in Eq. (12–15), our microscopic reaction will go either way with the same strength—that is, microscopic reversibility applies—the constant $C$ is the same for both processes, and we are left with the condition that

$$n_1 n_2 (1 - n_3)(1 - n_4) = n_3 n_4 (1 - n_1)(1 - n_2), \qquad (12\text{–}36)$$

or

$$\frac{n_1}{1 - n_1} \frac{n_2}{1 - n_2} = \frac{n_3}{1 - n_3} \frac{n_4}{1 - n_4}.$$

As we did in the classical case, we take the logarithm of the last equation:

$$\ln\left(\frac{n_1}{1 - n_1}\right) + \ln\left(\frac{n_2}{1 - n_2}\right) = \ln\left(\frac{n_3}{1 - n_3}\right) + \ln\left(\frac{n_4}{1 - n_4}\right).$$

This is in the nature of a conservation equation. In turn that means that each side of the equation can involve only quantities that are conserved in the collision, such as the energy. For the energy we have

$$E_1 + E_2 = E_3 + E_4.$$

Thus the logarithm $\ln[n/(1 - n)]$ must be linear in the energy. We write this linearity in a way that will be convenient later, namely

$$\ln\left(\frac{n(E)}{1 - n(E)}\right) = \beta(\mu - E). \qquad (12\text{–}37)$$

Note that here we have shifted our notation away from the quantum number designation to a dependence on energy. Such a shift is possible because the energy depends on the quantum numbers. For the moment we shall leave $\beta$ and

$\mu$ unspecified, except to insist that they do not depend on $E$. We can take the exponential of this equation to find that

$$\frac{n}{1 - n} = \exp[\beta(\mu - E)].$$

Finally we solve for $n$:

$$n = \frac{1}{\exp[\beta(E - \mu)] + 1}. \qquad (12\text{–}38)$$

So far in this discussion there has been no mention of spin. It enters in the following way: In equilibrium each spin direction is equally probable. For, say, spin-1/2 particles there are two such directions for each value of the energy, so to find the actual average number $N(E)$ of spin-1/2 particles at the given energy $E$ we must double the $n(E)$ we found before. In other words,[8]

$$N(E) = 2n(E) = \frac{2}{\exp[\beta(E - \mu)] + 1}. \qquad (12\text{–}39)$$

### Identification of the Constants

To identify the constants $\beta$ and $\mu$, we can use some known limits. In the limit of high energies, we can use the fact that the energy distribution must reduce to the classical Boltzmann distribution. That is because at very high energies the average occupation number in any given state will be so small that the effects of a second particle—and therefore the exclusion principle—are negligible. But for large $E$, Eq. (12–39) behaves as $N(E) \propto \text{const.} \times \exp(-\beta E)$. This proportionality can be the Boltzmann distribution only if $\beta = 1/kT$, just as in the classical case.

Once we know that $\beta = 1/kT$, we can plot the energy distribution for a variety of temperatures (•Fig. 12–13). Doing so leads us to our second constant, $\mu$, which we find by considering low temperatures, or equivalently $\beta \to \infty$. In this limit $N(E)$ in Eq. (12–39) can take on only two values. If $E > \mu$ then $N(E) = 0$, while if $E < \mu$ then $N(E) = 2$. This is just the form we expect for a collection of fermions in their ground state; there will be two electrons for each possible

■ See Chapter 10 for a discussion and calculation of the Fermi energy.

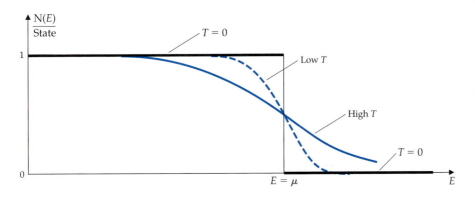

• **Figure 12–13**   A plot of the Fermi–Dirac distribution for various values of the temperature $T$. For a collection of spin-1/2 particles, there are two states. At $T = 0$, $N(E)/2 = 1$ for $E < \mu$, and for $E > \mu$ the distribution vanishes. For small $T$, the deviation from the case where $T = 0$ (dotted line) is concentrated near $E = \mu$. For higher temperatures there is no dramatic change near $E = \mu$.

---

[8] The factor 2 applies only for $s = 1/2$. The number of spin states is $2s + 1$ for a spin-$s$ object, which in this case could be $1/2, 3/2, 5/2, \dots$. But the case $s = 1/2$ is by far the most important one.

energy value up to the Fermi energy $E_F$, and none above $E_F$. Thus we identify $\mu$ with $E_F$. Summarizing, we have found the **Fermi–Dirac distribution**,

$$N(E) = \frac{2}{\exp\left[(E - E_F)/kT\right] + 1}.$$  (12–40)

Note that this distribution is not yet normalized; ultimately, we will want the sum over all energies of $N(E)$ to be the total number of particles $N$.

**Example 12–7**   As the temperature rises the Fermi–Dirac distribution spreads away from the steplike form it takes at $T = 0$. More particles in a gas of fermions take on energies greater than $E_F$, and fewer take on energies less than $E_F$. Consider a system with $E_F = 5.0$ eV. For what temperature is the number of particles at $E_F/2$ equal to 90% of the zero-temperature number at $E_F/2$?

**Solution**   According to the Fermi–Dirac distribution, the number of particles at $E < E_F$ for $T = 0$ is 2. (The normalization does not concern us here since we are going to compare the values of the distribution at two different values of $T$.) Thus we want the value of $T$ for which

$$\frac{2}{\exp\left[\dfrac{(E_F/2 - E_F)}{kT}\right] + 1} = \frac{2}{\exp\left[-\dfrac{E_F/2}{kT}\right] + 1} = (0.9)2 = 1.8.$$

This means that

$$\exp\left(-\frac{E_F}{2kT}\right) = \frac{2 - 1.8}{1.8} = 0.11.$$

We can now take the logarithm of this equation to find $T$. We obtain

$$\ln\left[\exp\left(-\frac{E_F}{2kT}\right)\right] = -\frac{E_F}{2kT} = \ln(0.11),$$

or

$$T = \frac{-E_F}{2k\ln(0.11)} = -E_F/\left[2k(-2.2)\right] = 0.23E_F/k.$$

Using $k = \left[1.38 \times 10^{-23}\ \text{J/K}\right]/\left[1.6 \times 10^{-19}\ \text{J/eV}\right] = 8.63 \times 10^{-5}$ eV/K, we have

$$T = \frac{(0.23)(5.0\ \text{eV})}{(8.63 \times 10^{-5}\ \text{eV/K})} = 1.33 \times 10^4 \text{K}.$$

## 12–6 The Bose–Einstein Distribution

In contrast to fermions, any number of identical bosons can occupy the same state. In fact, as we shall show in our treatment of lasers (Chapter 13), there is in some sense "encouragement" for there to be more than one boson in a particular state. Einstein had already demonstrated how this might operate in his 1917 work on blackbody radiation, many years before the Schrödinger equation and before the issue of the symmetry of wave functions arose. Einstein showed that the transition rate to a state already containing $n$ photons is *enhanced* by a factor $n + 1$. Thus the rate is $R(1 + 2 \rightarrow 3 + 4) = Cn_1n_2(n_3 + 1)(n_4 + 1)$, where $C$ is independent of the quantum labels.

Now we apply the equilibrium condition $R(1 + 2 \rightarrow 3 + 4) = R(3 + 4 \rightarrow 1 + 2)$, just as we did for nonidentical (classical) particles and for fermions. In the case of identical bosons, the equilibrium condition reads

$$n_1 n_2 (n_3 + 1)(n_4 + 1) = n_3 n_4 (n_1 + 1)(n_2 + 1). \qquad (12\text{–}41)$$

We can rearrange this equation to the form

$$\left( \frac{n_1}{n_1 + 1} \right) \times \left( \frac{n_2}{n_2 + 1} \right) = \left( \frac{n_3}{n_3 + 1} \right) \times \left( \frac{n_4}{n_4 + 1} \right).$$

Again, we take the logarithm:

$$\ln \left( \frac{n_1}{n_1 + 1} \right) + \ln \left( \frac{n_2}{n_2 + 1} \right) = \ln \left( \frac{n_3}{n_3 + 1} \right) + \ln \left( \frac{n_4}{n_4 + 1} \right).$$

When we compare this equation with that for the conservation of energy, $E_1 + E_2 = E_3 + E_4$, we again see that the logarithms must be linear in the energy:

$$\ln \frac{n}{n + 1} = \beta(\mu - E).$$

We take the exponential of this equation and solve the resulting linear equation in $n$ to find

$$n(E) = \frac{1}{\exp[\beta(E - \mu)] - 1}. \qquad (12\text{–}42)$$

Recall that for the Fermi–Dirac distribution [Eq. (12–39) or Eq. (12–40)], there was a factor 2 in the numerator that reflected the possibility that a spin-$1/2$ particle has two spin states ("up" and "down"), each with the same energy. Likewise, since our bosons have (an integer-valued) spin label $s$, there are $2s + 1$ spin states that all have the same energy. Thus the distribution takes the form

$$N(E) = \frac{2s + 1}{\exp[\beta(E - \mu)] - 1}. \qquad (12\text{–}43)$$

Finally we must deal with the thus-far unknown constants $\beta$ and $\mu$. We can identify $\beta$ just as we did in the Fermi–Dirac case: $N(E)$ must reduce to the Boltzmann distribution for high energies. Because $N(E) \propto \exp(-\beta E)$ in the large $E$ limit, we see immediately that $\beta = 1/kT$, as before. What is the parameter $\mu$? We can find it from the normalization—that is, from the requirement that the total number of particles is the fixed number $N$. Hence we would find $\mu$ from the requirement that

$$\sum N(E) = N, \qquad (12\text{–}44)$$

where the sum is over all states that have energy $E$.

Without actually trying to apply this condition, we can learn something important about $\mu$. The quantity $N(E)$ is never negative—it is basically a count of the number of particles with energy $E$. That means that the exponential factor must be larger than unity, which in turn means that $\beta(E - \mu)$ is positive and, since $\beta = 1/kT$ is positive, that $E - \mu$ is positive also. This must be true for all values of $E$. But the lowest possible value of the kinetic energy $E = p^2/2m$ is zero, so that $\mu \leq 0$.

When a system consists of relativistic particles, a new element enters: *It is not always true that the number of particles is fixed.* Photons represent the most important example. Their variability in number can be seen on a microscopic level in processes in which photons can be absorbed or emitted, such as $e + e \rightarrow \gamma + e + e$. In a blackbody distribution (see Section 12–8), the average number of photons, and therefore the total number of photons, changes with the temperature. Without a constraint on the number of photons there is nothing to fix the normalization and, consequently, nothing to fix $\mu$. It is perfectly consistent—in fact, necessary—to set the parameter $\mu = 0$ for photons.

We have one additional remark to make about photons: Because they are massless, special relativity imposes some special requirements. One familiar one is that photons always move with speed $c$. A second, less familiar, property is that even though they have spin 1, photons possess only two, not three, spin degrees of freedom.[9] Thus the factor $2s + 1$ in the numerator should be 2, not 3.

To summarize, when the boson number is conserved, the number distribution takes the form of Eq. (12–43) with $\beta = 1/kT$; that is,

$$N(E) = \frac{2s + 1}{\exp\left[(E - \mu)/kT\right] - 1},$$ (12–45)

where $\mu$, which is never positive, is determined from the requirement that the sum of $N(E)$ over all energies is the total number of bosons. This distribution is the **Bose–Einstein distribution** (•Fig. 12–14).

For the special case of photons the distribution is

$$N_\gamma(E) = \frac{2}{\exp\left[E/kT\right] - 1}.$$ (12–46)

We shall see in Section 12–8 that Eq. (12–46) is associated with the Planck blackbody distribution.

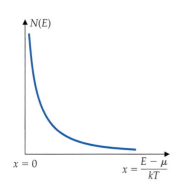

• **Figure 12–14** A plot of the Bose–Einstein distribution as a function of the variable $x = (E - \mu)/kT$. Note that at $x = 0$ the distribution is infinite. This infinity is canceled out when we integrate over the distribution to find average quantities.

## 12–7 Transition to a Continuum Distribution and the Calculation of Averages

Once we have our distributions, we can compute expectation values. To do so the distribution in question must be properly normalized. This is done by dividing the distribution by its sum over all quantum numbers. Such a process ensures that the sum over the newly normalized distribution is unity. For example, the average energy of a gas of particles confined to a box is

$$\langle E \rangle = \frac{\sum EN(E)}{\sum N(E)},$$ (12–47)

where again the sums are taken over energies and where $N(E)$ is given by Eq. (12–40) for fermions or Eqs. (12–45) and (12–46) for bosons other than photons and the special case of photons, respectively.

In a real experiment on bulk systems we measure the number of particles in a range of energies from $E$ to $E + \Delta E$, not at a precise energy $E$. This difference is relevant to us because, when $L$—the characteristic size of the container for the particles—is large compared to characteristic quantum wavelengths, the separation between allowed energies is small, and there are many levels in

[9] Classically this is the requirement that electromagnetic waves are transverse. The polarization directions are always perpendicular to that of the momentum.

the range from $E$ to $E + \Delta E$. Accordingly, we must count the states in the interval from $E$ to $E + \Delta E$ and thereafter treat the energy as a continuous variable. Let us see how this works for the case of the cubical box.

Earlier (see the subsection "Three Dimensions" of Section 10–5, together with •Fig. 10–9), we studied the problem of counting the number of box states in our energy interval. It will be useful to repeat the argument here. In Chapter 10 we saw that the number of states in the range from $E$ to $E + \Delta E$ is determined by finding the number of positive integers $n_x, n_y, n_z$ in the volume of a thin shell that forms the octant of a sphere. (The fact that the integers are all positive restricts us to the octant.) The radius $R$ of the sphere is given by

$$R^2 = n_x^2 + n_y^2 + n_z^2 = 2mL^2E/(\pi^2\hbar^2),\tag{12–48}$$

or

$$R = \sqrt{2mE}\,\frac{L}{\pi\hbar}.\tag{12–49}$$

Here we have used the relation $E = [\hbar^2\pi^2/2mL^2](n_x^2 + n_y^2 + n_z^2)$.

The thickness of the shell is found by taking the differential of each side of Eq. (12–48): $2R\Delta R = 2mL^2\Delta E/(\pi^2\hbar^2)$. Including a factor $1/8$ for the positive-integer octant, we find that the volume of the shell—which is equal to the number of states in the range from $E$ to $E + \Delta E$—is

$$\frac{1}{8}4\pi R^2\Delta R = \frac{\pi}{4}R(2R\Delta R) = \frac{\pi}{4}\left(\frac{\sqrt{2mE}L}{\pi\hbar}\right)\left(\frac{2mL^2\Delta E}{\pi^2\hbar^2}\right) = \frac{(2m)^{3/2}L^3}{4\pi^2\hbar^3}\sqrt{E}\,\Delta E.\tag{12–50}$$

Although we have done our counting for a cubical box, it applies to a container of any shape if the factor $L^3$ is replaced by the container's volume $V$. Note also that the counting of levels we have done here applies without distinction for identical particles of any kind—fermions or bosons.

The preceding discussion was based on the energy having the form $\vec{p}^2/2m$. If the particles are relativistic, as is the case for photons or for massive particles at very high temperatures, then we must modify the result. We shall see how this works in Section 12–8.

## The Transition to the Continuum

At this point we can see how the foregoing sums become integrals. Consider any of the distributions for noninteracting particles that we have discussed: Boltzmann, Fermi–Dirac, Bose–Einstein, or Planck. In the continuum limit we count the number of levels in the energy range from $E$ to $E + dE$ and then replace the sum over the intervals of "width" $dE$ by an integral over energy:

$$\sum_E f(E) \to \int_{E_{min}}^{E_{max}} f(E)\,\frac{(2m)^{3/2}V}{4\pi^2\hbar^3}\sqrt{E}\,dE = \frac{(2m)^{3/2}V}{4\pi^2\hbar^3}\int_{E_{min}}^{E_{max}} f(E)\sqrt{E}\,dE.\tag{12–51}$$

The quantity $f$ could be the distribution itself, or it could be associated with an average we want to find. We give explicit examples next.

## Finding Averages

At this point we are ready to use the distributions we have found for quantum systems of identical particles, whether fermions or bosons. Let us first work with the Fermi–Dirac distribution, Eq. (12–40). We consider spin-1/2 fermions such as electrons, so that the numerator factor in the distribution, which counts the number of spin states with the same energy, is indeed 2. The number distribution $N(E)\,dE$ is the number of fermions in the range from $E$ to $E + dE$. Using the counting factor of Eq. (12–50) with $L^3 = V$, along with Eq. (12–40), we obtain

$$N(E)\,dE = \frac{V}{2\pi^2}\left(\frac{2m}{\hbar^2}\right)^{3/2}\frac{\sqrt{E}\,dE}{\exp[\beta(E - E_F)] + 1}. \tag{12–52}$$

The total number of particles, $N$, is given by the integral of $N(E)$ over all possible energies (0 to $\infty$):

$$N = \int N(E)\,dE = \frac{V}{2\pi^2}\left(\frac{2m}{\hbar^2}\right)^{3/2}\int_0^\infty \frac{\sqrt{E}\,dE}{\exp[\beta(E - E_F)] + 1}. \tag{12–53}$$

It is useful to make the variable of integration dimensionless, and the natural quantity to use for the task is $kT$. Thus we change variables from $E$ to $x \equiv E/kT$. (This is a technique that is quite generally applicable to calculations involving statistical systems.) For this change of variable, $dE = kT\,dx$. Dividing by the integral prefactor in Eq. (12–53), we then have

$$2\pi^2\frac{N}{V}\left(\frac{\hbar^2}{2m}\right)^{3/2} = (kT)^{3/2}\int_0^\infty \frac{\sqrt{x}\,dx}{\exp[x - \beta E_F] + 1}$$

$$= (kT)^{3/2}\exp(E_F/kT)\int_0^\infty \frac{\sqrt{x}\,dx}{\exp(x) + \exp(E_F/kT)}. \tag{12–54}$$

We will use this expression shortly.

For Fermi–Dirac particles, the total number of particles is fixed and independent of the temperature. But the integrand of Eq. (12–54), and hence the integral, depends both on $\beta = 1/kT$ and on $E_F$. That means that the Fermi energy $E_F$ has to depend on temperature in such a way as to keep $N$ constant. In effect, the Fermi energy is the energy required to add one electron to the system—this is clear at $T = 0$. As the temperature rises, electrons are promoted from the low-lying states, and places open up for new electrons. Thus the Fermi energy decreases as the temperature increases.

We can use the Fermi–Dirac distribution to find the average energy per particle. We first divide by $N$ in order to ensure that we are using a probability distribution $n(E)/N$ rather than a number distribution $n(E)$. The resulting expression is

$$\langle E \rangle = \frac{1}{N}\int EN(E)\,dE = \frac{V/N}{4\pi^2}\left(\frac{2m}{\hbar^2}\right)^{3/2}\int_0^\infty E\frac{\sqrt{E}\,dE}{\exp[\beta(E - E_F)] + 1}$$

$$= \frac{V/N}{2\pi^2}\left(\frac{2m}{\hbar^2}\right)^{3/2}\int_0^\infty \frac{E^{3/2}\,dE}{\exp[\beta(E - E_F)] + 1}. \tag{12–55}$$

We could use the same formalism to compute the energy density $\rho$ of this system. Instead of dividing the energy-averaging integral by $N$, we would divide it by $V$. We will have occasion to compute $\rho$ in some of the problems at the end of the chapter.

---

**Example 12–8**    Show that in a gas of noninteracting fermions the average energy per particle in the high-temperature limit is the classical result, $\langle E \rangle = 3kT/2$.

**Solution**    We can use our expression for the average energy, Eq. (12–55), to write the integral as

$$\exp\left(\frac{E_F}{kT}\right)\int_0^\infty \frac{E^{3/2}\,dE}{\exp(E/kT) + \exp(E_F/kT)}.$$

The $\exp(E_F/kT)$ term in the denominator is small at high temperatures compared to the first exponential, so we drop it. We are then left with

$$\langle E \rangle = \frac{V/N}{2\pi^2}\left(\frac{2m}{\hbar^2}\right)^{3/2}\exp(E_F/kT)\int_0^\infty \frac{E^{3/2}\,dE}{\exp(E/kT)}$$

$$= \frac{V/N}{2\pi^2}\left(\frac{2m}{\hbar^2}\right)^{3/2}(kT)^{5/2}\exp(E_F/kT)\int_0^\infty e^{-x}x^{3/2}\,dx.$$

We have changed the variable of integration from $E$ to $x = E/kT$. This is far less complicated than it appears to be. We found the quantity $N/V$ in Eq. (12–54), and the high-temperature form of that equation is

$$2\pi^2\frac{N}{V}\left(\frac{\hbar^2}{2m}\right)^{3/2} = (kT)^{3/2}\exp\left(\frac{E_F}{kT}\right)\int_0^\infty e^{-x}\sqrt{x}\,dx.$$

When we use this quantity in the expression for $\langle E \rangle$, we have

$$\langle E \rangle = kT\,\frac{\displaystyle\int_0^\infty e^{-x}x^{3/2}\,dx}{\displaystyle\int_0^\infty e^{-x}x^{1/2}\,dx}.$$

The calculation is thus reduced to 2 integrals. The integral in the numerator is $3\pi^{1/2}/4$, while that in the denominator is $\pi^{1/2}/2$. Thus in the high-temperature limit $\langle E \rangle = 3kT/2$.

## 12–8  Systems of Relativistic Particles and the Blackbody Distribution

Two situations of interest to us require the inclusion of relativistic effects. The most important is in blackbody radiation, which is a system of *massless* particles—photons. In a very hot material such as might appear in astrophysical contexts, we may also have to include relativistic effects for massive particles. Here, however, we concentrate on blackbody radiation.[10] How does the relativistic nature of such radiation enter into the situation? The most important change has to do with the altered relation between momentum and energy.

For photons (or other massless particles) the relation between momentum and energy is $E = pc$, quite different from the nonrelativistic relation $E = p^2/2m$. Recall from Chapter 6 that the conditions for free particles in a well determine the allowed *wavelengths* associated with the particles, and because momenta are proportional to the inverse of these wavelengths, so are the allowed momenta. Thus the allowed energies will change if the relation between $E$ and $p$ changes. For massless particles, the allowed energies in a cube become

$$E_n^2 = c^2\frac{\pi^2\hbar^2}{L^2}(n_1^2 + n_2^2 + n_3^2), \tag{12–56}$$

or $E_n = (\pi\hbar c/L)R$, where as before $n_1$, $n_2$, and $n_3$ are positive integers and $R$ is defined as $R^2 = n_1^2 + n_2^2 + n_3^2$. (This result should be contrasted with its analogue for nonrelativistic particles.) We again use the fact that the number of states in the range from $E$ to $E + dE$ is $4\pi R^2\,dR/8$, which in this case is

$$\frac{\pi}{2}R^2\,dR = \frac{\pi}{2}\left(\frac{L}{\pi\hbar c}\right)^3E^2\,dE = \frac{V}{2\pi^2}\frac{E^2}{(\hbar c)^3}\,dE. \tag{12–57}$$

[10]See also Example 10–6 for a discussion of another ultrarelativistic system.

Here we have replaced $L^3$ by $V$, the volume of the cube. Considering that we can make up any shape out of a superposition of cubes, our result holds for a volume of any shape.

When we apply this equation to photons, we must multiply by a factor of 2 to account for the two possible polarization states of the photons, states with the same momentum and hence the same energy. When we form the number distribution of photons, we set $\mu = 0$, because there is no constraint on the total number of photons. Thus, using Eq. (12–57) in place of the counting factor of Eq. (12–50), we have, for the number of photons in the energy window $dE$ around $E$,

$$N(E)\, dE = 2\, \frac{V}{2\pi^2}\, \frac{1}{(\hbar c)^3}\, \frac{E^2}{\exp(E/kT) - 1}\, dE. \tag{12–58}$$

Equation (12–58) is associated with the Planck distribution law we studied in Chapter 4. In that chapter we concentrated on the energy density rather than the number density of photons—which is what Planck did. He did not know about photons and was interested in the energy content of a volume containing black-body radiation. The energy density $\rho$ is related to the number density $n$ by $\rho = CkTn$, where $C$ is a constant of proportionality determined by the necessary integrals in the two cases. (See Problem 36 *et seq.* at the end of the chapter.)

---

**Example 12–9**    Use Eq. (12–58) to find an expression for $n$, the number density of photons—the number of photons per unit volume—in a blackbody at temperature $T$.

**Solution**    From Eq. (12–58) we see that

$$n = \frac{1}{\pi^2(\hbar c)^3} \int_0^\infty \frac{E^2}{\exp(E/kT) - 1}\, dE = \frac{1}{\pi^2}\left(\frac{kT}{\hbar c}\right)^3 \int_0^\infty \frac{x^2}{e^x - 1}\, dx,$$

where we have used the integration variable $x = E/kT$. The density $n$ increases as the third power of the temperature. This fact can be understood from dimensional analysis. Planck's constant has the dimensions of energy $\times$ time, so that $(\hbar c)^3$ has the dimensions of (energy)$^3$ $\times$ volume. Since $n$ must have the dimensions of $1/V$ we must have a factor of $(kT)^3$, which has the dimensions of (energy)$^3$, in the numerator. The massless character of the photon has allowed us to factor out the physical dimensions from the integral, yielding a simple temperature dependence for $n$. For particles of finite mass this factoring out is not possible, and the temperature dependence is complicated. The integral goes all the way back to Planck; its value, 2.404, can be found only numerically. (See Problem 37 at the end of the chapter.)

---

## 12–9 Some Applications

### The Specific Heat of Electrons in Metals

In Section 12–4 we discussed the role of equipartition in determining the specific heat $C$ of a substance, defined as

$$C \equiv \frac{\partial U}{\partial T},$$

where $U$ is the thermal energy of the substance. Equipartition states that for one mole of the substance $U = N_A(skT/2) = sRT/2$, where $s$ is the number of degrees of freedom and $R = N_A k$ is the universal gas constant. We saw in our discussion of the law of Dulong and Petit that for a three-dimensional lattice such as that which forms a solid we expect $s = 6$. This leads to the prediction

$$C = 3R,$$

which is just the law of Dulong and Petit. We discussed how this law is realized experimentally for high enough temperatures.

An immediate question that arises is that in a metal we also have a large number of "free" electrons—about one per ion. Counting degrees of freedom, we would expect from classical considerations that these electrons would contribute an additional amount $3R/2$ to the specific heat. Why is there no effect? The answer lies in the nonclassical distribution of the electrons.

If we look at the expression for the average energy of an electron, Eq. (12–55), we see that for $E - E_F$ large compared with $1/\beta = kT$, the integrand is very small. Thus we have significant contributions to the average energy only from the region around $E = E_F$. There, however, the exponential does not differ much from unity, so we expect little temperature dependence and hence little contribution to the specific heat. Rather than sharpening this mathematical argument, let us instead think about a more physical argument.

Most of the electrons fill the levels up to near the Fermi energy. Thus it is not on the whole possible to "store" thermal energy by increasing the kinetic energy of the electrons, because that would mean moving an electron into a higher energy level. Such levels are for the most part occupied. As the temperature increases from zero, the distribution is described by saying that a few electrons with energy near the Fermi energy have spilled over into energy levels above $E_F$, leaving a few vacancies in the energy levels below $E_F$. The number of electrons that do this is relatively small until the temperatures involved are comparable to or higher than the Fermi temperature $T_F$, defined as $E_F = kT_F$. Indeed, if Eq. (12–55) for the average energy is integrated, we find that for $T \ll T_F$

$$\langle E \rangle \cong \frac{\pi^2}{4} \frac{k^2 T^2}{E_F}. \tag{12–59}$$

The total energy of a mole is $N_A$ times this quantity, and when we differentiate with respect to $T$ and use the definition of the Fermi temperature, we find for the electrons' contribution to the specific heat

$$C = \frac{\pi^2}{2} R \frac{T}{T_F}. \tag{12–60}$$

A typical Fermi energy in metals is on the order of 5 eV. If we recall that 1 eV corresponds to a temperature of 12,000K, then $T_F \cong 60,000$K. At ordinary temperatures, then, and even at temperatures for which the law of Dulong and Petit is well satisfied, $T/T_F$ is completely negligible, and the electrons contribute little to the total specific heat.

## The Specific Heat of Molecules

The existence of a hierarchy of three types of molecular excitation energies—rotational, vibrational, and electronic, in order of energy differences—manifests itself macroscopically in the form of the specific heat of molecules at a constant volume as a function of temperature. The equipartition theorem specifies that each degree of freedom associated with a gas molecule implies a contribution of $kT/2$ to the average energy per molecule $\langle E \rangle$, and as mentioned earlier, this means that if $s$ is the number of degrees of freedom, then the total internal energy $U = N_A \langle E \rangle$ is $sRT/2$ and the molar heat capacity $c'_V$ at constant volume is

$$c'_V = \frac{\partial U}{\partial T}\bigg|_V = \frac{sR}{2}.$$

In reality the internal energy of a molecular gas depends on the temperature in a more complicated way than our classical expressions suggests. Experiments show that the degrees of freedom somehow "turn on" as the temperature is increased. As we have seen, the three translational degrees of freedom are always present, even at very low temperatures, so that there is always a contribution of $3R/2$ to the specific heat. But at low temperatures, neither the two rotational nor the two vibrational degrees of freedom that we expect to be present for a diatomic molecule are evident in the specific heat.

To see how quantum mechanics explains this behavior, let us use the Boltzmann distribution to find the average rotational energy. The possible rotational energies are $E_{rot} = \ell(\ell + 1)\hbar^2/2I$. Thus the probability that the molecule is in the rotational state labeled $\ell$ is

$$P(\ell) = \frac{(2\ell + 1)\exp\left(-\dfrac{\ell(\ell + 1)\hbar^2}{2IkT}\right)}{\sum (2\ell + 1)\exp\left(-\dfrac{\ell(\ell + 1)\hbar^2}{2IkT}\right)}. \tag{12–61}$$

The denominator, which we denote henceforth by $Z$, ensures that the sum of the probabilities is unity. Note the factor $(2\ell + 1)$ which accounts for the $2\ell + 1$ degenerate states corresponding to different values of the quantum number for the z-component of angular momentum. Using this distribution we can compute various averages; for example, the average rotational energy is

$$\langle E \rangle = \frac{\sum \dfrac{\ell(\ell + 1)\hbar^2}{2I}(2\ell + 1)\exp\left(-\dfrac{\ell(\ell + 1)\hbar^2}{2IkT}\right)}{Z}. \tag{12–62}$$

In classical physics we have no discrete index $\ell$, and instead of sums we have integrals over continuous values for the energy, starting from an *arbitrarily small* rotational energy of zero. This is the difference: In quantum mechanics, there is a spacing—an energy gap—to the first rotational energy level that appears. How big is this gap? It is given by the difference between the rotational energy for $\ell = 1$ and that for $\ell = 0$. The rotational energy for $\ell = 0$ is 0, so the gap energy $\Delta E = \ell(\ell + 1)\hbar^2/2I$ for $\ell = 1$, or $\Delta E = \hbar^2/I$. We can estimate $I$ for $H_2$ by using a reduced mass of $m_p/2$ and a nuclear separation of $2a_0$, which gives

$$I = \left(\frac{m_p}{2}\right)(2a_0)^2 = 2m_p a_0^2.$$

Thus

$$\Delta E = \frac{\hbar^2}{4m_p a_0^2} \approx 6.0 \times 10^{-22}\,\text{J} = 3.7 \times 10^{-3}\,\text{eV}.$$

We can convert this energy to a temperature by dividing by the Boltzmann constant $k = 1.38 \times 10^{-23}$ J/K. ($kT$ has the dimensions of energy.) We find that

$$\frac{\Delta E}{k} \approx 43\text{K}.$$

Since the argument of the exponential in the Boltzmann distribution is the factor $-\Delta E/kT$, the exponential factor in Eq. (12–62) suppresses the average energy for values of $T$ less than 40K or so, and the equipartition theorem does not hold. In effect, the presence of the gap means that the rotational energies do not get excited until we get to temperatures on the order of 100K.

We can apply the same reasoning to the vibrational energies. The gaps for these are some 100 times greater still than the gaps for the rotational energies. Thus we do not expect these degrees of freedom to come into play until the gas of diatomic molecules reaches temperatures in the range of several *thousand* degrees.

In this way, the contributions $kT/2$ for each degree of freedom are activated one at a time. According to our estimates, for $H_2$ we would have $c_V' = 3R/2$ up to about 100K, $5R/2$ from 100K to about 1,000K, and $7R/2$ for higher energies. These values are in agreement with observation (•Fig. 12–11). With another factor of about 100—bringing the temperature up to about $10^5$K—the electronic levels enter the picture. Hotter than that, and the hydrogen molecule ionizes—we have a plasma. Let us emphasize that the steplike behavior of the specific heat is a quantum mechanical effect, inexplicable by classical means.

## Bose–Einstein Condensation

Recall from the discussion following Eq. (12–45) that the parameter $\mu$ is determined for massive bosons[11] by the requirement that the sum of the numbers of bosons with all energies add to the total number $N$ of bosons. This number is independent of the energy, so $\mu$ must vary with energy in just such a way as to keep $N$ fixed. The volume $V$ is also fixed, so that the number density $n = N/V$ is fixed. Using Eq. (12–50) for the counting factor, we integrate Eq. (12–45) to obtain

$$n = \frac{1}{4\pi^2}\left(\frac{2m}{\hbar^2}\right)^{3/2}\int_0^\infty \frac{\sqrt{E}\,dE}{\exp\left[(E-\mu)/kT\right]-1}. \tag{12–63}$$

With the usual change of integration variable from $E$ to $x \equiv E/kT$, we have

$$n = \frac{1}{4\pi^2}\left(\frac{2mkT}{\hbar^2}\right)^{3/2}\int_0^\infty \frac{\sqrt{x}\,dx}{\exp\left[x-\mu/kT\right]-1}. \tag{12–64}$$

Now recall that $\mu \le 0$—otherwise the argument of the exponential, $x - \mu/kT$, could become negative, and the integral itself could go negative, giving us a negative density. Let us lower the temperature of this boson gas. The left-hand side of Eq. (12–64) is the fixed density. As $T$ decreases, the prefactor on the right decreases, so that we require the integral itself to increase. Since the quantity $-\mu/kT$ is positive, it must decrease in magnitude as $T$ decreases. In other words, $\mu$ must increase towards $\mu = 0$ as $T$ decreases. But now we have a problem: At some point as $T$ decreases, $\mu$ arrives at zero. Let us call $T_c$ the value of $T$ for which $\mu$ reaches zero. The numerical value of $T_c$ is given by

$$n = \frac{1}{4\pi^2}\left(\frac{2mkT_c}{\hbar^2}\right)^{3/2}\int_0^\infty \frac{\sqrt{y}\,dy}{\exp y - 1} = \frac{1}{4\pi^2}\left(\frac{2mkT_c}{\hbar^2}\right)^{3/2}\times 2.31.$$

But with $\mu \le 0$ this is as large as the integral in Eq. (12–64) is going to get. If we decrease $T$ beyond $T_c$, we appear to be in real trouble: We will no longer be able to compensate for the decrease in the prefactor by increasing the integral.

Einstein went though this reasoning in 1924. He conjectured that the way out would be to separate the ground state with $E = 0$ from the rest of the sum in Eq. (12–64). Thus Eq. (12–63) actually applies only to particles that are *not* in the ground state. The number of particles not in the ground state is less than the

---

[11] We are assuming that the number of these bosons, whose prototype would be helium atoms with two protons and two neutrons in their nuclei, is fixed and that the temperatures are low enough to avoid the necessity of a relativistic formulation.

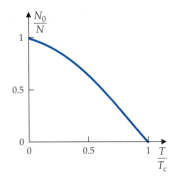

• **Figure 12–15**   The fraction of bosons in the Bose–Einstein condensate as a function of temperature. Note that the fraction approaches unity as the temperature drops to absolute zero.

total number $N$, and the "problem" is resolved by sending any excess particles that cause the problem into the ground state. There would then be two distributions of particles with relative numbers $N_0$ and $N_e$ for the ground state and the excited states respectively. The total would always add up to $N$. As the temperature drops below $T_c$, a larger and larger fraction "condenses" into the ground state. As the temperature approaches absolute zero, all the particles condense into the ground state. This phenomenon, which occurs even in the absence of direct interactions between the bosons, is called **Bose–Einstein condensation**.[12] •Figure 12–15 shows how the fraction $N_0/N$ of condensed bosons behaves as a function of temperature.

To be specific, let us consider 1 liter of liquid composed of helium atoms below a temperature of 2.18K, the transition temperature at which the condensation begins. The mass density of liquid helium is $\rho = 0.145$ kg/m$^3$, so there are on the order of $10^{23}$ atoms in the 1-liter container. A sizable fraction of these, perhaps 10%, will be found in the single-particle ground state. If we consider the same system at a temperature above, say, $T = 3$K, an estimate of the average number of atoms per energy level will show that there fewer than 10 atoms per energy level. This fact emphasizes the dramatic change that occurs due to the condensation.

Note that the fraction of particles that are condensing into the ground state are *all in the same quantum state*. The **condensate** (as it is called) is a **macroscopic quantum system**—a quantum system that makes its quantum properties known through large-scale observable effects. It is interesting that although Einstein proposed the condensation phenomenon for a gas of noninteracting particles in 1924, it took until 1995 actually to observe the phenomenon visually. The reason is that the condensation temperature is very low and at these temperatures most materials solidify. When atoms (if they are bosons) solidify to form a crystal lattice, the fact that they are identical bosons is irrelevant. In that case, the atoms are distinguished by their positions.

In 1995 a group of physicists headed by E. Cornell and C. Weiman in a joint program of the University of Colorado and the U. S. Commerce Department's National Institute of Standards and Technology performed experiments that showed the Bose–Einstein condensation in the most direct way possible. A gas of $^{87}$Rb atoms, which are bosons, was cooled to a temperature about 300 times lower than any material had ever been cooled—about 20 billionths of a degree above absolute zero. The atoms at this temperature have average speeds of only about one meter per hour! The experimenters used a set of ingenious schemes, in which laser light played an important role. (See the discussion of cooling by lasers in Chapter 13.) At a critical temperature $T_c = 190 \times 10^{-9}$ K, a Bose–Einstein condensation occurred. A study of the speed distribution showed a transition from the Maxwell–Boltzmann distribution above $T_c$ to one in which there was a pronounced peak around $v = 0$ as the temperature was lowered beyond $T_c$. The beautiful thing about this experiment is that the condensate and the rest of the atoms could be separated spatially, and one could study the pure condensate for extended periods of time—up to fifteen seconds in the initial experiments (•Fig. 12–16). The experiment has been reproduced elsewhere on other elements such as $^7$Li. A new state of matter explicable only by the quantum theory has been created.

---

[12] The Bengali physicist Satyendra Nath Bose had sent Einstein a manuscript in which he had derived the Planck distribution by using what we now call Bose–Einstein statistics. Einstein was so impressed that he translated the manuscript into German and had it published. But Bose had nothing to do with what we now call Bose–Einstein condensation.

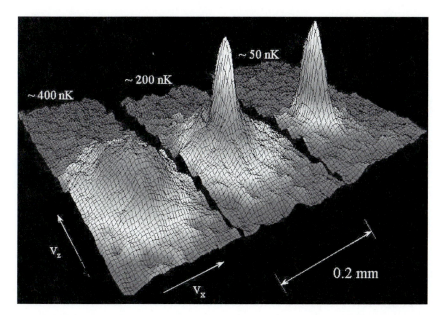

• **Figure 12–16** Recent experimental results on Bose–Einstein condensation. The plots shown are computer-generated images of the velocity distribution of rubidium atoms undergoing Bose–Einstein condensation. One should read the photos from left to right. Just prior to condensation there is no concentration in the $v = 0$ ground state. As the condensation proceeds by lowering the temperature, the atoms condense into that state, which means that there is a sharp spike at $v = 0$. These experiments, which were done at the University of Colorado, were the first to show an actual condensed state.

### Liquid Helium and Superfluidity

Liquid helium is another system in which the Bose–Einstein condensate occurs, although the fact that the helium atoms interact with one another—because the atoms are closely spaced—modifies the properties of the condensate substantially. The effects associated with the condensate in liquid helium were observed long before the noninteracting condensate just described was.

The story of the discovery of what was later learned to be a condensate in liquid helium is an interesting one. The French physicists Louis Cailletet and Raoul Pictet had succeeded in liquefying air in 1877, and it was with an improved version of their apparatus that the Dutch physicist Heike Kamerlingh Onnes, working at Leiden, in the Netherlands, first succeeded in liquefying helium in 1908. The transition from gas to liquid occurs[13] at 4.2K. Kamerlingh Onnes's success opened up a new range of low temperatures at which the behavior of materials could be studied. Indeed, he studied properties of helium down to 2K. In 1928 the Dutch physicist Willem Hendrik Keesom, working in the Kamerlingh Onnes Laboratory[14] in Leiden, discovered that at the temperature of $T_c = 2.17$K liquid helium makes a transition to a new phase. Below this critical temperature liquid helium has some surprising properties. We list a few that are easily demonstrated in the laboratory:

• When liquid helium at a temperature above $T_c$ is cooled by letting it evaporate and pumping the vapor away, the liquid beneath the vapor boils. As the temperature drops below $T_c$, the boiling stops very suddenly.

• Above $T_c$, liquid helium behaves like any other viscous liquid; for example, it cannot flow through channels that are too narrow. Below $T_c$, the flow through a narrow channel is completely unobstructed, to the point that the hellium appears to have no viscosity whatever.

---

[13] In contrast to other substances, helium does not solidify at very low temperatures. The zero-point energy of the atoms overcomes the bonding energy that usually leads to the formation of a solid, and helium remains liquid at temperatures as low as we can go, at least at normal pressures.

[14] Interestingly, Kamerlingh Onnes had been one of the physicists who had not given Einstein a job when he graduated from the Polytechnical Institute.

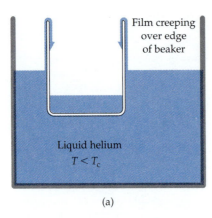 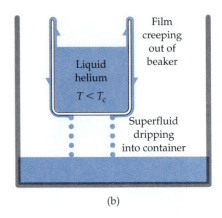

Film creeping
over edge
of beaker

Film
creeping
out of
beaker

Liquid
helium

$T < T_c$

Superfluid
dripping
into container

Liquid helium

$T < T_c$

(a)                    (b)

• **Figure 12–17** A superfluid
creeps up a glass surface
against gravity. The vapor that
condenses on the surface of the
glass acts as a siphon for the
superfluid.

• When a beaker is placed inside a container of liquid helium below $T_c$, liquid
helium from the larger container creeps up the side of the beaker and down
into it, as shown in •Fig. 12–17a. Similarly, if a beaker containing liquid helium below $T_c$ is placed in the container and the level of helium in the beaker
is above that of the container, then liquid helium creeps up the side of the
beaker from the inside and goes over the top into the larger container
(•Fig. 12–17b).

• When liquid helium below $T_c$ is put in a bottle with a stopper containing a
channel too narrow for normal fluids to pass through, and the bottle is heated (e.g., by infrared radiation), the liquid helium in the bottle produces a jet-
like fountain as it passes through the stopper channel (•Fig. 12–18).

These phenomena, together with accurate measurements of some others,
lead to the following picture of what is happening to the liquid helium: The
material consists of a set of bosons. At $T_c$, the helium atoms begin to form a
condensate—a single quantum state analogous to the ground state in the
Bose–Einstein condensate. Just at $T_c$ the fraction of the atoms in the condensate
is small, and as $T \rightarrow 0$ this fraction increases to unity. The condensate becomes
a **superfluid**.

Because below $T_c$ a fraction of the liquid helium is a superfluid, it is useful
to describe helium at these temperatures as consisting of a mixture of two fluids—the normal fluid and the superfluid—with the property that the relative
density of normal fluid to superfluid changes with temperature between $T = 0$
and $T = T_c$. The special properties of liquid helium described in the foregoing
list are a reflection of the properties of the superfluid.

What are the properties of the superfluid that could explain the phenomena that are observed? Because the helium atoms interact, even weakly, with
each other, the phenomenon is different from Bose–Einstein condensation. The
condensate wave function of the superfluid involves *all* the atoms of the superfluid. It is not possible for a small subset of atoms of the superfluid to leave
the ground state and get into an excited state. As a consequence it takes a macroscopic amount of energy to create excited states of the superfluid. Since viscosity—a kind of internal friction—is just a manifestation of energy dissipation
in a fluid as a result of its flow, the inability of the condensate to lose arbitrarily small amounts of energy manifests itself in the absence of viscosity. Thus the
superfluid component will flow unimpeded through narrow capillaries that
would not, due to viscosity, let the normal fluid get through.

The absence of viscosity also is responsible for the creeping of the film of
superfluid up the side of the beaker. In liquid helium, as indeed with all fluids,

• **Figure 12–18** The fountain
effect in liquid helium below
the critical temperature. The
liquid jetting through a narrow
channel in the beaker's stopper
is a pure superfluid. This effect
is a consequence of the capa-
bility of the superfluid to pass
through very narrow capillaries, the inability of normal liquid helium to do so, and the
incompressibility of liquid
helium.

a thin layer of fluid gets formed on the side of the container by the condensation of some of the vapor. In the absence of viscosity, the only forces that act on the fluid are gravity and the molecular force between the helium atoms and the molecules that form the beaker. For those helium atoms that lie in a thin film near the surface, the latter is stronger than the force of gravity, allowing the film to creep upwards. Effectively, the film acts as a siphon.

The cessation of boiling at $T_c$ is also explained by the presence of the superfluid. In a normal liquid, boiling occurs because local "hot spots" develop, which cause the expansion of gas contained in tiny bubbles. In the superfluid the thermal conductivity is huge (about a million times larger than in normal liquid helium), so that no hot spots can develop. In the superfluid heat is conducted in a different way than it is in ordinary fluids. Usually thermal energy propagates in a fluid by a diffusion mechanism: Faster molecules from a hotter region collide with slower molecules in the cooler region and thus transfer energy. This mechanism is not available to the superfluid, within which the motion of the molecules is closely correlated in the form of the single wave function. There the only mechanism for the transfer of thermal energy is an acousticlike wave, except that it is a wave of heat. This phenomenon, known as *second sound*, has been studied in detail, as has the rapid rise in thermal conductivity that occurs when the temperature drops through the critical point.

The fountain effect can be explained as follows: When liquid helium below $T_c$ absorbs some heat, there is a local decrease in superfluid: The number of atoms in the quantum condensate falls. To make up for the local decrease, superfluid from neighboring regions flows into the region in which the local decrease exists. This influx increases the density of that region and hence the pressure, and since only the superfluid can penetrate the narrow capillary stopper the fountainlike jet is pure superfluid.

Liquid helium is far from a closed subject. Still, there is little doubt that its exotic properties are a consequence of the boson character of the helium atoms and of a mechanism that bears a close resemblance to the Bose–Einstein condensation. It is worth noting that $^3$He (helium with a nucleus consisting of two protons and a neutron and thus existing as a fermion) does not exhibit any of these properties. It has its own set of strange properties that are somewhat akin to superconductivity, a phenomenon to be discussed in the next chapter.

## SUMMARY

A statistical treatment of matter is an absolute necessity given the large numbers of particles involved; for that same reason, the predictions of such a treatment are absolutely reliable. A gas of pointlike noninteracting atoms provides the most direct example of a classical gas. Quantum mechanical effects associated with identical particles are also extremely important in everyday situations. We can extract the following ideas:

- For a classical gas, the temperature is a measure of the average kinetic energy of the molecules of the gas. A connection is made between the temperature and other thermodynamic variables such as pressure or volume through kinetic theory.
- The distribution of energies in a large system is described by the Boltzmann distribution, which contains the pertinent dynamical factor $\exp(-E/kT)$. This factor, the Boltzmann factor, describes a great deal of the behavior of the physical world. When energies are purely kinetic, as in the classical gas of

pointlike atoms, the Boltzmann distribution becomes a distribution in velocity called the Maxwell distribution.

- When the constituents of a gas can have degrees of freedom other than those of center-of-mass motion in space, the Boltzmann distribution leads to the equipartition theorem, which gives a more general form for the heat capacity.

- Equipartition fails for gases made of nonpointlike molecules, and this failure is important evidence for the quantum theory, which correctly describes the departure from equipartition as due to a finite spacing between possible energies for the constituents of the gases.

- When the particles making up the statistical system are identical, quantum mechanics has a profound effect, even if the system again behaves as a classical system when the temperature is high enough. The energy distribution for identical fermions—the Fermi–Dirac distribution—and that for identical bosons—the Bose–Einstein distribution—both occur in common situations, the former describing electrons in a metal and the latter photons (which constitute the special case of *massless* bosons) in a blackbody. In the case of fermions, the exclusion principle acts to limit the number of spin-1/2 fermions that can go into the same state to two, while in the case of bosons, the probability that a given state contains a given number of bosons is enhanced over the classical expression. Both of these quantum mechanical systems reduce properly to the Boltzmann distribution at high enough temperatures.

## QUESTIONS

**1.** Why does quantum mechanics suggest that absolute zero is not attainable? Think in terms of the uncertainty principle.

**2.** In our derivation of the various number distributions we used the principle of micro-reversibility. In other words, we assumed that if we reversed all the initial and final velocities in a given collision we would have a collision that was equally as probable as the original collision. This assumption gave rise to a puzzle that first appeared in the 19th century, when work on these number distributions was being created. If this principle is true, how can any macroscopic system evolve? How is aging possible? Here are some questions that may give you an idea of how the puzzle might be solved: **(a)** Consider a molecule of some dye suspended in water with water molecules colliding with it from all sides with equal probability. What happens to that molecule? Why does it not just stay in one place? **(b)** Suppose we have a large number of such molecules and we place them initially in the corner of the container. What will happen? **(c)** If you took a series of photographs of the process you come up with in Part b, how could you tell, between two photographs, which one was taken earlier and which one later?

**3.** When does a gas fail to be "ideal"? Give an example.

**4.** In what sense does it make sense—if, indeed, it does at all—to talk about the temperature of a single molecule? Would it make sense to talk about the temperature of a football that sits in a room somewhere? If there is a difference between these situations, what is it?

**5.** A calculation would show you that the mean free path for collisions in a gas may be quite a bit larger than the mean intermolecular spacing of the gas. Does this make sense?

**6.** The Fermi temperature of, say, a metal is not the same as the metal's actual temperature. What would it mean if the actual temperature were much less than the Fermi temperature? Much greater than the Fermi temperature?

**7.** The cross section for a low-energy photon passing through a plasma—a gas of free electrons (moving in a background of positive charges, so that the total system is electrically neutral)—is denoted by $\sigma$, whose numerical value happens to be $0.67 \times 10^{-28}$ m$^2$. How would you roughly estimate the mean free path of a low-energy photon near the center of the Sun? What other numbers do you need to know?

**8.** Consider the "gas" of microscopic needles described on page 337 of the text. Qualitatively sketch the average energy of the system as a function of the temperature, given what you have learned about the system. Also, make a sketch of the specific heat of the system as a function of energy.

**9.** A cloud of CN molecules in the interstellar medium has been identified by the spectra of the molecules. A detailed study of the spectra allows one to find the number of molecules in the $\ell = 0$ state relative to the number in the $\ell = 1$ state. How could you use this ratio to determine the temperature of the cloud?

**10.** In Example 12–5 we considered a gas of hydrogen atoms, but we didn't treat them as identical particles and use a quantum mechanical distribution. Why was our approach a valid way of treating the gas?

# PROBLEMS

**1.** ▮ A water droplet 1 micron $(10^{-6}$ m$)$ in diameter is in equilibrium with air molecules at room temperature (295K). What is the root-mean-square (rms) speed of the droplet?

**2.** ▮ Dust grains of diameter $5 \times 10^{-7}$ m and density $10^3$ kg/m$^3$ are in equilibrium with air $(A = 28.9)$ at a temperature of 273K. What is the rms speed of the dust particles?

**3.** ▮▮ A certain gas is composed of a mixture of all the inert gases and thus consists exclusively of monatomic molecules. The temperature of the gas is 200K. Find the rms speed of the molecules. For masses, use the following values: helium $(A = 4)$, neon $(A = 20)$, argon $(A = 40)$, krypton $(A = 84)$, xenon $(A = 131)$, and radon $(A = 222)$.

**4.** ▮ A gas obeys the ideal-gas equation of state $pV = NkT$, where $N = nN_A$ is the number of molecules in the volume $V$ at pressure $p$ and temperature $T$ and $n$ is the number of gm-moles of the gas. Calculate the volume occupied by 1 gm-mole of the gas at atmospheric pressure $(p = 1.01 \times 10^5$ N/m$^2)$ and $T = 0°$ C = 273K.

**5.** ▮ Given that the molecular weight of air is 28.9, what is the density of air under the conditions specified in the previous problem?

**6.** ▮▮ The average molecular weight of air is 28.9, meaning that one mole of air, composed of $N_A = 6.02 \times 10^{23}$ molecules, has a mass of 28.9 g. Assume that air obeys the ideal-gas law in order to estimate **(a)** the number of molecules in your lecture room; **(b)** the mass density of helium gas $(A = 4)$ at standard temperature and pressure (STP; $T = 273$K and $p = 1.01 \times 10^5$ N/m$^2)$; and **(c)** the amount of helium needed to lift a mass (payload plus balloon) of 100 kg at STP.

**7.** ▮ Neon gas $(A = 20)$ obeys the ideal-gas law. What is the average of the square of the velocity of this gas for $T = 300$K? for $T = 3{,}000$K? for $T = 30$K?

**8.** ▮ Hydrogen gas is composed of molecules with a molecular weight of 2. A sample of this gas obeys the ideal-gas law. What must the temperature be so that the average of the square of the velocity is $(4 \text{ km/hr})^2$? (This figure corresponds roughly to walking speed.)

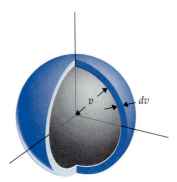

• **Figure 12–19** Problem
12–10.

**9.** ▮▮ Consider the Maxwell distribution for a gas of pointlike molecules that can move in only one dimension. Write the appropriate formula for this distribution and calculate the average energy of a molecule in terms of the temperature. Your result should be $kT/2$, a special case of equipartition for a single degree of freedom.

**10.** ▮▮ Distributions can often be interpreted in terms of more than one variable. Use the Maxwell distribution to describe a distribution in energy—that is, the number of molecules with energies in the range form $E$ to $E + dE$. (*Hint*: write $f$ in terms of $E = mv^2/2$. When the integrand does not depend on directions, the integral over the volume $d^3\vec{v}$ can be written as an integral over shells of a given speed $v$. (See •Figure 12–19.) For a given speed, these shells have a volume given by (surface area) × (thickness) $= 4\pi v^2\,dv$. This step is equivalent to having done the angular part of the integration. Then convert the integration over $v$ to one over $E$ by making use of the fact that $E = mv^2/2$.)

**11.** ▮▮▮ *The Speed Distribution.* The Maxwell distribution $f(\vec{v})$, Eq. (12–11), describes molecular velocities. Since the equation does not depend on direction, it can be used to find a distribution of speeds. We do this by summing over the probabilities of motion in all directions. In the limit, we have

$$\int_{\text{directions}} f(\vec{v})\,d^3\vec{v} \equiv g(v)\,dv,$$

where $g(v)\,dv$ is the probability that a molecule has a speed between $v$ and $v + dv$. The quantity $g(v)$ defined by this equation is the *speed distribution*. **(a)** Demonstrate by integrating over directions that

$$g(v) = 4\pi\left(\frac{m}{2\pi kT}\right)^{3/2} v^2 \exp\left(-\frac{mv^2}{2kT}\right).$$

**(b)** Verify that $g(v)$ is properly normalized; that is, show that $\int_0^\infty g(v)\,dv = 1$.

**12.** ▮▮ The speed distribution $g(v)$ (see Problem 11) has a maximum. The speed at which this maximum occurs is the *most likely speed* $v_{\text{ml}}$. **(a)** Find $v_{\text{ml}}$. **(b)** What is the numerical value of $v_{\text{ml}}$ for air ($A = 28.9$) at 273K?

**13.** ▮▮ Calculate $\langle v\rangle^2$ and $\langle v^2\rangle$ using the speed distribution of Problem 11. Is there a difference? Under what conditions will a distribution in $x$ give $\langle x\rangle^2 = \langle x^2\rangle$, assuming that $x$ is restricted to the range from 0 to $\infty$? Give an example of such a distribution.

**14.** ▮▮ Find the relative probability that a molecule in a sample of nitrogen gas ($A = 28$) at 373K has a speed twice the rms speed compared with a speed equal to the rms speed. (See Example 12–1.) Repeat the exercise for a speed 10 times the rms speed.

**15.** ▮▮ Consider the distribution of velocities in a beam emerging from a hole in a hot oven perpendicular to the hole, as given in Eq. (12–13). **(a)** Calculate $\langle v_z\rangle$. (*Caution*: Only positive values of $v_z$ are permitted.) **(b)** A gas of $UF_6$ consists of molecules made up of an atom of uranium and six fluorine atoms. Fluorine has atomic weight $A = 19$. Uranium comes in two varieties, with atomic weights 238 and 235, respectively. Use the results of Part (a) to calculate the difference between the values of $\langle v_z\rangle$ for the two varieties of uranium for $T = 850$K and for $T = 1{,}200$K. Your answer has applications to the separation of these two varieties of uranium, a problem that was of great importance in the World War II Manhattan Project. The separation used a diffusion technique that concentrated $^{235}U$ by utilizing the fact that $\langle v_z\rangle$ for $^{235}U$ and that for $^{238}U$ differ from one another.

In the text we considered only situations in which there was no dependence on space of the Boltzmann factor. In the next four problems we discuss situations into which spatial dependence enters.

**16.** ▮▮ Consider a general spatially dependent potential $V(\vec{r})$. **(a)** Find an expression for the distribution function $n(\vec{r})$ in this case. **(b)** Specialize to $V(\vec{r}) = mgz$, where $m$ is the mass of some molecule, $g$ is the local gravitational acceleration, and $z$ is the height above some reference point. Find an expression for $n(z)/n(0)$. The resulting expression for the variation in density with height is the **barometric formula. (c)** Use the barometric formula to find the height above sea level at which the air pressure is one-half its value at sea level, assuming that air can be treated as an ideal gas that is 80% $N_2$ with an atomic weight of 28 and 20% $O_2$ with an atomic weight of 32. You can take $T = 273K$.

**17.** ▮ The air inside a building is held at a constant temperature. A measurement shows that the pressure at the top floor is 99% that at the ground floor. What is the building's height? You will need the barometric formula of Problem 16.

**18.** ▮▮ By using the barometric formula (Problem 16), estimate the total number of molecules in Earth's atmosphere. For this purpose, you will want to know that the mean molecular weight $A = 28.9$; the acceleration due to gravity, $g = 978$ cm/s$^2$; and the mean temperature $T = 270K$. Earth's radius is 6,370 km. The number density at the surface can be found from the ideal-gas law with pressure $10^6$ dynes/cm$^2$ and temperature 293K.

**19.** ▮▮ When a particle of charge $q$ is placed in a uniform magnetic field $\vec{B}$ that points in the z-direction, it will move in a circle of radius $r = mv/qB$ as long as it has no initial motion in the z-direction. Consider a gas of charged particles, restricted to move in the $xy$-plane, in equilibrium at temperature $T$. Calculate $\langle r^2 \rangle$ as a function of temperature. What is the value of $\sqrt{\langle r^2 \rangle}$ if $B = 5$ tesla, $q = 1.6 \times 10^{-19}$ C, the mass of the particles (deuterons) is $3.3 \times 10^{-27}$ kg, and $T = 10^8K$. What is the value for electrons, for which $m = 0.9 \times 10^{-30}$ kg, under the same conditions?

**20.** ▮▮ Consider the two-state model presented in Section 12–3. Suppose that $E_1 < E_2$. Show that there is no temperature for which the fraction of needles in State 2 exceeds the fraction in State 1. As we shall see later, a laser requires the equivalent of $\langle n_2 \rangle > \langle n_1 \rangle$, where these are the average numbers of needles in the respective states, so that in a laser it is necessary to find a way around the Boltzmann distribution.

**21.** ▮▮ Consider the two-state model presented in Section 12–3. Find an explicit expression for the average energy of the system. Show that your result is a sensible one in the limit $T \to 0$ as well as in the limit $T \to \infty$.

**22.** ▮▮ Consider a system that has only three possible energies: $-E$, 0, and $+E$. (This system can actually be realized for certain atomic levels acting as magnets in an external magnetic field.) **(a)** Obtain an expression for the probability that, when the system is in thermal equilibrium at temperature $T$, the highest level is occupied. **(b)** Find the average energy $\langle E \rangle$ of the system. Plot $\langle E \rangle$ as a function of $T$. **(c)** If the heat capacity is given by $C = \partial \langle E \rangle / \partial T$, plot $C$ as a function of $T$ on the same graph as your plot of $\langle E \rangle$.

**23.** ▮▮ The "needle" of the two-state model presented in Section 12–3 has an energy of $-2$ eV when it is aligned with a certain exterior magnetic field and an energy of $+2$ eV when it is aligned opposite the direction of the field (antialigned). At what temperature are there twice as many needles aligned as antialigned?

**24.** ▮▮ In quantum mechanics a one-dimensional harmonic oscillator has discrete energies given by $E_n = \varepsilon(n + 1/2)$, with $n = 0, 1, 2, 3, \ldots$. Consider a gas of one-dimensional oscillators in equilibrium at a temperature $T$. Calculate the relative probability of finding an oscillator in the $n = 1$ state compared with finding one in the $n = 0$ state if $\varepsilon = 1.6 \times 10^{-21}$ J and the temperature is **(a)** 20K and **(b)** 300K.

**25.** ▮▮ Consider the vibrational motion of a diatomic molecule for which the interatomic forces produce a "spring constant" $\kappa$. (See Chapter 11 for a discussion of the vibrational levels of diatomic molecules.) **(a)** What, according to the Boltzmann distribution, is the probability that at temperature $T$ a given molecule from a gas is in the vibrational state labeled by the quantum number $n_V$? **(b)** Let $\kappa = 3.0 \times 10^9$ J/m². Estimate the temperature for which the excitation of vibrational modes become significant. **(c)** Find an expression for the average vibrational energy of the gas.

**26.** ▮▮ Consider a gas of $O_2$ molecules, for which $\hbar^2/2I = 1.78 \times 10^{-4}$ eV. What is the probability that a molecule is found in the $\ell = 1$ state, compared with the $\ell = 0$ state, at a temperature $T = 70$K? (See Chapter 11 for a discussion of the rotational levels of diatomic molecules.)

**27.** ▮▮ Assuming that the forces between atoms that form molecules can be approximated by elastic restoring forces, count the number of degrees of freedom, including those for rotations, for (see •Fig. 12–20) **(a)** $O_2$ molecules, **(b)** $H_2O$ molecules (bonds between hydrogen and oxygen only), **(c)** A hypothetical triangular molecule consisting of three identical atoms with identical bonds between each pair, and **(d)** the ammonia molecule $NH_3$ (with bonds between each pair of atoms).

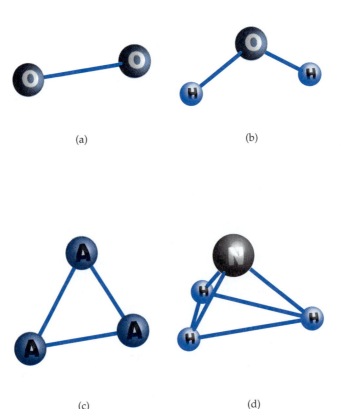

(a)

(b)

(c)

(d)

• **Figure 12–20** Problem 12–27.

**28.** ▮▮ A gas of one-dimensional oscillators for which the energy is $E = mv^2/2 + \kappa x^2/2$ is in equilibrium at a temperature $T$. If the position of a particular oscillator is given by $x(t) = A\cos(\omega t + \delta)$ and $v = dx/dt$, use the equipartition theorem to calculate $\langle A^2 \rangle$ in terms of $\kappa$, $m$, and $T$.

**29.** ▮▮ In order for a gas or, in the case of a metal, electrons to be classical, the mean separation $R$ of the molecules must be much larger than the de Broglie wavelength $\lambda = h/p$, where $p$ is the mean momentum. **(a)** Use the equipartition theorem to show that the preceding statement boils down to

$$ n^{-1/3} \gg \frac{h}{\sqrt{3mkT}}, $$

where $n$ is the particle density. **(b)** Does neon behave as a classical gas at a temperature of 100K and 1 atm pressure? At 10K and 1 atm pressure? At 1K and 1 atm pressure? **(c)** Treat electrons in a metal, one per ion, as a gas. Use estimated values for the density to learn the mean spacing. Is this gas classical at room temperature? At what temperature would it become classical?

**30.** ▮▮ As the temperature rises the Fermi–Dirac distribution spreads away from the steplike form it takes at $T = 0$. More particles in a gas of fermions take on energies greater than $E_F$, and fewer take on energies less than $E_F$. Show, however, that this spread is limited by demonstrating that there is no temperature for which the number of particles at $E_F/2$ is equal to half of the zero-temperature number at $E_F/2$.

**31.** ▮▮ What is the probability of finding an electron occupying the ground state in a box of length 0.1 nm on each side, at temperatures of **(a)** 10K, **(b)** 273K, and **(c)** 24,000K.

**32.** ▮▮ Consider a free-electron gas at a temperature $T$ such that $kT \ll E_F$. Write down an expression for the electron number density $N/V$ for electrons that have an energy in excess of $E_F$. Show, by making the change of variables $(E - E_F)/kT = x$, that the number density is proportional to $T$. Calculate an expression for $N/V$ under these circumstances, making use of the fact that

$$ \int_0^\infty \frac{dx}{(e^x + 1)} = \ln 2. \quad [\textit{Hint}: \text{In working out the integral over } E \text{ the integrand is} $$

such that $(x + E_F)^{1/2} \cong E_F^{1/2}.]$

**33.** ▮▮ Consider a free-electron gas in a metal for which the Fermi energy is 6.2 eV. What is the number density of electrons with energy above $E_F$ at **(a)** 4.0K, **(b)** 300K, and **(c)** 3,000K. Use the result of the previous problem.

**34.** ▮▮ By using the relation $\langle p \rangle = -(\partial \langle E \rangle/\partial V)_T$ at $T = 0$, find the pressure as a function of volume for an ideal gas of fermions at $T = 0$. (Here the subscript means that the temperature is held fixed as the derivative with respect to volume $V$ is taken.)

**35.** ▮▮ Consider a gas of nonrelativistic electrons confined to move in two dimensions. Obtain an expression for $N$ and for $\langle E \rangle$. Calculate these in the high-temperature limit. (*Hint*: The only change involves the counting of states, which alters the details of going to the continuum limit.)

**36.** ▮▮▮ Consider a gas of spin-1/2 massless Fermi–Dirac neutrinos in equilibrium at a temperature $T$. This gas can be described by a blackbody distribution, Eq. (12–58), in which the only change is to let the term $\exp(E/kT) - 1$ go into $\exp(E/kT) + 1$. Here we have assumed that the neutrino's chemical potential is zero, which is the case for the only situation that we know of in which the equilibrium just described was actually attained, namely the very early universe. (See Chapter 18.) Suppose that the neutrino has two spin orientations, so that the factor of 2 in the numerator of Eq. (12–58) remains the same. Compute the neutrino number density $n$ and the energy density $\rho$ for this distribution. Begin by arguing, on the basis of dimensional analysis, that these quantities will have the same dependence on $T$ as in the blackbody case. But the constant factors will be different. These can be found by doing numerical integrals, which, for this problem and for the blackbody, are of the form

$$\int_0^\infty dx \, x^n \frac{1}{\exp(x) \pm 1} = \int_0^\infty dx \, x^n \exp(-x) \frac{1}{1 \pm \exp(-x)},$$

where $n$ is 2 for the number density and 3 for the energy density. A rather accurate approximation to these integrals can be gotten by expanding the denominator in a geometric series and then integrating the elementary exponential integrals that result. Carry this calculation out for $n$ and for $\rho$ in this case—choosing the plus sign in the denominator—for the first three terms in the expansion and compare your result with the numerical values you can find in a standard table. Express $n$ and $\rho$ in terms of these numbers and the other dimensional and nondimensional factors in the expressions.

**37.** ▮▮ To compute the average photon number density in a blackbody at temperature $T$, we needed to integrate the expression

$$\int_0^\infty dx \, \frac{x^2}{e^x - 1}.$$

**(a)** Using the technique of Problem 36, show that the integral evaluates to approximately 2.404. **(b)** The universe, as we will discuss in more detail in Chapter 18, began with a "big bang" that has left a remnant in the present universe of a blackbody spectrum of radiation corresponding to a temperature of 2.73K. Find how many such blackbody photons there are on the average in any cubic centimeter of the present universe.

**38.** ▮▮ The power emitted by the radiation from a blackbody for an angular frequency interval $d\omega$ is given by

$$P(\omega) \, d\omega = \frac{\hbar}{4\pi c^2} \frac{\omega^3}{\exp(\beta\hbar\omega) - 1} \, d\omega.$$

The total power emitted per unit area of a blackbody is found by integrating this expression from zero to infinity. In Chapter 4 we showed by dimensional analysis that the power was proportional to $\sigma T^4$. Find $\sigma$. (You will need the results of a numerical integration; use the methods of Problem 36 to do this.)

**39.** ▮▮ Shortly after exploding, an atomic bomb produces a temperature of about $10^6$K—brighter than a thousand Suns!—over a sphere of about 10 cm in radius. What is the radiative power that this bomb would produce at a distance of 10 km?

**40.** ▮▮▮ The energy density of photons in a blackbody is given by the integral

$$\rho = \frac{1}{\pi^2(\hbar c)^3} \int_0^\infty dE \, \frac{E^3}{\exp(E/kT) - 1}.$$

By changing variables from $E$ to $x = E/kT$, show that $\rho$ is proportional to $T^4$. Use the technique of Problem 36 to evaluate the dimensionless integral that results when you change variables approximately, and compare your answer with what you find in a table of integrals. Express $\rho$ in terms of $n$, and find the energy density for photons left over from the big bang, which have a temperature of 2.73K.

**41.** ▮▮ If we call $P$ the pressure of a photon gas at equilibrium at temperature $T$ and $\rho$ the corresponding energy density, then it is possible to show, by using arguments from kinetic theory, that for this gas $P = \rho/3$. Express $\rho$ in terms of the photon number density $n$ and thereby derive the equation of state of the gas. How does that equation differ from the perfect-gas law for classical particles?

**42.** ▮▮▮ What is the density of a gas of $^{87}$Rb atoms (bosons) when it undergoes condensation at a critical temperature $T_c = 190 \times 10^{-9}$K. [*Hint*: Use the expression

$$\frac{N}{V} = \frac{2.31}{4\pi^2\hbar^3} \left(2mkT_c\right)^{3/2}.$$

relating the critical temperature at which a noninteracting boson gas undergoes condensation to the number density of atoms in the gas.

**43.** ▮▮ Consider a cube 10 cm on a side containing liquid helium, with a mass density $\rho = 0.145$ kg/m$^3$. **(a)** Given that a helium atom has a mass of roughly $6.7 \times 10^{-27}$ kg, find the possible energy levels for individual helium atoms in the cube. **(b)** Using the Boltzmann distribution, and assuming that you can neglect any interaction between the atoms, estimate the average occupation number at $T = 3$K for the two lowest single-atom levels in the cube. Note that there is some degeneracy to be taken into account for all but the lowest (ground) state.

# CHAPTER

# 13

# Decays, Radiation from Atoms, and Lasers

The Bohr postulates deal with the existence of states of quantized energy in the atom. We now understand that these are described by eigenfunctions of the Schrödinger equation and that the energies of the Bohr model are the associated energy eigenvalues. The Bohr postulates also state that energy conservation holds in the transition of an electron from one state to another with the spontaneous emission of radiation—so-called radiative transitions. This means that the emitted photon carries off the change in the electron's energy, which in turn specifies the wavelength of the photon. However, the Bohr postulates do *not* give the rate at which these spontaneous transitions occur.

In this chapter we approach this issue by making use of the correspondence principle, which leads to a quantum mechanical description of radiative transitions and of the radiation itself. We will also discuss induced transitions, which are transitions that take place in the presence of external electromagnetic fields. We finish with the most important practical application of these ideas: lasers.

## 13–1  Decay Rates and Exponential Decay

When a large number of quantum systems not in their quantum mechanical ground states—atoms or nuclei in excited states provide excellent examples—are observed during a given time interval, a certain number of them decay. The number $N(t)$ that remain in their excited states at any given time decreases with time, and the rate $dN(t)/dt$ of decrease is proportional to the number present at that moment. The proportionality constant is the **decay rate** $R$, and the rate of decrease is

$$\frac{dN(t)}{dt} = -RN(t). \qquad (13–1)$$

The minus sign appears because the number of electrons in state $n$ is *decreasing*. The solution of Eq. (13–1) is

$$N(t) = N_0 \exp(-Rt). \qquad (13–2)$$

The consequences of this **exponential decay law**, pictured in •Fig. 13–1, can be viewed in a number of ways. In particular, the time $\tau$ required for the number of particles in the excited state to decrease from any initial value by a factor of

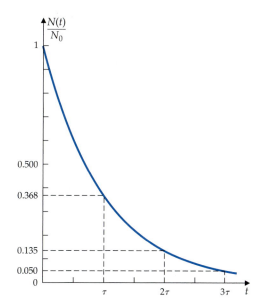

• **Figure 13–1** A plot of $N(t)/N(0)$ as a function of time in units of the lifetime $\tau$. $N(t)$ is the number of undecayed states at time $t$.

$e \cong 2.72$ is called the **lifetime**, or *mean life*, of the state. This occurs when the magnitude of the exponent in Eq. (13–2) is unity—that is, when $R\tau = 1$, so that

$$\tau = 1/R. \qquad (13\text{–}3)$$

Rather than the lifetime, one also uses the *half-life* $\tau_{1/2}$, the time for half of the sample to decay. As you can easily show (see Problem 1 at the end of the chapter), the half-life is related to the lifetime according to $\tau_{1/2} = (\ln 2)\tau$.

## Exponential Decay

The exponential decay law [Eq. (13–2)] for the members of a population appears in many fields. It describes, for example, the decrease in the number of fish in a lake when there is a decrease in the supply of nutrients, as well as the number of survivors of a population ill with tuberculosis. There are, however, some important differences between these classical examples and examples in quantum systems such as electrons in excited states or radioactive nuclei. First, in the classical examples, the decay rate can be manipulated. The rate depends on the behavior of individuals: Smokers or motorcycle riders without helmets will have a higher rate of dying than the population as a whole, and a change in behavior can change that rate. For the electrons in an excited atomic state, we can change the rate only in very limited ways; if, for example, the atom is accompanied by many other atoms, then collisions can hasten the rate. If the atom is isolated, there is nothing we can do to change the rate. We may phrase this notion differently: We believe that, *in principle*, a sufficiently detailed analysis of an individual's health could tell us when and for what reason that person would die without intervention, because *there are physiological mechanisms within individuals that make a kind of internal clock*. In contrast, we have found no evidence that there is an internal clock that determines when a *particular* electron in an excited state will decay. In fact, according to quantum mechanics *there is no such clock*. All we can know is the *probability* that the electron will decay.

There is a second major difference as well. The exponential decay law states that if we start with a set of excited atoms and then look after a time $\tau = 1/R$, the set will have only $1/2.7$ as many excited atoms. *Each of the atoms that remains after this time will have a lifetime of $\tau = 1/R$.* If we look after 10 lifetimes, we will see only a fraction $e^{-10} \cong 4.5 \times 10^{-5}$ of the original batch of excited atoms, but those excited atoms that are left continue to have just the same lifetime as before; in other words, they have the same chance to survive for another lifetime as those that decayed earlier. This situation is quite different from the description of how humans age. The average person's lifetime may be 75 years, and some people will live for 100 years. But once we have identified a small subpopulation that has lived 100 years, we can be sure that they only have a few more years to go; they do not have an additional average lifetime of 75 years!

This peculiarity of quantum mechanics still disturbs some people who believe that quantum mechanics is not a complete theory—that what is missing is a knowledge of some **hidden variables** that tell us about individual electrons. Such variables would supposedly describe the "clock" or other mechanisms that "explain" the peculiar features of quantum mechanics. With the continued absence of evidence for hidden variables and, indeed, the spectacular success of quantum mechanics in domains that were not accessible to experiment in the days of its discovery, the belief in hidden variables has largely disappeared among practicing physicists. There remains, however, an active discussion of these matters in the literature, especially among philosophers.

**Example 13–1**   The element uranium has isotopes—nuclei with differing numbers of neutrons. The fact that it is the chemical element uranium means that the number of protons is the same, 92, for all these isotopes. This number we label as Z. In addition, individual isotopes are labeled with the mass number A: the number of neutrons plus protons. The number of neutrons is different from one isotope to another. You can figure out how many there are in a given isotope by taking the difference between A and Z. Uranium isotopes are unstable and decay according to quantum mechanical rules. The lifetime of the isotope $^{238}$U is $6.0 \times 10^9$ yr, while that of $^{235}$U is $1.0 \times 10^9$ yr. Given that there is presently 140 times as much $^{238}$U as $^{235}$U, and assuming that when Earth was formed these isotopes were equally abundant, estimate Earth's age.

**Solution**   If we let $t = 0$ denote the time of Earth's formation, then the numbers of the two isotopes as a function of time are

$$N_{238}(t) = N_0 \exp\left(-\frac{t}{\tau_{238}}\right)$$

and

$$N_{235}(t) = N_0 \exp\left(-\frac{t}{\tau_{235}}\right),$$

where $N_0$ is the same for both species. If we label Earth's age as $T$, then the current conditions give

$$N_{238}(T) = N_0 \exp\left(-\frac{T}{\tau_{238}}\right) = 140 N_{235}(T) = 140 N_0 \exp\left(-\frac{T}{\tau_{235}}\right),$$

or

$$\exp\left(-\frac{T}{\tau_{238}}\right) = 140 \exp\left(-\frac{T}{\tau_{238}}\right).$$

If we now take the natural logarithm of each side, we get

$$-T/\tau_{238} = \ln(140) + (-T/\tau_{238}),$$

which is a linear equation for $T$. Solving, we have

$$T = \frac{\ln(140)}{\dfrac{1}{\tau_{235}} - \dfrac{1}{\tau_{238}}} = \frac{\tau_{235}\tau_{238}\ln(140)}{\tau_{238} - \tau_{235}} = \frac{(1.0 \times 10^9 \text{ yr})(6.0 \times 10^9 \text{ yr})(4.9)}{(6.0 \times 10^9 \text{ yr}) - (1.0 \times 10^9 \text{ yr})} = 6.0 \times 10^9 \text{ yr.}$$

$^{235}$U has a shorter lifetime than $^{238}$U, decaying more quickly, and as time passes there is relatively more $^{238}$U than $^{235}$U.

---

## 13–2 The Ingredients of a Quantum Calculation

Rather than giving a full quantum mechanical treatment of the rate of transitions between atomic states, we outline here a classical calculation and then indicate how things change in moving to quantum mechanics. We will be using the correspondence principle discussed in Section 5–4.

In classical physics, charges that accelerate emit radiation. Consider an electron, charge $-e$ and mass $m_e$, subject to a central spring force. Let us write the spring constant in the form $m_e\omega^2$. Then the equation of motion of the electron in the absence of an external electric or magnetic field takes the familiar form

$$\frac{d^2\vec{r}}{dt^2} + \omega^2\vec{r} = 0. \tag{13–4}$$

This equation represents a variety of motions; for example, the solution $x = r_0\cos\omega t, y = r_0\sin\omega t, z = 0$ represents circular motion in the $xy$-plane with radius $r_0$ and angular speed $\omega$.

An electron that undergoes that motion subject to a central spring force will radiate light with frequency $\omega$. More generally, any accelerating charge emits radiation, the instantaneous rate of which, or *instantaneous power P*, is given by a formula due to Joseph Larmor. This formula, which is studied in courses on advanced electricity and magnetism, is

$$P = \frac{2}{3}\frac{e^2}{4\pi\varepsilon_0}\frac{a^2}{c^3}, \tag{13–5}$$

where $a$ is the instantaneous acceleration of the charge. For the harmonic motion described in the previous paragraph, Eq. (13–5) becomes

$$P = \frac{2}{3}\frac{e^2}{4\pi\varepsilon_0}\frac{\omega^4}{c^3}\vec{r}^2. \tag{13–6}$$

We now apply the correspondence principle in the following form: Consider the emission of a single photon of energy $\hbar\omega$ in an average time $\tau$. The power emitted in this situation is the energy emitted per unit time, $\hbar\omega/\tau$. We equate this power to the classical expression for the power given by Eq. (13–6). This immediately yields an expression for the inverse of the average emission time $\tau$:

$$\frac{1}{\tau} = \frac{P}{\hbar\omega} = \frac{2}{3}\frac{e^2}{4\pi\varepsilon_0\hbar}\frac{\omega^3}{c^3}\vec{r}^2. \tag{13–7}$$

We can interpret the quantity $1/\tau$ as the decay rate $R$. It is convenient to write Eq. (13–7) in a form that uses the fine-structure constant $\alpha = e^2/(4\pi\varepsilon_0\hbar c)$, so that

$$R = \frac{2}{3}\alpha\frac{\omega^3}{c^2}\vec{r}^2. \tag{13–8}$$

Another form for this rate can be obtained by using the fact that the radiation emitted has a wavelength $\lambda = c/f = 2\pi c/\omega$, whereupon

$$R = \frac{2}{3}\alpha\left(\frac{2\pi c}{\lambda}\right)^3\frac{\vec{r}^2}{c^2} = \frac{16\pi^3}{3}\alpha f\frac{\vec{r}^2}{\lambda^2}. \tag{13–9}$$

Although the expression for $P$ in Eq. (13–6) and the expression for the rate in Eq. (13–9) were obtained by considering a radiating electron in harmonic motion, both expressions are generally true, provided that we use the $\vec{\mathbf{r}}$ and $\omega$ (or $\lambda$) applicable to whatever system we are considering. Thus, we may use Eq. (13–9) to estimate the decay rate for electrons in hydrogen.

---

**Example 13–2**   Estimate the rate of transition for an electron making the jump from the $2p$ to the $1s$ state in hydrogen by using the orders of magnitude $\vec{\mathbf{r}}^2 = a_0^2$, where $a_0$ is the Bohr radius.

**Solution**   The rate is given by Eq. (13–8), into which we substitute the Bohr radius

$$a_0 = \frac{\hbar}{m_e c \alpha}.$$

In addition, the angular frequency is given by

$$\omega = \frac{E_2 - E_1}{\hbar} = \frac{m_e c^2 \alpha^2}{2\hbar}\left[\left(-\frac{1}{4}\right) - (-1)\right] = \frac{3 m_e c^2 \alpha^2}{8\hbar}.$$

In both of these equations, we have used the results of Chapters 5 and 9. We can now substitute into Eq. (13–8) to obtain

$$R = \frac{2}{3}\,\alpha\,\frac{(\hbar\omega)^3}{\hbar^3 c^2}\,a_0^2 = \frac{2}{3}\,\alpha\,\frac{(3 m_e c^2 \alpha^2/8)^3}{\hbar^3 c^2}\left(\frac{\hbar}{m_e c \alpha}\right)^2 = \frac{2}{3}\left(\frac{3}{8}\right)^3 \alpha^5 \frac{m_e c^2}{\hbar}$$

$$= \frac{2}{3}\left(\frac{3}{8}\right)^3\left(\frac{1}{137}\right)^5 \frac{(0.9 \times 10^{-30}\ \text{kg})(3 \times 10^8\ \text{m/s})^2}{(1.05 \times 10^{-34}\ \text{J} \cdot \text{s})} = 5.6 \times 10^8\ \text{s}^{-1}.$$

---

## Quantum Mechanical Expression for the Transition Rate

The most important change in going from classical to quantum physics in Eq. (13–8) is a rather deep one. Since in quantum physics our electron has no position in the classical sense, we must find a quantity that replaces the position in Eq. (13–8). One way to do this would be with the use of an expectation value—that is, $\vec{\mathbf{r}} \to \int \psi^*(\vec{\mathbf{r}})\vec{\mathbf{r}}\psi(\vec{\mathbf{r}})d^3\vec{\mathbf{r}}$, where $\psi$ is either the initial or the final wave function. But this expression cannot be quite right, because the transition should somehow involve *both* the initial and the final wave function. A rigorous treatment in quantum mechanics shows that the correct way to proceed is by the replacement

$$\vec{\mathbf{r}} \to \int \psi_f^*(\vec{\mathbf{r}})\vec{\mathbf{r}}\psi_i(\vec{\mathbf{r}})d^3\vec{\mathbf{r}}. \tag{13–10}$$

Here $\psi_f$ is the wave function of the final state of the decaying system (e.g., the $1s$ state of the electron in the hydrogen of Example 13–2), and $\psi_i$ is the wave function of the initial state of the decaying system (e.g., the $2p$ state of the electron in the hydrogen of Example 13–2). The quantum mechanical expression for the transition rate is given by Eq. (13–8) with $\vec{\mathbf{r}}$ replaced by Eq. (13–10). We can continue to refer to $\vec{\mathbf{r}}$, but we must calculate it in a quantum mechanical fashion.

In Eq. (13–8) $\vec{\mathbf{r}}$ occurs in combination with $e$. The product $e\vec{\mathbf{r}}$ in classical electromagnetism is associated with a dipole moment, and the radiation emitted is called *electric dipole radiation*, both classically and quantum mechanically.

## Selection Rules

In our discussion of the Bohr atom in Chapter 7, we used the correspondence principle to arrive at a **selection rule**. The rule stated that in atomic transitions the angular-momentum quantum number could change by one unit—in the language of quantum mechanics, that $\Delta\ell = 1$, or that the angular momentum of the electron changes by one unit of $\hbar$. This selection rule emerges quite naturally from Eq. (13–10). The wave functions have an angular dependence, through the spherical harmonics (see Chapter 9), and some of the angular integrals are automatically zero.

We can easily see that Eq. (13–10) implies that there are no transitions between two $\ell = 0$ states. The wave functions of atomic states with $\ell = 0$ have no angular dependence. Then, picking out the $x$-component of $\vec{r}$—$r \sin\theta \cos\varphi$—we see that the integral in Eq. (13–10) in this case takes the form

$$\int_0^\infty r^2\, dr \int_0^\pi \sin\theta\, d\theta \int_0^{2\pi} d\varphi\, R_f^*(r) R_i(r)(r \sin\theta \cos\varphi).$$

The integration over $\varphi$ is zero. The same type of reasoning applies for the other components of $\vec{r}$.

---

**Example 13–3** Show that there are no transitions between an initial state of $\ell = 0$, where $\psi_i(\vec{r}) = R_i(r)$, and a final state of $\ell = 2$, where $\psi_f(\vec{r}) = R_f(r)(3\cos^2\theta - 1)$.

**Solution** The relevant integral is

$$\int_0^\infty r^2\, dr \int_0^\pi \sin\theta\, d\theta \int_0^{2\pi} d\varphi\, R_f^*(r) R_i(r)(3\cos^2\theta - 1)$$

$$(\vec{i} r \sin\theta \cos\varphi + \vec{j} r \sin\theta \sin\varphi + \vec{k} r \cos\theta).$$

The pieces of the integral associated with the $x$- and $y$-components vanish, because the $\varphi$ part of the integration gives zero, as in the text above. As for the $z$-component, the $\theta$ part of the integration is of the form $\int_0^\pi d\theta \sin\theta (3\cos^2\theta - 1)\cos\theta$, which, with the change of variable $\cos\theta = u$, becomes

$$\int_{-1}^{+1} u(3u^2 - 1)\, du = \int_{-1}^{+1} (3u^3 - u)\, du = \left[\frac{3u^4}{4} - \frac{u^2}{2}\right]_{-1}^{+1} = 0.$$

Thus the entire integral vanishes and there is no transition.

---

More generally, by using general properties of the spherical harmonics we can show that the only nonzero integrals involving $\vec{r}$ are those with

$$\Delta\ell = |\ell_{final} - \ell_{initial}| = 1. \tag{13–11}$$

In particular, the only transitions to the ground state of hydrogen, which is the state $n = 1$, $\ell = 0$, are from states with $n > 1$ and $\ell = 1$.

We can also note that what appears between $\psi_f$ and $\psi_i$ does not involve the spin. Thus the initial and final spin states must be the same: The electron spin cannot change its orientation as the electron emits radiation.

This selection rule is not, generally speaking, absolute: Transitions can still occur. When the integral in Eq. (13–10) vanishes, small correction terms that involve higher powers of $(r/\lambda)^2$ become important. A radiative transition with $\Delta\ell = 2$ is possible, but the rate is a factor $(Z\alpha)^2$—some $10^{-4}$ times smaller for hydrogen and very much smaller than for a transition with $\Delta\ell = 1$. The reason is that Eq. (13–10) is only approximate and has correction terms that change what is sandwiched between $\psi_f$ and $\psi_i$. These terms are of magnitude $r/\lambda$, where $\lambda$ is the wavelength of the emitted radiation.

Alternatively, the selection rule may be violated because the integral of Eq. (13–10) is zero for the initial and final states $A$ and $B$, while it is nonzero between states $A$ and $C$ and between states $C$ and $B$. Thus a transition $(n = 3, \ell = 0)$ to $(n = 1, \ell = 0)$ could occur through the chain $(n = 3, \ell = 0)$ $\rightarrow (n = 2, \ell = 1) \rightarrow (n = 1, \ell = 0)$. In this chain, two photons are emitted. Such rates are approximately equal to the square of the rate for an allowed single-photon rate.

## 13–3 Induced Transitions

Albert Einstein wrote an important paper on radiation in 1916, before the discovery of quantum mechanics, but after the Bohr model and the Planck hypothesis had taken root. Einstein pointed out that in addition to the **spontaneous** emission of radiation that accompanies the transition of an electron from a higher energy level to a lower one—the radiation we discussed in the first two sections of this chapter—there is **induced** or **stimulated** radiation, in which radiation already present in the vicinity of the atom enhances the transition rate. We adapt his beautiful, simple, and powerful approach to calculate the rate of induced radiation.

Consider a cavity containing blackbody radiation, of the type described in Chapter 4. The atoms of the cavity walls interact with the radiation; we make the simplifying assumption that these atoms have just two nondegenerate energy levels, $E_1$ and $E_0$, with $E_1 > E_0$, as in •Fig. 13–2. Now, how many atoms are there at each level? That is a question of temperature. According to Chapter 12, if the system is in equilibrium at a temperature $T$, the number $N_1$ of atoms with energy $E_1$ and the number $N_0$ with energy $E_0$ are related by the Boltzmann factor

$$\frac{N_0}{N_1} = \frac{\exp(-E_0/kT)}{\exp(-E_1/kT)} = \exp[(E_1 - E_0)/kT] = \exp(hf/kT). \quad (13\text{–}12)$$

The thermal equilibrium is dynamic: We have transitions *up* from $E_0$ to $E_1$, in which a photon with energy $hf$ is absorbed from the field in the cavity, and we have transitions *down* from $E_1$ to $E_0$, in which a photon with energy $hf$ is emitted. Let's see how these transitions match up. The rate $R_{up}$ of transitions *up* from $E_0$ to $E_1$ is proportional to the number of atoms in the state $E_0$ and also to

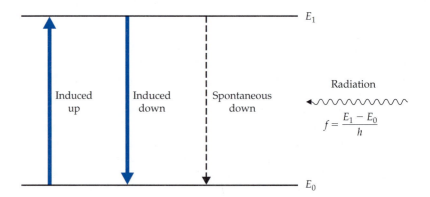

• **Figure 13–2** A two-level system with transitions between the levels. We refer to the two levels as the "up" and "down" states.

the intensity of the radiation in the cavity with the frequency of the light that must be absorbed—equivalently, to the number of photons present with frequency $f$. We thus write

$$R_{\text{up}} = B_{\text{up}} N_0 u(f, T). \tag{13–13}$$

Here, as in Section 4–4, $u(f, T)$ is the energy density for radiation of frequency $f$ in the cavity, and $B_{\text{up}}$ is a proportionality constant—assumed to be independent of temperature—that we want to determine. (Shortly we'll see that the assumption about the temperature independence is justified.)

The rate $R_{\text{down}}$ of transitions *down* from $E_1$ to $E_0$ is a sum of two parts. Each part will be proportional to the number of atoms $N_1$ in the state with energy $E_1$. One part is the rate $R_{\text{spon}}$ of spontaneous transitions, the rate we discussed in the first part of this chapter, which we shall here call $N_1 A$. This term is known already. The second part of $R_{\text{down}}$ is the part that Einstein postulated, a term in the overall rate that is **induced** by the presence of the radiation in the cavity and so is proportional to $u(f, T)$. We label this second part $R_{\text{induced}} = N_1 B_{\text{down}} u(f, T)$. Thus the overall rate for transitions from $E_1$ to $E_0$ is

$$R_{\text{down}} = R_{\text{induced}} + R_{\text{spon}} = N_1 \big( B_{\text{down}} u(f, T) + A \big). \tag{13–14}$$

Again, we are trying to find $B_{\text{down}}$, and again, we assume temperature independence.

In equilibrium there are as many transitions up as down, so that $R_{\text{up}} = R_{\text{down}}$. Then if we solve Eq. (13–13) for $N_0$ and Eq. (13–14) for $N_1$ and take the ratio of the two, we find that

$$\frac{N_0}{N_1} = \frac{u(f, T) B_{\text{down}} + A}{u(f, T) B_{\text{up}}}. \tag{13–15}$$

Equation (13–12) shows that the left-hand side is equal to $\exp(hf/kT)$. Then we can rearrange terms to obtain

$$A = u(f, T) \big[ B_{\text{up}} \exp(hf/kT) - B_{\text{down}} \big]. \tag{13–16}$$

At this point we can make several inferences. First, as we saw in our discussion of the Planck formula in Section 4–4, in the limit of high temperatures $u(f, T) \to (8\pi f^2/c^3)kT$, while the exponential factor $\exp(hf/kT) \to 1$ in this limit. Thus the right-hand side of Eq. (13–16) has the leading large-$T$ behavior $T(B_{\text{up}} - B_{\text{down}})$. It also has a nonleading term that behaves as a constant at large $T$. But $A$ is independent of temperature, so that the leading term must be equal to zero, and that can happen only if

$$B_{\text{up}} = B_{\text{down}} \equiv B. \tag{13–17}$$

Accordingly, we now need to determine only the single term $B$. We next use Eq. (13–17) in Eq. (13–16) to find $A = Bu(f, T)\big[\exp(hf/kT) - 1\big]$ or, equivalently,

$$\frac{B}{A} = \frac{1}{u(f, T)\big(e^{hf/kT} - 1\big)}. \tag{13–18}$$

Now we can use the full Planck formula, Eq. (4(19)):

$$u(f, T) = \frac{8\pi f^2}{c^3} \frac{hf}{\exp(hf/kT) - 1}.$$

This gives us immediately

$$\frac{B}{A} = \frac{c^3}{8\pi hf^3}. \tag{13–19}$$

This equation, which shows that $B$ is independent of temperature, justifies our earlier assumption—in other words, the result is internally consistent. And, given that $A$ is known already, it means that we have determined $B$.

With this result in hand, we can compute the rate $R_{\text{down}}$ per atom of the transition from the level $E_1$ to $E_0$ with the emission of a photon. From Eq. (13–14), we have

$$\frac{R_{\text{down}}}{N_1} = Bu(f, T) + A = A\left(1 + \frac{1}{e^{hf/kT} - 1}\right). \qquad (13\text{–}20)$$

Now the last factor is related to the average number of photons of frequency $f$ per unit volume in a blackbody cavity. Indeed, we calculated this quantity in Chapter 4; it is given by Eq. (4–16), $\langle n \rangle = \left(\exp(hf/kT) - 1\right)^{-1}$. Therefore we can rewrite the rate of induced emission per atom as

$$\frac{R_{\text{down}}}{N_1} = A\left(1 + \langle n \rangle\right) = \left(\frac{R_{\text{spon}}}{N_1}\right)\left(1 + \langle n \rangle\right), \qquad (13\text{–}21)$$

where we have interpreted the quantity $A$ as the rate of spontaneous decay per atom, $R_{\text{spon}}/N_1$. Equation (13–21) shows that the transition rate for emission is *enhanced* by the number of photons with the transition frequency already present, plus one.

The same type of argument gives the rate of absorption of photons from the cavity, viz.,

$$\frac{R_{\text{up}}}{N_0} = \left[R_{\text{spon}}/N_1\right]\langle n \rangle. \qquad (13\text{–}22)$$

Perhaps this result is less surprising than the result for $R_{\text{down}}$; we would expect the rate at which photons are absorbed to depend on the number of photons available to be absorbed.

Although Einstein's calculation dealt with blackbody radiation in a cavity, the proportionality of $R_{\text{down}}$ to $n(f) + 1$ generally holds even if we do not have a photon gas in equilibrium. Indeed, the factor of $n + 1$ appeared in our derivation of the Bose–Einstein distribution in Section 12–6. As we shall see in the next section, this fact has some powerful consequences for technology.

**Example 13–4**    As a result of the evolutionary history of the universe, it has a background of blackbody radiation in all directions whose characteristic temperature is 2.7K. In Chapter 4 we looked at this radiation and found that its distribution has a maximum at a certain frequency. Suppose a certain molecule has two levels whose energies differ by an amount corresponding to that frequency. **(a)** What is the maximum frequency and the corresponding energy? **(b)** By what factor does the rate of transitions between the postulated levels of our molecule differ from the rate of spontaneous transition from the upper to the lower level?

**Solution**    **(a)** In the discussion following Eq. (4–20) we found that the blackbody spectrum has a maximum at a frequency

$$f_{\text{max}} = \left[5.93 \times 10^{10}/\text{K}\cdot\text{s}\right]T.$$

(This is the Wien displacement law.) In this case $T = 2.7$K, so that

$$f_{\text{max}} = \left[5.93 \times 10^{10}/\text{K}\cdot\text{s}\right](2.7\text{K}) = 1.6 \times 10^{11} \text{ Hz}.$$

The corresponding energy is

$$E = hf_{\text{max}} = \left(6.63 \times 10^{-34} \text{ J}\cdot\text{s}\right)\left(1.6 \times 10^{11} \text{ Hz}\right) = \left(1.06 \times 10^{-22} \text{ J}\right)/\left(1.6 \times 10^{-19} \text{ J/eV}\right)$$

$$= 6.63 \times 10^{-4} \text{ eV}.$$

**(b)** The rate of induced transitions up and down is $B$, while that of spontaneous transitions is $A$. From Eq. (13–19), we have

$$\frac{B}{A} = \frac{c^3}{8\pi h f^3} = \frac{(3 \times 10^8 \text{ m/s})^3}{8\pi(6.64 \times 10^{-34}\text{J}\cdot\text{s})(1.6 \times 10^{11} \text{ s}^{-1})^3} = 3.95 \times 10^{23}.$$

Even at the low temperature of 2.7K, the enhancement due to the presence of the blackbody radiation is considerable.

## 13–4  Lasers

The **laser**—the name is an acronym for **l**ight **a**mplification by **s**timulated **e**mission of **r**adiation—is a device that takes advantage of the existence of stimulated emission to emit a coherent, monochromatic (i.e., single-frequency) beam of electromagnetic radiation. Here is how such a device could work: Consider a standing wave of frequency $f$ of electromagnetic field along the axis of a cavity also containing a gas (•Fig. 13–3). Since as we know (see Chapter 1) a standing wave can be decomposed into two oppositely moving traveling waves, this situation can be translated into photon language to say that we have a coherent beam of photons of energy $hf$ moving back and forth along the axis. If the gas is composed of atoms with two energy levels[1] $E_1$ and $E_0$, with $E_1 > E_0$, such that $E_1 - E_0 = hf$, then one could imagine that the rate of stimulated decay of the atoms from the state $E_1$ to the state $E_0$ is greatly enhanced by the factor $n(f)$, the number of photons of frequency $f$. This would effectively *amplify* the field in the cavity, with the amplification taking the form of an avalanche: Once one atom has decayed to produce another photon, that photon is available to stimulate the decay of the next atom even more strongly (•Fig. 13–4). Indeed, one could start with no radiation at all in the cavity, and the first atom that decays from the state $E_1$ would provide the first step in the avalanche. The amplification involves two important side benefits:

1. The amplified field (or collection of photons) has the same frequency and directionality as the stimulating field (photons); that is, it is along the axis of the cavity.
2. The amplified field is in phase with the stimulating field, so that we can speak of a *coherent beam* of electromagnetic radiation in the cavity. (See Section 1–4 for more on the meaning of coherence. A coherent beam of photons is quite different from the set of photons emitted by a lightbulb, or even by a lightbulb with a filter so that only a fixed frequency escapes.)

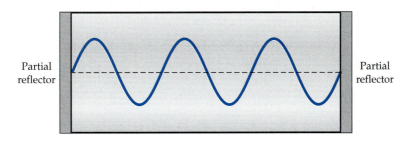

Partial reflector                    Partial reflector

• **Figure 13–3**  A standing wave is set up in a tubular cavity. The wave can be interpreted as the coherent—meaning that the two waves have fixed relative phase—superposition of two traveling waves, one moving to the right and the other to the left. The relation between the tube length $L$ and the wavelength $\lambda$ is $L = n\lambda/2$.

---

[1] Atoms do, of course, have many levels, but we need only consider the two levels for which the energy difference matches the photon frequency of interest to us.

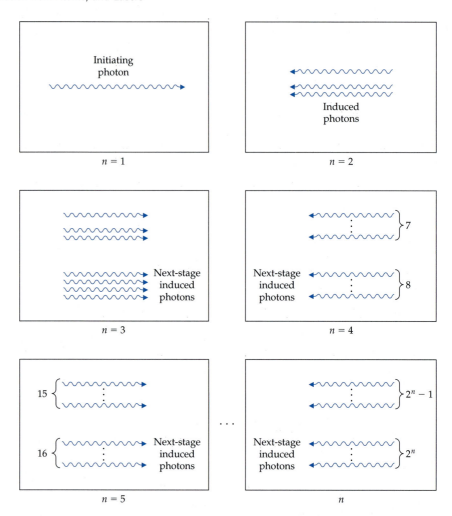

• **Figure 13–4** How a single photon in a cavity can, in one atomic lifetime, induce the emission of two photons of the same frequency and phase. These three photons induce the emission of four photons in a lifetime, and the resulting seven photons induce eight more, and so on, until after $n$ lifetimes there are $2^n - 1$ photons present.

Note that lasers can work either with a collection of atoms in a gas such as we have described, in which case the levels in question are atomic levels, or with the ensemble of atoms in a solid, in which case the levels in question are those of the solid. (We have not yet described how the energy levels of a solid are structured, but that information is not crucial for the processes described in this chapter.) In what follows, we concentrate mainly on atoms in gases.

The only problem with the scheme we have described is that it won't work! We learn from the Boltzmann formula that in thermal equilibrium the number of atoms in the states $E_1$ and $E_0$ are related by

$$\frac{N_1}{N_0} = \exp\left(-\frac{hf}{kT}\right). \tag{13–23}$$

[This equation is just the reciprocal of Eq. (13–12).] Equation (13–23) shows that in thermal equilibrium there are more atoms in the ground state than in the excited state, a result you should find hardly surprising. Now the lower states absorb photons of frequency $f$ (leading to transitions to the upper state), and the rate of absorption is proportional to the number $N_0$ of atoms in the ground state. The emission rate, on the other hand, is the same per atom, but is proportional to the number $N_1$ of excited atoms. This means that there will be more absorption than emission and hence that there will be a net attenuation of the field. To have more photons created than absorbed, we must have $N_1 > N_0$; that is, we must create a **population inversion**, in which atoms in the higher energy state are present in greater numbers than atoms in the lower energy state. Once this is achieved, we have the possibility of creating a laser.

## Creating a Population Inversion

To create a population inversion, we must construct a way of working outside of the equilibrium conditions. The simplest, although not necessarily the most efficient, way to evade the equilibrium condition of Eq. (13–23) is to use a three-level system, with energies $E_0$, $E_1$, and $E_2$ such that $E_2 > E_1 > E_0$ (•Fig. 13–5), and to put an excess of atoms into the level with $E_1$. The state with energy $E_2$ must have the twin properties that ($i$) it can be excited through a "pumping" process (to be described shortly) from the initially heavily occupied state $E_0$ and ($ii$) it can decay *rapidly* into the state $E_1$. The photons of frequency $f_{21} = (E_2 - E_1)/h$ are allowed to escape—a window can be provided that is transparent to this frequency—so that there is not too much absorption with accompanying transitions back up to the $E_2$ level. Furthermore, one chooses the state with energy $E_1$ to be *metastable*; that is, it is a state with a long lifetime. This allows the possibility of accumulating many atoms with energy $E_1$ before a spontaneously emitted photon of frequency $f_{10} = (E_1 - E_0)/h$ starts the rapid avalanche of coherent photons that form the laser with frequency $f_{10}$. Because the transitions between $E_1$ and $E_0$ are suppressed by selection rules (this is what makes the state with energy $E_1$ metastable), not many atoms in the lowest state are excited back to the state with $E_1$, and thus a population inversion is possible.

The problem with a three-level setup is that if the final state with energy $E_0$ is the ground state, then ultimately all the decays lead to an increase in $N_0$, and a lot of power must be expended to keep $N_0$ smaller than $N_1$. Another way of putting this is that every photon emitted by atoms in the state with energy $E_1$ reduces the imbalance between $N_1$ and $N_0$ by two: $N_1$ decreases by one and $N_0$ increases by one.

A way of getting around this problem is to use *four* atomic levels with appropriate properties (•Fig. 13–6). Atoms from the ground state $E_0$ are excited into the topmost level $E_3$. There is also a lower lying level $E_2$ such that the transitions from $E_3$ to $E_2$ are *rapid*. The photons of frequency $f_{32} = (E_3 - E_2)/h$ are allowed to escape. In this way, there is an efficient mechanism for putting atoms in the state $E_2$. The transitions from the level $E_2$ to the next lower level $E_1$, as well as to the ground state $E_0$, should be strongly inhibited by selection rules, so that there is a buildup of atoms in the state $E_2$. The level $E_1$ should have quantum numbers such that the transitions from it to the ground state $E_0$ are very rapid. The laser transition occurs between levels $E_2$ and $E_1$, which "empties" rapidly by transitions to the state $E_0$. Thus a population inversion between $E_2$

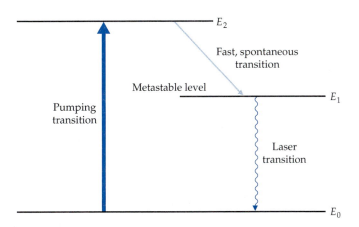

• **Figure 13–5**  A three-level scheme for creating a population inversion for a laser. The metastable level lies between the ground state and an excited state of energy $E_2$, which has been populated with a "pumping" mechanism from the ground state.

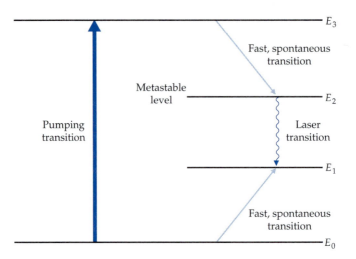

• **Figure 13–6** A four-level laser. The states with energies $E_2$ and $E_1$ between which the laser transition takes place lie between the ground state $E_0$ and the state $E_3$ to which there is pumping from the ground state. The state labeled $E_2$ is metastable. This scheme eliminates the buildup of filled states in the level $E_1$ at the bottom of the laser transition. This buildup makes the three-level scheme inefficient.

and $E_1$ is easily created and maintained: $N_1$ is always very small. In effect, every photon emitted with frequency $f_{21} = (E_2 - E_1)/h$ decreases $N_2$, but does not increase $N_1$, since level $E_1$ empties very rapidly.

### Pumping Schemes

The schemes described in the preceding section all involve what is referred to as *pumping*. This is some process that allows us to work around the thermal equilibrium situation by exciting atoms from the ground state to, for example, the highest levels in the three- or four-level methods of creating population inversions and, hence, laser transitions. Commonly used pumping techniques fall into two classes:

*Optical Pumping* In this scheme, the atoms that are to be excited are subject to a powerful source of light, which is absorbed by the atoms in the ground state. The source is generally a strong (incoherent) lamp. This method works best when the lasing medium is a solid—for example, in the so-called ruby laser.

*Electrical Pumping* In this scheme, the atoms are excited by nonoptical means. For instance, one can send an electrical discharge through the gas of atoms to produce ions and electrons. The electrons are accelerated by the electric field and, in collisions with the neutral atoms in the tube, excite those atoms through the reaction

$$e + A \rightarrow e' + A^*,$$

where the prime indicates that the electron has lost energy and the asterisk indicates that the atom is excited. The loss of kinetic energy of the electron balances the excitation energy of the atom. A closely related process, one that is used in the helium–neon (He–Ne) laser, is known as *resonant energy transfer*. Excited metastable helium atoms are produced by inelastic collisions with electrons that have been accelerated as just described. The excited state of helium is close in energy to a group of excited states of neon (•Fig. 13–7). Collisions between the excited atoms of helium and the ground-state atoms of neon result in a transfer of energy; the helium atoms end up in the ground state and the neon atoms in the excited states, from which laser transitions to intermediate states then occur. The intermediate states subsequently decay to the ground state.

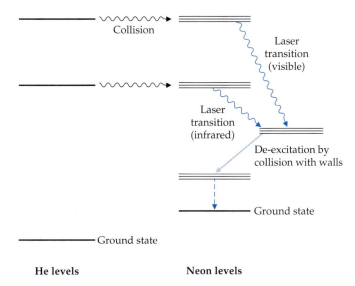

He levels**

**Neon levels**

• **Figure 13–7** In the resonant energy transfer method of pumping, helium atoms are electrically excited to levels that are near levels of interest in neon. The neon is easily excited to these levels by collisions with the excited helium atoms, with a transfer of excitation energy from helium to neon. The upper neon levels decay to make the laser transition.

## The Cavity

The collection of atoms or the material that actually undergoes the laser transitions (e.g., the gases in a He–Ne laser) are enclosed in a cavity, which is usually in the form of a cylinder of length $L$ with mirrors at the ends [•Fig. 13–8(a)]. The photons moving back and forth the length of the cylinder form the standing wave, something like sound within an organ pipe. The cavity will maintain

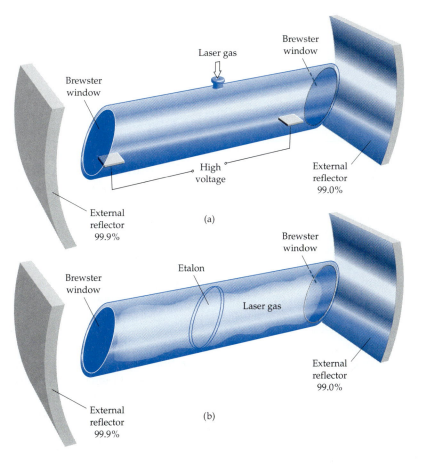

• **Figure 13–8** (a) Schematic of a cavity for a He–Ne laser. The 99%- and 99.9%-reflecting mirrors at the ends are placed beyond Brewster windows—windows that allow the transmission only of light that is polarized perpendicular to the axis of the cylinder. Only certain wavelengths can be supported, as the application of standing wave boundary conditions shows. (b) When an etalon is inserted in the cavity, its thickness determines the wavelengths that can be supported.

standing waves with wavelengths $\lambda_n$ given by $L = n\lambda_n/2$, where $n$ is an integer; that is,

$$\lambda_n = \frac{2L}{n}, \text{ where } n = 1, 2, 3 \ldots . \qquad (13\text{–}24)$$

In other words, only certain wavelengths can form standing waves. We would like $\lambda_n$ to match, for some value of $n$, the wavelength of the atomic transition in question. Since $L$ is on the order of a meter, while the atomic transition wavelength is around 500 nm, the value of $n$ that makes the match is in the range $4 \times 10^6$.

At finite temperatures, the particular wavelength just mentioned is not the only one supported. That is because at finite temperatures there is a so-called *Doppler broadening* of the emission wavelength. This phenomenon corresponds to the fact that the sources of the light—the atoms themselves—move with a variety of speeds at finite temperatures. Thus a number of adjacent $n$-values will be allowed, with the frequency spacing between modes given by $c/2L$, where $c$ is the speed of light *in the medium*. In other words, the light will not be truly monochromatic.

---

**Example 13–5**   Consider a He–Ne gas laser for which the laser radiation wavelength is 633 nm.
(a) Calculate the $n$-value for a standing wave in a cylindrical cavity 80 cm long. (b) Given that the spread in frequencies, $\Delta f$, of the neon atoms at room temperature is $1.7 \times 10^9$ Hz, what is the range of $n$-values covered by $\Delta f$?

**Solution**   (a) We have, from Eq. (13–24),

$$n = \frac{2L}{\lambda_n} = \frac{(1.6 \text{ m})}{(633 \times 10^{-9} \text{ m})} = 2.53 \times 10^6.$$

(b) From $n\lambda_n = nc/f_n = 2L = $ a constant, we see that $n = \text{const} \times f_n$. This in turn implies that $(\Delta n)/n = (\Delta f_n)/f_n$. Thus

$$\Delta n = \frac{n}{f_n} \Delta f_n = \frac{2L}{c} \Delta f_n = \frac{1.6 \text{ m}}{3 \times 10^8 \text{ m/s}} (1.7 \times 10^9 \text{ Hz}) = 9.$$

---

If it is important that the laser beam be truly monochromatic, one can effectively create a second cavity that allows only one frequency as follows: Insert a thin *etalon* consisting of a block of transparent material with the two faces made of reflectors [•Fig. 13–8(b)]. The thickness of the etalon now determines the spacing between the modes, and since that thickness can be made very small, the spacing between allowed modes becomes large, and only a single-frequency standing wave will be set up. Because the mirrors are not perfectly reflecting, some of the light will escape to be used as the laser beam, a highly collimated, coherent beam with a high degree of monochromaticity.

## Pulses

The laser that we have described is a steady source of power. It is often useful to have laser pulses of a very short duration (measured in nanoseconds) with a corresponding high peak power (on the order of megawatts). This combination can be achieved by means of a shutter that does not allow any radiation to escape. With the shutter closed, the population inversion can become very large. If the shutter is opened suddenly, the stored energy will be released in the form of a short, intense light pulse. The shutter can be mechanical—one of the end mirrors can rotate—or electrical—using material that becomes transparent when a voltage is applied. These methods to produce pulsing by altering the characteristics of the cavity are known as *Q-switching*.

Another method to produce pulses takes advantage of the fact that Doppler broadening leads to the excitation of more than one mode within the cavity. This technique is known as **mode locking**. We have seen that the light supported by the cavity is a series of frequencies that are integer multiples of a fundamental frequency; that is, $\omega_n = n\omega$. This is rather like the situation in which harmonic waves are passed through diffraction gratings, and, as in that situation, it is possible to show that when these different frequencies are added together, they make a resulting wave in the form of a series of regularly spaced pulses—short wave trains—that bounce back and forth within the cavity, in part passing through the mirror to become available for use. Such pulses are very much like the sharp spikes observed when light passes through a diffraction grating, except that in the case of the grating the pulses are separated in space (via an angle), whereas with mode locking they are separated in time. In practice, mode-locked lasers have produced pulses as short as 10 femtoseconds $(10^{-14}\,s)$, which corresponds to a length of only $3 \times 10^{-6}$ m.

We can divide the uses of such pulses into two categories. In the first, we can concentrate on the short-time feature. These pulses have been used to observe molecular processes that occur on time scales comparable to the pulse lifetime itself. For example, organic molecules can change their physical configuration on these scales in response to biological signals. In the second category, we can use the extraordinary large fields present in the pulses.[2] In the short pulses we have described, the field in the pulse is as much as 100 times the electric field in the vicinity of atomic electrons. Fields of this size permit the testing of ideas about quantum mechanics in completely unexplored regions, those where extreme relativistic phenomena such as pair production occur.

## Varieties of Lasers

Since the construction of the first laser in 1960, the variety of lasers has grown enormously. We list a few of them:

- *Solid-state lasers* consist of a cylindrical solid crystal with the ends polished optically flat. A thin layer of metal or dielectric is evaporated onto the ends, forming the mirrors, one of which must be slightly transparent to let out the beam of photons. The ruby laser is an example that uses a variant of the three-level system described earlier, while the YAG laser, made of neodymium ions in an yttrium–aluminum–garnet crystal, utilizes a four-level system. The optical pump is often made of a helical lamp surrounding the crystal cylinder. (•Fig. 13–9).

- *Gas lasers*, such as the He–Ne laser discussed before, are usually excited by means of an electrical discharge.

- *Dye lasers* use fluorescent organic dyes as the material that produces the laser light. Recall from our discussion of molecular spectra that each electronic

---

[2] The fact that the fields are large in the pulse is simply a consequence of the conservation of energy. There is a weak field or no field at all in between the peaks, where the interference is destructive. Only in the sharp pulse is there a substantial field. But the conservation of energy requires the total energy in the pulse then match the energy that would be present without interference, and that energy is very large since it is spread out over all space.

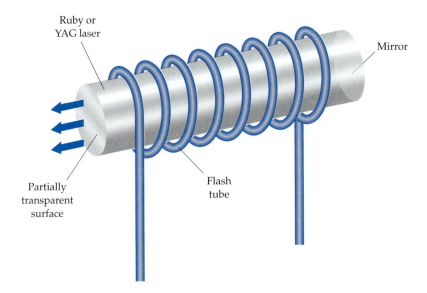

• **Figure 13–9** A flash tube provides an optical pump for populating the upper laser level in the ruby and other solid-state lasers.

level has vibrational levels built upon it. Optical pumping excites electrons to the upper set of states (•Fig. 13–10), and these electrons quickly cascade to the bottom of that set of states. The laser transitions to a variety of vibrational excited states built on a low-lying electronic level give a whole spectrum of potential frequencies. It is possible to pick out a particular transition and thus *tune the laser* by replacing one of the mirrors with a diffraction grating, whose angle will pick out any one of the closely lying lasing frequencies.

• *Semiconductor lasers* that use the levels of a solid rather than atomic levels will be briefly discussed in Chapter 14.

The technological uses of lasers are enormous, and their development is but one example of many of the interplay between basic science and technology. In the next three subsections we discuss some recent applications of lasers to physics and biology.

• **Figure 13–10** In dye lasers, the transitions occur between the vibrational levels of molecules, here those within the electronic ground state and those within the first excited electronic state. These levels are quite closely spaced, so that a wide variety of laser frequencies is possible.

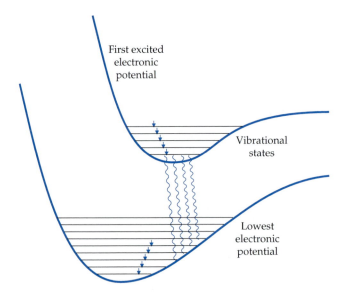

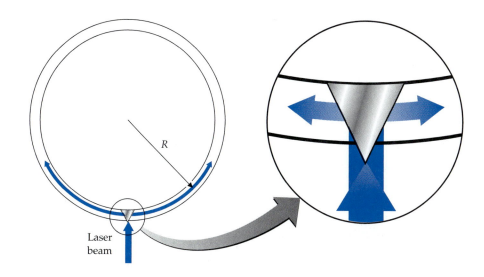

• **Figure 13–11** The ideal gyro laser shown here has a cavity that forms a ring. Two laser beams enter and are directed in opposite directions.

## The Gyro Laser

The gyro laser (•Fig. 13–11) detects slow rotational motions, down to $\mu rad/s$ and less. In one type, a hollow tube forms a ring of radius $R$ mounted on an object rotating with angular speed $\Omega$. A beam splitter on the ring divides an incoming laser beam of wavelength $\lambda$ into two pieces, one traveling counterclockwise (CCW) and the other clockwise (CW) around the ring.[3] These beams recombine at the beam splitter after one circuit and are removed. Without rotation, the beams travel the same distance $2\pi R$ around the ring in a time $t = 2\pi R/c$ and are in phase when they recombine. If, however, there is rotation in the CW direction, then as the CW-moving beam travels a circumference the beam splitter advances by a distance $\Omega R$. The beam must travel a total distance (approximate—the distance $\Omega R$ is very small)

$$L_{CW} = 2\pi R + (\Omega R)t = 2\pi R(1 + \Omega R/c).$$

Similarly, the path length for the CCW-moving beam is decreased:

$$L_{CCW} = 2\pi R(1 - \Omega R/c).$$

The two recombined beams are, in general, no longer in phase. Their phase difference is $\Delta\varphi = 2\pi\Delta L/\lambda$, where $\Delta L$ is the path length difference,

$$\Delta L = L_{CW} - L_{CCW} = 2\pi R^2 \Omega/c.$$

Thus $\Delta\varphi$ is proportional to the rotation speed. $\Delta\varphi$ will manifest itself as a reduction in the intensity of the recombined beam. As $\Omega$ changes, so does the interference.

In practical situations variants on this scheme are used. In one such scheme a laser medium that can send beams in both directions is put *into* the ring. In this case the frequencies supported in the cavity differ for the two directions. Suppose first that there is no rotation. Then the two beams will be maintained only if the travel distance $2\pi R$ is an integer $N$ times a wavelength; that is, $2\pi R = N\lambda = Nc/f_0$, or $f_0 = Nc/2\pi R$. (The subscript on $f_0$ signals "no motion.") This condition follows because without the integer requirement the beam will interfere destructively with itself when it makes a full circle.

---

[3] Below we describe how in practice the light actually travels in straight sections.

With rotation, the length $2\pi R$ for which the beams "bite their tails" is changed to $L_{CW}$ and $L_{CCW}$ for the CW and CCW paths, respectively. Thus the maintained frequencies *differ* for the two directions, taking the values

$$f_{CW} = \frac{Nc}{L_{CW}} = \frac{Nc}{2\pi R\left(1 + \dfrac{\Omega R}{c}\right)} = \frac{f_0}{\left(1 + \dfrac{\Omega R}{c}\right)} \cong f_0\left(1 - \frac{\Omega R}{c}\right)$$

and

$$f_{CCW} = \frac{Nc}{L_{CCW}} = \frac{Nc}{2\pi R\left(1 - \dfrac{\Omega R}{c}\right)} = \frac{f_0}{\left(1 - \dfrac{\Omega R}{c}\right)} \cong f_0\left(1 + \frac{\Omega R}{c}\right).$$

When the two beams are extracted and combined, their resultant contains easily detectable beats. The beat frequency is

$$\Delta f = f_{CCW} - f_{CW} = 2f_0\Omega\frac{R}{c} = 2\Omega\frac{R}{\lambda}. \tag{13–25}$$

Measurement of the beat frequency thus translates into measurement of $\Omega$, and by this method one can measure rotational velocities too small to be detected by any other means.

In practice the "ring" is not a circular cylinder, but rather a triangular path with mirrors at the corners (•Fig. 13–12); the triangle rests centered on the axis of the rotating object. The preceding formula for $\Delta f$ remains the same if one replaces $R$ by the ratio of twice the area enclosed by the path to the length of the path.

**Example 13–6**    Consider a circular gyro laser of radius 1.0 m that is to be used to detect a rotational velocity of 1 degree/day. What resolution of beat frequency is needed to achieve this level of detection if the wavelength of the light used is 440 nm?

**Solution**    The angular velocity to be detected is

$$\omega = 1\ \text{deg/day} \times (2\pi/360\ \text{rad/deg}) \times \left[1/(24 \times 3{,}600)\ \text{day/s}\right] = 2.0 \times 10^{-7}\ \text{rad/s}.$$

Using Eq. (13–25) we have immediately

$$\Delta f = 2 \times (2.0 \times 10^{-7}\ \text{rad/s})(1\ \text{m})/(440 \times 10^{-9}\ \text{m}) = 0.9\ \text{Hz}.$$

This is a frequency that is easily measurable.

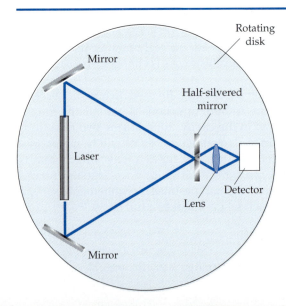

• **Figure 13–12**    In a real gyro laser the beams form a triangle, with mirrors that can also be used to extract the beams and detect any beats that might result from the rotation of the platform.

## Cooling and Trapping of Atoms

At room temperatures, a gas of atoms moves with speeds that we can estimate from equipartition to be given by $mv_{rms}^2/2 = (3/2)kT$. For sodium atoms, for example, $m = 22 \times 1.7 \times 10^{-27}$ kg, and then we find from this formula that $v_{rms} \approx 0.6 \times 10^3$ m/s. Here we describe techniques involving lasers that can stop atoms and allow us to study their properties at leisure.

Slowing an atom down is equivalent to cooling it. One way to cool atoms is to hit them head-on with a laser beam—if the photons of the beam have the right energy to be absorbed, then conservation of momentum implies that the atom slows. In 1985, Steven Chu devised a way to use this idea: Aim two oppositely directed laser beams aligned with, say, the x-axis at an atom, one from the right and one from the left. The frequency $f$ of these beams is chosen just a bit below the frequency for absorption; that is, $hf$ is slightly smaller than the energy difference between the first excited state and the ground state of the atom. Because of the uncertainty relation $\Delta E \, \Delta t \gtrsim \hbar$, an unstable excited state will have a finite width, and the laser frequency is picked to correspond to an energy on the lower side of the peak associated with the excited state. Accordingly, if the atom is at rest, a photon from the laser has only a small probability of being absorbed.

Now suppose that the atom is moving with speed $v$ in the negative x-direction. Then it "sees" the incoming light from the laser on the left shifted up in frequency to $f(1 + v/c)$. This slightly increased frequency is strongly absorbed, and the atom will be slowed down. As for the laser on the right, the atom "sees" its frequency moved down to $f(1 - v/c)$, and as this is even further off the correct frequency for absorption, that beam will have no effect. Similarly, if the atom is moving to the right, the right-side laser will slow it down, but the left-side laser will have no effect. The net effect is to slow the atom. Since atoms move in three dimensions, not one but six lasers are needed, two aligned along each axis. The electromagnetic field in which the atoms move has been described as "optical molasses."

Atoms that are trapped at low temperatures may be used for a variety of purposes. Trapped atoms can be allowed to fall freely. Their excitations can be measured as they are subsequently hit with a series of short laser pulses of different frequencies. These excitations change with the motion of the atoms, because there is a Doppler shift that is determined by the speed of the atoms. In this way the weakening of gravity due to a rise of as little as 3 cm has been measured! Such an improvement in the measurement of local variations in $g$ promises to be of great use in oil and mineral exploration.

## Optical Tweezers and Scissors

Lasers are now being used in basic biological research. Following a suggestion of Arthur Ashkin, then at AT&T Bell Laboratories, the radiation pressure of a laser beam has been made to act on viruses or bacteria as well as on atoms, but in a different way. As •Fig. 13–13 shows, the electric dipole forces due to a split laser beam act on a neutral particle (e.g., a bacterium or a bead) in a way that pushes the particle into the brightest area. These forces act to stabilize the target and bring its center towards the focus of the microscope lens through which the laser beam is directed. The forces are tiny, on the order of $10^{-12}$ N (piconewtons), but they are adequate when acting on such small objects. The bacterium or bead can thus be fixed and also pushed or pulled with a pair of laser beams, which thus act as *optical tweezers*. Another laser beam operated with short powerful pulses can be used to cut the target. One of the interesting developments in this area is the study of the forces that curl up long molecules such as DNA.

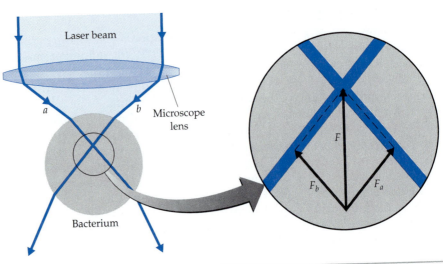

• **Figure 13–13**   The electric fields in a light beam can act on small objects. If the objects are neutral, the forces are due to dipoles that are induced in them (by the separation of charges). By focusing a laser beams through a lens, these forces act as restoring forces that bring the object to the focal point.

## SUMMARY

The question of how transitions between energy levels of atoms or molecules occur is central both to our understanding of the physical world and to engineering applications. Some important points are as follows:

- Quantum mechanical instability is a question of probability. Although it is not possible to say when an unstable state will decay, it is possible to describe the probability of the decay as a function of time in terms of the lifetime of the state, which governs the falling exponential form of this probability. The sequential decay of several unstable states can also be described in terms of the fundamental exponential decay law.
- The calculation of the lifetime itself is also possible within the constraints of quantum mechanics if one knows enough about the dynamics of the unstable state. An example is the dipole radiation of atomic states.
- An important feature of decays is the presence of selection rules; the dynamics may be such that certain decays are allowed while others are not. In the case of dipole radiation from atomic states, the angular momentum of the atomic state must change by unity.
- Because photons are bosons, atomic decays accompanied by the emission of radiation are enhanced if there are already other photons present that are identical to the emitted photon. This fact is the basis for the operation of the laser, which also involves being able to sidestep the Boltzmann factor for the number of atoms in various atomic states as a function of temperature.

## QUESTIONS

**1.** In Example 13–2, we estimated the rate of transition from the $2p$ to the $1s$ state in the hydrogen atom. In our treatment there, these states were given definite energies. Was this correct? Explain your answer.

**2.** If your answer to the previous question was "no," how would the fact that one or both of these states do not have a definite energy manifest itself in the observed spectrum of hydrogen?

**3.** Following the reasoning of Example 13–3, and without doing any serious calculations, can you see why the $2s$ state of hydrogen would not readily decay into the ground state?

**4.** Use our discussion of lasers to think of as many obstacles as you can to the construction of an X-ray laser, one in which the photon energy is in the range of kiloelectron volts.

**5.** An exponential decay law tells us how many individual members of a large collection of identical unstable objects decay in a small interval around some given time. How are we to interpret this law if there is just one unstable particle present?

**6.** The lifetime of a collection of at-rest unstable particles—muons, say—is different from the lifetime of a collection of moving muons. Why is that, and is the lifetime of the moving muons larger than or smaller than that of the at-rest muons?

**7.** We have stated that when thermal equilibrium is established in a system of atoms with two energy levels, as many atoms make transitions up by the absorption of photons as make transitions down by the emission of photons. But how does the system "know" the particular temperature at which it happens to be in equilibrium?

**8.** Single atoms can be confined in a *Penning trap* and observed. Suppose you shine a strong laser beam with a frequency that matches a transition between the ground state with energy $E_0$ and a level with energy $E_1$. If the transition is an allowed one, so that it takes place in a time on the order of $10^{-8}$ s, what do you expect to see when you look at the atom?

**9.** Suppose the confined atom of the previous question also has a level of energy $E_2$ for which the transition to the ground state is almost, but not totally, forbidden (so that it takes place in a time on the order of 0.1 s, say). If you shine a less intense laser beam tuned to have its frequency match that of the transition between the level $E_2$ and $E_0$, what do you expect to see when you look at the atom? Suppose you now turn on both of the laser beams. What do you expect to see when you look at the emission of radiation by the atom? Draw a figure of the intensity of the radiation as a function of time.

## PROBLEMS

**1.** ▌▌ Show, by finding the time for half of an initial sample of radioactive material to decay according to Eq. (13–2), that the half-life $\tau_{1/2}$ is related to the lifetime $\tau$ according to $\tau_{1/2} = (\ln 2)\tau$.

**2.** ▌ The half-life of a radioactive isotope is $1.4 \times 10^{11}$ years. What is the decay rate $R$? You may use the results of Problem 1.

**3.** ▌▌ In an experiment with a sample of a radioactive isomer (a nucleus with definite $A$ and $Z$ values), 10% of the original amount is found to have decayed by a specific mode after exactly one week. What is the lifetime for this particular mode of decay? How much time will pass after the beginning of the experiment before 50% of the original amount will have decayed?

**4.** ▌ Using the data in Appendix A, find the lifetime and the decay rate for the neutron. Do the same for tritium. Suppose you are given a mole of tritium. How many atoms will you have a thousand years later?

**5.** ▌ Suppose you were able to prepare a gram of $^{15}$O. How long would it take for the sample to be reduced to half a gram? A milligram? Use the data from Appendix A.

**6.** ▌▌ An unstable system may have two (or more!) decay modes, with a rate $R_1$ for decay through mode 1 and $R_2$ for decay through mode 2. (For example, a radioactive nucleus might decay with the emission of an alpha particle or, alternatively, with the emission of a photon.) Use the meaning of the rate [see, for example, Eq. (13–1)] to show that the total lifetime $\tau$ of the system is given by

$$\frac{1}{\tau} = \frac{1}{\tau_1} + \frac{1}{\tau_2},$$

where $\tau_1$ and $\tau_2$ are the lifetimes corresponding to the rates $R_1$ and $R_2$, respectively.

**7.** ▌▌ A nucleus decays into two channels with probabilities 0.62 and 0.38, respectively. Its lifetime is 20 hours. What are the decay rates into each of the two channels? (*Hint*: See Problem 6.)

**8.** ▌▌ Suppose that a system has many decay modes (see Problem 6) with very different lifetimes. Using a generalization of the results of Problem 6, show that the total lifetime of the system is approximately given by the shortest among the lifetimes of the different modes.

**9.** ▌▌ A particle called the $\pi^0$ decays into two energetic gamma rays with a mean life of $(8.4 \pm 0.6) \times 10^{-17}$ s. A recent table quotes a rest-mass energy of $134.9764 \pm 0.0006$ MeV for the particle. Is there any contradiction between this very accurate determination of mass and the limit imposed by the uncertainty principle involving $\Delta E$?

**10.** ▌ Which of the following atomic transitions is allowed according to the selection rules developed in this chapter?

$$^2S_{1/2} \rightarrow {}^2P_{3/2}, \qquad\qquad {}^2D_{3/2} \rightarrow {}^2D_{5/2},$$

$$^3D_2 \rightarrow {}^1P_1, \qquad\qquad {}^4D_{1/2} \rightarrow {}^4P_{1/2},$$

$$^2D_{3/2} \rightarrow {}^4S_{3/2}, \qquad\qquad {}^3P_0 \rightarrow {}^1S_0.$$

(Recall the spectroscopic notation for an atomic state, $^{2S+1}L_J$, where $L = 0, 1, 2, \dots$ is represented by $S, P, D, \dots$, respectively.

**11.** ▌▌ Consider the $3p \rightarrow 2p$ transition in hydrogen. By concentrating on the angular part show that this transition is forbidden according to Eq. (13–10).

**12.** ▌▌ Perform a dimensional analysis to find the form of the instantaneous power radiated by an accelerating charge $e$ in classical electromagnetism. Use the following facts: (1) According to special relativity, a uniformly moving charge cannot radiate, so that the only quantities available to make up the power are the acceleration $a$; the speed of light $c$; and the electric charge. (2) The electric charge always appears in the combination $e^2/4\pi\varepsilon_0$ in MKS units.

**13.** ▌▌ An excited state of a nucleus of radius $6 \times 10^{-15}$ m undergoes a radiative decay in which a photon of energy 1.5 MeV is emitted. Assuming that the dipole radiation formula is applicable, use it to estimate the lifetime of the excited state.

**14.** ▌▌▌ The $\ell = 1 \rightarrow \ell = 0$ transition between the two lowest rotational states in CO has a wavelength of 2.603 mm. Given that this represents dipole radiation, estimate the rate at which the transition takes place. (*Hint*: You need to calculate the internuclear separation first, using material from Chapter 11.)

**15.** ▌▌ Muonic hydrogen is ordinary hydrogen with the electron replaced by a *muon*, an unstable particle very much like the electron. The major difference is that the muon is some 207 times more massive than the electron. Estimate the lifetime of the $2p \rightarrow 1s$ transition in muonic hydrogen, given that the lifetime for this transition in hydrogen is $1.7 \times 10^{-9}$ s.

**16.** ▮▮ What is the ratio of the width $\Delta\lambda$ to the wavelength $\lambda$ itself of the line emitted by muonic hydrogen in the transition $2p \rightarrow 1s$? (*Hint*: The uncertainty in the energy of the photon that is emitted is related to the lifetime of the decaying state by the uncertainty principle, the energy of the photon is related to its frequency, and from $\Delta f$ one can find $\Delta\lambda$.)

**17.** ▮▮ Compare the rates at which induced transitions, both up and down, occur if the atoms undergoing the transitions are in equilibrium at the two temperatures $T_1 = 50K$ and $T_2 = 500K$. Make your calculation for the following frequencies: **(a)** a fixed frequency $f$; **(b)** the frequencies at which the energy density distribution of the radiation is a maximum; **(c)** the frequencies at which the number of photons in the cavities are a maximum. Assume in every case that suitable atomic levels are present.

**18.** ▮▮ The rate for transitions up and down at a frequency that corresponds to two levels differing in energy by 0.25 eV is found to be a certain value $R$ when the atom is placed in a blackbody cavity at a temperature of 640K. At what temperature(s) would the rate of the same transition be exactly 10 times $R$? Is it necessary to go to a higher temperature to get an increased rate of induced transitions? Why or why not?

**19.** ▮▮ The angular spread of a laser beam is ideally diffraction limited; that is, the angular spread is given by $\Delta\theta = 1.22\lambda/D$, where $\lambda$ is the wavelength of the light and $D$ is the diameter of the aperture of the source—the diameter of the laser rod. What is the size of a spot projected by the beam at a distance of 1 km for a ruby laser ($\lambda = 700$ nm) with diameter 1 cm? If the beam has a flux corresponding to $10^{18}$ photons emitted per second, what is the energy deposited per square centimeter per second at a target 1 km away?

**20.** ▮▮ A laser beam of wavelength 500 nm is directed at the moon through a 48-inch telescope (i.e., the effective diameter of the aperture is 48 inches). Estimate the area of the beam by the time it reaches the moon. You may want to look at Problem 19.

**21.** ▮▮ A laser beam with $\lambda = 600$ nm is 2 mm in diameter. If the beam delivers $10^{17}$ photons per second, what is the power of the laser? What is the radiation pressure this beam can exert? If we think of the beam as a target made of photons, what is the number of photons per cubic centimeter in that target?

**22.** ▮▮ Consider a monochromatic beam of photons with a wavelength of 460 nm, reflected back and forth between two mirrors 1.2 m apart. One of the mirrors is perfectly reflecting, and the other reflects 98.5% of the radiation incident on it. Ignore the fact that there may be a gas in the cavity through which the laser travels that reduces the speed of light to $c/n$, where $n$ is the index of refraction of the gas for these particular photons. **(a)** How long does it take a photon to make one trip back and forth between the mirrors? **(b)** What fraction of the photons will remain in the cavity after $N$ round-trips through the cavity? **(c)** How long will it take before half of the photons will have left the cavity? (This is the half-life of the cavity.)

# Conductors, Semiconductors, and Superconductors

The behavior of materials depends critically on quantum mechanical effects. In the years since the 1920's, the range of applications of quantum mechanics to this area—*solid-state physics*—has become so large that even a brief survey of all of the applications would by itself fill a book. Limited as we must be here, we concentrate mainly on one subject: the conduction of charge in materials—metals and semiconductors. We also describe superconductivity, an interesting phenomenon whose possible technological consequences are enormous.

## 14–1  The Classical Theory of Conductivity

We'll begin by discussing the classical treatment of currents in metals—the phenomenon of electrical conductivity. When we're done, we'll see that the evidence forces us to reconsider conductivity and to treat it from a quantum mechanical point of view.

When an electrical potential $V$ is applied across a metal, a current $I$ results. Ohm's law, an empirical rule that holds for many metals, states that the resistance, defined by the ratio $V/I$, does not vary with $V$. This is equivalent to saying that the ratio $E/j$ does not vary with $E$, where $E$ is the magnitude of the electric field across the metal and $j$ is the magnitude of the resulting current density. This ratio is the *resistivity* of the metal. In 1900 the German physicist Paul Drude developed a model of conductivity that led to Ohm's law. This model was based on the assumption that metals contain electrons that are free to move in response to an electric field. These electrons are the very loosely bound *valence* electrons. When an electric field is applied, the valence electrons in the metal accelerate in the opposite direction. Collisions with ions interrupt the acceleration, so that the ions act to produce a *drag force*. The motion is analogous to a parachute's fall being slowed down by collisions with air molecules;

as with the parachute, a simple model assumes that the drag force is proportional to the speed of the electron. In that case the equation of motion is

$$m \frac{dv}{dt} = eE - \frac{m}{\tau} v, \tag{14–1}$$

where $m$ is the electron mass and for dimensional reasons we have written the coefficient of $v$ in the drag force as this mass divided by a parameter $\tau$ with the dimensions of time. (See Problem 1 at the end of the chapter.) This parameter is known as the *Drude relaxation time*. As the electron accelerates, its speed increases until the drag term becomes large enough to cancel the acceleration. At that point the electron is at a terminal speed, known in this context as the **drift speed**. This quantity is given by the condition that the right side of Eq. (14–1) vanishes; that is,

$$v_d = eE \frac{\tau}{m}. \tag{14–2}$$

In turn, the current density for a gas containing $n_e$ electrons/m$^3$ is

$$j = en_e v_d = \frac{e^2 n_e \tau E}{m}.$$

The result that $j$ is linearly proportional to $E$ is a statement of Ohm's law, $E/j = \rho$, where

$$\rho = \frac{m}{e^2 n_e \tau} \tag{14–3}$$

is the resistivity of the gas.

What is the Drude relaxation time $\tau$? The only quantity with the dimensions of time that is relevant in this context is the time between the collisions that the electrons make with the impeding ions. But one could imagine that this quantity depends in turn on such variables as the density of the ions with which the electrons collide and how fast the electrons are moving. It turns out to be convenient for a discussion of this type of random collision to introduce some new quantities. Accordingly, in place of $\tau$, we think in terms of the *average distance* that an electron travels between collisions. This quantity is also known as the **mean free path** $\ell$. The mean free path plays such an important role in the motion of particles through a gas of other particles that it is worthwhile to digress briefly and discuss the concept in detail.

### Mean Free Path and Collision Cross Sections

Consider an electron moving through a gas of ions. We pose the following question: *How far will a given electron travel before it undergoes a collision?* This query is somewhat analogous to the question of how far you can run across a crowded field before bumping into someone. That depends on (a) how crowded the field is and (b) how big the people you might be bumping into are—a dependence that becomes more obvious if there were elephants instead of people on the field.

Let us address the issue of the size of the obstacle first. If we think of an electron as a little billiard ball of radius $r_1$ colliding with a larger billiard ball (an ion) of radius $r_2$, then, as •Fig. 14–1 shows, whenever the centers of the two objects are less than a distance $r_1 + r_2$ apart, a collision takes place. Thus the effective area that an ion presents to an electron is the area $\pi(r_1 + r_2)^2$ of a circle. This area is denoted by $\sigma$ and is generally called the **collision cross section**. The concept

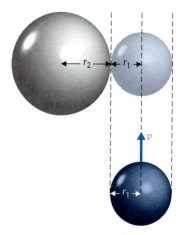

• **Figure 14–1** The maximum distance between the centers of colliding spheres is $r_1 + r_2$; this is the distance when there is a glancing collision. Thus the effective area blocked by one sphere in collisions with the other is $\pi(r_1 + r_2)^2$.

has meaning even though electrons and ions cannot really be viewed as sharply defined objects. We shall see shortly that the cross section can be found by measuring how many electrons are taken out of an incident beam of electrons in a given distance.

Now we look at the role of the density of target ions. We'll suppose that the density of target ions is $n$ ions/m$^3$. We'll also assume that the ions are small enough so that the likelihood of one target "masking" another is negligible. Consider, then, an electron beam with cross-sectional area $A$ traversing a gas of target ions. If the incoming beam contains $N(x)$ electrons at a depth $x$ into the gas, then at a depth $x + dx$ there will be fewer electrons in the beam because of collisions. The change $dN(x)$, which is negative, is minus the product of (*i*) the number of electrons present in the beam, $N(x)$, (*ii*) the number of target ions in the cylinder of area $A$ and depth $dx$, viz., $n \times A dx$, and (*iii*) the probability of a collision. The last of these is just the fraction of the area blocked out by the ions, so that the third factor is $\sigma/A$. Thus

$$dN(x) = -N(x) \times (nA dx) \times (\sigma/A) = -N(x)n\sigma \, dx. \qquad (14\text{–}4)$$

Note that the beam area $A$ has canceled out, and we have $dN/dx = -n\sigma N(x)$, a familiar equation with a familiar solution, namely

$$N(x) = N(0)\exp(-\sigma n x). \qquad (14\text{–}5)$$

We note the following points:

- A measurement of the attenuation of an incoming beam as it passes through a material of known density allows us to measure $\sigma$. This is indeed how cross sections are frequently measured.

- The ratio $N(x)/N(0) = \exp(-\sigma n x)$ is proportional to the probability that an electron in the beam has survived to a depth $x$. With proper normalization, this ratio forms a probability distribution, and it is possible to use it to find the average distance of penetration into the medium, $\langle x \rangle$. The latter quantity has no dependence on the density of the incoming beam or, indeed, on whether the beam was a real one. In other words, it is the average distance of travel before collision of any electron moving through the gas—what we referred to earlier as the mean free path $\ell$. A simple calculation (the denominator factor in the first line of Eq. (14–6) normalizes the distribution and makes it a true probability distribution) gives

$$\ell \equiv \langle x \rangle = \frac{\displaystyle\int_0^\infty x \exp(-\sigma n x)\,dx}{\displaystyle\int_0^\infty \exp(-\sigma n x)\,dx}$$

$$= \frac{1}{n\sigma}. \qquad (14\text{–}6)$$

Equation (14–6) is the answer to the question we posed at the beginning of this subsection.

The mean collision time $\tau$ is related to the mean free path by the fact that the mean free path is traversed by the electron in a time $\tau$ when it moves with an average speed $v_{av}$ relative to the ions. Thus we have

$$\tau = \frac{\ell}{v_{av}}. \qquad (14\text{–}7)$$

We can now rewrite Eq. (14–3) for the resistivity in terms of the more fundamental quantities $\sigma$ and the density $n_i$ of ions as

$$\rho = \frac{mv_{\text{av}}}{e^2 n_e \ell} = \frac{mv_{\text{av}}\sigma}{e^2}\left(\frac{n_i}{n_e}\right). \tag{14–8}$$

Equation (14–8) is the **Drude formula**.

### The Classical Drude Formula

What values does one take for the parameters of the Drude formula? In a *classical* picture of metals, the valence electrons are treated as if they form a classical gas at the temperature $T$ of the metal. Thus one replaces $v_{\text{av}}$ by the mean speed of a classical Maxwell–Boltzmann gas at temperature $T$. This speed is (see Example 12–3) $\sqrt{8kT/\pi m}$, where $k$ is Boltzmann's constant. (Aside from the factor $(8/\pi)^{1/2}$, this result for the average speed is dictated by dimensional arguments.) The cross section $\sigma$ can also be replaced by the effective area of an atom, $\pi a^2$, where $a$ is on the order of the magnitude of the Bohr radius. Apart from the origin of the size of the atom, these choices are classical, and when we substitute into the general Drude formula we find the *classical Drude formula*,

$$\rho = \frac{m\pi a^2}{e^2}\sqrt{\frac{8kT}{\pi m}}.$$

Here we have made the additional assumption that there is one valence electron per atom—that is, that $n_i = n_e$.

Although the preceding formula for $\rho$ gives a numerical result for the resistivity at room temperatures that is close—in retrospect, fortuitously—to the right magnitude, the temperature dependence $\sqrt{T}$ is totally wrong. Experimentally, the resistivity depends *linearly* on temperature. An example is shown in •Fig. 14–2.

What has gone wrong? We already know part of the answer: The electrons do not form a classical gas, but rather have a Fermi–Dirac distribution. This changes the speed that should be used in the Drude formula. The second thing that has gone wrong is in the description of the mechanism by which electrons are retarded in their passage through matter: When the ions are arranged in a periodic lattice, a quantum mechanical treatment shows that the mean free path for electron scattering becomes infinite! We'll explain this surprising statement in the next section. Thus the cross section $\sigma$ that is needed must be based on the departure of the periodic lattice from perfect periodicity, and this is not described by an atomic size such as $\pi a^2$.

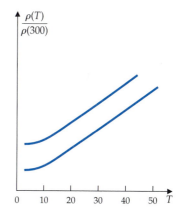

• **Figure 14–2**   Sketch of data on resistivity $\rho$, normalized to its value at 300K, for two particular samples of potassium. The linear dependence for $T > 30$K should have (and does have) a universal slope. The intercept depends on the purity of the sample.

## 14–2 The Quantum Mechanical Free-electron Model

The two German physicists Arnold Sommerfeld and his student Hans Bethe were the first to point out how valence electrons behave and how the exclusion principle affects that behavior. Solids consist of atoms which in turn consist of electrons bound to nuclei. In metals, the least strongly bound electrons are in unfilled shells. These are the valence electrons. When two atoms are brought close together, the valence electrons find it easy to tunnel to the sites of the neighboring atoms. Because the lower lying levels are filled with other electrons, the migrating electrons have no levels to fall to and cannot lose energy. Thus there is nothing to impede their mobility. These valence electrons must still satisfy the exclusion principle among themselves. Typically, the number of valence electrons per positive ion is of order 1, but it can be as large as 5 (for Bi and Sb).

The exclusion principle requires that at "low" temperatures—we'll see subsequently that for real materials room temperature qualifies as low—the valence electrons fill the available energy levels of the box, two electrons at a time, starting from the lowest level and working up to the Fermi energy $E_F$. In Chapter 10 we found that the Fermi energy for the box is given by Eq. (10–50), which, recalling that $n_e$ is the density of valence electrons, we can write in the form

$$E_F = \frac{\hbar^2 \pi^2}{2m} \left( \frac{3n_e}{\pi} \right)^{2/3}. \tag{14–9}$$

Typical Fermi energies for metals are in the range of electron volts, as Example 14–1 shows.

---

**Example 14–1**    Estimate the Fermi energy in copper. Assume that there is one free electron per ion in that metal. The atomic weight of copper is 63.5 and its density is $8.95 \times 10^3 \, \text{kg/m}^3$.

**Solution**    The only quantity we need to calculate to find $E_F$ is the density $n_e$ of free electrons. With one valence electron per ion, $n_e$ is the same as the number density of ions, which we estimate as follows: The atomic weight is the number of nucleons in an atom,[1] and since the mass of the atom is, to a good approximation, the mass of the nucleus, the mass of one copper atom is the mass $m_p$ of a nucleon multiplied by 63.5. The mass of a single nucleon is $m_p = 1.67 \times 10^{-27}$ kg. Thus the number of atoms per cubic meter is

$$n = \frac{8.95 \times 10^3 \, \text{kg/m}^3}{(1.67 \times 10^{-27} \, \text{kg})(63.5)} = 8.44 \times 10^{28} \, \text{atoms/m}^3.$$

Using this value for the number density of electrons, we have

$$E_F = \frac{(1.05 \times 10^{-34} \, \text{J} \cdot \text{s})^2 (3.14)^2}{2(9.11 \times 10^{-31} \, \text{kg})} \left( \frac{3(8.44 \times 10^{28} \, \text{m}^{-3})}{3.14} \right)^{2/3}$$

$$= 1.12 \times 10^{-18} \, \text{J} \cong 7.0 \, \text{eV}.$$

---

We can use the results of Example 14–1 to show that room temperature qualifies as a low temperature for the purpose of working with the exclusion principle. We can define a "Fermi temperature" $T_F$ by $E_F = kT_F$. For copper,

$$T_F = \frac{E_F}{k} = \frac{1.1 \times 10^{-18} \, \text{J}}{1.38 \times 10^{-23} \, \text{J/K}} \cong 8 \times 10^4 \text{K}.$$

The comparison of this quantity with the temperature of the metal determines how "rounded" the edge of the Fermi distribution is. (See •Fig. 12–13.) We see that $T_F$ is much greater than room temperature $T \cong 300$K. This fact justifies the use of a sharp Fermi distribution (corresponding to $T = 0$) in treating metals at room temperature.

### The Quantum Mechanical Speed for the Drude Formula

By using the energy–speed relation $E = mv^2/2$, we can find the speed $v_F$ of the electrons with energies near the Fermi energy:

$$v_F = \sqrt{\frac{2E_F}{m}} = \frac{\pi \hbar}{m} \left( \frac{3n_e}{\pi} \right)^{1/3}. \tag{14–10}$$

---

[1] The number for copper is not integral because copper nuclei come in two stable varieties, one with 63 nucleons and one with 65 nucleons; their abundances are such that 63.5 is the average.

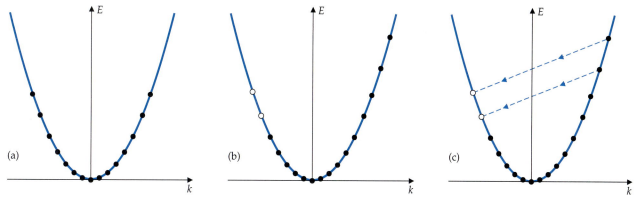

**• Figure 14–3**    (a) Relation between energy and momentum for electrons in a one-dimensional gas of free electrons. (b) Acceleration of electrons in the gas due to the imposition of an electric field leads to the promotion of electrons into higher energy states for positive momentum and a depletion of states containing negative-momentum electrons, which have moved into unoccupied states. (c) Collisions with ions reverse the directions of the electrons with the largest positive momenta, sending them back into the unoccupied negative-momentum levels.

*It is this speed, not the average of a classical velocity distribution, that is the relevant speed in the Drude formula, Eq. (14–8), for the resistivity.*

To see why, we consider a simple case in which electrons can only move in one dimension in a box of length $L$. •Figure 14–3a shows a plot of the energy as a function of the momentum of the electron. The momenta in a large box approximate a continuum, but for clarity we plot them as integer multiples of $\hbar\pi/L$. Momenta can be positive or negative, with the largest possible magnitude $p_F = mv_F$. When an electric field is applied to the left, the rightmost electron—the one whose speed is $v_F$—increases its momentum and jumps to the next available site. This allows the next electron to increase its momentum, and the next one after that, so that after a while the picture looks as shown in •Fig. 14–3b. But the jumping does not go on forever, because of collisions with ions. An electron that scatters can lose energy only by going into a lower lying unoccupied slot, and the only such slots that are available are at the back end of the distribution. Thus what the collisions do is replenish the slots by the transitions shown in •Fig. 14–3c. Now—and this is the crucial point—the skewing of the distribution is very slight. That means that the only electrons that can lose energy through collisions are grouped very closely around those with the Fermi speed. But this is precisely what we wanted to show: It is the electrons with $v_F$ that determine the collision frequency. The electrons so scattered provide empty spaces, and these get filled again because of the action of the electric field. The net result is that the distribution is displaced by a certain amount $mv_0$; in other words, the entire set of electrons moves with $v_0$, and this is what we mean by the drift speed: $v_d = v_0$. Since $v_d$ is on the order of millimeters per second, the displacement is indeed tiny. When the field is turned off, the collisions return the electrons to the original situation of •Fig. 14–3a. This picture can be generalized to three dimensions, where it is a bit more complicated, but the end result is the same: It is the electrons moving with the Fermi speed that determine the collision frequency.

The difference between $v_F$ and a classical form such as $\bar{v} = \sqrt{8kT/\pi m}$ is quite significant. Using the numbers we found for copper, we have

$$v_F = \sqrt{\frac{2E_F}{m}} = \sqrt{\frac{2(1.1 \times 10^{-18}\,\text{J})}{9.1 \times 10^{-31}\,\text{kg}}} = 1.5 \times 10^6 \text{ m/s},$$

whereas at room temperature,

$$\bar{v} = \sqrt{\frac{8kT}{\pi m}} = \sqrt{\frac{8(1.4 \times 10^{-23}\,\text{J/K})(300\text{K})}{(3.14)(9.1 \times 10^{-31}\,\text{kg})}} = 10^5\,\text{m/s}.$$

We have thus found that the velocity that goes into Eq. (14–8) is many times larger than the classical calculation indicates. If this were the only effect, the resistivity would be much too large. But as we shall see next, another quantum mechanical effect reduces the cross section to values below the $\pi a^2$ that was based on a classical particle scattering picture.

### Scattering from a Regular Lattice

In a metal—or, for that matter, in any solid—the ions form a more or less perfect crystal lattice. In the passage of electrons through a perfectly regular crystal lattice it is the wave character of the electrons that is operative. And it is generally true that, *for a perfectly rigid lattice with perfect periodicity, there is never any electron scattering.* Scattering takes place only in the presence of departures from perfect periodicity in the lattice, as caused by faults, impurities, or thermal effects that displace ions from their "rest" positions.

The mathematical expression of this result is **Bloch's theorem**, which concerns solutions of the Schrödinger equation for periodic potentials. The theorem is simple, although the proof, which we won't attempt here, is not. Consider a one-dimensional periodic potential with period $a$, for which

$$V(x + a) = V(x). \tag{14–11}$$

•Figure 14–4 shows an example of such a potential, which is the one-dimensional analogue of the type of potential one would expect an electron to "see" in a crystalline solid. (With very rare exceptions, all solids are crystalline, even if their structure is broken up into small grains.) Bloch's theorem, which is due to the Swiss physicist Felix Bloch, states that for such potentials solutions of the Schrödinger equation have the form

$$u_k(x) = e^{ikx}\varphi_k(x), \tag{14–12}$$

where $k$ is a (real) parameter that depends on the energy and $\varphi_k(x)$ has the spatial periodicity of the potential, namely

$$\varphi_k(x + a) = \varphi_k(x). \tag{14–13}$$

For $k > 0$, the solution represents a wave propagating in the positive $x$-direction. If the potential is zero, the solution to $d^2u/dx^2 + k^2u = 0$ which has the property that $|u(x + a)| = |u(x)|$ is uniquely the linear combination of $\cos kx$ and $\sin kx$ that leads to $C\exp(\pm ikx)$, where $C$ is a constant. (See Problem 11 at the end of the chapter.) Furthermore, $E = \hbar^2 k^2/2m$.

Equation (14–12) shows explicitly that there is no attenuation for the probability of finding an electron moving through the lattice. That is because the

• **Figure 14–4** A one-dimensional periodic potential of the type that might be encountered by electrons in a crystal of regularly spaced ions.

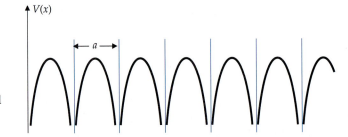

probability is $|u_k(x)|^2 = |\varphi_k(x)|^2$, and the latter does not systematically decrease as its argument increases. Rather, its periodicity shows that it regains its value when one moves from $x$ to $x + a$. This implies that *electrons do not scatter out of the incident beam in a perfectly periodic crystal lattice*, a result that is generalizable to three dimensions.

In saying that there is no attenuation of the probability of finding an electron as it passes through an ideal lattice, we are saying that *the mean free path of the electron in the ideal lattice is infinite*. Under these conditions, the resistivity of the ideal lattice, according to Eq. (14–8), is zero! Does this mean that the mean free path of an electron in a real metal is infinite? Not quite, because there are a number of ways that real systems deviate from perfect periodicity. One of them has to do with **impurities**—foreign atoms in the lattice—that destroy the crystal's perfect periodicity. While the impurities do scatter electrons, if the fraction of impurities is in the range $f \cong 10^{-5}$—a not untypical fraction—this is not a huge effect. The effective area of the collision-producing centers is around $f \times (\text{atomic area} \cong \pi a_0^2) \cong 10^{-25} \text{m}^2$.

There is a second and typically much more important way in which the mean free path becomes finite. When the temperature is not zero, the ions in the lattice will undergo thermal vibrations, whereupon the lattice is no longer perfectly regular and the electrons can scatter. This effect, in contrast to the effect due to impurities, is dependent on the temperature. We can make an estimate of its magnitude by using a classical picture of the ionic motion.

The area presented by the moving lattice points is proportional to the *square* of the amplitude of the vibration of the ions ($\bullet$ Fig. 14–5). Let us denote this amplitude by $R_0$; as an estimate in what follows, we'll take the cross section for electron scattering from the vibrating ions to be $\pi R_0^2$.

We can use equipartition to estimate the amplitude $R_0$. The total energy of a vibrating mass $M$—in this case the ionic mass—is given by the sum of the potential and kinetic energy, and this sum takes the form $M\omega^2 R_0^2$, where $\omega$ is the angular frequency of the vibrational motion, determined by the interionic spring forces. (We simplify matters by assuming that there is only one oscillation frequency in the region of temperatures that are of interest.) But there are two degrees of freedom in this system, one for the kinetic energy and one for the potential energy, so by equipartition the sum is given by $2(kT/2) = kT$, or

$$M\omega^2 R_0^2 = kT. \qquad (14\text{–}14)$$

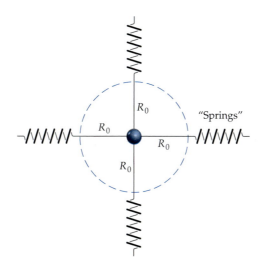

$\bullet$ **Figure 14–5**   A two-dimensional representation of the oscillations of an ion about its equilibrium position. The amplitude of the oscillations is $R_0$.

**Table 14–1    Useful Parameters of Common Metals**

| Element | $A$ | $n_e/n_i$ | $n_e$ (in units of $10^{28}$ m$^{-3}$) | $T_D$ (in K) | Resistivity at 273K (in units of $10^{-8}$ $\Omega \cdot$ m) |
|---|---|---|---|---|---|
| Li | 7 | 1 | 4.70 | 400 | 8.6 |
| Na | 23 | 1 | 2.65 | 150 | 4.2 |
| Mg | 24 | 2 | 8.60 | 318 | 3.9 |
| Al | 27 | 3 | 18.1 | 394 | 2.4 |
| K | 39 | 1 | 1.40 | 100 | 6.1 |
| Fe | 56 | 2 | 17.9 | 420 | 8.9 |
| Cu | 63.5 | 1 | 8.42 | 315 | 1.56 |
| Zn | 65 | 2 | 13.10 | 234 | 5.5 |
| Ag | 108 | 1 | 5.86 | 215 | 1.51 |
| Sb | 122 | 5 | 16.5 | 200 | 39 |
| Au | 197 | 1 | 5.90 | 170 | 2.04 |
| Pb | 207 | 4 | 13.2 | 88 | 19.0 |
| Bi | 209 | 5 | 14.1 | 120 | 107 |

This shows that the effective area of the ionic obstacles to the electron motion is proportional to the temperature.

The angular frequency $\omega$ of the lattice vibrations is a matter of the forces that bind the ions into a crystal. It is customary to use another parameter in its place, the **Debye temperature** $T_D$, defined by

$$\hbar\omega = kT_D. \tag{14–15}$$

The forces binding the ions together are such that $T_D$ lies in the range from 100K to 400K. (See Table 14–1.) In terms of this quantity, Eq. (14–14) gives

$$\pi R_0^2 = \frac{\pi kT}{M\omega^2} = \frac{\pi\hbar^2}{MkT_D^2}T. \tag{14–16}$$

When this equation is evaluated at room temperature we find a cross section[2] $\sigma = \pi R_0^2$ on the order of $10^{-22}$ m$^2$. This is much larger than the $10^{-25}$ m$^2$ that is the typical contribution from impurities in relatively pure metals.

With our modified value of speed and our modified value of $\sigma$, Eq. (14–16), one finds from Eq. (14–8) a resistivity at room temperature that is within an order of magnitude of the correct value, a satisfactory result given our crude estimates. In addition, the cross section, and hence the resistivity, is proportional to the temperature, in accordance with experiment. Table 14–1 gives some of the relevant parameters for metals.

**Example 14–2**    Calculate, according to the model just discussed, the resistivity of silver ($A = 108$, mass density $\rho_m = 10.5 \times 10^3$ kg/m$^3$) at $T = 300$K, given that the Debye temperature is $T_D = 215$K and that there is one valence electron per ion. The experimental value is $\rho = 1.66 \times 10^{-8}$ $\Omega$-m at a temperature of 300K.

**Solution**    We start with Eq. (14–8) for the resistivity. We then use $n_i = n_e$, the expression given in Eq. (14–10) for the Fermi speed in terms of the valence electron density, and the expression $\sigma = \pi R_0^2 = \pi\hbar^2 T/MkT_D^2$ [Eq. (14–16)]. Together, these give

$$\rho = \frac{m\sigma v_F}{e^2} = \frac{m}{e^2}\left(\frac{\pi\hbar^2 T}{MkT_D^2}\right)\left(\frac{\pi\hbar}{m}\left(\frac{3n_e}{\pi}\right)^{1/3}\right) = \frac{\pi^2\hbar^3 T}{e^2 MkT_D^2}\left(\frac{3n_e}{\pi}\right)^{1/3}.$$

---

[2]It is interesting to compare this number with the typical atomic area $\pi a_0^2 \cong 7 \times 10^{-21}$ m$^2$, where $a_0$ is the Bohr radius.

As in Example 14–1, the ion density (equal to the electron density) is

$$n_i = \frac{\text{mass density}}{Am_p} = \frac{10.5 \times 10^3 \text{ kg/m}^3}{108 \times 1.67 \times 10^{-27} \text{ kg}} = 5.82 \times 10^{28} \text{ m}^{-3}.$$

Using this value for $n_e$ in our expression for $\rho$, we have

$$\rho = \frac{\pi^2(1.05 \times 10^{-34} \text{ J} \cdot \text{s})^3(300\text{K})}{(1.6 \times 10^{-19} \text{ C})(108 \times 1.67 \times 10^{-27} \text{ kg})(1.38 \times 10^{-23} \text{ J/K})(215\text{K})^2}$$
$$\times \left[\frac{3(5.82 \times 10^{28} \text{ m}^{-3})}{\pi}\right]^{1/3}$$

$$= 0.44 \times 10^{-8} \ \Omega \cdot \text{m}.$$

This is about four times smaller than the experimental value.

---

As Example 14–2 shows, our theory is rather crude. Our description of the interaction of electrons with the lattice is far from a proper quantum mechanical description of that interaction; furthermore, the description of the valence electrons under the influence of an external electric field as a gas of free electrons is also oversimplified. Despite all this, one can claim for our picture *the successful prediction of a linear temperature dependence for the resisitivity and a very rough prediction of its magnitude.* The theory as we have presented it does not give an accurate numerical prediction of the resistivity.

## 14–3  Band Structure

The Drude model with its necessary quantum mechanical modifications does grossly describe the difference between conductors and insulators: Insulators have no free electrons. But there exist materials such as silicon and germanium that are insulators at low temperatures and conductors at high temperatures. Moreover, it is not true that all atoms with loosely bound valence electrons make equally good conductors. To understand thoroughly the electrical (and other!) properties of materials, we must bring in a remarkable—and highly nonclassical—feature of the allowed energies in periodic potentials. This feature is the existence of a series of essentially continuous energy levels separated by regions of energies in which there are no allowed energy levels. We refer to the energy regions with a continuum of levels as **bands** and to the regions where there are no allowed levels as **gaps**.

We want to see how bringing together a set of identical atoms can produce a picture with something like bands and gaps. In our discussion of these properties, we concentrate on motion in one dimension. Let's start by thinking about two such atoms. If the atoms are far apart, each atom has its own energy levels: a 1s level, a 2s level, and so on. And since the atoms are identical, any given energy level is repeated twice. In other words, each level is at least twofold degenerate (in addition to any degeneracy that is present in the single atom). When the two atoms are brought closer together, the state in which one of the electrons is localized around, say, ion 1 now includes a small probability for localization around ion 2 and similarly for the other electron localized around ion 2. This effect increases as the ions get closer, and the correct description involves a single electron moving in a potential due to the two ions together. •Figure 14–6 represents the potential the electron "sees," and for states that are localized about one of the force centers the wave function will be peaked at that center. But as •Fig. 14–7 shows, the wave functions of the electrons localized about one or the other ion may be combined into wave functions that are

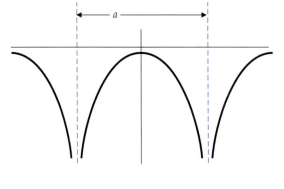

• **Figure 14–6**   Potential energy of an electron in the field of two positively charged ions a distance $a$ apart.

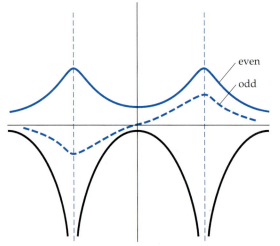

• **Figure 14–7**   Even and odd superpositions of wave functions localized about each of two ions. The even solution is the sum of the wave functions centered about the two ions, and the odd solution is the difference of these wave functions.

even or odd about the midpoint of the line joining the two ions. This can be seen by adding or subtracting the two wave functions.

Now—and this is the important point—the energies of the odd and even wave functions are slightly different, with the odd state having more energy. The reason for this can be most easily understood as follows: The odd state in •Fig. 14–7 has more curvature than the even state, and the curvature, which is measured by the derivative, is proportional to the energy because the Schrödinger equation relates the second derivative of the wave function to the energy times the wave function. This effect must be very small when the ions are far apart, and that is the statement that there are degenerate energies. When the ions move closer together, *the degeneracy is broken.*

Now we imagine $N$ ions, with a set of $N$ degenerate energy levels when they are far apart. When the ions come together to form a solid, the degeneracy is again broken. As •Fig. 14–8 shows, as long as the energies do not change too much, a kind of band structure is formed, with most of the levels surrounding the location of the old degenerate energies and few levels in the region between the old locations where there were no energy levels.

The preceding argument merely shows that there will be a breaking of the degeneracy, with a larger number of states in the general vicinity of the old degenerate states. We now argue that the bands occur within well-defined limits. In other words, for $N$ atoms each initially $N$-fold degenerate level breaks its degeneracy within distinct upper and lower energy limits. This phenomenon is sufficiently universal, depending only on the identity of the atoms—that is, on the periodicity of the potential—that we can see the same effect in a set of

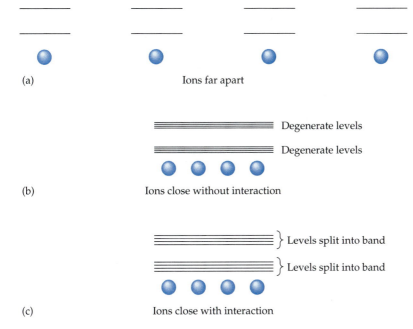

(a) Ions far apart

(b) Ions close without interaction

Degenerate levels
Degenerate levels

(c) Ions close with interaction

Levels split into band
Levels split into band

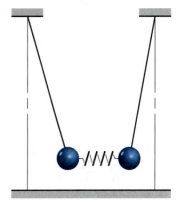

• **Figure 14–8** (a) Ions that are very far apart, with the two lowest energy levels for each atom shown. (b) In the absence of all interactions, bringing the $N$ ions closer together changes nothing; we have two $N$-fold degenerate energy levels in the system. (c) Because of interactions, each of the $N$-fold degenerate levels splits, leading to a pair of bands. If the interactions are strong enough, the bands can overlap.

• **Figure 14–9** Two identical pendulums coupled by a spring, moving in the plane of the paper. Without the coupling, the pendulums move with the same frequency—that is, they are "degenerate."

coupled pendulums. Suppose that, to start, we have two identical pendulums swinging in the same place. The frequency of each pendulum is a measure of its energy. When the pendulums are far apart, they have identical frequencies $\omega_0$ and hence identical energies; this common energy is to be thought of as the initial degenerate energy of one of the atomic levels. Now suppose the pendulums are brought together and that there is a weak coupling—a weak spring, say—between the masses (•Fig. 14–9). In this situation, there will be two *normal modes*—two natural frequencies (•Fig. 14–10). In one mode, the masses move together, and in that case the spring neither stretches nor compresses, and the frequency is that of the uncoupled oscillators. In the other mode, the masses move in opposite directions. We expect the frequency for this second mode to be higher, because in this case the spring provides an additional force bringing the masses together or forcing them apart. In other words, the energy is increased. In this way, the two degenerate frequencies $\omega_0$ separate into $\omega_0$ and $\omega_1$. (See Problem 16 at the end of the chapter.)

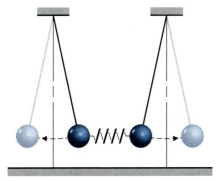

(a) Mode in which masses move in opposite directions.

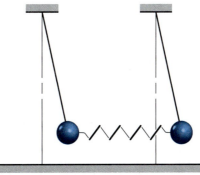

(b) Mode in which masses move together.

• **Figure 14–10** The normal modes of two coupled pendulums. (a) A mode in which the masses move in opposite directions, characterized by a frequency higher than that of either pendulum alone. (b) A mode in which the masses move together (in phase), in which the connecting spring never stretches or compresses. This motion has the frequency of the original degenerate frequency.

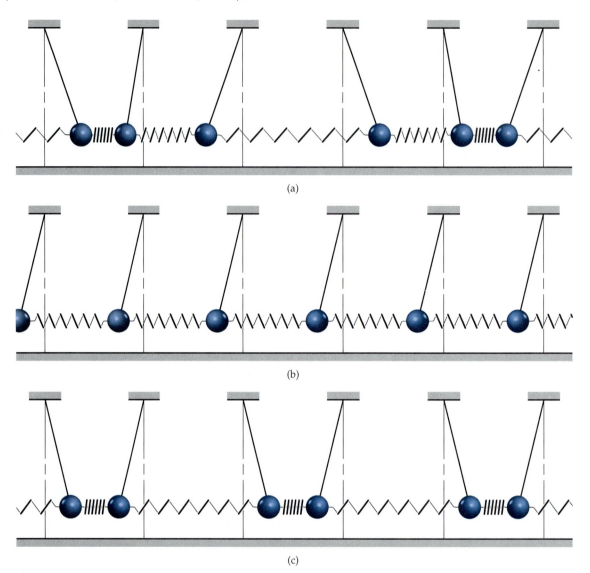

(a)

(b)

(c)

• **Figure 14–11** (a) $N$ identical pendulums coupled to their neighbors by identical springs. (b) The lowest frequency mode of the coupled motion is the one in which all the pendulums move together, with no stretching or compressing of the springs. (c) The mode with the highest frequency is the one in which successive pendulums move in opposite directions, stretching half the springs and compressing the other half to the maximum extent possible.

Now instead of two pendulums we bring many of them together, in a line, to form a model of a one-dimensional crystal.[3] The pendulums interact through weak springs joining neighbors (•Fig. 14–11a). Of all the complicated motions that can occur, there is a simple one in which the pendulums swing together as a unit (•Fig. 14–11b). The frequency of this state is $\omega_0$, because the springs joining the bobs neither stretch nor compress and hence play no role in the motion. And there is a maximum frequency, $\omega_{max}$, the motion in which adjacent pendula move in opposite directions[4] (•Fig. 14–11c). This frequency is a maximum be-

---

[3] Explicit treatments of this situation can be found in many books; see for example, H. Georgi, *The Physics of Waves*, (Englewood Cliffs, NJ: Prentice Hall, 1993).

[4] This frequency is actually that of $\omega_1$ in the two-pendulum case, although the actual value of the maximum frequency is not important to our argument.

cause it is the motion in which the maximum restoring force works on each pendulum bob. In all the other complicated motions one can imagine—the other normal modes—there are some bobs moving together, or at least more nearly together, so that the spring forces acting between them are smaller and the frequency is smaller. But of course it is never smaller than $\omega_0$, the frequency that occurs when there may as well be no springs at all; that is, $\omega_0$ is a minimum.

This classical system is indeed a good analogue of the quantum mechanical problem of the degeneracy breaking that occurs when a solid forms. In sum, when an $N$-fold degenerate energy $E$ is modified by bringing the atoms together to form a solid, the degeneracy breaking produces $N$ separated levels within a fixed range. When $N$ is large, there is, for all purposes, a continuum of energies between fixed limits, which we refer to as bands.

### The Connection between Bands and Propagation in a Lattice

The existence of energy bands and their associated energy gaps has an impact on the relationship between energy and momentum. The label $k$ that appears in Eq. (14–12) would have a simple interpretation if there were no ions: The solutions would be those of free particles, $u_k(x) = \exp(\pm ikx)$, plane waves with electron momentum $\pm \hbar k$ and electron energy $E = \hbar^2 k^2 / 2m$.

The interaction of the electrons with the ions changes this state of affairs. We can understand the nature of the change by the following argument: When the lattice ions are present, a would-be electron traveling wave is reflected and transmitted at each lattice point. In the special case that the waves reflected from successive lattice points differ in phase by an integral multiple of $2\pi$—that is, when $ka = \pm n\pi$—a pattern of *standing* waves will be set up in the form of a sum or a difference of $\exp(+i\pi x/a)$ and $\exp(-i\pi x/a)$. These combinations are the two (properly normalized) wave functions

$$\psi_{\text{even}} = \frac{1}{\sqrt{2}} \cos\left(\frac{\pi x}{a}\right)$$

and

$$\psi_{\text{odd}} = \frac{1}{\sqrt{2}} \sin\left(\frac{\pi x}{a}\right). \tag{14–17}$$

Now these two standing waves would be degenerate in energy in the absence of any further interactions of our electrons with the lattice. Once the interaction is taken into account, however, the energies will differ. The reason is that the ions are located at $x = ra$, where the $r$-values are integers. The $\psi_{\text{even}}$'s peak at these locations, while the $\psi_{\text{odd}}$'s peak in between them (•Fig. 14–12). Since the electron–ion interaction is attractive, and since the electron is more likely to be in the vicinity of the ions for the even wave functions, the energy of the even wave functions is lowered relative to that of the odd wave functions. Thus at the values $ka = \pm n\pi$ we expect to see a splitting in the energy levels. Such splits show up as gaps in the energy when it is plotted as a function of $ka$. From this point of view, energy bands of finite width are allowed because not all the

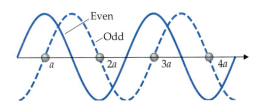

• **Figure 14–12**   Harmonic wave functions with wavelength $2a$. The even ones peak at the location of the ions, and the odd ones vanish there.

electrons have precisely the energies that will give the perfect conditions for reflection and transmission.

The details of how $E$ varies with $k$ depend on the electron–ion interaction, and we shall discuss two simple models shortly. Before doing that, however, we can give a more quantitative treatment of the splitting of energy levels by estimating the change in energy of electrons due to the ionic potential. A rough estimate is obtained by evaluating the average of the ionic potential in the electronic state,

$$\Delta E = \int \psi^{(0)*}(\vec{r})V(r)\psi^{(0)}(\vec{r})\,d^3r. \tag{14–18}$$

Here the superscript "0" on the wave functions emphasizes that these are the wave functions that apply in the absence of the potential—that is, Eq. (14–17). For simplicity, we'll work with a one-dimensional lattice with ionic separation $a$. Since everything repeats itself over the distance $a$, we limit ourselves to staying within a single "cell" of width $a$. If we choose the interval to lie between 0 and $a$, then the integral in Eq. (14–18) runs from 0 to $+a$. Moreover, the wave functions themselves must be normalized such that

$$\int_0^a \psi^{(0)*}(\vec{r})\psi^{(0)}(\vec{r})\,d^3r = 1. \tag{14–19}$$

You can check that this normalization is realized with the wave functions of Eq. (14–17).

We can now find the energy shift from the original even and odd levels by looking at the integration in Eq. (14–18), using either $\psi_{\text{even}}$ or $\psi_{\text{odd}}$ respectively for $\psi^{(0)}$. We have

$$\Delta E_{\text{even}} = \frac{1}{2}\int_0^a \cos^2(\pi x/a)V(x)\,dx = \frac{1}{4}\int_0^a V(x)\big[1 + \cos(2\pi x/a)\big]\,dx;$$

$$\Delta E_{\text{odd}} = \frac{1}{2}\int_0^a \sin^2(\pi x/a)V(x)\,dx = \frac{1}{4}\int_0^a V(x)\big[1 - \cos(2\pi x/a)\big]\,dx. \tag{14–20}$$

From these expressions we can conclude that not only are both levels shifted, but they are in addition *pushed apart by equal and opposite amounts* (•Fig. 14–13). This is true for all the even and odd levels.

Just how the energy depends on $k$ depends on details of the periodic potential. A simple example of a one-dimensional crystal is provided by a series of square wells of depth $V_0$ and width $b$, with the centers of the potentials separated by the periodicity length $a$, as in •Fig. 14–14. This representation of a crystal is the *Kronig–Penney model*. The bound-state energies can be calculated in closed form. The result is simplest in the limit that $V_0 \to \infty$ and $b \to 0$ such that the product $bV_0$ is finite; in particular,

$$bV_0 = \frac{\hbar^2}{2ma}\xi, \tag{14–21}$$

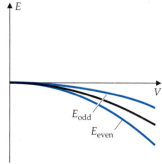

**• Figure 14–13** Energies of the even and odd wave functions, shifted and split by the potential. The energies of the even ones lie lower because, for them, the electrons and ions, which attract each other, are more likely to be near one another.

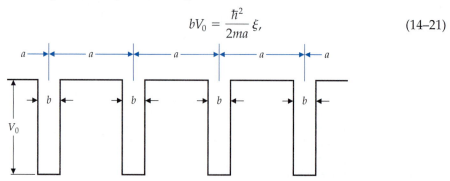

**• Figure 14–14** In the Kronig–Penney model, the periodic potential energy consists of deep, narrow square wells.

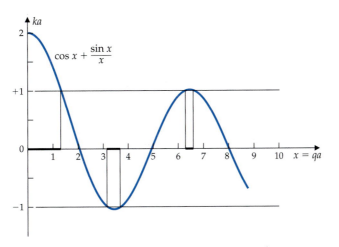

• **Figure 14–15**  A plot of the values of $\cos ka$ as a function of $qa$ for the parameter $\xi = 2$. Since $|\cos ka| < 1$, the values of $qa$ corresponding to the dark band are forbidden. These values are those for which there are energy gaps in the Kronig–Penney model.

where $\xi$ is some fixed number and the factor that multiplies it has the dimensions of energy. If we define $q$ by writing the energy as

$$E = \frac{\hbar^2 q^2}{2m},$$  (14–22)

we find the relation between $k$ and $q$—or equivalently between $k$ and $E$—by solving the eigenvalue equation

$$\cos ka = \cos qa + \frac{\xi}{2}\frac{\sin qa}{qa}.$$  (14–23)

(This equation is algebraically more complicated than, say, a counterpart for the square-well potential, Eq. (10–42), but is of the same qualitative type.)

The eigenvalue equation can best be solved graphically. If $\xi = 0$, meaning that there is no potential, then $q = k$ and the energy is indeed $(\hbar k)^2/2m$, the free-particle energy associated with a free-particle momentum $\hbar k$. If $\xi \neq 0$, we have something more complicated. •Figure 14–15 is a plot of the right side of Eq. (14–23) as a function of $qa$. The only allowed values of the left side of that equation are those that lie between $-1$ and $+1$. This condition is also plotted on •Fig. 14–15 and the combination of the two curves reveals the presence of the allowed **energy bands.** We see on the figure that the first one starts at $\cos ka = 1$ and ends at $\cos ka = -1$. There is then an **energy gap**—a continuous range of excluded $k$-values—and at that point the second energy band starts. This behavior is further reflected in •Fig. 14–16, a plot of $E(k)$ vs. $k$.

• **Figure 14–16**  Plot of $E(k) = q^2\hbar^2/2m$ as a function of $k$. The energy gaps occur at $ka = \pm n\pi$.

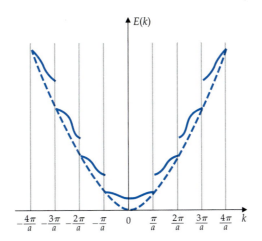

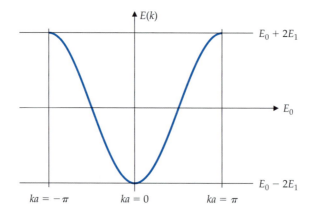

• **Figure 14–17** Plot of $E(k)$ as a function of $k$ for the model of Eq. (14–24), restricted to the principal region $-\pi/a \leq k \leq +\pi/a$.

Although the Kronig–Penney model is somewhat artificial, the bands and gaps it predicts are in fact the principal feature of solids. A second model, one that leads to a more transparent relation between $E$ and $k$, is based on the quantum mechanical version of a series of pendulums coupled to their nearest neighbors by springs. In this model the relation between $E$ and $k$ is

$$E = E_0 - 2E_1 \cos ka. \tag{14–24}$$

In order to have a single-valued relation between $E$ and $k^2$, we limit ourselves to the range $-\pi \leq ka \leq \pi$ (•Fig. 14–17). This model gives only a single energy band, since $E$ must lie between $E_0 + 2E_1$ and $E_0 - 2E_1$, but in spite of this failing it has a number of other, realistic properties. In particular, for small $k^2$ the relation between $E$ and $k$ is

$$E = E_0 - 2E_1 + k^2\left(a^2 E_1\right) = E_0 - 2E_1 + \frac{\hbar^2 k^2}{2m^*}, \tag{14–25}$$

where the quantity

$$m^* = \frac{\hbar^2}{\left(2a^2 E_1\right)}. \tag{14–26}$$

Thus in this model (as well as in the Kronig–Penney model, which is algebraically more complicated), the energy at the bottom of the band is of the form of a free-electron energy, but with an **effective mass** $m^*$. This observation will prove important when we discuss semiconductors.

If the lattice is not perfectly periodic, two things can happen. On the one hand, irregularities in the lattice that are not too disruptive maintain its conductivity. Sodium represents a very good example of this. The band structure of sodium in the solid form—that is, when it forms a periodic lattice—is very well defined, and solid sodium is a good conductor. But sodium remains a good conductor even when it is melting—that is, when the periodic structure is breaking down but there is still some maintenance of order. On the other hand, if the irregularities in the lattice structure become too large, so that one approaches a condition of random potentials, then electrons cease to propagate. They are trapped at the sites of the breakdown of periodicity. The material has changed from a conductor to an insulator.

### The Differences between Conductors and Insulators

Given the band structure, one can understand qualitatively why some materials, even if they have a periodic structure, conduct differently than others. Remember, the bands represent the *possible* energy levels of electrons in the

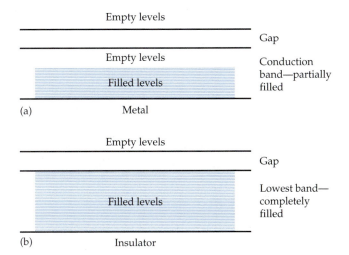

• **Figure 14–18** (a) In a conductor, the highest occupied band is partially filled, with empty energy levels within that band available to electrons. (b) In an insulator, the highest band with electrons within it is filled, and the nearest empty levels are separated from the filled band by a finite gap.

material. Whether those levels are filled or not is a question of how many electrons are around to fill them. When the number of electrons is such that a band is partially filled, then *nearby* unoccupied energy levels can be filled by electrons that acquire energy by being accelerated in an external electric field (•Fig. 14–18a). Such a material is a **conductor**. (Another way to realize a conductor is via a situation in which the energies of the most energetic electrons are in a region where bands fully overlap, so that there are always unoccupied levels just above the energy level of the most energetic electron.) The topmost, partially filled, band is called the **conduction band**. When, on the other hand, the number of available electrons is such that all the possible states in a band are filled, the exclusion principle forbids electronic acceleration by an external field, and we have an **insulator** (•Fig. 14–18b). In this situation, we could imagine imposing a large external field, in which case the lattice will be distorted and electrons can tunnel through to the next band: The insulator breaks down.

Depending on the particular atoms making up the solid, the separations between the bands may be quite narrow. If the most energetic electrons lie just at the top of a band, filling it, the material is technically an insulator. However, if the energy separation to the next (empty) band—the energy gap—is small enough, the material will become conducting at higher temperatures, because the electrons can be thermally excited into the empty conduction band above the gap. In this case we have a material that conducts more or less well according to the temperature: a **semiconductor**. These materials are of great technological importance and form the subject of the next few sections. The degree to which their behavior can be controlled is quite remarkable, and because of this, semiconductors have become the principal tools of modern electronics.

## 14–4 Semiconductors

The energy gap between a highest filled band and an empty band above it can vary quite a bit from material to material. For example, whereas the gap for carbon in the form of diamond is 5.4 eV wide, the gap width is 1.17 eV for silicon, 0.74 eV for germanium, and only 0.24 eV for InSb (indium antimonide). Although the last three materials are technically insulators—they contain a filled band with the next band above it empty—the small size of the energy gap allows them to conduct when the temperature is not zero. That is because at finite temperatures the picture of electrons filling all the levels up to a sharply

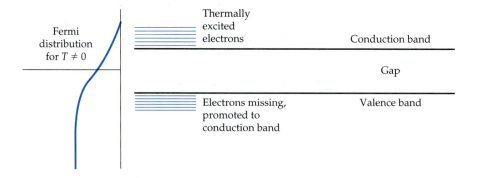

• **Figure 14–19**   At finite temperatures, the Fermi–Dirac distribution does not have a sharp edge. Some electrons from the valence band are accordingly promoted to the conduction band. The number depends on the temperature and the width of the energy gap.

defined Fermi energy is not correct. Electrons have some probability of having energy that lifts them into the next band, especially if the gap is not too large. Whereas at zero temperature we have a completely filled **valence band**, at finite temperatures there are some electrons in the band above it—the **conduction band** (•Fig. 14–19), so named because all but the lowest levels in it are empty. Materials with relatively narrow gaps are called **semiconductors**. Table 14–2 on page 416 gives the gap width for a selection of semiconductors, as well as some other data to be discussed later.

We can estimate the number of electrons that are thermally excited across the gap by noting that they are distributed in energy according to the Fermi distribution function [Eq. (12–40)]. In particular, the distribution depends on the temperature. We have

$$N(E) = \frac{2}{\exp\left[(E - E_F)/kT\right] + 1}. \tag{14–27}$$

We can simplify this expression by estimating the values of the variables that are relevant to us. The first question that arises concerns the value of $E_F$: What exactly is the Fermi energy in this situation? For an electron gas at $T = 0$, $E_F$ is well defined as the energy above which there are no electrons. But with a gap present, $E_F$ could lie anywhere within it. Later we'll show an important result: $E_F$ *lies at the midpoint of the gap*. Next we ask what values of $E$ enter into the situation. To have conduction—and this is all we care about here—it is necessary only to promote electrons to the very bottom of the "empty" band, the conduction band. Thus we are interested in Eq. (14–27) at $E = E_c$, the value of the energy at the bottom of the conduction band (•Fig. 14–20). Hence for us the factor $E - E_F = E_c - E_F = E_g/2$, where $E_g$ is the gap width. At room temperature, say 300K, Eq. (14–27) simplifies because the gap width is generally much larger than $kT$. For 300K, the value of $kT$ is roughly $1/40$ eV (a good number to remember!). For germanium, a typical semiconductor, with $E_g = 0.74$ eV, the value of $E_g/(2kT) \cong (0.74 \text{ eV})/(1/20 \text{ eV}) \cong 15$. Thus the exponential factor

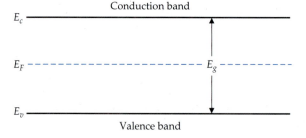

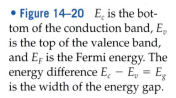

• **Figure 14–20**   $E_c$ is the bottom of the conduction band, $E_v$ is the top of the valence band, and $E_F$ is the Fermi energy. The energy difference $E_c - E_v = E_g$ is the width of the energy gap.

in the denominator of Eq. (14–27) dominates, and this allows us to approximate the distribution by

$$N(E) \cong 2 \exp\left[-\frac{(E - E_F)}{kT}\right]. \tag{14–28}$$

The exponential factor in Eq. (14–28) is $\exp(-15) \cong 3 \times 10^{-6}$. It is worth comparing this figure to the corresponding factor for carbon in diamond form—an insulator by virtue of the fact that its gap width is more than 5 eV; for diamond the factor is $\exp(-100) \cong 10^{-44}$!

### Electrons and Holes

When finite temperature effects promote some electrons from the valence band to the conduction band, these electrons leave **holes** behind them in the valence band. What is the significance of the holes? A useful analogy is to think of a bubble in a fluid as a space in which there is no fluid. Under the influence of gravity, fluid fills the space, leaving a space above it, fluid from above fills that space, leaving an empty space above it, and so on. The effect is that the bubble moves *upward* under the force of gravity, as if it had *negative mass*. A hole can move under the influence of an electric field in the same sense that a bubble in a fluid moves in the presence of gravity. As •Fig. 14–21 shows, a field acting in the positive x-direction, for example, will allow an electron in the same band as the hole to move in the negative x-direction, filling the hole and leaving a new hole behind it. This hole is in its turn filled by another electron moving in the negative x-direction, and so on, with the hole effectively moving to the right: It acts *as if it were a positively charged particle*. The motion of holes in the valence band thus contributes to the conductivity of the material.

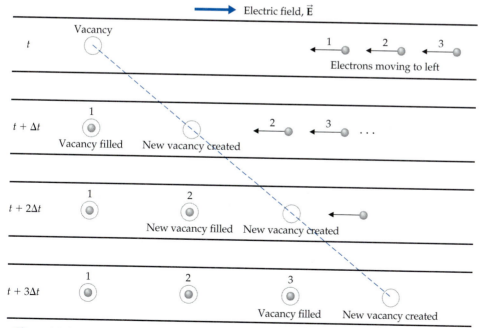

• **Figure 14–21** The motion of a hole under the influence of an electric field. Time increases from the top to the bottom of the figure. The vacancy is marked with an empty circle. The movement of the hole produces a current that adds to the electron current.

Since the holes act as positively charged particles, they are called **p-carriers** (p for *positive*), while the electrons that carry current are called **n-carriers** (n for *negative*). As we have just seen, the exponential falloff of the distribution favors the presence of the n-carriers at the bottom of the conduction band, and the number density n of n-carriers will, according to Eq. (14–28), have a temperature dependence of the form

$$n = N_c \exp\left[-\frac{(E_c - E_F)}{kT}\right], \tag{14–29}$$

where $N_c$ is a normalizing factor that we shall determine later. For now we need only state that compared to the exponential factor, $N_c$ has a weak temperature dependence, so that the temperature dependence of n is dominated by that of the exponential factor. The exponential factor in Eq. (14–28) shows that it is easiest to promote an electron from the very top of the valence band, at $E \cong E_v$ (•Fig. 14–20), so that is where we will find the holes. Now the probability of finding a hole with a given set of quantum numbers is one minus the probability of finding an electron with those quantum numbers. Therefore

$$N_p(E) = 2\left(1 - \frac{1}{\exp[(E - E_F)/kT] + 1}\right) = 2\frac{\exp[(E - E_F)/kT]}{\exp[(E - E_F)/kT] + 1}$$

$$= 2\frac{\exp[-(E_F - E)/kT]}{\exp[-(E_F - E)/kT] + 1}. \tag{14–30}$$

The relevant energy here is $E \cong E_v$, and $E_F - E_v$ is, again, approximately $E_g/2$. We have already seen that $E_g$ is much larger than $kT$, so that we can drop the exponential in the denominator compared to 1. Thus the density of p-carriers will take the general form

$$p = N_v \exp\left[-\frac{(E_F - E_v)}{kT}\right], \tag{14–31}$$

where, again, we leave the factor $N_v$ to be determined later.

We see from Eqs. (14–29) and (14–31) that the product of the densities of the two types of current carriers is

$$np = N_c N_v \exp\left[-\frac{(E_c - E_v)}{kT}\right] = N_c N_v \exp\left(-\frac{E_g}{kT}\right). \tag{14–32}$$

In other words, the temperature dependence of the product np is largely controlled by the width of the gap. This relation is quite general, and we shall have occasion to make use of it later.

Equation (14–32) allows us to show that the Fermi energy lies at the center of the energy gap. Each hole is created by the promotion of an electron, so that

$$n = p. \tag{14–33}$$

We can accordingly equate n (or p) with the square root of the product np. The relation $n = (np)^{1/2}$ reads

$$N_c \exp\left[-\frac{(E_c - E_F)}{kT}\right] = \sqrt{N_c N_v} \exp\left(-\frac{E_g}{2kT}\right). \tag{14–34}$$

We can rearrange Eq. (14–34) to the form

$$\sqrt{\frac{N_c}{N_v}} = \exp\left(-\frac{E_g}{2kT}\right)\exp\left[\frac{(E_c - E_F)}{kT}\right].$$

By taking the logarithm of both sides, we find that

$$\ln\sqrt{\frac{N_c}{N_v}} = \frac{E_c - E_F - E_g/2}{kT} = \frac{\frac{1}{2}(E_c + E_v) - E_F}{kT},$$

or

$$\frac{1}{2}(E_c + E_v) - E_F = kT \ln\sqrt{\frac{N_c}{N_v}}. \tag{14-35}$$

Now unless the ratio $N_c/N_v$ is singular at $T = 0$—and we'll see later that there is nothing at all ill behaved about that ratio—the $T = 0$ limit of this result shows that at $T = 0$ $E_F$ is exactly in the middle of the gap. For finite $T$ there is a small correction: $N_c$ and $N_v$ are not terribly different, as we shall see, so that the logarithm of their ratio is small; moreover, by assumption, $kT$ is small compared to the energies.

We have one last task left: to calculate $N_c$ and $N_v$. To do so we can look back to Eq. (12–50). That expression, which takes into account the fact that there are many different quantum states, each with the same energy, gives the number of free-electron states within a window $dE$ around energy $E$ in a volume $V$:

$$N(E)\, dE = \frac{V}{2\pi^2}\left(\frac{2m}{\hbar^2}\right)^{3/2} \frac{\sqrt{E}\, dE}{\exp\left[(E - E_F)/kT\right] + 1}$$
$$\cong \frac{V}{2\pi^2}\left(\frac{2m}{\hbar^2}\right)^{3/2} \exp\left[-\frac{(E - E_F)}{kT}\right]\sqrt{E}\, dE. \tag{12-50}$$

Note that, as usual, in the last step we have dropped the factor of 1 compared to the exponential in the denominator. We are interested in the *density*, so we divide out the factor $V$. How would this form change when the electrons are within a material rather than free? Recall from Chapter 12 that the factor $(E)^{1/2}\, dE$ came from a factor $k^2\, dk$ in the counting of states together with the free-particle relation $E = p^2/2m = \hbar^2 k^2/2m$. Two changes are necessary here. First, we measure the energy from the bottom of the conduction band; therefore we replace $(E)^{1/2}$ by $(E - E_c)^{1/2}$. Second, there is a change in the relation between $E$ and $k$ inside the crystal lattice. As we noted in the discussion following Eq. (14–24), for $k$-values near the bottom of a band, the effect of the lattice is to change the free-electron mass to an effective mass $m^*$. For rather complicated reasons, $m^*/m_e$ is quite small whenever the gap energy $E_g$ is small. Denoting the effective mass of the negative carriers by $m_n^*$, we see that the net result is that for negative carriers Eq. (12–50) can be transformed to an expression for the density that takes the form

$$n(E)\, dE \cong \frac{1}{2\pi^2}\left(\frac{2m_n^*}{\hbar^2}\right)^{3/2} \exp\left[-\frac{(E - E_F)}{kT}\right]\sqrt{E - E_c}\, dE. \tag{14-36}$$

Equation (14–36) is a density, whereas Eq. (14–29), whose coefficient $N_c$ we want to find, is a total number of carriers. Therefore we want to integrate Eq. (14–36) over the energy $E$ to make a direct comparison between the two. We reserve this exercise for Problem 26 at the end of the chapter; the result is

$$n = \int n(E)\, dE \cong \frac{1}{4}\left(\frac{2m_n^*}{\pi\hbar^2}\right)^{3/2} \exp\left[-\frac{(E_c - E_F)}{(kT)^{3/2}}\right]. \tag{14-37}$$

Comparing this equation with Eq. (14–29) gives

$$N_c = \frac{1}{4}\left(\frac{2m_n^* kT}{\pi\hbar^2}\right)^{3/2}. \tag{14-38}$$

We forego a similar derivation for the holes. We need only remark that holes also behave like free particles, this time with an effective mass $m_p^*$, and

$$N_v = \frac{1}{4}\left(\frac{2m_p^* kT}{\pi\hbar^2}\right)^{3/2}. \tag{14-39}$$

**Example 14-3**   A certain semiconductor such as we have just described has $m_n^* = 0.015m_e$ and $m_p^* = 0.39m_e$ and a gap width of $E_g = 0.18$ eV. What is the number density of negative (or positive) carriers at 293K?

**Solution**   With a semiconductor of this type, $n = p$, so we need to find only one of them. We use Eq. (14-32), from which it follows that

$$n^2 = np = N_c N_v \exp\left(-\frac{E_g}{kT}\right).$$

From Eqs. (14-38) and (14-39), we have

$$N_c N_v = \frac{1}{16}\left(\frac{2kT}{\pi\hbar^2}\right)^3 (m_n^* \, m_p^*)^{3/2},$$

so that

$$N_c N_v = \frac{1}{16}\left[\frac{2(1.38 \times 10^{-23}\,\text{J/K})(293\,\text{K})}{\pi(1.05 \times 10^{-34}\,\text{J·s})^2}\right]^3 (0.015 \times 0.39)^{3/2}(0.9 \times 10^{-30}\,\text{kg})^3$$

$$= 2.7 \times 10^{47}\,\text{m}^{-6}.$$

In addition, with $E_g = (0.18\,\text{eV})(1.6 \times 10^{-19}\,\text{J/eV}) = 1.8 \times 10^{-20}$ J, we find that

$$\frac{E_g}{kT} = \frac{(1.8 \times 10^{-20}\,\text{J})}{[(1.38 \times 10^{-23}\,\text{J/K})(293\text{K})]} = 7.1.$$

Pulling things together,

$$n = (5.2 \times 10^{23}\,\text{m}^{-3})\exp(-3.55) = 1.5 \times 10^{22}\,\text{carriers/m}^3.$$

Table 14-2 gives the effective masses for some $n$- and $p$-carriers. As we noted earlier, the effective masses of electrons at the bottom of the conduction band—that is, the $n$-carriers—are frequently an order of magnitude smaller than the electron masses. For gallium arsenide (GaAs), for example, $m_n^*/m_e = 0.067$. In fact, the table shows that the smaller the gap width, the smaller is the effective mass. The $p$-carriers generally have $m_p^*$ much closer to the regular electron mass.

| Table 14-2 | Data for Some Semiconductors | | | | | |
|---|---|---|---|---|---|---|
| Material | $E_g$ (eV) | $N_c$(m$^{-3}$) | $N_v$(m$^{-3}$) | $m_n^*/m_e$ | $m_p^*/m_e$ | $\kappa$ |
| Si | 1.14 | $2.7 \times 10^{25}$ | $1.1 \times 10^{25}$ | 1.06 | 0.58 | 11.7 |
| Ge | 0.67 | $1.0 \times 10^{25}$ | $5.2 \times 10^{24}$ | 0.56 | 0.35 | 15.8 |
| GaAs | 1.43 | $4.6 \times 10^{23}$ | $1.5 \times 10^{25}$ | 0.07 | 0.71 | 13.1 |
| InP | 1.35 | $4.9 \times 10^{23}$ | $6.9 \times 10^{24}$ | 0.07 | 0.42 | 12.4 |
| InSb | 0.18 | $4.6 \times 10^{22}$ | $6.2 \times 10^{24}$ | 0.015 | 0.39 | 17.9 |

Incidentally, you might wonder whether $m_n^*$ and $m_p^*$ are just fudge factors to make things come out right or whether the electrons and holes behave in other circumstances as if they had the modified masses. This can be checked by imposing an external magnetic field and looking for the frequency $\omega$ (the *cyclotron frequency*) with which the electrons in the semiconductor orbit about the field lines. A simple application of Newton's second law, $F = ma$, gives

$$qvB = \frac{m^* v^2}{r},$$

so that $m^* = qBrv/v^2 = qBr^2\omega/r^2\omega^2 = qB/\omega$. One uses what are known as cyclotron resonance techniques[5] to measure $\omega$ and hence the effective mass. It is *this* effective mass that also appears in our formulas.

---

**Example 14–4**   In a cyclotron resonance experiment, resonances occur at $B = 0.12$ T and at 0.45 T for the given fixed frequency $f = 2.4 \times 10^{10}$ Hz. Assuming that the first resonance corresponds to $n$-carriers and the second to $p$-carriers, what are the effective masses $m_n^*$ and $m_p^*$ as multiples of $m_e$?

**Solution**   With the relation given in the text between the frequency and the mass, we solve the equation to find the masses involved, namely $m/m_e = eB/(m_e\omega_{res}) = eB/(2\pi f_{res}m_e)$. Then

$$\frac{m_n^*}{m_e} = \frac{(1.6 \times 10^{-19}\,\text{C})(0.12\,\text{T})}{2 \times 3.14(2.4 \times 10^{10}\,\text{s}^{-1})(0.9 \times 10^{-30}\,\text{kg})} = 0.14$$

and

$$\frac{m_p^*}{m_e} = \left[(0.45\,\text{T})/(0.12\,\text{T})\right]\frac{m_n^*}{m_e} = 0.53.$$

---

**Example 14–5**   Assume that the only difference between the resistivity of a semiconductor and that of a metal resides in the density of $n$- (and $p$-) carriers. The resistivity of a semiconductor for which $n = p$ at $T = 400$K is $9.5 \times 10^{-2}\,\Omega \cdot$m, while at $T = 700$K it is $1.5 \times 10^{-3}\,\Omega \cdot$m. Assuming that the $T^{3/2}$ dependence of the density $n(= p)$ can be ignored, use this information to calculate the gap width in the semiconductor.

**Solution**   The resistivity $\rho_n$ due to the $n$-carriers is inversely proportional to $n$. The $p$-carriers also transport charge in parallel with the $n$-carriers. Recalling that for parallel circuits it is the inverse resistances that add, we have

$$\rho^{-1} = \rho_n^{-1} + \rho_p^{-1} = (\text{constant}) \times (n + p) = (\text{constant}) \times 2n.$$

Ignoring the $T^{3/2}$ dependence, we have $\rho = C \exp(E_g/2kT)$, where $C$ is another constant. In turn,

$$\frac{\rho(T_1)}{\rho(T_2)} = \exp\left[\frac{E_g}{2k}\left(\frac{1}{T_1} - \frac{1}{T_2}\right)\right].$$

We can solve this equation for $E_g$ by taking the logarithm:

$$E_g = 2k\frac{T_1 T_2}{T_2 - T_1}\ln\left(\frac{\rho(T_1)}{\rho(T_2)}\right).$$

In our case $T_1 = 400$K and $T_2 = 700$K. Thus

$$E_g = 2(1.38 \times 10^{-23}\,\text{J/K})\frac{(400\text{K})(700\text{K})}{700\text{K} - 400\text{K}}\ln\left(\frac{9.5 \times 10^{-2}}{1.5 \times 10^{-3}}\right) = 1.1 \times 10^{-19}\,\text{J} = 0.67\,\text{eV}.$$

---

[5] These are of the type discussed in connection with *nuclear magnetic resonance* in Chapter 9.

## 14–5 Intrinsic and Extrinsic Semiconductors

The system we described in the previous section, in which each hole near the top of the valence band is matched with an electron at the bottom of the conduction band, is called an **intrinsic semiconductor**. In this kind of semiconductor, the number of $n$-carriers is identical to the number of $p$-carriers. It turns out that one can get an excess of positive or negative carriers by **doping**—that is, adding *impurities* to an intrinsic semiconductor. Materials changed in this way are called **extrinsic semiconductors**.

Consider, for example, a crystal lattice of silicon atoms to which some—typically one part in $10^5$ or so—arsenic atoms have been added (•Fig. 14–22). With such a small admixture, the crystal structure will be unchanged, with the occasional arsenic ion replacing a silicon ion. However, the electron structure will change, because arsenic has *five* valence electrons, while silicon has *four*. Four of the valence electrons supplied by the arsenic atom play the same role in the chemical structure of the crystal as do the four silicon valence electrons—they form the chemical bonds that hold the crystal together. But one more electron is present, and it is very weakly bound to the ion, so that it is easily ionized and becomes a carrier of current. Arsenic is thus called a **donor impurity**, and since an $n$-carrier has been added, the doped material is called an **$n$-type semiconductor**.

The arsenic atom contributes significantly to the current-carrying ability of the material. To see just how weakly the extra electron is bound, we assume that it is bound to the screened nucleus $(Z_{\text{eff}} = 1)$ with the atomic binding energy of the Bohr model, namely

$$E_B = -\frac{1}{2}\frac{e^4 m_e}{(4\pi\varepsilon_0\hbar)^2} = -13.6 \text{ eV}.$$

However, because $m_e$ is replaced by $m_n^*$, which is generally much smaller than $m_e$, the binding is reduced. Furthermore, in the Bohr model the orbital radius is

$$a_0 = \frac{\hbar}{m_e c \alpha},$$

so that $a_0$ is inversely proportional to the mass. Thus when $m_e$ is replaced by $m_n^*$, the orbital radius of the last electron is increased enough to put the electron well into the full environment of the material. Since the semiconductor has a dielectric constant $\kappa = \varepsilon/\varepsilon_0$, the effective binding is of an order of magnitude

$$E_d = -\frac{1}{2}\frac{e^4 m_n^*}{(4\pi\kappa\varepsilon_0\hbar)^2}. \tag{14–40}$$

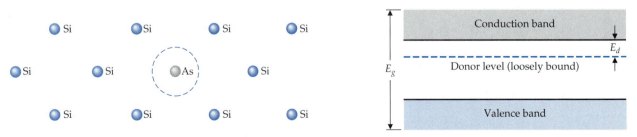

• **Figure 14–22** An impurity As atom inserted into a lattice of Si atoms, with four valence electrons, does little to distort the lattice. But, since As has an extra valence electron, that electron is very loosely bound, and only an energy $E_d$ (donor energy gap) is needed to ionize it.

We have used the notation $E_d$ for the ionization energy of the donor levels. Typically, $\kappa$ is on the order of 10 (see Table 14–2), and with $m_n^* = m_e/10$, this binding energy is 1,000 times smaller than the naive value. Equation (14–40) is only suggestive and does not rigorously apply to real semiconductors. Nevertheless, the order of magnitude is right: This binding energy is typically some thousand times smaller than the normal binding energy; that is, it is on the order of 10 millivolts.

## Fermi Energies in Doped Semiconductors

The changes in the binding energy we have described have some important consequences. We concentrate here on an understanding of the locations of the Fermi energies for $n$-type semiconductors, which, as we'll see a little later, allows us to understand important properties of modern electronic devices.

The donor electrons have energy levels $E_d$ that lie *just below the conduction band* (•Fig. 14–23a). The donor electrons in these levels are therefore easily excited into the conduction band. The donor levels thus replace the top of the valence band as the source of $n$-carriers. In fact, for an $n$-type semiconductor, almost all of the conduction electrons come from the donors. Thus $E_d$ plays the role that $E_v$ played for an intrinsic semiconductor. The same reasoning we used in our discussion of the Fermi energy shows that for an $n$-type semiconductor $E_F$ lies approximately midway between $E_d$ and $E_c$, very close to the bottom of the conduction band. A repeat of the discussion that gave us this relation for intrinsic semiconductors shows that there is a small correction to $E_F$ that is linear in the temperature:

$$E_F = \frac{E_c + E_d}{2} + kT \times (\text{small factor}).  \qquad (14\text{–}41)$$

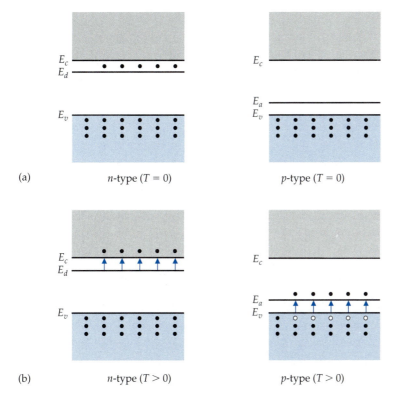

(a)  $n$-type ($T = 0$)  $p$-type ($T = 0$)

(b)  $n$-type ($T > 0$)  $p$-type ($T > 0$)

• **Figure 14–23**  (a) At $T = 0$, an $n$-type semiconductor will have donor levels, filled with extra valence electrons, just below the conduction band. A $p$-type semiconductor will have unfilled acceptor levels just above the valence band. (b) At finite temperatures, the electrons in the donor level of an $n$-type semiconductor are easily promoted into the conduction band. In a $p$-type semiconductor, electrons near the top of the valence band are easily promoted into the nearby acceptor levels, leaving holes near the top of the valence band. (From *Physics for Scientists and Engineers*, 2nd ed., by Paul Fishbane, Stephen Gasiorowicz, and Stephen T. Thornton. Copyright © 1996 by Prentice-Hall, Inc., Upper Saddle River, NJ.)

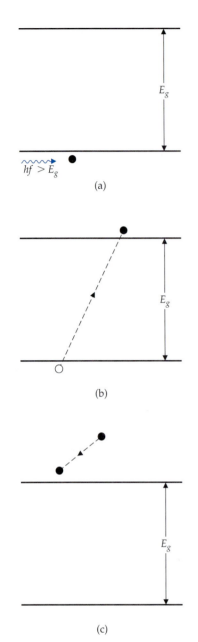

(a)

(b)

(c)

(d)

An **acceptor impurity** has *fewer* valence electrons than the bulk of atoms in the semiconductor. With such an impurity, a surplus of *p*-carriers will be created. Boron is an example when it is used to dope germanium; boron has three valence electrons. At the site where there is a boron atom instead of a germanium atom, only three electrons are available for making the chemical bonds necessary for the crystal structure. One of the electrons from the top of the valence band becomes attached to the boron atom. That electron leaves behind a hole. Thus each boron atom adds a hole or, equivalently, a *p*-carrier.

Semiconductors with a surplus of holes are called *p*-**type semiconductors**. Here too the Fermi energy is displaced. The lowest lying acceptor levels, with energies that we label as $E_a$, are the ones that electrons from the valence band go to. Those levels therefore act like levels near the bottom of the conduction band in intrinsic semiconductors, and we expect $E_F$ to lie approximately halfway between $E_v$ and $E_a$ (•Fig. 14–23b).

## 14–6 Engineering Applications of Semiconductors: Present and Future

### Optical Effects in Semiconductors

Materials are transparent to light, or to electromagnetic radiation of a given frequency, when they don't absorb the radiation that impinges on them. Absorption occurs when the electrons in the material can be excited to a higher level, so that the probability that a photon will be absorbed by a material generally depends on the frequency of the photon. In a semiconductor, only for a frequency $f$ such that $hf$ is larger than the gap energy $E_g$ can an electron in the valence band absorb the associated photon and make a transition to an unoccupied level in the conduction band as a result. When this absorption occurs, we say that the electron has undergone **photoexcitation**. The existence of the threshold frequency, reminiscent of the threshold for the photoelectric effect discussed in Chapter 6, implies transparency below the threshold frequency and opacity above the threshold frequency. Electrons that are photoexcited into the conduction band add to the *n*-carriers and thus enhance a current when a given voltage is employed. In other words, the current increases when light of high enough frequency is employed. This is how CdS cells that operate exposure meters work. With an energy gap of 2.42 eV, the maximum wavelength for photoabsorption is approximately 438 nm.

The photoexcited electrons will lose energy and eventually fall into a hole in the valence band (**recombination**). The process produces light (**photoluminescence**) as described in •Fig. 14–24.

In intrinsic semiconductors electron–hole recombination is generally rapid, with a lifetime on the order of $10^{-8}$ s, typical of atomic processes. In extrinsic semiconductors there will be impurity levels in the energy gap, and electrons may undergo transitions to these levels. Under certain circumstances such levels may be *metastable*; that is, further transitions from the impurity levels to levels in the valence band are strongly suppressed by selection rules of the kind discussed in Chapter 13. In that case, the electrons are said to be *trapped*, and depending on the degree of metastability, the time for a transition can be measured in seconds or even minutes. The transition from the metastable state is as-

• **Figure 14–24** Photoluminescence. (a) A photon excites an electron from the valence band to the conduction band. (b) A hole is left behind. (c) After many ionic collisions the excited electron has the energy of the bottom of the conduction band. (d) The electron falls back into the hole, emitting a photon with frequency given by $hf = E_g$.

sociated with **phosphorescence**. The frequency—and thus the color—of the light emitted depends on the impurity levels in the gap, which can be controlled by the choice of impurities. This effect is used in the design of color television tubes.

---

**Example 14–6** An LED device has an energy gap of 1.82 eV. What is the wavelength of the light emitted during recombination?

**Solution** We set the energy of the photon equal to the gap energy $E_g$; that is, $hf = E_g$. Thus

$$\lambda = \frac{c}{f} = \frac{hc}{E_g} = \frac{(6.63 \times 10^{-34}\,\text{J·s})(3 \times 10^8\,\text{m/s})}{(1.82\,\text{eV})(1.6 \times 10^{-19}\,\text{J/eV})} = 683\,\text{nm}.$$

---

## The *p–n* Junction

Many of the important properties of electronic devices, including the digital devices that lie behind computers, have to do with how various semiconductors act when they are brought together. The simplest illustration occurs when *n*-type and *p*-type semiconductors are brought into contact, making what is called a *p–n* **junction**. Moreover, we shall see shortly that the *p–n* junction appears as a basic building block in more complicated semiconductor devices.

Consider two pieces of the same intrinsic semiconductor, one doped to *n*-type and one to *p*-type. The hole densities in the *n*-type and *p*-type semiconductors will be denoted by $p_n$ and $p_p$, respectively; the electron densities in the semiconductors will be respectively denoted by $n_n$ and $n_p$. We bring the two semiconductors together. We call the side with the *n*-type material the *n*-side and the other side the *p*-side.

The energy-level structures of the two types of semiconductors before and after contact are shown in •Fig. 14–25. The energy gap is the same for both, because the same intrinsic semiconductor is used. But after contact, the conduction and valence bands will be distorted. This happens because $E_F$ must be the same for two systems in contact and in equilibrium.[6] As a result the energy levels are

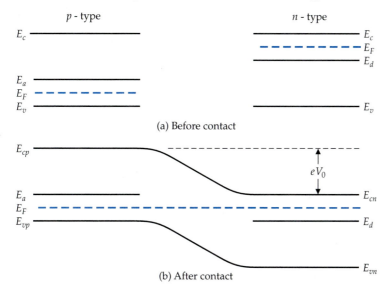

(a) Before contact

(b) After contact

• **Figure 14–25** The energy structure of the *p–n* junction. (a) Energy levels in *p*- and *n*-type semiconductors before these materials are put in contact with each other. (b) Upon contact the levels must distort because, once equilibrium is established, the Fermi energies must match. (After B. G. Streetman, *Solid State Electronic Devices*. Copyright © 1980 by Prentice-Hall, Inc., Upper Saddle River, NJ.)

---

[6] A technical proof, which is fundamentally a question of thermodynamics, can be found on page 1196 in *Physics for Scientists and Engineers*, 2d ed., by Fishbane, Gasiorowicz, and Thornton. Copyright © 1996 by Prentice-Hall, Inc., Upper Saddle River, New Jersey.

lowered on the *n*-side and raised on the *p*-side. In particular, as the figure shows, the energies of the bottoms of the conduction bands on the two sides differ after the junction is formed. If $E_{cn}$ and $E_{cp}$ are the energies of the bottom of the conduction bands in the *n*- and *p*-type pieces, respectively, then their difference is conventionally written in the form

$$E_{cp} - E_{cn} = eV_0; \tag{14–42}$$

a similar expression holds for the differences in energies of the tops of the valence bands. The quantity $V_0$, which is an electric potential, is called the **contact potential**. Its precise value is a property of the materials making up the junction.

Knowing the energy levels, we can use our previous work to learn about the density of the carriers. First we give a quantitative description; then we present a qualitative explanation of the results. We have for the densities of the *n*-carriers on the two sides

$$n_n = N_c \exp\left[-\frac{(E_{cn} - E_F)}{kT}\right]; \qquad n_p = N_c \exp\left[-\frac{(E_{cp} - E_F)}{kT}\right]. \tag{14–43}$$

By taking the ratio of these equations we find

$$\frac{n_p}{n_n} = \frac{\exp[-(E_{cp} - E_F)/kT]}{\exp[-(E_{cn} - E_F)/kT]} = \exp\left[-\frac{(E_{cp} - E_{cn})}{kT}\right] = \exp\left(-\frac{eV_0}{kT}\right). \tag{14–44}$$

In exactly the same way,

$$\frac{p_n}{p_p} = \exp\left(-\frac{eV_0}{kT}\right). \tag{14–45}$$

The physical interpretation of these relations is as follows: As •Fig. 14–26 shows, the *p*-side has an excess of mobile holes, while the *n*-side has an excess of mobile electrons. The densities need not be the same, but each material must be electrically neutral. This is assured on the *p*-side by the fixed negative acceptor ions (negative because they have captured an electron) and on the *n*-side by the fixed positive donor atoms (positive because they have given up an electron). When the two sides are put together, some holes from the *p*-side will diffuse to the *n*-side, and some electrons from the *n*-side will diffuse to the *p*-side. Ultimately equilibrium will be achieved because the fixed positive charges remaining on the *n*-side and the fixed negative charges remaining on the *p*-side set up an electric field that stops the flow. The *n*-side will then be at a higher potential—just the potential $V_0$—than the *p*-side. Note that since electrons have a negative charge, the energy of electrons on the *n*-side is *reduced* by $eV_0$.

• **Figure 14–26**   Physical mechanism for the distortion of energy levels shown in •Fig. 14–25. Negative ions, with their holes, diffuse from the *p*-type region into the intermediate *depletion region*. They diffuse because they are electrically neutral. Positive ions, with their electrons, also electrically neutral, diffuse from the *n*-type region into the depletion region. Within the depletion region, the electrons recombine with the holes. The leftover negative and positive ions create a charge separation that gives rise to an induced electric field that limits further diffusion by exerting dipole forces on additional neutral combinations.

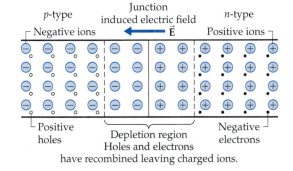

Equation (14–44) shows that the concentration of the electrons on the $p$-side is less than the concentration on the $n$-side by a factor $\exp(-eV_0/kT)$. You might think this means that to achieve equilibrium the $n$-side electrons would migrate to the $p$-side. However, electrons on the $n$-side must climb a potential hill of height $eV_0$ to get to the $p$-side. Only a number reduced by the Boltzmann factor $\exp(-eV_0/kT)$ will be able to make the climb. In that way equilibrium is established. (This, indirectly, is the ultimate reason for the equality of the Fermi energies.) Similar arguments can be made about the holes.

*The p–n Junction as Diode*   A $p$–$n$ junction acts as an important electronic device when an external potential $V_{ext}$ is applied across it, as in •Fig. 14–27. The system is no longer in equilibrium, and a net charge flow is possible. Note that since the system is no longer in equilibrium, the Fermi energies on the two sides no longer have to match. Consider first the case in which the external potential is higher on the $p$-side than on the $n$-side. This situation is that of a **forward bias**. The energy levels for the electrons (which, remember, have charge $-e$) on the $n$-side are raised relative to the energy levels on the $p$-side, and the potential difference $|V_0|$ is *reduced* to $(|V_0| - |V_{ext}|)$. The electrons that migrate from the $n$-side to the $p$-side have less of a hill to climb.

The holes that might tend to migrate from the $p$-side to the $n$-side also have a lesser hill to climb: The potential difference is still $(|V_0| - |V_{ext}|)$, but holes have a charge opposite that of electrons. The net result is an increased flow of negative charge from the $n$-side to the $p$-side, or a net current from the $p$-side to the $n$-side. This current increases as the forward bias increases and $(|V_0| - |V_{ext}|)$ decreases.

We could, on the other hand, apply a **reverse bias**: an external potential higher on the $n$-side than on the $p$-side. Then the energy levels on the $n$-side are lowered relative to the energy levels on the $p$-side. The potential difference is now *increased* to $(|V_0| + |V_{ext}|)$. Both electrons from the $n$-side and holes from the

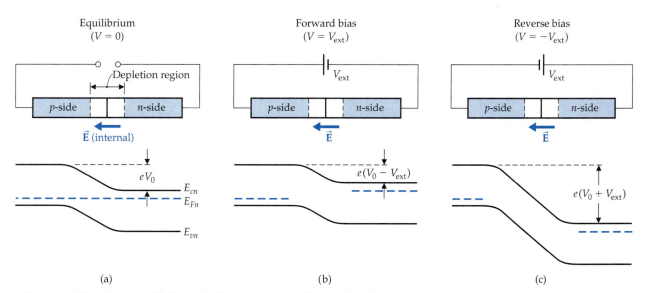

**• Figure 14–27**   (a) At equilibrium, the Fermi energy on the $p$-side is the same as that on the $n$-side, as in •Fig. 14–25. (b) With a forward bias, an external potential $eV_{ext}$ opposing the contact potential $eV_0$ reduces the internal electric field, making the flow of charge from the $n$-side to the $p$-side easier. (c) With a reverse bias, the external potential adds to the contact potential, thus increasing the internal electric field and making the flow of charge from the $n$-side to the $p$-side more difficult. (After B. G. Streetman, *Solid State Electronic Devices*. Copyright © 1980 by Prentice-Hall, Inc., Upper Saddle River, NJ.)

*p*-side have an even harder time climbing the hill, so that there is very little current due to the charge carriers.

In sum, a forward bias induces a current, while a reverse bias does not. This is the behavior of a **diode**, an electronic device that passes current in one direction only. In effect the diode is a switch that turns off or on according to whether the bias is forward or reverse. And switches are the key elements in digital circuits.

*Current in a p–n Junction*   Here we describe the behavior of the *p–n* junction as diode in a more quantitative manner. We suppose that the biasing voltage $V_{\text{ext}}$ will be either a forward or reverse bias according to whether $V_{\text{ext}}$ is positive or negative, respectively. Let's consider explicitly the current due to the motion of the holes. The potential barrier $e(V_0 - V_{\text{ext}})$ (see •Fig. 14–27) inhibits the positive charge carriers from the *p*-side, whose density is $p_p$, from flowing to the *n*-side, and only the fraction

$$p_p \exp\left[-e(V_0 - V_{\text{ext}})/kT\right] \tag{14–46}$$

gets through. Positive carriers from the *n*-side do not encounter such a barrier, but their density is already reduced by the factor $\exp(-eV_0/kT)$. The same sort of reasoning holds for the *n*-carriers.

Now we think about the current. Let $I_0$ be the current composed of *p*-carriers moving to the *p*-side and *n*-carriers traveling to the *n*-side. In the absence of any biasing potential, $I_0$ will be canceled by the current of *p*-carriers going to the *n*-side and *n*-carriers going to the *p*-side, both of which are proportional to $\exp(-eV_0/kT)$. When the biasing voltage is turned on, the latter current—holes moving to the *n*-side and electrons to the *p*-side—is modified by the factor $\exp(+eV_{\text{ext}}/kT)$, as we see from Eq. (14–46). Thus the net current traveling to the *n*-side is

$$I = I_0\left[\exp(+eV_{\text{ext}}/kT) - 1\right]. \tag{14–47}$$

The relation between the current and the external potential expressed by this equation is shown in •Fig. 14–28. We see that the junction acts as a rectifier, with the net current behaving quite differently according to whether $V_{\text{ext}}$ is positive or negative.

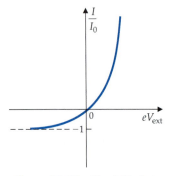

• **Figure 14–28**   The *I–V* plot for a *p–n* junction, showing how the junction acts as a rectifier.

## Transistors

Digital circuits involve the sensitive and rapid control of switches; in turn this suggests the need for controlled amplification, so that small currents can have large effects. The transistor is a device that amplifies. Today there are a variety of transistors in use, and it would take us too far afield to describe all of them. We can, however, examine one transistor, the *bipolar junction transistor*, which illustrates well how the *p–n* junction is used as a building block. This transistor was the first one, so it is of historical interest as well. It was developed at the end of the 1940s by John Bardeen, Walter Brattain, and William Shockley, working at Bell Laboratories (•Fig. 14–29). One could argue that, since it was the device that made digital electronics possible,[7] it is the single most important invention of the century.

We describe the *n–p–n* bipolar junction transistor, so called because of an arrangement that looks, at least schematically, like an $n - p$ junction adjoined

• **Figure 14–29**   John Bardeen, Walter Brattain, and William Shockley, developers of the transistor. The device lies at the base of all of modern electronic technology.

---

[7] Before the transistor, computers and other digital devices were constructed by using vacuum tubes, a particularly awkward technique. The teams that ran the earliest programmable computers spent a large part of their time changing tubes!

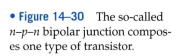
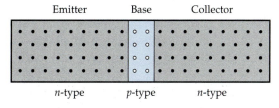

Emitter    Base    Collector

• **Figure 14–30** The so-called *n–p–n* bipolar junction composes one type of transistor.

*n*-type    *p*-type    *n*-type

to a *p–n* junction (•Fig. 14–30). The *n*-type material on the left is the *emitter*, and the *n*-type material on the right is the *collector*, while the *p*-type material in between is the *base*. This arrangement differs most importantly from a simple connection of two junctions in that the base is very narrow. By "narrow," we mean that an electron entering the *p*-region (the base) from the emitter can traverse that region with only a small probability of recombination with a hole before it reaches the collector.

As we showed in our discussion of the *n–p* junction, the *n*-region is at a higher potential than the *p*-region, so that the contact potential for the bipolar junction has a step form, as shown in •Fig. 14–31(a). Because the electron is negatively charged, its potential energy in the *n*-region is *lower* than in the *p*-region, and its potential energy is the inverse of the contact potential, as shown in •Fig. 14–31(b).

Suppose we now put a biasing potential $V_{eb}$ between the emitter and the base such that the potential energy of an electron in the base region is lowered from $eV_0$ to $e(V_0 - V_{eb})$. An additional biasing potential between the base and the emitter, $V_{bc}$, chosen to be much larger than $V_{eb}$, further lowers the electron potential energy at the collector, and the electron potential-energy curve now has the shape shown in •Fig. 14–32. The potential barrier for the electrons has been lowered to the point that electrons will flow from the emitter to the base. Because the base is very narrow, the electron density is only slightly reduced by recombination with the holes in the base, and most of the electrons enter the collector. The emitter current is large, because $V_{eb}$ is positive—we lie on the steep part of the curve of •Fig. 14–28. There is also a very small current $I_b$ entering the base, since some electrons do leave the region, and current conservation gives

$$I_c = I_e + I_b. \tag{14–48}$$

Keep in mind that $I_b$ is much less than both $I_e$ and $I_c$.

Suppose that we now make a small change in $V_{eb}$. Because we are on the very steep part of the $I/I_0$ curve, a small change in $V_{eb}$ makes a large change in $I_e$ and therefore in $I_c$. In this way the transistor makes a very good amplifier.

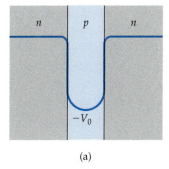

(a)

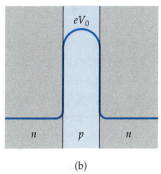

(b)

• **Figure 14–31** (a) The contact potential in an *n–p–n* bipolar junction. (b) The potential energy of an electron in an *n–p–n* bipolar junction.

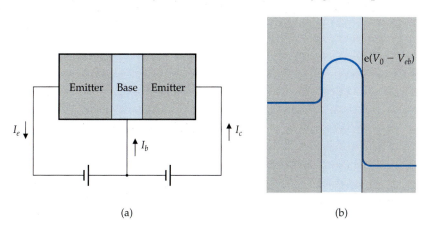

(a)

(b)

• **Figure 14–32** (a) Biasing potentials and current flows in a bipolar junction transistor. (b) The change in the electron potential barrier in the presence of biasing potentials. (From *Physics for Scientists and Engineers*, 2nd ed., by Paul Fishbane, Stephen Gasiorowicz, and Stephen T. Thornton. Copyright © 1996 by Prentice-Hall, Inc., Upper Saddle River, NJ.)

Note that holes flowing from the base to the emitter can counter the large current $I_e$, so, to avoid this effect, the doping of the emitter region is much larger than that of the base region, which in its turn is also larger than that of the collector region.

### Semiconductor Lasers

The juxtaposition of $n$–$p$ junctions, as in $n$–$p$–$n$ and $p$–$n$–$p$ junctions—these latter form the transistor—is by now a routine procedure. One of the structures that can be created by such a juxtaposition is the **semiconductor laser**. The semiconductor within which the laser light will be produced—GaAs, for example—is sandwiched between two layers of semiconductors that are $p$-doped on one side and $n$-doped on the other. •Figure 14–33 shows the energy diagram for this arrangement. When a voltage is applied so that the $p$-side is at a higher potential, electrons flow towards the $p$-side at the bottom of the conduction band, but they accumulate in the transition region because of the potential barrier that is visible in the figure. Similarly, the holes flow towards the $n$-side, but the barrier keeps them in the transition region. Thus in the transition region electrons accumulate on top and holes at the bottom, and a population inversion is built up. The electrons and holes recombine by the emission of photons, which very rapidly stimulate further recombination. This is the laser mechanism described in Chapter 13.

### Nanostructures and Integrated Circuits

Beginning in the 1980s engineers learned to fabricate combinations of semiconductors and other materials with a structure whose dimensions are on the order of nanometers—called **nanostructures**. These structures involve the deposition of thin layers of one material on top of another by means of molecular beams. A bottom layer composed of a single crystal, the *substrate*, ensures a flat surface. By controlling the beams, the thickness of the layers can be controlled, and donors or acceptor impurities can be added in precisely controlled amounts. Finally, further structure can be included by chemical etching, in which patterns are made by removing material from the layers.

Why are nanostructures interesting? The speed of a circuit is limited, quite literally, by its size, with the speed of light providing an absolute limit. Thus it is advantageous for computers to have circuits that are as small as possible. Circuits that are printed on substrates using the techniques thus far described in this section may contain millions of transitors and other types of logic gates. These **integrated circuits** are reliable and cheap; the days of applying solder at each junction of a circuit are over. To the extent that the structures are as small as possible, the circuit will be as fast as possible. The semiconductor lasers we have described have been produced in large numbers as nanostructures and

• **Figure 14–33**  Enegy levels of a semiconductor laser. An applied voltage causes electrons to flow into a region of lowered energy in the conduction band and holes to flow into a region of raised energy in the valence band. A recombination of a hole and an electron is accompanied by the emission of a photon, which in turn stimulates more recombination and a buildup of photons. (From *Physics for Scientists and Engineers*, 2nd ed., by Paul Fishbane, Stephen Gasiorowicz, and Stephen T. Thornton. Copyright © 1996 by Prentice-Hall, Inc., Upper Saddle River, NJ.)

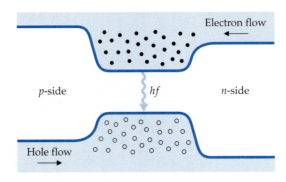

could transmit signals from circuit element to circuit element with optical waves. Integrated circuits with a structure on the scale of nanometers will provide the basis for new electronic advances in the next century. The next subsection describes one type of structure that may well prove important.

## Artificial Atoms

A particularly important use of the nanostructure techniques described at the end of the previous subsection is the construction of **heterojunctions**—junctions involving two different semiconductors. Interesting properties emerge from the fact that different semiconductors have different gaps. For example the junction in •Fig. 14–34a is made of two different semiconductors, each doped as an *n*-type material, with the semiconductor on the right having a wider gap than the semiconductor on the left. In each case $E_F$ will lie closer to $E_c$ than to $E_v$. When the materials are joined to form a heterojunction, charges flow until the Fermi energy is the same on both sides. This moves the levels around, but since the gap widths on the two sides do not change—they are an intrinsic property of each semiconductor—we may develop kinks in the potential, as shown in •Fig. 14–34b. A potential well in one dimension has been created; electrons in the conduction band at the interface find themselves within a potential well that is aligned perpendicular to the plane of the figure, and their motion is thus channeled in that direction. Since layers can be deposited so that the donor ions are relatively far from the potential kink, the electrons do not find any impurity ions in their way, and they are very mobile. This configuration is useful in devices for which speed is important, such as switches in supercomputers.

When several heterojunctions are put together, it is possible to create what have been called **artificial atoms**. This term refers to bound systems in which, like atoms in nature, quantum mechanical behavior dominates. Consider the following structure, pictured in •Fig. 14–35: A layer of silicon-doped GaAs is deposited on a substrate; silicon-doped GaAs is an *n*-type semiconductor that in the application to be described will act as one metal plate of a capacitor. A thin (10 nm) GaAs layer is deposited on top of this and will act as a layer

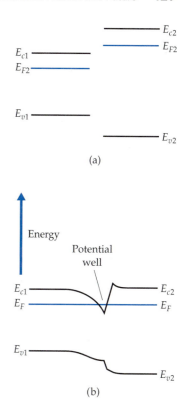

(a)

(b)

• **Figure 14–34**   (a) Two different *n*-type semiconductors are brought into contact to form a heterojunction. (b) As a result, the levels distort, with the edge of the conduction band forming a kink that acts as a one-dimensional potential well for electrons. (From *Physics for Scientists and Engineers*, 2nd ed., by Paul Fishbane, Stephen Gasiorowicz, and Stephen T. Thornton. Copyright © 1996 by Prentice-Hall, Inc., Upper Saddle River, NJ.)

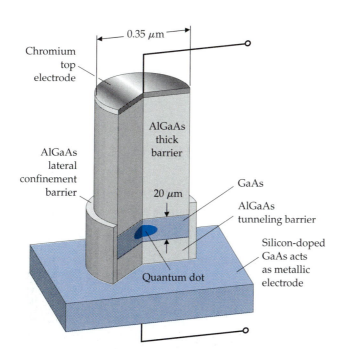

• **Figure 14–35**   The construction of an artificial atom, analogous to a three-dimensional well within which are discrete electronic levels.

through which electrons can tunnel. Above this layer, one constructs a three-dimensional **quantum dot**—a three-dimensional box—by creating an "island" of GaAs, surrounded by an AlGaAs barrier. Above this layer, a layer of Al-GaAs too thick to tunnel through is deposited, and above that chromium is deposited, making the second plate of the capacitor. Effectively, we have a box of GaAs between two plates of a capacitor, with barriers that allow electrons to pass between the lower plate and the box, but not between the upper plate and the box. This box is, as we shall see shortly, our artificial atom. Note that the box is not empty: With dimensions of, say, 10 nm × 100 nm × 100 nm and an average interatomic distance of 0.2 nm, the box contains many atoms. These atoms, however, form the usual lattice, and any additional electrons that enter the box *move freely with an effective mass*, just as they do in a larger crystal.

If a voltage is applied between the plates of the capacitor so that the top plate is made positive compared with the lower one, then electrons will be attracted from the bottom plate towards the box. One can then pull electrons into the box one by one. Each additional electron will be harder to get in because of the repulsion of the other electrons in the box and because of the exclusion principle. There will be a sequence of potential differences between the bottom plate and the box as the electrons are pulled into the box. In this way one can study the effects of electron–electron repulsion in these artificial atoms. The energy levels are separated by (see Chapter 8)

$$\Delta E = \frac{\hbar^2 \pi^2}{2m_n^* a^2},$$  (14–49)

where $a \cong 100$ nm and the effective electron mass $m_n^*$ is on the order of a tenth of the electron mass. Thus the level splittings $\Delta E$ are in the range $4 \times 10^{-5}$ eV $\cong k(0.5\text{K})$. These spacings are small, and as a result it is necessary to perform the experiments at low temperatures, say $T \cong 0.1\text{K}$. The existence of such levels has been confirmed by the selective absorption of laser light at frequencies that correspond to the differences between the levels in the well, $hf = \Delta E$.

One can construct "artificial molecules" by placing two quantum dots close to each other, and there has been work on constructing a "crystal" made up of regularly spaced quantum dots or wells. Since all this must be done at low temperatures, we are as yet far from any practical application. Nevertheless, the ingenuity with which research and development have been carried out in this field is so impressive that one can look forward to applications in the near future. These applications will make use of the ability to control energy spacings through control of the box size.

## 14–7 Superconductivity

In 1911, working at Leiden, Holland, Heike Kamerlingh-Onnes, who had been able to reach new lows in temperature after his success in liquifying helium, discovered in a routine study of the properties of mercury that the resistivity of the element drops very sharply to zero at 4.2K. This phenomenon, called superconductivity, is not confined to mercury: About half of the metallic elements exhibit it, as do many alloys, with the transition temperature varying from material to material. Some other metallic superconductors with their critical temperatures are niobium $(T_c = 9.5\text{ K})$, lead $(T_c = 7.5\text{ K})$, and $Nb_3Ge$ $(T_c = 23.2\text{ K})$. (Later we'll discuss so-called high-temperature superconductors.)

It should be stressed that there really is *no* resistance below the transition temperature. We know from standard circuit analysis that if a current is set up in a ring of inductance $L$ and resistance $R$ the current decays according to $I(t) = I(0)\exp(-Rt/L)$. The circulating current sets up a magnetic field, and by measuring the field one can measure the current or its decay. Recent experiments have shown that the decay time $L/R$ is greater than $10^5$ years(!), compatible with the notion that $R = 0$.

The transition to the superconducting state is very sharp, taking place for very pure metals over a range of only $10^{-5}$K. This strongly reinforces the identification of the transition to the superconducting state with a sharp transition analogous to the freezing of water at $0°$ C, to the Bose–Einstein condensation (Chapter 12), or to the transition to superfluidity in liquid $^4$He (Chapter 12). If we think of conductivity as associated with the flow of electrons through a lattice, then we can recognize some features in common with superfluidity. In particular, in both superfluid $^4$He and superconducting metals, a flow occurs with no energy loss. We shall see that the real similarity between these phenomena is that in both cases there is a Bose–Einstein condensation: a collective quantum behavior of a large number of particles.

## Magnetic Properties of Superconductors

In 1933, the German experimental physicists M. Meissner and R. Ochsenfeld discovered that if a superconductor is cooled below $T_c$ in a magnetic field, then the magnetic flux lines are *expelled* from the material (•Fig. 14–36), leaving $B = 0$ inside it, except for a thin surface layer ranging from 1 to 50 nm. This phenomenon is known as the **Meissner effect**. The word "expel" is important. The mere fact that the resistivity in a superconductor is zero would imply that a magnetic field could not penetrate a superconductor, because any buildup of magnetic field inside the sample would result in a change of flux, giving rise to an electric field that, in its turn, would produce a current opposing the magnetic field (Lenz's law). With zero resistance, this current could be arbitrarily large, and it would totally cancel the imposed magnetic field. By the same token, the prohibition of a change in flux in a perfect conductor would mean that if a magnetic field is already present in a sample that is then cooled below $T_c$, the magnetic field would be frozen in the sample. The Meissner effect implies that a superconductor is more than a zero-resistance conductor.

If a material in a superconducting state is placed in an external magnetic field that is then increased beyond a critical field $B_c$, superconductivity is lost. The critical field, which typically ranges from 0.01 to 0.08 T at the lowest end of the temperature spectrum, depends on temperature, and drops as the temperature is increased, as we can see in •Fig. 14–37. This phenomenon has some technological applications, since it is possible to switch a sample back and forth between the normal and the superconducting state by quickly varying the magnetic field.

For certain materials, the magnetic field, instead of being expelled, is concentrated into filaments where the field is large and the material is in a "normal," not a superconducting, state. Such materials are called Type II superconductors.

## Specific Heat and the Superconducting Energy Gap

The specific heat of metals is due primarily to the heat capacity of their lattices. We saw in Chapter 12 that the electronic contribution is very small, since only the small fraction of electrons outside the filled levels below the Fermi energy contribute. As we shall see, superconductivity arises from the electrons in the metal, so that the lattice contribution to the specific heat is independent of whether or not the material is in its superconducting state. We can isolate the

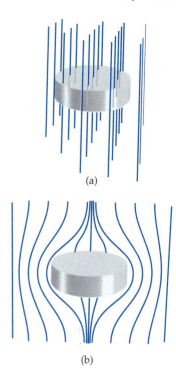

**• Figure 14–36** (a) Magnetic field penetrating a superconductor above the critical temperature $T_c$. (b) The magnetic field is expelled when the superconductor is cooled below $T_c$.

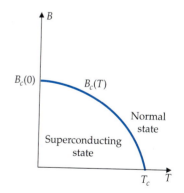

**• Figure 14–37** The critical magnetic field as a function of temperature. Superconductivity is maintained in the region below the curve. The sample ceases to be superconducting in the region outside the curve.

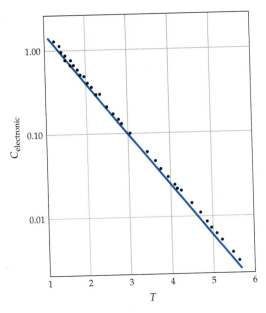

• **Figure 14–38** Electronic component of the specific heat as a function of $T_c/T$. The $y$-axis is drawn on a logarithmic scale, so that the straight line represents the behavior $\exp(-\Delta/kT)$.

lattice contribution by using a magnetic field to turn the superconductivity on or off and measuring the specific heat. Any difference is due to the electronic contribution, and when one performs the experiment, one finds that the temperature dependence of the *electronic* component is (•Fig. 14–38)

$$C_{\text{electronic}} \propto T, \quad \text{for } T > T_c$$

and

$$C_{\text{electronic}} \propto \exp(-\Delta/kT), \quad \text{for } T < T_c.$$

The quantity $\Delta$ has the dimensions of an energy and is typically on the order of several $10^{-4}$'s eV. It provides a significant clue to the behavior of superconductors.

We know [see, for example, Eq. (1–42)] that the specific heat of a material is generally given by a relation of the form $C = \partial\langle E\rangle/\partial T$, where $\langle E\rangle$ is some average energy (in our case of the electron). This means that if the specific heat contains an exponential factor such as $\exp(-\Delta/kT)$, the average energy must also contain such a factor. By writing the average energy over a distribution, it becomes possible to ascribe that factor to the existence of an *energy gap* in the energy spectrum of the electrons.[8] This implies that the probability of exciting an electron beyond the gap is the Boltzmann-like factor $\exp(-\Delta/kT)$, and that factor is responsible for a corresponding decrease in the number of electrons that can contribute to the heat capacity of the material. To put it in other terms, *there is an energy gap between the lowest state of the system and the levels that allow the system to store energy.* Moreover, since the effect on the specific heat is a macroscopic one, *large numbers of electrons—a fraction of the total number of valence electrons—must be unable to participate in storing energy.*

This gap, which can be confirmed by a variety of further experiments, is *different* from the gap between energy bands. It is essentially a modification of the Fermi surface and not a characteristic of the forces between ions. The gap width $E_g$ is in fact two times the factor $\Delta$. The gap lies just above the lowest state—that

---

[8] In fact there exist exceptional superconductors without gaps, so-called gapless superconductors. We do not deal with these here.

is, just above the Fermi energy, and it is realized by pushing up the levels within a range $\Delta$ above $E_F$ and pushing down the levels within a range $\Delta$ below $E_F$.

The presence of a gap in a superconductor implies the existence of an energy state that lies below the normal ground state of electrons in a metal. Moreover, as we have mentioned, the state must be populated with large numbers of electrons. This leaves us with two crucial questions:

1. What is the attractive force that gives rise to a lowering of the ground-state energy by the small amount $\Delta$?

2. How is it possible to get a large number of electrons into this new ground state, given that electrons are fermions?

The answers to these questions were provided in 1957 by three Americans: John Bardeen, Leon Cooper, and Robert Schrieffer.

### The Bardeen–Cooper–Schrieffer (BCS) Theory

Bardeen, Cooper, and Schrieffer were able to identify an indirect energy-lowering interaction between electrons by making use of the lattice structure. An important hint that the lattice was involved was given by experiments which showed that the transition temperature depended on the atomic mass of the lattice ions even in cases where the chemistry of the lattice (i.e., the Z-value of the atoms) did not change. This so-called *isotope effect* is measured by changing the percentage of atoms with a given $A$ in the lattice.

The indirect electron–electron interaction is subtle: An electron interacts with the lattice and because of the electrostatic attraction with the ions the electron deforms the lattice by pulling the neighboring positive ions towards itself. A second electron "sees" the first electron with its entourage of positive ions as a net positive charge and is attracted to it. Leon Cooper carried out a detailed analysis of the interaction and showed that the attraction gives rise to a bound state of the electrons—we call the bound pair a *Cooper pair*—if their energy is in the neighborhood of the Fermi energy. In particular, binding is favored for electrons with opposite spins (forming a spin-zero state) and equal and opposite momenta. It is this binding that lowers the energy, solving the first problem raised at the end of the previous subsection.

The existence of Cooper pairs also solves the second problem: How can individual electrons, which are fermions, occupy the new quantum mechanical ground state in large numbers? A Cooper pair acts as a single boson, even if the two electrons making up the pair may be many lattice sites away from each other. We can then have a Bose–Einstein condensation, in which the Cooper pairs are the bosons. The Cooper pairs can all fall into the same state, described by a single wave function. (We say that the pairs form a *coherent state*.)

Zero resistivity follows. As we described earlier, when a normal current flows, electrons scatter off the lattice fluctuations, giving rise to ohmic heating. In a superconductor, the current is formed by a large number of pairs *that move together*. In order to interrupt the flow of charge, the total state would have to be broken up. Although the gap energy is on the order of $10^{-4}$ eV, the number of pairs is huge (e.g., $10^{22}$), and the energy required to break up this system is simply unavailable. This accounts for the persistent currents observed in superconductors. We may visualize the situation by a diagram of the kind shown in •Fig. 14–39. In this case, we connect electrons by a kind of rod. When a potential difference is imposed, we get a situation very much like that shown in •Fig. 14–3b. Now, however, the slowing-down mechanism due to collisions can no longer work, since *each* of the two electrons connected by the rod must get into empty slots and, as the figure shows, there aren't any.

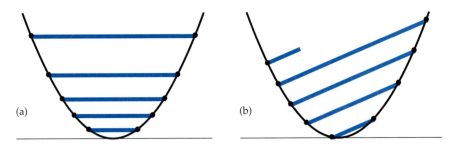

• **Figure 14–39** An analogue of •Fig. 14–3. An electric field accelerates electrons. Here, however, because the pairs of equal and opposite momenta are "bound" together, a return to the situation with no field must involve the shift of all the electrons in the collective state.

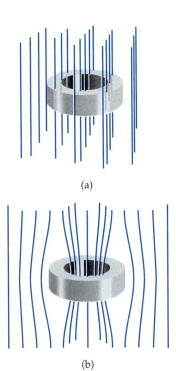

• **Figure 14–40** When a superconducting ring is cooled below the critical temperature $T_c$, the flux through the material of the ring is expelled and must go either to the outside of the ring or through the hole in the middle. The flux through the hole is an integral number of flux units $\Phi_0$.

We used the same reasoning to explain why an object moving through a superfluid does not lose energy: It would have to give up enough energy to break up the entire coherent state. In ${}^4$He, the atoms themselves can form the condensate, since they are spin-zero objects and therefore bosons. ${}^3$He atoms are more like electrons in a superconductor because ${}^3$He atoms are fermions (two electrons together with a nucleus with an *odd* number of nucleons). The ${}^3$He atoms actually form pairs analogous to the Cooper pairs, and it is these pairs that form the requisite condensate. The mechanism is different from the one that applies in superconductors, because there is no lattice. The ${}^3$He atoms do attract each other weakly, and this is enough to lead to the Cooper-pair formation. Since the masses of the ${}^3$He atoms are so much larger than those of electrons, all of this happens at much lower temperatures than it does for superconductors.

The Meissner effect also follows from the coherence of the state formed by the condensed Cooper pairs. If one were to try to increase the magnetic field inside a superconductor, then, by Faraday's law, an electromotive force (emf) would be induced within the material. This emf would then give rise to an induced current to oppose the change in magnetic flux. Because all Cooper pairs act together, the smallest change in magnetic flux will immediately give rise to a current sufficiently large to cancel the flux within.

## Magnetic Flux Quantization

The coherence of the wave function has another interesting consequence. Suppose we place a ring made of a superconductor in its normal state (i.e., $T > T_c$) in a magnetic field and then cool the metal to a temperature below $T_c$ (•Fig. 14–40). The magnetic field within the material will be expelled, and some of the expelled magnetic field lines will be trapped *inside* the ring. One can show that the magnetic flux inside the ring is quantized, taking on the values

$$\Phi = n\frac{h}{2e} = n\frac{\pi\hbar}{e} \equiv n\Phi_0, \tag{14–50}$$

where the $2e$ in the denominator refers to the charge on the electron pair that forms the basic unit. This result is a consequence of the requirement that the coherent wave function of the superconducting state be single valued. The magnitude of the flux quantum $\Phi_0$ is very small, approximately $2.1 \times 10^{-16}$ T·m².

*Josephson Effects*  When two superconductors are separated by a very thin strip of insulator—as little as a single nanometer—they form a *Josephson junction*, named after the English physicist Brian Josephson, who predicted the phenomena to be described here in 1962. Single electrons can tunnel from one superconductor to the other, but, it turns out, so can Cooper pairs. If the (coherent) wave functions of the superconducting state on the two sides of the junction are described by $\psi_1$ and $\psi_2$, respectively, then the effect of pair tunneling is to provide a mixing of the two wave functions. When this mixing is analyzed, a number of experimentally observable phenomena emerge.

First, as a consequence of the fact that Cooper pairs tunnel across the insulator, there is a net current across the junction, known as a *Josephson DC current*, even in the absence of a potential difference. The largest value of the current is called the *critical current*; its magnitude depends on the width of the insulator, on the material, and on the temperature.

Second, if there is a potential difference $V$ between the two sides of the junction, there will be an oscillation of the tunneling current with angular frequency $\omega = (2e)V/\hbar$. Since frequencies can usually be measured to a high degree of accuracy, this presents a new method of high-precision measurements of voltages $V$. Or, if the potential is known, the measurement of $\omega$ provides a way to measure the fundamental ratio $e/\hbar$.

Third, two Josephson junctions connected in parallel in a superconducting ring (•Fig. 14–41a) form what is known as a superconducting quantum interference device, or **SQUID**. When a current is sent into this arrangement, the current splits, flowing across the two opposite arcs. The now separate currents can interfere with each other. The two junctions in effect act like two slits, but whereas the interference pattern for two slits depends on the distance between the slits, the interference pattern for the two junctions depends on the magnetic flux $\Phi$ enclosed by the junctions. This is a quantum mechanical effect caused by the fact that in the superconducting state the flow of Cooper pairs is described by a single wave function. The resulting interference gives a net current that depends harmonically on the enclosed flux; that is,

$$I = I_0 \cos\left(\frac{e\Phi}{\hbar}\right), \tag{14–51}$$

where $I_0$ depends on the characteristics of the junctions. A change in the flux threading the loop manifests itself in a change in the current and hence in the voltage $V$ across the device, and this can be measured to high precision. •Figure 14–41b is a typical trace illustrating the dependence of $V$ on $\Phi$. The voltage can be measured with enough accuracy to be translated into a measurement in change of flux to an accuracy[9] of $10^{-15}$ T!

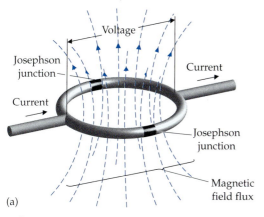

(a)

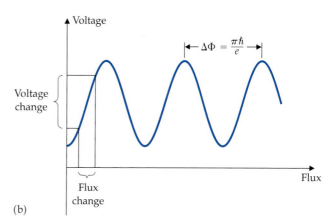

(b)

• **Figure 14–41**   (a) A superconducting quantum interference device (SQUID) consists of a pair of Josephson junctions arranged so that they are in parallel. The current that flows, and hence the voltage across the SQUID, is sensitive to the magnetic flux enclosed by the loop. (b) The dependence of the voltage on the enclosed flux.

---

[9] Since Earth's magnetic field is $10^{-4}$ T and fluctuations due to auroral currents are 1% of that, high-precision measurements must be made in heavily shielded enclosures.

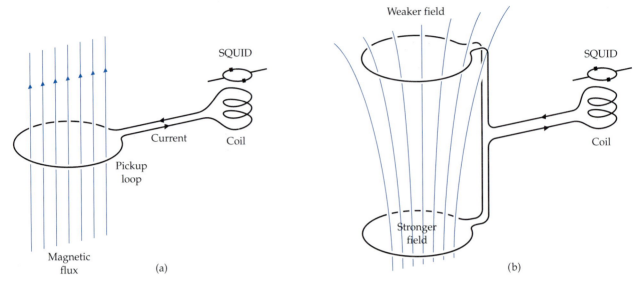

**• Figure 14–42**   (a) The use of a SQUID to measure magnetic flux. (b) The use of a SQUID to measure gradients in magnetic fields. (From "SQUIDS" by John Clarke, *Scientific American*, vol. 271, no. 2, p. 47, 1994. Illustration by Ian Warpole.)

In practice a SQUID is used as follows. A pickup loop in the region of the magnetic field to be measured is connected to a coil placed beneath the SQUID (•Fig. 14–42a). A change in flux through the pickup loop induces an EMF in the loop, which causes a current to flow through the wire and the coil. This current effects a change in flux through the SQUID, which in turn changes the current $I$ in Eq. (14–51). An estimate illustrates the parameters involved. Equation (14–51) shows that the distance between peaks in the current is given by $eBA/\hbar = \pi$, where $A$ is the area enclosed by the parallel junctions. With an enclosed area of 0.01 cm$^2$, if $B$ changes by as little as $\pi(1.05 \times 10^{-34}\,\text{J}\cdot\text{s})/\left[(0.01 \times 10^{-4}\,\text{m}^2)(1.6 \times 10^{-19}\,\text{C})\right] = 2.1 \times 10^{-9}$ T, two peaks will be seen in the current. With a measurement of the distance between peaks only 10% accurate, one gets a sensitivity on the order of $10^{-10}$ T.

The SQUID can accordingly be used to measure the time changes at a particular point in space if the currents producing the magnetic field are not static. If the currents producing the magnetic field are static, then by moving the pickup loop through space and varying its orientation it can measure the variation in the magnetic field over space. Changes in the direction of magnetic fields can be measured with a double pickup coil arranged such that the change in magnetic flux in one of the coils gives a current flowing in the direction opposite that of the current flowing in the other coil (•Fig. 14–42b). Thus the SQUID measures the difference between the fluxes—that is, the gradient of the magnetic field. Without going into details, there is enough information in the current through the SQUID to measure the magnetic field as well as any changes in its magnitude or direction.

SQUIDs are playing an important and growing role in noninvasive medical procedures. For example, they can be used to locate the site of an arrhythmic electrical discharge in heart muscle and in locating the sites of "electrical storms" in the brain during epileptic seizures. Another application is in the field of archeology. When pottery is fired in a kiln, the clay out of which the pottery is made loses whatever weak magnetization it had, and when it is cooled, the permanent magnetic dipoles align themselves with Earth's local magnetic field. Measurement of the magnetization in old kilns can be used to determine

Earth's field at the time the last firing took place, and since the variation in the field over the past several thousand years is known by other methods, this provides a means of dating the kilns and thus their products.

### High-Temperature Superconductors

The primary technological advantages of superconductors center around the fact that no energy is lost when currents run through them. It was always recognized that these advantages would be enormously enhanced if one could fabricate compounds whose critical temperature was higher than the typical values of $T_c$, which range up to around 15K. It is expensive to cool materials to very low temperatures, and the mechanical properties of cold materials are not always a help. Thus it was with great excitement that Georg Bednorz and Alexander Mueller announced in 1986 their discovery of a new class of materials with a superconducting transition temperature $T_c \cong 30K$. Within the next two years this group of *high-temperature superconductors* was expanded to include materials with $T_c \cong 100K$, and as of 1999 the record high temperature was around 150K. Although these are not yet "room-temperature" superconductors, it is now possible to think of technological devices that can be cooled with liquid nitrogen rather than liquid helium. This represents an enormous reduction in the cost of cooling, and it was hoped that the discovery would immediately translate into technological breakthroughs. Unfortunately the new materials are not easy to work with—it is hard to make wires with them, for example—and the extensive use of superconducting power lines still lies in the future. It is safe to say that the tremendous technological payoff of further research in this field will maintain interest in the fundamental explanation of high-temperature superconductivity for many years. At this time, the only clear thing is that the conventional BCS mechanism does not explain the new materials.

## S U M M A R Y

The behavior of matter in the solid state, in which there is an underlying crystalline structure, is very substantially controlled by quantum mechanical effects. This is revealed phenomenologically by, among other things, the failure of a classical picture of electrical conductivity in metals, in which classical (valence) electrons move through a medium that is studded with obstacles (ions). When quantum mechanics is applied, the following features appear:

- The valence electrons in a material have a behavior dominated by the exclusion principle, severely limiting the possibilities of their scattering, because the states into which they could classically be scattered are largely already occupied. At room temperatures, the Fermi–Dirac distribution of these electrons is such that nearly all the states are filled, with only electrons near the Fermi energy being able to scatter.
- Scattering takes place from the periodic lattice formed by the ions, and the solution of the Schrödinger equation in this circumstance shows that, to the extent that the lattice is perfect and rigid, there is *no* scattering whatsoever. Scattering occurs only when there are impurities in the lattice or when, because of finite temperature effects, the lattice ions vibrate about their equilibrium positions.
- The periodic lattice structure has one other important effect: Interactions of the electrons with the lattice distort the structure of the possible energy levels of the electrons into a series of very closely spaced levels called bands, separated by energies for which there are no allowed levels, called gaps.

Moreover, the electrons behave as if they have an effective mass that can differ quite substantially from their "isolated" mass.

- The band structure and the number of electrons available to fill the allowed states of the bands determine the differences between conductors, insulators, and semiconductors: In conductors, electrons do not fill any given band, so that there are nearby empty states for electrons with the highest energies; these states are available for the electrons to scatter into, and conductivity results. In insulators, an allowed band is filled, and the nearest empty level is in the next allowed band, separated from the most energetic electrons by the gap energy. This energy is so large that scattering into the first empty level is unlikely. A semiconductor has a filled band (the valence band), with the next band (the conduction band) empty, but the gap is small and the Boltzmann factor puts some electrons into the conduction band at finite temperatures. These few electrons, as well as the holes they left behind in the valence band, can conduct.

- By doping—that is, selectively introducing various impurities into a material that contribute additional electrons or that remove them, leaving additional holes—the properties of semiconductors can be very finely tuned. Also, semiconductors of various types can be put into contact with one another, and in these cases the equality of the Fermi energy across the contact points allows one to control the placement of the bottom of the conduction band and the top of the valence band. The capacity to perform these manipulations has led to a vast set of electronics applications, including diodes, transitors, many optical effects, semiconductor lasers, and so forth.

- The interaction of electrons with a background lattice can also lead to superconductivity if the electrons can pair properly. Superconductivity is a macroscopic quantum effect that involves not only an absence of resistance, but also some special magnetic phenomena, including the existence of a quantum of magnetic flux. The applications here include the lossless conduction of electricity, as well as very precise detection techniques for magnetic fields.

## QUESTIONS

**1.** We argued that for the electrons in a metal the speed that goes into the Drude formula is not the classical average speed but something that works out to be many times larger—the Fermi speed. Shouldn't this decrease the mean time between the collisions and therefore increase the resistivity, in contradiction to what is observed?

**2.** Take a $p$–$n$ junction and put a reverse bias on it. This device detects light. What is the mechanism?

**3.** You have perhaps seen a demonstration in which a permanent magnet floats suspended above a superconductor. What is the mechanism behind this phenomenon? (*Hint*: It helps for the superconductor to be in the shape of a concave dish.)

**4.** When two pieces of the same metal are separated by a very thin insulator, symmetry suggests that there will be no current crossing the insulator. This symmetry is manifested in the fact that the Fermi energies of the two pieces are identical. A current is possible if an electric field is applied, so that the energy levels on one side are slightly lowered by the potential. Given what you have learned about tunneling, can you find how the current depends on the thickness of the insulator separating the two metal pieces? How would it depend on the work function that you learned about in our discussion of the photoelectric effect?

**5.** According to Ohm's law, the current in a metal is proportional to the imposed potential, $I = V/R$, where $R$ is a constant. Keeping in mind the previous question, how does the picture of electrons tunneling from occupied energy levels on one side of the metal to empty levels on the other side fit in with a proportionality to $V$? It will be helpful to draw a figure of the energy levels for the two adjacent "boxes."

**6.** Suppose that the two pieces of metal in the previous two questions are cooled and become superconducting. Can you make a prediction as to how the relation $I = V/R$ would be changed?

**7.** In a monoenergetic beam of electrons, the electrons penetrate through matter with different path lengths, something that is known as "straggling." In a general way, how would you explain this phenomenon?

**8.** A photon beam passing through a medium is attenuated, with its intensity falling off exponentially as a function of the mean path length of the beam. What atomic processes are responsible for this?

**9.** The intensity of X rays as they pass through matter falls off as $\exp(-\alpha x)$, where $\alpha$ is characteristic of the absorbing medium and $x$ is the thickness of the medium. For lead $\alpha = 55$ cm$^{-1}$. If in the dentist's office you are given a lead apron to wear that is 3 millimeters thick, should you worry about being irradiated when you are given an X-ray examination?

## PROBLEMS

**1.** ▌ Show that for a drag force proportional to the speed $v$ of the moving object the coefficient of $v$ has dimensions of mass/time.

**2.** ▌ Assuming that an electron under the influence of a drag force starts from rest at $t = 0$, show that the solution of Equation (14–1) is

$$v(t) = \frac{eE\tau}{m}\left[1 - \exp\left(-\frac{t}{\tau}\right)\right].$$

The drift speed of Eq. (14–2) is the large-time limit of this speed.

**3.** ▌▌ In the space between the arms of a galaxy, a gas composed of atomic hydrogen is present in the amounts of 0.5 atom per cm$^3$. Assuming that atomic hydrogen acts like a sphere of radius $10^{-10}$ m insofar as its interaction with other atoms in the interstellar gas is concerned, what is the mean free path of a hydrogen atom in the gas?

**4.** ▌ What is the collision cross section for the interaction of a shotgun pellet of diameter 2 mm and a ball of diameter 25 cm? for the interaction of a hydrogen atom (radius $10^{-10}$ m) and the ball? Is it reasonable that your answers are so similar?

**5.** ▌▌▌ Use the results of this chapter to calculate the probability that a molecule in a gas will have just one collision in a time interval $\Delta t$. Repeat the exercise to find the probability that the molecule will have exactly two collisions in the same time interval. Generalize your answer to the case of $n$ collisions in the interval, and then use the resulting distribution to find the average number of collisions in the interval. The answer you obtain should be equal to the time interval divided by the average collision time.

| **The following table gives some properties of metals at room temperature that are useful in the next four problems:** | | | | | | |
|---|---|---|---|---|---|---|
| | Metal (atomic weight) | | | | | |
| Property | Li (6.9) | Na (23.0) | Cu (63.5) | Mg (24.3) | Fe (55.8) | Al (27.0) |
| Drude relaxation time $\tau$ (units of $10^{-14}$ s) | 0.88 | 3.2 | 2.7 | 1.1 | 0.24 | 0.80 |
| Fermi velocity (units of $10^6$ m/s) | 1.29 | 1.07 | 1.57 | 1.58 | 1.98 | 2.03 |
| Mass density (units of $10^3$ kg/m³) | 0.53 | 0.97 | 8.93 | 1.74 | 7.87 | 2.70 |

6. ▮ What are the Fermi energies of the metals in the table?

7. ▮ For each of the metals shown use the data in the table to calculate the ratio $n_e/n_i$—that is, the number of valence electrons per ion.

8. ▮▮ Assuming that somehow the classical Drude model were applicable, use the Maxwell–Boltzmann value of $v_{av}$ to calculate the mean free path of electrons in the metals listed in the table. Assume room temperature.

9. ▮▮ The quantum mechanical Drude model correctly predicts the mean free path $\ell$ of electrons in the metals of the table. Find the respective values of $\ell$.

10. ▮▮ A current of 1 A flows in a copper wire 2 mm in diameter. The density of valence electrons in copper is roughly $9 \times 10^{28}$ m$^{-3}$. Find the drift speed of these electrons.

11. ▮▮▮ Consider the most general solution of the differential equation $d^2u/dx^2 + k^2u = 0$, and show that the requirement that $|u(x + a)| = |u(x)|$ singles out the form $\exp(\pm ikx)$. (*Hint*: Work with $A \cos kx + B \sin kx$.)

12. ▮▮ The atomic weight of copper is $A = 63.5$ and its Debye temperature is 343K. Find the temperature at which the cross section for electron scattering from crystal vibrations, given by Eq. (14–14), is roughly the area presented by an atom.

13. ▮▮ Magnesium ($A = 24.3$) has two valence electrons and a density of $1.74 \times 10^3$ kg/m³. The resistivity at 295K is $4.30 \times 10^{-8}$ $\Omega \cdot$m. Estimate the Debye temperature. The experimental value is 400K.

14. ▮▮ Use Table 14–1 to calculate the resistivity of Li, Mg, K, Cu, Ag, Au, and Bi, and compare your answers with the experimental values also given in the table.

15. ▮▮ When an electron moves in a crystal lattice, one effect of all of the forces is to change the value of the electron mass that appears in formulas to an effective mass $m_{eff} = \kappa m_e$. Calculate the values of $\kappa$ needed to make the quantum mechanical Drude formula give the experimental value of resistivity for Na, Al, Fe, Zn, Sb, and Pb. You can use the data in Table 14–1.

16. ▮▮▮ Consider the coupling of two identical pendulums, with energy given by

$$E = \frac{1}{2} m \left( \frac{dx_1}{dt} \right)^2 + \frac{1}{2} m\omega^2 x_1^2 + \frac{1}{2} m \left( \frac{dx_2}{dt} \right)^2 + \frac{1}{2} m\omega^2 x_2^2 + \frac{1}{2} m\lambda(x_1 - x_2)^2;$$

the last term contains the coupling between the pendulums. **(a)** Use energy conservation, $dE/dt = 0$, to obtain the equations of motion for the two masses. **(b)** Show that the combinations $X \equiv (x_1 + x_2)/2$ and $x \equiv x_1 - x_2$ are involved in simple uncoupled equations, and find the associated frequencies. What are the motions like that correspond to $x = 0$ and $X = 0$, respectively?

**17.** ▌▌ Consider the two curves for the temperature dependence of the resistivity shown in •Fig. 14–2. Which curve represents the purer sample of potassium, and why?

**18.** ▌▌ Consider a two-band problem defined by

$$E = E_0 - 2E_1 \cos ka, \qquad 0 \le ka \le \pi;$$

$$E = E_0^* + 2E_1^* \cos ka, \qquad \pi \le ka \le 3\pi.$$

**(a)** Under what condition is there an energy gap—that is, the bands do not overlap? **(b)** What is the effective mass at the bottom of the second band?

**19.** ▌▌▌ Consider the Kronig–Penney model, which, with the definitions $ka \equiv x$ and $qa \equiv y$, consists of the relations $2mE/\hbar^2 = y^2$ and $\cos x = \cos y + (\xi/2) \sin y/y$. **(a)** Suppose $(\xi/2) = 0.96$. Find the value of $y$ for which $ka = \pi$. **(b)** Use the definition of the effective mass $m^*$ of the previous problem, and calculate the effective mass at $k$ such that $ka = \pi$ for $(\xi/2) = 0.96$. You will find, in contrast to the previous problem, that the effective mass can be much lighter than the free-electron mass.

**20.** ▌▌ The gap width for undoped Silicon is 1.14 eV. Given that the effective masses of the $n$- and $p$-carriers are 1.06 and 0.58 times the electron mass, respectively, calculate the density of $n$-carriers at **(a)** $T = 300K$ and **(b)** $T = 77K$.

**21.** ▌▌ An intrinsic semiconductor has $m_n^*/m_p^* = 0.94$ and $m_n^*/m_e = 0.067$. The $n$-carrier density at 300K is $6.74 \times 10^{11}$ per cubic meter. What is the gap width for this semiconductor?

**22.** ▌▌ Use the result of Problem 21 to calculate the $n$-carrier density in the semiconductor of that problem at 250K.

**23.** ▌▌ Given that $N_c = 2.8 \times 10^{25} \text{ m}^{-3}$ and $N_v = 0.48 \times 10^{25} \text{ m}^{-3}$ at $T = 300K$, calculate $m_n^*$ and $m_p^*$.

**24.** ▌▌ The statement that the Fermi energy lies in the middle of a gap has a correction to it that is linear in $kT$. Calculate the coefficient of $kT$ for the five semiconductors listed in Table 14–2.

**25.** ▌▌ Use the result of Problem 23 to calculate the values of $T$ for which $[E_F - (E_c + E_v)/2]/\Delta E = 0.20$ for the five semiconductors listed in Table 14–2.

**26.** ▌ Fill in the steps leading from Eq. (14–36) to Eq. (14–37). The integration of Eq. (14–36) is over an energy range from $E = E_c$ to the top of the conduction band. Use the fact that, because the falloff of the integrand is exponential, it is safe to put the upper limit of the integration at $E = \infty$. (*Hint*: Change variables from $E$ to $E - E_c$ so that the integral over the latter variable runs from 0 to $\infty$. Make a change of scale so that the exponent is defined to be $x$, and look up the integral, to check that you indeed get Eq. (14–37).

**27.** ▌▌ Calculate the ionization energy for the donor electron in gallium arsenide, assuming that Eq. (14–40) applies. Use the data given in Table 14–2.

**28.** ▌ What is the maximum wavelength of light required to excite an electron from the top of the valence band to the bottom of the conduction band in InP?

**29.** ▌ A CdS solar cell produces current if light with frequency of at least $5.8 \times 10^{14}$ Hz falls on it. What is the gap width?

**30. ▮▮▮** Calculate the ratio of the currents through an $n$–$p$ junction for positive and negative biasing voltages of magnitudes 0.10 V and 0.15 V at a temperature of 77K.

**31. ▮▮▮** A one-dimensional quantum well of width 12 nm is created, and measurements show that the energy difference between the ground state and the first excited state is 0.16 eV. What is the effective mass of the electron in the well?

**32. ▮▮▮** For a *quantum wire* the energies corresponding to the two confined dimensions are quantized. If the dimensions of the two-dimensional box are $a_1$ and $a_2$, the energy is given by

$$E = \frac{\hbar^2 \pi^2 n_1^2}{2m^* a_1^2} + \frac{\hbar^2 \pi^2 n_2^2}{2m^* a_2^2} + \frac{p_z^2}{2m^*}, \quad \text{with } n_1, n_2 = 1, 2, 3, \dots.$$

If the motion is restricted so that $p_z = 0$, $a_1 = 10$ nm, and $a_2 = 30$ nm, and if $m^* = 0.3\, m_e$, what are the minimum photon frequencies needed to excite the first two excited states?

**33. ▮▮▮** Consider a *quantum dot*, a three-dimensional box with dimensions 40 nm × 30 nm × 10 nm. If $m^* = 0.067\, m_e$, what are the energies of the first four excited states?

# The Atomic Nucleus

In the study of atoms, molecules, and the solid state, it is a very good approximation to treat the nucleus of the atom as a point particle—a source of the coulomb field that binds electrons. Atomic structure is greatly simplified by the fact that the nucleus is thousands of times more massive than the orbiting electrons. The electrons therefore move in a (very nearly) fixed potential created by the nuclear charge, modified only by screening from other electrons. In contrast, there is no neat division between light and massive particles in the study of nuclei; all the nuclear constituents (protons and neutrons—collectively, nucleons) have about the same mass. Moreover, the internuclear force—the force that holds the nucleus together—is much more complicated than the coulomb force and the small corrections to it that arise from relativistic and spin effects. The nuclear force is strong and does not have a simple $r$-dependence; it also depends on the relative velocity of the interacting nucleons, on the directions of their spins, and on their relative orbital angular momenta. In its complexity, the nuclear force is reminiscent of molecular forces. This fact suggests that nucleons are not the "elementary particles" that many had hoped them to be. We shall learn in Chapter 16 that nucleons are made of more basic building blocks: quarks and gluons. In spite of the difficulties, nuclear physics has become a mature field with consequences for technology. Because the nucleus is a many-body quantum system, it shares a number of collective effects with solids. Our understanding of nuclear structure is based on a number of complementary pictures: The ground state of nuclei and the energy levels that characterize their states of excitation exhibit semiclassical features as well as familiar single-particle quantum features. In this chapter we discuss these complementary pictures in a mostly qualitative way.

## 15–1  Neutrons and Protons

The notion of a massive atomic center, or **nucleus**, was introduced by Ernest Rutherford (•Fig. 15–1) in 1911, but it is fair to say that nuclear physics really began with the discovery of the neutron by the Englishman James Chadwick (•Fig. 15–2) in 1932. The neutron has nearly the same mass as the proton—the neutron is a little bit more massive—and no net electric charge. Like protons, neutrons have spin 1/2, so that they, too, are fermions. This fact has great importance for nuclear structure. When Chadwick first discovered the neutron, he

• **Figure 15–1** Ernest Rutherford was one of the great experimental physicists of the 20th century. He carried out basic work in radioactivity and was largely responsible for making nuclear physics a field of research. Here he is shown in his laboratory, where, in an unsuccessful attempt to get him to modulate his booming voice, his assistants put up the notice "TALK SOFTLY, PLEASE."

thought it was a strongly bound electron–proton system. Such a system would have an integer angular momentum, because the spins of two spin-1/2 particles add to an integer and because orbital angular momentum is always an integer. However, evidence from rotational spectra of molecules suggested that the $^{14}$N nucleus had to be a boson, which meant that neutrons had to be fermions. Later, direct measurements showed that the neutron, like the proton, has spin 1/2.

With the discovery of the neutron, a proper explanation of the nature of the nucleus became possible. A given species of atom has $Z$ electrons—as we discussed in Chapter 11, the number $Z$ determines the chemistry of the element. The nucleus of an atom is said to have **atomic number** $Z$ and **atomic weight**[1] $A$, where $A$ is the integer nearest to the ratio $M_{nucleus}/M_{hydrogen}$. Such a nucleus consists of $Z$ protons and $N \equiv A - Z$ neutrons. Because the electric charge of the proton is $+e$, exactly opposite to that of the electron, the electric charge of the nucleus is $+Ze$ and the total charge of the atom is zero. Nuclei with a given $Z$ and $A$ are **nuclides**, of which there are a great variety, both stable and unstable.

There is a standard notation for nuclides (except for the deuteron, which has the special symbol $d$): A nuclide with a given $Z$ takes the chemical symbol of the atom to which it corresponds, and the atomic weight $A = Z + N$ is added as a superscript. Thus $^4$He is the helium nucleus, with $Z = 2$ and $A = 4$, while $^3$H is the triton, which has $Z = 1$ and $A = 3$.

Nuclei with the same $A$ but different $Z$'s and $N$'s are called **isobars**. Examples are $^{15}$N and $^{15}$O. Nuclei with a given $Z$ but different $N$'s are **isotopes**. Examples are $^{15}$O and $^{16}$O. We mention in particular hydrogen, for which there are three isotopes: the standard one whose nucleus consists of a single proton; *deuterium*, whose nucleus, the *deuteron*, consists of a stable bound state of a neutron and a proton; and *tritium*, whose nucleus, the triton, consists of two neutrons and one proton.

• **Figure 15–2** James Chadwick, who was a student of Rutherford, discovered the neutron. He is shown here on the right, along with Peter Kapitza, a great Russian experimentalist who made major contributions to low-temperature physics.

**Example 15–1**    The following table lists various nuclei and their charges $Z$:

| Nucleus: | He | Cl | Br | Pb | U |
|---|---|---|---|---|---|
| Charge Z: | 2 | 17 | 35 | 82 | 92 |

**(a)** List the number of protons and neutrons in each of the following nuclides: $^4$He, $^{37}$Cl, $^{81}$Br, $^{208}$Pb, and $^{235}$U. **(b)** For which nuclides is the angular momentum an integer? **(c)** Which of the neutral *atoms* with these nuclei are fermions?

**Solution**    **(a)** We use the fact that the number of neutrons $N$ is $A - Z$. Then

$^4$He has $Z = 2$ and $N = 4 - 2 = 2$.
$^{37}$Cl has $Z = 17$ and $N = 37 - 17 = 20$.
$^{81}$Br has $Z = 35$ and $N = 81 - 35 = 46$.
$^{208}$Pb has $Z = 82$ and $N = 208 - 82 = 126$.
$^{235}$U has $Z = 92$ and $N = 235 - 92 = 143$.

**(b)** The nuclides with $A$ even will have an even number of spin-1/2 particles, and therefore the addition of angular momenta will give an integer-valued angular-momentum quantum number. Thus the nuclei of $^4$He and $^{208}$Pb will have integral total angular momentum. The others will each have a half-integer total angular momentum. **(c)** The neutral atom consists of a nucleus with $A$ spin-1/2 constituents and $Z$ electrons, each of which also has spin 1/2. Hence the total number of spin-1/2 particles is

---

[1] See Example 14–1 for another use of the atomic weight.

$A + Z = N + 2Z$. Since $2Z$ is always even, it is $N$ that determines the character of the total system. Among our examples, only the neutral atom with $^{235}$U as its nucleus has $N$ odd, and that atom is the only fermion in this group.

---

The proton and neutron masses have been measured to great accuracy. To a first approximation the nearly equal masses are about $1.67 \times 10^{-27}$ kg. Since we will work here in MeV as a convenient energy unit, we give the more precise results for the proton and neutron masses in those units:

$$m_p c^2 = 938.38 \text{ MeV and } m_n c^2 = 939.57 \text{ MeV.}$$

For comparison, you might recall that the electron mass is roughly $0.51$ MeV$/c^2$. Because the electron is some 2,000 times less massive than the nuclear constituents, it is a good approximation to say that the entire mass of an atom is contained in its nucleus.

Although the neutron has no net electric charge, it does have electromagnetic properties. For example, it has a magnetic moment, so you can deflect a neutron in a magnetic field. This fact provided the first suggestion that a neutron has structure—a subject that we shall cover in Chapter 16. For the time being we treat protons and neutrons as "fundamental" particles, just as we treat electrons and nuclei as "fundamental" when we talk about atomic physics.

The proton and the neutron have so many properties in common that they are together known as **nucleons**.

## 15–2  Nuclear Size and Mass

We begin our discussion of the gross structure of nuclei with their sizes and masses. In the process of understanding these properties we can learn quite a bit about the forces that hold nuclei together. We start with a discussion of nuclear radii. Rutherford found the first evidence that nuclei had radii when he observed a deviation from the pure coulomb scattering form for especially energetic alpha projectiles—projectiles that had enough energy to be able to penetrate *within* the nucleus. In interpreting the original experiments in which he and his younger colleagues discovered the atomic nucleus, Rutherford assumed that the charge in the nucleus was concentrated at a "point," so that a pure coulomb potential with a $1/r$ dependence described the scattering of the alpha particles. If the charge is spread out, this alters the force that the alpha particle feels. A distributed charge deflects energetic alpha particles less than a point charge does. Inverse square forces such as the coulomb force have the property that once the particle being scattered—here the alpha particle—penetrates within the radius of a spherically symmetric charge distribution, that particle "feels" only the effect of the charge inside its radial distance. How far the alpha particle has penetrated into the nucleus can be estimated by knowing the angle with which the alphas are scattered by the nucleus. Small deflections are due to glancing collisions, and for those the nucleus acts as a point source of charge. Large deflections are due to close collisions, and for those the alpha comes close to and may penetrate the nucleus. In fact, Rutherford's first experiment already showed that, for large angle scattering, the measured distribution of scattered alpha particles deviated from that predicted for a point nucleus. The implication was that the nucleus was small, but not a point particle. And there is one more complication: The deviations from those expected for a pointlike nucleus can be attributed to nuclear forces that act at short distances as well as a distributed electric charge distribution.

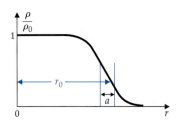

• **Figure 15–3**  For large enough nuclei ($Z > 10$, say) the nuclear density is well described by the formula

$$\rho(r) = \frac{\rho_0}{\{1 + \exp[(r - r_0)/a]\}},$$

where $\rho_0 = 0.17$ nucleon/fm³, $r_0 = (1.2A^{1/3} - 0.48)$ fm, and $a = 0.55$ fm. This formula is obtained from electron–nucleus collision experiments, on the assumption that the nuclear density is the same as the electric charge density—that is, that the protons and neutrons are distributed in the same way. It is the charge density that is actually measured by the electron-scattering experiments.

The two effects—that of the distributed electric charge and that of the nuclear force—can be better separated by using as projectiles particles that are not influenced by nuclear forces. Indeed, the best measurements of the charge density of nuclei are obtained by scattering high-energy *electrons* from them. We have a good theoretical understanding of the scattering of electrons[2] by point charges, so the scattering of electrons from a known spread-out charge is just given by adding the contributions of the various parts of the charge. In the case of electrons scattering from nuclei, we have a kind of inverse problem, in which information on how the electron is scattered is used to extract the charge distribution of the nucleus. A series of experiments pioneered by Robert Hofstadter at Stanford University in the 1950s led to a picture of the nuclear charge density shown in •Fig. 15–3.

A serviceable description of Hofstadter's results is that for $Z > 10$, the nuclear density has a constant value of about 0.17 nucleon/fm³ out to a radius of about

$$R = R_0 A^{1/3} = (1.1 \text{ fm})A^{1/3} = (1.1 \times 10^{-15} \text{ m})A^{1/3} \qquad (15\text{–}1)$$

and is zero beyond this radius. If we were to pack $A$ nucleons inside a sphere of radius $R$, with volume $4\pi R^3/3$, a rough measure of the spacing between the nucleons would be $a$, with $a^3 = 4\pi R^3/(3A) = (1.8 \text{ fm})^3$. The nucleons in a nucleus, therefore, are very close together, with a spacing about $10^5$ times smaller than the typical atomic distances of $10^{-10}$ m.

We can convert the number density just quoted into a nuclear mass density by using the nucleon masses:

$$\frac{(1.67 \times 10^{-27} \text{ kg})A}{\left(\dfrac{4\pi}{3}\right)[(1.1 \times 10^{-15} \text{ m})A^{1/3}]^3} = 3.0 \times 10^{17} \text{ kg/m}^3. \qquad (15\text{–}2)$$

This is a huge density compared to that of any material you can mention, and the reason is that in ordinary matter the nuclei occupy very little of the apparent volume. The neutron stars mentioned at the end of Chapter 10 have about this density. In that sense, a neutron star is something like a gigantic nucleus!

The fact that the mass density of nuclei beyond the lightest ones does not depend on the number of nucleons tells us something about nuclear forces: It implies that the addition of a nucleon to a nucleus, a change from $A$ to $A + 1$, does not increase the density, as might be the case if the additional nucleon were attracted by *all* the other $A$ nucleons. (If the force were similar to the electrostatic force, the attraction of the extra nucleon to the nucleus would be $A$ times the two-nucleon force. Nuclei would then share the property of atoms that after a certain point they do not increase in size. In fact, atoms with $Z$ up to 100 still have atomic radii on the order of a few times $10^{-10}$ m.) This suggests that *nuclear forces have a short range*. Beyond a certain distance—some 1.0–1.2 fm, about the size of a nucleon–a nucleon no longer feels the presence of other nucleons.

Nuclear forces are also very strong. It takes 13.6 eV to remove an electron from hydrogen, while it takes several MeV (as we shall see) to remove a nucleon from a nucleus. By this measure the nuclear force between closely packed nucleons is some $10^6$ times stronger than the electric force between an electron and a proton at atomic distances. Another important property of nuclear forces deserves mention here: Once the coulomb repulsion between protons is taken into account, the forces between protons and protons are the *same* as those be-

---

[2] An electron is a better choice than an alpha particle, because the electron, unlike the alpha particle, has no structure of its own and furthermore does not respond to nuclear forces.

tween protons and neutrons and between neutrons and neutrons. Crudely speaking,[3] the nuclear potential energies have the property that for two nucleons in the same state,

$$V_{nn} = V_{np} = V_{pp}. \tag{15-3}$$

We shall study some consequences of this fact later.

# 15-3  The Semiempirical Mass Formula

Many different nuclei are found in nature, and because of their large binding energies nuclear masses are not found simply by adding the masses of their constituents. In this section we describe a formula that gives a very good, though not perfect, fit to nuclear masses as a function of $A$ and $Z$: the semiempirical mass formula. The structure of the formula reveals some classical aspects as well as some clearly quantum mechanical aspects of nuclear structure. We'll imagine here that we are working with relatively large nuclei—those with, say, more than 10 nucleons. We'll also use energy units rather than mass units—we write the mass in terms of $Mc^2$ rather than $M$. In this way we calculate the *energy* of the nucleus rather than its mass. We can enumerate the terms that contribute to the energy of a nucleus as follows:

1. The nucleon masses themselves provide a kind of zeroth-order term. The nucleus consists of $Z$ protons and $N$ neutrons, so that the rest mass contribution is

$$(Zm_p + Nm_n)c^2. \tag{15-4}$$

2. The nuclear forces have a short range; thus each nucleon "feels" a force that depends only on the (fixed) number of its immediate neighbors rather than on the total number of nucleons. This force acts on *each* nucleon—there must be a factor of $A$—and acts to bind the nucleon, leading us to expect a negative contribution to the nuclear energy of the form

$$-Ab_{vol}^{(1)}. \tag{15-5}$$

   The constant $b$ here and similar constants to be discussed are always taken as positive. All the $b$'s have dimensions of energy, and it turns out to be practical to use MeV as the scale. We use the subscript "vol" here because, for a nucleon density that is the same from nuclide to nuclide, the total number $A$ of nucleons is proportional to the nuclear volume. The superscript "(1)" is present because, as we shall see, we must take into account a second contribution of this type.

3. A correction to Eq. (15-5) comes from the fact that nucleons on the surface of the nucleus have fewer neighbors holding them in than do nucleons in the interior, so we expect a reduction in the binding energy that is proportional to the surface area; in turn, because the density of nucleons does not vary with the nucleus, the surface area is proportional to $A^{2/3}$. This represents a positive contribution of the form

$$+A^{2/3}b_{surf}. \tag{15-6}$$

4. The protons in the nucleus *repel* each other through the coulomb force, and this repulsion increases the nuclear energy. If one assumes that the positive

---

[3] The reason for this reservation is that for two identical nucleons (e.g., $nn$ or $pp$), there are certain states that are not permitted. The orbital angular-momentum zero state with spins parallel is an example.

charge is uniformly distributed throughout a volume of radius $R$, then a calculation of the energy associated with a coulomb force gives an energy $(3/5)Z^2e^2/(4\pi\varepsilon_0 R)$, which translates into

$$\frac{0.7Z^2}{A^{1/3}} \text{ MeV.} \tag{15-7}$$

5. If this were all, then the energy would be lowered by lowering $Z$, and the most stable nuclei would have many neutrons and few protons. You might imagine stable isotopes of hydrogen with dozens of neutrons! But in fact the most stable nuclei have only a slight excess of neutrons, and there are no stable nuclei with huge neutron (or proton) excesses. Thus the departure from this symmetry—a symmetry in which $N = Z$—increases the energy, and a possible term describing that fact has the form

$$+ \frac{(N - Z)^2}{2A} b_{\text{symm}}. \tag{15-8}$$

You could also imagine trying higher powers of $(N - Z)^2$, but one wants to have the smallest number of arbitrary parameters needed to fit the data. This effect is purely quantum mechanical and is due to the Fermi nature of the nucleons. As we shall see later, we can think of each nucleon moving in a central nuclear potential formed by the effect of all the other nucleons. Each energy level associated with this potential can be filled by two neutrons and two protons, in accordance with the exclusion principle. If there is an excess of neutrons, for example, then a larger number of energy levels get filled up by neutrons, and the neutrons in the higher energy levels will have more energy than any of the protons. Under these circumstances it is favorable for neutrons to decay by a process—to be discussed further in Chapter 16—known as **beta decay**, given by

$$n \rightarrow p + e^- + \bar{\nu}. \tag{15-9}$$

(A neutron decays into a proton, an electron, and an antineutrino.[4]) Beta decay decreases the number of neutrons and increases the number of protons. In the same way, if there were an excess of protons, the protons in the highest energy levels would decay into a neutron, a positron (an antielectron), and a neutrino; that is,

$$p \rightarrow n + e^+ + \nu. \tag{15-10}$$

In free space only the first of these processes can occur, because a neutron has more mass than a proton and an electron together. (The neutrino is massless, or at least nearly so compared to any other particle with which we are familiar.) However, both processes can occur as long as energy can be conserved, and this is possible within the nucleus because the nucleus that is produced in the decay can have a different mass from the original nucleus.

The preceding argument describes why there should be a term that has a minimum when $N = Z$. To see how the precise dependence $(N - Z)^2/2A$ in Eq. (15-8) emerges, one can ignore the nucleon–nucleon interactions and consider $N$ neutrons and $Z$ protons in a spherical container whose volume is that of the nucleus, viz., $V = (4/3)\pi R^3 = (4/3)\pi R_0^3 A$, where we have used Eq. (15-1). The neutrons and protons each form a collection of identical fermions. The calculation of the total energies of the neutrons and protons that

---

[4] The antineutrino is more specifically of a type associated with electrons. More details about neutrinos and their antiparticles are presented in Section 15–5 and Chapter 16.

fill the lowest levels in the spherical container is identical to a calculation carried out in Chapter 10. [See Eq. (10–57).] The details of the procedure are treated in Problems 6 and 7 at the end of the chapter. If you do those problems, you will find that when the difference between the number $N$ of neutrons and the number $Z$ of protons is small compared to $A$, the sum of the energies of the neutrons and protons takes the approximate form

$$E_p + E_n \approx K \times \left( A + \frac{10}{9} \frac{(N - Z)^2}{2A} \right), \tag{15–11}$$

where we have neglected terms of higher order in the small ratio $(N - Z)^2/A^2$. The constant $K$ that appears here is also calculated in the same problems, and its value is

$$K = \left( \frac{9}{8\pi^2} \right)^{5/3} \frac{4\pi^4\hbar^2}{15mR_0^2}. \tag{15–12}$$

We have thus found an expression for the energy associated with the fact that the nucleons are confined to the volume of the nucleus; this expression takes into account the fact that the nucleons obey Fermi–Dirac statistics. The formula for the energy contains two terms. If we compare the second term on the right-hand side of Eq. (15–11) with Eq. (15–8), then we see that we have precisely the symmetry term, with

$$b_{symm} = \frac{10}{9} K = \left( \frac{9}{8\pi^2} \right)^{5/3} \frac{4\pi^4\hbar^2}{15mR_0^2} \frac{10}{9} = \left( \frac{3}{\pi} \right)^{1/3} \frac{\pi\hbar^2}{4mR_0^2}. \tag{15–13}$$

Numerical evaluation of the coefficient gives $b_{symm} = 33$ MeV, about two-thirds of the best fit to the data, as we shall subsequently see. Given the crudeness of our approach, we should not be too disappointed.

The first term on the right-hand side of Eq. (15–11)—the term proportional to $A$—is *positive*. This term represents the kinetic energy of the nucleons confined in the sphere of radius $R$. It must be combined with the term proportional to the coefficient $b_{vol}^{(1)}$ in Eq. (15–5) when we group the energy terms.

6. A last contribution must be added to give a better fit to the masses: the so-called *pairing term*, a term that increases the binding of "even–even" nuclei—those with even numbers of protons and neutrons—and decreases that of "odd–odd" nuclei—those with odd numbers of each type of nucleon. This term is of the form

$$\Delta mc^2 = \begin{cases} + \dfrac{b_{pair}}{\sqrt{A}} & \text{for } N \text{ and } Z \text{ even} \\ 0 & \text{for } N \text{ even and } Z \text{ odd or for } N \text{ odd and } Z \text{ even.} \\ - \dfrac{b_{pair}}{\sqrt{A}} & \text{for } N \text{ and } Z \text{ odd} \end{cases} \tag{15–14}$$

The origin of this term—the cause of the pairing effect—is purely quantum mechanical, with an origin that is believed to be the same as that leading to superconductivity. (See Chapter 14.) The term reflects what is observed about the relative stability of the nuclides.

We can bring together all these contributions to find the **Weizsacker–Bethe semiempirical mass formula**, here given in terms of energy:

$$E(A, Z) = (Zm_p + Nm_n)c^2 - Ab_{vol} + A^{2/3}b_{surf}$$
$$+ \frac{0.7Z^2}{A^{1/3}} \text{ MeV} + \frac{(N - Z)^2}{2A} b_{symm} + \Delta mc^2. \tag{15–15}$$

A best fit to the nuclear masses is obtained with

$$b_{vol} = 16 \text{ MeV}, \, b_{surf} = 17 \text{ MeV}, \, b_{symm} = 47 \text{ MeV}, \, \text{and} \, b_{pair} = 12 \text{ MeV}, \quad (15\text{–}16)$$

with a charge distribution radius of $R_c = 1.24 A^{1/3}$ fm.

We may note here that $-b_{vol}$ is the sum of the two terms $-b_{vol}^{(1)}$ [see Eq. (15–5)] and $(9/10)b_{symm}$ [see Eq. (15–11)]. Using the numbers of Eq. (15–16), this sum gives $b_{vol}^{(1)}$, which represents the interaction potential energy per nucleon, a numerical value of some 60 MeV. Compare this to the interaction energy per electron in atoms, which is on the order of eV. No wonder that the nuclear force is called a "strong" force!

While Eq. (15–16) gives a very good fit to the general run of nuclear masses, it does not take into account variations among neighboring nuclei associated with the nuclear shell structure to be discussed later in the chapter. The shell structure explains why certain nuclei are particularly stable. These nuclei—those with the strongest binding energy per nucleon—are those for which $N$ or $Z$ is 2, 8, 14, 20, 28, 50, 82, or 126,[5] the so-called *magic numbers*.

With the semiempirical mass formula, Eq. (15–15), we can easily find the energy per nucleon, $B(A, Z)/A$, required to disassemble the nucleus into $A$ separate nucleons:

$$\frac{B(A, Z)}{A} = -\frac{E(A, Z) - (Zm_p + Nm_n)c^2}{A}$$

$$= b_{vol} - \frac{b_{surf}}{A^{1/3}} - \frac{(0.7 \text{ MeV})Z^2}{A^{4/3}} - \frac{(N - Z)^2}{2A^2} b_{symm} - \frac{\Delta mc^2}{A}. \quad (15\text{–}17)$$

A plot of the experimental value of this quantity—the so-called *curve of binding energy*—is given in •Fig. 15–4. The figure shows that the binding energy rises rapidly, with some structure, as $A$ increases, peaks in the vicinity of $A$ around

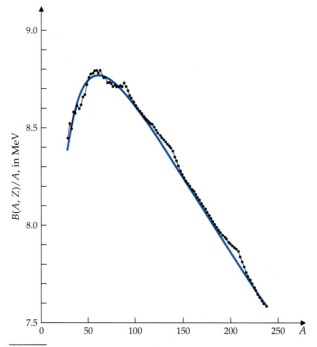

• **Figure 15–4** The comparison of the experimental binding energy per nucleon, $B(A, Z)/A$, given by the line connecting the dots, with the value of the binding energy per nucleon obtained from the semiempirical mass formula, shown by the smooth curve. The peaks in the experimental distribution signal closed-shell effects not included in the mass formula.

---

[5] There are no naturally occurring nuclei for which $Z = 126$, although $N = 126$ is realized.

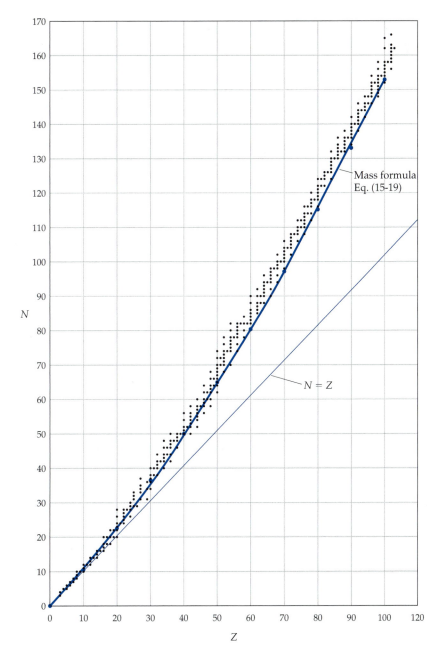

60—the iron region of the periodic table—and then drops slowly as $A$ increases further. The figure also shows the result of using Eq. (15–17).

The symmetry term suggests that there should be a tendency for $Z$ to equal $N$. The data that best illuminate this possibility are shown in •Fig. 15–5; the curve on which the known nuclides lie is known as the *stability curve*, and it does indeed follow a line in which $Z = N$. Indeed, as the next example shows, the stability curve follows well the more general prediction of the full semiempirical mass formula.

↓**Example 15–2**   Use the semiempirical mass formula to calculate the value of $Z/A$ that minimizes the energy of Eq. (15–15). Use the fact that $m_n c^2 - m_p c^2 = 1.3\,\text{MeV}$, and ignore the pairing term. (*Hint*: Express the relevant terms in terms of $A$ and $Z$, and minimize.)

**Solution**   When we drop the pairing term $\Delta m c^2$ and replace $N$ with $A - Z$ in Eq. (15–15), we see that an equation for the terms that depend on $Z$ is

$$Z(m_p - m_n)c^2 + \frac{(0.7\,\text{MeV})Z^2}{A^{1/3}} + \frac{(A - 2Z)^2}{2A}b_{\text{symm}}$$

$$= \left[ -1.3Z + \frac{0.7Z^2}{A^{1/3}} + \frac{25(A - 2Z)^2}{A} \right]\text{MeV}.$$

Hence the minimum occurs when

$$\frac{\partial E}{\partial Z} = 0 = \left[ -1.3 + \frac{1.4Z}{A^{1/3}} - \frac{100(A - 2Z)}{A} \right]\text{MeV}. \tag{15–18}$$

We solve Eq. (15–18) for $Z/A$:

$$\frac{Z}{A} = \frac{101.3}{200 + 1.4A^{2/3}}. \tag{15–19}$$

This quantity is plotted in •Fig. 15–5, along with the experimental stability curve. Experiment indeed gives $Z/A \cong 1/2$ for the lighter nuclei, in agreement with the prediction of the symmetry term alone; the ratio decreases somewhat for heavy nuclei, in good agreement with the prediction of Eq. (15–19).

### Nuclear Decays

The decay of nuclei goes under the very general heading of **radioactivity**. The term was introduced by Marie Curie (•Fig. 15–6), who, together with her husband Pierre, was the first person to identify a new radioactive element; in 1898, they isolated from the mineral pitchblende an element they named polonium and then another they called radium. Ernest Rutherford and Frederick Soddy were the first to understand that radioactivity is a property of individual nuclei and the first to measure their exponential decay rates. (See Chapter 13.) Nuclei can decay in a variety of ways that we call **decay modes**. A common class of decays leads to the emission of one of the following particles:

- a neutron.
- an alpha ($\alpha$) particle, which is the nucleus of helium, made up of two neutrons and two protons; for this nucleus, $Z/A = 1/2$. From what we have said earlier, we expect the $\alpha$ particle to be particularly stable.
- an electron. In the context of radioactive decays, the electron is called a beta ($\beta$) particle and the decay mode is known as **beta decay**. (When these decays were first studied, it was not clear what the charged particle was. Only later was it identified with the electron. It was also not clear at the time what the alpha particle was.) Moreover, as we shall see later, a second particle, one of the so-called *neutrinos*, is also emitted in this type of decay.
- a photon. This mode is known as **gamma** ($\gamma$) **decay**, with the photon called a $\gamma$-particle or $\gamma$-ray.

In each of these cases we refer to the original nucleus as the **parent** and the nucleus that remains after the decay as the **daughter**. Such decays, which we'll treat in more detail in Section 15–5, can occur only if the energetics allow. The semiempirical mass formula can give some guidance with the energetics, but the safest thing is to use a table of masses of the nuclei such as that found in

• **Figure 15–6**   Marie Curie, one of the handful of people to have won two Nobel prizes.

Appendix A. Thus, if $E(A, Z)$ is the total energy of a given nuclide, we learn whether the nuclide will spontaneously emit a neutron by calculating

$$Q_n \equiv E(A, Z) - [E(A - 1, Z) + m_n c^2].$$

If $Q_n > 0$, then the decay can occur, with the energy $Q_n$ going into the kinetic energy of the daughter nucleus and the neutron.

A different $Q$-value is necessary if we want to see whether decay by $\alpha$ emission is possible. In this case we want

$$Q_\alpha \equiv E(A, Z) - [E(A - 4, Z - 2) + E(4, 2)].$$

If $Q_\alpha$ is positive, then the decay is allowed.

---

**Example 15–3**   Obtain a formula for the kinetic energy $K$ of an $\alpha$-particle emitted in the at-rest decay of a nucleus of mass $M^*$ to a daughter nucleus of mass $M$ and an $\alpha$-particle. What is this energy in the decay $^{212}\text{Po}(Z = 84) \rightarrow {}^{208}\text{Pb}(Z = 82) + \alpha$? You are given that $(\Delta M)c^2 = 8.95$ MeV, where $\Delta M$ is the difference in mass between that of $^{212}\text{Po}$ and the sum of the masses of $^{208}\text{Pb}$ and $\alpha$.

**Solution**   Because the parent nucleus is at rest when it decays, the energy available as kinetic energy for the daughter nucleus and the $\alpha$-particle is $(\Delta M)c^2 = M^*c^2 - Mc^2 - m_\alpha c^2$. If we denote the momenta of the daughter nucleus and the alpha as $-\vec{p}$ and $\vec{p}$, respectively, then

$$\Delta M c^2 = \frac{p^2}{2M} + \frac{p^2}{2m_\alpha}.$$

Hence, solving for $p^2$, we get

$$p^2 = 2\Delta M c^2 \frac{M m_\alpha}{M + m_\alpha}$$

and

$$K = \frac{p^2}{2m_\alpha} = \frac{M}{M + m_\alpha} \Delta M c^2 = \frac{\Delta M c^2}{\left(1 + \dfrac{m_\alpha}{M}\right)}.$$

Substituting the numbers, we find that $K \cong (8.95 \text{ MeV})/(1 + 4/208) = 8.78$ MeV.

---

Incidentally, you can easily demonstrate from the preceding example a general feature of decays of this type. The $\alpha$-particle and the daughter come off with equal and opposite momenta, but the mass of the daughter is much greater than that of the $\alpha$-particle so the latter moves much faster. In turn, the $\alpha$-particle carries off the lion's share of the energy available for motion. This is true for all the single-particle decay modes we have described.

In the beta decay of a nucleus, a neutron decays according to Eq. (15–9), and the proton remains within the daughter nucleus. Because the antineutrino has little or no mass, the appropriate $Q$-value is

$$Q_e \equiv E(A, Z) - [E(A, Z + 1) + m_e c^2].$$

If $Q_e$ is positive, this energy is available for the kinetic energy of the daughter nucleus, the electron, and the antineutrino. If, instead, a proton decays according to Eq. (15–10)—a process that cannot occur outside the nucleus because of the energetics—with the consequent emission of a positron from the nucleus, then the $Q$-value that must be positive is

$$Q_{e+} \equiv E(A, Z + 1) - [E(A, Z) + m_e c^2].$$

Another type of nuclear decay process is **spontaneous fission**, in which a heavy nucleus splits into two much lighter nuclei and, often, some additional smaller pieces. Again, energetics poses some limitations. For the straightforward split into two smaller nuclei, we must have a positive value for the quantity

$$Q_f \equiv E(A, Z) - \left[E(A_1, Z_1) + E(A - A_1, Z - Z_1)\right], \qquad (15\text{--}20)$$

where $A_1$ and $Z_1$ refer to either of the final nuclei.

---

**Example 15–4**    **(a)** Use the semiempirical mass formula to estimate the energy released when a single $^{235}$U (uranium; $Z = 92$) nucleus fissions into two nuclei of equal size and charge. Neglect the contribution of the (small) pairing term. **(b)** Estimate the energy, in joules, released in the fissioning of 1 kg of $^{235}$U nuclei.

**Solution**    **(a)** We are here interested in the application of Eq. (15–20) when $A_1 = A/2$ and $Z_1 = Z/2$. For the energies of all the nuclei concerned we use the semiempirical mass formula, Eq. (15–15). Then you can verify quickly that in the difference $E(A, Z) - 2E(A/2, Z/2)$ only the coulomb and surface terms survive:

$$E(A, Z) - 2E(A/2, Z/2) = \left[A^{2/3} - 2\left(\frac{A}{2}\right)^{2/3}\right]b_{surf} + \left[\frac{Z^2}{A^{1/3}} - 2\frac{(Z/2)^2}{(A/2)^{1/3}}\right](0.7 \text{ MeV})$$

$$= (1 - 2^{1/3})A^{2/3}b_{surf} + (1 - 2^{-2/3})\frac{0.7Z^2}{A^{1/3}} \text{ MeV}.$$

Using $b_{surf} = 17$ MeV, $A = 235$, and $Z = 92$, we numerically evaluate this formula to obtain $E(A, Z) - 2E(A/2, Z/2) = 187$ MeV. It is interesting to note that the contribution of the surface term is $-168$ MeV while that of the coulomb term is about $+355$ MeV. It is the (positive) coulomb term that makes a breakup of the nucleus energetically favorable.

The experimental number for the energy release in this fission reaction is some 170 MeV per fission. This number is millions of times larger than the typical chemical—that is, atomic—energy of several eV.

**(b)** Using 1 MeV $= 10^6$ eV $= 10^6(1.6 \times 10^{-19}$ J$) = 1.6 \times 10^{-13}$ J, we find that each fission produces $170 \times 1.6 \times 10^{-13}$ J $= 2.7 \times 10^{-11}$ J. Now we find the number of nuclei in 1 kg. Each $^{235}$U has a mass a little less than $235m_{nucleon} = 235 \times 1.67 \times 10^{-27}$ kg $\cong 4 \times 10^{-25}$ kg ("a little less" because of the binding energy, typically a 1% effect in nuclei). Thus the energy release per kg of $^{235}$U is $(2.7 \times 10^{-11}$ J$)/(4 \times 10^{-25}$ kg$) \cong 7 \times 10^{13}$ J/kg. For comparison, about $4 \times 10^6$ J/kg is released in a TNT (chemical) explosion; equivalently, the energy released in the fission of 1 kg of $^{235}$U is that of the explosion of some 20,000 metric tons of TNT.

---

## 15–4  Aspects of Nuclear Structure

The nucleus consists of many neutrons and protons. Each of them interacts strongly with its neighbors, and on these grounds one would certainly expect a structure quite different from that seen in atoms. A nucleon traveling inside a nucleus would be expected to have a very short mean free path, approximately equal to the internucleon spacing. This suggests that it is better to think about collective behavior in the nucleus rather than single-particle behavior. Some features of nuclear structure bear out that picture. On the other hand, the exclusion principle implies that when a nucleon strikes another nucleon and recoils into another state, that state must have been unoccupied. Thus scattering is inhibited inside nuclear matter. A nucleon still is acted upon by nuclear forces, but these play more of a background role as a potential inside which a given nucleon is free to move. This motion represents a form of single-particle behavior, analogous to the motion of electrons in an atom, and is associated with the shell structure in nuclei. We give a brief and qualitative discussion of these two contrasting aspects of nuclear behavior in this section.

## The Liquid-drop Model

An emphasis on the strong interaction between a nucleon and its nearest neighbors suggests an analogy to a droplet of a liquid. The so-called **liquid-drop model** of the nucleus in effect smoothes over the individuality of the nucleons in a nucleus. In fact, the leading terms of the semiempirical mass formula can be associated with an almost classical picture of the nucleus: a picture of a uniformly charged fluid held together by surface tension in spite of the repulsive electrostatic interaction that tends to break up the fluid. Such a picture leads to some definite predictions.

We start by asking about the dominant excitations of an electrically charged liquid drop. These will be vibrational excitations of the spherically symmetric equilibrium shape. From the semiempirical mass formula, we see that the nucleus is virtually incompressible, and therefore any deformations will preserve its volume. To show incompressibility, we show that the energy, as given by the formula, changes by a large amount when the volume changes. (See Problem 5 at the end of the chapter.)

Incompressibility, however, does not mean that the nucleus cannot deform—only that the volume must remain the same under deformation. •Figure 15–7 shows a series of possible oscillations of an **incompressible liquid drop**, the basis for the liquid-drop model. The dynamics behind this model consists of a *kinetic term* having to do with the motion of the parts of the drop and a *potential-energy term* having to do with the energy costs of a distortion. In particular, we can cite a type of "surface tension" that tends to make the spherical shape stable—this term is proportional to the surface area and hence, for constant density, to $A^{2/3}$. It can be seen in the surface term, Eq. (15–6), of the semiempirical mass formula. This term acts to bring the nuclei back to a spherical shape. However, for large enough $Z$, the coulomb repulsion will be larger than the surface tension, and any oscillations become unstable. This accounts for the phenomenon of spontaneous fission that we described in the previous section, in which the nucleus breaks up into two large pieces. In Example 15–3 we assumed that the fission split the nucleus into two *equal* parts. This rarely happens, however; the most likely division is into unequal mass fragments (•Fig. 15–8).

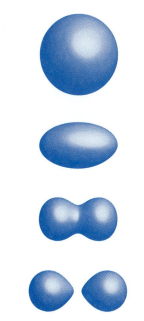

• **Figure 15–7** The sequence of shapes assumed by a charged oscillating liquid drop. The last stage is at the point of fission into two smaller drops.

*Reactions in the Liquid-drop Model* Suppose a low-energy neutron enters a nucleus. With many nucleons in its way, it undergoes collisions. It quickly loses enough energy so that it no longer has enough kinetic energy to escape the nuclear forces. Thus the capture of the particle becomes probable. (We can make an analogy with a fast-moving marble that enters a bowl of marbles; the energy of the initial marble is quickly shared with other marbles, and none of the marbles have enough energy to emerge rapidly from the bowl.) The nucleus, together with the additional particle, forms a **compound nucleus**, a term coined by Niels Bohr.

One expects the compound nucleus to be long lived in comparison with the time it takes a particle to cross a nuclear diameter. This is because the excitation energy, shared by all the nucleons, has to be concentrated in a single particle if that particle is to leave the nucleus. Furthermore, with the general sharing of the energy one expects that the compound nucleus "forgets" how it was formed, and the final breakup of the nucleus should not depend on the formation mechanism. In other words, the production of the compound nucleus and its decay are independent processes. This translates into the following prediction: Consider the reaction $n_1 + N_1 \rightarrow n_2 + N_2$, where $n_1$ and $n_2$ respectively stand for

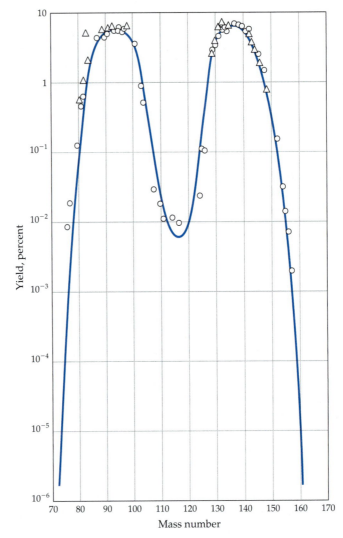

• **Figure 15–8**  The mass distribution of fission products of $^{235}U$ when the fission is induced by thermal neutrons. The result of a theoretical calculation is drawn as a solid line.

initial- and final-state particles and $N_1$ and $N_2$ for initial- and final-state nuclei. Then the rate of this reaction must have the form

$$R(n_1 + N_1 \rightarrow n_2 + N_2) = R(n_1 + N_1 \rightarrow N^*)R(N^* \rightarrow n_2 + N_2), \quad (15\text{–}21)$$

where $N^*$ is the unstable compound nucleus. For example, a number of reactions such as

$$\alpha + {}^{60}Ni \rightarrow n + {}^{63}Zn,$$

$$\alpha + {}^{60}Ni \rightarrow 2n + {}^{62}Zn,$$

$$\alpha + {}^{60}Ni \rightarrow n + p + {}^{62}Cu,$$

$$p + {}^{63}Cu \rightarrow n + {}^{63}Zn,$$

$$p + {}^{63}Cu \rightarrow 2n + {}^{62}Zn,$$

and

$$p + {}^{63}Cu \rightarrow n + p + {}^{62}Cu,$$

all go through the same compound nucleus $^{64}Zn$. If the incident $\alpha$-particle and $p$ energies are chosen so that in the compound nucleus the same energies are excited, then the rates for the reactions are simply related to each other. (See Problem 14 at the end of the chapter.) Experiments support these relations.

## The Shell Model

A picture that is complementary to the liquid-drop model is one in which we emphasize the independent motion of a single nucleon in a potential created by the environment in which that nucleon moves. At the beginning of this section we argued that the exclusion principle would suppress the scattering of nucleons inside nuclei, so that a single nucleon would move as if it were in a potential due to all the other nucleons. This picture is reminiscent of the single-particle description of electron motion in atoms—except for the presence of a dominant potential, the nuclear coulomb potential. Electron motion in atoms shows a shell structure that we discussed in Chapter 11. Clear evidence comes from a plot of the ionization potential as a function of $Z$, as shown in •Fig. 15–9. One might similarly expect to see a shell structure in nuclei, and one does. As an example, a plot of the relative abundance of different nuclides as a function of $A$ shows strong peaks associated with the set of magic numbers (•Fig. 15–10) referred to in Section 15–3. As we described there, the nuclei in which either $Z$ or $N$ had the values 2, 8, 20, 28, 50, 82 and, for $N$ alone, 126, are particularly stable.

Why these numbers rather than others is a matter of detail. For example, a simple three-dimensional harmonic oscillator, for which the energy eigenvalues are $\hbar\omega(2n_r + \ell + 3/2)$, where $n_r = 0, 1, 2, \ldots$ and $\ell$ is the orbital angular-momentum quantum number, does *not* give this set of magic numbers. (See Problem 15 at the end of the chapter.) Potentials of various shapes were proposed in a search for a central potential that would have a sequence of bound states that matched the magic numbers, but none seemed suitable. The German-born theoretical physicists Maria Goeppert-Mayer and Hans Jensen were the ones to solve the problem. They pointed out that a strong *spin-orbit interaction*, of the form $V_{\text{s.o.}} = -(2\beta/\hbar^2)\vec{\mathbf{L}} \cdot \vec{\mathbf{S}}$, could reshuffle the levels of a simple harmonic oscillator potential and account for the magic numbers. The actual ordering of the shells is given in •Fig. 15–11. The orbital angular momenta are denoted by the usual letters $s(\ell = 0)$, $p(\ell = 1)$, $d(\ell = 2)$, $f(\ell = 3)$, and so on, and the subscripts indicate the $j$ values. The energy gaps make for the particularly stable shells that give rise to the magic numbers.

The shell model makes unambiguous predictions about the spins of nuclei that consist of a closed shell plus a single nucleon or a closed shell that is missing a single particle (that is, a closed shell plus a hole). For example, since the

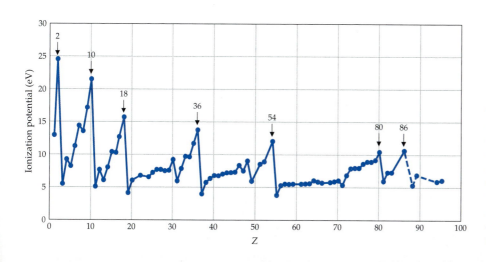

• **Figure 15–9** Atomic ionization potentials as a function of $Z$, showing that the atom has a shell structure, from data compiled by C. E. Moore (1970). (From *Subatomic Physics*, 2nd ed., by Hans Frauenfelder and Ernest Henley. Copyright © 1991 by Prentice-Hall, Inc., Upper Saddle River, NJ.)

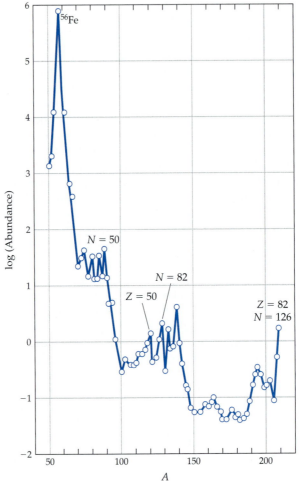

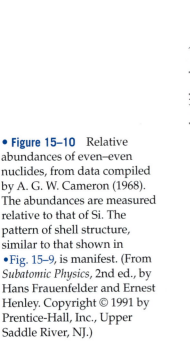

• **Figure 15–10** Relative abundances of even–even nuclides, from data compiled by A. G. W. Cameron (1968). The abundances are measured relative to that of Si. The pattern of shell structure, similar to that shown in •Fig. 15–9, is manifest. (From *Subatomic Physics*, 2nd ed., by Hans Frauenfelder and Ernest Henley. Copyright © 1991 by Prentice-Hall, Inc., Upper Saddle River, NJ.)

closed shell has a total angular momentum of zero, the angular momentum of the nucleus in the first case is that of the next shell that the single particle resides in. In this way we can use the $j$ values and the occupation number $(2j + 1)$ for the nuclear shells given in •Fig. 15–11 to make predictions. As an example, consider $^{13}C$, which has six protons and seven neutrons. The six protons and six of the neutrons fill the first two shells, and there is one neutron left, which must occupy the second $j = 1/2$ shell. Thus the predicted spin of the $^{13}C$ nucleus is $1/2$, in accord with experiment.

**Example 15–5**    Consider the nucleus $^{133}Sb$, which has 51 protons and 82 neutrons. Use the shell assignments in •Fig. 15–11 to predict the angular momentum of that nucleus.

**Solution**    •Figure 15–11 shows that 82 neutrons form a closed shell, as do 50 protons. Thus the additional proton must lie in the shell above the closed shell corresponding to 50 protons, which is a $g_{7/2}$ shell. Therefore $^{133}Sb$ has an angular momentum $j = 7/2$.

Nuclear spins are measurable because they are proportional to the magnetic moment of the nucleus. The shell model also makes some definite predictions regarding the magnetic moments of nuclei that consist of a closed shell plus a single particle or hole. While a detailed discussion of these would take us too far afield, we can say that for light nuclei the predictions are in good

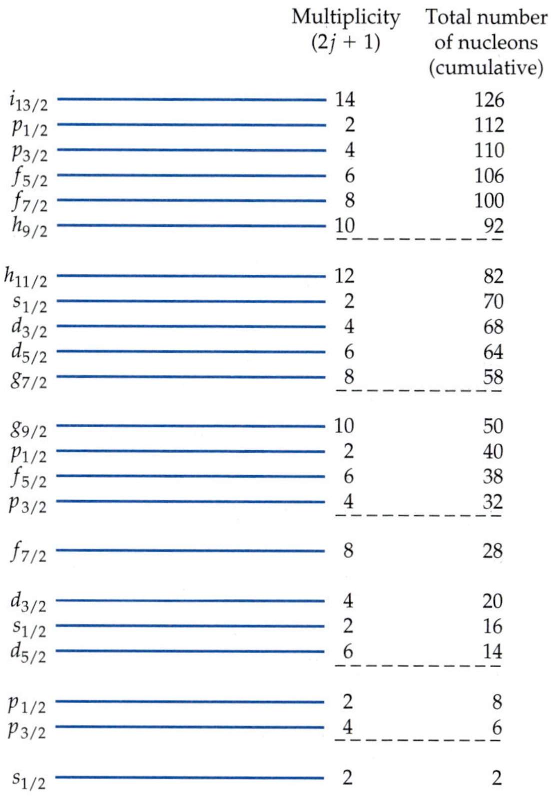

| | Multiplicity $(2j + 1)$ | Total number of nucleons (cumulative) |
|---|---|---|
| $i_{13/2}$ | 14 | 126 |
| $p_{1/2}$ | 2 | 112 |
| $p_{3/2}$ | 4 | 110 |
| $f_{5/2}$ | 6 | 106 |
| $f_{7/2}$ | 8 | 100 |
| $h_{9/2}$ | 10 | 92 |
| $h_{11/2}$ | 12 | 82 |
| $s_{1/2}$ | 2 | 70 |
| $d_{3/2}$ | 4 | 68 |
| $d_{5/2}$ | 6 | 64 |
| $g_{7/2}$ | 8 | 58 |
| $g_{9/2}$ | 10 | 50 |
| $p_{1/2}$ | 2 | 40 |
| $f_{5/2}$ | 6 | 38 |
| $p_{3/2}$ | 4 | 32 |
| $f_{7/2}$ | 8 | 28 |
| $d_{3/2}$ | 4 | 20 |
| $s_{1/2}$ | 2 | 16 |
| $d_{5/2}$ | 6 | 14 |
| $p_{1/2}$ | 2 | 8 |
| $p_{3/2}$ | 4 | 6 |
| $s_{1/2}$ | 2 | 2 |

• **Figure 15–11** The experimental ordering of nuclear shells for protons. The levels are labeled by $(\ell, j)$. The energy gaps, characterized by the "magic numbers" 2, 8, 28, 50, 82, and 126, are impressionistically drawn. The shell structure for neutrons differs slightly from that for protons at the higher levels. We ignore the difference.

agreement with experiment. For heavier nuclei, the interaction between the particle (or hole) and the nuclei in the closed shells has a significant effect, and the simple picture sketched here no longer works.

## 15–5 Nuclear Reactions

Earlier we listed a variety of nuclear decays. Decays represent one class of reactions that nuclei can undergo. In this section we discuss some of the different classes of possible reactions and their properties.

### Time Dependence in Quantum Mechanical Decays

In our discussion of quantum mechanics in this chapter, we noted the existence of states that are not "stationary"—states that undergo decay. The excited states of atoms and molecules provide examples, but perhaps the most interesting are nuclear decays, some of which were discussed in Section 15–3. The decays of unstable nuclei are just like the decay of a generally unstable state described in Chapter 13. The rate at which $N$ unstable nuclei undergo decay is proportional to $N$ itself, and the proportionality constant $R$ is the *decay rate* [Eq. (13–1)]:

$$\frac{dN(t)}{dt} = -RN(t). \tag{15–22}$$

This rate does not depend on any external factors such as the temperature of the nuclei. We do not even have to know whether the nuclei are in a solid or a liquid. As we saw in Chapter 13, the solution of the differential equation (15–22) is

$$N(t) = N_0 \exp(-Rt), \tag{15–23}$$

where $N_0$ is the number of unstable nuclei at the time $t = 0$. In writing the equation this way, we have fixed the arbitrary constant that always arises when you integrate a derivative by demanding that the solution reduce to $N_0$ at $t = 0$. This equation is frequently written in the form

$$N(t) = N_0 \exp(-t/\tau), \qquad (15\text{-}24)$$

where the constant $\tau = 1/R$ is the *mean life* or *lifetime* of the decaying nucleus.

If the nucleus can decay in a variety of ways, then there are a variety of *partial decay rates* $R_1, R_2, \ldots$. The nucleus is said in this case to decay into various **channels**. Under those circumstances Eqs. (15–23 and 15–24) still hold, but with

$$R = R_1 + R_2 + \ldots. \qquad (15\text{-}25)$$

Decays frequently occur in **series**. An example among the heavy elements is the chain leading from $^{238}$U to $^{206}$Pb (lead; $Z = 82$), shown in •Fig. 15–12. Equation (15–23) can be generalized to describe such a chain. The rate of change of a particular species in some chain, such as $^{231}$Pa (protactinium; $Z = 91$), contains a negative term associated with a loss—the decay to a daughter nucleus $^{227}$Ac (actinium, $Z = 89$)—as well as a positive term associated with a gain— the decay of $^{231}$Th (thorium, $Z = 90$) into $^{231}$Pa. When there is a series, there is a set of coupled equations with positive and negative terms describing the numbers of the different species involved. There may also be branches, as in •Fig. 15–12, and in that case the equations are more complicated.

Let us look at an example. We suppose—see •Fig. 15–13a—that a nucleus 1 decays into a daughter nucleus 2 with a decay constant $R_{12}$, while the daughter 2 can further decay into a nucleus 3 with a decay constant $R_{23}$. Nucleus 3 is stable. The equations describing this pair of reactions are

$$\frac{dN_1(t)}{dt} = -R_{12}N_1(t)$$

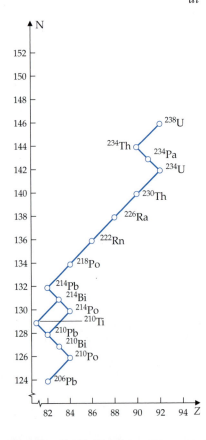

• **Figure 15–12** Decay series leading from $^{238}$U to $^{208}$Pb by steps involving alpha decay (a drop in $Z$ and $N$ by two units each) and beta decay (a drop in $N$ by one and an increase in $Z$ by one).

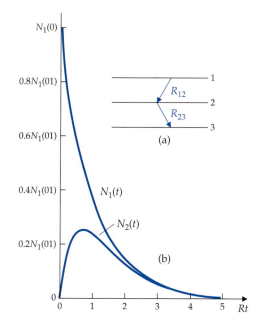

• **Figure 15–13** (a) Schematic picture of the nuclear decay sequence $N_1 \rightarrow N_2 \rightarrow N_3$ with rates $R_{12}$ and $R_{23}$, respectively. (b) Plot of Eq. (15–27) in arbitrary time units for the parameters $R_{23} = 2R_{12}$.

and

$$\frac{dN_2(t)}{dt} = R_{12}N_1(t) - R_{23}N_2(t).$$

In the second of this set of equations, the first term on the right-hand side is the rate at which decays of nucleus 1 are adding to the numbers of nucleus 2. This set of equations can be solved. We'll assume as an initial condition that we have a pure sample of $N_1(0)$ nuclei of type 1 at $t = 0$. The first equation then has the familiar solution

$$N_1(t) = N_1(0) \exp(-R_{12}t). \tag{15–26}$$

This can be inserted into the second equation, which then becomes an equation for $N_2(t)$ alone. Direct substitution will confirm that this new equation has the solution

$$N_2(t) = N_1(0)R_{12} \frac{\exp(-R_{23}t) - \exp(-R_{12}t)}{R_{12} - R_{23}}. \tag{15–27}$$

Note that $N_2(0) = 0$, a result consistent with our assumption that initially there were no nuclei of type 2. •Figure 15–13b illustrates the evolution of these numbers. As the figure shows, for short times there is a buildup of daughter nuclei as the parent decays. Then the daughter decays. The way the equation has been written, the implicit assumption has been made that $R_{12} > R_{23}$, so that $N_2 > 0$. For very long times, then, the daughter decays with sensibly a single exponential determined by $R_{23}$.

## Nuclear Decay Modes

Here we make further comments about nuclear decay modes.

*Alpha Decays* In Chapter 8 we gave a brief presentation of how alpha decay is explained as the tunneling of an $\alpha$-particle—a helium nucleus—through the barrier formed by the well produced by the nuclear forces and the coulomb repulsion of the positively charged protons in the daughter nucleus (•Fig. 15–14).

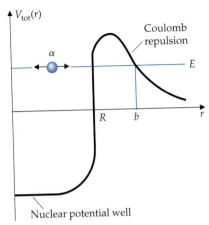

• **Figure 15–14**  The total potential energy includes both a nuclear part and a coulomb repulsion part. An α-particle can tunnel through the coulomb barrier. The energy of the emitted α-particle is $E$, $R$ is the nuclear radius, and $b$ is the distance at which the α-particle acquires a positive kinetic energy.

We can think of the α-particle as preexisting in the nucleus and rattling back and forth between the sides of the barrier until it gets out. The most important feature of this (or any other) tunneling phenomenon is the exponential factor of Eq. (8–19), which we shall here label by $S$. This factor is sensitive to the height and width of the barrier potential energy $V(r)$ and provides the dominant energy dependence in the tunneling probability $|T|^2$, which is proportional to $\exp(S)$. According to Eq. (8–19),

$$S = -2 \int dr \sqrt{\left(\frac{2m}{\hbar^2}\right)(V(r) - E)}. \qquad (15\text{–}28)$$

As •Fig. 15–14 shows, in the α-decay problem the integration in the exponent is from the onset of the barrier, which we may regard as the nuclear radius $R$, to the radius $b$ at which the kinetic energy of the α-particle is zero. Since the outer part of the barrier is purely coulomb, the quantity $b$ is determined by the condition that $V_{\text{coulomb}}(b) = E$, or

$$b = \frac{2(Z - 2)e^2}{4\pi\varepsilon_0 E}. \qquad (15\text{–}29)$$

(We have noted that the charge of the alpha particle is 2, so that the charge of the daughter nucleus is $Z - 2$.) We must also set $m = m_\alpha$ in Eq. (15–28), which now reads

$$S = -\frac{2\sqrt{2m_\alpha}}{\hbar} \int_R^b \sqrt{\frac{2(Z - 2)e^2}{4\pi\varepsilon_0 r} - E}\, dr$$

$$= -\frac{2\sqrt{2m_\alpha}}{\hbar} \sqrt{\frac{2(Z - 2)e^2}{4\pi\varepsilon_0}} \int_R^b \sqrt{\frac{1}{r} - \frac{1}{b}}\, dr.$$

In the last step we have substituted $E$ in terms of $b$ from Eq. (15–29). While this last integral can be done exactly, we can get a more understandable result by assuming that the energy $E$ of the α-particle is small compared to the height of the barrier at $r = R$. In that case, •Fig. 15–14 shows that $b \gg R$. We can then approximate

$$\int_R^b \sqrt{\frac{1}{r} - \frac{1}{b}}\, dr \cong \int_R^b \sqrt{\frac{1}{r}}\, dr = 2(\sqrt{b} - \sqrt{R}) \cong 2\sqrt{b}.$$

This gives the approximate result

$$S \cong -\frac{4\sqrt{2m_\alpha}}{\hbar} \frac{2(Z - 2)e^2}{4\pi\varepsilon_0} \frac{1}{\sqrt{E}} = -4 \frac{\sqrt{2m_\alpha c^2}}{\sqrt{E}} 2(Z - 2) \frac{e^2}{4\pi\varepsilon_0 \hbar c}, \qquad (15\text{–}30)$$

where we have again used Eq. (15–29), this time to substitute back for $b$ in terms of $E$. In Eq. (15–30) the last factor is the fine-structure constant, roughly $1/137$. Finally we can make a numerical approximation by using the fact that $m^\alpha c^2 \cong 3{,}750$ MeV. Then if we agree to measure $E$ in MeV a good numerical approximation is

$$S \cong -4 \frac{Z - 2}{\sqrt{E}}. \qquad (15\text{–}31)$$

We can now make use of our calculation to estimate the lifetime of a nucleus subject to $\alpha$-decay. We picture the $\alpha$-particle moving back and forth within the well with speed $v = \sqrt{2E/m}$ until it escapes. As an estimate of the number of times $N$ that it bounces before escaping, we suppose that $N \times$ (probability of escape with one bounce) $= 1$. But the probability of an escape with one bounce is just $|T|^2$, so that we estimate $N = |T|^{-2} \propto \exp(-S)$. (We use proportionality because we are only going to compare predicted $E$-dependence with experiment.) Now that we have an estimate for $N$, we can ask how long it takes the $\alpha$-particle to collide $N$ times with the barrier. This is the lifetime $\tau$. In our picture, the time of one round-trip is $t_1 = 2R/v = 2R[m_\alpha/(2E)]$, and $\tau = Nt_1$. Thus, using our numerical estimate for $S$, Eq. (15–31), we find that the lifetime is proportional to

$$\tau = 2R\sqrt{\frac{m_\alpha}{2E}}\, N \propto \sqrt{\frac{1}{E}}\exp(-S) = \sqrt{\frac{1}{E}}\exp\left(4\frac{Z-2}{\sqrt{E}}\right). \qquad (15\text{–}32)$$

This result can be tested by taking the logarithm; we would expect a plot of $\ln \tau$ against $(Z - 2)/(E)^{1/2}$ to be linear with coefficient 4. •Figure 15–15 shows

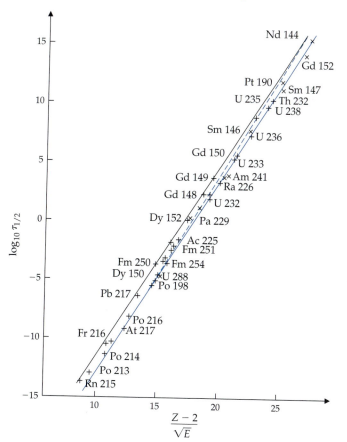

• **Figure 15–15** The logarithm of the half-life of a large number of different alpha emitters as a function of $(Z - 2)/(E)^{1/2}$.

just such a plot for a wide range of alpha emitters, with lifetimes ranging from microseconds to billions of years. Amazingly, the nuclear lifetimes *all* lie on a curve that is given numerically as

$$\ln \tau \propto 3.7 \frac{Z - 2}{\sqrt{E}}.$$

Despite the general crudeness of the model—the mixture of quantum and classical ideas—the agreement is truly remarkable and shows us that the treatment of alpha decay as a tunneling phenomenon is basically correct.

*Beta Decays*    These decays have played an important part in the evolution of our knowledge of the subatomic world. In beta decays two charged particles are observed to emerge: the daughter nucleus $^A(Z + 1)$ and an electron $e^-$. If this was all there was to it—if the decay were what we call a two-body decay—the electron's energy and momentum would be fixed uniquely by the kinematics in terms of the masses of the particles involved. (See Problem 21 at the end of the chapter.) But this is not what is observed. Instead the data show that the electron's kinetic energy ranges[6] from zero to a maximum value of $m_Z c^2 - m_{Z+1} c^2 - m_e c^2$, where $m_Z$ stands for the mass of the nuclide $^AZ$.

In order to explain the variation in the electron's energy some puzzled scientists—including Niels Bohr—went so far as to postulate the failure of energy conservation! In 1931, Wolfgang Pauli proposed a solution. He postulated that the decay is actually a *three*-body decay, in which the new third particle is electrically neutral and, to fit the observed peak of the electron kinetic energy measurements, has a mass less than that of the electron. Since the new particle was not observed Pauli assumed that it barely interacted with anything. He was so "embarrassed" by the strange properties of this object that his suggestion was made in a letter rather than a published paper. The hypothetical particle was dubbed the **neutrino** (the "little neutral one" in Italian) by Fermi and was given the symbol $\nu$. To explain why it had not been detected, the neutrino had to have very weak interactions with matter. Indeed it can pass through thousands of *light years* of lead without a single interaction! The nuclear beta decay process described here is actually a neutron decaying within the nucleus according to the more basic process of Eq. (15–9)—that is why the Z-value of the nucleus increases by 1. Incidentally, the particle emitted in conjunction with the electron is actually the *antineutrino*, symbol $\bar{\nu}$. The reason that this particle is the one that is emitted will be given in our discussion in Chapter 16 of the general class of fundamental interactions called the **weak interactions**.

In addition to the fundamental kinematic restrictions laid out in Section 15–3, the rates for beta decay processes in nuclei depend on the spins of the parent and daughter nuclei, with selection rules that are dictated by the conservation of total angular momentum and on how much orbital angular momentum must be carried off by the electron and neutrino to make up for the difference in spin. Beta decay rates in nature have lifetimes ranging from about 1 s to more than $10^{13}$ yr. The process of free neutron decay, given by Eq. (15–9), occurs with a mean lifetime of some 15 min.

*Gamma Decays*    The mechanism that gives rise to the emission of a photon from an unstable nucleus is the same as that which occurs in atoms and is there-

---

[6] Actually, this range is realized when the parent nucleus decays at rest and the daughter nucleus is so massive compared to the electron that it is a good approximation to ignore the daughter's momentum.

fore well understood. This makes γ-decays particularly useful for the study of nuclear levels. Because the energy level differences are much larger in nuclei than in atoms—typically, thousands of times larger—the frequencies of γ-decay photons are thousands of times larger than the frequencies of atomic photons, but the same selection rules apply. By using a series of decays such as $A^{**} \rightarrow A^* + \gamma$ and then $A^* \rightarrow A + \gamma$, it is possible to measure the *correlation* in the directions of the two photons. Such correlations help to determine properties such as the angular momentum of the excited states $A^{**}$ and $A^*$ of the nucleus $A$.

In another process—also electromagnetic in origin—known as **internal conversion**, an excited state $A^*$ of a nucleus can decay electromagnetically to the ground state $A$, this time without the emission of a photon. In this process the energy difference $E(A^*) - E(A)$ is used to eject an electron from one of the inner atomic shells of the atom containing the nucleus. In atomic language the process reads

$$A^* \rightarrow A^+ + e^-,$$

where $A^+$ is a singly ionized atom with $A$ as its nucleus. This is a two-body decay, and its kinematic properties make it clearly distinguishable from beta decay, which is a three-body process.

## Collision Reactions

We have not been very explicit about the existence of a complex energy-level structure in the nucleus. It is true that the notion of a shell model, with the lowest shells occupied by protons and neutrons, carries with it the implication that there are higher shells that are unfilled and that can be excited by promoting one or more nucleons into them. We also mentioned the existence of energy levels in our discussion of γ-decays. The complementary aspects of nuclear structure, independent particle motion and collective effects, can be unraveled only by a precise study of the energy levels. The procedure is somewhat analogous to the study of atomic or molecular spectroscopy, in which levels are excited by electronic, radiative, or thermal effects, and their subsequent decays provide the basic information on their energies, angular momenta, and widths. In nuclei the structure is more like that of molecules, in that not only are there single-particle excitation levels (the kind we expect from a shell structure), but in addition there are collective effects leading to rotational and vibrational excitations.

To excite nuclei one employs reactions of the form

$$a + A \rightarrow b + B, \tag{15–33}$$

where $A$ is a target nucleus and $a$ is a projectile; typically the projectile, which may be an electron, a proton, a neutron, a photon, a deuteron, an alpha particle, or even a heavy nucleus, is brought to an appropriate energy by an accelerator. The end products of the reaction are an emergent projectile $b$—any one of the particles labeled by $a$, but also, possibly a pair of protons, an electron and a nucleon, a neutrino and one or more nucleons, and so on—and a residual nucleus $B$. The reaction must obey conservation laws for energy, momentum, and charge, as well as other conservation laws that will be discussed in the next chapter.

In Chapter 14 we described the effective area $\sigma$ that a target particle presents to an incoming projectile. This quantity is the *collision cross section*. One can measure a cross section for every available energy and, indeed, for every available incoming and outgoing projectile. The sum of all of this information can then be analyzed using quantum mechanics, and in this way one can learn about the wave functions of the excited energy levels that are accessed by the

• **Figure 15–16** A sketch of the relative number of gamma rays seen in the reaction discussed in the text, as a function of energy. The nearly-equal spacing between the peaks strongly suggests that these spectral lines arise from transitions between rotational levels; that is, $j + 2 \rightarrow j$. (Based on data of R. M. Diamond and F. S. Stephens, *Arkiv For Fysik*, **36**, 221, © 1987 by permission.)

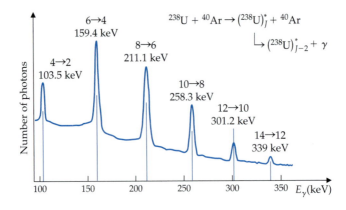

combination of projectile and target. We give just one example that illustrates this process, namely

$$^{238}\text{U} + {}^{40}\text{Ar} \rightarrow {}^{238}\text{U} + {}^{40}\text{Ar} + \gamma.$$

The outgoing gamma rays (photons) have a spectrum shown in •Fig. 15–16. The interpretation is superimposed on the figure: The $^{40}$Ar comes close to the uranium nucleus and, through its coulomb field, excites different states in the uranium. The pattern of gamma-ray energies shows peaks that are more or less equally spaced, which is characteristic of rotational spectra. In this case the different peaks are associated with $j = 4 \rightarrow j = 2$ transitions, $j = 6 \rightarrow j = 4$ transitions, and so on, up to $j = 14 \rightarrow j = 12$ transitions. From the data one can deduce the rotational inertia, and an analysis of the rotational inertia then leads to a more accurate picture of the different ways in which nuclei can be excited.

Of more practical interest are reactions in which neutrons collide with heavy nuclei and cause fission—a splitting of the target nucleus. This phenomenon is sometimes referred to as induced fission, to distinguish the process from spontaneous fission. Neutrons are particularly effective in destabilizing nuclei that are on the borderline of undergoing spontaneous fission, because spontaneous fission is actually a type of tunneling process and the neutron may bring enough energy to put the tunneling fragments nearer to the top of the barrier or even over its top. These neutrons are even more effective if they are very slow: Fast neutrons are more likely to knock out particles from the nucleus, but slow (thermal) neutrons are captured on entering the nucleus (recall our discussion of the compound-nucleus model) and thus add about 6 to 8 MeV of binding energy to the system. This energy will cause the "liquid drop" to undergo large oscillations that in turn cause the nucleus to "neck off" (see •Fig. 15–7) and break up into two daughter nuclei of smaller mass. Thus, if one wants to induce fission efficiently, one must slow the neutron projectiles. Neutrons are slowed down by passing them through material that contains a preponderance of light nuclei (e.g., paraffin, which contains a large number of hydrogen atoms), so that in sucessive elastic collisions the neutrons give up kinetic energy to the almost stationary light nuclei. After many collisions, the neutrons come into thermal equilibrium and, by equipartition, have kinetic energies on the order of $3kT/2$, where $T$ is room temperature. Using the equipartition relation $\langle mv^2/2 \rangle = 3kT/2$ as a way to measure the average speed, we would find that at room temperature ($T = 294$K), the speed would be around 3,000 m/s, very slow by microscopic standards.

It is more likely than not that in the fission process one or more additional neutrons are emitted. The reason is that heavy nuclei have relatively more neutrons than protons—for example, $^{235}$U is 60% neutrons by number, but palladium, which has half the Z-value, is only 57% neutrons by number. Indeed, it is most likely that the neutrons from fission will be produced by the daughter nuclei as they shed neutrons to produce a more nearly equal neutron-to-proton ratio—a property that tends toward nuclear stability. A great deal of kinetic energy is generated in fission. Most of it is in the kinetic energy of the heavy daughter nuclei, with the rest carried off by neutrons and possibly gamma rays. In Example 15–3 we used the semiempirical mass formula to estimate the energy released when a heavy nucleus splits into two equal pieces. That energy is typically about 200 MeV, of which the fission fragments take off something like 167 MeV in, for example, the case of $^{235}$U.

Another important class of reactions are those of **fusion**. These reactions, which occur when light nuclei combine to make heavier ones, are of particular astrophysical interest, as we shall see in the next section.

# 15–6 Applications

### Geological and Archeological Dating

Alpha decay has found an application in geology in the determination of the time elapsed since rocks were formed. When the minerals that form rock are in the liquid state, any helium they may contain boils off. However, when the rock crystallizes the helium may remain trapped. Suppose now that the rock contains a certain amount of uranium, primarily $^{238}$U. When this nuclide decays along the path shown in •Fig. 15–12, it ultimately ends up as $^{206}$Pb plus eight $\alpha$-particles. If these particles do not escape from the rock, the ratio between the remaining uranium present and the helium in the rock is a function of the age of the rock. Since some of the helium may diffuse out of the rock, an alternative is to measure the ratio of lead to uranium. The oldest rocks encountered on the surface of the earth are found in this way to be approximately 3.7 billion years old. The oldest meteorites are found in the same way to be 4.5 billion years old, an age that corresponds to the formation of planets, earlier than the time of formation of surface rocks.

The decay of potassium into argon $\left(^{40}\text{K} \rightarrow {}^{40}\text{Ar}\right)$ with a mean life of $1.8 \times 10^9$ yr has been used to obtain the ages of rocks both on Earth and on the Moon. Potassium is a very common element in igneous rocks, which initially contain no $^{40}$Ar. Argon produced in the decay of potassium escapes if the rocks are still molten, but is trapped once the rocks solidify. In this way the ratio of $^{40}$Ar to $^{40}$K in solid rock can be used to estimate the age of the rocks. The method has been checked on lava beds whose age is known in other ways.

**Carbon dating** is a similar application used with a time scale of 20,000 years or less. The basic idea is that cosmic rays colliding with molecules in the atmosphere form the unstable isotope $^{14}$C of carbon, which then combines with oxygen and enters into organic matter such as trees. In a living plant $CO_2$ undergoes a continuous exchange with atmospheric $CO_2$, so that the ratio $^{14}$C/$^{12}$C is the same as in the atmosphere. When the plant dies, however, the exchange stops. Since $^{14}$C decays with a lifetime of 8,270 yr, while $^{12}$C is stable, the ratio $^{14}$C/$^{12}$C in the dead plant diminishes with time. A measurement of this ratio is equivalent to a measurement of the time elapsed since the plant has died.

**Example 15–6**   Climbers on Mount Ararat in Turkey discover some wooden timbers that they think may be the remnants of Noah's ark. They compare the radioactivity of the timbers with that of a tree they find at their base camp and discover that, for the same size blocks of wood, the timbers' $^{14}$C activity—the number of decays of $^{14}$C per unit time—is just half of that of the tree. What is the age of the timbers?

**Solution**   The difference between the wood the climbers think is from the ark—the timbers—and that of the living wood is that the living wood is constantly having its $^{14}$C replenished, so that the activity of the tree is the same as the activity of the trees out of which the timbers were made when those trees were living. Once they were cut down, the $^{14}$C was no longer being replenished and the activity decreased. To proceed we need to find a formula for the activity $A$. We have Eq. (15–23), which expresses the number of radioactive nuclei $N(t)$ as a function of time. The number of decays per unit time is the number at time $t$ minus the number at $t + \delta t$, divided by the time $t$ change $\delta t$, where $\delta t$ is a small time interval:

$$A(t) = \frac{N(t) - N(t + \delta t)}{t - (t + \delta t)} = \frac{N_0 \exp(-Rt) - N_0 \exp(-R(t + \delta t))}{\delta t}$$

$$= N_0 \exp(-Rt) \frac{1 - \exp(-R\delta t)}{\delta t}.$$

We expand this expression for small $\delta t$; to $O(\delta t)$ the numerator factor is $+R\delta t$, so that

$$A(t) = RN_0 \exp(-Rt).$$

The activity decreases with time, as we would expect.

Now let $T$ be the time elapsed since the timbers were cut. Because the blocks of wood from the two trees are identical, presumably they start with the same amount of $^{14}$C, so that $N_0$ is the same. Thus the only difference between the activity of the blocks is that for the timbers that were cut $t = T$, whereas for the tree $t = 0$. Since the activity of the timbers is half that of the block cut from the tree, it follows that

$$\frac{1}{2} = \exp(-RT), \text{ or } -RT = \ln\left(\frac{1}{2}\right) = -0.69.$$

For $^{14}$C, $R = (8{,}270 \text{ yr})^{-1}$, so that $T = (0.69)/R = (0.69)(8{,}270 \text{ yr}) = 5{,}730 \text{ yr}$. This number does indeed put the piece of wood back in biblical times.

## Nuclear Chain Reactions (Fission)

Induced fission accompanied by the emission of neutrons can give rise to a **chain reaction**—a series of fissions that are either very rapid[7] (as in an atomic bomb) or controlled (as in a nuclear reactor). A nuclear reactor consists of a core made of cells of natural uranium embedded in a *moderator*, a substance that slows neutrons, usually graphite. (Earlier we spoke of paraffin, but graphite is more suitable as a construction material.) Natural uranium consists of 99.3% $^{238}$U and 0.7% $^{235}$U. Only the latter undergoes induced fission with slow neutrons. These fissions produce fast neutrons, with energies in the MeV range, and the moderator slows them down enough to be useful for inducing fission in other $^{235}$U nuclei. Thus a chain reaction is generated. Measurements show that on the average 2.47 neutrons are produced in a single fission of $^{235}$U. A small fraction of these immediately cause fission in $^{238}$U, so that the number of neutrons per fission of $^{235}$U that enter the moderator is approximately 2.5. Some 10% of these are captured in the moderator, leaving 2.24 neutrons available, and in a typical design 88% of these are captured by $^{235}$U. But only half of those

---

[7] In a rapid chain reaction, the interval between successive fission reactions is on the order of $10^{-8}$ s, a number derived from the fact that the neutrons which are produced move with speeds on the order of $10^9$ cm/s. (This is the speed that a neutron with a kinetic energy of 1 MeV would have.) Thus the neutron does not have to have much of a share of the total available kinetic energy to move rapidly.

actually induce a fission. In effect, in the design cited a single neutron from the first fission reaction produces 1.07 neutrons for the next reaction. We call this number the *reproduction factor k*. As long as $k$ is larger than unity, we have a chain reaction. Indeed, the problem may be to keep the number of neutrons from becoming too large. Cadmium control rods, which strongly absorb very low energy neutrons—often called thermal neutrons because their kinetic energy written as $K = (3/2)kT$ corresponds to room temperature—can be inserted into the uranium whenever the number of neutrons begins to grow beyond the accepted bounds. In the normal operation of a reactor, $k$ is kept close to unity. In fact some of the neutrons needed to keep $k$ at that value come from the decay of the fission fragments, and that decay may take place seconds after the fission. These so-called delayed neutrons are crucial to the safety of nuclear reactors. If something goes wrong, one has seconds rather than microseconds to insert the control rods. Fission reactors are actually very sophisticated devices, and we have described only their bare essentials.[8]

*The Okla Reactor*   A remarkable example of a reactor that is too interesting not to mention is the so-called fossil reactor making up a mineral deposit in Okla, Gabon, in Africa. As we have mentioned, in natural uranium the percentage of $^{235}U$ is 0.7202%. But in one part of the deposit in Okla some of the $^{235}U$ was discovered to have been depleted: The concentration was only 0.7171%, a difference of four parts per thousand. Examination of the rest of the deposit revealed that in some samples the depletion was as much as half. This meant that something had been depleting the $^{235}U$ beyond the depletion that is due to spontaneous fission and that is present in all natural $^{235}U$. Now since the half-life of $^{235}U$ is 700 million years, the concentration of $^{235}U$ two billion years ago would have been about 3.5%. This is about the same concentration you find in a standard reactor. Moreover, two billion years ago was about the time that plants appeared in Earth's newly formed atmosphere of oxygen. This organic matter in the sea precipitated uranium in the form of pitchblende in geological traps, where it became concentrated. If the concentrated uranium was in a swamp, then the water would act like a moderator, and all the conditions for a natural reactor would be present. So far no other natural fossil reactors have been found, but many must have existed.

## Fusion Reactions

Fusion reactions play a critical role in the evolution of stars, and the understanding of these reactions has allowed astrophysicists to explain, in detail, both the abundances of elements in stars and the histories of different classes of stars. Our sun "burns" through a series of nuclear reactions that can be summarized as the combination of two protons and two neutrons forming an $\alpha$-particle: $p + p + n + n \rightarrow {}^4He$. The process does not actually occur this way, because the probability of four particles getting simultaneously close to one another is very small. Instead, the process occurs as the end result of a series of other processes that form cycles. Some of these processes are very slow, and some are fast. The reaction times, or equally their reciprocals—the rates—are sensitive to the environment in which the reactions occur. For example, the fusion reactions must overcome the coulomb barrier that keeps the charged nuclei apart, so they are sensitive to the temperature of the environment, which determines the average speed of the nuclei. Moreover, all the reactions depend on the

[8] A very good description of the building of the first reactor, as well as information about the atomic bomb, may be found in Richard Rhodes, *The Making of the Atomic Bomb* (New York: Simon & Schuster, 1986).

density of the materials in which they take place. One of these cycles is (the mean reaction times are given on the right of each reaction and are appropriate to the conditions in the Sun, whose central temperature is about $1.4 \times 10^7$K)

$$p + p \rightarrow d + e^+ + \nu + 0.41 \text{ MeV} \qquad (7 \times 10^9 \text{ yr});$$

$$d + p \rightarrow {}^3\text{He} + \gamma + 5.51 \text{ MeV} \qquad (4 \text{ s});$$

$${}^3\text{He} + {}^3\text{He} \rightarrow {}^4\text{He} + p + p + \gamma + 12.98 \text{ MeV} \qquad (4 \times 10^5 \text{ yr}).$$

(Recall that $d$ represents the deuteron—the nucleus with one proton and one neutron.) We have included the kinetic energy released in each reaction, energy that, in part, arrives on Earth in the form of sunshine. This set of reactions effectively amounts to $6p \rightarrow 2p + 2e^+ + {}^4\text{He} +$ photons and neutrinos. Note that the initiating reaction is much slower than the others; as we describe in Chapter 16, reactions involving neutrinos are slow.

A second important cycle for many stars is the **carbon cycle**, in which ${}^{12}\text{C}$ acts as a catalyst:

$${}^{12}\text{C} + p \rightarrow {}^{13}\text{N} + \gamma + 1.93 \text{ MeV} \qquad (10^6 \text{ yr});$$

$${}^{13}\text{N} \rightarrow {}^{13}\text{C} + e^+ + \nu + \gamma + 1.20 \text{ MeV} \qquad (10 \text{ min});$$

$${}^{13}\text{C} + p \rightarrow {}^{14}\text{N} + \gamma + 7.60 \text{ MeV} \qquad (2 \times 10^5 \text{ yr});$$

$${}^{14}\text{N} + p \rightarrow {}^{15}\text{O} + \gamma + 7.39 \text{ MeV} \qquad (< 3 \times 10^7 \text{ yr});$$

$${}^{15}\text{O} \rightarrow {}^{15}\text{N} + e^+ + \nu + \gamma + 1.71 \text{ MeV} \qquad (2 \text{ min});$$

$${}^{15}\text{N} + p \rightarrow {}^{12}\text{C} + {}^4\text{He} + 4.99 \text{ MeV} \qquad (10^4 \text{ yr}).$$

This cycle, which is effectively equivalent to $4p \rightarrow {}^4\text{He} + 2e^+ +$ photons + neutrinos, can act only in stars that are mature enough to have accumulated sufficient carbon, which is produced through still another set of nuclear reactions.

The reaction times given in the foregoing equations show the large variety of rates involved in nuclear reactions. Many of these reactions are inhibited by the presence of a barrier produced by coulomb repulsion. Quantum mechanical tunneling through this barrier plays a role, and the rates contain a typical tunneling factor $\exp(-Z_1 Z_2 \alpha c/v)$, where $Z_1$ and $Z_2$ are the atomic numbers of the interacting particles, $\alpha = 1/137$, and $v$ is the relative velocity of the colliding particles. Thus large values of $v$—high energies—favor a reaction. On the other hand, we know from our work on statistical mechanics that the number of particles with a large velocity falls off as $\exp(-E/kT)$, where $E$ is the energy. In stars the temperatures lie in the range from $10^7$ to $2 \times 10^7$ K. All this makes for a rather complicated set of calculations. It is a tribute to the ingenuity of nuclear experimentalists and theorists who have measured the reaction rates under terrestrial laboratory conditions and then calculated the rates under stellar conditions that not only the current burning of stars, but also their evolution from the earliest stages to their ends as novae or supernovae, is reasonably well understood. A supernova observed in 1987—the first one studied in great detail by astronomers and particle physicists who detected the arrival of antineutrinos in underground laboratories—behaved in many ways as the experts in this field had predicted.

Our sun and other stars may be viewed as gigantic thermonuclear reactors, with gravity keeping the ingredients together for energy generation. Projects to create a terrestrial thermonuclear reactor, in which electromagnetic fields hold the ingredients together, have been under way since the 1950s. Although progress has been slow, our energy needs make this an important research project. Indeed, while it is not our intention to enter into debates about "clean" and not-so-clean energy, it is clear for many reasons that at some point humankind will need to go beyond the burning of fossil fuels. Both fission and fusion must remain as serious candidates for future energy sources.

### Effects of Radiation

When natural radioactivity was discovered at the turn of the century—the name "radioactivity" is due to Madame Curie—there was total ignorance as to its biological effects. The conditions under which, for example, the Curie's worked and that certainly were directly or indirectly reponsible for their deaths would not be tolerated in any modern physics laboratory. Nonetheless, until fairly recently radioactivity was even thought to have a tonic effect. Labels on bottles of mineral water from Europe would sometimes mention the amount of radioactivity in the water, with the idea that this was somehow beneficial. It is, notwithstanding, very difficult to specify the exact amount of radiation that is actually biologically harmful, although as a general rule one can say less is better. This raises the question of what is more and what is less—what are the units? We turn to that question next.

The simplest thing to quantify is the **activity** of a radioactive sample—the number of decays per second of the sample. If we know the decay rate $\lambda$ in terms of which the number $N(t)$ of decaying particles is given as a function of the time $t$, then, as we showed in Example 15–6, the activity $A = N(t)\lambda$. The common unit for $A$ is the *curie*, symbolized Ci:

$$1 \text{ Ci} = 3.7 \times 10^{10} \text{ decays/s.}$$

Notice that no reference is made to what the decays are. Typical laboratory samples have activities that are in the microcurie to the millicurie range.

The curie is a useful unit for comparing the strengths of different amounts of the same radioactive sample—say, 1 gram versus 10 grams of tritium. But for biological purposes we often want to compare the effects of different kinds of radioactivity, say an alpha versus a gamma ray emitter. Toward that end we use the fact that products of radioactive decays ionize the medium in which they travel. For example, a gamma ray traveling in air will knock the electrons out of some of the neutral atoms it encounters, rendering them charged. Indeed, the way in which such radiation is detected is to measure an effect of this ionic charge. The measurement leads to the definition of the so-called *roentgen* (symbolized R), after Wilhelm Conrad Roentgen, who discovered X rays at the turn of the century. The roentgen is a measure of how much ionizing charge is produced per unit mass of whatever it is that is being ionized. To normalize things, the roentgen refers to the ionization of 1 cm$^3$ of air under standard conditions, namely 0°C and 760 mm of pressure. By definition,[9] 1 R produces $3.3 \times 10^{-10}$ C of charge in that amount of air under standard conditions, and this translates to

$$1 R = 2.58 \times 10^{-4} \text{ C/kg.}$$

**Example 15–7**    Suppose that in air each ion that is produced by an ionizing radiation of 1 R has the charge of one electron, that is, $1.6 \times 10^{-19}$ C. How many ions are produced in a kilogram of air under standard conditions? Given that it takes about 34 eV to form an ion, how much energy per kilogram is being dumped into the air?

**Solution**    The total number of ions per kilogram formed is

$$N = (2.58 \times 10^{-4} \text{ C/kg})/(1.6 \times 10^{-19} \text{ C}) = 1.6 \times 10^{15} \text{ ions/kg.}$$

---

[9] The roentgen is most certainly not an SI unit!

Each ion has taken up 34 eV = $34 \times 1.6 \times 10^{-19}$ J. Thus 1 kg has taken up

$$(1.6 \times 10^{15} \text{ ions/kg})(34 \times 1.6 \times 10^{-19} \text{ J/ion}) = 8.8 \times 10^{-6} \text{ J.}$$

We are, of course, not made out of air. What we want is a measure of the *dose* of radiation a given part of our body absorbs. This is given in terms of a unit called the *rad*—the *radiation absorbed dose*. The rad is a measure of the amount of energy per gram that the radiation is dumping into the material— our bones, for example. It is defined as 1 rad = 100 erg/gm of absorbed energy. Given that 1 erg = $10^{-7}$ J, you can calculate from Example 15–7 that, for air, 1 $R$ = 0.87 rad.

But different kinds of radiation have different effects. An electron and an alpha particle passing through matter at the same initial kinetic energies will lose energy in matter quite differently. The unit that meaures these relative effects is called the *quality factor*, QF. It is a pure number that indicates how much energy a type of radiation produces over a given path length. X rays, for example, typically have a QF of 1, while alpha particles have a QF of 20. Putting QFs and rads together, one can define something called the *dose equivalent* as DE = rad × QF. This quantity is measured in *rems*, the units that you usually read about when you are warned of what doses you can tolerate in, say, a year. At sea level at a latitude of about 50°, the natural background radiation to which we are all exposed is about 0.1–0.2 rem per year. This radiation comes in part from the heavens and in part from the Earth. The heavenly part—cosmic radiation—consists of a large variety of particles ranging from electrons and muons to gamma rays and neutrinos. The earthly part consists of decays of radioactive isotopes such as the uranium and thorium series and $^{40}$K. The recommended maximum absorbed amount per year has gone down over the years, but for the general public it is now set at 0.5 rem/yr. A dental X ray produces about 0.002 rem, while a chest X ray produces about 0.05 rem. What effect such small doses have on your health is not predictable, but if you break your arm it is clearly better to have an X ray than to worry about the additional fraction of a rem.

## SUMMARY

The problem of understanding the atomic nucleus is more complicated than the problem of understanding the atom itself, for several reasons: The constituents of the nucleus—protons and neutrons—interact so much more strongly than electrons interact with the nucleus that techniques appropriate for atomic interactions are not suitable for nuclear interactions. In fact, we are not even sure of the internucleon forces, nor can such forces be simply written in terms of a potential energy, because special relativity plays an important role in nuclei. The structure of nuclei is, moreover, a many-body problem with all of the constituents of roughly equal mass. In spite of this complexity, some simple pictures, appropriate to different aspects of nuclear physics, have been developed:

- Nuclear sizes are measured by scattering electrons or other projectiles from nuclei and by corrections to atomic structure. The observations show that the nucleus can be regarded as a closely packed collection of nucleons.
- The forces that bind nucleons are so strong that the mass of a nucleus is only some 99% of the sum of the masses of its constituents. A good account of the masses of nuclei, especially for $Z$ values greater than 10, can be made with

the semiempirical mass formula, which takes into consideration the various aspects of nuclear structure.

- Details of nuclear structure are best described by models that emphasize its different aspects. The most "collective" view is the liquid-drop model, which treats the nucleus as a continuous fluid and can explain nuclear collisions in which intermediate compound nuclei form and decays in which the nucleus fissions. The most "individualistic" view is the shell model, in which separate nucleons move in a kind of collective potential formed by the others. This model successfully accounts for nuclear energy levels, spins, and electromagnetic properties of nuclei, as well as the systematics of stable nuclei.

- According to the respective numbers of neutrons and protons, the masses of many nuclides permit decays to other nuclei through various modes, including neutron decay, beta decay, gamma decay, and alpha decay, in which the particle emitted is a neutron, an electron, a photon, and a helium nucleus, respectively. Large nuclei can also undergo spontaneous fission into smaller fragments. The mechanism for beta decays is understood in terms of the more fundamental beta decay of the neutron. The alpha decay mechanism is a quantum tunelling process.

- Certain collision processes are of special interest, including induced fission, which has important applications in nuclear reactors, and fusion, which is crucial for our understanding of stellar processes and may eventually be realized in some application.

## QUESTIONS

**1.** The half-life of a certain unstable nuclide is 5 min. Does this mean that after 10 min, all the members of a collection of this nuclide will have decayed? If not, how long will it be before all the members will have decayed?

**2.** In many fission reactors a large number of antineutrinos are emitted. Why is that?

**3.** When a nucleus undergoes fission, the large fragments are never stable. Why?

**4.** Given that it takes about 8 MeV of energy to remove a neutron from a nucleus and that the radius of a nucleus such as lead is about 6.6 fm, how would you go about estimating the depth of the potential in which the neutron sits?

**5.** The masses of nuclei are measured by taking ions of a particular atom and passing them through simultaneous electric and magnetic fields, appropriately oriented. What is an example of "appropriately oriented" electric and magnetic fields? Does your configuration make the determination of the nuclear masses depend on the charge-to-mass ratio of the nucleus or on the mass alone? Ions generally carry some electrons attached to the nucleus. With what accuracy is the nuclear mass determined if you ignore the mass associated by relativity with the binding of these electrons?

**6.** Experiments show that when neutrons are made to collide with heavy nuclei, the number of neutrons absorbed per target nucleus has an energy dependence of the form of a series of very sharp, narrow spikes with widths of fractions of electron volts separated by large gaps. We are used to thinking about nuclear energies in the realm of MeV. If you think of the spikes as the analogue of spectral lines (which they are), what do these numbers tell you about heavy nuclei? Which of the models that we have discussed in the text would get support from these experimental results?

7. We have seen that the $\alpha$-decay of a nucleus is a quantum tunneling effect. We visualize the phenomenon as some of the nucleons in the nucleus combining into an $\alpha$-particle in the nucleus. The $\alpha$-particle subsequently tunnels through the coulomb barrier created by the remaining protons. Until Fermi came up with the correct theory of $\beta$-decay, it was thought that perhaps the process was of the same kind as $\alpha$-decay, viz., electrons tunneling through a coulomb barrier. The following considerations, however, will show you why this picture does not work: **(a)** If we imagine the $\alpha$-particle confined to a space of nuclear dimensions, it will have a de Broglie wavelength on the order of the nuclear radius. How big is this? **(b)** The momentum of the $\alpha$-particle is determined approximately by the wavelength from Part (a). How are the kinetic energy of the $\alpha$-particle and the de Broglie wavelength related? **(c)** The observed kinetic energies of the $\alpha$-particles in these decays are on the order of a few MeV. Is this consistent with the momentum given by the de Broglie wavelength? It has to be if the tunneling picture works. **(d)** But now you will see the problem with the electron in $\beta$-decay. It too has energies on the order of MeV. What does this say about its de Broglie wavelength? How does this wavelength compare with that of an $\alpha$-particle with the same energy? How can this electron then possibly fit into a nucleus? It doesn't, so the tunneling picture does not work. In the Fermi theory the electron is created in the decay!

8. Why are light nuclei, but not heavy nuclei, suitable as moderators for reactors? What sort of collisions slow down the neutrons that are being moderated? The best moderator would be hydrogen, except that it "eats" neutrons in the reaction $n + p \rightarrow \gamma + d$.

9. In the process described in the preceding question, the gamma ray carries off energy. How is this possible? After all, the deuteron's nucleus consists of a neutron and a proton.

10. The neutron and the proton form a bound state—the deuteron, symbol $d$. The average separation of the constituents is on the order of 1 fm. What can you deduce about the size of the potential energy from the fact that there is no $np$ bound state lower in mass that the deuteron, nor is there an $np$ bound state with higher mass?

## PROBLEMS

1. ▮ Show that if the nuclear density is constant the volume of the nucleus is proportional to $A$, the radius is proportional to $A^{1/3}$, and the surface area is proportional to $A^{2/3}$.

2. ▮▮ Given that it may take as much as 10 MeV to remove a nucleon from a nucleus, estimate the difference between the mass of a nucleus and the sum of the masses of the nucleons that compose it. How big is the corresponding effect in atoms if we regard the atomic constituents as the nucleus as a whole plus the electrons?

3. ▮ Noting that there are just $3.6 \times 10^6$ J in a kW-hr, make an estimate of how many 100-watt lightbulbs could be lit for an hour if you could fission one kg of uranium. (See Example 15–4 for the critical number.) Electricity costs around 10 cents per kW-hr. What would you have to pay for the usage described here?

4. ▮▮ Consider the nuclei $^{13}$N ($Z = 7$) and $^{13}$C ($Z = 6$). The difference in mass between them is measured to be 1.19 MeV/$c^2$. Which terms in the semiempirical mass formula contribute to that difference? Use your answer in conjunc-

tion with the expression for the coulomb energy in the semiempirical mass formula, $(3/5)Z^2e^2/(4\pi_0 R_0 A^{1/3})$, to calculate $R_0$.

**5.** ▮▮ In this problem you will use the semiempirical mass formula to conclude that the nucleus is, to a good approximation, incompressible. Start by noting that the volume term in the formula dominates for, say, $A$ greater than 100. We can call the resulting energy $E_{vol} = -b_{vol}A$. **(a)** Express $E_{vol}$ as a function of radius. **(b)** By how much is the energy altered for a change in volume that results from an overall change in the radius $\Delta R$? In addition to providing an analytic answer, evaluate your result numerically by finding the energy change for a 1% change in the radius. Your numbers should be on the order of 50 MeV per nucleon—such a large number that you can conclude that the nucleus is incompressible.

**6.** ▮▮ Calculate the total energy of the ground state of a collection of $Z$ protons within a sphere whose volume is that of the nucleus, $(4/3)\pi R_0^3 A$. [See Eq. (10–57).] Repeat the calculation for a collection of $N$ neutrons.

**7.** ▮▮▮ In Problem 6 you will have found that the ground-state energy of a collection of $Z$ noninteracting protons placed within the volume of a nucleus with $A$ nucleons is

$$E_p = \left(\frac{9}{4\pi^2}\right)^{5/3} \frac{2\pi^4 A\hbar^2}{15mR_0^2} \left(\frac{Z}{A}\right)^{5/3},$$

and similarly, the ground-state energy of $N$ noninteracting neutrons is

$$E_n = \left(\frac{9}{4\pi^2}\right)^{5/3} \frac{2\pi^4 A\hbar^2}{15mR_0^2} \left(\frac{N}{A}\right)^{5/3}.$$

In these formulas we have neglected the small neutron–proton mass difference and used a common mass $m$. **(a)** Express the total ground-state energy of a nucleus made of $N$ neutrons and $Z$ protons confined within the nuclear volume—the sum of $E_p$ and $E_n$—in terms of the variables $A = N + Z$ and $t = N - Z$. **(b)** Assuming that $t/A \ll 1$, make an appropriate expansion of your result in Part (a), using the fact that, for $x$ small, $(1 + x)^n \approx 1 + nx + n(n - 1)x^2/2 + \ldots$. Show that the terms which are linear in $t$ cancel, leaving the first nontrivial term proportional to $t^2$. You will thereby have shown that Eq. (15–11) is correct. **(c)** Using $mc^2 = 940$ MeV and $R_0 = 1.24 \times 10^{-15}$ m, find a numerical value for the constant $K$ in Eq. (15–12). Express your answer in MeV.

**8.** ▮▮▮ Consider the semiempirical mass formula for strontium ($Z = 38$). There are isotopes of strontium that have values of $A$ from 95 down to 80. **(a)** Which is the stablest of these isotopes? **(b)** What is the smallest value of $A$ for which beta decay, $^A Z \rightarrow {}^A(Z + 1) + e^- + \bar{\nu}_e$, becomes possible if $m_e c^2 = 0.51$ MeV?

**9.** ▮▮ Use the semiempirical mass formula to calculate the kinetic energy of the $\alpha$-particle produced in the $\alpha$-decay of $^{235}$U. Check your results by using the measured masses given in Appendix A.

**10.** ▮ A beam of deuterons incident on $^{29}$Si causes nuclear reactions in which a particle is emitted, leaving a recoil nucleus. Give the mass number and the atomic number of the recoil nucleus if the emitted particle is **(a)** an $\alpha$-particle, **(b)** a proton, and **(c)** a neutron.

**11.** ▮▮ Consider the collision of a projectile of mass $m$ with a target particle of mass $M$ that is at rest. What is the maximum energy loss of the projectile in the collision? If you wanted to slow down neutrons with collisions such as this, would you use a target made of heavy or light nuclei?

**12.** ▮▮ The process $n + p \rightarrow \gamma + d$ is known as *radiative capture*. Given the masses involved, what is the energy of the gamma ray if the capture takes place with the neutron and the proton at rest?

**13.** ▮▮ The binding energy of the deuteron is the minimum energy needed to break it up into its component parts. Using the table of isotopes in Appendix A, find the binding energy of the deuteron in Mev and in joules. What would the energy be of a photon that would just photodisintegrate the deuteron? Why is this number different from the result you obtained in the preceding problem?

**14.** ▮▮ Assuming that the factorization of the rates given by Eq. (15–21) holds, prove that, for the physically measurable rates $R$ shown,

$$\frac{R(\alpha + {}^{60}\text{Ni} \rightarrow n + {}^{63}\text{Zn})}{R(p + {}^{63}\text{Cu} \rightarrow n + {}^{63}\text{Zn})} = \frac{R(\alpha + {}^{60}\text{Ni} \rightarrow 2n + {}^{62}\text{Zn})}{R(p + {}^{63}\text{Cu} \rightarrow 2n + {}^{62}\text{Zn})}$$

$$= \frac{R(\alpha + {}^{60}\text{Ni} \rightarrow n + p + {}^{62}\text{Cu})}{R(p + {}^{63}\text{Cu} \rightarrow n + p + {}^{62}\text{Cu})},$$

provided that the energies are such that all the reactions go through the same intermediate compound nucleus ${}^{64}\text{Zn}^*$.

**15.** ▮▮ The energy levels for a particle in a three-dimensional harmonic oscillator are given by the formula $\hbar\omega(2n_r + \ell + 3/2)$ with $n_r$ taking the values 0, 1, 2..., and $\ell$, the angular-momentum quantum number, taking the values 0, 1, 2, .... Plot the energy spectrum going up to $n_r = 3$ and $\ell = 6$. Label each state by the degeneracy, and write down the degeneracy for each value of the energy. Give the magic numbers (i.e., the cumulative values of the states forming closed shells) for this model of the nucleus.

**16.** ▮▮ What is the ${}^{14}\text{C}$ activity, defined to be the number of decays per unit time in a sample of the material, expected in a timber structure that is 800 years old? 1,200 years old? Give your results as a ratio of the activity of a sample of the timber to that of the same-sized sample of modern wood. (*Hint*: See Example 15–6 for a discussion of the activity.)

**17.** ▮ A long-lived isotope of polonium listed in Appendix A has an $\alpha$-decay. Identify the daughter nuclide.

**18.** ▮ A long-lived isotope of actinium listed in Appendix A has a $\beta$-decay. What is the daughter nuclide?

**19.** ▮▮▮ In our treatment of radioactive decays—see especially the discussion following Eq. (15–25)—we said that the daughter decayed into a "granddaughter," but we did not include that in our equations. Suppose the "granddaughter" is stable. Find $N_3(t)$.

**20.** ▮▮ Consider the decay series described between Eqs. (15–25) and (15–26). Assuming that $R_{12} = 0.0040\,\text{s}^{-1}$ and $R_{23} = 0.0017\,\text{s}^{-1}$, plot the activity as a function of time for the case in which we initially ($t = 0$) have $10^{23}$ nuclei of type 1 and none of type 2 or type 3. Assume that the type 3 nucleus is stable. (For a definition of the activity, see Example 15–6.)

**21.** ▮▮ Using relativistic kinematics, find how the kinetic energy is shared when a parent nucleus decays into a daughter nucleus and a neutron. (*Hint*: Work in the rest frame of the parent nucleus.)

**22.** ▮▮ Using relativistic energy and momentum conservation and assuming that beta decay is of the two-body form ${}^{A}Z \rightarrow {}^{A}(Z + 1) + e^-$, find the momentum and the kinetic energy of the electron as a function of the masses of the particles involved. Relativistic kinematics is essential here, since in the actual decays the electrons move with appreciable fractions of the speed of light. You may want to use the hint of the previous problem.

**23.** ▮▮ In the case of a typical fission of $^{235}$U the fission fragments have a mass ratio of 95/140. Assuming that the total kinetic energy they have between them is 167 MeV, with what percent of the speed of light is each fragment traveling? To do this problem you can use the approximation that the mass of a nucleus is given by $Zm_p + Nm_n$ in any calculation concerning the motion of the nuclei.

**24.** ▮▮ *Mirror nuclei* are pairs of nuclei in which the numbers of protons and neutrons are exchanged. $^{11}$B(boron; $Z = 5$) and $^{11}$C(carbon; $Z = 6$) are examples. The mass difference of this pair of mirror nuclei is 1.980 MeV$/c^2$, with carbon being the heavier. Will one of these nuclei decay into the other? Which one? By what process?

**25.** ▮▮ Consider the mirror nuclei described in the previous problem. Show that if the *nn* nuclear force differed from the *pp* nuclear force, the masses of the mirror nuclei would have a non-zero contribution from this effect. Such a term is explicitly ruled out in the semiempirical mass formula. Compare a calculation of the mass difference from the semiempirical mass formula with the measured mass difference of 1.980 MeV$/c^2$. Some 20 pairs of mirror nuclei are known and these strongly suggest the equality of the nuclear forces between nucleons.

**26.** ▮▮ A nucleus decays into two channels with relative probabilities 0.62 and 0.38, respectively. The half-life of the nucleus is 200 s. What are the partial decay rates into the two channels?

**27.** ▮▮ Consider the $\gamma$ spectra shown in •Fig. 15–16. Assume that energy levels can be described by the formula

$$E = Wj(j + 1) + K[j(j + 1)]^2,$$

where $W$ and $K$ are constants. Given the data of the figure, calculate $W$ and $K$.

**28.** ▮▮▮ The parameter $W$ of the previous problem has the usual form $\hbar^2/2I$, where $I$ is the rotational inertia of the nucleus. Estimate $I$ by assuming that the $^{238}$U nucleus (92 protons and 146 neutrons) consists of a core of a doubly magic nucleus of 82 protons and 126 neutrons and a spherical nucleus containing the the remaining 30 nucleons. Treat these two nuclei as two point objects with masses characterized by $A = 208$ and $A = 30$, respectively, separated by a distance that is the "radius" of $^{238}$U. How does your number compare with the result of the previous problem?

**29.** ▮▮ What is the minimum energy that can be released in the reaction $n + {}^{235}$U $\rightarrow {}^{93}$Rb $+ {}^{141}$Cs $+ 2n$? In addition to the masses listed in Appendix A, you will need the mass of $^{93}$Rb, an isotope of rubidium (92.92172 amu), and the mass of $^{141}$Cs (140.91949 amu).

**30.** ▮▮ Suppose that at the present time, $t = 0$, two (unrelated) radioactive isotopes, labeled 1 and 2, are present in numbers $N_1(0)$ and $N_2(0)$, respectively, so that their present relative abundance is $N_1(0)/N_2(0)$. **(a)** What is their relative abundance at any previous (negative) time, given that the decay constants for their respective decays are $R_1$ and $R_2$. **(b)** Apply your result from Part (a) to find the relative abundance on Earth of the two uranium isotopes $^{235}$U and $^{238}$U two billion years ago given that their relative abundance today is 0.00725. You will need to find the principal decay modes for these isotopes from Appendix A.

**31.** ▮▮ In his experiment to identify the neutron Chadwick used the reaction

$$\alpha + {}^{11}\text{B} \rightarrow n + {}^{14}\text{N}.$$

He measured the kinetic energies of the particles involved, which were 5.26 MeV for the $\alpha$-particle, 3.26 MeV for the neutron, and 0.57 MeV for the nitrogen nucleus; the boron nucleus was at rest. Using these data and the masses of the boron nucleus, the nitrogen nucleus, and the $\alpha$-particle listed in

Appendix A, find the mass of the neutron. How does your result compare to the value for the neutron mass given in that Appendix?

**32.** ▮▮ If the density of uranium is 18.97 grams per cubic centimeter, what is the radius of a sphere that contains 56 kilograms of uranium? This is the so-called critical mass of uranium—the minimum amount which, if assembled, will spontaneously explode. If all of this material were to fission how much energy would be released, given that the number of nuclei per kilogram is $2.58 \times 10^{24}$ and that each fission produces $2.7 \times 10^{-4}$ erg? In a nuclear weapon about 2% of the nuclei are fissioned. If a gram of TNT produces an explosion of $4 \times 10^{10}$ ergs, how much TNT would you need to produce an explosion equivalent to 2% of the fissioning of a critical mass?

# Frontiers

All the areas of physics remain active fields of research. In mechanics, current research includes the delicate dependence of the solutions of Newton's equations on initial conditions, a topic closely associated with the notion of *chaos*. In the field of fluid flow, research is focusing on the complex problem of *turbulence*. In the area of solid-state physics, two current fields of great interest are high-temperature superconductivity and the quantum properties of systems that are considerably larger than atoms—the field of mesoscopic physics. In each of the preceding fields, one is building on a platform that has its own laws and principles, as well as the particular variables that describe the physics of that field. Ultimately though, all of this research rests on the notion of the atomic structure of matter, and a study of atoms necessarily leads us to nuclei, the nuclear constituents—protons and neutrons—and then to the question of what *these* are made of.

These considerations lead us to an important property of physical laws: At the particular **scale** of size at which a given set of phenomena is described, the detailed physical laws of the scale below that do not have to be known; the effect of the "smaller scale" physics can be described by a few constants. For example, to understand atoms, we don't need to know much about their nuclei, other than that they are almost pointlike on the scale of atomic radii, that their spins are half-integral if $A$ is odd or integral if $A$ is even, that they are characterized by definite masses, which we take from experiment, and that they have certain electromagnetic properties (e.g., magnetic moments), which we also take from experiment. Our discussion of nuclear physics, though cursory, indicated that we did not need to know much about the internal structure of protons and neutrons to get a good description of nuclear structure. The structure of nuclear particles and of electrons and neutrinos is a field of research and speculation that may be called the *small-scale frontier of physics*, a topic discussed in Chapter 16.

At *large* scales, gravity is sufficiently well understood in terms of Newton's law to explain the motions of planets, solar systems, and even galaxies with high precision. But when the scale gets larger and masses increase still further, new laws of physics make their appearance. The general theory of relativity of Einstein contains and supersedes Newton's law of gravity, and general relativity, too, is a field in which one is approaching the frontiers of physics. A brief introduction to the subject appears in Chapter 17.

In an interesting convergence of the largest scales and the smallest scales, the field of *cosmology*—the study of the evolution of the universe—draws heavily both on the subatomic scale and on general relativity for an understanding of what has been observed. A brief introduction to cosmology in Chapter 18 concludes our survey of these frontiers of physics.

# Elementary Particle Physics

The subject of this chapter is the underlying structure of the elementary particles that make up matter and the forces between these particles. The field is also known as **high energy physics**. The reason for this name is that to probe distances of dimension $\Delta x$, one needs projectiles with a de Broglie wavelength $\lambda$ smaller than $\Delta x$, and this implies projectile momenta larger than $\hbar/\Delta x$. For small $\Delta x$ one needs large momenta and therefore large energies. At the same time, high energies imply high speeds, and thus an understanding of the physics in this domain necessitates a fusion of quantum mechanics and special relativity. Although the field originally developed through an attempt to understand the nuclear forces—the nuclear potential $V(r)$ that was to be inserted in a Schrödinger equation—we now know that the nuclear force problem is peripheral to the much deeper understanding opened up by the development of relativistic quantum mechanics.

We begin with a bit of historical background to the main content of this chapter. While all of physics consists of a continual dialogue between theory and experiment, elementary particle physics is particularly striking in this respect, representing a kind of leapfrog game in which theory may get ahead temporarily, to be leapt over by experiment, in its turn to be leapt over by theory again, and so on. Sometimes the theorists' imagination surpasses the technical possibilities for experiment. Sometimes there is a technological breakthrough that leads to new discoveries—some anticipated by the theorists and some not. We can see this pattern in elementary particle physics from the beginning if by the beginning we take the year 1897, when the Englishman J. J. Thomson first identified the electron. But Thomson's discovery had been preceded by theoretical speculation that electric currents were carried by charged particles and by a mid-19th century technological breakthrough: The German glassblower and mechanic Johann Geissler had invented the first glass vacuum tube. Within these tubes vacua were maintained—without pumping—better than anyone had been able to achieve before. Following Geissler, the Englishman William Crookes created vacuum tubes in which electrical "rays" cast sharp shadows. This led Crookes to believe that these rays were made up of charged corpuscles. In 1897 both Thomson and the German Walter Kaufmann showed that the ratio $e/m$ of the charge $e$ to the mass $m$ of the particles making up the rays was much larger than that of hydrogen, and Thomson speculated correctly that $m$ was small compared to the mass of hydrogen.

Thus by the turn of the century there was one known elementary particle: the *electron* as it came to be called. But were atoms themselves elementary particles? Here there was great controversy. Maxwell argued that they were indivisible but nonetheless had complex internal motions such as vibrations. Thomson himself believed that atoms were made up of electrons and *protons*, as the positively charged component of the hydrogen atom came to be called. In Thomson's picture there were *two* elementary particles—the electron and the proton. But as we described in Chapter 15, the features of nuclei demand the existence of neutrons and that the neutron be a spin-1/2 particle, ruling out the possibility that it could be a bound state of an electron and a proton. Chadwick's 1932 discovery of the neutron gave us a third elementary particle.

The understanding of the rules that govern the interactions of these particles begins with the introduction of relativity into quantum mechanics.

## 16–1 Relativistic Quantum Mechanics and Antiparticles

Attempts to combine quantum mechanics with relativity date back to the early days of quantum mechanics. Schrödinger looked at the relativistic energy–momentum relation, which can be written in the form

$$E - V = K + mc^2 = \sqrt{p^2c^2 + m^2c^4}$$

or, equivalently,

$$(E - V)^2 = p^2c^2 + m^2c^4. \tag{16–1}$$

He made the association $E \leftrightarrow i\hbar\, \partial/\partial t$, $p_x \leftrightarrow -i\hbar\, \partial/\partial x$ to get the equation

$$\left(i\hbar \frac{\partial}{\partial t} - V\right)^2 \psi(\vec{r}, t) =$$

$$-c^2\hbar^2\left[\left(\frac{\partial}{\partial x}\right)^2 + \left(\frac{\partial}{\partial y}\right)^2 + \left(\frac{\partial}{\partial z}\right)^2\right]\psi(\vec{r}, t) + m^2c^4\psi(\vec{r}, t). \tag{16–2}$$

Writing $\psi(\vec{r}, t) = u(\vec{r}) \exp(-iEt/\hbar)$, one obtains an eigenvalue equation. When the coulomb potential is used in the equation, the energy is to leading order the Bohr result, Eq. (5–11), which is of order $O(\alpha^2)$. But the new equation also gave $O(\alpha^4)$ terms in the energy, and these corrections disagreed with the accurate experimental data available at the time. Schrödinger realized that the absence of electron spin in his equation was important.

The credit for a correct equation that describes a relativistic spin-1/2 particle is due to Dirac, in 1928. This equation led to totally new insights into the nature of matter. Dirac decided that, to maintain the probability interpretation, an appropriate equation would have to involve only the first derivative $\partial/\partial t$ (i.e., it should contain only the first power of $E$). To maintain the relativistic relation $E = \sqrt{p^2c^2 + m^2c^4}$ which holds in the absence of a potential, he needed to know how to deal with the operator $\sqrt{-\hbar^2c^2\nabla^2 + m^2c^4}$. (Note that $\sqrt{-\hbar^2c^2\nabla^2 + m^2c^4}\,\psi(r)$ *cannot* be calculated as $\sqrt{-\hbar^2c^2\nabla^2\psi(r) + m^2c^4\psi(r)}$; the wave function cannot simply be brought under the square-root sign.) This could be done only by replacing a single equation for $\psi(\vec{r}, t)$ by a set of four coupled equations. The electron spin fell right out of his equations, as did a good approximation to the electron magnetic moment, something that until then had no real theoretical foundation. You could solve the equations in the presence of a coulomb potential and thereby calculate relativistic corrections to the energy levels of hydrogen that were in excellent agreement with the experimental values known at the time. It was wonderful, but there was a catch that seemed so large that many physicists thought that the Dirac equation was crazy.

The catch is that when you take the square root of something you always have a sign ambiguity. The square root of $E^2$ is plus or minus $E$. This meant that the Dirac equation had solutions corresponding to *negative energies*! Why not simply throw them out? In quantum mechanics you can't do this, since in the presence of a potential the positive-energy solutions and the negative-energy solutions could be connected to each other by quantum mechanical tunneling. So what to do? The solution appeared once one coupled a Dirac electron to an electromagnetic field. One then saw that it was possible to rewrite the negative-energy solutions of the Dirac equation as positive-energy solutions, but for "electrons" of the *opposite charge*! Hence one got two particles for the price of one—particles with equal and opposite charges. And one got some new predictions. For example, the electron and its **antiparticle**, as the new particle came to be called, could meet and *annihilate*, turning all their rest energy plus any kinetic energy they may have had into electromagnetic radiation in the form of gamma rays. It is in such processes that the Einstein relation $E = mc^2$ is most visibly realized. We'll come back to these processes later.

But what was the antiparticle of the electron? When the concept was first understood the only positively charged elementary particle known was the proton. Accordingly, the assumption was made that this must be the antiparticle. In 1930, however, the American theorist Robert Oppenheimer showed that that was impossible. He argued that the annihilation of an electron and a proton in a hydrogen atom would take place so rapidly that all matter would be totally unstable. Finally in 1931 Dirac proposed that the antiparticle of the electron was something "unknown [then] to experimental physics." It would be a particle of exactly the same mass and spin as the electron but with opposite (positive) charge.

At the turn of the century the Englishman C. T. R. Wilson invented a particle detector that became known as a *cloud chamber*. This device contains, say, air saturated with moisture. It can be suddenly cooled, at which point the air contains more moisture than is needed to saturate it. The air has become *supersaturated*. If a charged particle passes through this supersaturated air, then water condenses around ions left behind and makes a track like the trail left by a high-flying jet, which can then be photographed. By the early 1930s, the Wilson cloud chamber was the principal method of detecting charged particles in **cosmic rays**—particles that continually bombard Earth from outer space. In 1930 Carl Anderson, a young American postdoctoral student at the California Institute of Technology, succeeded in building a cloud chamber that was placed in the highest magnetic field ever attained in such a device. Why was this important? Recall from Chapter 1 that a particle of charge $q$ and momentum $p$ moving perpendicularly to a constant magnetic field $B$ follows a circular path of radius $R$ given by

$$p = qBR. \tag{16–3}$$

By measuring the curvature of trajectories in known fields and combining this observational data with information on energy—something that can be found by measuring the degree of ionization in the chamber, which is in turn related to how much condensation there was around the track—one can learn a great deal about the particles. In particular, knowing both the energy and momentum one could determine the mass. Anderson soon detected long tracks due to particles with the mass of an electron, but that curved in the direction opposite

• **Figure 16–1** Photograph from a cloud chamber, with charged particle tracks on both sides of the central lead plate. The larger curvature in the upper half reveals that the particle had less momentum above than below, indicating that the particle lost energy in the plate and therefore went from below to above. This fact, combined with the known direction of the magnetic field, shows that the particle has positive charge.

that which an electron coming from outer space would go. There were two choices: Either the particles that produced the tracks were ordinary electrons moving upward from Earth—a possibility that seemed highly unlikely—or they had a positive charge. How to distinguish the two possibilities? Anderson had the ingenious idea of putting a quarter-inch-thick lead plate in the middle of the chamber. That would slow down these mysterious particles, and he could then tell whether they were coming from above or below, depending on how the tracks were bent on either side of the plate. He soon persuaded himself that the particles were indeed positively charged and that apart from the reverse bending their tracks were exactly like those of electrons. In short he had discovered a *positive electron*! This particle, the antiparticle of the electron, is the **positron**, symbol $e^+$ (•Fig. 16–1). Later on Anderson would note that while he had vaguely heard of the Dirac equation it had no influence on his actual discovery, which was a pure accident. In any event, the subject of *antiparticle physics* had been born.

The consequences of the existence of antiparticles are manyfold. In particular, when enough energy was available, electromagnetic radiation could form particle–antiparticle pairs in a kind of inverse annihilation. In other words, *a relativistic quantum theory had to be a theory that involved infinitely many particles.* More than any other feature, this distinguishes the world of elementary particles from all that had come before.

**Example 16–1**    Because they are oppositely charged, an electron and a positron attract each other and can form a hydrogenlike atom. This atom, called **positronium**, is unstable because there is a finite probability that the wave functions of the $e^+$ and $e^-$ overlap. Once this happens the $e^+$ and $e^-$ can annihilate. Given that $m_e c^2 = 0.51$ MeV, find the energy and frequency of each photon in the annihilation process

$$e^+ + e^- \rightarrow \gamma + \gamma.$$

If instead the reaction is $e^+ + e^- \rightarrow 3\gamma$, can you predict the energy of the photons?

**Solution**    To the accuracy given, we can completely ignore the effects of the positronium binding energy, which is of order $O(\text{eV})$, and treat the $e^+$ and $e^-$ as if they were free and at rest. The initial energy is $2m_e c^2 = 1.02$ MeV. Because of momentum conservation, the two photons come off with momentum $\vec{p}$ and $-\vec{p}$, so that their energies are equal at one-half of 1.02 MeV; that is,

$$hf = 0.51 \text{ MeV} = (0.51 \times 10^6 \text{ eV})(1.6 \times 10^{-19} \text{ J/eV}) = 0.82 \times 10^{-13} \text{ J}.$$

Hence the frequency is $f = (0.82 \times 10^{-13} \text{ J})/(6.6 \times 10^{-34} \text{ J} \cdot \text{s}) = 1.2 \times 10^{20}$ Hz.

When there are three photons, we can use the momentum of two of the photons to define a plane, and then by momentum conservation the momentum of the third photon must lie in that plane. Since $\vec{p}_1 + \vec{p}_2 + \vec{p}_3 = 0$, the three momenta form a triangle. The energy conservation equation $p_1 c + p_2 c + p_3 c = 2m_e c^2$ determines the length of one side of this triangle in terms of the length of the other two, but that is not enough to find the angle between two of the momentum vectors. The best we can do is find the maximum value of any one of the momenta.

## 16–2  Conservation Laws

The conservation of energy, momentum, and angular momentum in isolated systems—laws familiar from classical mechanics—also hold in quantum mechanics. We have mentioned a number of instances in which these apply throughout this book. Relativistic quantum mechanics contains the possibility of converting energy into particles and antiparticles, so that the number of particles is no longer fixed. Even so, the presence of conservation laws can restrict the possibilities. The law of charge conservation, first seen in classical electromagnetism, is, as far as is presently known, an exact law. This law certainly provides a constraint on the processes that can occur. It means, for example, that when two electrons collide the final state can consist of four electrons and two positrons but not of six electrons. Similarly a reaction like $e^- + e^- \rightarrow 2\gamma$ is explicitly forbidden by the conservation of electric charge.

The question of conservation laws is relevant to a general rule that appears to hold in particle physics: *Anything that is not explicitly forbidden will occur.* We must therefore have a conservation law that forbids the (unobserved) annihilation of an electron and a proton in a reaction like $e^- + p \rightarrow 2\gamma$. Such a reaction is allowed by electric charge conservation.

All relativistic quantum theories have the property that particles have antiparticles associated with them. In some cases the particle and antiparticle are identical. For example, the antiparticle of the photon is the photon itself, and no new particle is involved. The photon is said to be its own antiparticle. On the other hand, the proton, like the electron, has an antiparticle that differs from it in that it has the opposite charge. (The mass of the antiproton is the same as that of the proton.) The proton's antiparticle is the **antiproton**, symbol $\bar{p}$. These particles were first produced in the laboratory in the mid-1950s in experiments designed expressly for their anticipated discovery. (See •Fig. 3–7.) Protons and antiprotons can annihilate each other and produce photons or pions—see Section 16–4 for more on pions—or other particles. Thus

$$p + \bar{p} \rightarrow a + b + \dots. \tag{16–4}$$

This annihilation process, like that of electrons and positrons, must satisfy the known conservation laws. Protons are much more massive than electrons, so the possible final states consistent with conservation of energy are much greater than in electron–positron annihilation. For example, the initial state has zero net electric charge. As a consequence of charge conservation, the sum of the charges of all of the particles in the final state must also be zero. Or, since both proton and antiproton have spin 1/2, the initial state has an integer value of angular momentum. Since angular momentum is conserved, the number of fermions in the final state of Eq. (16–4) must be even. Thus when one of the particles on the right side of that equation is a fermion, a second spin-1/2 (or 3/2, or 5/2, ...) particle must be produced.

What distinguishes a neutron from an antineutron? Both are electrically neutral, yet neutrons are definitely not their own antiparticles. If they were, they would rapidly annihilate in all nuclei, and the only surviving nuclei would have either no neutrons or one neutron. Following a suggestion of the Hungarian-born Eugene Wigner, who was one of the pioneers in the development of quantum mechanical symmetries, physicists have resolved the question by defining a quantity called the **baryon number**. This quantity is assigned to both neutrons and protons, in analogy to electric charge. The neutron is given a baryon number $B = +1$, so that the antineutron has baryon number $-1$, and similarly for the proton. An antineutron is thus distinguished from a neutron by its baryon number, a quantity conserved in nuclear processes. Note that this does not mean that the neutron by itself is stable; it is not, decaying by beta decay when it is free [see Chapter 15 and Eq. (16–17), presented later], with a lifetime of about 15 minutes. In this decay the final state has net baryon number +1, accounted for by the single proton. Could a free proton decay by some process with the emission of a single neutron? As far as baryon number conservation is concerned, yes; but the masses are such that the energy would not be conserved, so the process is forbidden. When the proton is bound in nuclei, energy can indeed be conserved in the decay, and as we shall see in our discussion of the weak interactions, protons and neutrons do turn into each other, at least temporarily. Similarly, the process represented by Eq. (16–4) has total baryon number $B = 0$ in the initial state, so that must be true for the final state as well. Incidentally, particles with baryon number 1—and we shall see that there are others beyond the proton and neutron—are called **baryons**.

At this point we can see why electrons and protons do not annihilate to radiation, something that would have turned all atoms into neutrons and photons. Electrons have a baryon number of zero, and the proton has $B = 1$. Radiation also has a baryon number of zero, so that the process $e^- + p \rightarrow$ radiation would violate the conservation of baryon number.

### Does the Proton Decay?

Baryon number conservation and electric charge conservation appear to be very similar laws. There is an important difference, however: The quantum theory of electricity and magnetism associates an *exact* conservation law with the $1/r$ form of the coulomb potential, and this is in turn associated with the fact that photons are massless. Measurements such as the $r$-dependence of Jupiter's magnetic field show that if the photon has any mass at all, it is less than $10^{-21}$ of the electron mass! The baryon number, however, is not connected with any such field, so it is perfectly possible to imagine that baryon number conservation is not exact. We do know that the lifetime of the proton, the lightest of the $B = 1$ particles, is very large: Measurements show that it is larger than $10^{25}$ yr and is longer than $10^{32}$ yr for certain decay modes. Because of this fact, it is a very good approximation to treat baryon number as an exactly conserved quantity for most purposes.

The ultimate stability of the proton is nevertheless of some importance, since current explanations of the excess of matter over antimatter in the visible universe require that $B$ not be conserved. These explanations are not in conflict with experiment because they suggest proton lifetimes well in excess of experimental limits. Teams of experimenters have noted that in order to test such ideas by detecting the decay mode that is predicted to occur most frequently, namely

$$p \rightarrow e^+ + (\text{other } B = 0 \text{ "stuff"}), \qquad (16\text{–}5)$$

one needs

1. a lot of protons (see Example 16–2),
2. a way of detecting the positron, and
3. a way of discriminating between the preceding reaction and other reactions that do not involve $\Delta B = 1$, but that still emit an energetic $e^+$. An example of such a reaction would be $\bar{\nu} + p \rightarrow e^+ + n + (B = 0 \text{ "stuff"})$, induced by a very high energy antineutrino associated with cosmic rays. You may not get many of these reactions, but the number of predicted proton decays is excruciatingly small.

---

**Example 16–2**  Suppose there was a proton decay that had a lifetime of $2 \times 10^{30}$ years. How many cubic meters of water would have to be observed if one wanted to have 100 events a year? Ignore neutron decays.

**Solution**  Since one proton decays in $2 \times 10^{30}$ yr, on the average, one needs $2 \times 10^{30}$ protons to get 1 decay/yr and therefore $2 \times 10^{32}$ protons for 100 decays. Now for $H_2O$, there are $6 \times 10^{23}$ molecules in 18 g of water, and since oxygen, with $Z = 8$, contains 8 protons, and H contains 1 proton, there are 10 protons per water molecule. Since the density of water is 1 g/cm$^3$, 18 g of water, or equivalently 18 cm$^3$ of water, contain $6 \times 10^{23} \times 10$ protons. Hence 1 m$^3$ of water contains $6 \times 10^{24} \times (10^6/18) = 0.33 \times 10^{30}$ protons. Thus the number of cubic meters of water is $(2 \times 10^{32})/(0.33 \times 10^{30}) = 600 \text{ m}^3$. This is a volume of $10 \times 10 \times 6$ m of water, much larger than your average swimming pool!

How is the positron in the decay of Eq. (16–5) detected? As long as the "stuff with $B = 0$" in the reaction has a total mass much less than the proton mass—which is what the speculations predicted—the positron will be emitted with an energy equivalent to about half the proton mass, on the order of 450 MeV. The emitted positron will be highly relativistic, and a charged particle such as the positron that moves through material with a speed greater than the speed of light *in the material* gives rise to an analogue of a sonic boom: Radiation is emitted in a sharp cone about the positron's direction. (This is known as Čerenkov radiation). Even though the positron itself has little chance of getting to a detector at the edge of the volume of water, the Čerenkov radiation does arrive at the edge. The tank's surface is covered with photomultipler tubes—devices that are very sensitive to incoming radiation and that can detect the Čerenkov radiation and determine its energy (•Fig. 16–2). Incidentally, as of 1997 experiments of this type showed that if the dominant decay mode of the proton is $p \rightarrow \pi^0 + e^+$, then the lifetime of the proton is greater than $10^{32}$ yr.

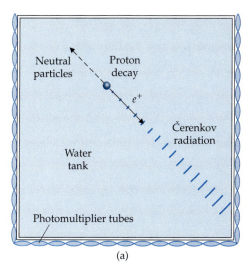

(a)  (b)

• **Figure 16–2** (a) A schematic plan of an experiment to detect possible proton decay into a positron $e^+$ and neutral particles. The $e^+$ is highly relativistic, with a speed greater than that of light *in water*. This produces Čerenkov radiation, which is detected by photomultiplier tubes on the walls of the water tank. (b) Photograph of a section of the wall in the Super-Kamiokande Laboratory, showing the array of photomultiplier tubes. The facility, which was originally built to look for proton decay, is currently being used for the study of neutrinos produced in the cosmos. The technicians in the rubber dinghy cleaning the tubes before the tank was completely filled with water give a sense of the scale of the facility.

# 16–3 Virtual Particles and a Pictorial Representation

In Chapter 7 we introduced the time–energy uncertainty principle $\Delta E \Delta t \cong \hbar$. [See Eq. (7–40).] This relation means that within a time interval $\Delta t$ we cannot specify the energy to a greater accuracy than $\Delta E \cong \hbar/\Delta t$. For microscopic times the uncertainty in energy can be large, and it can accommodate the "energy-violating" creation of particles for short times. Consider, for example, the appearance of a positron–electron pair and its disappearance again after a time interval $\Delta t$. The relation now reads

$$\Delta t\,(2m_e c^2) \cong \hbar; \qquad (16\text{–}6)$$

that is, such a pair can appear only for the brief time interval given by this equation. Any particle or pair of particles that appear and disappear within time intervals constrained by the uncertainty principle are called **virtual particles** or **virtual states**. Since there is no corresponding uncertainty for charge or angular momentum, these quantities maintain their values all along.

Suppose now that a virtual pair appears in the vicinity of an electron in the hydrogen atom. During the appropriate time interval $\Delta t$, the pair can interact with the atom: The $e^-$ in the virtual pair is repelled by the electron, and the $e^+$ is attracted to it. Since the creation of virtual pairs can take place repeatedly (with each occurrence limited to a time interval $\Delta t$), these pairs act to provide some screening of the electron charge and thus reduce the magnitude of the binding energy of the electron by a tiny amount. In spite of the size of the energy change (it is actually $1.8 \times 10^{-8}$ eV), it has been measured. This means that we can no longer talk about an atom with a single electron: A more accurate description of a wave function has to include terms that involve the atom + one pair, atom + two pairs, and so on. By the same token, virtual photons can exist, and these too will participate in a precise wave function. As the tiny number just quoted indicates, ordinary atomic physics hardly needs these corrections, even for fairly precise numerical calculations, but they are nevertheless present.

### Feynman Diagrams

The American physicist Richard Feynman (•Fig. 16–3) was one of many people who studied the interactions between electrons and positrons and the electromagnetic field. He devised some simple rules for making calculations that led to a nice pictorial representation of particles and virtual particles. We draw a box. We may think of time as increasing as we go from the bottom edge to the top edge of the box. Any particle in the initial state enters the box at its bottom edge or, if the particle is present in the final state, leaves at the top edge; for example, in •Fig. 16–4a an electron is drawn. A positron (or any antiparticle) is also drawn in the same way, except that there is a convention that the arrow points downwards[1] (•Fig. 16–4b). A photon is drawn as in •Fig. 16–4c. A photon that is emitted by an electron and that appears in the final state starts on the electron line—we refer the place where this happens as a *vertex*—and then goes on to the top of the frame (•Fig. 16–4d); an initial-state photon absorbed by an electron comes from the bottom edge of the frame and ends somewhere on the electron line (•Fig. 16–4e). The process in which a photon converts to an electron–positron pair is described by •Fig. 16–4f, and the inverse process, in which an electron–positron pair annihilate to produce a photon, is described by •Fig. 16–4g.

Because of energy and momentum conservation a free electron cannot absorb or emit a real photon. (See Problem 20 at the end of the chapter.) Thus in all of these diagrams at least one particle must be virtual and attached to something else. Equivalently, virtual particles cannot be in initial or final states: They can last only a finite time and thus appear only in the middle of the box.

• **Figure 16–3** Richard P. Feynman was one of the leading theoretical physicists of the second half of the 20th century. His range of interests was wide, but he is best known for his reformulation of quantum mechanics and for his work on quantum field theory. He is shown here in discussion with Paul Dirac at a gravitation conference held in Warsaw in 1963.

---

[1] It is this convention that has led to the colorful description of a positron as "an electron traveling backward in time."

(a) Electron     (b) Positron     (c) Photon     (d) Real photon emission     (e) Real photon absorption     (f) Real pair creation     (g) Real pair annihilation

• **Figure 16–4** Feynman graphs involving electrons, positrons, and photons. Time flows upwards in these graphs. (a) The up arrow indicates that this particle is an electron. (b) The down arrow indicates that this particle is a positron, the antiparticle of the electron. (c) A photon. In the next four diagrams, one of the particles does not leave the "box" since, in three-body reactions involving electrons, positrons, and photons, energy and momentum conservation make it impossible for all particles to be real. (d) Vertex in which an electron emits a photon. (e) Vertex in which an electron absorbs a real photon. (f) Vertex indicating the conversion of a photon into an electron–positron pair. (g) The annihilation of a real electron and real positron into a virtual photon.

An electron that emits a virtual photon and then reabsorbs it is represented by •Fig. 16–5a. A photon creating a virtual pair for an instant, with the pair annihilating to give back the photon, is shown in •Fig. 16–5b. These diagrams, which we call **Feynman diagrams**, are of some visual help. Keep in mind that the real aim of Feynman's method is to facilitate actual numerical calculation.

The consideration of electrons and photons leads one to an important extension of the notion of virtual particles. Because photons are associated with electric and magnetic fields, we may turn this around and ask about the meaning of "virtual" electric and magnetic fields. It so happens that one can translate the coulomb force between two charged particles into a picture of an exchange of a virtual photon between the two particles, as shown in •Fig. 16–5c.

Actually, the exchange of a virtual photon and the calculation of the process using Feynman's rules resolve some features that are very difficult to deal with in any other way. Because no signal can propagate with a speed greater than light, *retardation effects* come into play. To illustrate this you can think about the repulsive potential electron 1 senses at time $t$, due to electron 2. This potential cannot be of the simple form $1/|\vec{r}_1 - \vec{r}_2|$. Instead, it must be of that form, *but with the proviso* that the position of electron 2 is to be evaluated at an earlier time, namely $\{t - |\vec{r}_1 - \vec{r}_2|/c\}$. In this way the arrival of a "signal" from electron 2 will arrive at the location of electron 1 at the time $t$. In Feynman's method *the exchange of a virtual photon automatically takes into account all of these retardation effects.*

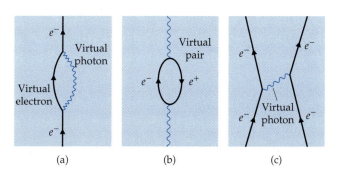

(a)              (b)              (c)

• **Figure 16–5** Feynman diagrams for processes in which virtual particles appear. (a) An electron emits and reabsorbs a virtual photon. (b) A photon creates a virtual electron–positron pair, which subsequently annihilates into a photon with the energy and momentum of the original photon. (c) Two electrons interact by means of the exchange of a virtual photon.

## Quantum Electrodynamics

In the late 1940s, a concerted and spectacularly successful attack on the fully relativistic theory of electrons interacting with photons, including all the effects of antiparticles, virtual pairs, and virtual photons—a theory known as **quantum electrodynamics**—was carried out by the American physicists J. Schwinger and R. P. Feynman, the Japanese physicist S. Tomonaga, and the British physicist F. J. Dyson. The theory involves a small parameter, the fine-structure constant $\alpha = e^2/(4\pi\varepsilon_0\hbar c) \cong 1/137$ first introduced in Chapter 5, and physical effects can be formulated as a series in this small quantity. Such a series allows for some very precise calculations.

Take the magnetic moment of the electron. The simplest theory of the magnetic moment, Dirac's union of quantum theory and relativity, predicts that the electron has a g-factor[2] of 2. The possibility of virtual pairs and of virtual photons modifies the electron's properties. If an electron encounters a magnetic field, its magnetic moment will appear different from the one predicted by Dirac's theory. The reason is that the virtual particles enter into the calculation. For example, when an electron emits and reabsorbs a virtual photon, it must share its initial momentum with the virtual particle and hence must recoil during this intermediate step. A moving charge is acted upon by a force $e(\vec{\mathbf{v}} \times \vec{\mathbf{B}})$ due to the magnetic field in addition to the force it undergoes due to the intrinsic magnetic dipole moment associated with the spin, and this changes the effective magnetic moment (•Fig. 16–6a). In addition, the magnetic field itself can create a virtual pair, which annihilates and produces a virtual photon or even three (•Fig. 16–6b, c), all of which act to alter the magnetic dipole moment.

The calculations carried out with quantum electrodynamics provide the most accurate predictions ever made. For example, the predicted value of the electron g-factor $g_e$ is not 2, as Dirac's theory would indicate, but is corrected by about 1 part in 1,000, according to the formula

$$\frac{g_e - 2}{2} = (1{,}159{,}652{,}201.4 \pm 27.1) \times 10^{-12}.$$

(The reason for the presence of an error in a theoretical calculation such as the one that produced this number is the uncertainty in some of the input constants—in particular the fine-structure constant.) The experimental value is

$$\frac{g_e - 2}{2} = (1{,}159{,}652{,}188.4 \pm 4.3) \times 10^{-12}.$$

Comparable accuracy would be achieved by measuring the distance between New York and Los Angeles to within the thickness of a hair! In all of science, this remarkable agreement between experiment and theory is matched only in some other calculations of quantum electrodynamics.

• **Figure 16–6** Feynman diagrams showing how the presence of virtual particles changes the interaction of an electron with an external magnetic field $\vec{\mathbf{B}}$, indicated by the thick wavy line. (a) The external magnetic field interacts with a virtual electron during the time that a virtual photon is present. (b) The external magnetic field creates a virtual pair that annihilates to a virtual photon which is then absorbed by the electron. (c) The external magnetic field creates a virtual pair that annihilates, creating three virtual photons, which are then absorbed by the electron.

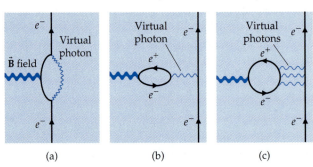

[2] The g-factor, or gyromagnetic ratio, is defined in Eq. (9–52).

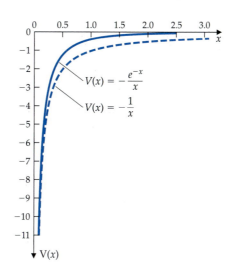

• **Figure 16–7**   Hideki Yukawa was the first Japanese physicist to win the Nobel prize.

• **Figure 16–8**   Comparison of the shapes of the Yukawa potential $e^{-x}/x$ and the coulomb potential $1/x$.

## 16–4  The Yukawa Hypothesis and Pions

In 1934, shortly after the discovery of the neutron, the Japanese physicist Hideki Yukawa (•Fig. 16–7) had the inspired idea that the nuclear forces could be described by an analogue of the electromagnetic field. This field would have quanta that are analogous to photons, and the nuclear force, with its retardation effects, could be generated by the exchange of the quanta of the field. He found that if the quantum had a mass $\mu$, then the proper analogue of the coulomb potential $e^2/(4\pi\varepsilon_0 r)$ would take the form (•Fig. 16–8)

$$V(r) = -g^2 \frac{\exp(-\mu c r/\hbar)}{r}, \qquad (16\text{–}7)$$

where $g$ is some coupling strength analogous to the electric charge $e$. This result reduces to the coulomb form when $\mu = 0$. The **range** $r_0$ of the potential is defined to be the distance where the exponential factor falls to $1/e$; that is,

$$r_0 = \frac{\hbar}{\mu c}. \qquad (16\text{–}8)$$

We can understand the association of a mass $\mu$ with the range if we use the energy–time uncertainty relation. A virtual quantum requires the "borrowing" of an energy of at least $\mu c^2$. This can be done for a time $\Delta t \cong \hbar/\mu c^2$ during which the virtual quantum can propagate for a distance of, at most, $c\Delta t = \hbar/\mu c$. That distance matches precisely $r_0$. If $\mu = 0$, the range is infinite.

The potential in Eq. (16–7) is known as the **Yukawa potential**. Yukawa predicted the associated quanta, subsequently named pi-mesons, or **pions** (symbolized by $\pi$). He estimated their mass by its association with an experimental value for the range that came from the known properties of nuclear forces (see Chapter 15), namely

$$m_\pi = \frac{\hbar}{r_0 c} = \frac{1.05 \times 10^{-34}\,\text{J}\cdot\text{s}}{(1.2 \times 10^{-15}\,\text{m})(3.0 \times 10^8\,\text{m/s})} = 0.3 \times 10^{-27}\,\text{kg} \cong 320 m_e.$$

In energy units, $m_\pi c^2 = 163$ MeV. Yukawa did not specify the spin of the meson, but the particle had to be a boson, since it was emitted by a nucleon, and a nucleon remains in the final state (•Fig. 16–9). Nucleons have spin 1/2, so the meson must have integer spin. Analogy with photons suggested that it would have spin 1, whereas simplicity suggested spin 0.

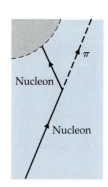

• **Figure 16–9**   Feynman diagram containing a vertex in which a nucleon emits a pion and becomes a virtual nucleon.

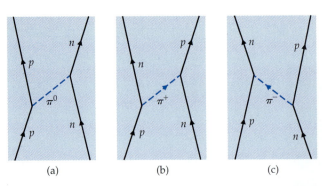

• **Figure 16–10** Feynman diagrams showing how the exchange of $\pi^+$, $\pi^-$, and $\pi^0$ give rise to proton–nucleon scattering. (a) In the exchange of a $\pi^0$ the nucleons do not change charge. In (b) and (c) there is an exchange of charge among the nucleons.

Any quantum that can be exchanged could also be expected to be produced as a physical particle in a reaction with sufficient energy. This condition set the stage for experimental efforts to prove the existence of pions by their direct production. When the pion was discovered in the late 1940s, experiments showed that the spin was 0. The mass turned out to be in the vicinity of 140 MeV/$c^2$, remarkably close to the predicted value. Pions come in three charge forms, $\pi^+$, $\pi^0$, and $\pi^-$, which allows them to mediate the forces between nucleons and antinucleons (•Fig. 16–10).

The Yukawa hypothesis has an importance that transcends its direct application to nuclear forces. Calculations in relativistic quantum mechanics—the calculations that use Feynman diagrams—show that

**(i)** any particle that is exchanged by two particles (•Fig. 16–10) gives rise to a force between the two particles and

**(ii)** the only consistent way to construct interactions between sets of particles is by postulating the exchange of *some* particle between them.

It is understood that the emission and absorption of the particle being exchanged must not violate any conservation laws. For example, $n \rightarrow p + \pi^-$ and $\pi^- + p \rightarrow n$, which occur in the interaction leading to the reaction $n + p \rightarrow p + n$ (•Fig. 16–10) are permissible, whereas $n \rightarrow p + \pi^+$ and $\pi^+ + p \rightarrow n$, each of which violates charge conservation, are not. If the latter reaction were allowed it would in turn allow the (forbidden) process $n + n \rightarrow p + p$.

There was a brief and interesting detour along the way to the discovery of pions: Particles for which $mc^2 \cong 110$ MeV were discovered in cosmic rays in 1936—after Yukawa's suggestion. Although their mass was about right for them to be pions, measurements of their absorption in matter showed that they did not interact strongly enough with nuclei to give rise to the nuclear force. They were a new and unexpected particle dubbed the **muon**, symbol $\mu$. When the physicist I. I. Rabi first heard of them he asked, "Who ordered that?" It turned out that muons were just like electrons, only heavier, and that they had very little to do with nuclear forces. We shall see later that they nevertheless play an important role in elementary particle physics.

**Example 16–3**    One possible particle that can be exchanged between two protons is the $\rho^0$-meson, whose mass is 770 MeV/$c^2$. The force due to this exchange is repulsive. What is your prediction of the range of the repulsive potential?

**Solution**    The virtual particle can exist for a time $\Delta t \cong \hbar/m_\rho c^2$, and during that time it can travel at most a distance $c\Delta t$. So that is the range of the repulsive force, namely

$$r_{rep} = \frac{\hbar c}{m_\rho c^2} = \frac{(1.06 \times 10^{-34} \text{ J·s})(3 \times 10^8 \text{ m/s})}{(770 \text{ MeV})(1.6 \times 10^{-13} \text{ J/MeV})} = 2.6 \times 10^{-16} \text{ m}.$$

This is about one-fifth of the range due to the pion exchange, as could, in fact, have been obtained by noting that $m_\rho \cong 5m_\pi$.

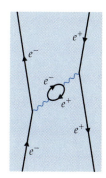

**Example 16–4**   In addition to the exchange of a photon between an electron and a positron (•Fig. 16–5c), there is a process in which the exchanged photon makes a virtual electron–positron pair that annihilates to a virtual photon, which is in turn absorbed by the positron (•Fig. 16–6). Although the probability of this happening is small, it does change the coulomb potential. Estimate how close the electron has to be to the positron in order to experience this change.

**Solution**   The virtual pairs can exist only for a time $\Delta t = \hbar/(2m_e c^2)$. In that time, the pairs can travel a distance of no more than $c\Delta t = \hbar/(2m_e c)$. This is the distance over which they have an effect. The Bohr radius is $a_0 = \hbar/(m_e c\alpha) \cong 137\,\hbar/(m_e c)$. Thus the distance over which the coulomb potential is changed extends to $1/274$ of the Bohr radius (•Fig. 16–11).

• **Figure 16–11**   Modification of the coulomb potential due to the creation of a virtual pair.

**Example 16–5**   If the photon had a tiny mass $\mu$, then instead of behaving as inverse powers of $r$, static electromagnetic fields would behave at large distances as inverse powers of $r$ times an exponential factor of the form $\exp(-\mu c r/\hbar)$. Suppose you are given the information that the result of a measurement of Earth's magnetic dipole field by space probes at a distance of 10,000 km agrees within 5% with what is expected from Maxwell's equations. What upper limit does this measurement set on the photon mass?

**Solution**   The presence of a photon mass means that the expected result is multiplied by a factor $\exp(-\mu c r/\hbar)$, and this factor is measured to be within 0.05 of 1 for $r = 10,000$ km. When an exponential factor is near unity, the argument must be small, so we can use the small-argument approximation $\exp(x) \approx 1 + x$. In this case, $|x| < 0.05$ and $x = -\mu c r/\hbar$. Including the absolute value—we know only that $x$ lies between $-0.05$ and $+0.05$—we have $\mu c r/\hbar < 0.05$, or

$$\mu < 0.05\,\frac{\hbar}{cr} = \frac{(0.05)(1.05 \times 10^{-34}\,\text{J}\cdot\text{s})}{(3.00 \times 10^8\,\text{m/s})(10^7\,\text{m})} = 1.75 \times 10^{-41}\,\text{kg}$$

$$= 1.75 \times 10^{-41}\,\text{kg}\,\frac{m_e}{0.91 \times 10^{-30}\,\text{kg}}$$

$$= 1.9 \times 10^{-11}\,m_e.$$

This mass is very small indeed.

# 16–5  The Particle"Zoo" and the Discovery of Quarks

The discovery of pions in the late 1940s coincided with a return of many physicists from defense-related work in World War II, and their successes brought with them a national commitment to the growth of science, including such esoteric fields as elementary particle physics. The construction of accelerators of ever-increasing energy allowed an exploration of the interactions of pions with nucleons. Already in 1952 the elastic scattering of $\pi^+$ by protons (an analogue of Compton scattering of light by electrons) yielded an interesting phenomenon: The incoming pion and the proton formed a new "particle" of mass 1,232 MeV/$c^2$, which then decayed back into a proton (or neutron) and a single pion (•Fig. 16–12). The "particle" was not stable; indeed, its lifetime was exceedingly short. One can think about such particles by analogy to spectral lines in atoms. For example a photon impinging on one particle (the ground state of hydrogen) "produces" an excited state of hydrogen, which then decays back to the ground state. If $hf$ is the excitation energy, the light that is emitted has a frequency $f \pm \Delta f$ and an energy spread $2h\Delta f$ (the width of the spectral line). A

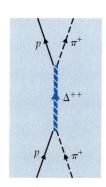

• **Figure 16–12**   Feynman diagrams for the formation of a $\Delta^{++}$ particle in the collision of a $\pi^+$ and a proton. The $\Delta^{++}$ bears some resemblance to the "compound nucleus" described in Chapter 15.

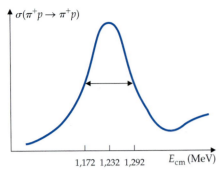

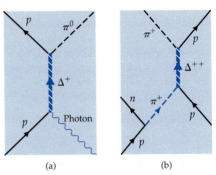

(a)          (b)

• **Figure 16–13**   The "width" of the $\Delta^{++}$ appears in the shape of the cross section for the reaction $\pi^+ + p \rightarrow \pi^+ + p$. The initial collision produces the $\Delta^{++}$ particle, which exists for a time $\Delta t$ determined by the energy uncertainty, which we call the "width" (here about 120 MeV) in this context.

• **Figure 16–14**   The existence of the $\Delta$ particle is independent of how it is produced, as shown by the fact that it can be produced (a) by photons impinging on a proton and (b) in nucleon–nucleon collisions.

measurement of the energy spread can be used, via the relation $\Delta t = \hbar/\Delta E$, to determine the lifetime of the excited state. For the 1,232-MeV/$c^2$ particle, subsequently named the $\Delta$ particle, a measurement of the energy of the outgoing pion showed that the width of the line was in the vicinity of 120 MeV/$c^2$ (•Fig. 16–13), so that the lifetime is

$$\tau_\Delta = \frac{\hbar}{\Delta E} = \frac{1.05 \times 10^{-34}\,\text{J·s}}{(120\,\text{MeV})(1.6 \times 10^{-13}\,\text{J/MeV})} = 5 \times 10^{-24}\,\text{s}.$$

This lifetime is so short—smaller than the time it takes a photon to traverse the diameter of a proton—that one is reluctant to use the term "particle." In fact, the object is more commonly called a **resonance**. Nevertheless, the description of the $\Delta^{++}$ and its nearly-equal-mass partners $\Delta^+$, $\Delta^0$, and $\Delta^-$ as particles has much justification. For example, the $\Delta$'s have a definite spin $S = 3/2$. Moreover, they can be produced in other ways—for instance, in photoproduction and in production by virtual pions (•Fig. 16–14). The characteristics of the $\Delta$ do not depend on how it is produced.

More experiments uncovered many other unstable particles. An example is the $\rho$-meson (see Example 16–3), which comes in three charge states ($\rho^+$, $\rho^0$, and $\rho^-$) and has spin 1 and mass $m_\rho c^2 \cong 770$ MeV. This particle decays primarily into a pair of pions with $\Delta E \cong 150$ MeV (•Fig. 16–15).

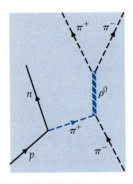

• **Figure 16–15**   The $\rho$ meson is produced in the reaction $\pi + \text{nucleon} \rightarrow \rho + \text{nucleon}$. The $\rho$ decays quickly to a pair of pions. The energy distribution of the two pions in the final state shows clearly that they are the decay products of a single particle.

**Example 16–6**   A $\rho^0$-meson decays into a $\pi^+$ and a $\pi^-$. **(a)** If the $\rho^0$-meson is at rest, what are the momenta of the two pions? Express your answer in the form $p/m_\pi c$, where $p$ is the magnitude of the desired momentum and $m_\pi$ is the pion mass. **(b)** If the pion momentum has a range $p \pm \Delta p$, with $\Delta p/p = 0.07$, use the result of part (a) to find the width of the $\rho$-meson resonance; that is, if the $\rho$-meson has rest energy $Mc^2 \pm \Delta Mc^2$, what is $\Delta Mc^2$?

**Solution**   **(a)** Since the $\rho^0$-meson is at rest, the two pions must come off back to back. Thus their momenta are $\vec{p}$ and $-\vec{p}$. The energy of each pion is therefore $E = \sqrt{p^2c^2 + m_\pi^2 c^4}$. By conservation of energy, the sum of the energies, $2E$, must equal the initial energy $Mc^2$, where $M$ is the $\rho^0$ mass. Thus

$$2\sqrt{p^2c^2 + m_\pi^2 c^4} = Mc^2,$$

or $p^2c^2 = (M^2c^4)/4 - m_\pi^2 c^4$. Solving for $p$ we have

$$p = \sqrt{\frac{M^2c^2}{4} - m_\pi^2 c^2}.$$

Hence

$$\frac{p}{m_\pi c} = \sqrt{\frac{M^2}{(2m_\pi)^2} - 1} = \sqrt{\left(\frac{770}{280}\right)^2 - 1} = 2.56.$$

**(b)** The uncertainty in the energy of the pion is

$$E \pm \Delta E = \sqrt{(p \pm \Delta p)^2 c^2 + m_\pi^2 c^4} = \sqrt{p^2 c^2 \left[\frac{(1 \pm \Delta p)}{p}\right]^2 + m_\pi^2 c^4}$$

$$= m_\pi c^2 \sqrt{\left(\frac{p}{m_\pi c}\right)^2 \left[\frac{(1 \pm \Delta p)}{p}\right]^2 + 1}$$

$$= (140\,\text{MeV}) \sqrt{(2.56)^2 (1 \pm 0.07)^2 + 1}.$$

This value ranges between 408 MeV (the plus sign) and 362 MeV (the minus sign). The value of $Mc^2$ is twice the energy, so we can think of the $\rho$-mass $M$ as having a width; that is, $Mc^2 \pm \Delta Mc^2$ runs from 816 MeV to 724 MeV. Therefore the spread, or width, is 92 MeV.

As if all this were not enough, a new class of particles, dubbed the **strange particles**, was discovered.[3] These particles were counterparts of the pions, the nucleons, and the other resonances involving pions and nucleons, but they differed in that their production patterns required the existence of a new label, or quantum number, something suggested by the American physicist M. Gell-Mann (•Fig. 16–16) and the Japanese physicist K. Nishijima. This new quantum label, the **strangeness**, was assigned so that its conservation would make the production and decay pattern consistent—in much the same way that baryon number is assigned. Particles with nonzero strangeness are termed **strange particles**. For example, a strange particle called the $\Lambda^0$ ($\Lambda$ is a capital Greek *lambda*), of mass $m_\Lambda = 1{,}115\,\text{MeV}/c^2$, decays in a pattern similar to that of the $\Delta$; that is,

$$\Lambda^0 \rightarrow p + \pi^-. \tag{16–9}$$

However, this decay is $10^{14}$ times slower than $\Delta$ decay. One assigns a nonzero strangeness to the $\Lambda^0$ and a zero strangeness to the pion and the proton and postulates that strangeness is conserved in the production reactions and *violated* in the decay reactions. This postulate allows for the systematic explanation of all the production and decay rates of strange particles.

As the number of particles, strange and nonstrange, grew, people began to look for ways of explaining their proliferation.

• **Figure 16–16** Murray Gell-Mann made enormous contributions to the development of elementary particle physics. For more than two decades he blazed trails in almost every area of that field.

### The Quark Model

What to do? One is confident in calling certain particles elementary only as long as there are not too many of them. The very existence of 100 chemical elements suggests that atoms have an underlying structure. The search for a pattern in the particles we have described was spearheaded by Gell-Mann. The end result was a picture in which all the hadrons—the strongly interacting states, all the states we have described thus far except the photon, the electron, and the close partners of the electron—are constructed from a set of basic building blocks: the **quarks** and their antiparticles.[4]

---

[3] A term such as this responds in a whimsical fashion to a necessity. Later in the book we'll meet others: color, flavor, charm, and so forth. Such terms refer, for the most part, to something that is either exactly or approximately conserved. You might say that we need a word to express a conservation law for something we had not earlier seen to be conserved; any word for the conserved quantity will serve the purpose.

[4] The word "quark" was taken by Gell-Mann from a phrase in James Joyce's novel *Finnegans Wake*.

**Table 16–1    The Light Quarks and Some of Their Properties**

| Name | Charge, in Units of $e$ | Baryon Number | Strangeness |
|---|---|---|---|
| $u$ (up quark) | +2/3 | +1/3 | 0 |
| $d$ (down quark) | −1/3 | +1/3 | 0 |
| $s$ (strange quark) | −1/3 | +1/3 | −1 |
| $\bar{u}$ | −2/3 | −1/3 | 0 |
| $\bar{d}$ | +1/3 | −1/3 | 0 |
| $\bar{s}$ | +1/3 | −1/3 | +1 |

The proposal that all the hadrons are composed of quarks and antiquarks, made independently by Gell-Mann and George Zweig in 1964, started out as a simple classification scheme—something like a periodic table for elementary particles. The quarks are spin-1/2 particles. Initially it was assumed that there were only three of them, along with their antiparticles. Because much later three more quarks were discovered with much more mass than the original three quarks, we refer to the original set as the *light quarks*. (See Table 16–1.)

As the table shows, the scheme works only if these particles are assigned fractional electric charges such as $\frac{1}{3}e$ and fractional baryon number. Prior to this suggestion, no one had ever proposed fractional electric charges for elementary particles, and the proposition was a startling one. The antiquarks, symbolized with a bar over a letter, have electric charge, baryon number, and strangeness opposite to those of the corresponding quarks.

The mesons, particles with integer spin, are combinations of a quark and an antiquark—for example,

$$\pi^+ = u\bar{d}; \pi^- = \bar{u}d; \quad \text{and} \quad \pi^0 = (u\bar{u} - d\bar{d}). \quad (16\text{–}10)$$

In the $\pi^0$ state, the quark–antiquark pairs are arranged with a total angular momentum of zero (a singlet spin wave function with zero orbital angular momentum). The (spin-1) $\rho$-mesons have a similar quark structure, but with the quark and antiquark in a triplet spin state with zero orbital angular momentum. The strange particles involve the $s$-quark.

**Example 16–7**    Pions are made of $u$ and $d$ quarks and their antiquarks. The quark model can also account for mesons with strangeness—the $K$-mesons, which are made up of a quark–antiquark pair in which one of the pair is an $s$ or $\bar{s}$ quark. List all of the possible combinations, give their charge, strangeness, baryon number, and lowest angular-momentum state.

**Solution**    If one of the quarks is an $s$-quark, then its partners must be antiquarks. We thus have the following two possibilities:

$s\bar{u}$: $Q = -\dfrac{1}{3} + \left(-\dfrac{2}{3}\right) = -1$; $S = -1$; $B = \dfrac{1}{3} - \dfrac{1}{3} = 0$. This is the $K^-$ meson.

$s\bar{d}$: $Q = -\dfrac{1}{3} + \left(\dfrac{1}{3}\right) = 0$; $S = -1$; $B = 0$. This is the $\bar{K}^0$ meson.

If the strange antiquark is involved, its partners must be quarks, so that we have the following possibilities:

$\bar{s}u$: $Q = \dfrac{1}{3} + \left(\dfrac{2}{3}\right) = 1$; $S = +1$; $B = -\dfrac{1}{3} + \dfrac{1}{3} = 0$. This is the $K^+$ meson.

$\bar{s}d$: $Q = \dfrac{1}{3} + \left(-\dfrac{1}{3}\right) = 0$; $S = +1$; $B = 0$. This meson differs from the $\bar{K}^0$ by the sign

of the strangeness, and it may be viewed as the antiparticle of the $\bar{K}^0$. We refer to it as the $K^0$ meson. All of these particles are bosons, since they are composites of two fermions.

The lowest total angular momentum is $J = 0$, for which $\ell = 0$ and the spins are in a spin-zero singlet state.

Baryons are constructed of three quarks, because that is the only way to make a state with $B = 1$. A look at the way three quarks can be combined to make the proper electric charges would suggest that

$$p = uud \quad \text{and} \quad n = udd. \tag{16–11}$$

As far as angular momentum is concerned, the three quarks are arranged in combinations with no orbital angular momentum and with one quark's spin aligned in the direction opposite to the spins of the two others. In this way the nucleons would have spin 1/2. From that point of view, the $\Delta$-states are easy to understand: They are the same sort of combination of up and down quarks as the nucleons, but with the spins aligned, so that the $\Delta$ is a spin-3/2 particle. For example,

$$\Delta^{++} = uuu. \tag{16–12}$$

•Figure 16–17 shows how production of the $\Delta^{++}$ state appears in the quark model. Strange baryons such as the $\Lambda^0$ would involve a single strange quark:

$$\Lambda^0 = usd. \tag{16–13}$$

The $\Lambda^0$ is also a spin-1/2 particle, so the same remarks about angular momentum that we made for the nucleons apply to it.

The quark model of the observed particles has turned out to be very useful both for classification and for making predictions. Entire series of mesons and baryons are explained by including orbital angular momentum in the wave functions. Or we can predict that any state with $B = 0$ can be made only with equal numbers of quarks and antiquarks, so that such states must have integer spin. Or we can understand the differences in mass among various particles in terms of the differences in mass among the quarks that constitute them.

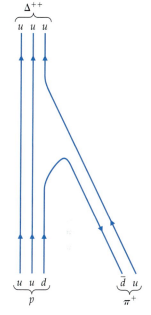

• **Figure 16–17** Quark model depiction of the production of $\Delta^{++}$ in a $\pi^+$–$p$ collision. The annihilation of the $d$ and the $\bar{d}$ is accompanied by the production of virtual gluons, not shown here.

---

**Example 16–8**   Consider the (strangeness-conserving) production reactions **(a)** $\pi^- + n \to \Lambda^0 + X$; **(b)** $p + \bar{p} \to K^0 + X$; and **(c)** $\pi^+ + \pi^0 \to \bar{K}^0 + X$. Assuming that the $X$ particles are single $K$-mesons, which reactions are possible, and if so, which $K$'s are the $X$'s?

**Solution**   **(a)** In the initial state we have $Q = -1$, $B = 1$, and $S = 0$. The quantum numbers are conserved. Charge conservation demands that $X$ have $Q = -1$, and since the $\Lambda^0$ has $S = -1$, the $X$ must have $S = +1$. None of the $K$'s satisfy this condition ($K^-$ has $S = -1$), so reaction (a) cannot happen as described. **(b)** In the initial state $Q = 0$, $B = 0$, and $S = 0$. The charge on $X$ must be zero, $B = 0$, and the strangeness is $S = -1$. The $\bar{K}^0$ particle fits the bill. **(c)** The initial state has $Q = +1$, $B = 0$, and $S = 0$. Since the $\bar{K}^0$ has strangeness $-1$, the particle must have $S = +1$ and $Q = +1$. $K^+$ is then a possible state.

---

## Quark Confinement and the Experimental Discovery of Quarks

As mentioned, quarks must have charges that are fractions of the electron charge. This fact sent the experimental community on a search. Fractional charge had never been observed, and one explanation was that quarks and antiquarks were extremely massive and bound together by such strong potentials that their bound states were *relatively* light. If that were the explanation, then quarks would be produced only in very high energy collisions, but then, in the absence of antiquarks, they would be around for a long time. This bit of reasoning suggested looking for quarks in cosmic rays, something that was done with no success. Unsuccessful searches were also made in the debris of the collisions produced by the highest energy accelerators. One experiment looked for quarks in oysters since, it was reasoned, most of the Earth's surface is water and oysters filter water and might thereby concentrate quarks! None were found.

The absence of any evidence for free fractional charge has led us to the conviction that the physics of hadrons as combinations of quarks bears little resemblance to that of atoms as combinations of electrons and nuclei. Atoms can readily be broken up into their constituent parts. In contrast to the interaction between electrons and nuclei, whose attraction becomes weaker as their separation increases, the attraction between quarks must *increase* as their separation increases. In this way it would take infinite energy to separate a quark from a bound state with other quarks. One says that quarks are **confined** in their bound states. Quark confinement is an entirely new feature of the physical world.

How, then, can we speak of the experimental discovery of quarks? The answer lies in the probe of the *internal* properties of nucleons. To take an example, even if it has no charge, the neutron does have some electromagnetic properties. It has a magnetic moment, something that a purely neutral system would not possess. The internal structure of the neutron is in principle explained by the quark model. The magnetic moment, for example, can be explained by the fact that both the up and the down quarks have magnetic moments determined by their charges and masses (taken to be $m_u \cong m_d \cong m_p/3$), together with the specific form of the wave function of the three-quark system.

The most convincing experiment to detect quarks used the technique pioneered by Rutherford to probe the atom: *scattering* from nucleons. The nucleon has a radius of about $10^{-15}$ m. To probe the interior of something so small by scattering, one needs high energies. The uncertainty principle suggests that if we want to use a scattering experiment to measure distances with a certain precision $\Delta x$, we need momentum transfers in the scattering on the order of $\Delta p = \hbar/\Delta x$. In turn, the momentum of the probing projectile must be of about this size. In the case we are examining, we might be interested in a precision on the order of $10^{-16}$ m. Thus we estimate that the energy of our projectile, which must be ultrarelativistic, is at a minimum

$$E = pc \cong \frac{\hbar c}{\Delta x} = \frac{(1.05 \times 10^{-34}\,\text{J·s})(3 \times 10^8\,\text{m/s})}{10^{-16}\,\text{m}}$$

$$= 3 \times 10^{-10}\,\text{J} = 2 \times 10^9\,\text{MeV} = 2\,\text{GeV}.$$

Accelerators that could provide projectiles of such energies were not available before the 1960s. In 1968, a group working at the Stanford Linear Accelerator Center (SLAC) scattered electrons from nucleons and discovered that an unexpectedly large number were scattered at large angles. Just as the unexpectedly large number of alpha particles scattering at large angles from atoms suggested to Rutherford that there was a dense center—the nucleus—so did the electron scattering results suggest to the SLAC experimenters that there were point objects within the nucleon. These points were the quarks, whose properties—including their fractional charges!—were verified by these and further experiments of this sort.

# 16–6  Interactions among the Quarks: Quantum Chromodynamics

Study of the wave function of the $\Delta$, and indeed of many of the quark-model wave functions of hadrons that we have discussed, reveals a problem. The $\Delta^{++}$ is composed of three $u$-quarks with total spin 3/2. If we pick the state with spin projection 3/2, then the spins are aligned. This means that the spin part of the wave function is symmetric under the interchange of any two quarks—having all the spins aligned means that all three spins are spin up or all three spins are spin down. Moreover, the $\Delta^{++}$ has a spatial wave function that is symmetric under the interchange of quarks. Therefore the total wave function, spatial and spin, is symmetric. Yet quarks are fermions. *This violates the exclusion principle.*

The correct solution to this dilemma was the introduction of a new quantum number that was dubbed the **color**. Each quark was to come in *three* versions. In terms of the color nomenclature there were for example three $u$ quarks, identical except for the color label—$u_R$, $u_B$, and $u_G$, where $R$ might stand for red, $B$ for blue, and $G$ for green. The three-quark state $uuu$ with all spins aligned can now be made totally antisymmetric by making it antisymmetric in the color labels. The color is a kind of charge, which in turn suggests the introduction of an analogue of the electromagnetic field. This field had to allow a reaction in which quarks change color—for example, a $u_R$ could turn into a $u_B$—so that the analogues of photons could not be color neutral in the sense that ordinary photons are electrically neutral. The new field was to be transmitted by colored equivalents of photons. These quanta became known as **gluons**, since their interactions "glued" together the quarks that were constituents of particles such as protons and neutrons.

Now consider a transition of the form

$$u_R \rightarrow u_G + g, \tag{16–14}$$

where the symbol $g$ stands for the gluon, the analogue of the photon $\gamma$ in the reaction $e^- \rightarrow e^- + \gamma$. Now if color, like electric charge, is exactly conserved, then *the gluon needs color indices*. Thus a more appropriate labeling for the gluon in Eq. (16–14) would be $g_{R\bar{G}}$, indicating that the gluon carries off color charges $R$ and $\bar{G}$. A labeling of this type distinguishes gluons from photons in a fundamental way. Photons themselves carry no electric charge, so that photons can scatter from photons only because each is coupled to electrons and positrons (•Fig. 16–18a). But gluons carry the color charge, so that they can scatter *directly* from other gluons (•Fig. 16–18b).

The account we have just given is a rough qualitative description of a more complete theory called **quantum chromodynamics** (QCD). The form of the direct gluon–gluon couplings in this theory leads to a theoretical structure that is

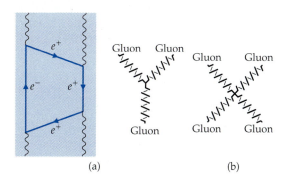

• **Figure 16–18**  (a) Photons can scatter from photons only through the creation and annihilation of virtual electron–positron pairs. (b) Gluons can scatter directly from other gluons, with no virtual particles taking part in the process.

• **Figure 16–19**  (a) Electric field lines in the interaction of an electron and a positron. (b) Conjectured configuration of gluonic field lines in the interaction of a quark and an antiquark. This type of configuration leads to confinement.

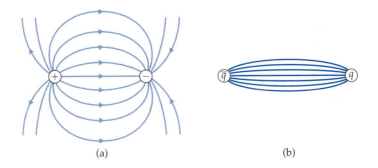

(a)                              (b)

quite different from quantum electrodynamics. The *hope* is that this basic difference in structure will account for the fact that not only quarks and antiquarks, but also gluons, have never been observed—that no particles carrying color charge *can* ever be observed and that only bound states which are colorless in a technical sense can escape and be observed in the laboratory. In this picture fractional electric charge is never isolated because all the states with fractional electric charge have color. Just how close is this hope to a reality? The basic problem is that at "low" energies—energies comparable to the spectrum of the low-lying hadrons—the equations are just too complicated to work with. Numerical calculations done on specially designed supercomputers reveal that there is indeed *color confinement* in the theory. At the time of this writing, the numerical calculations are also beginning to give reliable values for some of the masses and transition rates involving baryons and mesons. What remains undone in this approach is a simple description of how the quarks, gluons, and $q\bar{q}$ pairs form the wave functions of the observed particles.

There are some very telling analogies with superconductivity that have produced a kind of folklore of confinement in QCD. The picture is that quarks and antiquarks, or triplets of quarks, are bound together by gluonic energy that is confined to thin tubes (•Fig. 16–19), in contrast to the fanned-out field lines that characterizes the coulomb situation. The energy in such a tube is proportional to its length, so that the potential energy of a pair is proportional to the distance between them. In a nonrelativistic approximation we would say, as a guess, that the potential energy has the form

$$V(r) = Kr. \tag{16–15}$$

(This is the interaction of increasing strength that we alluded to in the previous section.) If we were to try to separate a $q\bar{q}$ pair we would have to put in more and more energy. Rather than stretching a tube indefinitely—something that requires an indefinitely large amount of energy—we break the tube with the creation of a $q\bar{q}$ pair at the break point. Thus a quark and an antiquark produced in a reaction such as $e^+ + e^- \rightarrow q + \bar{q}$ will generate, through the successive breaking of the thin tube connecting them, many $q\bar{q}$ pairs, as suggested by •Fig. 16–20a. A useful analogy is what happens when a bar magnet is cut in two:

• **Figure 16–20**  (a) A magnet broken in two gives two magnets rather than a north and south "magnetic monopole." (b) When a quark and an antiquark are pulled apart, energy has to be put into the system. At some point it is energetically favorable for a new $q\bar{q}$ pair to be created, resulting in two such pairs, or, equivalently, in two bosons.

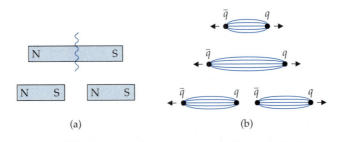

(a)                              (b)

We do not end up with two isolated magnetic charges; rather, we are left with two bar magnets (•Fig. 16–20b).

Fortunately quantum chromodynamics has another property: At high energies the strength of the quark–gluon couplings becomes small. This makes it possible to use the same calculational techniques that have been so successful in electrodynamics. Hence at high energies one can predict what happens when a quark–antiquark pair is produced, as in the reactions

$$e^- + e^+ \rightarrow \text{photon} \rightarrow q + \bar{q}$$

and                                                                    (16–16)

$$e^- + e^+ \rightarrow \text{photon} \rightarrow q + \bar{q} + g.$$

The theory describes successfully not only the rate of these reactions but also how the energy carried by the quark, the antiquark, and the gluon manifests itself in the flow of energy of **jets** of ordinary particles such as pions and nucleons. The experimental study of these reactions is very complicated, since energies and directions have to be measured with high precision. Furthermore since many events occur in a short time, time discrimination is important. Thus the detectors are large and complex. The results are collected with high-speed electronics, and displayed in a form ready for analysis. (•Fig. 16–21).

### Heavy Quarks and Quarkonium

More support for QCD comes from another source, the study of *heavy* quark systems. In 1974 a new quark, the **charm quark** ($c$) was discovered simultaneously by two groups, one at SLAC and the other at the Brookhaven National Laboratory. This particle, dubbed the $J/\psi$ (for the separate naming by the Brookhaven/SLAC groups), was spectacularly different from the run of hadronic resonances. It was rather massive, with $mc^2 = 3,097$ MeV, compared to 938 MeV for the proton, but that was not in itself unusual. Its width, however, was only 88 *keV*, rather than the 100 MeV or more that was the rule for hadronic resonances as massive as 3,000 MeV. Further experimentation revealed an entire set of states in the same energy region. The ensemble of $J/\psi$ states were interpretable as bound states of a new quark, $c$, and its antiparticle. (Actually, a number of theorists anticipated the finding of another quark on grounds that we'll discuss later, so that the zeroing in on the $c$ quark was a quick and fairly natural process.) These bound states make **charmonium**, an "atom" similar to positronium. The large energy scale implies that the $c$ quark is heavy, in contrast to the "light" quarks $u$, $d$, and $s$.[5]

Positronium decays into two or three photons when the $e^-$ and the $e^+$ annihilate, and charmonium would decay by an analogous process. The difference is that if, like the light quarks, the $c$ quarks carry color, then the products of the annihilation process could be gluons as well as photons. The gluons would then produce pairs of light quarks, which would appear as pions and other mesons. But why the small width? A small width, remember, means a long lifetime and hence a relatively feeble decay interaction. This could be due to some new conservation laws that were violated in the decay process, an explanation that worked for the strange decays. But in fact the correct explanation does not require the introduction of another new concept. Recall the remark we made in the previous section: "At high energies the strength of the quark–gluon couplings becomes small." The large $c$ mass implies that the gluons produced in

---

[5]To be more precise, $mc^2$ is around 1,500 MeV for the $c$ quark, in contrast to masses on the order of 300 MeV for the light quarks. The value of 1,500 MeV is compatible with the notion that for heavy quarks, most of the mass of a quark–antiquark bound state is the rest mass of its components.

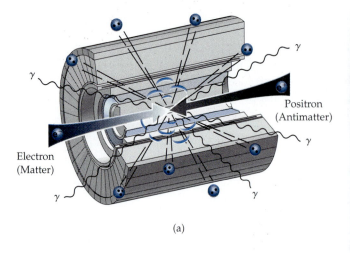

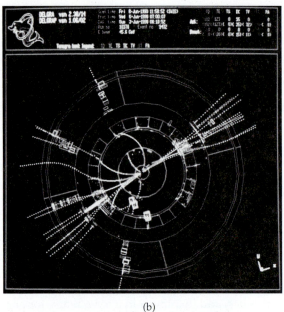

(a)

(b)

(c)

• **Figure 16–21**    (a) In $e^+$–$e^-$ colliders, a detector takes the form of a cylinder with the colliding beams along the axis. These detectors measure the collision products, their energies, and their directions. The results are displayed in the form shown in (b), in this case a two-jet event with two isolated muons. (c) An experimental area at the LEP accelerator at CERN, near Geneva, Switzerland, illustrates the complexity and size of such detectors.

annihilation will have a high energy, so the couplings will indeed be small. More quantitative analysis shows that the width of the $J/\psi$ will lie somewhere in between electrodynamic widths such as that of positronium—in which the coupling is proportional to $\alpha = 1/137$ and thus the width is very small—and the widths characteristic of the hadronic resonances. This analysis successfully predicts the properties of the $J/\psi$.

In the last 20 years, an extensive charmonium spectroscopy has developed. All of the aspects of atomic structure are there. The difference is that in addition to the coulomblike potential between $c$ and $\bar{c}$ due to the exchange of a gluon, there is the confinement potential $V = Kr$. The energy-level structure of charmonium is shown in •Fig. 16–22. In addition to the $c\bar{c}$ bound states, other mesonic states such as $c\bar{u}$, as well as baryon number 1 states like $cdd$, $csd$, ..., have been identified and their decays measured.

Within three years of the discovery of charmonium, another narrow state much like the $J/\psi$ was found at Fermilab, near Chicago, this time with $mc^2 = 9{,}460$ MeV and a width of 53 keV. It was apparent that the $c$-story was repeating. The underlying quark was named the **bottom quark** ($b$), with a mass around 4.5 GeV/$c^2$. The spectroscopy of these states provides still another window into the dynamics of QCD. The bottom–antibottom bound states—"bottomonium"—that are the analogue of charmonium are still another example of what is now know generically as **quarkonium**.

Finally, for theoretical reasons, a pairing $(u, d)$, $(c, s)$ demanded a quark to pair up with the $b$ quark. The expected quark is the **top quark** ($t$), discovered in 1995. Its energy-equivalent mass lies in the vicinity of 175,000 MeV! By itself the top quark is more massive than most atoms. There is little that experimentalists have measured about top–antitop systems (toponium), but theorists have a lot of predictions based on their work with charmonium and bottomium, so that QCD can be vigorously tested.

In sum, we have three pairs of quarks. In each pair there is one quark with electric charge $+(2/3)e$ and one with charge $-(1/3)e$. The quarks and the

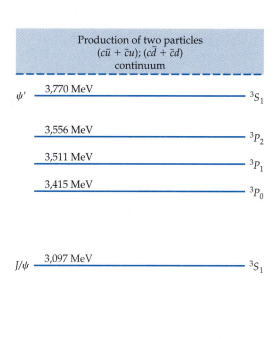

• **Figure 16–22** Selected energy levels in $c\bar{c}$ bound states (charmonium), with their masses, spin states, orbital angular momentum, and total angular momentum identified.

underlying theory describing their interactions account for the interactions among the hadrons.

## 16–7 Weak Interactions and Leptons

Enrico Fermi identified the underlying **weak interaction** responsible for the beta decay of nuclei. He was the first person to make predictions based on Pauli's conjectured neutrino. Prior to Fermi it had been assumed that in nuclear beta decay the electron was *inside* the nucleus waiting to tunnel out. In Fermi's theory the electron is created at the time of decay, in the reaction

$$n \rightarrow p + e^- + \bar{\nu}_e. \tag{16–17}$$

The reason for labeling the third particle on the right as an antineutrino will become clear later, when we discuss leptons and their conservation. The weak interactions govern a variety of phenomena. The rules of relativistic quantum mechanics allow the reversal of the arrow in a reaction, as well as the transfer of particles from one side of an equation like Eq. (16–17) to the other, provided that the particle is replaced by its antiparticle. Thus the same interaction leads to the process

$$p \rightarrow n + e^+ + \nu_e. \tag{16–18}$$

While the reaction in Eq. (16–18) is kinematically impossible for a free proton, whose mass is smaller than that of the neutron, it can and does occur in nuclei—for example in $^{66}\text{Ga} \rightarrow {}^{66}\text{Zn} + e^+ + \nu_e$. That is because binding-energy effects give us nuclear masses for which reactions such as this become kinematically possible.

The neutrino or antineutrino is never directly observed as part of the measurement of these processes. As we show in Example 16–9, the interaction of neutrinos with matter is so weak that a neutrino would likely pass through *light years* of lead. If, however, many neutrinos are available, then we may see an occasional interaction. It was in this way that F. Reines and C. L. Cowan first directly observed antineutrinos in 1956. These two researchers set up a large scintillator counter near the Savannah River nuclear reactor. The reactor produces enormous numbers of antineutrinos from the nuclear reactions within: A flux on the order of $10^{13}$ antineutrinos/cm$^2$/s arrived at their detector, and a few antineutrinos interacted inside the detector through the reaction

$$\bar{\nu}_e + p \rightarrow e^+ + n. \tag{16–19}$$

The reason that antineutrinos rather than neutrinos are produced is that they arise from the decay of the fission fragments in the reactor. The fission fragments are isotopes with many more neutrons than protons. Hence most of the decays are of neutrons. The annihilation of the $e^+$ in Eq. (16–19) with an atomic electron in the scintillator gives rise to two energetic gamma rays that produce a characteristic scintillation. The process is so weak that Cowan and Reines observed only about three such events an hour in spite of the enormous flux of antineutrinos.

---

**Example 16–9**  Low-energy neutrinos traverse a piece of solid iron, for which the density is $\rho = 7.9 \times 10^3 \, \text{kg/m}^3$. At low energies the cross section for neutrino–nucleon scattering is $\sigma \cong 10^{-47} \, \text{m}^2$. Estimate the mean free path of the neutrinos in the iron.

**Solution**  We learned in Chapter 14 that the mean free path is given by

$$\lambda = \frac{1}{n\sigma},$$

where $n$ is the number of targets per cubic meter and $\sigma$ is the cross section for scattering from the target. In this case each nucleon present in the iron is a target, and their number density is

$$n = \rho \times (\text{number of nucleons/unit mass})$$

$$= (7.9 \times 10^3 \, \text{kg/m}^3)(\text{number of nucleons/kg}).$$

Now

$$\text{number of nucleons/kg} = \frac{1}{\text{nucleon mass}} = \frac{1}{1.67 \times 10^{-27} \, \text{kg}} = 0.60 \times 10^{27} \, \text{kg}^{-1};$$

hence

$$\lambda = \left[(7.9 \times 10^3 \text{kg/m}^3)(0.60 \times 10^{27} \, \text{kg}^{-1})(10^{-47} \, \text{m}^2)\right]^{-1} = 2.1 \times 10^{16} \, \text{m}.$$

Since 1 light-yr $= (3.15 \times 10^7 \, \text{s/yr})(3.0 \times 10^8 \, \text{m/s}) \cong 10^{16} \, \text{m}$, it follows that $\lambda = 2$ light-yr!

The basic weak interaction process as described by Fermi is in fact an interaction between quarks involving the following reactions ($\bullet$Fig. 16–23):

$$
\begin{aligned}
(i) & \quad d \rightarrow u + e^- + \bar{\nu}_e; \\
(ii) & \quad u \rightarrow d + e^+ + \nu_e; \\
(iii) & \quad \nu_e + d \rightarrow u + e^-; \\
(iv) & \quad e^- + u \rightarrow d + \nu_e.
\end{aligned}
\qquad (16\text{–}20)
$$

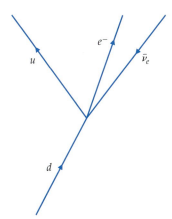

$\bullet$ **Figure 16–23** The quark weak decay that underlies the reaction of Eq. (16–17).

These reactions are crossover equivalents of one another, meaning that a particle on one side can be replaced with an antiparticle on the other. Examples in which each of these reactions governs physical processes are ($i$) beta decay of the neutron, ($ii$) the decay of the $\pi^-$ meson, ($iii$) the absorption of electron neutrinos by matter; and ($iv$) $K$-capture, a process in which an electron in orbit around a nucleus $(A, Z)$ is absorbed by the nucleus to give a new nucleus $(A, Z - 1)$ with the emission of a neutrino.[6] (For the processes governed by the preceding weak interactions to be realized physically, masses must be such that the processes are kinematically possible.)

The neutrino $\nu_e$ has spin $1/2$, just like the electron, and it is electrically neutral. Measurements of the maximum energy of the electron in reactions like that of Eq. (16–17) suggest that if the neutrino has any mass at all, that mass is very small—the best experiments to 1998 of this type set an upper limit of around $10^{-5} m_e$. Still, there is recent solid evidence from experiments of a different type that the neutrino cannot be absolutely massless. As with the neutron, we may ask, What distinguishes the neutrino from the antineutrino? We relied on the baryon number $B$ to distinguish between the neutron ($B = 1$) and the antineutron ($B = -1$). With regard to neutrinos, we must introduce an analogous quantum number called the **lepton number**. The electron and neutrino are assigned lepton number $N_L = +1$, and the positron and antineutrino are assigned $N_L = -1$. Particles with nonzero lepton number are generally referred to as **leptons**; hadrons have lepton number 0. Lepton number conservation implies that a reaction like

$$\bar{\nu}_e + {}^{37}\text{Cl} \rightarrow {}^{37}\text{Ar} + e^-$$

---

[6] The name of this process comes from the fact that the absorbed electron most often comes from the atomic $K$ shell, a historical name for the atomic ground state $n = 1$.

does not occur—the lepton number would have to change by 2. In fact in a subsequent experiment Cowan and Reines looked for this reaction and could not find it. On the other hand, the reaction

$$\nu_e + {}^{37}\text{Cl} \rightarrow {}^{37}\text{Ar} + e^-$$

does occur, because the lepton number is the same on both sides. Lepton number conservation has been confirmed in many experiments.

### Yukawa Hypothesis for the Weak Interactions

We suggested in our discussion of the Yukawa hypothesis that the exchange of a particle lies behind particle interactions. With the addition of the fact that there are several leptons that can contribute, Equation (16–20) underlies the weak interactions of "ordinary" particles such as protons and neutrons. How can the Yukawa interaction be brought in? Extensive experimental studies of the process represented by Eq. (16–20) shows that the range of the interaction has to be very short, so that the exchanged particle or particles must be very massive. Moreover, the exchanged particle must have spin 1, like the photon of electrodynamics or the gluon of quantum chromodynamics. Finally, as •Fig. 16–24 shows, the exchanged particle (labeled $W^-$) must carry a negative charge. The corresponding antiparticle $W^+$ will have positive charge, and since, in analogy with the pion, particles come in sets with charges +1, 0, and −1, there should be a neutral spin-1 partner (•Fig. 16–25), dubbed the $Z^0$. In 1983 a successful search was made for these particles, primarily at the European Nuclear Research Center CERN. The particles were produced in proton–antiproton collisions and identified through decays such as $W^+ \rightarrow e^+ + \nu_e$ or $Z^0 \rightarrow e^+ + e^-$. The $W^\pm$ has mass $m_W c^2 = 80$ GeV (80 times the mass of the proton!), while the $Z^0$ has mass $m_Z c^2 = 91$ GeV.[7] The W and Z particles are the **intermediate vector bosons**. (The word "vector" is a term often applied to particles with spin 1, an angular momentum with three possible values of the z-component of the spin.)

### More Leptons

Why did we bother to include the subscript "$e$" on the neutrino in the interaction given by Eq. (16–17)? The addition of the subscript is not fussy pedantry on our part; we did it because in fact there are other neutrinos. The subscript "$e$" is attached to neutrinos involved in reactions with electrons. We mentioned in Section 16–3 that in the search for Yukawa's meson the muon was discovered. This particle behaves much like an electron, only it is some 207 times more mas-

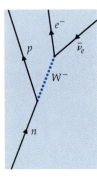

• **Figure 16–24** The underlying mechanism at the nucleonic level for the neutron decay $n \rightarrow p + e^- + \bar{\nu}_e$ is the exchange of the W intermediate boson between baryons and leptons.

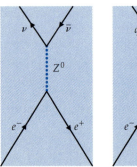 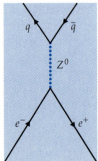

• **Figure 16–25** The neutral $Z^0$ must decay into neutral final states. Examples are
$Z^0 \rightarrow \nu + \bar{\nu}$ and
$Z^0 \rightarrow$ quark + antiquark.

[7] Actually, as we shall see shortly, the $Z^0$ is closely related to the photon, a relation that accounts for the fact that the Z has a mass very different from the W.

sive than the electron. Just like the electron, it has an antiparticle $\mu^+$ of opposite charge. *And there is a special neutrino with which it is involved: the $\nu_\mu$.* Here are some of the salient facts:

**1.** The muon is unstable via the decay

$$\mu^- \rightarrow \nu_\mu + e^- + \bar{\nu}_e. \qquad (16\text{–}21)$$

In fact all that a detector actually observes here is a muon abruptly turning into an electron. But kinematic analysis of many such observations would reveal that the muon decays into three particles, two of them nearly massless. We also remark here that the fact that the muon is unstable while the electron is stable is due only to the fact that the electron is the lightest charged lepton. The muon's lifetime is $2.2 \times 10^{-6}$ s. This number, as well as various rates for other reactions involving it, can be explained if the strength and the form of the muon weak interaction is the *same* as that of the electron. Only the labels are different.

**2.** The evidence that $\nu_\mu \neq \nu_e$ involves our knowledge of pion decay and scattering experiments. Almost from the moment of their discovery, charged pions were known to decay into muons and neutrinos, via the processes

$$\pi^+ \rightarrow \mu^+ + \nu_\mu \qquad \text{and} \qquad \pi^- \rightarrow \mu^- + \bar{\nu}_\mu.$$

The pion can also decay into electrons, but the rate for $\pi^+ \rightarrow e^+ + \nu_e$ is about 10,000 times smaller than the rate for $\pi^+ \rightarrow \mu^+ + \nu_\mu$. Thus we can be sure that the debris of pion decay is mainly "muonic." Now high-energy accelerators produce very large numbers of pions through $p$–$p$ collisions. These pions can be focused into highly collimated beams, and when they decay their debris rests within the beam. The charged muons within the debris-laden beam can be blocked out, and what remains is a neutrino beam. In a beautiful experiment carried out in 1962, a beam of this type was used, and although a reasonable number of reactions of the type neutrino $+ n \rightarrow p + \mu$ were observed, none were found of the type neutrino $+ n \rightarrow p + e$. Hence the neutrinos in the beam were of the muon type, and these do not associate with electrons. The results of this experiment necessitate the restatement of lepton conservation in a stricter form: *There are separate conservation laws for e-leptons and for $\mu$-leptons.*

**3.** All the reactions that occur for the electron and its neutrino occur for the muon and its neutrino. In particular, there is a reaction with quarks that is the analogue of the first reaction of Eq. (16–20), namely

$$d \rightarrow u + \mu^- + \bar{\nu}_\mu. \qquad (16\text{–}22)$$

In the language of the intermediate vector bosons, the muon and its neutrino couple to the $W$ and $Z$ exactly as the electron and its neutrino.

In the mid-1970s a *third* lepton, the $\tau$, was discovered. The $\tau$ appears to be just another heavy electron, with $mc^2 = 1{,}777$ MeV. It has its associated neutrino $\nu_\tau$ that appears to have properties similar to those of the other neutrinos and that reacts only with the $\tau$, just as $\nu_e$ reacts only with electrons or $\nu_\mu$ reacts only with muons.

In sum, there appear to exist three lepton doublets, $(e, \nu_e)$, $(\mu, \nu_\mu)$, and $(\tau, \nu_\tau)$, matching the three quark doublets, $(u, d)$, $(c, s)$, and $(t, b)$. There are good theoretical reasons for the match, even if no one knows why there are *three and only three* of these sets.

---

**Example 16–10**    The analogue of Eq. (16–22) holds for the $\tau$ lepton. **(a)** What is the simplest decay mode of the $\tau^+$ into a $\pi$-meson? **(b)** Given that in part (a) a neutrino must appear in the final state, calculate the neutrino's energy. Give your answer in the frame in which the $\tau$ is at rest. **(c)** Can the $\tau^+$ decay into a final state containing one or more protons?

**Solution**    **(a)** The rearrangement of Eq. (16–22) that starts with a single $\tau$ in the initial state is

$$\tau^+ \rightarrow \bar{\nu}_\tau + u + \bar{d}.$$

In turn the combination $u\bar{d}$ is the $\pi^+$ [Eq. (16–10)], so that the decay $\tau^+ \rightarrow \bar{\nu}_\tau + \pi^+$ is directly possible from the quark model point of view; it is kinematically possible as well. **(b)** The neutrino and the $\pi$-meson have equal and opposite momenta when the $\tau$ decays at rest. Let these momenta have magnitude $p$. Energy conservation then implies that

$$m_\tau c^2 = pc + \sqrt{(pc)^2 + \left(m_\pi c^2\right)^2}.$$

We have used the fact that the neutrino mass is very small; that is also the case for the pion mass compared to the $\tau$ mass; to order $\left(m_\pi/m_\tau\right)^2$ (that is, to better than 1%) we can neglect $m_\pi$. Then the right-hand side of the foregoing equation is $2pc$, and hence

$$E_\nu = pc = m_\tau c^2/2 = 839 \text{ MeV}.$$

**(c)** If a proton is contained in the decay products of a $\tau$, then by baryon conservation an antiproton must be present as well. Since the total proton–antiproton mass of 1,878 MeV/$c^2$ is larger than the mass of the $\tau$, this $\tau^+$ decay cannot be realized physically.

---

# 16–8  Pulling Things Together

## Internal Conservation Laws Revisited

Let us briefly summarize the basic conservation laws in the strong, electromagnetic, and weak interactions in terms of quarks and leptons.

*The Strong Interaction*    Only the quarks, which come in the six **flavors** $u, d, s, c, b, t$, and not the leptons, participate. The basic interactions involve gluons, which will generally change the "color" content of the quark, but which do not change the flavor labels of the quarks; for example, a $u$ quark of one color may change its color with the emission of an appropriate gluon, but it will remain a $u$ quark. This means that gluons carry color labels but not flavor labels. The physical particles—the ones that can be observed traveling through detectors, even if they are not stable—are *colorless* combinations of quarks. The fact that quarks do not change their flavor imposes certain restrictions. For instance, in a reaction like $\pi^- + p \rightarrow \Lambda^0 + X$, we have the following quark contents:

$$(d\bar{u}) + (uud) \rightarrow (uds) + X.$$

Now a quark and an antiquark could annihilate, but unless this occurs a quark in the initial state must appear in the final state. This means that $X$ must contain an $\bar{s}$ and a $d$. Thus in the simplest reaction $X = (d\bar{s})$, and $X$ is a $K^0$ particle. In other words the reaction

$$\pi^- + p \rightarrow \Lambda^0 + K^0 \tag{16–23}$$

is expected. In addition, the final state can contain any number of $(u\bar{u})$, $(d\bar{d})$, $(s\bar{s})$, ... pairs, so this will conserve the net number of quarks of a definite flavor. Thus we expect that at high enough energies we will see reactions like

$$\pi^- + p \rightarrow \Lambda^0 + K^0 + n\pi^0 + \pi^+\pi^- + \dots. \tag{16–24}$$

Since the number of $s$-quarks is conserved, so is *strangeness*.

*The Electromagnetic Interaction*   The basic interaction here is the emission of a photon by a charged particle, without the alteration of the nature of that particle. The basic conservation law is that of electric charge, and since the photon does not carry charge, the basic interactions are $q$(charge $-e/3$) $\rightarrow$ $q$(charge $-e/3$) $+ \gamma$ and $q$(charge $+2e/3$) $\rightarrow$ $q$(charge $+2e/3$) $+ \gamma$. In addition, the photon does not carry flavor, so that there are no terms of the type $d \rightarrow s + \gamma$. (Recall that both $d$ quarks and $s$ quarks have charge $-e/3$.) Similarly, photons do not carry color. We say that both flavor and color are conserved in the electromagnetic interaction.

*The Weak Interaction*   The rules governing the weak interaction are best understood by reducing our discussion to the quark level. Thus the Fermi interaction $n \rightarrow p + e^- + \bar{\nu}_e$ can be expressed in quark language as $d \rightarrow u + e^- + \bar{\nu}_e$, the first reaction of Eq. (16–20). (See •Fig. 16–26.) We notice one major difference from the strong and electromagnetic interactions: *Flavor changes.* (Since no gluons are involved, color does not change, and electric charge is, as always, conserved.) Earlier we described how the Yukawa hypothesis applied to the weak interaction restates the basic process as the emission and reabsorption—the exchange—of the intermediate vector bosons, particles which mediate the weak interaction in much the same way that photons mediate electromagnetic ones. In this picture the first reaction of Eq. (16–20) consists of a pair of virtual reactions, namely (•Fig. 16–26)

$$d \rightarrow u + W^- \quad \text{and then} \quad W^- \rightarrow \ell^- + \bar{\nu}_\ell, \quad (16\text{–}25)$$

where the symbol $\ell$ stands for any of the three types of leptons.

We have mentioned that the decay $\Lambda^0 \rightarrow p + \pi^-$ is slow. Expressed in terms of quarks, this decay reads

$$(uds) \rightarrow (uud) + (d\bar{u}).$$

The slowness of the reaction implies that it is governed by weak interactions, in this case of the two-step form

$$s \rightarrow u + W^- \quad \text{and then} \quad W^- \rightarrow \bar{u} + d. \quad (16\text{–}26)$$

Recall that we can transfer a particle from one side of the reaction to the other by changing the particle to its antiparticle. Thus we can regard the first step in Eq. (16–26) as the reaction $\bar{u} + s \rightarrow W^-$. This way of thinking about things helps clarify how the $W^-$, as well as its partners with other charges—we have not yet discussed these—couples to other particles.

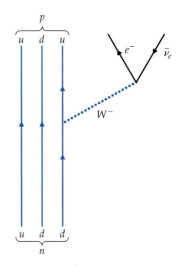

**• Figure 16–26**   The underlying mechanism at the quark level for neutron decay is the transition $d \rightarrow u + W^-$, with the virtual $W^-$ decaying into $e^- + \bar{\nu}_e$.

---

**Example 16–11**   The negative cascade particle, symbol $\Xi^-$, has quark content $(ssd)$. Show, as in Eqs. (16–25) and (16–26), the steps that occur in the observed decay (1) $\Xi^- \rightarrow \Lambda^0 + \pi^-$, and show that the (unobserved) decay (2) $\Xi^- \rightarrow n + \pi^-$ cannot occur with a single intermediate $W^-$.

**Solution**   The observed reaction (1) is, in quark form,

$$(ssd) \rightarrow (uds) + (d\bar{u})$$

We may take one $sd$ pair in the initial and final states as "uninvolved bystanders." The other changes can be achieved with

$$s \rightarrow u + W^- \quad \text{and then} \quad W^- \rightarrow \bar{u} + d.$$

Reaction **(2)** involves $(ssd) \rightarrow (udd) + (d\bar{u})$, so that two $s$ quarks must change flavor, and hence two $W^-$'s must be involved. This turns out to be highly unlikely.

We emphasize here that some conservation laws hold for all interactions. Conservation of electric charge is one such law, and we have reason to believe that it is exact. Other conserved quantities may not be exactly conserved. For example, in each of the three kinds of interaction—strong, electromagnetic, and weak—there is no transformation of a quark into an antiquark. This guarantees conservation of baryon number. There *could* exist a superweak interaction in which $q \rightarrow \bar{q}$ does occur, in which case baryon number would not be exactly conserved, but to date there is no evidence for such an interaction. Similarly, lepton number is conserved in the three interactions.

## The Standard Model

The strong, electromagnetic, and weak interactions share many properties. In particular, in all cases the interaction is mediated by the exchange of spin-1 bosons, and the ways in which these particles—photons, gluons, and intermediate vector bosons $(W^{\pm}, Z^0)$—are absorbed or emitted by the fundamental fermions—quarks and leptons—are similar. Indeed, the way in which we understand the intermediate vector bosons has come from an effort to understand how to unite these basic reactions. A significant clarification that truly unified the weak and electromagnetic interactions was achieved some 30 years ago in work by, among others, the Americans Steven Weinberg and Sheldon Glashow and the Pakistani Abdus Salam. Their fundamental unification is based on the Yukawa hypothesis applied to quarks and leptons, together with a kind of mixing that takes place between the photon and the $Z^0$. This mixing implies a true unification of the weak and electromagnetic interactions. The work made precise predictions for the masses of the intermediate vector bosons, and these predictions proved to be right on the money. Today we refer not to the weak and electromagnetic interactions but to the **electroweak interaction**. Its similarity to quantum chromodynamics brings us close to a further unification. Differences remain, however. For example, consider the manner in which the spin-1 mesons manifest themselves in reality: Photons are massless, gluons are massless, carry color, and are confined, while the $W^{\pm}$ and the $Z^0$ differ from both of these by not carrying color and being massive. (The $W^{\pm}$ also carry electric charge.)

The combination of the electroweak interaction with quantum chromodynamics is known as the **standard model**. (•Figure 16–27) It is hoped that the two parts of the standard model represent two branches of a "grand unified theory." The standard model has a number of parameters that have to be put in "by hand"—the masses of the leptons and the quarks, for example. The fundamental origin of mass represents just one of the important questions that remain to be answered. Perhaps the grand unified theory will give compelling—and experimentally testable—reasons why the numbers are what they are or why there are three sets of quarks and leptons, to name just two of several issues. Such a theory will undoubtedly predict some new phenomena. It is even possible that gravity will have to play a role. This is an active area of research. Many interesting speculations have been made, and physicists await data from new accelerators for clues. And physicists are also looking elsewhere: outward, towards the distant heavens, where, as we'll see in Chapter 18, some important evidence awaits.

# SUMMARY

The process of dismantling matter to explore the nature of its constituents continues. We are currently exploring the constituents of nucleons, which exist at a level parallel to that of electrons, photons, and neutrinos. A number of features of this exploration bring us to the edge of unexplored territory:

- Special relativity is all important in any discussion of subnuclear physics. This importance has several consequences: The idea of a simple Newtonian force between particles no longer applies, and the (nonrelativistic) Schrödinger equation must be modified; particle production and the presence of antiparticles become necessary; and the uncertainty principle, combined with particle production, permits us to deal with virtual particles.

- Where we do understand the dynamics, as in quantum electrodynamics, the relativistic theory of electrons and electromagnetic radiation, a set of extremely accurate and entirely verified predictions emerges. Some of the principles that emerge can be extended to situations in which we are not entirely certain of the dynamics or of how to handle it. Thus the exchange of virtual particles is at the heart of fundamental interactions. Indeed, the application of the particle exchange idea to nuclear forces leads to verified predictions about the $\pi$-meson.

- The interactions of elementary particles are restricted by conservation laws, the old ones with which we are familiar, such as conservation of energy, momentum, and electric charge, and new ones, such as conservation of strangeness, baryon number, and lepton number. Many of these conserved quantities are associated with an underlying symmetry, an important guidepost for our current ideas about physical law. Some of the new conservation laws are not exact, and one of the pressing experimental tasks is the verification of the extent to which they are or are not exact. The detection of proton decay, for example, would test to what extent the conservation of baryon number is exact.

- The set of hadrons—nucleons, pions, and their short-lived resonant partners—are in fact composed of a much smaller set of fermionic consituents—quarks—interacting by means of a set of bosons, the gluons. For reasons having to do with the fundamental nature of the the interaction, quarks and gluons are never seen isolated.

- At present three fundamental forces are seen in nature: the electroweak force, which consists of an intimate mixture of the weak interaction, responsible for the beta decay of the neutron, and the electromagnetic forces; the strong force, governing the behavior of quarks and gluons; and gravity. Together the first two describe the world of quarks, gluons, leptons, and photons. Just as electricity and magnetism were united in the 19th century, one hopes that these separate forces can be united under a single banner.

# QUESTIONS

**1.** We have remarked several times in the text that while the neutron is electrically neutral it does have an electromagnetic structure. Indeed it has both a magnetic moment and a charge distribution, each of which can be measured in electron–neutron scattering. Try to draw a few Feynman diagrams involving neutrons, protons, $\pi$-mesons and $\gamma$'s that would take into account such a structure.

# Standard Model of Fundamental Particles and Interactions

The Standard Model summarizes the current knowledge in Particle Physics. It is the quantum theory that includes the theory of strong interactions (quantum chromodynamics or QCD) and the unified theory of weak and electromagnetic interactions (electroweak). Gravity is included on this chart because it is one of the fundamental interactions even though not part of the "Standard Model."

## FERMIONS
matter constituents
spin = 1/2, 3/2, 5/2, ...

### Leptons  spin = 1/2

| Flavor | Mass GeV/c² | Electric charge |
|---|---|---|
| $\nu_e$ electron neutrino | <1×10⁻⁸ | 0 |
| e electron | 0.000511 | -1 |
| $\nu_\mu$ muon neutrino | <0.0002 | 0 |
| μ muon | 0.106 | -1 |
| $\nu_\tau$ tau neutrino | <0.02 | 0 |
| τ tau | 1.7771 | -1 |

### Quarks  spin = 1/2

| Flavor | Mass GeV/c² | Electric charge |
|---|---|---|
| u up | 0.003 | 2/3 |
| d down | 0.006 | -1/3 |
| c charm | 1.3 | 2/3 |
| s strange | 0.1 | -1/3 |
| t top | 175 | 2/3 |
| b bottom | 4.3 | -1/3 |

## BOSONS
force carriers
spin = 0, 1, 2, ...

| Unified Electroweak spin = 1 | | | Strong (color) spin = 1 | | |
|---|---|---|---|---|---|
| Name | Mass GeV/c² | Electric charge | Name | Mass GeV/c² | Electric charge |
| γ photon | 0 | 0 | g gluon | 0 | 0 |
| W⁻ | 80.4 | -1 | | | |
| W⁺ | 80.4 | +1 | | | |
| Z⁰ | 91.187 | 0 | | | |

## Structure within the Atom

Electron Size < 10⁻¹⁸ m

Quark Size < 10⁻¹⁹ m

Neutron & Proton Size = 10⁻¹⁵ m

Nucleus Size = 10⁻¹⁴ m

Atom Size = 10⁻¹⁰ m

**Spin** is the intrinsic angular momentum of particles. Spin is given in units of $\hbar$, which is the quantum unit of angular momentum, where $\hbar = h/2\pi = 6.58 \times 10^{-25}$ GeV s $= 1.05 \times 10^{-34}$ J s.

**Electric charges** are given in units of the proton's charge. In SI units the electric charge of the proton is $1.60 \times 10^{-19}$ coulombs.

The **energy** unit of particle physics is the electron volt (eV), the energy gained by one electron in crossing a potential difference of one volt. **Masses** are given in GeV/c² (remember $E = mc^2$), where 1 GeV $= 10^9$ eV $= 1.60 \times 10^{-10}$ joule. The mass of the proton is $0.938$ GeV/c² $= 1.67 \times 10^{-27}$ kg.

**Color Charge**  Each quark carries one of the three types of "strong charge," also called "color charge." These charges have nothing to do with the colors of visible light. There are eight possible types of color charge for gluons. Just as electrically-charged particles interact by exchanging photons, in strong interactions color-charged particles interact by exchanging gluons. Leptons, photons, and W and Z bosons have no strong interaction and hence no color charge.

**Residual Strong Interaction**  The strong binding of color-neutral protons and neutrons to form nuclei is due to residual strong interactions between their color-charged constituents. It is similar to the residual electrical interaction which binds electrically neutral atoms to form molecules. It can also be viewed as the exchange of mesons between the hadrons.

If the protons and neutrons in this picture were each 10 cm across, then the quarks and electrons would be less than 0.1 mm in size and the entire atom would be about 10 km across!

**Quarks Confined in Mesons and Baryons**  One cannot isolate quarks and gluons; they are confined in color-neutral particles called hadrons. This confinement (binding) results from multiple exchanges of gluons among the color-charged constituents. As color-charged particles (quarks and gluons) move apart, the energy in the color force between them increases. This energy eventually is converted into additional quark-antiquark pairs. The quarks and antiquarks then combine into hadrons; these are the particles seen to emerge. Two types of hadrons have been observed in nature: mesons qq̄ and baryons qqq.

# Properties of the Interactions

### Baryons qqq and Antibaryons q̄q̄q̄
Baryons are fermionic hadrons.
There are about 120 types of baryons.

| Symbol | Name | Quark content | Electric charge | Mass GeV/c² | Spin |
|---|---|---|---|---|---|
| p | proton | uud | 1 | 0.938 | 1/2 |
| p̄ | Anti-proton | ūūd̄ | −1 | 0.938 | 1/2 |
| n | neutron | udd | 0 | 0.940 | 1/2 |
| Λ | lambda | uds | 0 | 1.116 | 1/2 |
| Ω⁻ | omega | sss | −1 | 1.672 | 3/2 |

**Baryons** are a type of hadron composed of three quarks (or three antiquarks).

### Mesons q̄q
Mesons are bosonic hadrons.
There are about 140 types of mesons.

| Symbol | Name | Quark content | Electric charge | Mass GeV/c² | Spin |
|---|---|---|---|---|---|
| $\pi^+$ | pion | ud̄ | +1 | 0.140 | 0 |
| K⁻ | kaon | s̄u | −1 | 0.494 | 0 |
| $\rho^+$ | rho | ud̄ | +1 | 0.770 | 1 |
| B⁰ | B-zero | d̄b | 0 | 5.279 | 0 |
| $\eta_c$ | eta-c | cc̄ | 0 | 2.980 | 0 |

**Matter and Antimatter** For every particle type there is a corresponding antiparticle type, denoted by a bar over the particle symbol (unless + or 0 charge is shown). Particle and antiparticle have identical mass and spin but opposite charges. Some electrically neutral bosons (e.g., $Z^0$, gamma, and $\eta_c = c\bar{c}$, but not $K^0 = d\bar{s}$) are their own antiparticles.

This chart has been made possible by the generous support of:
US Department of Energy
Lawrence Berkeley National Laboratory
Stanford Linear Accelerator
American Physical Society, Division of Particles and Fields
Burle Industries, Inc.

| Property | Gravitational | Weak (Electroweak) | Electromagnetic (Electroweak) | Strong — Fundamental | Strong — Residual |
|---|---|---|---|---|---|
| Acts on: | Mass - Energy | Flavor | Electric charge | Color Charge | See Residual Strong interaction note |
| Particles experiencing: | All | Quarks, Leptons | Electrically charged | Quarks, Gluons | Hadrons |
| Particles mediating: | Gravitation (not yet observed) | W⁺ W⁻ Z⁰ | $\gamma$ | Gluons | Mesons |
| Strength relative to electromag? for two u quarks at: $10^{-18}$ m | $10^{-41}$ | 0.8 | 1 | 25 | Not applicable to quarks |
| $3 \times 10^{-17}$ m | $10^{-41}$ | $10^{-4}$ | 1 | 60 | |
| for two protons in nucleus | $10^{-36}$ | $10^{-7}$ | 1 | Not applicable to hadrons | 20 |

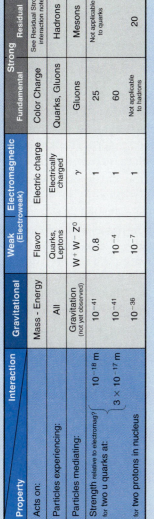

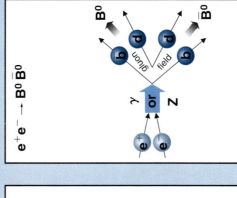

$$n \rightarrow p\,e^- \,\bar{\nu}_e$$

A neutron decays to a proton, an electron, and an antineutrino via a virtual (mediating) W boson. This is neutron $\beta$ decay.

$$e^+e^- \rightarrow B^0\,\bar{B}^0$$

An electron and positron (antielectron) colliding at high energy can annihilate to produce $B^0$ and $\bar{B}^0$ mesons via a virtual Z boson or a virtual photon.

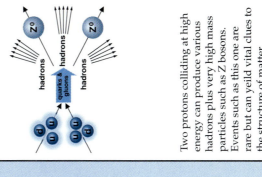
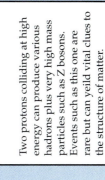

$$p\,p \rightarrow Z^0\,Z^0 + \text{assorted hadrons}$$

Two protons colliding at high energy can produce various hadrons plus very high mass particles such as Z bosons. Events such as this one are rare but can yeild vital clues to the structure of matter.

• **Figure 16–27** A summary chart for many of the phenomena and properties discussed in this chapter. (Copyright 1999 by the Contemporary Physics Education Project, Lawrence Berkeley National Laboratory, Berkeley, CA. Reprinted by permission.)

2. The Cowan–Reines experiment detected antineutrinos that came from the reactor that was used. We said that this was because the fission fragments are neutron rich—they have many more neutrons than protons. Why is that?

3. In a recent experiment performed in Japan which seems to show that neutrinos have a tiny mass some of the neutrinos were alleged to have penetrated the entire Earth before being detected. How is this possible? Could electrons penetrate the entire Earth? Why not?

4. Consider a world just like ours, with the single exception that the neutron and proton masses are identical. What would this world look like?

5. Positronium is the atom formed by an electron and a positron. Would you expect the mass of positronium to be equal to, larger than, or smaller than the sum of the masses of the electron and positron (which are equal to each other)?

6. Positronium, the atom formed by an electron and a positron, is unstable because the electron and positron can annihilate to photons. In which atomic state will positronium annihilate more quickly, a 2s state or a 2p state?

7. Even though no fractionally charged particle has ever been observed to pass through a particle detector, we say we have discovered quarks. Does this discovery stand up to the "usual" criteria for scientific discovery? Argue your case.

---

## PROBLEMS

1. ▮▮ Consider the time-independent relativistic equation obtained from Eq. (16–2) by inserting $\psi(\vec{r}, t) = u(\vec{r})\exp(-iEt/\hbar)$ into it. If we write $E = mc^2 + \varepsilon$ with $\varepsilon \ll mc^2$, and if $V \ll mc^2$, what is the form of the eigenvalue equation? (*Hint*: The equation that emerges from the relation $(E - V)^2 = K^2 = m^2c^4 + p^2c^2$ is similar to that of the hydrogen atom if it is recognized that $V^2$ just changes the $\ell(\ell + 1)$ coefficient of $1/r^2$ in the centrifugal term of the radial wave equation.)

2. ▮ Suppose a particle of mass $m = 140$ MeV/$c^2$ decays at rest into an $e^+$ and an $e^-$. If this occurs in a magnetic field of magnitude 2 T, what is the radius of the orbits that these charged particles travel in? (Assume that the charged particle paths curl in a plane perpendicular to the magnetic field.)

3. ▮▮ Consider a free electron with momentum $\vec{p}$ that emits a free photon with momentum $\vec{q}$. Given that momentum is conserved in the process, show that energy cannot be conserved, which means that the final electron must be virtual.

4. ▮▮ Energy and momentum are conserved. Use this fact to decide whether the annihilation $e^+ + e^- \rightarrow \gamma$ is allowed.

5. ▮▮ The $\omega^0$ is a particle that decays into three pions: $\omega^0 \rightarrow \pi^+ + \pi^- + \pi^0$. Given that the mass of the $\omega^0$ is 782 MeV/$c^2$ and that of the pions is 138 MeV/$c^2$, on the average, what is the largest possible momentum that a single pion can have? Assume that the $\omega^0$ decays from rest.

6. ▮▮ In 1987 astronomers saw light from a supernova in the Large Magellanic Cloud, which is a galaxy about $1.8 \times 10^5$ lt-yr away from us—relatively nearby. In addition to the light, antineutrinos were detected in large underground detectors. One of the detectors was in Kamioka, Japan. This detector found 11 antineutrinos that arrived in a time interval of about 12.4 seconds. (Because of the detection method the detector was sensitive only to antineutrinos.) The energies of these antineutrinos ranged between 7 and 50 MeV. In this problem, we use these facts to find an approximate bound on the mass $m$ of the antineutrinos and hence on neutrinos. **(a)** It is known from terrestrial experiments that

$mc^2 \ll 1$ MeV. Use this approximation to show that if neutrinos have a mass and therefore move less rapidly than the speed of light then

$$\frac{v}{c} \cong 1 - \frac{1}{2}\frac{m^2 c^4}{E^2}.$$

**(b)** Suppose that two neutrinos of the same mass but different energies $E_1$ and $E_2$ arrive at the detector, which is at a distance $d$ from the supernova, with a time separation of $\Delta t$. Assuming that $E_1 > E_2$, show that

$$m^2 c^4 \cong \frac{2c\Delta t}{d}\frac{E_1^2 E_2^2}{E_1^2 - E_2^2}.$$

**(c)** Put in the relevant parameters from the Kamioka experiment to find $m$. This is really an upper bound on $m$, since some of the time difference could be the time difference at which the different neutrinos were emitted from the super-nova. If they were all emitted at exactly the same time but they arrived at different times this would fix $m$.

**7.** ▌ In low-energy nuclear physics reactions the only particles that are relevant are the neutron $n$; the proton $p$; the electron $e^-$; its antiparticle, the positron $e^+$; the antineutrino $\bar{\nu}_e$; and the photon $\gamma$. Working only with these particles, and neglecting energy conservation, consider the neutron decay process

$$n \rightarrow p + X.$$

**(a)** What must be the quantum numbers of $X$ (i.e., charge, baryon number, spin, and so forth)? **(b)** Can $X$ be a single particle? If not, what is the fewest number of particles that will work?

**8.** ▌ Take the conditions of the previous problem and consider the reaction

$$e^- + p \rightarrow \nu_e + X.$$

**(a)** What must be the quantum numbers of $X$? **(b)** Can $X$ be a single particle? If not, what is the fewest number of particles that will work?

**9.** ▌ Take the conditions of Problem 7 and consider the reaction

$$e^- + e^+ \rightarrow \bar{\nu}_e + X.$$

**(a)** What must be the quantum numbers of $X$? **(b)** Can $X$ be a single particle? If not, what is the fewest number of particles that will work?

**10.** ▌▌▌ In Example 16–1 we suggested that one could find the maximum value of the energy of any one photon in the three-photon decay of positronium. **(a)** Find the value of the photon energy and the magnitude of its momentum for the third photon in the case where two photons go in one direction and the third goes off in the opposite direction. **(b)** Find the value of the energy and the magnitude of the corresponding momentum of any one photon for the sym-metric decay, wherein the three photons go off making a $120°$ angle with each other.

Case (a) is the maximum-energy case.

**11.** ▌▌ When a $\pi^-$ is absorbed in liquid hydrogen, it can be captured by a pro-ton before decaying, whereupon it forms a hydrogenlike "atom." What is the energy of the photon emitted when this "atom" makes a transition from the $n = 2, \ell = 1$ state to the $n = 1, \ell = 0$ ground state.

**12.** ▌ Sketch a Feynman diagram or diagrams for each of the following processes: **(a)** $e^- + \mu^- \rightarrow e^- + \mu^-$. **(b)** $e^- + e^- \rightarrow e^- + e^- + \gamma$.

**13.** ❚ As we have seen, the weak interactions are mediated by intermediate bosons. The charged intermediate boson is the $W$, and its mass has been measured as $m_W c^2 = 81$ GeV ($1$ GeV $= 10^9$ eV). What is the range of the weak force?

**14.** ❚❚ Suppose that the photon had a mass of $0.001$ eV/$c^2$, so that the coulomb potential is modified to the form of Eq. (16–7). *Estimate* the effect of this mass on the binding energy of monatomic hydrogen. (*Hint*: Estimate $\mu r/\hbar c$ for the given value of $\mu c^2$ and for $r$ on the order of the Bohr radius, and show that $r^{-1}\exp(-\mu r/\hbar c) \cong r^{-1} - \mu/\hbar c$, which should be taken as the potential in the Schrödinger equation.)

**15.** ❚❚ Strangeness is conserved in strong interactions, while in weak interactions it changes by one unit. Furthermore, electric charge, baryon number, lepton number, and angular momentum are, as far as we can tell, always conserved. Which of the following reactions will be **(a)** fast, **(b)** slow, or **(c)** totally forbidden: $(i)$ $\pi^- + p \rightarrow n + K^0$; $(ii)$ $\bar{p} + p \rightarrow p + \pi^-$; $(iii)$ $e^- + p \rightarrow \bar{\nu}_e$; $(iv)$ $\Delta^{++} \rightarrow \pi^+ + \pi^+ + n$; $(v)$ $\Delta^{++} \rightarrow K^+ + K^+ + n$; $(vi)$ $\pi^- + p \rightarrow \Lambda^0 + K^0$.

**16.** ❚ Baryon number is conserved in the reaction $\pi^+ + n \rightarrow K^0 + \Sigma^+$ and the decay $K^0 \rightarrow \pi^+ + \pi^-$. Given that the baryon number of the neutron is $+1$ and that of all pions is zero, find the baryon numbers of the $K^0$ and the $\Sigma^-$.

**17.** ❚ Using the fact that the baryon number of the pions is zero, deduce the baryon number of the photon from the (baryon-number-conserving) reaction $\pi^0 \rightarrow \gamma + \gamma$. From what you have learned from this result and the previous problem, does the decay $\Sigma^+ \rightarrow K^+ + \gamma$ conserve baryon number?

**18.** ❚ **(a)** What must be the net strangeness of the final state in the collision process $\pi^- + p$? **(b)** What must be the net strangeness of the final state in the collision of a $K^-$ and a proton?

**19.** ❚❚ There is a set of 10 particles that may be viewed as belonging to a single family. They all have spin 3/2 and $B = 1$. Their masses and strangeness are given in the following table:

| Particle | Strangeness | Approximate Mass |
|---|---|---|
| $\Delta^{++}, \Delta^+, \Delta^0, \Delta^-$ | 0 | 1,232 MeV/$c^2$ |
| $\Sigma^{*+}, \Sigma^{*0}, \Sigma^{*-}$, | −1 | 1,385 MeV/$c^2$ |
| $\Xi^{*0}, \Xi^{*-}$, | −2 | 1,530 MeV/$c^2$ |
| $\Omega^-$ | −3 | 1,672 MeV/$c^2$ |

**(a)** What is the quark content of each of the members of this family? **(b)** Estimate the masses of the $u$ and $s$ quarks from three of the mass assignments, assuming that the binding energy is independent of strangeness and that the $u$ and $d$ quarks have the same mass. Check for consistency in your assumptions by using the fourth mass.

**20.** ❚ A certain particle—call it the $X^{++}$—is observed to decay rapidly into two $\pi^+$ mesons. What can you say about the particle's mass, its strangeness, and whether it is a boson or a fermion?

**21.** ❚❚ Consider the $X^{++}$ particle of the previous problem. Draw quark diagrams to investigate the possible quark content of the $X^{++}$. If the interactions between the quarks involved are independent of whether they are $u$ or $d$ quarks, can you predict the existence of other $X$-like particles with more or less the same mass? You may treat the $u$ and $d$ quarks as having the same mass.

**22.** ▮▮ Consider a quark–antiquark pair, each member of which has mass $m$, bound by a central potential $V(r) = Kr$ and moving in a circular orbit. Use the Bohr quantization rules of Chapter 5 to find an expression for the energy levels in terms of $K$, $m$, and a principal quantum number $n$. Suppose that $m = 1{,}500$ MeV/$c^2$ and $E(n = 1) - E(n = 0) = 150$ MeV. What is $K$?

**23.** ▮▮ The difference in the mass of the $\pi$ and that of the $\rho$ lies in the fact that the one is a spin singlet and the other a spin triplet, so that an analogue of the hyperfine interaction (see Section 9–6) causes the split. Assume that the masses of the singlet states containing quarks and antiquarks of masses $m_1$ and $m_2$ are of the form

$$M_{\text{singlet}} = m_1 + m_2 - \frac{3a}{m_1 m_2},$$

while the masses of the triplet states take the form

$$M_{\text{triplet}} = m_1 + m_2 + \frac{a}{m_1 m_2}.$$

Use the masses of the $\pi$ and the $\rho$ together with the assumption that the $u$ and $d$ quarks have the same mass $m$, to obtain the values of $m$ and $a$. (Note that the extra term in the mass of the singlet and of the triplet is associated with a magnetic dipole–magnetic dipole interaction; the magnetic moment is inversely proportional to the mass. This, as explained in Chapter 9, is why the factors $3a$ and $a$ appear in the expressions for the respective masses.)

**24.** ▮▮ The lowest state of the $c\bar{c}$ system is an $\ell = 0$ state of mass 2,980 MeV/$c^2$ and total spin 0. The first excited state is an $\ell = 0$ state of mass is 3,097 MeV/$c^2$ and total spin 1. Assume that the energy can be written in the form

$$E = 2m_c c^2 + \left( \frac{\vec{S}_1 \cdot \vec{S}_2}{\hbar^2} \right) \frac{(m_c c^2 \beta^2)}{n^2},$$

where $\vec{S}_1$ and $\vec{S}_2$ are the spins of the $c$ and anti-$c$, respectively. Using the data given and the fact that the $c$ has spin 1/2, estimate $m_c c^2$ and $\beta$. Use these estimates to calculate the energy of the $n = 2$, $S = 0$, $\ell = 0$ state and the $n = 2$, $S = 1$, $\ell = 0$ state. (*Hint*: Recall that $(\vec{S}_1 + \vec{S}_2)^2 = \vec{S}_1^2 + \vec{S}_2^2 + 2\vec{S}_1 \cdot \vec{S}_2$, and use the fact that quarks have spin 1/2.)

**25.** ▮▮ Suppose the energies of the low-lying $b\bar{b}$ states have the same form as given in Problem 24, with the same value of $\beta$. Given that the lowest lying state, the singlet $\ell = 0$, $n = 1$ state, has energy 9,460 MeV, what is the mass of the $b$ quark? What are the energies of the $n = 1$ triplet state and the $n = 2$ singlet and triplet states?

**26.** ▮▮ As the energy of a neutrino goes up, the cross section it presents as a target for collisions with nucleons goes up. For the most energetic neutrinos that can be produced in accelerators the cross section is $10^{-42}$ m$^2$, which can be compared to the cross section $10^{-47}$ m$^2$ for low-energy neutrinos. Estimate how many of these high-energy neutrinos are necessary to cause 100 reactions when they are sent into a slab of iron 10 m long.

**27.** ▮ Draw the Feynman graphs for the reaction $n \to p + e^- + \bar{\nu}_e$ in terms of quarks and $W$ exchange. Do the same for the weak decay $\Lambda^0 \to p + \pi^-$.

**28.** ▮▮ Use the fundamental interactions described by Eq. (16–20) to draw Feynman diagrams for the processes **(a)** $\nu + p \to \nu + p$ and **(b)** $e^+ + e^- \to \mu^+ + \mu^-$. The last reaction can also take place with electromagnetic interactions; draw the Feynman diagram for that process.

# General Relativity

When Einstein published his paper on the special theory of relativity in 1905, only two kinds of forces were known: electromagnetism and gravity. The atomic nucleus had yet to be discovered, so the question of what held the nucleus together—the strong interaction described in Chapter 16—had not yet arisen. Although radioactivity had first been observed the previous decade, it would not be until the 1930s before it was clearly realized that many of these decays involved still another force—the weak interaction.

The question that came up and that preoccupied Einstein above all was how to bring the theory of gravitation under the umbrella of relativity. A model for this blending of ideas was what Einstein had achieved with his unification of electricity and magnetism under special relativity's umbrella. While this unification was implicit in Maxwell's equations, Einstein emphasized that electricity and magnetism were manifestations of the same force and that which manifestation one observed depended on one's state of motion. It was natural for Einstein to go on and think about gravitation. His relativistic theory of gravity of 1915—what is referred to as **general relativity**—is the subject of this chapter.

## 17–1  The Equivalence Principle

Maxwell's equations for the electric field $\vec{E}$ and magnetic field $\vec{B}$ satisfy the principle of special relativity, as Einstein already noted in the great 1905 paper in which he proposed the theory. The notion of a field is very important; if we want to study the effect of one moving charge on another, then, because of the finite speed of propagation of a signal from one charge to another—this was briefly mentioned in Chapter 16 under the name of *retardation*—a treatment without fields would require knowing the full history of the charge's motion. The idea of a field simplifies this. A charge in motion gives rise to an electromagnetic field that propagates with the speed of light and affects all other charges around it.

In 1907 Einstein undertook the study of the gravitational effect of one mass on another in a way that was consistent with the special theory. The natural procedure was to look for a gravitational analogue of the electromagnetic field. Ultimately the search for the analogue to Maxwell's equations led to equations for a gravitational field. In looking for these equations, Einstein arrived at some deep physical insights.

Because of the connection of gravity with acceleration, Einstein found it natural to turn his attention from inertial frames to accelerated frames. The special theory of relativity states that the laws of physics are the same in all inertial systems. If one tries to extend this principle to accelerated frames, one meets an immediate difficulty: Observers in accelerating frames perceive "forces" that have no identifiable sources, in contrast to, say, the way the force on an electron is related to the electric charges in its neighborhood. These forces, which are termed "fictitious forces"—the "force" that "pushes" you into the door of your car when the car takes a curve is an example—are artifacts of the observer's acceleration. An inertial observer would properly attribute the contact force on you due to the car door as the force responsible for your acceleration as you move along a circular path. Whereas you, the accelerating observer, seems to observe an *outward* force, the inertial observer sees an inward force. In November of 1907 Einstein realized that there was a connection between the "fictitious" force produced by a uniform acceleration and the force of gravity. Eventually this turned out to be the key that unlocked his theory of gravitation.

This extraordinary realization, he later recalled, "was the happiest thought of my life." As he reminisced, "I was sitting in a chair in the patent office in Bern [the Swiss National Patent Office, where Einstein was still working as a patent examiner] when all of a sudden a thought occurred to me. 'If a person falls freely he will not feel his own weight.' [If such a person is standing on a scale, and the person and the scale are together falling freely, the scale will read zero.] I was startled. This simple thought made a deep impression on me. It impelled me toward a theory of gravitation."

Let us analyze this idea of Einstein in steps. First, why does a person falling freely in a uniform gravitational field[1] of the kind that is found, approximately, near Earth's surface become weightless? To understand this phenomenon, recall Newton's law of gravitation. If we call the distance above Earth's surface $h$ and if $h$ is much smaller than Earth's radius $R$, Newton's law in this situation is approximately

$$m\frac{d^2h}{dt^2} = -\frac{GmM}{R^2},$$ (17–1)

where $M$ is Earth's mass and $m$ is the person's mass. (We are ignoring the fact that the gravitational force varies with the height above Earth.) Canceling the mass $m$ on both sides of Eq. (17–1), we learn that $d^2h/dt^2$ is a constant equal to $-GM/R^2$. Thus, neglecting such effects as air resistance, we see that all objects at Earth's surface fall freely with the same acceleration. This is the reason the scale will read zero when the person falls freely under the influence of gravity. The scale and the person will be falling with the same acceleration. The relative acceleration between them is zero, so there is no force between them, and since the function of a scale is to register the force between scale and person, the scale in this case registers nothing! If the person is wearing blinders so that he or she cannot see the unfortunate approach of Earth's surface, it will seem to that person as if the gravitational field has been switched off.

That all objects fall with the same acceleration in the gravitational field at Earth's surface was something that scientists had appreciated since the days

---

[1] Here we are using the word "field" in a loose way; we simply mean that the force of gravity is constant when we say that the field is uniform. All of the discussion in this section is restricted to that situation. What took Einstein nearly a decade to figure out after he had his "happiest thought" was how to generalize these ideas to situations where the potential varied with position—gravitational fields that are *not* uniform.

when Galileo allegedly dropped weights from the leaning tower of Pisa to see how rapidly they fell. It is something that we routinely demonstrate in introductory physics by evacuating a tube containing a coin and a feather and then watching them fall. But Einstein recognized that an unspoken assumption has been slipped into this argument: We have assumed that the masses $m$ on the left and right sides of Eq. (17–1) are the same. This is not obvious, however; the mass on the left side of the equation is a measure of the *inertia* of the object—how difficult it is to accelerate the object when a force acts on it, whether the force is gravitational or not. There is no *a priori* reason why this mass has anything to do with gravitation. Following Einstein, we can call this the **inertial mass** and designate it by $m_i$. On the other hand, the mass on the right side of the equation is a measure of the strength by which our mass couples *gravitationally* to Earth's mass. This mass Einstein called the **gravitational mass**—to be designated by $m_g$. Thus the assumption that underlies the weightlessness of objects falling in a uniform gravitational field is that these two masses are identical—that is, that $m_i = m_g$. Any experiment, such as that involving the coin and the feather, that tests whether all masses fall at the same rate demonstrates the equality of the inertial and gravitational masses, and experimental physicists have been ingenious in their efforts to make precision tests. The Hungarian nobleman Lorand von Eötvös showed in a series of experiments carried out early in the century that $m_i = m_g$ to one part in $10^9$. Over the years, the accuracy of the experiments has been steadily improving. To quantify it, let us define the quantity $\rho$ for a given substance as $\rho \equiv m_g/m_i$. Then[2]

$$|\rho(\text{wood})/\rho(\text{Pb}) - 1| < 1.0 \times 10^{-9} \quad \text{Eötvös (1922);}$$

$$|\rho(\text{Au})/\rho(\text{Al}) - 1| < 1.0 \times 10^{-11} \quad \text{Dicke–Roll–Krotkov (1964);}$$

$$|\rho(\text{Pt})/\rho(\text{Al}) - 1| < 1.0 \times 10^{-12} \quad \text{Braginsky–Parov (1974).}$$

But even before any of these experiments were carried out, Einstein was certain that this remarkable identity between gravitational and inertial mass was the key to any theory of gravitation.

The experiments that Baron Eötvös began in 1889 and continued for the next 30 years are the prototypes for all the precision experiments made to verify the equivalence of inertial and gravitational mass. Eötvös used a torsion balance with weights at either end. The arms of the balance pointed east–west. Since the experiment took place in Hungary and not at the North Pole, Earth's rotation would produce a force—indeed, a torque—that would turn the balance if the two weights had different inertial masses (•Fig. 17–1). Recall that the degree of acceleration depends on the inertial mass. Now what Eötvös did was to use as weights different materials, but in amounts that weighed the same. Since the weight of an object depends on its *gravitational* mass, his choice amounted to using weights of identical gravitational mass. With these weights at the ends of his torsion balance he looked to see if it turned through an angle, however tiny. To the accuracy of his measurements, which kept improving, he never saw such an angular shift. As different sorts of masses have been discovered this question has been restudied using them. For example when nucleons are bound in a nucleus the bound object is less massive than the nucleons that make it up by an amount $\Delta m$. To break up the nucleus we have to supply an amount of equivalent energy $E = (\Delta m)c^2$. Does this "binding-energy mass" show the same characteristics as other sorts of masses? Experiment shows that it does.

---

[2]You might worry that the ratios that follow would be unity even if $m_g$ were, say, twice $m_i$ for all materials. But in fact, that is good enough: All that matters is that $\rho$ be the *same* for all materials—we can always redefine $G$ so that the ratio is unity.

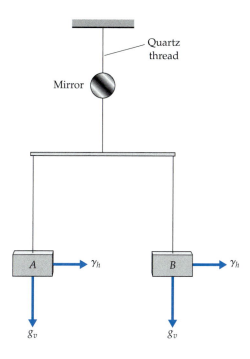

• **Figure 17–1**   Schematic design of Eötvös's experiment. The vertical forces on the masses $A$ and $B$ include Earth's gravity and vertical components of the Sun's gravity, both proportional to $m_g$ and leading to an acceleration $g_v$. Horizontal forces contain horizontal components of the Sun's gravity, proportional to $m_g$, as well as centripetal (noninertial) forces proportional to $m_i$, all leading to an acceleration $\gamma_h$. There will be no rotation about a *horizontal* axis if $A$ and $B$ have the same gravitational mass. If $m_i \neq m_g$, then the horizontal noninertial forces will give rise to a net torque about the vertical quartz thread from which the rod is suspended, and the resulting rotation can be detected through the deflection of a ray of light from the mirror.

To make the issue as vivid as possible, consider what has come to be known as the "Einstein elevator" (•Fig. 17–2). This is an imaginary box placed in empty space with a rope attached to the top of it. We imagine that there are no other masses anywhere, so that no gravitational force acts on the elevator or its contents. Now suppose there is what Einstein referred to as a "being" who can pull the elevator up so that it accelerates uniformly. That is, the "being" applies a constant force on the rope. Then the floor of the elevator will rise with a constant acceleration. An object released inside the elevator will appear to "fall" to the floor with a uniform acceleration, since the floor is rising to meet it. But if inertial and gravitational masses are exactly equal, this situation is indistinguishable from one in which the elevator is at rest—there is no "being" pulling on it—but is suspended in a uniform gravitational field that is just strong enough to make an object that is released in it accelerate just as the "being"

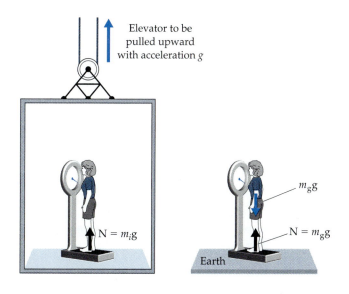

• **Figure 17–2**   (a) The Einstein elevator, a closed box in empty space away from all masses, is accelerated upward with magnitude $g$ by an external "being" pulling with some constant force on the rope attached at the top. The upward force on the passenger is $m_i g$, and that is what the scale reads. (b) For comparison, we have the same passenger standing on a scale on Earth. The scale reads $m_g g$.

made it do so. The equality of inertial and gravitational masses thus led to a new sort of relativity principle, what Einstein called the **principle of equivalence**:

*No experiment can distinguish between the behavior of a system accelerating uniformly and the behavior of that system in a uniform gravitational field. The two situations are equivalent.*

If we take this principle as a law of physics, then we have an explanation of why the inertial and gravitational masses are the same. As a matter of fact, the equivalence principle has extraordinary consequences, and we shall discuss its implications for electromagnetic radiation in Sections 17–2 and 17–3.

**Example 17–1**   Suppose you attach a mass of 10 kg to the end of a massless cord 1.5 m long whose other end is tied to the roof of an Einstein elevator. Now you straighten the cord so that it is perpendicular to the elevator roof, displace the cord through a small angle—say less than 1°—and accelerate the elevator with an acceleration $g$ of 9.8 m/s². The system will act as a pendulum. What is its period? What would it be if the mass were 20 kg?

**Solution**   By the principle of equivalence, we can replace this situation by one in which the system is in a uniform gravitational field of sufficient strength to produce the same acceleration, which is that at Earth's surface. Since the initial displacement is small we can use the small-angle approximation to the period, which is

$$T = 2\pi\sqrt{\frac{\ell}{g}} = 2\pi\sqrt{\frac{1.5 \text{ m}}{9.8 \text{ m/s}^2}} \cong 2.5 \text{ s},$$

where $\ell$ is the length of the pendulum and $g$ is the acceleration. This number is independent of the mass that is suspended. In fact we can turn the argument around and say that this fact about pendulums, apparently first noted by Galileo, is a confirmation of the principle of equivalence.

## 17–2  The Gravitational Redshift

Suppose (•Fig. 17–3a) that we have a device S (for "source") that emits radiation of proper frequency $f$ and that we place this emitter a height $h$ above a detector; we have in mind that these devices are oriented vertically on Earth's

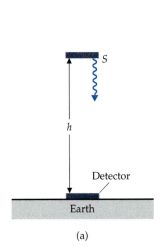

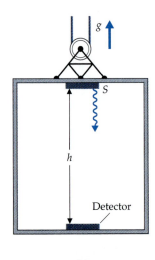

• **Figure 17–3**   (a) A source S emits radiation of frequency $f$ in the direction of a detector a distance $h$ below it. Both source and detector are located on Earth, where the gravitational field strength is $g$. (b) The same experimental setup is placed in an elevator in a gravity-free region. The elevator is accelerating upward with magnitude $g$.

(a)                    (b)

surface. The gravitational potential difference between the emitter and detector is $\Delta\varphi = gh$, with the emitter at the higher potential. According to the equivalence principle, the physics of this situation should be the same as if, instead of being on Earth's surface, the entire system were accelerated in an upward direction with acceleration $g$ (•Fig. 17–3b). If during this acceleration a pulse of radiation is emitted at time $t = 0$, then in the time $t = h/c$ that the radiation has reached the detector, the detector will have acquired a speed $v = gt = gh/c$. Thus the detector "sees" the radiation as having a blueshift to an increased frequency $f'$, which for small values of $v/c$ is approximately

$$\frac{f'}{f} = 1 + \frac{v}{c} = 1 + \frac{gh}{c^2} = 1 + \frac{\Delta\varphi}{c^2}. \qquad (17\text{–}2)$$

Consequently light moving towards Earth—or indeed any mass—acquires a blueshift; similarly, light rising away from Earth—or any mass—acquires a redshift (a decreased frequency).

In Problem 3 at the end of the chapter we discuss some numbers for experiments of the type we have described performed at Harvard University by R. V. Pound and his collaborators starting in 1960. These researchers were able to measure the frequency shift of light that was detected at an apparatus 74 feet below an emitter; they found exactly the frequency shift predicted.

When we observe light from a star, the light must first rise from the star and then fall to Earth. The star is far more massive than is Earth, so the change in potential is essentially all due to the star. (See Problem 8.) In effect, we are at zero potential while the star's surface is at the potential $-GM/R$, where $R$ is the radius of the star and $M$ is its mass. Thus the light reaching us is redshifted by an amount

$$\frac{\Delta f}{f} = \frac{f' - f}{f} = \frac{\Delta\varphi}{c^2} = -\frac{GM}{Rc^2}. \qquad (17\text{–}3)$$

Note the sign: The potential difference between emitter and detector is negative, so the frequency decreases, the expected redshift. For the Sun, $M = 2 \times 10^{30}$ kg and $R = 7 \times 10^{8}$ m. The fractional change in frequency, Eq. (17–3), is then $2.1 \times 10^{-6}$. Recent measurements carried out on the sodium D-line in the solar spectrum have confirmed the predicted redshift to an accuracy of 5%.

While we have expressed the effect of gravity in terms of the frequency shift, we can alternatively express it in terms of elapsed time. The frequency $f$ of a periodic system is related to the system's period $T$ by $f = 1/T$. Then $\Delta f = -\Delta T/T^2$, so that $\Delta f/f = -\Delta T/T$, or, with the observation that the fractional shift in period is the same as the fractional shift in elapsed time,

$$\frac{\Delta t}{t} = -\frac{\Delta\varphi}{c^2} = \frac{GM}{Rc^2}. \qquad (17\text{–}4)$$

An observer lower in a potential sees a periodic source higher in the potential oscillating faster and sees the same periodic source lower in the potential oscillating more slowly.

A periodic source that is a beam of light makes a clock, with its frequency marking off the ticks. As we saw in Chapter 2, its properties must be shared by *any* clock. Thus the behavior of light in a gravitational potential is representative of any clock in that potential. Because the frequency of light increases as the light descends towards a mass, that will also be true of any clock. Hence all other things being the same an observer at a given gravitational potential will

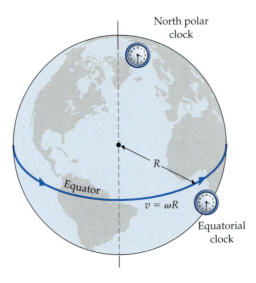

North polar
clock

• **Figure 17–4**   One clock is placed at the North Pole (the northern end of the axis of rotation of Earth) and the other is placed at rest with respect to Earth on the equator, so that it moves with respect to an inertial observer with speed $v = \omega R$.

Equator

$v = \omega R$

Equatorial clock

$R$

see a clock at a higher gravitational potential running more quickly and a clock at a lower gravitational potential running more slowly. Of course, we are considering only the effects of gravity; in computing a *net* effect we must consider *both* the effect of gravity and the effect of motion.

As we noted in Chapter 2, Einstein predicted in his 1905 paper on special relativity that two clocks, one at Earth's pole and one at the equator ( •Fig. 17–4), will not tick with equal frequencies. However—and remember, this was 12 years before general relativity—he took into account only the effect of the fact that a moving clock runs slowly, and the clock on the equator moves relative to that on the pole. Accordingly he predicted a fractional frequency shift down of magnitude $v^2/2c^2$ for the equatorial clock, where $v$ is the equatorial speed; that is, when the numbers are put in, a fractional shift of about $10^{-12}$ occurs. We now know—and he subsequently learned—that Einstein's prediction was wrong, because it does not take into account the gravitational shift.

Referring back to Chapter 2, we see that the time dilation due to the motion together with a shift due to any difference $\Delta\varphi$ in gravitational potential between equator and pole gives a net shift

$$\frac{\Delta f}{f} = \frac{\Delta\varphi}{c^2} - \frac{v^2}{2c^2}. \qquad (17\text{--}5)$$

Now although a clock at the pole and a clock on the equator might at first appear to be under exactly the same gravitational potential, that is not quite right. The clock on the equator is subject to an acceleration of $v^2/R = \omega^2 R$ due to the axial rotation, where $R$ is Earth's radius. The clock is fixed, in other words, to an accelerating frame, and because of that an observer in that frame would say that there is an additional force acting on the clock. Let us call this force $F_{\text{centrif}}$ and label the corresponding acceleration $a_{\text{centrif}}$; keep in mind that this is strictly a frame-dependent effect, not a true force with a source. This force can be derived from a potential energy, so the associated acceleration (force per unit mass) can be derived from a potential (potential energy per unit mass). What exactly do we mean by "deriving an acceleration from a potential"? A force is given in terms of a potential energy as the negative of the space derivative of the potential energy. Thus the acceleration should be given as the negative of the space derivative of the potential. Let us label this potential as $\varphi_{\text{centrif}}$ to emphasize that it is associated with $F_{centrif}$. In the case under consideration you can quickly verify that the potential which gives the acceleration is

$$\varphi_{centrif} = -\frac{1}{2}\omega^2 r^2 = -\frac{1}{2}v^2. \tag{17-6}$$

For accelerating clocks this potential must be included along with the gravitational potential $\varphi_{grav}$ to make a net potential $\varphi_{net}$. *The equivalence principle requires that we do this.*

For the clock at the equator we must include $\varphi_{centrif}$ evaluated at $r = R$; for the clock at the pole there is no acceleration and hence $\varphi_{centrif}$ is zero. Thus if the net potential for the clock at the pole is $\varphi_{net,pole} = \varphi_{grav,pole}$, then that on the equator is $\varphi_{net,equator} = \varphi_{grav,equator} - v^2/2$. Now—and here is the important point— *the net potential at the equator must equal the net potential at the pole.* If it did not, an observer sitting on the rotating Earth would see an object on the equator accelerating with respect to another object at the pole. The oceans would fly off into space—or, more soberly, the oceans, as well as Earth itself, would move to adjust things so that the net potentials did equalize at the surface. In fact, the mean ocean level is a surface of constant net potential. Mathematically, the equality of $\varphi_{net}$ at the two points reads $\varphi_{grav,pole} = \varphi_{grav,equator} - v^2/2$, or

$$\varphi_{grav,equator} = \varphi_{grav,pole} + v^2/2. \tag{17-7}$$

That is, the gravitational potential at the equator is higher than that at the pole by an amount $v^2/2$. Physically, this happens because Earth is not spherical. It bulges at the equator; a point on the equator is a little farther from the center than a point on the pole, so the potential is a little higher at the equator. In Problem 12 at the end of the chapter, we verify this fact numerically.

At this point we can return to the frequency shift, Eq. (17–5). The potential difference that appears there is the gravitational potential difference. Thus, using Eq. (17–7) for the gravitational potential difference, we obtain

$$\frac{f_{equator} - f_{pole}}{f_{pole}} = \frac{\varphi_{grav,equator} - \varphi_{grav,pole}}{c^2} - \frac{v^2_{equator} - v^2_{pole}}{2c^2} \tag{17-8}$$

$$= \frac{\varphi_{grav,equator} - \varphi_{grav,pole}}{c^2} - \frac{v^2_{equator}}{2c^2} = +\frac{v^2_{equator}}{2c^2} - \frac{v^2_{equator}}{2c^2} = 0.$$

In other words, *there is no frequency difference at all* between the signals sent out by the two clocks. Einstein was certainly wrong in 1905, and we might wonder what his reaction would have been had there been clocks accurate enough to test his prediction at that time. But as we shall see momentarily, present-day atomic clocks readily reach the accuracy necessary to show that the 1905 prediction was wrong.

The combination $\varphi_{grav} - v^2/2$, a quantity that takes both gravity and rotation into account, is known as the **geopotential**. We have argued that the geopotential must be the same at equator and pole, because otherwise there would be a relative acceleration between a point on the pole and a point on the equator. This argument holds for any two points at different latitudes (latitude is the measure of the distance from equator to pole) but the same height above the surface. The geopotential must have the same value at all points on Earth's surface. This prediction is borne out by experiment; experiments that test dependence on latitude have never observed an effect, even with clocks that are more than accurate enough to pick out the presence of either of the two terms.

In the 1970s some remarkable experiments were carried out that first confirmed these various effects in detail. In one experiment that was performed in late 1976 a C141 "Starlifter" transport plane was flown from Washington, DC, to Thule, Greenland, and back. Aboard were an array of cesium and rubidium atomic clocks as well as eight physicist "clock-watchers." These clocks keep time to nanosecond accuracy. An identical array of clocks remained on the ground. In concept this experiment is a little different from Einstein's 1905 "experiment." In that thought experiment the clocks were simply stationed at two positions on Earth's surface without taking into account how they got there. In the 1976 experiments one set of clocks flew in an airplane while the other remained on the ground. All measurements are made with respect to a nonrotating coordinate system fixed to Earth's center.

In addition to the effects of the different values of the gravitational potential due to the change in position of the airplane, the effects of the motion of the flight also have to be taken into account. For example there is one that is sometimes called the east–west effect. Because of the direction of Earth's rotation, an airplane flying west to east moves more rapidly with respect to the fixed system than one does that flies from east to west. The size of this effect decreases as the latitude increases towards the poles. Another effect involves the change in gravitational potential associated with changes in the airplane's altitude. Einstein's theory was fully confirmed in these experiments.

## Some Practical Matters

Today we routinely use clocks and optical techniques that are more than sensitive enough to register all the effects we have described. For instance, when sufficiently sensitive clocks are mounted in airplanes, altitude-dependent effects can serve as altimeters. To take another and much more important example, satellites function at altitudes that are substantially larger than Earth's radius. In a typical communication satellite time dilation effects are as large as $10^{-5}$ s a day. These are substantial corrections for an atomic clock and must be taken into account in the operation of the satellites. The NAVSTAR Global Positioning System is based on a set of orbiting clocks, and the effects of general relativity are very important to its operation. If someone tries to argue that general relativity has no practical applications, these are good counterexamples to point to.

**Example 17–2**    The NAVSTAR Global Positioning System consists of 18 satellites that orbit Earth every 12 hours at an altitude of 14,000 kilometers above Earth's surface, a height that makes the system's speed 2.96 km/s. Each satellite is outfitted with a highly stable and precise atomic clock. For the system to work properly these clocks must be synchronized with the corresponding identical Earth clocks. How would you have to adjust the satellite clock to make it synchronous with an Earth clock located at the North Pole?

**Solution**    We are interested in finding how the satellite clock runs compared to the Earth-based clock. By placing the Earth clock at the pole we eliminate any effects, either from special relativity or from the centripetal acceleration, of the Earth clock. From Eq. (17–5), the satellite-based clock differs from the Earth-based one according to the formula

$$\frac{\Delta t_{sat}}{t} = -\frac{\varphi_{sat} - \varphi_{gnd}}{c^2} + \frac{v_{sat}^2}{2c^2},$$

where $\varphi_{sat}$ and $\varphi_{gnd}$ are the gravitational potentials at the satellite and on the ground, respectively. Using the given satellite speed, we find that the special-relativistic contribution to the fractional time shift of the satellite clock is $v_{sat}^2/2c^2 = 5 \times 10^{-11}$. The sign is positive, indicating that the satellite clock runs slowly compared to the Earth-based clock.

Gravitation produces a much larger effect that is of the opposite sign. The gravitational potential of an object a distance $d$ from Earth's center is $-GM/d$. If $h$ is the height of the satellite above Earth's surface, then the gravitational contribution to the fractional time shift is

$$-\frac{GM}{c^2}\left(\frac{1}{R_E} - \frac{1}{R_E + h}\right) = -\frac{GMh}{R_E^2 c^2}\left(1 + \frac{h}{R_E}\right)^{-1} = -\frac{gh}{c^2}\left(1 + \frac{h}{R_E}\right)^{-1}.$$

Here we have reexpressed the gravitational constant in terms of the gravitational acceleration at Earth's surface, $g = 9.8$ m/s$^2$. Numerical evaluation gives for the gravitational contribution

$$-\frac{gh}{c^2}\left(1 + \frac{h}{R_E}\right)^{-1} = -\frac{(9.8\ \text{m/s}^2)(1.4 \times 10^7\ \text{m})}{(3 \times 10^8\ \text{m/s})^2}\left(1 + \frac{14{,}000\ \text{km}}{6{,}347\ \text{km}}\right)^{-1} = -4.8 \times 10^{-10}.$$

Note the sign, which indicates that the satellite clock runs fast compared to the Earth-based one. The gravitational speeding up is an order of magnitude larger than the slowing down due to the motion. At the end of one day the satellite clock will have gained 44,000 ns compared to the Earth-based clock. To adjust for this, the clocks that are to go into orbit are set to run more slowly in their own rest frames by this amount, so that once in orbit they are synchronized with the Earth clocks.

# 17–3  The Bending of Light

Not only can masses alter the frequency of light, they can also bend it. That is, light follows what we would instinctively call a *curved path* when it passes in the vicinity of masses. To see how this works we return to Einstein's elevator and use it to discuss the propagation of light in a gravitational field. •Figure 17–5 illustrates our point. In Part (a), light enters one side of the elevator and goes out the other while the elevator is accelerating uniformly upward. The floor of the elevator moves up to meet the light. Hence to an observer in the elevator, it will appear as if the light is "falling." Its trajectory will be curved [Part (b)]. But now we can replace the uniformly accelerating elevator by an elevator in a uniform gravitational field, as in Part (b). By the equivalence principle the two situations are equivalent. We conclude that *gravity bends light.*

The idea that gravity bends light precedes Einstein by a good deal: Newton believed that light was some kind of particle phenomenon and that particles of light had mass and therefore could be acted on by gravity.

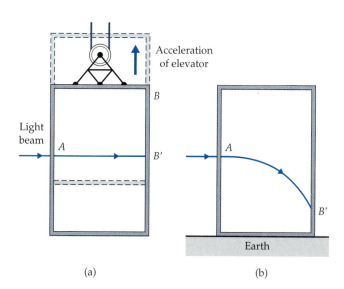

(a)                                    (b)

• **Figure 17–5**  (a) Light enters an upward-accelerating elevator at right angles through hole $A$. If the elevator were not accelerating, the beam would strike point $B$ on the other wall of the elevator, and as this wall moved farther away the spots $B$ would align to form a straight line. But with a non-zero acceleration, the elevator has moved further upwards as the beam crosses it and strikes $B'$, lying below $B$. As the distance to the far wall varies, and if the acceleration is constant, the spots $B'$ would form a parabola. (b) The equivalent picture of an elevator in a constant gravitational field must have the beam strike at $B'$. The light beam must "fall" in the field.

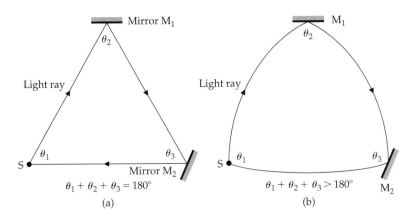

**• Figure 17–6** (a) Light travels in a closed triangular path in a gravity-free region of space. The angles add to 180°. (b) Light travels in a closed triangular path in a region of space with masses present. Because light falls in the presence of a gravitational field, the angles do not add to 180°.

To get a feeling for Einstein's approach, let us imagine that we construct a giant triangle in outer space made up of light beams (•Fig. 17–6). The vertices of the triangle are a source S and two mirrors $M_1$ and $M_2$. If there is no gravitational field around, the sum of the interior angles of this imaginary triangle will be 180° (•Fig. 17–6a). This is one of Euclid's theorems, so we can say that the geometry of our imagined space is "Euclidean." Now let us suppose that there is a uniform gravitational field in the region of our triangle. In this case, the light beams will be curved. Direct measurement of the curvature would be difficult, but there is one simple and direct way to verify it: We remeasure the interior angles of our triangle (•Fig. 17–6b). They will no longer add up to 180°. We can think of other situations in which the sum of the interior angles of triangles is not 180°. For example, if we draw a triangle made up of great circles on Earth's surface—think of an ordinary globe—the sum of the interior angles will be greater than 180° (•Fig. 17–7a), and this is true for any triangle we draw on our globe. We say that the space formed by the globe's surface is **non-Euclidean** and, in particular, when the sum of the interior angles of a triangle on the space is greater than 180°, that the space has a **positive curvature**. On the other hand, if we draw a triangle on the surface of a figure that looks like a saddle, the sum of its interior angles will be less than 180° (•Fig. 17–7b). Such a space is said to have a **negative curvature**.

A remarkable intellectual leap took place regarding these issues once general relativity was fully formulated. When the triangle of light beams is placed in a region where there is mass, it deforms as we have described. But what is it that is deforming? One commonly says that light follows straight line paths. Indeed, this *defines* what we mean by a straight line. But if masses are present these straight lines will no longer make Euclidean triangles. This is what we

**• Figure 17–7** (a) A triangle drawn on the surface of a sphere. The angles add up to a number larger than 180°. (b) A triangle drawn on a saddle-shaped surface has angles that add to less than 180°.

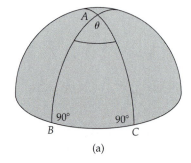

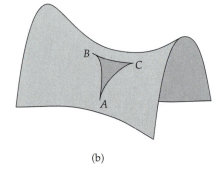

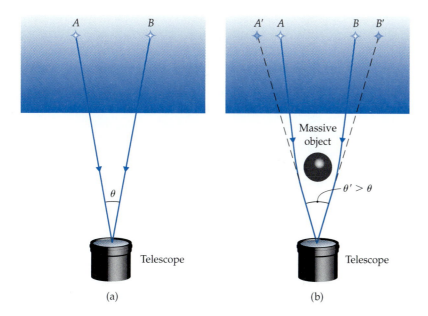

• **Figure 17–8** (a) The rays from two stars $A$ and $B$ subtend an angle $\theta$ at a telescope. (b) When the two rays straddle a massive body, they are deflected. An observer at $T$ will see the apparent positions of the stars at $A'$ and $B'$ and measure the subtended angle as larger than $\theta$.

mean by saying that space is positively curved or negatively curved when masses are present. "Straight lines" no longer obey the axioms of Euclidean geometry. For example, through a point external to a given "straight line" it may be possible to draw many lines or no lines parallel to the given line. The actual nature of the curvature depends on the configuration of the masses present. In other words, what general relativity describes is the way in which the presence of masses bends space. As we have seen, both space and time are affected. The geometry involved is four dimensional, with three space dimensions and one time dimension. The "straight lines" that are involved are really trajectories of light in this four-dimensional world. We are far from Newton's idea of a force rather mysteriously associated with mass. In a real sense, there are no forces in Einstein's theory—only modifications of space and time due to the presence of masses.

While Einstein did not understand the full implications of his idea at the time he enunciated the equivalence principle, he understood enough to propose a test. According to general relativity, when the light from a star skirts the Sun it is very slightly bent by the Sun's gravitation. Hence Einstein proposed that astronomers photograph a known field of stars near the Sun at a time of total solar eclipse, when the otherwise blinding light from the Sun is shielded. A comparison with a photograph of the same stars in the absence of the Sun would show a shift in apparent positions (•Fig. 17–8). When Einstein first made this proposal in 1911, he argued for it in very much the way we have here. In other words, he considered only, in effect, the curvature of space. He left out the fact that gravity also alters time. Thus he predicted a bending that was, in fact, only half the actual bending.

For various reasons—the First World War broke out, for one—this early prediction was never tested. In 1916 Einstein published the full theory, which included the effects of gravity on both space and time. The theory also allowed one to treat nonuniform gravitational fields and arbitrary motions. At every point in space–time there is a different geometry that is determined by the gravitational fields. In fact, in very strong gravitational fields space and time get mixed up. It was this complication that took several years of intense effort for

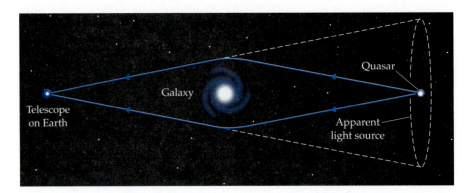

• **Figure 17–9** The lensing effect of a distant galaxy, which distorts the image of a point source behind it into an apparent ring shape.

Einstein to overcome. In his 1916 paper he predicted that the gravitational field of the Sun would bend the light from a star whose light "grazed" the Sun by 1.74 seconds of arc. In 1919 a solar eclipse occurred that allowed Einstein's prediction to be confirmed. The result of the measurement made Einstein an international celebrity. The theory has been confirmed many times since.

The bending of light around masses has become a tool of modern astronomy through what are known as **gravitational lenses**. If we have a distant source with an intermediate mass intervening between us and the source, as in •Fig. 17–9, then light from the source is bent around the intermediate mass, which thereby acts as a lens. Depending on how the source and the intermediate mass are aligned, and depending on the masses and distances involved, the light from the source will appear to us as a ring, an arc (a portion of a ring), or a multiple image. These effects are today routinely observed; we show a series of examples in •Fig. 17–10, including an example of a full ring. The images

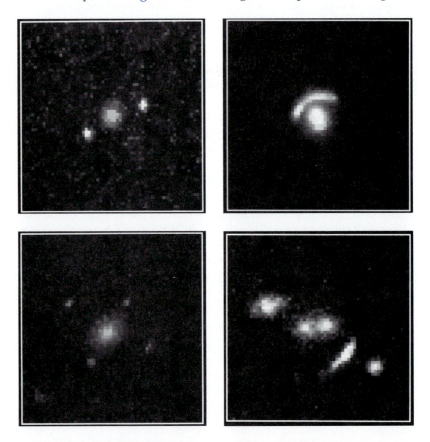

• **Figure 17–10** Some examples of gravitational lensing.

give us information about the distances and masses of the objects that produce the distorted images, as well as about the sources themselves. A large-scale experiment is now underway to observe the presence of dim intergalactic stars by observing their lensing effect on the light from stars in nearby galaxies. As the intermediate stars pass between us and the source stars the light from those stars flares up as it becomes focused more strongly on our telescopes.

We can make a simple estimate of the angle of deflection of light due to a mass. We take the "force of gravity" on a photon of energy $E$ a distance $r$ from a mass $M$ to be of Newtonian form $F = (GM/r^2)\mu$, where $\mu$ is an effective photon mass that we suppose is equal to $E/c^2$. If the closest distance of approach of the photon is $R$, then the time that the photon spends nearest the mass, which is where we expect the strongest influence, is $t \cong 2R/c$ and hence the impulse, or change of momentum, given to the photon is

$$\Delta p \cong Ft = \frac{GM}{R^2} \frac{E}{c^2} \frac{2R}{c}.$$

The corresponding deflection is given by

$$\Delta p = p(1 - \cos\theta) \cong \frac{p\theta^2}{2}.$$

In the last step we have assumed that the deflection is small. If we set $p = E/c$ and solve for $\theta$, we obtain

$$\theta^2 = \frac{4GM}{Rc^2}$$

Finally we can convert this equation to one involving a distance $D$ to an observer—someone who sees the photon arrive as in •Fig. 17–11. Geometry gives $R/D = \theta$, or $R = D\theta$. When we substitute this we find

$$\theta^3 = \frac{4GM}{Dc^2}.$$

This result shows that if we have an independent means of learning the distance $D$, then measurement of the deflection angle $\theta$ allows us to estimate the mass $M$.

## The Precession of the Perihelion of Mercury

We would be remiss not to mention one other historically important test of general relativity, one that is closely connected to the distortion of space associated with large masses. This is an effect that leads to a small but significant difference from the predictions of Newtonian gravitation, according to which the force due to the (spherical) Sun on another spherical body such as Mercury is an inverse square law. The orbits followed by a planet such as Mercury under the influence of a perfect inverse square law have a very special feature: They are closed. You cannot see such an effect in a circular orbit, but you will if the orbit is elliptical. The perihelion of a planet's (elliptical) orbit is the point of

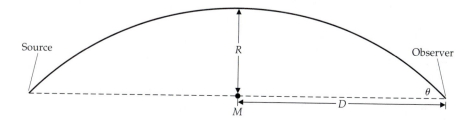

• **Figure 17–11**   The relation between the angle $\theta$ subtended by an image of a ring created by lensing, the distance $D$ to the lensing galaxy, and the mass $M$ of the lensing galaxy.

closest approach to the Sun, and the closure of the orbit means that once a planet has passed perihelion and come all the way back to that point, the curve closes, retracing itself over and over. On the other hand, if the force is not a pure inverse square law, then the orbit does not quite close—it is not quite a perfect ellipse. The perihelion shifts a little, and this shift is systematic, so that if the perihelion shifts by, say, 0.1° with each orbit, then after 100 orbits, the perihelion will have shifted by an angle of $100 \times 0.1° = 10°$. Generally, over many orbits, the perihelion will slowly rotate about the sun in a phenomenon called the precession of the perihelion.

Now there are many influences on a real planet—including the gravitational influences of other planets—that, in effect, modify the force on the planet from the pure inverse square law due to the Sun. The observed precession rate for the perihelion of Mercury is 574″ of arc per *century*. In one of the heroic efforts of 19th-century physics, the French astronomer Urbain Le Verrier had systematically accounted for all the possible classical reasons for this precession. He ended up with a tiny amount left over that he could not account for, some 43″ of arc per century. Only when general relativity was formulated could one account for this discrepancy. In effect, general relativity provides corrections to the Newtonian inverse square law and hence a precessing perihelion. The effects of general relativity on Mercury are relatively large because of Mercury's proximity to the Sun. This test was one that general relativity, which accounted precisely for Le Verrier's discrepancy, passed with a perfect grade.

## 17–4  Gravitational Radiation

Einstein predicted other effects, one of the most interesting of which is **gravitational radiation**. According to general relativity, when masses accelerate, disturbances propagate in space. The phenomenon is something like the emission of electromagnetic radiation that occurs when electric charges accelerate, except that in the case of an accelerating mass what propagates are distortions in space itself. The disturbances form periodic *gravitational waves* if the acceleration is periodic. The waves can carry energy away, and indeed astronomers have seen orbiting double neutron stars slow down as they lose energy from the emission of gravitational waves in just the way that general relativity would predict. This slowdown provides indirect evidence for gravitational waves.

We can get some idea of the size of these effects by quoting a formula that comes from the equations of general relativity: The rate at which energy is radiated (the power) by two point masses $M_1$ and $M_2$ separated by a distance $d$ and rotating about each other in circular orbits with an angular frequency $\omega$ is

$$P = \frac{32}{5} \frac{G}{c^5} \omega^6 d^4 \mu^2, \qquad (17–9)$$

where $\mu = M_1 M_2 / (M_1 + M_2)$ is the reduced mass. This formula is an analogue of the Larmor formula for the power radiated by a rotating charge [Eq. (13–6)].

To illustrate a typical application, we consider a system of two neutron stars, each of mass $1.4 M_0$, where $M_0$ is the mass of the Sun, in circular motion about their center of mass with a measured period of $T = 7.75$ hr. If the system is $1.5 \times 10^{20}$ m (15,000 lt-yr, part way across the galaxy) from Earth, we can find the energy that falls in one second on 1 m² of Earth's surface from the gravitational radiation emanating from the system. To apply Eq. (17–9) we must know the separation $d$ of the two neutron stars. Because the motion is nonrelativistic, we can use Newton's second law, which for two objects in circular motion about their center of mass takes the form

$$\mu\omega^2 d = \frac{GM\mu}{d^2},$$

where $\mu$ is the reduced mass (which cancels from both sides of this equation), $M$ is the sum of the masses, and $\omega$ is the angular frequency of the motion. If we use the relation $\omega = 2\pi/T$, we can solve the equation for $d$, giving

$$d^3 = \frac{GM}{\omega^2} = \frac{GMT^2}{(2\pi)^2}$$

$$= \frac{(6.67 \times 10^{-11}\text{ m}^3/\text{kg}\cdot\text{s}^2)(2 \times 1.4 \times (2 \times 10^{30}\text{ kg}))(7.75\text{ hr} \times 3{,}600\text{ s/hr})}{(2\pi)^2}$$

$$= 7.4 \times 10^{27}\text{ m}^3.$$

The cube root of this result is $d = 1.95 \times 10^9$ m. With these numbers, Eq. (17–9) gives us a total radiated power $P = 6.44 \times 10^{23}$ J/s. While this is a lot of power, our square meter on Earth is far away from the source. If we assume that the power is radiated uniformly in all directions, then the energy per unit time falling on an area $A$ a distance $D$ away is

$$\frac{PA}{4\pi D^2}.$$

With $A = 1$ m$^2$ and $D = 1.5 \times 10^{20}$ m, we find an energy per unit second of $3.4 \times 10^{-18}$ J/m$^2$/s. Numbers such as this suggest why the detection of gravitational waves is difficult.

The American astronomers J. H. Taylor and R. A. Hulse discovered a binary neutron star system ("a binary pulsar") with properties like those we have described. They observed that $\omega$ changes steadily with time. As the system radiates, the distance between the neutron stars decreases with time and the angular frequency increases steadily. The measurement of the change in $\omega$ over a period of years is in agreement with what is to be expected if the energy loss is due to gravitational radiation, and this observation may be viewed as a confirmation of the existence of that radiation.

**Example 17–3**   The binary system whose physical parameters we have just described represents roughly the binary pulsar studied by Taylor and Hulse. What is the rate of change of the rotational period $T$ due to the power radiated by gravitational radiation for this system? (*Hint*: The total energy of a pair of stars rotating about their center of mass is $E = \mu v^2/2 - GM\mu/d$. When the Newtonian equations of motion are used, the total energy takes the more compact form $-GM\mu/2d$. As we have seen, the distance $d$ is given by $d^3 = GM/\omega^2$.)

**Solution**   If we equate the rate of change of $E$ to the radiated power $P$ and then in turn express $E$ in terms of $T$, we can find the rate of change of $T$. First of all, using $\omega = 2\pi/T$ we have

$$d = \left(\frac{GM}{\omega^2}\right)^{1/3} = \left(\frac{GMT^2}{4\pi^2}\right)^{1/3},$$

and we can eliminate $d$ from the equation for $E$ in favor of $T$. We have

$$E = -\frac{GM\mu}{2d} = -\frac{1}{2}(2\pi GM)^{2/3}\mu T^{-2/3}.$$

Thus the rate of change of $E$ is

$$\frac{dE}{dt} = -\frac{1}{2}(2\pi GM)^{2/3}\mu\left(-\frac{2}{3}T^{-5/3}\frac{dT}{dt}\right) = \frac{1}{3}(2\pi GM)^{2/3}\mu T^{-5/3}\frac{dT}{dt}.$$

But from Eq. (17–9),

$$P = \frac{32}{5} \frac{G}{c^5} \omega^6 d^4 \mu^2 = \frac{32}{5} \frac{G}{c^5} \left(\frac{2\pi}{T}\right)^6 \left(\frac{GMT^2}{4\pi^2}\right)^{4/3} \mu^2 = \frac{(32)(2\pi)^{10/3}}{5} \frac{G^{7/3} M^{4/3} \mu^2 T^{-10/3}}{c^5}.$$

Equating $P$ and $dE/dt$, we can solve for the rate of change of $T$:

$$\frac{dT}{dt} = \frac{(96)(2\pi)^{8/3}}{5} \frac{G^{5/3} M^{2/3} \mu}{c^5} T^{-5/3}.$$

Every term on the right side is known, and when we insert the numerical values on that side we obtain

$$\frac{dT}{dt} = 2.0 \times 10^{-13}.$$

This rate corresponds to a change in the period of about $6 \times 10^{-6}$ per year. Remarkably, it is possible today to measure such changes in orbital periods of binary pulsars with Doppler shift techniques and to verify that these systems lose energies at a rate that is in agreement with what is predicted by general relativity.

### The LIGO Project

The direct detection of gravitational waves presents a tremendous challenge to experimentalists. First of all, the overall acceleration caused by a gravitational wave is not observable, since the laboratory and Earth are accelerated by the same amount. Thus it is only the "tidal" effects—the differences between gravitational effects at different locations—that are in principle measurable. A gravitational wave traveling along a z-axis will stretch or compress a bar that has been set up as a type of antenna along the corresponding x- or y-axis. This effect is somewhat analogous to that of an electromagnetic wave on an antenna, within which the wave causes charges to move in the direction transverse to the wave's propagation. An important difference between the two situations is that instead of inducing a dipole field, as is the case for electromagnetic waves, the gravitational field induces a quadrupole field. In effect, this means that a wave propagating in the z-direction squeezes matter along the x-direction and stretches it in the y-direction and half a period later reverses this process. There is another difference as well: The most dramatic sources of gravitational radiation—collisions of neutron stars or of black holes (see Section 17–5)—generate waves of frequencies below 100 Hz, corresponding to a wavelength larger than $\lambda = c/f = 3{,}000$ km. Finally, the power reaching Earth—even from the spectacular events we have considered—is so tiny that the induced strains in a typical antenna bar give a fractional stretching $\Delta L/L \cong 10^{-18}$!

The pioneering attempts by the American physicist Joseph Weber, who is rightly called the father of gravitational wave detection, to measure the strains in a one-ton piece of aluminum in the 1960s proved unsuccessful. Continued work has led to the current Laser Interferometer Gravitational-wave Observatory (LIGO) project.

In the LIGO project, a measurement of $\Delta L$ is to be carried out with a device like a Michelson interferometer. The basic idea (•Fig. 17–12) is that free-hanging masses at the ends and at the corner of an L-shaped structure move relative to one another when a gravitational wave hits the structure. The movement is measured by using interference between a precisely controlled laser beam that is split and sent through both arms. In order to get a measurable $\Delta L$, it is necessary to make $L$ as large as possible, so the length of the arms is 4 km, a length that is optimized for financial and geological reasons. The effective length is

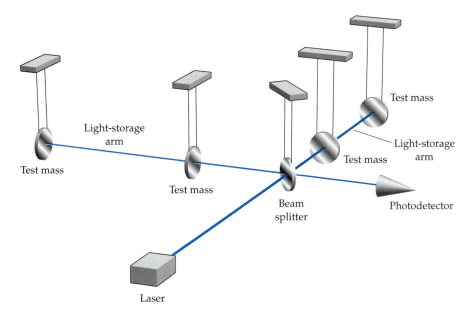

• **Figure 17–12**   Schematic drawing of the interferometer that forms the basis of the LIGO project detector. (From Weiss, Rainer. "Gravitational Radiation." *Reviews of Modern Physics* **71** (1999): Fig. 1, p. S187.)

increased manyfold by using mirrors to make the light bounce back and forth between the corner points and the endpoints hundreds of times. With the mirrors, the effective length is on the order of a quarter of a typical gravitational wavelength; this gives a maximum interference effect. The laser light is confined to vacuum tunnels 1 m in diameter, in order to minimize scattering. To avoid spurious effects caused by local gravitational disturbances (microearthquakes, trucks, etc.), two sites are under construction, one in Washington state and one in Louisiana (•Fig. 17–13). Both detectors will have to register an event in order to establish that it has an extraterrestrial origin. The device should be precise enough to detect a change in arm length of a *single nuclear radius*, and this should be enough to measure many predicted effects.

The LIGO facilities, which are to start collecting data in 2002, are designed to operate continuously, since big bursts of gravitational radiation are expected to be rather rare. Long-range plans include a third detector, and in the long run the system of detectors should be able to differentiate between different causes of the gravitational radiation. The project pushes many different technologies to their limits: Lasers whose frequencies can be stabilized over long periods, large vacuum systems, and mirrors with very small losses are just some of these devices. A 40-m-long prototype instrument has already been constructed. Similar interferometers are planned in Germany, Japan, and Italy, and all the projects involve large international collaborations. These projects will open a whole new window to our universe.

• **Figure 17–13**   Photograph of the LIGO facility near Hanford, Washington.

## 17–5   Black Holes

Although Einstein's theory is now a part of all discussions of black holes, there is a remarkable history behind them. Something like a black hole first appeared in Newtonian gravity. Several scientists after Newton asked what would happen if one increased the mass $M$ of a star from which light is emanating while holding its radius $R$ fixed. (Alternatively, one could decrease the radius while holding the mass fixed.) If, as Newton's followers might have done, we think of light as consisting of massive particles of mass $m$

that move, at least when they are emitted, with the speed of light, then the total energy $E$ of a "light particle" would be

$$E = \frac{1}{2}mc^2 - \frac{GmM}{R}. \tag{17–10}$$

We can see from this equation that we can increase the size of the potential-energy term either by increasing the stellar mass or decreasing the stellar radius (or both). If we change this term enough, the total energy would become negative, meaning that the light cannot escape. The case $E = 0$ is a critical one. We can set $E = 0$ and solve for the star's radius:

$$R_S = \frac{2GM}{c^2}. \tag{17–11}$$

This radius, which depends on $M$, is known as the *Schwarzschild radius*, after the German astronomer Karl Schwarzschild, who, while serving in World War I, was the first person to solve Einstein's equations of general relativity for a spherically symmetric mass distribution. The subscript on $R$ acknowledges Schwarzschild's work.

The Newtonian picture cannot be correct, because it assumes that light rises and then falls like a rock thrown from Earth's surface. The correct picture, which emerges from general relativity, is quite different. Schwarzschild was able to calculate the rate at which clocks tick in a gravitational potential of the form $\varphi(r) = -GM/r$, as would be produced[3] by a mass $M$ at the origin or more generally by a spherically symmetric distribution of matter with total mass $M$ inside the radius at which we observe the potential. One can translate his findings into the following description:

A clock at a point $P$ a distance $r$ from the mass $M$ may, by the equivalence principle, be treated as if it were in a gravity-free region but undergoing an instantaneous radially outward acceleration $g = GM/r^2$. Schwarzschild effectively calculated how the tick of the clock—call it $\Delta t_P$—varies with $r$ compared to some "standard clock." The latter will be at rest at point $P$ in this gravity-free region (and therefore is in an inertial frame). If the tick of this standard clock is $\Delta t_0$, then, according to Schwarzschild,

$$\Delta t_P = \Delta t_0 \sqrt{1 - \frac{2GM}{rc^2}}. \tag{17–12}$$

In other words, the accelerating clock runs slowly compared to the standard clock, and by the equivalence principle, the clock in the gravitational field runs slowly as well. Note that we can just as well consider $\Delta t_P$ to be the elapsed time on the $P$-clock if we regard $\Delta t_0$ as the elapsed time on the standard clock.

We can check a known limit here: If the quantity $2GM/rc^2$ is small, we can expand the square root in Eq. (17–12) to find an approximate formula for the time dilation, namely

$$\Delta t_p \cong \Delta t_0 \left(1 - \frac{2GM}{2rc^2}\right) = \Delta t_0 \left(1 - \frac{GMr}{r^2c^2}\right) = \Delta t_0 \left(1 - \frac{gr}{c^2}\right).$$

We can now use this equation to compute the approximate fractional difference between the ticks of our two clocks:

---

[3] We are assuming that this classical result holds in general relativity. It can in fact be proven with the full machinery of the theory.

$$\frac{\Delta t_0 - \Delta t_P}{\Delta t_0} = \Delta t_0 \left(1 - \left[1 - \frac{GM}{rc^2}\right]\right) / \Delta t_0 = \frac{GM}{rc^2}.$$

This is exactly Eq. (17–4), a result that was also derived under an approximation in which the effects of relativity were small.

Let us return to the correct formula for the time of the $P$-clock as seen from outside, Eq. (17–12). We see from this equation that something peculiar happens when

$$r = R_S = \frac{2GM}{c^2}, \tag{17–13}$$

where the Schwarzschild radius $R_S$ [Eq. (17–11)] has reappeared. At this point the $P$-clock looks to the outside observer as if it had stopped ticking or, equivalently, as if the frequency of the ticks had gone to zero. The clock has become frozen, and in fact these stars used to be known as "frozen stars." If the ticks represent the frequency of radiation emitted by the clock, then that radiation has been redshifted all the way to zero frequency. We no longer see such radiation, and that is why we use the word "black hole." (This term was coined by the American physicist John Wheeler [•Fig. 17–14], who has made many contributions to all areas of modern physics, but whose deepest influence has been in the revival of gravitational theory and the search for ever deeper implications of Einstein's work.) Any radiating system that crosses $R_S$ becomes invisible, although it is worth keeping in mind that because of the "freezing" we will never see any system actually cross $R_S$.

• **Figure 17–14** John Archibald Wheeler, one of the important figures in American physics of the 20th century.

**Example 17–4**   Find the Schwarzschild radius $R_S$ for **(a)** a mass of 12 solar masses and **(b)** a 1-kg mass.

**Solution**   **(a)** For $N$ solar masses, we have

$$R_S(N) = N \times \frac{2GM_{Sun}}{c^2} = N \times \frac{2(6.67 \times 10^{-11} \text{ m}^3/\text{kg} \cdot \text{s}^2)(2 \times 10^{30} \text{ kg})}{(3 \times 10^8 \text{ m/s})^2}$$

$$= N \times 3 \times 10^3 \text{ m}.$$

For $N = 12$, the result is 36 km. **(b)** For a mass of 1 kg, elementary substitution gives $R_S = 1.5 \times 10^{-27}$, only $10^{-12}$ times the proton radius!

How would an outside observer describe things *inside* the black hole? Our equations become meaningless for $r < R_S$. It is possible that this technical matter will be handled by noting that a system of coordinates $(t, \vec{r})$ may not be useful inside the horizon. But to an observer falling into the black hole, there is nothing special about $R_S$. It is not a particularly pleasant place to be, because there are important Newtonian tidal effects (see Example 17–5) and it is not very satisfying to know that no report to the outside world would be possible, but there is no "discontinuity" at the horizon.

**Example 17–5**   A remote probe is sent to investigate the vicinity of a black hole of 12 solar masses. The probe sends information back as it falls towards the horizon. (Of course the probe sends the information more and more slowly as it approaches the horizon.) The probe is 2 m long, and it falls lengthwise in—refer to the part closest to the hole as the base and the part farthest from the hole as the top. What is the difference between the $g$-value at the base and that at the top at a distance of 10 times the Schwarzschild radius?

**Solution**   The $g$-value is the force per unit mass. If we let $M$ be the mass of the black hole and $L$ be the length of the probe, and if we suppose that the center of the probe is at a distance $10R_S$, then ordinary Newtonian physics states that the $g$-values at top and base are, respectively,

$$g_{t,b} = \frac{GM}{(10R_S \pm L/2)^2} \cong \frac{GM}{100R_S^2}\left(1 \mp \frac{L}{10R_S}\right).$$

In the last step we have made an expansion based on the fact that $L \ll R_S$. The difference is

$$\Delta g = g_b - g_t \cong \frac{GML}{500R_S^3}.$$

For a numerical evaluation, we can use the result of Example 17–4, which shows that for 12 solar masses $R_S = 36$ km. Thus

$$\Delta g = \frac{(6.7 \times 10^{-11}\, \text{m}^3/\text{kg}\cdot\text{s}^2)(12 \times 2 \times 10^{30}\, \text{kg})(2\, \text{m})}{500 \times (3.6 \times 10^4\, \text{m})^3} = 2 \times 10^5\, \text{m/s}^2.$$

The acceleration due to gravity on Earth is 10 m/s²; by comparison, the value $\Delta g$ that we have calculated is so large that no probe will withstand the difference in acceleration between base and top. And the problem is even worse closer to the black hole!

## Astronomical Black Holes

Do black holes exist? Einstein didn't think so, but he never developed any method for testing his view. We now believe not only that black holes exist, but that they are a common feature of the cosmos. While mathematically a black hole can have any mass, the cases of most interest are astronomical objects. In Chapter 10 we discussed the formation of white dwarfs and neutron stars as end products of stellar evolution, forming after nuclear burning stops. White dwarfs form for stars of masses up to about 1.5 solar masses (the Chandrasekhar limit); these stars do not collapse further because the degeneracy pressure of electrons is sufficient to hold back a gravitational collapse. Up to a mass of approximately 3.5 solar masses—we do not know enough about the interaction between neutrons to calculate this number with an accuracy of better than about half a solar mass—neutron stars form, with the degeneracy pressure due to their neutrons preventing a total collapse. Beyond this mass, nothing can hold back a gravitational collapse. That is not to say that all stars with an initial mass larger than 3.5 solar masses will form black holes, because in any collapse following the end of burning a lot of matter can be blown off, leaving neutron stars. Nevertheless, any compact object with a mass greater than six solar masses is generally thought to be a black hole. It is believed that huge black holes lie at the centers of galaxies, including our own; such black holes may have the mass of one to 100 *billion* Suns. Note that for $10^{11}$ solar masses, $R_S = 3 \times 10^{14}\, \text{m} \cong 3 \times 10^{-2}\, \text{lt-yr}$, a very small size on a galactic scale.

How many black holes are there in our galaxy? Aside from a huge one at the center, whose formation would involve the accretion of matter over a long period of time, stellar black holes that are associated with collapsed stars are most likely formed in supernovas. A supernova is a violent collapse, followed by an explosion, of a star that has finished burning, and these events leave behind either a neutron star or a black hole, depending on the mass of the collapsing star. There may be several such events per century in the galaxy, and since they may have been happening for $10^{10}$ yr, there may be as many as $10^8$ black holes of this type in our galaxy.

## Seeing Black Holes

Even though no signal can emerge from a black hole, it still acts as a source of gravitational potential. This means that a black hole and a star can form a planetary system like Earth and the Sun and that the same Keplerian laws would apply provided only that the speeds are nonrelativistic. For purposes of illustration, consider a circular orbit with a black hole of mass $M$ and a star of mass $m \ll M$. Then the star orbits the (nearly) stationary black hole with a period $T$ determined from the formula

$$ma = m\omega^2 R = m\left(\frac{2\pi}{T}\right)^2 R = \frac{GMm}{R^2},$$

so that

$$T = 2\pi\sqrt{\frac{R^3}{GM}}. \qquad (17\text{--}14)$$

As long as the orbital plane of this system is not oriented perpendicular to us, the period can be determined by measurements of the Doppler shift of the luminous star—in one period it moves away from and then towards us. If in addition the diameter of the orbit can be resolved by a high-resolution telescope, Eq. (17–14) shows us that $M$ can be determined. If this mass is dark and in addition is very massive, then it is a good candidate for a black hole.

It is not necessary that the matter orbiting the black hole be in the form of a single star; masses of luminous gas can orbit the black hole in the form of an *accretion disk*. Measurements of the Doppler shift and the radius of the disk also allow the determination of the mass of the black hole. As an example, consider the galaxy M87, which was studied with the Hubble Space Telescope. This galaxy contains a disk of luminous matter at its center. Doppler shift measurements indicate that the outer part of the disk rotates at a speed 500 km/s about the center. This part of the disk is 60 lt-yr, or about $6 \times 10^{17}$ m, from the center about which it rotates. The mass about which the disk orbits is therefore determined by the equation $v^2/R = GM/R^2$, or

$$M = \frac{v^2 R}{G} = \frac{(5 \times 10^5 \text{ m/s})^2 (6 \times 10^{17} \text{ m})}{(6.7 \times 10^{-11} \text{ m}^3/\text{kg} \cdot \text{s}^2)} = 2.2 \times 10^{39} \text{ kg},$$

or about $10^9$ solar masses. Observations show that the region in the center contains very few visible stars, so that the most plausible conclusion is that this huge amount of matter is a black hole.

We can do better than the indirect evidence just described. Because of the powerful tidal effects described in Example 17–5, matter from a star in orbit about a black hole will be stripped off. This matter, which is fully ionized, will emit radiation, a strong component of which will be X rays. The field of X-ray astronomy, which began in the 1970s, has been essential to the study of black holes. In fact the first convincing candidate for a black hole was an X-ray source in the constellation Cygnus known as *Cygnus X1*. Its X-ray emissions are variable, with a time scale on the order of 1 s or less. This is significant because it means that the radiating source cannot be larger than 1 lt-s, or $3 \times 10^8$ m, which in turn means that the region from which the X rays come is very compact— remember that the radius of the Sun is about $10^9$ m. There is a visible star at Cygnus X1, and using the techniques we have described astronomers were able to show that it is in orbit about a dark companion with a mass of more than eight solar masses.

Here is what happens to make the X-ray source at Cygnus X1. The matter pulled from the visible star by the dark companion—at this point, given the large mass and compact size of the system, we'll call the companion a black hole—goes into orbit about the black hole, due to the fact that it already has some orbital motion associated with the star. The matter then forms an accretion disk. The interaction between the charged particles that make up the disk heats the disk to the point where it emits radiation strongly in the X-ray part of the spectrum. This hypothesis can be verified with calculations using ordinary classical physics.

You might worry that the evidence for the black hole remains circumstantial. A "smoking gun" for black holes would be the observation of a horizon, or of matter disappearing into the hole. While observing this is difficult, there are some hints of just such observations. Neutron stars with visible companions also give rise to compact accretion disks. But when matter from the accretion disk loses enough energy to fall onto the surface of the neutron star there should be some signal of heating at the surface of the star—radiation of some kind. In contrast, matter falling on a black hole gives no signal of any kind. A comparison of known neutron stars with companions and of conjectured black holes with companions appears to show differences that point to the existence of a horizon.

### Are Black Holes Forever?

• **Figure 17–15**   Stephen Hawking has made many contributions to the theory of gravitation. He is one of the leading researchers into the difficult and as yet unsolved problem of the quantum theory of gravitation.

Until the 1974 work of the English astrophysicist Stephen Hawking (•Fig. 17–15), it was said that any light—or, indeed, anything else—that was trapped at or inside the Schwarzschild radius would never get out. But by bringing relativistic quantum mechanics into play Hawking noted that black holes can radiate energy from the Schwarzschild radius. He proposed a mechanism in which virtual particle–antiparticle pairs are created in the strong gravitational field of the black hole (•Fig. 17–16). One member of the pair can fall into the black hole before the pair has time to annihilate. That particle is never seen again, while its partner remains outside the black hole and gives rise to radiation. Hawking showed that the radiation—we refer to it as *Hawking radiation*—is of blackbody form, characterized by a temperature given by

$$kT = \frac{\hbar c^3}{8\pi GM}. \tag{17–15}$$

Over the course of time the hole would evaporate by this mechanism. We can use Eq. (17–15) to estimate the lifetime of a black hole. We use the blackbody expression for power radiated at a temperature $T$ from a surface of radius $R_S$ (and area $4\pi R_S^2$):

$$P = (4\pi R_S^2)\sigma(kT)^4. \tag{17–16}$$

Here $\sigma = \pi^2/(60c^2\hbar^3)$ is the Stefan–Boltzmann constant. (See Chapter 4.) But this power also describes the rate at which the black hole loses mass: $P = -d(Mc^2)/dt$. If we use Eq. (17–10) for $R_S$ and Eq. (17–15) for temperature in Eq. (17–16), we find that

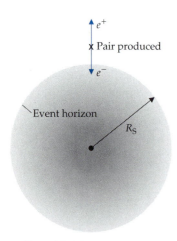

$e^+$

✕ Pair produced

$e^-$

Event horizon

$R_S$

• **Figure 17–16**   Mechanism for black-hole radiation. A virtual $e^+$–$e^-$ pair is produced by the strong gravitational field, analogous to the production of a virtual pair by an external magnetic field. One member of the pair is trapped in the black hole, and its partner emerges as a real particle.

$$-\frac{dM}{dt}c^2 = 4\pi\left(\frac{2GM}{c^2}\right)^2\frac{\pi^2}{60c^2\hbar^3}\left(\frac{\hbar c^3}{8\pi GM}\right)^4, \tag{17–17}$$

or

$$\frac{dM}{dt} = -\frac{C}{M^2}, \tag{17–18}$$

where the constant $C = 0.4 \times 10^{16}\ \text{kg}^3/\text{s}$.

Equation (17–18) is a differential equation that describes how $M$ changes with time. You can verify by direct substitution that the solution is of the form

$$M(t) = \{M(0)^3 - [0.4 \times 10^{16} \text{ kg}^3/\text{s}]t\}^{1/3}. \qquad (17\text{–}19)$$

With this result, we can find the time it takes for a black hole that starts with the value $M(0)$ to shrink to $M = 0$, namely

$$t = \frac{M(0)^3}{0.4 \times 10^{16} \text{ kg}^3/\text{s}}. \qquad (17\text{–}20)$$

For example, for $M(0) = 10^{31}$ kg (five solar masses), $t = 2.4 \times 10^{77}$ s. This is a long time; the age of the universe is some 10 billion yr, about $3 \times 10^{17}$ s. The mass of a black hole would have to be only $M(0) \cong 10^{11}$ kg, a very small mass by astronomical standards, to be presently on the point of evaporation.

## SUMMARY

The remarkable experimental fact that inertial mass and gravitational mass are one and the same leads to Einstein's general relativity, which is based on a restatement of this fact called the equivalence principle: The behavior of a system accelerating uniformly and the behavior of that system in a uniform gravitational field are indistinguishable. Some testable—and, in the case of the first two items, well-tested—consequences are as follows:

- Light moving away from a mass is redshifted; light moving toward a mass is blueshifted. Since this is also a shift in frequency, an observer will see a clock run more slowly when it moves closer to a mass and more rapidly when it moves away from a mass. In addition, because an accelerating clock cannot be distinguished from a clock in a gravitational potential, a uniformly accelerating clock also runs more quickly or slowly according to whether the acceleration is smaller or larger.

- Light is deflected in the presence of a mass. It's path is the straightest line possible in the distorted space whose curvature is produced by the presence of mass.

- Nonuniformly accelerating masses radiate energy in the form of gravitational radiation.

- The amount of the redshift observed when light falls towards a mass depends on the value of the mass and the radius within which the mass sits. If the mass is large enough or the radius is small enough, the redshift goes all the way to infinite wavelength (zero frequency) and no light is observed at all. In that case one has a black hole. The conditions for stars to evolve to black holes are entirely possible, as are the conditions for very massive black holes to form.

## QUESTIONS

**1.** A small space probe finds itself close to the horizon of a black hole. Is it possible for the difference between the gravitational forces at the two ends of the probe (the tidal force!) to be small enough so that the probe is not ripped apart? You might find it helpful to do some dimensional analysis.

**2.** One subject that actively engages theorists is the problem of a quantum theory of gravity. In such a theory gravitational attraction would be described as the exchange of a gravitational quantum—the graviton—in the same spirit that the exchange of photons leads to the coulomb law describing the attraction of two charged particles. From what you know of the spatial dependence of the gravitational potential, what would you predict the mass of the graviton to be?

For the next two questions you will find it useful to look at a globe. The questions explore the non-Euclidean geometry of a spherical surface.

**3.** Take two cities on the globe—say New York and Tokyo—and find the shortest distance between them assuming that you remain on the surface. Are you surprised by the route?

**4.** You can construct "right triangles" on the surface of a sphere by drawing the shortest lines between the poles and the equator. What do you notice about the sums of the interior angles of the triangles as you let their area get larger? Does anything like this happen if you draw right triangles in "flat" Euclidean space?

**5.** Discuss the gravitational redshift of light by thinking of light in terms of photons.

**6.** Suppose an experiment discovered that not all masses accelerate in the same way under the sole influence of gravity, in the sense that a brick accelerated differently than half a brick. Could Newtonian gravitation still be viable? How about Einstein's general relativity? Suppose a second experiment found that not all masses accelerate in the same way under the sole influence of gravity, in the sense that a 1-kg inertial-mass brick accelerated differently than a 1-kg inertial mass of feathers. Could Newtonian gravitation still be viable? How about Einstein's general relativity?

**7.** In Section 17–3, we argued that a light beam coming in perpendicular to the side wall of an elevator and passing through a hole to the interior would hit the opposite wall at a point below where one would expect it to because the elevator had moved up in the time it takes for the light to cross the space. But wouldn't this be true for a uniformly moving elevator? What distinguishes the two cases?

**8.** The period of a pendulum of length $L$ undergoing small oscillations is $(L/g)^{1/2}$, and the fact that this result contains no mass factors is evidence that gravitational mass and inertial mass are identical. How much of the preceding statement is true of a pendulum that undergoes oscillations of large amplitude?

**9.** Black holes are formed by the gravitational collapse of massive stars or collections of stars. Can you think of a reason that black holes might be rotating? If a black hole does rotate, can you give a qualitative estimate of the largest angular velocity that it could have? In making your estimate, ignore the fact that a rotating body tends to bulge out in the equatorial plane.

## PROBLEMS

**1.** ▮▮ Newton made a direct test of the equality of inertial and gravitational masses. He constructed two pendulums with identical little boxes at the ends and filled those boxes with identical gravitational masses of two different materials. (The boxes guaranteed that air resistance would be the same for the two pendulums.) He then observed that the pendulums oscillated in unison over a long period. **(a)** How could Newton ensure that the gravitational masses of the two materials were the same? **(b)** Suppose Newton could observe the two pendulums remain together to within an accuracy of 1/20 of an oscillation period during 500 oscillations. With what accuracy could he say that the gravitational and inertial masses are equal? [*Hint*: Find an expression for the period of the pendulum assuming that the gravitational and inertial masses are *not* equal.]

**2.** ▮ Suppose you want to use a pendulum to test the equivalence of inertial and gravitational masses on the Moon. (See the preceding problem.) Derive the period for small oscillations of the pendulum in terms of the Moon's mass $M$, its radius $R$, and Newton's constant $G$. Point out precisely where in your derivation you have used the equivalence between the two sorts of masses.

**3. ∎** In the 1960s Professor R. V. Pound of Harvard and his collaborators performed some celebrated experiments on the gravitational shift of the wavelength of light. There was a tower 74 feet high down which they could send radiation of frequency $f$. How much of a change in frequency should they have expected if they placed a detector at the bottom of the shaft?

**4. ∎** How much slower will a clock at the base of Mount Everest go than at the summit? The top of Mount Everest is 29,028 feet above sea level.

**5. ∎∎** Obtain the formula for the gravitational redshift by treating a photon as a particle that falls in a gravitational field just like a particle of mass $m$. The photon has no rest mass, but if it has an energy $E$ then Einstein teaches us that it will respond to gravity with an effective mass $E/c^2$. [*Hint*: The energy of the photon is related to its frequency by Planck's formula $E = hf$.]

**6. ∎∎** In addition to the general relativistic effect described in the text, there is a special relativistic effect in the measurement on Earth of radiation emitted by the Sun. The detector on Earth is in motion relative to the source of the radiation. This motion is a combination of Earth's axial rotation about its axis and Earth's orbital motion around the sun. **(a)** Find the speed due to the axial motion. You will want the information that Earth's radius is about $6.4 \times 10^6$ m and there are about $8.6 \times 10^4$ s in a day. **(b)** Find the speed due to the orbital motion. Here you can use the facts that the orbital radius is $1.44 \times 10^{11}$ m and that there are $3.15 \times 10^7$ s in a year. **(c)** The answers to parts (a) and (b) allow you to make an approximation for the speed of a point on Earth relative to the Sun, as well as to show that the motion is to a good approximation *transverse* to the direction from which light arrives. This means that there is an additional transverse Doppler shift that must be taken into account, and as we showed in Chapter 2, the measured frequency $f'$ differs from $f$ according to $f' = f\sqrt{1 - v^2/c^2}$, or

$$\frac{\Delta f}{f} = \frac{f' - f}{f} = \sqrt{1 - \frac{v^2}{c^2}} - 1 \cong \left(1 - \frac{v^2}{2c^2}\right) - 1 = -\frac{v^2}{2c^2}.$$

Here we have used the fact that $v \ll c$ to make our approximation. Find the numerical value of this contribution to $\Delta f/f$ and compare it with the numerical value due to general relativity.

**7. ∎∎∎ (a)** Find the gravitational redshift, given by Eq. (17–3), for light reaching us from the surface of a neutron star whose radius is 10 km and whose mass is $4 \times 10^{30}$ kg. Such stars rotate quite rapidly, leading to a spread in the redshift. **(b)** A particular neutron star rotates with a period of 1.3 s. Estimate the contribution to the shift due to the motion at the star's surface.

**8. ∎∎** Equation (17–3) describes the gravitational redshift of the light from a star, assuming that Earth's surface is at a gravitational potential of zero, an approximation justified by the fact that a star is far more massive than is Earth. Write a formula for the fractional frequency shift that takes into account the fact that the light not only rises from the star but also falls in arriving at Earth's surface. *Estimate* the size of the effect for light from the Sun.

**9. ∎∎∎** The gravitational redshift [Eq. (17–3)] for light reaching us from the surface of the Sun is $\Delta f/f = 2.1 \times 10^{-6}$. Estimate the temperature at which the Doppler spreading due to the motion of the gases at the Sun's surface would mask this effect. (*Hint*: You want the percentage width of the frequency peak for a given line due to Doppler spreading to exceed $2 \times 10^{-6}$. Recall that a gas at a temperature $T$ has an rms speed $v_{rms} = \sqrt{\langle v^2 \rangle} = \sqrt{3kT/m}$, where $m$ is the molecular mass. Make your estimate on the assumption that the gas consists of hydrogen atoms.)

**10.** ▮▮ The acceleration $g$ due to gravity is given by a radial space derivative of the geopotential. Verify that this derivative gives

$$g = \frac{GM}{R^2} - \frac{v^2}{R}.$$

The first term is the effect of the gravitational force, while the second is a term associated with the fact that a given point on Earth is accelerating centripetally and $g$ is observed in this accelerating frame. (*Hint*: Note that it is only the angular speed $\omega$, and not the speed $v$, that is a constant when $r$ varies.)

**11.** ▮▮ In this problem we verify the consistency of the constancy of the geopotential with the data on the acceleration due to gravity at Earth's surface. The value of g at sea level at the equator is 9.78 m/s$^2$, while that at the pole is 9.83 m/s$^2$. Given that the Earth rotates once every day and that its mean radius is $6.37 \times 10^6$ m and its mass is $5.98 \times 10^{24}$ kg, use the expression for g in the preceding problem and verify that the geopotential takes on the same value at equator and pole.

**12.** ▮▮ Using the constancy of the geopotential at Earth's surface, find the difference in Earth's mean radius at the equator from that at the pole. You may want to use the fact that the mean radius is approximately $6.37 \times 10^6$ m and that Earth's mass is $5.98 \times 10^{24}$ kg.

**13.** ▮▮▮ Example 17–5 gives the tidal force—the difference between the gravitational force on the top and the bottom of an object of length $L$—near the Schwarzchild radius $R_s$. Show that this tidal force is larger if the mass $M$ of the source of the gravitational force is smaller. Show also that the force varies as $1/M^2$.

**14.** ▮ Find the Schwarzschild radii of the Sun, Earth, and a proton.

**15.** ▮▮ What is the magnitude of the Schwarzschild radius for an object with a mass of $2.5 \times 10^9$ Sun? Assuming that we can use the Newtonian formula for the force of gravity, calculate at what distance from this mass, assuming it forms a spherical distribution, will an object of length 2 m, pointing in a radial direction, experience a difference of $1,000g$ in the $g$-value between the top and bottom of the object. Assume that the object is entirely outside the distribution.

**16.** ▮▮ What is the initial mass of a black hole whose lifetime is about that of the age of the universe, $3 \times 10^{17}$ s. What is the initial value of the Schwarzschild radius for such an object? In principle, such objects could have been created in the big bang (see Chapter 18), although we have no idea whether their demise would be detectable.

**17.** ▮▮ How long would it take for a black hole of mass 1 kg, perhaps produced in the big bang (see Chapter 18), to disappear because of Hawking radiation?

**18.** ▮▮ A star of one solar mass $M_0$ orbits about a black hole of mass $10M_0$ at a radius $2R_S$. Calculate the speed of the orbiting star. Use Eq. (17–16) to calculate the power radiated in the form of gravitational radiation.

**19.** ▮▮ Show, using dimensional analysis, that the Hawking blackbody temperature given in Eq. (17–15) is, aside from the factor $8\pi$, the only possible combination involving gravitationally relevant quantities and the quantum parameter $\hbar$.

**20.** ▮▮▮ In 1916, one year after Einstein presented his account of general relativity, Karl Schwarzschild calculated the invariant interval between two events a radial distance $r$ from a spherically symmetric static mass distribution of total mass $M$ to be

$$ds^2 = gc^2 dt^2 - dr^2/g - r^2 d\theta^2 - r^2 \sin^2\theta d\varphi^2,$$

where

$$g = 1 - \frac{2GM}{c^2 r}$$

summarizes the effect of the presence of the mass. ($M$ is the mass inside the radius $r$.) This expression should be compared with that for the interval in the absence of mass, viz.,

$$ds^2 = c^2 dt^2 - dr^2 - r^2 d\theta^2 - r^2 \sin^2\theta d\varphi^2 = c^2 dt^2 - x^2 - y^2 - z^2,$$

a quantity with which we are familiar from our discussion of special relativity. The Schwarzschild formula for the "lightlike" separation $ds^2 = 0$ describes the trajectory of a beam of light in the vicinity of the mass—in other words, it summarizes the bending of space–time due to the mass. The case $g = 1$ corresponds to flat space, in which the trajectory of the beam follows what we conventionally think of as a straight line with speed $c$. Calculate how close to flat space is **(a)** at the surface of the Moon, **(b)** at Earth's surface, and **(c)** at the Sun's surface.

**21.** ∎∎∎ Use the first two equations in Problem 20 to compute the speed of light passing outside the Schwarzschild radius of a spherical mass—indeed, outside the mass itself—for trajectories that are **(a)** purely radial and **(b)** purely transverse. Show that in both cases the speed is less than $c$—in other words, gravity slows the light! This makes the passage of light in such a gravitational field resemble its passage through a medium with an index of refraction. What would you expect in a Newtonian picture in which light is a massive particle? Incidentally, this effect has been tested by bouncing radar signals off of planets and seeing that the presence of the Sun delays the round-trip.

# CHAPTER

# 18

# Cosmology

Cosmology—the study of the origin and evolution of the universe as a whole—not only is of obvious and long-standing interest to humanity, but is one of the most active topics in modern physical science. One of the reasons for the current level of interest is that many of the technological advances the understanding of quantum mechanics has made possible—in particular the solid-state devices that have fostered new types of detectors and large-scale computing—have made that part of the cosmos which is accessible to us vastly larger than it has been. With new computer-corrected telescopes and the routine use of space as an observing platform, we can measure with unprecedented accuracy and in heretofore invisible wavelengths. In addition, our ability to handle large amounts of data has improved; indeed, some scientists have estimated that the rate at which we can gather information relevant to cosmological questions will grow by a factor of *billions* over the next few years!

Work on cosmology is guided by some basic assumptions. First, our understanding of the universe must be grounded in observation, and what we propose should have testable consequences—this is a central tenet of science as a whole. Second, the same laws of physics apply everywhere in the universe, and these are the laws that we could in principle test here on Earth—we refer to this assumption as the *central dogma*. Finally, we adopt the *cosmological principle*: There is no preferred center of the universe. The cosmological principle extends the Copernican principle that *we* are not the center of the universe. In its more technical form given in the next section, and together with the laws of physics we already know, we can go remarkably far in understanding what the universe is and how it behaves, and we can confront some new physical ideas.

## 18–1  The Expanding Universe

Modern cosmology has its origins in Einstein's 1915 discovery of the general theory of relativity and gravitation, in which the nature of space and time is dynamically determined. Newton, of course, also had a cosmology, even if it was a crude one. In his picture, individual stars moved in accordance with the laws of Newtonian gravity; the space in which the movement took place was like an infinite fixed stage.

Newton implicitly assumed—or nearly so—what is today a central feature of cosmology, the **cosmological principle**:

- On the average the universe is *isotropic*; that is, any observer will see the same general features no matter in which direction the observer looks.

- On the average the universe is *homogeneous*; that is, an observer stationed anywhere will, on the average, see the same thing as any other observer.

The experimental evidence for a high degree of isotropy is compelling and will be discussed later. Homogeneity holds only on the average. There are structures (clusters of clusters of galaxies and regions with very few galaxies) the size of $10^8$ lt-yr, but these are still small compared to the size of the universe, which we shall see is on the order of $10^{10}$ lt-yr. In any case, these structures appear uniformly throughout the universe, so that as long as you don't look too closely, homogeneity applies.

There is a problem in Newton's picture: Any region that has a higher mass density than that of a neighboring region will act to attract the neighboring matter to it and increase the disparity. We say there is a *gravitational instability*— or more simply, matter tends to clump. If the universe is infinitely large and static, the matter will have an infinitely long time to clump, and the clumps will become infinitely large. Or if the universe is finite and static, all the matter will eventually gather into a single large clump.

There is an additional problem with an infinitely large and static homogeneous universe. Although the problem can be traced back to Kepler in the early 17th century, we know it today as **Olbers' paradox**, after the 19th century German physician and astronomer Heinrich Olbers.[1] Let us assume—and this is well borne out by the data—that most stars have more or less the same *intrinsic luminosity L*; that is, the total radiant energy emitted per unit time by a star is more or less the same for all stars. A single star a distance $r$ from us will then shine on us with an *observed* power per unit area of $L/(4\pi r^2)$, a quantity that we denote by $l_{obs}$. Under the assumption of homogeneity, the number of stars that can be found in a shell of thickness $\Delta r$ at a distance $r$ from us is $\Delta N = 4\pi r^2 \Delta r\, n$, where $n$ is the number of stars per unit volume and $4\pi r^2 \Delta r$ is the volume of the shell. (Homogeneity implies that $n$ is a constant that is independent of $r$.) The observed luminosity of *all* the stars in the shell is thus

$$L_{obs} = 4\pi r^2 \Delta r\, n l_{obs} = 4\pi r^2 \Delta r\, n \left(\frac{L}{4\pi r^2}\right) = \Delta r\, nL, \qquad (18\text{--}1)$$

which is *independent* of $r$. Thus, if we add up the effects of all the shells and if the universe is infinitely large and began infinitely long ago, so that the light from all the stars has had a chance to arrive at Earth, we get an infinite luminosity! But of course the night sky is not ablaze with light—fortunately for us, because that would be equivalent to living in a large oven, with a temperature of thousands of degrees! We conclude that a homogeneous universe cannot be both infinitely large and static.

When Einstein wrote his first paper on the cosmological consequences of relativity, Olbers' paradox was irrelevant to him, since all the evidence he possessed suggested that the universe was finite in space, with only a finite number of stars. Moreover, the general theory of gravitation—general relativity—gave Einstein a powerful tool to think about what it would mean for the universe to be finite. Without the geometrical insight provided by general relativity, one has a hard time thinking about what a finite universe could possibly mean. After all, if the universe has an "edge," you might think, Why can't you just go

---

[1] The first person to point out that, given a finite speed of light, Olbers' paradox is resolved with a finite age for the stars was none other than Edgar Allen Poe, in 1848.

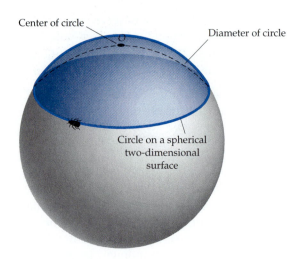

Center of circle

Diameter of circle

Circle on a spherical two-dimensional surface

• **Figure 18–1** An (intelligent) insect living on the surface of a sphere measures the ratio of the circumference to the diameter of a large circle (here a circle of fixed latitude) drawn on the sphere. The diameter is given by the shortest distance between opposite points on the circle.

beyond the edge and see what is on the other side. And if you include "the other side," wouldn't you in fact be thinking about an infinite space? But in general relativity the nature of space is determined by the matter in it.

It is possible to consider different kinds of space—flat and curved, finite and infinite. We described these possibilities in Chapter 17. In particular, although a curved three-dimensional space is difficult for us to visualize, such spaces have *measurable* properties that reveal the nature of the space. To take a two-dimensional analogue, a bug that lives on a two-dimensional space could, by measurements of angles or distances, determine the kind of space it lives on, even without leaving the surface. For example, a bug on a sphere could measure the ratio between the circumference of a circle and its diameter—in this case the diameter is the shortest distance between two points 180° apart. As • Fig. 18–1 illustrates, the bug would determine that this ratio is smaller than $\pi$ and thereby deduce the *curvature* of the space it lives on. Note that a bug on a sphere, which would be an example of a closed two-dimensional space, cannot go to an edge and peek over it to learn what is on the "other side"; there is no edge, and there is no other side. Note also that there is no special point in this space—no "center." This caution is worth emphasizing. It is very tempting to think of the universe as a large balloon that expands or contracts. But any balloon we are familiar with expands into *something*, and it is only natural to think of ourselves outside the balloon watching it expand. However we are never *outside* the universe.

Einstein was able to interpret what was meant by a finite universe and was accordingly able to visualize what it meant for such a universe to be homogeneous. While Einstein did not think that the universe was infinite, he had a strong prejudice for a *static* universe. A static finite universe has the same problem of gravitational instability that exists in an infinite static universe, and prejudice towards a static universe led both Newton and Einstein to create models with this same fatal difficulty. Because of his insistence on a static universe, Einstein missed predicting one of the most significant discoveries of 20th-century science—that the universe is evolving in time. In fact, it is expanding! We shall review the experimental evidence for the expansion of the universe in Section 18–2. On the theoretical side, the Russian Alexsandr Friedmann[2] showed

[2] Alexsandr Alexandrovich Friedmann had been a meteorologist and a mathematician and had even supervised the production of airplanes. But having learned of Einstein's theory, he taught it to himself and in 1922 published the first of two papers in which he derived the version of the expanding universe that we still use today.

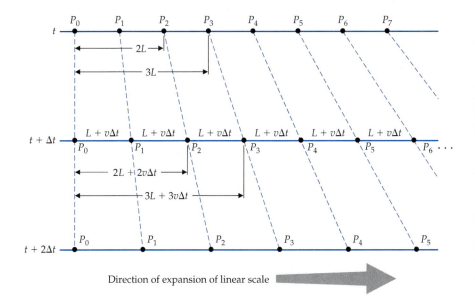

• **Figure 18–2** Points on a line undergoing uniform expansion. One can see that the recession speed of each point is proportional to its distance from the point of observation. With $P_0$ the location of the observer, $P_1$ is a distance $L$ away at time $t$ and a distance $L + v\Delta t$ at time $t + \Delta t$. The distance from $P_1$ to $P_2$ is again $L + v\Delta t$, so that the distance from $P_0$ to $P_2$ is $2L + 2v\Delta t$, and so on. Thus the speed at which $P_1$ moves away from $P_0$ is $v$, while the speed at which $P_2$ moves away from $P_0$ is $2v$, and so on.

that the equations of general relativity lead to nonstatic solutions, with one that is of particular interest being a solution in which space (the homogeneous universe) expands, but is finite in extent.

There is an important relation between the relative velocity of points—think of galaxies—and the separation of those points in an expanding homogeneous universe. Consider a number of initially equidistant points on a straight line in this expanding universe (•Fig. 18–2). Since there is no center to the universe, we may position ourselves at any one of the points, say $P_0$. The point $P_1$ is a distance $L$ away from us, the point $P_2$ is $2L$ away, and so on. At time $\Delta t$ later, the universe has expanded somewhat and the points have separated. The point $P_1$ will be $L + v_1\Delta t$ away from us, where $v_1$ is the instantaneous velocity of $P_1$. The point $P_2$ will be a distance $2L + v_2\Delta t$ from us, where $v_2$ is the instantaneous velocity of $P_2$, and so forth. But the universe must remain homogeneous. Therefore the distance $P_2P_0$ must remain twice the distance $P_1P_0$ (or $P_2P_1$). In other words,

$$2L + v_2\Delta t = 2(L + v_1\Delta t); \qquad (18\text{–}2)$$

that is,

$$v_2 = 2v_1.$$

In the same way, the distance $P_nP_0 = nL + v_n\Delta t = n(L + v_1\Delta t)$, so that quite generally

$$v_n = nv_1. \qquad (18\text{–}3)$$

In this derivation we have implicitly assumed that all the speeds are much smaller than the speed of light $c$. Otherwise we would have to use the relativistic velocity addition theorem, so that no speed would exceed $c$.

Now the factor $n$ is a measure of how far the $n$th galaxy is from us. In other words, Eq. (18–3) states that the farther a galaxy is from us, the more rapidly it recedes, in exact proportion to its distance from us. Put another way, if we station an observer at some point in the universe—*any* point—the rest of the universe will appear to expand away from this point at a rate that increases in proportion to the distance between the observer and any remote point. There is no preferential origin in space for this expansion. There *can't* be, since we are assuming our evolving universe to remain always homogeneous and isotropic. It is customary to rewrite Eq. (18–3) in the form $v(t) = Hr(t)$, where $v(t)$ is the velocity with which a given point (or galaxy) recedes with respect to any fixed point and $r(t)$ is the separation between the two points. Finally, we have one further refinement to make here: We must add the possibility that the proportionality "constant" $H$ can in fact be a function of time—shortly, we'll see more precisely how. Thus

$$v(t) = H(t)r(t). \tag{18–4}$$

This relation is known as **Hubble's law**, and $H$ is **Hubble's "constant"**—we'll explain the role of Edwin Hubble in all this in the next section.

In our discussion we have neglected the effects of special relativity. Hence we expect the linear relation between recession speed and distance of Eq. (18–4) to hold only for "nearby" distances. The reason is that if we let $r(t)$ get arbitrarily large then $v$ will eventually exceed the speed of light, and that cannot happen. That means that the true law must be a nonlinear one for which Eq. (18–4), even with a constant value of $H$, is a nonrelativistic approximation. The precise form of this law will not concern us here. We also remark that there is no limitation on the magnitude of $H$, since it measures the scale of expansion of space itself. In fact in some models $H$ increases indefinitely with time. This may even be the case for our actual universe.

There is another form of Hubble's law that we'll find convenient later. The distance between any pair of galaxies has the form

$$r(t) = r_0 R(t), \tag{18–5}$$

where $r_0$ is the separation between the galaxies *now* and $R(t)$ is a *universal* scaling factor, one that applies in every corner of the universe. (If it differed from point to point we would not be able to say that the universe is homogeneous.) If *now* is at $t = t_0$, then $R(t_0) = 1$. We can differentiate Eq. (18–5) with respect to time—we employ the notation $dx/dt = \dot{x}$—to find the relative speed of separation any pair of galaxies:

$$v(t) = \dot{r}(t) = \frac{d}{dt}(r_0 R(t)) = r_0\dot{R}(t) = \frac{r(t)}{R(t)}\dot{R}(t) = \left[\frac{\dot{R}(t)}{R(t)}\right]r(t). \tag{18–6}$$

By comparison with Eq. (18–4), we can identify the quantity in square brackets with Hubble's constant:

$$H(t) = \frac{\dot{R}(t)}{R(t)}. \tag{18–7}$$

## 18–2 The Discovery of the Expanding Universe

What does observation tell us about the matters discussed in the previous section? We concentrated there on homogeneity and relative motions. Equation (18–4) is the central feature of the expanding (or contracting) homogeneous and isotropic universe. To test it, we need two principal ingredients: We must be able to measure the speed with which matter is receding from us, and we must know the distance that that matter lies from us. The speed with which a galaxy moves away from us can be determined by studying the spectra of elements in the galaxy. All the galaxies contain hydrogen, whose spectrum is well known. We are familiar with the Doppler shift, which describes the observed wavelength of radiation emitted from a source moving with respect to the observer. (See Chapter 1.) When a source emits light of wavelength $\lambda$ in its own rest frame and moves away from the observer with relative speed $v$, the observed wavelength is given by

$$\frac{\lambda_{\text{obs}}}{\lambda} \cong 1 + \frac{v}{c},$$

where the approximation holds for $v/c \ll 1$. In other words, the observed wavelength increases above the value a stationary observer sees. We can define $Z$ by

$$\frac{\lambda_{\text{obs}} - \lambda}{\lambda} = \frac{\Delta\lambda}{\lambda} \equiv Z \cong \frac{v}{c}. \tag{18–8}$$

The quantity $Z$ represents the fractional redshift due to the expansion of the universe.[3] Note that even when $v/c$ is not small, so that the linear Hubble law must fail, $Z$ can still be used as a measure of the full expansion law. In the expanding universe, the farther the object, the larger is the value of $Z$.

The relative speed of a galaxy is comparatively easy to measure in the way we have just decribed. But to test Eq. (18–4) we must also know the distance of the galaxy, and measurements of the distances of galaxies that are far from us are difficult to carry out. For nearby stars, measurements of how the star moves against the background of more distant stars as Earth moves in its solar orbit can provide information on distance. But for more distant objects such as galaxies such methods are hopeless. One can measure the *apparent* brightness of a galaxy (effectively counting the number of photons per unit time that enter the aperture of a telescope), but unless one knows the *intrinsic* brightness of the galaxy, which measures the total number of photons that it emits per unit time, one cannot use the inverse square law to deduce the galaxy's distance from us.

Astronomers have often relied on **standard candles** to measure distances to galaxies. A standard candle is an object whose intrinsic brightness we think we know, so that its apparent brightness measures its distance. The first standard candle was established through the work of Henrietta Leavitt, working at the beginning of this century as an assistant at Harvard. She discovered that a class of powerful variable stars called the *Cepheids* had a variation period—typically tens of days—that was closely correlated to their intrinsic brightness. These stars can be observed in individual nearby galaxies and enable us to measure distances to those galaxies.[4] Once one measures Cepheids that are close enough to establish a distance by more direct means one has a way to measure the distance to nearby galaxies. Because Cepheids are useful just for nearby galaxies, we can make the jump to more distant ones only by using the now-measured distance of nearby galaxies and seeking other types of stars or features within those galaxies whose intrinsic properties are known. As long as these stars or other features can be measured from further away than single Cepheids, they can be used to extend the distance scale.

A new standard candle having just this property has been established. When certain white dwarfs collapse, the collapse is followed by an explosion

---

[3] Since galaxies have random velocities that are not connected to any expansion of the universe, they also have random Doppler shifts—both blueshifts and redshifts. In the experiments that established the expansion, Hubble was persuaded that he could subtract out this effect by averaging over many galaxies—the random motions are about the same for all galaxies. This claim is now very well confirmed.

[4] Indeed Hubble established the fact that galaxies are distant collections of stars by measuring the presence of Cepheids with a low apparent luminosity in what were at the time thought to be clouds of glowing gas within our galaxy.

• **Figure 18–3** Edwin Hubble.

called a *supernova*. Supernovas of a particular type are highly uniform. The astronomer Allan Sandage has studied nearby supernovas (ones whose distance can be independently found) as they occur and has found that their peak luminosity varies from a fixed number—about $10^{10}$ times that of the Sun—by only $\pm18\%$. The uniformity and the huge luminosity, which allows supernovas in very distant galaxies to be seen, makes them ideal candidates for standard candles that can extend our distance measurements considerably. The Cepheids and supernovas together establish a distance "ladder." But doing that is a difficult process and that is why distance measurements continue to form the weak link in our knowledge of the expanding universe.

One of the most important figures in 20th-century astronomy, and the man who laid the experimental foundation of cosmology, was Edwin Hubble (•Fig. 18–3). Born in Missouri in 1889, Hubble was a great college athlete as well as an outstanding scholar. He won a Rhodes scholarship to Oxford after graduating from the University of Chicago. He served briefly in the army in the First World War and then accepted a job at what was then the largest telescope in the world, at the Mount Wilson Observatory near Pasadena, California. In the 1920s Hubble made two major discoveries:

1. Certain fuzzy patches in the sky (*spiral nebulae*) were not clouds of gas within our galaxy, but were distant galaxies as large as our own. Our galaxy is some 100,000 light years in radius, while Andromeda, the nearest outside galaxy, is two *million* light years away. Thus Hubble established the fact that the universe was far larger than had ever been envisioned. Furthermore, Hubble found the distant galaxies to be, on the average, uniformly distributed.

2. Hubble measured the velocities of the galaxies by using the Doppler shift, as we have described. When he plotted his data for recession speed versus distance for a limited number of relatively nearby galaxies (•Fig. 18–4), he found that to a good approximation the galaxies were receding from us with a speed

$$v = H_0 r, \tag{18–9}$$

where $H_0$ was more or less the same for all galaxies. This is of course Hubble's law, Eq. (18–4), just the relationship we expect for an expanding ho-

• **Figure 18–4** Velocity–distance relation for distant galaxies and clusters of galaxies. The black dots represent galaxies, and the solid line represents a best straight-line fit to these points. The hollow circles represent clusters of galaxies, and the dashed line is the best straight-line fit to these.

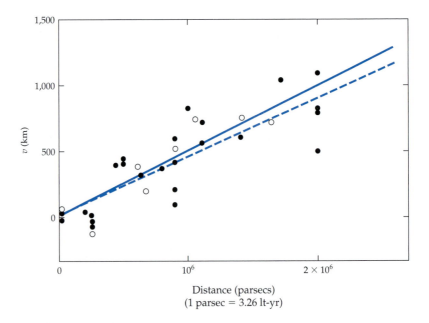

mogeneous universe. Hubble was the first person to state clearly that this law applies to our actual universe.

The measurement of $H_0$—$H(t)$ at the present time—has been a great challenge to astronomers. The uncertainties in distance measurements lead to an uncertainty in the value of $H_0$. At the end of 1999 a generally agreed value was

$$\frac{1}{H_0} = (14 \pm 3) \times 10^9 \text{ yr.} \qquad (18\text{–}10)$$

When we want to make estimates we'll sometimes just use $10 \times 10^9$ yr. Since as we shall see momentarily it is more than reasonable to assume that $H$ varies with time, we prefer to call $H_0$, the value of $H$ at *this time*, **Hubble's parameter** rather than Hubble's constant.

---

**Example 18–1**   Estimate the distance in light-years at which the naive Hubble's law breaks down due to special relativity. Assume that Hubble's parameter is a constant with the central value of Eq. (18–10).

**Solution**   The naive form of Hubble's law will break down when the speed becomes relativistic. We'll suppose for purposes of estimation that this happens for $v/c = 0.5$ and solve for the distance from the naive law:

$$r = \frac{v}{H_0} = (0.5c)(13 \times 10^9 \text{ yr}).$$

Because 1 light-year is the distance light travels in 1 year, which is equal to $c \times (1 \text{ yr})$, solving for $r$ is simple:

$$r = (0.5)(13 \times 10^9) \text{ lt-yr} = 6.5 \text{ billion lt-yr.}$$

This is a distance that we have seen and measured. Note that the value of $Z$ measured at these distances remains correct, but it can no longer be interpreted as $v/c$ except as an approximation—one that will become worse as $Z$ increases.

---

If the universe evolves, then it has a future and a past, and it might also have an age. In particular, if the universe is expanding, then it must have started expanding at some time, and one possibility is that it started expanding from something small. This is the hypothesis of the **big bang theory**. Olbers' paradox is no longer an issue once we accept the big bang idea: If the universe has a finite lifetime, there are only a finite number of stars and Olbers' argument breaks down.

In the big bang picture, Hubble's parameter provides us with a scale for the age of the universe, some $10^{10}$ yr; this estimate should be reasonable if the parameter doesn't change too rapidly with time. We can similarly estimate the size of the universe by noting that the speed of light provides a measure on how far things could have gone since the big bang. By this reasoning the size of the universe is $c/H_0$, roughly $10^{10}$ lt-yr. The distance to the Andromeda galaxy is about a million light years, so there is plenty of room in the visible universe. Our estimate for the age is a satisfactory one. A lifetime much shorter than this would conflict with data on the age of the galaxy and the solar system. Example 13–1 illustrates how we estimate such ages and gives an age of $6 \times 10^9$ yr, a number that fits in nicely with our ideas about the evolution of the universe.

One of the most exciting aspects of cosmology is that it has allowed us to learn some things—indeed, a great deal—about the future and past of the universe. We deal with these in the next two sections.

## 18–3 The Future of the Universe and the Critical Density

Given the universe's current expansion, we can envision three possibilities (•Fig. 18–5):

1. The universe can continue to expand forever; the expansion could slow down more and more, but never really end. In that case, we say that the universe is **open**.

2. The expansion can continue, but slow down enough so that in the limit of infinite time it stops. The universe is **flat**.

3. The universe can expand up to a point, but the expansion slows and stops in a finite amount of time, and the universe collapses back to a point. In this case the universe is **closed**.

By studying the Friedmann solution we know that which of these alternatives actually happens depends on how much matter—or, to be more specific, *energy*—there is in the universe. The more energy there is, the more the gravitational attraction tends to pull the matter together. In a period of expansion, this energy tends to slow down the expansion. With ideas about how the universal scale factor $R$ changes in time, we can even make the expansion of the universe quantitative. Moreover, because of our simplifying assumptions about the universe—that the matter is distributed uniformly—we can arrive at the same results as those of Einstein's theory by using rather simple classical arguments.

We want to find an equation for $R(t)$. Since according to our assumptions $R(t)$ is the same everywhere in the universe, if we find it anywhere we will have found it everywhere. Let us then consider at some time $t$ a sphere of radius[5] $r_0 R(t)$ containing mass distributed uniformly with density $\rho(t)$. (This same density applies throughout the universe.) The total mass $M$ in our sphere is then $(4\pi/3)r_0^3 R^3 \rho$. This mass never changes as the sphere expands or contracts. If we have a point mass $m$ at the surface of the sphere, we know from our study of Newtonian mechanics that $m$ is affected by gravity in this highly symmetric situation only by the mass within the sphere and only as if there were a point mass $M$ concentrated at the center of the sphere. Thus the whole surface of the sphere is—depending on initial conditions—expanding or contracting

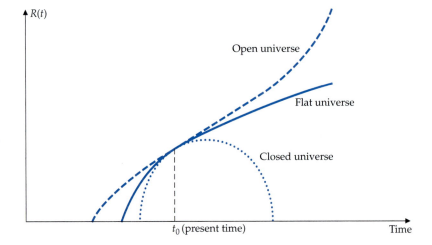

• **Figure 18–5** The three possible fates of the universe, as characterized by the radius $R$ as a function of time. The different evolution possibilities start at different times because we have chosen then to have the same radius now ($t = t_0$), and they each take different times to evolve to $R(t_0)$.

---

[5] The scale factor $R$ is dimensionless, so we have included a length factor $r_0$ that contains the dimensional information.

as if it were under the influence of this one central mass. Now our typical point mass $m$ on the surface obeys Newton's law,

$$m r_0 \ddot{R}(t) = -\frac{GmM}{r_0^2 R(t)^2}.$$

This is our equation for $R(t)$. We also know that the conservation of energy is applicable and yields

$$\frac{1}{2} m r_0^2 (\dot{R}(t))^2 - \frac{GmM}{r_0 R(t)} = E,$$

where $E$ is a constant with the dimensions of energy. The first term in this equation is a "kinetic energy" while the second term is a "potential energy." We see that the sign of $E$ depends on which of these terms dominates. All of this is very Newtonian. So are the next steps, which are to divide through by $m$ and to replace $M$ by its expression in terms of $R(t)$, namely $M = (4\pi/3) r_0^3 R^3 \rho$. Thus

$$\frac{1}{2} (\dot{R}(t))^2 - \frac{4\pi}{3} G (R(t))^2 \rho(t) = E', \qquad (18\text{--}11)$$

where $E' = E/(mr_0)^2$ has the same sign as $E$. (Only the sign will be of interest here.) The next step would not have occurred to Newton. The universe is made up of more than just massive Newtonian objects such as stars and planets. There is also radiation. Because of Einstein's relation $E = mc^2$, radiation interacts gravitationally, with an effective mass that is found by dividing the radiative energy by $c^2$. We can—and should—include its effect in our equation for $R$. This effect will be manifest in Eq. (18–11) in a contribution to the quantity $\rho$ given by the radiation energy density divided by $c^2$. Because of the universality of $R$ that equation holds everywhere in our highly symmetric universe, and we can now forget how we derived it in terms of our little sphere.

One more manipulation of Eq. (18–11) is useful: We divide by the positive quantity $(4\pi/3) G R^2$, leaving

$$\frac{3(\dot{R}(t)/R(t))^2}{8\pi G} - \rho(t) = \frac{3(H(t))^2}{8\pi G} - \rho(t) = E'', \qquad (18\text{--}12)$$

where again $E''$ has the sign of $E$ and where the density $\rho(t)$ contains the radiation energy density divided by $c^2$. Note the presence of the Hubble parameter $H(t)$ defined by Eq. (18–7). Note also that $E''$ varies with time.

What is the significance of the sign of $E$? If a mass is thrown from the solar system with a total energy of zero, it will come to rest in the infinite future infinitely far from us; if its energy is positive, it will be moving with finite kinetic energy even when it is infinitely far, and if its energy is negative, then it will fall back into the solar system. Similarly, positive, zero, and negative $E$ correspond in our descriptions at the beginning of this section to an open, a flat, and a closed universe respectively. A look at Eq. (18–12) shows us that for a closed universe (negative $E$), $H$ is limited to lie *below* a critical value. For an open universe (positive $E$), $H$ lies *above* a critical value. The dividing lines between these cases is determined by the flat universe ($E = 0$). This dividing line can be expressed in terms of a *critical density* defined by

$$\rho_c = \frac{3H^2}{8\pi G}. \qquad (18\text{--}13)$$

*If $\rho$ exceeds $\rho_c$, then the universe is closed; if $\rho$ is less than $\rho_c$, then the universe is open.* Note that since the Hubble parameter in general depends on time, so does the critical density. At present, the critical density is roughly

$$\rho_c\big|_{now} = \frac{3\left[(1.3 \times 10^{10}\ \text{yr})(3 \times 10^7\ \text{s/yr})\right]^{-2}}{8\pi(6.67 \times 10^{-11}\ \text{m}^3 \cdot \text{kg}^{-1} \cdot \text{s}^{-2})} = 1.2 \times 10^{-26}\ \text{kg/m}^3. \qquad (18\text{--}14)$$

The striking thing about this number is how small it is. If the actual density of the universe is of this order of magnitude—and we believe it is—then the universe is all but empty space. In fact, the number in Eq. (18–14) corresponds to only about seven hydrogen atoms per cubic meter!

**Example 18–2**    Suppose that the universe is flat (i.e., the density is the critical density). What was the value of the Hubble parameter when the density of the universe was the density of ordinary matter?

**Solution**    The density of Earth, which we may take as a measure of the density of ordinary matter, is $5.5 \times 10^3\ \text{kg/m}^3$. We can find the value of the Hubble parameter $H$ at the time when $\rho_c$ had this value from Eq. (18–13):

$$H = \sqrt{\frac{8\pi G \rho_c}{3}} = \sqrt{\frac{8\pi(6.67 \times 10^{-11}\ \text{N} \cdot \text{m}^2/\text{kg}^2)(5.5 \times 10^3\ \text{kg/m}^3)}{3}}$$

$$= 1.8 \times 10^{-3}\ \text{s}^{-1} = 5.5 \times 10^4\ \text{yr}^{-1}.$$

When we compare this result with the current value of some $8 \times 10^{-11}\ \text{yr}^{-1}$, we see that it is indeed misleading to think of $H$ as a constant. We can also estimate that if the universe is flat it would have the density value we just found at an age of $1/H$, or about 10 min (See Section 18–4).

What do the data on density tell us about whether the universe is open or closed? It is customary to discuss the magnitude of the density in terms of the ratio

$$\Omega = \rho/\rho_c. \qquad (18\text{--}15)$$

Astronomers estimate that there are about $10^{11}$ galaxies within the visible universe, each with about $10^{11}$ stars. Typically these galaxies are about 60,000 light years in size. Astronomers have various techniques for determining the mass of galaxies. For example, some of them rotate around a central mass, and one can use Kepler's laws of motion to deduce information about their mass. A convenient mass unit is that of the Sun, $M_\odot = 1.989 \times 10^{30}$ kg. Many galaxies are grouped in "clusters"—hundreds, if not thousands, of individual galaxies attracted to each other gravitationally. The mass of a typical galactic cluster is about $10^{15} M_\odot$. But measurements of the galactic motion within clusters show that most of this mass cannot be identified with matter that is actually "shining." There must be *dark matter*—matter whose nature is under debate among physicists—even if it is invisible to telescopes. The existence of dark matter in the regions surrounding galaxies—galactic halos—is determined by studying the orbits of stars at the edge of the visible galaxy. If the only matter affecting these stars were visible matter, their orbits would obey the relation $v^2/R = GM/R^2$, where $R$ is the radius of their (galactic) orbit and $M$ is the visible mass of the galaxy. This relation suggests that the speed of the stars in question should fall as $1/R^{1/2}$. But observations show that these speeds increase with $R$ before finally tapering off. This suggests the presence of a considerable amount of dark matter outside the visible domain of galaxies. (See Problem 5 at the end of the chapter.) Nevertheless, as of 1999, this dark matter does not appear to be enough to close the universe. The best estimate, including both luminous and nonluminous matter, is that $\Omega$ is about 0.2. If only the luminous matter had been included then $\Omega$ would have been some two orders of magnitude smaller. The implication appears to be that the universe is open.

The question of the nature of the dark matter is an interesting one, because it may give us some new insight into physical law. We are confident that we know what stars are made of: protons; some nuclei of light elements such as alpha particles; some heavier nuclei; and electrons to balance the electrical charge. We are less confident that we know the nature of the dark matter that lies in the halos that surround the visible galaxies or that lies within the visible galaxies themselves. Among the candidates that have been suggested are such things as *brown dwarfs*—small, very massive stars that, like the planet Jupiter, are not hot enough to radiate enough energy to make them visible—and elementary particles such as the elusive neutrino. (See Chapter 16.) Neutrinos may have tiny masses, as has been suggested by some recent experiments. Neutrinos with mass can get trapped gravitationally in the galactic halos and might contribute to the dark matter. Or—and here is where new physics may come in—the dark matter may be of a form we have not yet observed on Earth.

It is actually quite amazing that $\Omega$ is as close to unity as it is. A different way of stating this fact is the following: In problems involving gravity the natural unit of time is the *Planck time*, a quantity that on dimensional grounds (see Problem 10) has the form

$$t_P = \sqrt{\frac{\hbar G}{c^5}} = 5.4 \times 10^{-44} \text{ s}$$

At 20 billion years, the universe is $10^{62}$ units of $t_P$ old. This is very old for a universe as structurally rich as ours. If $\Omega$ had been much greater than unity, the universe would have collapsed on itself early on and never arrived at our current ripe age. By contrast, if $\Omega$ had been much less than unity, the expansion would have been so rapid that there would have been no time for matter to have lumped and formed galaxies. Many cosmologists feel that the only way to understand this is to suppose that for reasons we do not at present understand $\Omega$ is *exactly* unity. Of course, that is really a prejudice. The final word will have to come from nature, and we will probably have a good idea of the answer sometime in the first decade of the 21st century. We'll come back to this in the last section.

## 18–4 The Early Universe and the Background Blackbody Radiation

The remarkably imaginative George Gamow[6] was the first to apply the notion that our current ideas about physics could shed light on the early universe and vice versa (•Fig. 18–6). It had occurred to the Belgian abbot–astronomer Georges Lemaître in 1927 that, if one could imagine running the history of the universe backwards in time, there must have been an initial time in which the temperature of the universe was essentially infinite! This is the time of what has come to be known as the **big bang**. We know what happens to matter when it is heated: If the matter is hot enough, atoms dissociate into electrons and nuclei; if it is still hotter, the nuclei dissociate into protons and neutrons, and at a still hotter temperature the nucleons dissociate into quarks and gluons. To the extent that we can understand how the temperature of the universe changes with

• **Figure 18–6** George Gamow made important contributions in many areas of science. He explained alpha decay as a manifestation of quantum mechanical tunneling, he applied nuclear physics to astrophysical problems, he played a critical and prophetic role in cosmology, and he even spent time working on the genetic code.

---

[6] Gamow, who was born in Odessa, Russia, had intended to study with Friedmann at the University of Petrograd, but Friedmann died in 1925 of typhoid fever at the age of 37. Gamow managed to escape from the Soviet Union, passed through Copenhagen at Niels Bohr's institute, and eventually landed at George Washington University in Washington, DC, where, in the late 1940s, he did his work on cosmology in collaboration with some younger students and research associates.

time—how "hotter" means "earlier"—we can single out a number of eras that are characterized by different regimes of matter.

Along with protons, electrons, and other particles, the early universe certainly contained electromagnetic radiation—photons. That is because photons take part in thermal processes just as material particles do. (See Chapter 12.) As long as the photons were in thermal equilibrium they would form a blackbody spectrum characteristic of the temperature. Gamow and his young collaborators Ralph Alpher and Robert Herman studied the implications of the presence of this cosmological radiation. They asked, What does this radiation look like now? They reasoned that in the early universe photons and charged particles such as electrons interacted with each other very rapidly compared to the expansion rate $H(t)$. This meant that the photons and the charged particles reached thermal equilibrium at a *common* temperature $T$. The shape of the photon spectrum could be used to measure the temperature.

Gamow and his colleagues argued that when the temperature dropped to about 3,000K—it turns out that this is about 100,000 years after the initial explosion—electrons and protons could combine into neutral hydrogen atoms without these atoms being ripped apart by energetic photons from the high-energy tail of the blackbody spectrum. When this *recombination*—cosmologists use this word even though prior to this epoch the protons and electrons were never in fact "combined"—takes place, the photons no longer find substantial numbers of charged particles to interact with and become basically free. The blackbody radiation has, to a good approximation, **decoupled** from matter. The time at which recombination takes place is called the **recombination time**.

Now as we have already emphasized, when we say that the universe expands, we mean that all length scales grow. This means that the wavelengths of the photons making up the now nearly free blackbody radiation grow. However, the blackbody nature of the spectrum is preserved—the relative amounts of radiation with different wavelengths remain the same—since there are no processes to change it. We can note that increasing the wavelengths of the radiation corresponds to decreasing the frequencies and hence the energy: The radiation will still be of blackbody form, but at a much lower temperature than when it disconnected from matter. Thus Gamow and his collaborators predicted in the late 1940s that *there should at present be cosmic background radiation at a temperature much lower than the 3,000K at which recombination took place and radiation ceased to interact with matter.*

Using their best knowledge of how much the universe must have expanded since the recombination time, the Gamow group conjectured that the temperature at which one would find this radiation would be a few degrees above absolute zero. Let us apply the Wien displacement law for blackbody radiation, which states that $\lambda_{max}T = 0.29 \times 10^{-2}\,\text{K} \cdot \text{m}$, where $\lambda_{max}$ is the wavelength at which the distribution reaches a maximum for a given temperature $T$. (See Chapter 4.) If we take $T$ to be, for example, 10K, then $\lambda_{max} \cong 3 \times 10^{-4}$ m, a wavelength in the microwave regime. Because this regime had been utilized in the Second World War in radar, the technology to perform experiments to detect that radiation was at hand. But the experiments were not carried out, and the work of the Gamow group was largely ignored or forgotten.

### The Experimental Discovery of the Blackbody Radiation

In 1964, Arno Penzias and Robert Wilson were engaged at Bell Telephone Laboratories to work part time on communications satellites and part time on radio astronomy. They were attempting to do the latter with what was known as a "horn-reflector" radio telescope—so called because it looked like a large horn

• **Figure 18–7**   A. A. Penzias, R. W. Wilson, and the horn antenna they used to discover the cosmic background blackbody radiation.

(•Fig. 18–7)—but they found that their telescope was unexpectedly "noisy." They tried for almost a year to get rid of the noise when they learned that a group from Princeton University, not far away, was in the process of building a radio telescope precisely to detect the "noise" that Penzias and Wilson had already discovered and that had been predicted, unbeknownst to any of them, some 20 years earlier by Gamow and his collaborators. The noise was formed by the cosmic background photons. Penzias and Wilson's measurements covered only a small frequency range of the spectrum, and it has since been confirmed that these background photons are distributed in a blackbody spectrum. The temperature corresponding to this spectrum is about 3K; today's best value, which was determined with the use of the Cosmic Background Explorer (COBE) satellite, is the astonishingly accurate number 2.726 ± 0.002K. •Figure 18–8, plots the COBE data; it resembles nothing more than a textbook drawing of a blackbody curve. In fact it is the best blackbody curve that has ever been measured! The discovery of the cosmic background blackbody radiation, together with the redshift observations of the expansion of the universe, vastly extended and refined since Hubble's measurements, give very strong evidence in favor of a universe evolving from a big bang.

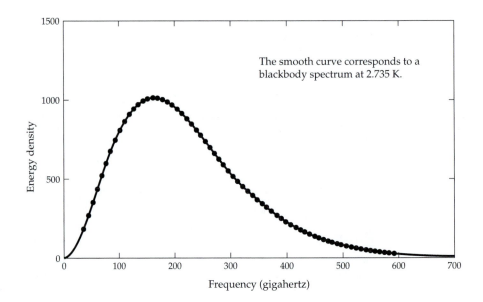

• **Figure 18–8** A plot of the intensity of the background blackbody radiation as measured by the COBE experiment.

In Example 12–4, we found that on average the number of photons per unit volume in a blackbody cavity is given as a function of temperature by

$$n_\gamma = \left( \frac{2.4}{\pi^2} \right) \left( \frac{kT}{\hbar c} \right)^3, \tag{18–16}$$

where $k = 1.381 \times 10^{-23}\,\mathrm{J \cdot K^{-1}}$ and the product $\hbar c = 3.162 \times 10^{-26}\,\mathrm{J \cdot cm}$. With $T = 2.73\mathrm{K}$ one finds that $n_\gamma = 4.13 \times 10^8/\mathrm{m^3}$. In other words, although we are quite unaware of their presence, there are, on the average, some 400 microwave photons in any cubic centimeter in the universe left over from the big bang. Incidentally, this number gives a ratio of photons to hydrogen atoms that is about

$$\frac{n_\gamma}{n_H} \cong 10^{10}. \tag{18–17}$$

Measurements of the background radiation in different directions show that it is not exactly isotropic. In the main this is not a violation of the premise of the cosmological principle that we started with. The explanation for the largest part of the anisotropy is simple and elegant: Our galaxy is in motion. It appears to be moving with respect to the background radiation in the general direction of the constellation Hydra with a speed of about 600 km/s. We know this because there is a direction-dependent Doppler shift of the cosmic background radiation that makes it appear as if the blackbody spectrum has a temperature that is higher in the direction of our motion and lower in the direction from which we have come. This is an effect of order $v/c$, in this case about $10^{-3}$—small, but readily observable. Once this effect is subtracted out, the remaining background radiation is remarkably, if not perfectly, isotropic. These remaining deviations of the temperature $\Delta T$ from the average of 2.73K are on the order of $\Delta T/T \cong 10^{-6}$. We'll come back to discuss why the deviations are important in the last section.

### The Relation between Time and Temperature in the Expansion of the Universe

The foregoing discussion makes it clear that temperature is a very important ingredient in any evolutionary development of the universe from the big bang. To approach the question of how the temperature might vary with time, recall the Stefan–Boltzmann law for the energy density (total energy per unit volume at a given temperature) of blackbody radiation, namely Eq. (4–22):

$$\rho_\gamma(T) = \frac{\pi^2}{15} \frac{(kT)^4}{(\hbar c)^3} \, \text{J/m}^3 = (8.46 \times 10^{-16})T^4 \, \text{J/m}^3. \tag{18–18}$$

This radiation contributes to the energy density that acts to slow down the expansion of the universe. If we put $T = 2.73$K into the equation, we find that at present $\rho_\gamma = 4.7 \times 10^{-14}$ J/m$^3$. We can divide by a factor of $c^2$ to find an equivalent in kg/m$^3$ and then compare that value with the critical density for closure, Eq. (18–14). When we do this, we find that $(\rho_\gamma/c^2)/\rho_c \cong 10^{-5}$. Therefore the radiation does not now have much effect on the rate of expansion of the universe; accordingly we describe the present epoch as being **matter dominated**.

However, the universe was not always matter dominated. To understand this idea we note that after the recombination time the number of photons remained constant. The photon density is the total number of photons divided by the volume, and since the volume is proportional to the cube of the scale factor $R(t)$, $n_\gamma$ is proportional to $[R(t)]^{-3}$. But $n_\gamma$ is, as we see from Eq. (18–16), also proportional to $T^3$. Thus the temperature must be inversely proportional to $R(t)$. In addition, $T(t_0) = 2.73$K ($t_0$ is now), and by definition $R(t_0) = 1$. Therefore

$$T(t) = \frac{2.73}{R(t)}. \tag{18–19}$$

We can now look back at Eq. (18–18) to see that the energy density in the blackbody radiation is proportional to $1/[R(t)]^4$. In other words, as we go back in time and $R(t)$ decreases, the energy density in the blackbody radiation increases.

We would also like to know how the mass density of particles scales with the expansion parameter $R(t)$. Most of this mass density is in nucleons, and even for times much earlier than the recombination time, there is not enough energy to make nucleon–antinucleon pairs. That means that the total number of nucleons (like the total number of photons) is fixed, and the mass density is inversely proportional to the volume of the universe—that is, to $1/[R(t)]^3$. This relationship is to be compared to the $1/[R(t)]^4$ behavior of the energy density in the radiation: As we go backwards to earlier times and smaller $R(t)$, the energy density of radiation increases *faster* than that of nucleons by a power of $1/R$. At some point in the compression (as seen in our "trip" to earlier times) the expansion rate is governed more by radiation than by matter; we say that at that point the universe becomes **radiation dominated**.

A numerical evaluation of the crossover point can be made once we know how $R(t)$ varies with time, a subject we'll discuss next. Once the numbers are put in, we find that the radiation energy density and the energy density in matter become equal at about $10^5$ yr after the big bang, which is about the recombination time. It is reassuring that this is also a time when we don't have to worry about the creation of nucleon–antinucleon pairs, an assumption we made in our discussion.

We now turn to the question of how the scale factor $R(t)$ varies with time. We can make an estimate of this without having to take into account relativity, either general or special. We'll suppose that the energy density of the universe is the critical density, the case where the expansion slows down and ends, but doesn't produce a subsequent contraction. In that case we can go back to Eq. (18–12) and set $E$ and hence $E''$ to zero:

$$0 = \left(\frac{\dot{R}}{R}\right)^2 - \frac{8\pi}{3} G\rho. \tag{18–20}$$

We can treat Eq. (18–20) as a differential equation for $R$. We can solve it only if we know $\rho(t)$, which differs according to whether we are in the radiation- or matter-dominated epochs. For the radiation-dominated epoch, we have

$$\rho(t) \cong \rho_\gamma(t) = \frac{\rho_\gamma(t_0)}{[R(t)]^4}. \tag{18–21}$$

Then Eq. (18–20) becomes

$$\frac{dR}{dt} = \sqrt{\frac{8\pi G\rho_\gamma(t_0)}{3}} \frac{1}{R(t)} = \frac{A}{R(t)}, \tag{18–22}$$

Which we may rewrite as

$$R\left(\frac{dR}{dt}\right) = A,$$

or

$$\frac{1}{2}\frac{d}{dt}R^2 = A.$$

This equation has the solution

$$R^2(t) - R^2(t_i) = 2A \times (t - t_i),$$

where $t_i$ is some initial time. If we take $R(t_i = 0) = 0$ as our initial condition, then we arrive at

$$R(t) = (2At)^{1/2}. \tag{18–23}$$

In turn, since the temperature varies inversely as $R(t)$, the temperature varies as $t^{-1/2}$ in the radiation-dominated epoch.

For the matter-dominated epoch, which starts at a crossover time $t_x$, the density is

$$\rho(t) \cong \rho_m(t) = \frac{\rho_m(t_0)}{[R(t)]^3}. \tag{18–24}$$

Then Eq. (18–20) takes the form

$$\frac{dR}{dt} = \sqrt{\frac{8\pi G\rho_m(t_0)}{3}} \frac{1}{\sqrt{R(t)}} = \frac{B}{R^{1/2}(t)}. \tag{18–25}$$

Since $R^{1/2}\, dR/dt = (2/3)\, dR^{3/2}/dt$, Eq. (18–25) can be rewritten as

$$\frac{2}{3}\frac{d}{dt}R^{3/2} = B.$$

To set the initial conditions we may take $R(t = 0) = 0$. (That is not quite right, but the error we make is a small one.)[7] With this initial condition, the solution of our equation for $R$ in the matter-dominated period is

$$R(t) = \left(\frac{3Bt}{2}\right)^{2/3}. \tag{18–26}$$

Again, since we know that the temperature is inversely proportional to $R$ in the matter-dominated period, we conclude that the temperature is currently proportional to $t^{-2/3}$.

---

[7] More properly, we should choose the crossover time $t_x$ as the time at which we set the initial conditions. That time is roughly $10^5$ yr, much less than the current value of order $10^9$ yr, and as Example 18–3 shows, $R(t_x)$ is very small compared to the present size of $R$.

We have seen that the temperature drops at a faster rate in the current matter-dominated epoch than in the radiation-dominated epoch. Keep in mind that the radiation-dominated epoch lasted only $10^5$ yr, as opposed to some 10 billion (and still counting) for the matter-dominated epoch. Nevertheless, as far as the standard cosmology is concerned, the matter-dominated epoch is pretty uneventful; all the action took place prior to $10^5$ yr, when the universe was radiation dominated.

**Example 18–3**   By what factor has the radius of the universe increased since the decoupling of radiation from matter (the recombination time)? Estimate the radius of the universe at the recombination time.

**Solution**   The single equation (18–26) describes the change in scale in the matter-dominated epoch. Since we are only asking for the factor by which the radius changes, we don't have to know $B$. The recombination time is $t = 10^5$ yr, while today we are at about $10^{10}$ yr. The factor by which the radius has increased is

$$\left[ \frac{(10^{10} \text{ yr})}{(10^5 \text{ yr})} \right]^{2/3} = 10^{10/3} \cong 2{,}000.$$

If we say the radius today is $10^{10}$ lt-yr, then at the recombination time the radius was $1/2{,}000$ of this number, $5 \times 10^6$ lt-yr, some 60 times the size of our galaxy. This number translates into roughly $5 \times 10^{22}$ m for the size of the universe at the recombination time.

# 18–5  Another Smoking Gun: Helium Abundance

One other piece of evidence points strongly to a big bang: the current abundance of helium and of some other elements in the heavens. To see how this fits in with a big bang, consider how elements are produced.

All the energy that stars emit is generated in nuclear fusion reactions. The Sun is a typical example of one of the cooler stars. Its central temperature is about $14 \times 10^6$K. Basically, what happens in the Sun is that, through a sequence of reactions, four protons are converted into a $^4$He nucleus with a net loss of mass. The difference in mass has been converted into radiant energy. Since the sequence is quite simple, we'll spell it out. The first step is the reaction[8]

$$p + p \rightarrow d + e^+ + \nu.$$

The next two reactions involve the production of the gamma rays that we eventually observe as sunlight. The first reaction is

$$d + p \rightarrow {}^3\text{He} + \gamma.$$

($^3$He is a light isotope of helium.) The final reaction is

$$^3\text{He} + {}^3\text{He} \rightarrow {}^4\text{He} + 2p.$$

Thus the overall reaction is

$$4p \rightarrow {}^4\text{He} + 2e^+ + 2\nu,$$

with a release of 26.7 MeV in kinetic and radiant energy. In the hotter stars there is a sequence that starts with carbon and a proton and ends up with carbon and helium. We see that the end result of these nuclear cycles is always to produce $^4$He. Moreover, the radiant energy produced in the cycles is known, a point that will be important in what follows.

---

[8] The solar neutrinos so produced have actually been detected, and their intensity has given us information about the nature of neutrinos.

Just how much helium is there in the universe? It is customary to introduce a quantity called $Y$ that measures the total mass of helium relative to the rest of the mass in the visible universe. Most of the matter in the visible universe is, in fact, in the form of hydrogen, both atomic and molecular. Between the stars, for example, there are giant clouds of hydrogen out of which new stars are continually forming. Accordingly the ratio $Y$ is defined as

$$Y = \frac{M_{\mathrm{He}}}{M_{\mathrm{H}} + M_{\mathrm{He}}}, \tag{18–27}$$

where $M_{\mathrm{H}}$ and $M_{\mathrm{He}}$ are the total masses of hydrogen and helium respectively. (Here we refer to $^4$He, which is by far the most abundant helium isotope.) By doing spectroscopic measurements on a variety of astronomical objects, astronomers have measured $Y$ to be approximately 0.25. In other words, $^4$He makes up about one-fourth of the nucleonic mass in the visible universe.

Can all this helium have been produced in stars? While scientists engage in lively debates as to how old the oldest galaxies are, they are certainly several billion years old and several billion years younger than the 12 to 20 billion years assigned to the age of the universe. If all the helium the astronomers observe had been produced in stellar processes, then we could calculate how much radiant energy had been produced along with the helium. It turns out that this would have been about *100 times* larger than the radiant energy we see coming from stars. In other words, not all of the observed helium can come from the stars. By itself, the preceding calculation did not prove that the observed helium has a cosmological origin, but it did suggest further study.

And indeed, the subject was investigated further by Gamow and his young colleagues in the early 1950s. Gamow realized that the conditions underlying the formation of the elements in the early universe and the conditions in the interior of stars were entirely different. In the first place all the element formation in the early universe had to take place in the first few minutes after the big bang, if for no other reason than the fact that neutrons were involved and the free neutron is unstable.[9] Stars, on the other hand, have billions of years during which fusion reactions can occur, because no free neutrons are needed.

Furthermore, the temperatures and densities in stars and the early universe are entirely different. Nuclear reactions take place at a scale of about 1 MeV or, using the fact that for $T = 300K$, $kT \cong (1/40)$ eV, at a temperature of about $10^{10}$K (10 billion kelvins). This number should be compared with the temperatures at the center of stars, some tens of millions of kelvins. Nuclear reactions are much slower at the latter temperature. Moreover, using the current photon density of $4 \times 10^8$ per m$^3$ (400 per cm$^3$) and the knowledge that the photon density varies as the cube of the temperature, we find that when the universe was at $10^{10}$K, the photon density was

$$n_\gamma(1 \text{ MeV}) = \frac{(10^{10}\text{K})^3}{(2.73\text{K})^3} \times (4 \times 10^8 \text{ m}^{-3}) \cong 10^{37} \text{ m}^{-3}.$$

To see how odd these conditions are, recall that in a mole of a standard gas there are "only" some $6 \times 10^{23}$ molecules.

Finally, over billions of years sequences of improbable fusion reactions in stars can build up elements as heavy as iron. The fact that metals like iron are created in the normal nuclear processes going on in the interior of stars gives

---

[9] It decays into a proton, an electron, and an antineutrino with a lifetime of about 15 minutes.

astronomers a method for distinguishing the "cosmological," or **primordial**, helium from the rest. They study groups of stars that exhibit little or no metallic content. But these stars do contain helium. Astronomers reason that this helium must be primordial since the processes that produce helium in stars inevitably also produce metals. The observations suggest that it is useful to work with an analogue to $Y$ that is called $Y_P$, where the subscript refers to primordial helium. By measuring a large variety of these metal-free sources, astronomers estimate that $Y_P = 0.235 \pm 0.003$, an astonishingly precise number and one which suggests by comparison with $Y$ that most helium is primordial rather than produced in stars. *But what is even more astonishing is that big bang cosmology is able to reproduce this number theoretically.*

## The Prediction of the Primordial Helium

The interaction that starts all cosmological element building is

$$p + n \rightarrow d + \gamma,$$

the capture of a neutron by a proton to form a deuteron with the emission of a gamma ray. At very low energies—meaning that the capture is made with the neutron and proton at rest—$E_\gamma = E_B$, where $E_B$ is the binding energy of the deuteron. $E_B$ is also the threshold $\gamma$-energy for the reverse process,

$$\gamma + d \rightarrow n + p,$$

that is, the photodisintegration of the deuteron. This energy is about 2 MeV, and if the temperature of the universe is above the equivalent of that value the background radiation will quickly break up any deuterons that have been formed in the capture process. Detailed calculations that take into account the photon distribution yield a different value for the temperature $T_c$ above which deuterons are broken up, namely $kT_c = 0.086$ MeV, well below the 2.23-MeV deuteron binding energy. It can also be shown from the equations that describe the big bang that this temperature is reached at about three minutes after the big bang.

Once the universe has been cooled below $T_c$, deuterons are stable and $^4$He is formed very rapidly by short sequences of nuclear reactions, the most significant of which is

$$d + d \rightarrow t + p$$

followed by

$$d + t \rightarrow {}^4\text{He} + n,$$

where $t$ is the triton, the isotope $^3$H. In a similar sequence, the triton is replaced by a $^3$He nucleus and the roles of the neutron and proton are interchanged. This is how the cosmological helium gets produced. To get the *amount* of helium produced requires further calculations involving the rates of these reactions. Suffice it to say that by using known physics the observed value of $Y_P$ is obtained.

Something emerges from the calculations that illustrates how one can turn things around and learn something about terrestrial physics from our astronomical observations. The equations governing helium production depend on the energy density $\rho$. In particular, helium production is increased when more particles contributing to $\rho$ are present. Neutrinos are likely candidates to put into $\rho$ in the period when nuclei form. Now we know from experiments done at large accelerators that there are at least three types of neutrinos. (See Chapter 16.) But are there more types than these experiments have discovered? The calculations of helium production in the big bang show that for each new type of neutrino $Y_P$ is increased by some six percent. But the experiments give, as we have seen, a very precise value for $Y_P$ that appears to be consistent with exactly three types of neutrino—the same number of neutrino types is found in the accelerator studies. This is another success of the standard big bang cosmology.

## 18–6 Beyond the Big Bang

The big bang picture has, as we have seen, testable consequences. Indeed, it is so successful that it is now referred to in the literature as the **standard model of cosmology**. However the standard model presents at best an incomplete picture. To start with, there is no mechanism that explains why the big bang occurred in the first place—the big bang itself is simply hypothesized. In addition, there are many issues, even given this hypothesis, that the big bang alone does not address. In what follows we discuss some of these issues.

Recall our earlier remark that there is a natural time unit, the **Planck time** $t_P$, that is obtained by constructing a quantity with the dimensions of time out of $G$, $\hbar$, and $c$, namely

$$t_P = \sqrt{\frac{\hbar G}{c^5}} \cong 5.4 \times 10^{-44} \text{ s}. \tag{18–28}$$

The fact that $G$ and $\hbar$ appear in the equation for $t_P$ tells us that this time has to do with quantum physics and gravity. For times earlier than $t_P$ in the universe's evolution we must combine quantum physics and gravity. But we have no convincing quantum theory of gravity at present, so we simply do not know how to deal with the physics that obtains for $t < t_P$, when the universe is packed into a very small space indeed.

Another issue not addressed by the standard model has to do with just how homogeneous the universe is. We know that if we look closely enough homogeneity fails: Earth, the Sun, the Milky Way galaxy—all these are really inhomogeneities. In fact, galaxies come in clusters and in clusters of clusters, and there are regions of vast emptiness where there are relatively few galaxies. These structures—the clusters and the holes—have lengths of about one-tenth the length scale of the universe as a whole. We will be learning much more about such structures and about whether there are even larger structures over the first decade of the 21st century. Just where did the universe's structure come from? A *perfectly* homogeneous blackbody background radiation leaves no place for structure.

In fact, this problem has a reverse side to it that is even more serious than the preceding paragraph would indicate. In the standard model, whatever we see now in the way of the background radiation evolved directly from what must have existed at the time of electron–proton recombination. It was at that time that the radiation was "released" to expand freely with the universe. We have already estimated, in Example 18–3, the factor by which the size of the universe at the recombination time differed from the present size of the universe, a factor of about 2,000. The recombination time was about $10^5$ years, or about $3 \times 10^{12}$ seconds. If we multiply this figure by the speed of light we find the size of a "connected domain," a region within which signals traveling no faster than the speed of light could travel. The numerical size of the connected domain was $(3 \times 10^{12} \text{ s})(3 \times 10^8 \text{ m/s})$, or about $10^{21}$ m. Events separated from each other by more than this amount cannot be causally connected to each other. In Example 18–3, we saw that the size of the universe at the recombination time was some $5 \times 10^{22}$ m, 50 times larger than the size of the connected domain. In other words, the universe at the recombination time consisted of

$(50)^3 \cong 10^5$ causally disconnected volumes. When this number is extrapolated to our time, we find that if we look at two points in the sky separated by more than one degree—corresponding to twice the Moon's diameter as seen from Earth—then these two points contain light (and in particular background black-body radiation) from distant parts of the universe that were causally disconnected at the recombination time.

But let's think what this means. How did all these unconnected domains "know" enough to produce blackbody distributions with the same temperature—the 2.73K that we see everywhere in the sky—to one part in a hundred thousand? To smooth out the temperature through the transmission of information across different domains would require signals moving faster than light! This possibility would violate our central dogma that the laws of physics are the same everywhere and were the same at that earlier epoch. This is the dilemma known as the **horizon problem.**

Another feature left unexplained by the standard model has to do with the density of the universe. In Section 18–3 we described the significance of the parameter $\Omega$, defined by Eq. (18–15). A value of $\Omega = 1$ corresponds to a universe with a flat geometry—one that slows to a halt but does not fall back on itself. The measured value of this parameter, which lies between 0.1 and 0.2, is the value of $\Omega$ *now*, $\Omega(t_0)$—if $\Omega$ were *exactly* unity, it would not change with time. But if $\Omega$ is not exactly unity, then *it too will evolve with the universe*. To reach a value of 0.2 today, the standard model finds that the value of $\Omega$ at the earliest time we feel confident about understanding the evolution mechanisms, about $10^{-35}$ s, had to differ from unity by *one part in $10^{57}$*! In other words, the history of our universe is very delicately balanced between the two cases of return and escape. We would very much like to know how to explain an initial condition with $\Omega = 0.99999...999$ (57 nines); in the standard model it is simply an unexplained constant. This is what is known as the **flatness problem**.

## Inflation

In recent years progress has been made towards realizing an idea that could resolve the three issues of inhomogeneity, the horizon problem, and the flatness problem. This compelling mechanism was suggested in 1980 by the American cosmologist Alan Guth, and it has become the basis of much recent work in theoretical cosmology. It goes under the name **inflation**. While no single dynamical model that can produce the desired effect has emerged as a favorite, at least there *are* dynamical models for it, unlike the big bang, for which we do not have any dynamical model. These different models differ in their predictions only for the fine details.

Here is a sketch of the idea of inflation: At a time of about $10^{-35}$ s, which is about as far back as we can go with our present understanding of short-distance physics, the universe was tiny; its radius was $10^{-27}$ m, $10^{12}$ times smaller than the radius of a proton. At that time, all of the universe was causally connected. The something happened—exactly what is not agreed on among the theorists, but there are candidate explanations—to put the universe into a very fleeting epoch of *exponential expansion*, wherein $R(t) \propto \exp(Kt)$. Such an expansion corresponds to a solution of Eq. (18–20) with a density that is independent of time. We can choose the constant $K$, which should eventually come from some theory, so that in a fraction of a second the universe expands enormously. This epoch in the universe's evolution is the "inflation" period; it lasts only a minuscule time and "shuts off" due to some dynamical mechanism, at which point the type of expansion we discussed earlier takes over.

The causal connectedness of the universe when inflation takes over permits a single characteristic temperature at that time, and because the inflation is uniform, there is a single temperature remaining even when the universe has broken into disconnected pieces, at the end of the inflation. In this way inflation resolves the horizon problem.

Inflation also addresses the flatness problem. The inflationary period is special in the way the radius of the universe evolves, and it is also special in the way the density parameter $\Omega$ evolves: *towards unity*. An analogue that can help here is to think of a balloon being blown up to a very large size. As the balloon expands, any surface irregularities or surface curvature smooths out. In the inflation picture the universe smooths out so much that it becomes flat, meaning that $\Omega = 1$ is a clear prediction of inflation. The data on the density parameter of the universe is based on studies of large clusters of galaxies and on extensions of the work on primordial helium to other light nuclei. These data yield a value of $\Omega = 0.3 \pm 0.1$, including dark matter that is associated with the orbital motion of the large clusters. So how do we get to $\Omega = 1$? One solution is to postulate the existence of dark matter distributed more uniformly over the universe—that is, dark matter not associated with galactic clusters. Particle theorists have potential candidates for exotic contributions to dark matter, and in addition some believe that quantum vacuum fluctuations may contribute significantly to the energy that could increase the value of $\Omega$. The search for contributions to $\Omega$ is a very active one.

Finally, the inhomogeneities in the universe are understood in the inflation scenario as being due to *quantum fluctuations* in the universe just before inflation starts. These fluctuations are blown up during inflation along with everything else, and they form places where matter can clump or where inhomogeneities in the blackbody radiation can form. This is a crucial observation, because it turns out to be the key way to test inflation against observation. Inflation does very well in its prediction of the size of inhomogeneities in the blackbody radiation. And inflation predicts some striking features having to do with the size and distribution of both galactic clusters and empty spaces. Observations about these structures that are extensive enough to test inflation represent a major target of astronomical work for the first decades of the 21st century.

There is one other loose end, having to do with the number of background photons compared to the number of hydrogen atoms, the $10^{10}$ factor of Eq. (18–17). One might expect the universe to have started out neutral in every possible way, with baryon number zero—that is, an equal number of protons and antiprotons or, more correctly, an equal number of quarks and antiquarks of each type. Why then should there be any protons left—protons that were not annihilated by antiprotons in the first second or so of the existence of the universe? Some current theories about elementary particles allow for this asymmetry and even for a numerical value on the order of $10^{10}$ for the ratio of background photons to hydrogen atoms. More data are needed, both from the search for dark matter and from accelerator physics.

Cosmology is a fascinating subject to end on. It combines all the fundamental fields of physics—quantum physics, particle physics, the physics of large collections of matter, and special and general relativity—in a search for something of the greatest importance to us: the history of our universe. In the next 10 years or so a profound change in this field will come in one way or another. Soon we will see clearly to the edge of the universe, as far back as it is possible to see. We will thus reach a goal that was set when our neolithic ancestors began to gaze at the sky.

# SUMMARY

The fact that the speed of light is finite means that observations of the distant universe are observations of how things were long ago. Several features dominate our observations of the universe as a whole:

- The universe is isotropic.
- The universe is on the average homogeneous.
- The universe is expanding according to Hubble's law; that is, it is expanding in a way that is consistent with the preceding two observations: Any part of the universe is moving away from us with a speed proportional to its distance from us. The most distant parts of the universe we can see are some 10 billion years in the past, and in this region things are moving away from us with speeds that are sizable fractions of the speed of light.
- An important additional question having to do with the future evolution of the universe is whether its geometry—its curvature—is flat, open, or closed. This question concerns the average mass density of the universe and is still under investigation.

Centered around the observations just described is a set of questions having to do with the early evolution of the universe. These questions bring in all the subjects of modern physics that we have studied in this book. The idea that the universe is expanding contains within it the question of how the expansion began and leads to the big bang model, which hypothesizes that the universe began from a single point at an infinite temperature some 10 to 15 billion years ago. As the universe expands in this picture it cools, passing through regimes of temperatures that correspond to those characteristic of particle physics, then nuclear physics, and then atomic physics. The evidence for such a picture is much stronger than the simple extrapolation of Hubble's law. It includes the following observations:

- The presence of a background blackbody radiation that, having evolved along with the universe, is currently at a temperature of around 3K. This radiation decoupled from matter some 100,000 years after the big bang.
- The fraction of helium present in the universe. This number corresponds to how nuclear processes would have produced helium in the "oven" of the universe when it was several minutes old.

Still earlier times in the universe correspond to higher energies. In effect the early universe provides us with a high-energy accelerator laboratory, with energies much higher than any terrestrial laboratory can afford. The laws of physics at these huge energies are constrained by the requirement that the early universe evolved into our present observable universe.

# QUESTIONS

**1.** Here is a way of imagining that you might beat Olbers' paradox: You could suppose that, like blackbodies, the nearby stars absorb all the light from the further-away stars, so that our night sky is not ablaze. Why does this not work? Think about the properties of a blackbody.

**2.** The star closest to us is known as $\alpha$ Centauri. As the Earth moves from one side of its orbit to the other this star appears to shift its position by a tiny amount. Why? The angle through which the star shifts is called its parallax.

For $\alpha$ Centauri the parallax is 0.760″. How would you use this number to find the distance from here to the star? For stars much further away this angle is too small to be measured, and the method won't work.

**3.** Why did the Planck system of natural units have to wait for Planck? Try to construct a set of units out of the charge on the electron, its mass, the speed of light, and $G$, and see what you can come up with. You will be able to find a distance, but this distance will not depend on constants that are independent of the properties of any particular particle. In that sense it is not "natural."

**4.** Some scientists have suggested that some of the constants of nature may not in fact be constant—that they vary slightly with time. Suppose the fine-structure constant $\alpha = e^2/4\pi\varepsilon_0\hbar c$ changed by some fraction over a period of $10^{10}$ yr. What kind of experiment could you suggest to test this hypothesis?

**5.** Our current ideas about particle physics allows us to explain, at least qualitatively, the observational result that $n_H/n_\gamma \cong 10^{-10}$. Some people argue that the observed value of $n_H$ is obtained by assuming that all the matter we see is in fact matter and that none of it is antimatter. On grounds of symmetry, however, one would like to assume that equal amounts of matter and antimatter were created in the big bang. What would an experimentalist have to look for to check whether half of the observed matter is really antimatter?

**6.** Cosmologists have suggested one way that would allow the universe to be closed yet still not have enough matter to reach the critical density would be to add a new force acting between matter. Would this force have to be an attractive or a repulsive force?

**7.** In the expanding universe, the temperature at which recombination took place and the radiation ceased to interact with matter was 3000K. So why is it that today we see this radiation at a much lower temperature?

**8.** Hubble's law is consistent with our position being equivalent to all other positions in the universe. Yet Hubble's law seems to pick *us* out quite precisely: A galaxy $n$ times as far from *us* as another galaxy moves with $n$ times the velocity with respect to *us* as the other galaxy. Is there an inconsistency here?

## PROBLEMS

**1.** ▮ Consider a one-dimensional model of the universe in which the galaxies, including our own, were equally spaced. Show that if all the galaxies moved away from us with equal speeds, that would single out our location as a special one (i.e., homogeneity would be lost).

**2.** ▮ Consider Example 13–1, and suppose that the present measured value of the ratio of the abundance of uranium isotopes, $^{235}U/^{238}U$, were 0.07 rather than 0.007. How would this modification change the time since the formation of the heavy elements? Similarly, supposing that the initial value of the ratio were 3.3 rather than 1.65, how does the result change?

**3.** ▮▮ Use Eq. (18–12) to show that if $E$ is negative, $H$ must be *below* a certain limit, whereas if $E$ is positive, $H$ must lie *above* a certain limit. In each case, what is the limit? How does your answer show that the universe expands forever for negative $E$ or eventually recollapses for positive $E$? [*Hint*: Consider Eq. (18–7).]

**4.** ▮▮ In Example 18–2 we found the value of the Hubble parameter when the density of the universe was that of Earth, assuming that the universe was flat. Estimate the value of the radius of the universe under those conditions.

**5.** ▮▮ Suppose that the galaxy is viewed as consisting of a central mass $M$ (the visible matter) and dark matter distributed with a density $\rho(r) = K/r$ for $r > R_0$. What is the speed of a star moving in a circular orbit of radius $R$ about the center, with $R > R_0$?

**6.** ▮▮ How much time will have elapsed since the big bang when the temperature of the background blackbody radiation is dominated by radiation with wavelengths around 1 m? At that time, what will be the value of the ratio $(\rho_\gamma/c^2)/\rho_c$? (Don't forget that the value of $\rho_c$ will have changed.)

**7.** ▮ Our galaxy moves relative to the blackbody background radiation with a speed of roughly 300 km/s. By how much is the frequency of this radiation shifted due to the movement in the direction towards the radiation at the peak frequency of the radiation? In the direction away from the radiation at this same peak frequency? It is possible to subtract the effect of the motion because its effect can be found for any direction.

**8.** ▮▮ In Eq. (18–27) we define the ratio $Y$. This quantity is measured to be about 0.25. Neglect the binding energy of the nuclei involved as well as the difference between neutron and proton masses in order to express $Y$ in terms of the proton mass and the numbers of helium and hydrogen nuclei in the universe; you should find that the proton mass cancels from the expression. Solve the equation you come up with in order to find the ratio of the number of helium nuclei to the number of hydrogen nuclei.

**9.** ▮▮ We stated that if somehow the mass density of the universe were constant with time, the radius of the universe would grow exponentially with time. Prove that this is the case by investigating the solution of Eq. (18–20) with a constant value of $\rho$.

**10.** ▮▮ **(a)** Use the physical constants $G$, $\hbar$, and the speed of light $c$, together with dimensional analysis to construct a quantity with the dimension of time. You will have found the *Planck time*. **(b)** Construct a natural mass and a natural length from these quantities—the *Planck mass* and *Planck length*, respectively. What are the numerical values of your constants?

**11.** ▮▮▮ Using only two-body collisions and strong interactions, find the sequences of collisions that can produce $^4$He starting with deuterium, protons, and neutrons. You may have occasion to produce the nucleus $^3$H or the nucleus $^3$He along the way. Note also that there are no $nn$ or $pp$ nuclei.

**12.** ▮▮▮ Starting from Eq. (18–11) show that the constancy of $E'$ implies that

$$H^2(t) - \frac{8\pi G}{3}\rho(t) = \left[ H^2(t_0) - \frac{8\pi G}{3}\rho(t_0) \right]\left( \frac{R(t_0)}{R(t)} \right)^2.$$

**13.** ▮▮▮ Use the result of the previous problem to derive the relation

$$\Omega(t) = \frac{\Omega(t_0)}{\Omega(t_0) + \left(1 - \Omega(t_0)\right)\left(R(t)/R(t_0)\right)^{n-2}},$$

where $n = 4$ for the radiation-dominated times and $n = 3$ for the matter-dominated times.

**14.** ▮▮ If $\Omega(t_0) = 0.2$ ($t_0$ is now), what was $\Omega(t_x)$, where $t_x$ is the time the matter dominated epoch started, about $10^5$ yr? Use the result of Example 18–3 that the radius of the universe has increased by a factor of 2,000 since then.

# Appendix A
## Tables

### Table A–1    Some Fundamental Constants

| Constant | Symbol | Value |
|---|---|---|
| Speed of light in vacuum | $c$ | $2.99792458 \times 10^8$ m/s |
| Elementary charge | $e$ | $1.60217738 \times 10^{19}$ C |
| Electron rest mass | $m_e$ | $9.1093897 \times 10^{-31}$ kg |
| Permittivity constant | $\varepsilon_0$ | $8.85418781762 \times 10^{-12}$ F/m |
| Permeability constant | $\mu_0$ | $1.25663706143 \times 10^{-6}$ H/m |
| Electron charge-to-mass ratio | $e/m_e$ | $1.75881961 \times 10^{11}$ C/kg |
| Proton rest mass | $m_p$ | $1.6726230 \times 10^{-27}$ kg |
| Neutron rest mass | $m_n$ | $1.6749286 \times 10^{-27}$ kg |
| Muon rest mass | $m_\mu$ | $1.8835326 \times 10^{-28}$ kg |
| Planck constant | $h$ | $6.6260754 \times 10^{-34}$ J·s |
| Planck constant/$2\pi$ | $\hbar$ | $1.0546 \times 10^{-34}$ J·s |
| Boltzmann's constant | $k$ | $1.380657 \times 10^{-23}$ J/K $= 8.617 \times 10^{-5}$ eV/K |
| Avogadro's number | $N_A$ | $6.0221367 \times 10^{23}$ mol$^{-1}$ |
| Universal gas constant | $R$ | $8.314510$ J/mol·K |
| Stefan–Boltzmann constant | $\sigma$ | $5.67050 \times 10^{-8}$ W/m²·K⁴ |
| Rydberg constant | $R_y$ | $1.0973731534 \times 10^7$ m$^{-1}$ |
| Gravitational constant | $G$ | $6.67260 \times 10^{-11}$ m³/s²·kg |
| Bohr radius | $a_0$ | $5.29177249 \times 10^{-11}$ m |
| Electron magnetic moment | $\mu_e$ | $9.2847700 \times 10^{-24}$ J/T |
| Proton magnetic moment | $\mu_p$ | $1.41060761 \times 10^{-26}$ J/T |
| Fine-structure constant | $\alpha = \dfrac{e^2}{4\pi\varepsilon_0\hbar c}$ | $1/137.035982$ |
| Bohr magneton | $\mu_B = \dfrac{e\hbar}{2m_e}$ | $9.2740154 \times 10^{-24}$ J/T |
| Solar mass | $M_{\text{sun}}$ | $1.989 \times 10^{30}$ kg |
| Solar radius | $R_{\text{sun}}$ | $6.9598 \times 10^8$ km |
| Earth mass | $M_{\text{earth}}$ | $5.977 \times 10^{24}$ kg |
| Earth radius | $R_{\text{earth}}$ | $6.3782 \times 10^5$ km |

### Conversion Factors

| | |
|---|---|
| 1 radian | $57.30°$ |
| 1 kilometer | 0.6214 mi |
| 1 angstrom | $10^{-10}$ m |
| 1 fermi | $10^{-15}$ m |
| 1 light year | $9.460 \times 10^{12}$ km |
| 1 electron volt | $1.602 \times 10^{-19}$ J |
| 1 tesla | $10^4$ gauss |

## Table A–2    Properties of Some Isotopes

In the table that follows, the masses are given in terms of the so-called atomic mass unit u. In these units, $^{12}C$ is defined to have a mass of exactly 12.0000, so that hydrogen has a mass of 1.0078 u. In kilograms, 1 u = $1.6605402 \times 10^{-27}$ kg. The masses are given for the neutral atom, so that the very small electron mass contribution of the Z electrons is included. The other entries in the table should be self-explanatory. Note that if an abundance is given as 100%, the other isotopes are unstable.

| Atomic Number Z, Mass Number A | Element | Symbol | Atomic Mass | Percent Abundance or Decay Mode | Half-life if Unstable |
|---|---|---|---|---|---|
| **Light Elements** | | | | | |
| 0, 1 | Neutron | $n$ | 1.008665 | $\beta^-$ | 10.4 min |
| 1, 1 | Hydrogen | H | 1.007825 | 99.985 | |
| 1, 2 | Deuterium | D | 2.014102 | 0.015 | |
| 1, 3 | Tritium | T | 3.016049 | $\beta^-$ | 12.33 yr |
| 2, 3 | Helium | He | 3.016020 | 0.000137 | |
| 2, 4 | | | 4.002602 | 99.999863 | |
| 3, 6 | Lithium | Li | 6.015121 | 7.5 | |
| 3, 7 | | | 7.016003 | 92.5 | |
| 4, 7 | Beryllium | Be | 7.016928 | $\gamma$ | 53.29 days |
| 4, 9 | | | 9.012182 | 100 | |
| 5, 10 | Boron | B | 10.012936 | 19.9 | |
| 5, 11 | | | 11.009305 | 80.1 | |
| 6, 11 | Carbon | C | 11.011433 | $\beta^+$ | 20.385 min |
| 6, 12 | | | 12.000000 | 98.90 | |
| 6, 13 | | | 13.003355 | 1.10 | |
| 6, 14 | | | 14.003242 | $\beta^-$ | 5,730 yr |
| 7, 13 | Nitrogen | N | 13.005738 | $\beta^+$ | 9.965 min |
| 7, 14 | | | 99.63 | | |
| 7, 15 | | | 0.37 | | |
| 8, 15 | Oxygen | O | 15.003065 | $\beta^+$ | 122.24 s |
| 8, 16 | | | 15.994915 | 99.76 | |
| 8, 18 | | | 17.999160 | 0.20 | |
| **Medium Elements** | | | | | |
| 17, 35 | Chlorine | Cl | 34.9688853 | 75.77 | |
| 17, 37 | | | 36.965903 | 24.23 | |
| 18, 40 | Argon | Ar | 39.962384 | 99.600 | |
| 19, 39 | Potassium | K | 38.963708 | 93.2581 | |
| 19, 40 | | | 39.96400 | $\beta, \gamma$ | $1.227 \times 10^9$ yr |

*continued*

| Atomic Number Z, Mass Number $A$ | Element | Symbol | Atomic Mass | Percent Abundance or Decay Mode | Half-life if Unstable |
|---|---|---|---|---|---|
| | | | **Medium Elements** | | |
| 28, 58 | Nickel | Ni | 57.935346 | 68.077 | |
| 28, 60 | | | 59.930789 | 26.233 | |
| 29, 63 | Copper | Cu | 62.929599 | 69.17 | |
| 29, 65 | | | 64.927791 | 30.83 | |
| 30, 64 | Zinc | Zn | 63.92914 | 48.6 | |
| 30, 66 | | | 65.926035 | 27.9 | |
| | | | **Heavy Elements** | | |
| 82, 206 | Lead | Pb | 205.974440 | 24.1 | |
| 82, 207 | | | 206.975871 | 22.1 | |
| 82, 208 | | | 207.976627 | 52.4 | |
| 82, 210 | | | 209.984163 | $\beta^-, \gamma, \alpha$ | 22.3 yr |
| 82, 211 | | | 210.9888734 | $\beta^-, \gamma$ | 36.1 min |
| 82, 212 | | | 211.991872 | $\beta^-, \gamma$ | 10.64 hours |
| 82, 214 | | | 213.999798 | $\beta^-, \gamma$ | 26.8 min |
| 84, 210 | Polonium | Po | 209.982848 | $\alpha, \gamma$ | 138.376 days |
| 84, 214 | | | 213.995177 | $\alpha, \gamma$ | 164.3 $\mu$s |
| 88, 226 | Radium | Ra | 226.025402 | $\alpha, \gamma$ | 1,600 yr |
| 89, 227 | Actinium | Ac | 227.027749 | $\beta^-, \gamma, \alpha$ | 21.773 yr |
| 90, 228 | Thorium | Th | 228.028716 | $\alpha, \gamma$ | 1.9131 yr |
| 90, 231 | | | 231.036299 | $\beta^-$ | 25.52 hours |
| 90, 232 | | | 232.038051 | 100%; $\alpha, \gamma$ | $1.405 \times 10^{10}$ yr |
| 91, 231 | Proactinium | Pa | 231.037131 | $\alpha, \gamma$ | $3.276 \times 10^4$ yr |
| 92, 232 | Uranium | U | 232.037131 | $\alpha, \gamma$ | 68.9 yr |
| 92, 233 | | | 233.039630 | $\alpha, \gamma$ | $1.592 \times 10^5$ yr |
| 92, 235 | | | 235.043924 | 0.720; $\alpha, \gamma$ | $7.038 \times 10^8$ yr |
| 92, 236 | | | 236.045562 | $\alpha, \gamma$ | $2.3415 \times 10^7$ yr |
| 92, 238 | | | 238.050784 | 99.2745; $\alpha, \gamma$ | $4.468 \times 10^9$ yr |
| 92, 239 | | | 239.054289 | $\beta^-, \gamma$ | 23.50 min |
| 93, 239 | Neptunium | Np | 239.052932 | $\beta^-, \gamma$ | 2.355 days |
| 94, 239 | Plutonium | Pu | 239.052157 | $\alpha, \gamma$ | 24,119 yr |

# Appendix B
## A Mathematical Tool Chest

### B–1 Vectors

Vectors are quantities with both magnitude and direction that obey simple addition rules. We'll assume that you have learned to represent, add, and subtract vectors in your first course in physics and restrict our discussion to a simple reminder of certain operations. We denote the vector $\vec{A}$ with boldface notation and an arrow above the symbol. The vector $\vec{A}$ is represented graphically by an arrow whose direction specifies the direction of $\vec{A}$ and whose length is the magnitude of $\vec{A}$. By setting down a coordinate system in the space where the vector is located we can characterize the vector by its coordinates in that space. Consider, for example, the simple set of Euclidean coordinates $(x, y, z)$. The components of $\vec{A}$ in this system are the projections of $\vec{A}$ along the three coordinate axes, $A_x, A_y$, and $A_z$. We can also define the dimensionless unit vectors $\vec{i}, \vec{j}$, and $\vec{k}$ along the axes $x, y$, and $z$, respectively. Then the vector $\vec{A}$ can be represented in the following equivalent ways:

$$\vec{A} = (A_x, A_y, A_z) = A_x\vec{i} + A_y\vec{j} + A_z\vec{k}.$$

The dimensions of $\vec{A}$ are contained in its components.

There are two ways two vectors can be multiplied. Suppose the vectors $\vec{A}$ and $\vec{B}$ make an angle $\theta$ when they are placed tail to tail (Fig. B–1). Then the scalar product (or dot product) of $\vec{A}$ and $\vec{B}$ is a scalar given by

$$\vec{A} \cdot \vec{B} = AB \cos\theta.$$

The vector product (or cross product) $\vec{A} \times \vec{B}$ is a third vector $\vec{C}$ whose magnitude is given by

$$C = AB \sin\theta.$$

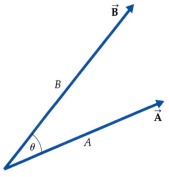

Fig. B–1

The direction of $\vec{C}$ is defined by a right-hand rule sketched in Fig. B–2: $\vec{C}$ is perpendicular to the plane defined by $\vec{A}$ and $\vec{B}$ with its orientation given by the thumb when the fingers of the right hand are aligned with $\vec{A}$ and curled towards $\vec{B}$. With this definition it is easy to convince yourself that

$$\vec{A} \times \vec{B} = -\vec{B} \times \vec{A}.$$

Keep in mind that the scalar product has maximum magnitude when the two vectors that form it are parallel to each other and vanishes when the two vectors that form it are perpendicular to each other, while the vector product has maximum magnitude when the two vectors that form it are parallel and vanishes when those vectors are perpendicular.

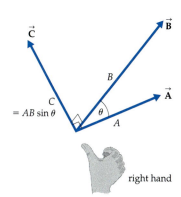

right hand

Fig. B–2

## B–2 Calculus

In this section we assume that you are familiar with the interpretations of derivatives and integrals and restrict ourselves to a few general results, as well as some special ones of interest for this course.

### Some General Rules

Fundamental rule of calculus:

$$f(x) = \frac{d}{dx}\left( \int^x f(x')\,dx' \right) \text{ or, equivalently, } f(x) = \int^x \frac{df}{dx'}\,dx'$$

chain rule:
$$\frac{d}{dx} F[g(x)] = \frac{dF}{dg}\frac{dg}{dx};$$

$$\frac{d}{dx}[f(x)g(x)] = g\frac{df}{dx} + f\frac{dg}{dx}.$$

### Some Simple Derivatives

$$\frac{d}{dx}(ax^n) = anx^{n-1};$$

$$\frac{d}{dx}(e^x) = e^x;$$

$$\frac{d}{dx}[\sin(x)] = \cos(x);$$

$$\frac{d}{dx}[\cos(x)] = -\sin(x).$$

### Taylor Expansion

This rule is useful when one knows the value of a function for some given argument $x_0$ and wants to find the function for arguments near $x_0$. We suppose that $x$ is small compared to $x_0$. Then

$$f(x_0 + x) = f(x_0) + x\frac{df(z)}{dz}\bigg|_{z=x_0} + \frac{x^2}{2!}\frac{d^2f(z)}{dz^2}\bigg|_{z=x_0} + \ \dots.$$

Some special cases of Taylor expansions follow. The first is the **binomial expansion**; the series terminates if $n$ is an integer:

$$(a + b)^n = a^n + na^{n-1}b + \frac{n(n-1)}{2!}a^{n-2}b^2 + \ \dots.$$

The following expansions are good to $O(x^3)$, where the magnitude of $x$ is less than unity:

$$e^x = 1 + x + \frac{x^2}{2!} + \frac{x^3}{3!} + \ \dots;$$

$$\sin x = x - \frac{x^3}{3!} + \ \dots;$$

$$\cos x = 1 - \frac{x^2}{2!} + \dots ;$$

$$\frac{1}{1 + x} = 1 - x + x^2 - x^3 + \dots ;$$

$$\ln(1 + x) = x - \frac{x^2}{2} + \frac{x^3}{3} - \dots ;$$

$$\sqrt{1 + x} = 1 + \frac{x}{2} - \frac{x^2}{8} + \frac{x^3}{16} + \dots .$$

## Some Simple Integrals

Using the fundamental rule and our simple derivatives, we can immediately deduce some simple integrals:

$$\int x^n \, dx = \frac{x^{n+1}}{n} ;$$

$$\int e^x \, dx = e^x ;$$

$$\int \sin(x) \, dx = -\cos(x);$$

$$\int \cos(x) \, dx = \sin(x).$$

## Partial Integration

This useful rule follows from the fundamental rule of calculus and the rule for the derivative of a product of functions:

$$\int g(x) \frac{df(x)}{dx} \, dx + \int f(x) \frac{dg(x)}{dx} \, dx = f(x)g(x).$$

For an example of how of this rule is used, suppose $g(x) = x$ and $f(x) = e^x$. Then $df/dx = e^x$, and the rule gives

$$\int xe^x \, dx = xe^x - \int e^x \left(\frac{d}{dx} x\right) dx = xe^x - \int e^x \, dx = (x - 1)e^x.$$

## Some Special Integrals

The following integrals are of direct use in the text:

$$\int_{-\infty}^{+\infty} \exp(-Ax^2) \, dx = \sqrt{\frac{\pi}{A}} \equiv I(A);$$

$$\int_{-\infty}^{+\infty} x^2 \exp(-Ax^2) \, dx = \frac{dI(A)}{dA} = \frac{\sqrt{\pi}}{2A^{3/2}}$$

$$\int_0^\infty \frac{x^3}{e^x - 1} \, dx = \frac{\pi^4}{15}.$$

Generally speaking, the evaluation of these integrals involves advanced methods. (The first two are so-called Gaussian integrals.)

## B–3 Probabilities and Probability Distributions

Probabilities enter physics when many outcomes of an experiment are possible. These outcomes could be discrete, as in the number of counts of a Geiger counter, or they could be continuous, as in the angular deflection of one object colliding with another. We'll treat these cases separately and give examples of each as we proceed.

### Discrete Distributions

To describe the possible results of an experiment in which the measurements are of $j_{max}$ different discrete quantities labeled with integers, we can imagine performing the experiment many times, say $N_{tot}$. Then $N(1)$ times the result can be labeled with the discrete integer 1, $N(2)$ with the discrete integer 2, and so forth. The collection $\{N(1), N(2), \ldots, N(j_{max})\}$ describes the results of our set of experiments. We must have the condition

$$\sum_{j=1}^{j_{max}} N(j) = N_{tot}. \tag{B.3–1}$$

We can make the passage from numbers to probabilities by dividing each $N(j)$ by $N_{tot}$. Then the probability $P(j)$ of having made a measurement with the result labeled by $j$ is the fraction of experiments that give the result labeled by $j$; that is,

$$P(j) = \frac{N(j)}{N_{tot}}. \tag{B.3–2}$$

The sum of these probabilities adds to unity, a result that in the language of probabilities represents a certainty. We interpret this state of affairs as meaning that the probability of measuring the result $j_1$ *or* the result $j_2$ is the sum of the probabilities of measuring each one separately [i.e., $P(j_1) + P(j_2)$], and the probability of measuring *some* result in an entire series of $N_{tot}$ experiments is unity. This result follows immediately from Eq. (B.3–1) and the definition of $P(j)$:

$$\sum_{j=1}^{j_{max}} P(j) = \sum_{j=1}^{j_{max}} \frac{N(j)}{N_{tot}} = 1. \tag{B.3–3}$$

We can also think about *joint probabilities*, in which two or more discrete quantities are measured. For example, we could imagine setting an array of Geiger counters around a radioactive source, with the discrete result registering the particular counter that lights up. It is possible that more than one lights up. The probability $P(j_1, j_2)$ that the discrete result labeled with $j_1$ *and* that labeled with $j_2$ occur at the same time is the *product* of the two probabilities $P(j_1)$ and $P(j_2)$;

$$P(j_1, j_2) = P(j_1) \times P(j_2). \tag{B.3–4}$$

When this equation holds we say the two measurements in question are *independent*. In determining joint probabilities, one is not restricted to the same type of measurement. We illustrate this statement in Example B.3–1.

Knowing all the probabilities we can compute averages for our series of measurements. These averages are sums weighted with the probabilities. Generally the average value (also called the *expectation value*) of any function of the discrete variable $f(j)$ is the weighted sum of $f(j)$:

$$\langle f(j) \rangle = \sum_{j=1}^{j_{max}} f(j)P(j). \qquad (\text{B.3–5})$$

Note the use of angle brackets.

**Example B.3–1**    Consider the city of Middletown, with $N_{tot} = 200$ buildings, each with a variety of stories ranging from 1 to 12. For this example, a complete description of Middletown is a set of data giving the number of stories in each of these buildings. Thus $N(1)$ is the number of buildings with one story, $N(2)$ the number with two stories, and so forth. The following table gives the data:

| Number of Stories, $j$ | Number of Buildings with $j$ Stories, $N(j)$ |
|:---:|:---:|
| 1 | 45 |
| 2 | 63 |
| 3 | 37 |
| 4 | 16 |
| 5 | 14 |
| 6 | 5 |
| 7 | 4 |
| 8 | 3 |
| 9 | 6 |
| 10 | 2 |
| 11 | 3 |
| 12 | 2 |

You can verify that Eq. (B.3–1) is satisfied; that is, when the entries in the right-hand column of the table are added, they give 200. The ratio $N(j)/N_{tot}$ is the fraction of buildings with $j$ stories, and the interpretation of this ratio is simple: If you select any of the 200 buildings at random, then $P(j)$ is the probability that you will pick a building with $j$ stories. Thus the probability of finding a building with 1 story is $45/200 = 0.23$, and that of finding a building with 12 stories is $2/200 = 0.01$. The sum of the probabilities is unity, since that is the probability of finding *some* number of floors in a building. The sum also may be decomposed; that is, the sum over, say, $P(1)$ and $P(2)$ is the probability that you will find a building with 1 *or* 2 stories.

Let us illustrate the joint probability for different quantities. For example, we could ask for the probability that a front door have certain allowed colors—red, blue, and yellow, say—with corresponding data. By the same kind of counting we have just outlined, we would find the probabilities $P_b$, $P_r$, and $P_y$ that satisfy the sum

$$P_b + P_r + P_y = 1.$$

At this point in our example, we can ask for *joint* probabilites: What is the probability that a building has nine stories *and* a blue door. If there is no prearranged color–height connection, we say that the probabilities for height and door color are *independent* quantities, meaning that the measurement of one does not affect the measurement of the other. In this case, the probability $P_b(9)$ that a building has nine stories *and* a blue front door is a product of the independent probability factors:

$$P_b(9) = P_b \times P(9).$$

Finally, let us look at an average. The average number of stories of all the buildings in Middletown is the weighted sum

$$\langle j \rangle = \sum_{j=1}^{12} jP(j).$$

A calculation using the numbers in the table would show you that the average number of stories in Middletown is $\langle s \rangle = 3.15$. Similarly, the square of the average number of stories is $\langle s^2 \rangle = 15.7$. Note that the square root of 15.3 is not 3.15; in general $\sqrt{\langle j^2 \rangle} \neq \langle j \rangle$.

**Example B.3–2**   In a physics class of 250 students the final grades are distributed as follows: 37 students get A, 62 students get B, 98 students get C, 30 students get D, and 23 students get F. If we assign four points to an A, three points to a B, etc., calculate the average grade $\langle g \rangle$.

**Solution**

$$\langle g \rangle = \frac{\sum gN(g)}{\sum N(g) = N} = \frac{\left[(37 \times 4) + (62 \times 3) + (98 \times 2) + (30 \times 1) + (23 \times 0)\right]}{250}$$

$$= 2.24,$$

a little above a C.

**Example B.3–3**   It is easy to check that exactly the same average grade as we found in Example B.3–2 would be obtained in a class with 140 A's and 110 F's. To distinguish between such cases, one is interested in the **standard deviation**, which is a measure of the how "widely" the grades depart from the average. The standard deviation is defined as

$$\sigma = \sqrt{\langle (g - \langle g \rangle)^2 \rangle}.$$

Calculate $\sigma$ for the distribution given in Example B.3–2 and for the distribution given at the beginning of this example.

**Solution**   For the distribution of Example B.3–2, we have the following values:

| | | |
|---|---|---|
| A: | $g - \langle g \rangle = 4.00 - 2.24 = 1.76;$ | $(g - \langle g \rangle)^2 = 3.10.$ |
| B: | $g - \langle g \rangle = 3.00 - 2.24 = 0.76;$ | $(g - \langle g \rangle)^2 = 0.58.$ |
| C: | $g - \langle g \rangle = 2.00 - 2.24 = -0.24;$ | $(g - \langle g \rangle)^2 = 0.06.$ |
| D: | $g - \langle g \rangle = 1.00 - 2.24 = -1.24;$ | $(g - \langle g \rangle)^2 = 1.54.$ |
| F: | $g - \langle g \rangle = 0.00 - 2.24 = -2.24;$ | $(g - \langle g \rangle)^2 = 5.02.$ |

Thus

$$\langle (g - \langle g \rangle)^2 \rangle =$$

$$\frac{\left[(37 \times 3.10) + (62 \times 0.58) + (98 \times 0.06) + (30 \times 1.54) + (23 \times 5.02)\right]}{250} = 1.27.$$

The standard deviation is the square root of this number, $\sigma = 1.13$.

For the second distribution, we note that the average grade is the same, so the same table applies for the different grades. Hence for this distribution

$$\langle (g - \langle g \rangle)^2 \rangle = \frac{\left[(140 \times 3.10) + (110 \times 5.02)\right]}{250} = 3.94,$$

or $\sigma = 1.99$. Thus the standard deviation is larger for the second distribution, corresponding to grades that, on the whole, differ from the average to a greater degree than they do for the first distribution. If every student in the class had a C, the average would be 2.00, and the standard deviation would be zero!

It is easy to show that $\sigma$ can also be expressed as $\sigma = \sqrt{\langle g^2 \rangle - \langle g \rangle^2}$.

## Continuous Distributions

Many physical quantities do not come in discrete packages, or if they do, the step from one value to the next is a small one. For example, although the number of stories in the buildings of Example B.3–1 is an integer, the actual height of the buildings is a continuous variable. With such a variable—call it $x$—it makes sense to ask only whether the measured value lies within some range. Suppose the overall range of the physical variable is $x_0$ to $x_f$. Then a complete set of data for $N_{\text{tot}}$ experiments consists of the set of numbers $N(x_i, \Delta x)$ in the range $x_i$ to $x_i + \Delta x$, where $x_i$ lies in the range from $x_0$ to $x_f$ and where every interval in the full range is covered. By making the interval widths $\Delta x$ small enough, the number of experiments given a result in the interval will be proportional to $\Delta x$, that is,

$$N(x_i, \Delta x) = n(x_i)\Delta x.$$

When we sum all these numbers together, we get the total number of experiments $N_{\text{tot}}$, or

$$\sum_i n(x_i)\Delta x = N_{\text{tot}}.$$

We can describe $n(x)$ as a **number density**.

If we divide $n(x_i)$ by $N_{\text{tot}}$, we deal with probabilities. Thus we define

$$p(x_i) \equiv \frac{n(x_i)}{N_{\text{tot}}}. \tag{B.3–6}$$

In terms of $p(x)$, the probability of finding a result for our physical variable in the interval from $x_i$ to $x_i + \Delta x$ is

$$P(x_i, \Delta x) = p(x_i)\Delta x,$$

and the normalization requirement is

$$\sum_i p(x_i)\Delta x = 1. \tag{B.3–7}$$

The quantity $p(x)$ is called a **probability density**.

At this point we can make a transition from sums to integrals. We have broken things up into slices, but we can take the limit in which these slices are infinitesimally thin, in which case the probability that our measurement lies between two very different values $x_1$ and $x_2$ is

$$P(x_1 < x < x_2) = \sum_{x_i=x_1}^{x_2} p(x_i)\Delta x \rightarrow \int_{x_1}^{x_2} p(x)\,dx,$$

where the arrow indicates the limiting process. The normalization condition represents the fact that *some* value of $x$ will be measured:

$$1 = \int_{x_0}^{x_f} p(x)\,dx. \tag{B.3–8}$$

If $x$ were by nature positive—a speed, say—we could in fact write the limits on this integral as 0 and $\infty$ without harm, since $p(x)$, for example, would vanish outside the allowed range.

Averages are expressed as before. For example, the average value of $x$ is

$$\langle x \rangle = \sum_{h_i=x_0}^{x_f} x_i p(x_i)\Delta x \rightarrow \int_{x_0}^{x_f} x p(x)\,dx. \tag{B.3–9}$$

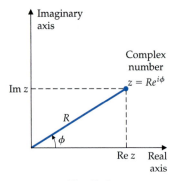

Fig. B–3

To perform the integration, we must of course know the probability density function $p(x)$.

# B–4 Complex Numbers

Complex numbers are based on the quantity $i = \sqrt{-1}$, or $i^2 = -1$. A **complex number** $z$ actually represents two ordinary numbers $x$ and $y$, where $z = x + iy$. We say that $x$ is the **real part** of $z$ and $y$ is the **imaginary part**, symbolized respectively by

$$x = \mathrm{Re}(z) \text{ and } y = \mathrm{Im}(z). \tag{B.4–1}$$

The use of the variables $x$ and $y$ suggests that a complex number can be associated with a point in a plane, as in Fig. B–3. We also define the **complex conjugate** of $z$, as $z^* = x - iy$.

When an equation between complex numbers holds, it is actually two equations, one for the real parts of both sides and one for the imaginary parts.

Transcendental functions of complex numbers are defined by power series in the usual way. This definition implies the remarkable relation

$$\exp(i\theta) = \cos\theta + i\sin\theta. \tag{B.4–2}$$

Equation (B.4–2) allows us to separate an arbitrary complex quantity $z$ in a second form, namely

$$z = R\exp(i\phi), \tag{B.4–3}$$

where

$$R \equiv |z| = \sqrt{[\mathrm{Re}(z)]^2 + [\mathrm{Im}(z)]^2} \tag{B.4–4}$$

is the **absolute value** of $z$ and $\phi$ is the **phase** of $z$, given by

$$\tan\phi = \frac{\mathrm{Im}(\phi)}{\mathrm{Re}(\phi)}. \tag{B.4–5}$$

Both $R$ and $\phi$ are real. Note that $\mathrm{Re}(z) = R\cos\phi$ and $\mathrm{Im}(z) = R\sin\phi$.

We can see immediately that $z^* = R\exp(-i\phi)$, and, because the phase cancels in the product of $z^*$ and $z$, that

$$R = \sqrt{z^*z}. \tag{B.4–6}$$

Trigonometric identities follow easily using complex numbers. For example, consider the sum of two angles, $\theta_1 + \theta_2$. We have

$$\exp[i(\theta_1 + \theta_2)] = \exp[i(\theta_1)] \times \exp[i(\theta_2)]$$
$$= \{\cos(\theta_1) + i\sin(\theta_1)\} \times \{\cos(\theta_2) + i\sin(\theta_2)\}$$
$$= [\cos(\theta_1)\cos(\theta_2) - \sin(\theta_1)\sin(\theta_2)] + i[\cos(\theta_1)\sin(\theta_2) + \sin(\theta_1)\cos(\theta_2)].$$

But the left side is $\cos(\theta_1 + \theta_2) + i\sin(\theta_1 + \theta_2)$. Matching real and imaginary quantities on both sides, we immediately have the identities for $\cos(\theta_1 + \theta_2)$ and $\sin(\theta_1 + \theta_2)$.

# Answers to Odd-Numbered Problems

(Problems asking for proofs or "show that" solutions are presented in the Solutions Manual)

## CHAPTER 2

**1.** $T = L/(c + v\cos\theta) + L/(c - v\cos\theta)$ **3.** 0.97 ns
**5.** 247.49 years **7.** $v = (1 - 1.5 \times 10^{-12})c$ **9.** The clocks differ by 62. ns: 5% of 62. ns = 3-ns accuracy needed
**11.** 2.1 Hz **15. (a)** 0.946$c$ **(b)** 2.6 $\mu s$ **17.** 3.3 $\mu s$
**19.** 335. nm; same (if emitter moved toward you)
**21.** $0.20 \times 10^{-15}$ Hz **23. (a)** $f' = f[(1 - u/c)/(1 + u/c)]^{0.5}$
**(b)** $f = f'[(1 - v/c)/(1 + v/c)]^{0.5}$ **(c)** $w = (u + v)/(1 + uv/c^2)$
**25.** $u'_y = u_y/(\gamma(1 - vu_x/c^2))$

## CHAPTER 3

**1.** $v = (1 - 32 \times 10^{-7})c = c - 96.$ m/s **3.** $f = 0.999933$
$\times 10^{15}$ Hz $= 10^{15}$ Hz $- 6.7 \times 10^{10}$ Hz **7.** $f = f'$
$\gamma(1 + v/c) = f'[(1 + v/c)/(1 - v/c)]^{1/2}$ **9.** $v = 0.90c$,
$T = 9.3$ years **11.** $1.73 \times 10^5$ N-s, relativistic; $1.50 \times 10^5$
N-s, nonrelativistic **13.** $1.04 \times 10^{14}$ J $= 6.5 \times 10^{32}$ eV
**15.** 0.93 kg **19.** $5.0058 \times 10^{-27}$ kg **21.** $v = 0.9999996c$
$= c - 96.$ m/s **23.** 130. MeV **25.** 4.05 years
**29.** 0.51 MeV **31.** 8.6 $\mu$m **33.** $p_0 = -p_+ = 357.$
MeV/$c$ **35. (a)** 45.6 GeV/$c$; **(b)** 4.44 TeV **39.** To the observer, the relative speed is $1.03c$, but $A$ sees $B$ moving $0.926c$ away and $B$ sees $A$ moving $0.926c$ away.

## CHAPTER 4

**1.** $1.2 \times 10^7$ meters $= 1.2 \times 10^4$ km **3.** $6.56 \times 10^{-22}$
MeV-s **5.** $3.87 \times 10^{21}$ photons/m²-s **7.** No. The number emitted/s (current) will increase **9.** $2.4 \times 10^{-19}$ J $= 1.5$ eV
**11.** $1.4 \times 10^8$ atoms **13.** 88 degrees **15.** $5 \times 10^5$ K
**17.** $\lambda T = (h/2\pi c/(4.96\ k) = 4.6 \times 10^{-4}$ mK **19.** 0.17 mm; $7.4 \times 10^{-3}$ eV **21.** $4.47 \times 10^{26}$ J/s radiated; about $10^3$ years if energy is chemical release; temperature would change as fuel became depleted. **23.** $4 \times 10^8$ photons/s
**25. (a)** $2.9 \times 10^2$ K **(b)** $2.7 \times 10^2$ K **27.** $8.7 \times 10^{27}$ Hz
**29.** $(\sqrt{2} - 1)mc^2$, or 41% of the rest energy **31.** $1.3 \times 10^{-34}$
meter; $2.6 \times 10^{-34}$ radian **33. (a)** $8.69 \times 10^{-18}$ J $= 54.3$
eV **(b)** 0.0294 eV $= 4.71 \times 10^{-21}$ J

## CHAPTER 5

**1.** $2.5 \times 10^{65}(h/2\pi)$ **3.** It takes 13.6 eV to set the electron free; the photon in question has only 6.9 eV, not enough to excite it. **5. (a)** 121.5 nm, 102.5 nm, 97.2 nm, 94.9 nm
**(b)** 91.13 nm **7.** $n = 3$ to 2 produces 656 nm, no $\Delta n = 1$ in that range **9.** 13.6 eV **11.** 137. **13.** 9.0042, assuming Li$^7$ **15. (a)** 5.1 eV **(b)** $1.06 \times 10^{-10}$ m
**17.** $\lambda = (4c/3)(m(h/2\pi)/u_0^2)^{1/3}(n^{2/3} - m^{2/3})^{-1}$ **19.** 66.41 nm
**21. (a)** $a = \alpha^3 c^3\ m/((h/2\pi)n^4) = 3.888 \times 10^{23}/n^4$ m/s²
**(b)** $p = (2/3)\alpha^7 c^4\ m^2/((h/2\pi)n^8)$ **(c)** $m\alpha^2 c^2/n^8 = 27.2$ eV/$n^8$, for $n \gg 1$ **(d)** $(3/2)(h/2\pi)n^5/(m\alpha^5 c^5)$ **(e)** 0.6 ns **23.** Transitions between rotational states of a molecule from $L = 2(h/2\pi)$ to $L = (h/2\pi)$ state: $\lambda = 2\pi MR^2 c/(h/2\pi)$
**25.** 0.262 eV vibration; same for Bohr rotation **29.** Hydrogen radius/spacing $= 0.015 = 1/66$ **31.** Maximum wavelength to photodisintegrate hydrogen is 91.4 nm; excess energy becomes kinetic energy for the electron

## CHAPTER 6

**1.** $C = a^{-1/2}\pi^{-1/4}$ **5. (a)** $1.5 \times 10^2$ eV **(b)** $6.0 \times 10^2$ eV
**7.** $3 \times 10^7$ **9.** $n = 18, 19, 20$ **11.** $\langle x \rangle = L/2$ **17.** 90%
**19.** 1/2 **23.** $E = E_x + E_y = (n_x^2 + n_y^2)(h/2\pi)^2\pi^2/(2mL^2)$
**25.** $E(\text{KeV}) = 0.30, 0.75, 1.19, 1.49, 1.94$

## CHAPTER 7

**1.** $\cos[\Delta px/(h/2\pi)]$ **5.** $v_g = \sqrt{\dfrac{2(E - V_0)}{m}}$ **7.** E $\geq 1.5$ KeV

**9.** Could resolve 5. pm, energy 32. eV **11.** $2 \times 10^{-5}$

radian spread **13.** $E_0 \approx \left(\sqrt{\dfrac{\left(\dfrac{h}{2\pi}\right)^2 (V_0)^2}{4ma^2}}\right)^3$, assuming

$E_0 \ll mc^2$ **15.** $1.6 \times 10^2$ MeV **17.** $1.3 \times 10^{-6}$ eV
**19.** $5.9 \times 10^{-24}$ second in lab frame; travels 0.59 fm

**21.** 0.7 fm **23.** $C = \sqrt{\dfrac{a}{2\pi\left(\dfrac{h}{2\pi}\right)^2}}$

## CHAPTER 8

**1.** 0.79% **9.** Reflected: $R' \exp(ikx)$, where $k^2 = 2m(E - V_0)$
$/(h/2\pi)^2$ transmitted: $T' \exp(-iqx)$, where $q^2 = 2mE/(h/2\pi)^2$

**13. (a)** $|T^2| \approx \exp\left[-\dfrac{16\pi a}{3h}\sqrt{2mA}\right]$

**(b)** $|T^2| = \exp\left[-\dfrac{8\pi a}{3h}\sqrt{mA}\left(\sqrt{8}-1\right)\right]$ **(c)** 0.25

**15.** $\varepsilon = 3.9 \times 10^{36}$ W$^{3/2}$ in SI units ($\varepsilon$: volts/m, W: joules) $\varepsilon = 2.5 \times 10^8$ W$^{3/2}$ ($\varepsilon$: volts/m, W: eV) **17.** 0 for $x \le 0$; $A\cos(kx) + B\cos(kx)$, where $V = 0, k^2 = 8\pi^2 m(E - V_1)/h^2$; $C[\exp(qx) - \exp(-qx)]$ for $0 < E < V_1, q^2 = 8\pi^2 m|(V_1 - E) |/h^2$; if $E > V_1$, $F\sin(qx)$

**25.** $R = \dfrac{i(q^2 - k^2)\sin(2qa)\exp(-2ika)}{2kq\cos(2qa) - 1(k^2 + q^2)\sin(2qa)}$,

$T = \dfrac{2kq\exp(-2ika)}{2kq\cos(2qa) - i(k^2 + q^2)\sin(2qa)}$, where

$q^2 = 8\pi^2 m(E - V_0)/h^2$ and $k^2 = 8\pi^2 mE/h^2$; classically, we expect $T = 1$ and $R = 0$, which we get for $|E_0| \gg V_0$
**27.** $E = 0.4$ eV; $E = (1.5n^2 - 1.1)$ eV

## CHAPTER 9

**1.** $-\dfrac{h^2}{8\pi m}\dfrac{d^2u}{dr^2} + \left[-\dfrac{Ze^2}{4\pi\varepsilon_0 r} + \dfrac{h^2}{8\pi m}l(l+1)\dfrac{1}{r^2}\right]u = Eu$, for

$0 \le r < \infty$ **7. (a)** $\approx 9.5 \times 10^{27}$ **(b)** $5.2 \times 10^{-13}$ eV
**9.** $2.5 \times 10^{15}$ Hz; reduced mass is 0.05% less than electron mass, so energy and frequency are reduced by 0.05% $\left(1.4 \times 10^{12}\text{ Hz less}\right)$ **11.** $\langle r^2\rangle = 3a_0^2$; $\langle r\rangle^2 = \left[(3/2)a_0\right]^2$

**13.** $\langle V\rangle = \dfrac{me^2}{16\varepsilon_0^2 h^2}$ **15. (a)** $5 \times 10^{-16}$ **(b)** $2 \times 10^{-57}$; the angular momentum keeps the electron further from the origin

**17.** $R_{10}$ is maximum at $r = 0$, but $\langle r\rangle_{10} = \dfrac{3}{2}\dfrac{a_0}{Z}$, $R_{21}$ is maximum at $r = 2\dfrac{a_0}{Z}$, but $\langle r\rangle_{21} = 5\dfrac{a_0}{Z}$

**19.** 15 lines if only $\Delta m = -1, 0, 1$ allowed; 11. tesla **21.** $n = 3$ to $n = 2$ lines not appearing: $m$ states $-2$ to 1; $-2$ to 0; $-1$ to $+1$; $+1$ to $-1$; $+2$ to 0; $+2$ to $-1$; three spectral lines of different frequencies in spectrum: $4.56 \times 10^{14}$ Hz and $4.56 \times 10^{14} \pm 5.6 \times 10^{10}$ Hz **23.** If in ground state, the orbital angular momentum is zero, so total angular momentum number is the spin number $s = 1/2$; gradient 0.18 tesla/meter; $K = (6.0\text{ eV/meter})(\text{field length})$; If 10-cm field, then $K = 0.6$ eV **27.** $j = 5, 4, 3, 2, 1$

**29. (a)** $B = \dfrac{\mu_0 eh}{8\pi^2 R^3 m}l$ **(b)** $1.3 \times 10^{-6}$ eV **31.** $l = 3$

**33.** 6. tesla

## CHAPTER 10

**3.** $N_A = 2(1 - |K_{12}|^2)$; $N_S = 2(1 + |K_{12}|^2)$ **5.** Ground-state $\Psi = \dfrac{\sqrt{2}}{a}\left[\sin\left(\dfrac{\pi}{2a}\right)x_1\sin\left(\dfrac{\pi}{2a}\right)x_2\right]$ first excited

$\Psi = \dfrac{1}{a\sqrt{2}}\left[\sin\left(\dfrac{\pi x_1}{a}\right)\sin\left(\dfrac{\pi x_2}{2a}\right) + \sin\left(\dfrac{\pi x_1}{2a}\right)\sin\left(\dfrac{\pi x_2}{a}\right)\right]$

**7.** $= \dfrac{1}{a\sqrt{2}}\left[\sin\left(\dfrac{\pi x_1}{a}\right)\sin\left(\dfrac{\pi x_2}{2a}\right) - \sin\left(\dfrac{\pi x_1}{2a}\right)\sin\left(\dfrac{\pi x_2}{a}\right)\right]$;

$\Psi = 0$, at $x = 0$, $L/2$, $L$ **9. (a)** $N_G = (2/L)^{3/2}$
**(b)** $N_F = (2/L)^{3/2}/\sqrt{3}$ **11.** $3/(m\pi)$ if $m = n$; 0 if $m^2 \ne n^2$ and both odd or even $\left(\dfrac{1}{\pi}\right)(1 + 1/3), -\left(\dfrac{1}{\pi}\right)(1/3 + 1/5)$, etc ... **13.** 7.0 eV for Cu, 5.9 eV for Pb **15.** 1 free electron/atom **19.** $E \approx V^{-2/3}h^2/m$ **23.** $6.4 \times 10^{10}$ N/m$^2$ **25.** 9.8 km **27.** 2-dimensions $\Delta N = \dfrac{4\pi mL^2}{h^2}\Delta E$;

1-dimensions $\Delta N = \dfrac{16mL^2}{Nh^2}\Delta E$; 3-dimensions

$\Delta N = \left(\dfrac{3N}{\pi^4}\right)^{1/3}\dfrac{4\pi^2 mL^2}{h^2}\Delta E$; 2-dimensions $\dfrac{dN}{dE} = \dfrac{4\pi L^2}{h^2}$;

1-dimensions $\dfrac{dN}{dE} = \dfrac{16mL^2}{h^2 N}$; 3-dimensions $\dfrac{dN}{dE} = \left(\dfrac{3}{\pi^4}\right)^{1/3}$

$\dfrac{4\pi^2 m}{h^2}L^2 N^{1/3}$

## CHAPTER 11

**1.** $1s^2 2s^2 2p^6 3s^2$; $J = 0$ **3.** $L = 2$; $S = 1/2$; possible $J = 5/2$, 2, 3/2; lowest energy for $J = 3/2$ **5.** 0.11 nm **7.** 0.22 nm, using $s = 3/2$ from Problem 6 **9.** Not compatible; $Z_e = 1.5$ **11.** 135. eV **13.** Assuming $R_0 \approx 2a_0$ and $\Delta l = 1$: 2.3 mm, 1.1 mm, 0.75 mm, 0.56 mm, ..., $2.3/n$ mm **15.** $I_1 = 1.415 \times 10^{-46}$ kg-m$^2$; $I_0 = 1.463 \times 10^{-46}$ kg-m$^2$; 0.114 nm, 0.116 nm **17. (a)** $a_0 = 0.14$ pm when $R = 0$, so expect $\approx 1/4$ to 1/2 pm **(b)** 4. nm **19.** $1.1 \times 10^{-4}$ K

## CHAPTER 12

**1.** 4.8 mm/s **3.** He, 1.1 km/s; Ne, 0.50; Ar, 0.35; Kr, 0.24; Xe, 0.19; Rn, 0.15 km/s **5.** 1.29 kg/m$^3$ **7.** $\langle v^2\rangle(m^2/s^2) = 3.72 \times 10^5$ @ 300 K, $3.72 \times 10^6$ @ 3,000 K, $3.72 \times 10^4$ @ 30 K **9.** In 1-dimension:

$$f(v) = \sqrt{\dfrac{m}{2\pi kT}}\exp\left(-\dfrac{mv^2}{2kT}\right); \langle K\rangle = \tfrac{1}{2}kT$$

**13.** $\langle v\rangle^2 = \dfrac{8}{\pi}\left(\dfrac{kT}{m}\right), \langle v^2\rangle = 3\left(\dfrac{kT}{m}\right), \langle v^2\rangle/\langle v\rangle^2 = \dfrac{3\pi}{8}$;

$\langle x^2\rangle = \langle x\rangle^2$ when there is no spread about the average. For example, if $g(v) = \delta(v - \langle v\rangle)$ (the Dirac delta function), then $g(v) = 0$ for $v \ne \langle v\rangle$ and $g(v) = \infty$ when $v = \langle v\rangle$

**15. (a)** $\langle v_z\rangle = \sqrt{\dfrac{\pi}{2}}\sqrt{\dfrac{kT}{m}}$ **(b)** $\Delta v_z = 0.76$ m/s @ 850 K, 0.90 m/s @ 1,200 K **17.** 80. meters (using $T = 273$ K)

**19.** $\langle r^2\rangle = 4.6$ mm; 77. $\mu$m

**21.** $\langle E\rangle = \dfrac{E_1\exp(-E_1/kT) + E_2\exp(-E_2/kT)}{\exp(-E_1/kT) + \exp(-E_2/kT)}$

**23.** $67. \times 10^3$ K   **25. (a)** $\exp\left(-\dfrac{n_v hf}{kT}\right)$ **(b)** $3 \times 10^6$ K

**(c)** $\langle E \rangle = \left(\dfrac{hf}{2}\right)\dfrac{1 + 3\exp(-hf/kT)}{1 + \exp(-hf/kT)}$   **27. (a)** 4 **(b)** 7 **(c)** 9

**(d)** 15   **29. (b)** classical @ 100 K and 10 K, nonclassical at 1 K **(c)** $T \gg 3$ K, classical   **35.** $N \approx \left(\dfrac{4\pi m A k T}{h^2}\right)\exp\left(\dfrac{E_F}{kT}\right);$

**35.** $N \approx \left(\dfrac{4\pi m A k T}{h^2}\right)\exp\left(\dfrac{E_F}{kT}\right); \langle E \rangle = kT;$ as expected, for 2-dimensions, $\langle E \rangle = 2(kT/2)$   **37. (b)** $4.2 \times 10^8$ photons/m$^3$   **39.** $5.7 \times 10^6$ watts spread over a sphere of radius 10. km is 1.4 mW/m$^2$   **41.** $PV = 0.900\, NkT$ for photon gas; $PV = NkT$ for classical gas

## CHAPTER 13

**3.** $\tau = 9.5$ weeks; $t = 6.6$ weeks   **5.** 122. s; $1.22 \times 10^3$ s
**7.** $R_1 = 0.031$ hr$^{-1}$ = 0.74 day$^{-1}$; $R_2 = 0.091$ hr$^{-1}$ = 0.46 day$^{-1}$   **9.** $\Delta E \Delta t = 0.0006$ MeV $\left(8.4 \times 10^{-17}\text{ s}\right) = 5 \times 10^{-20}$
MeV $- s > \dfrac{h}{2\pi} = 6.6 \times 10^{-22}$ MeV $- s$, no contradiction
**13.** $\tau = 1/R = 4.3 \times 10^{-17}$ second   **15.** $\tau = 8 \times 10^{-12}$ s
**19.** 9.5 cm diameter spot; 39.9 J/cm$^2$-s   **21.** 33. mW; $3.5 \times 10^{-5}$ N/m$^2$; $1.06 \times 10^{-14}$ photons/cm$^3$

## CHAPTER 14

**1.** $F_d = kv$;   $[k] = [F]/[v] = [\text{kg-m-s}^2/(\text{m/s})] = [\text{m/s}]$
**3.** $1.59 \times 10^{13}$ m = 0.61 light-day   **7.** $n_e/n_i$: Li, 1; Na, 1; Cu, 1; Mg, 2; Fe, 2; Al, 3   **9.** Li, 11. nm; Na, 34. nm; Cu, 42 nm; Mg, 17. nm; Fe, 4.8 nm; Al, 16. nm   **13.** $1.1 \times 10^2$ K, estimate   **15.** Na, 1.4; Al, 3.4; Fe, 30.; Zn, 7.4; Sb, 67.; Pb, 11   **17.** The lower curve represents the purer sample.
**19. (a)** $y = qa = \pi$   **21.** 1.4 eV   **23.** $m_n^* = 9.72 \times 10^{-31}$ kg = $1.1 m_e$; $m_p^* = 3.00 \times 10^{-31}$ kg = $0.33 m_e$   **25.** Si, $1.1 \times 10^3$ K; Ge, $0.64 \times 10^3$ K; GaAs, $1.4 \times 10^3$ K; InP, $1.3 \times 10^3$ K; InSb, $0.17 \times 10^3$ K   **27.** $-5.6 \times 10^{-3}$ eV   **29.** 2.4 eV
**31.** $2.2 \times 10^{-32}$ kg = $0.024\, m_e$   **33.** 19. eV; 22. eV; 24. eV; 27. eV

## CHAPTER 15

**3.** $1.9 \times 10^8$ bulbs; $1.9 million   **5. (a)** $-b_{\text{vol}}\left(\dfrac{4\rho\pi}{3m_p}R^3\right)$

**(b)** $\Delta E_v = -b_{\text{vol}}\left(\dfrac{4\pi\rho R^2}{m_p}\right)\Delta R = -b_{\text{vol}}\left(\dfrac{36\pi}{m_p}\right)^{1/3}\rho A^{2/3}\Delta R$

**7. (a)** $E_p + E_n = \left(\dfrac{9}{4\pi^2}\right)^{5/3}\dfrac{\pi^2 h^2}{30mR_0^2}\left[A\left(\dfrac{A-t}{2A}\right)^{5/3}\right.$

$\left. + A\left(\dfrac{A+t}{2A}\right)^{5/3}\right]$ **(b)** $\approx$ constant $*\left[A + \dfrac{5t^2}{(9A)}\right]$ **(c)** 19. MeV
**11.** Maximum energy loss is 25% of the original kinetic energy. Target of light nuclei would better slow the neutrons.

**13.** 15.42 MeV. The $\gamma$ must supply momentum and KE to the neutron and proton.   **17.** $^{205}$Pb

**19.** $N_3 t = \dfrac{N_1(0)}{R_{12} - R_{23}}\left[R_{12}(1 - \exp(-R_{23}t))\right.$
$\left. - R_{23}(1 - \exp(-R_{12}t))\right]$   **23.** $v/c = 0.047$ (smaller mass), 0.032 (larger mass)   **25.** Mass formula gives 2.2 MeV compared to measured 1.98 MeV   **27.** W = 7.5 KeV; $K = -3.7$ eV   **29.** 180.8 MeV   **31.** $1.010369u$, which is 0.17% too large

## CHAPTER 16

**1.** $\varepsilon u(r) = -\dfrac{h^2}{8\pi m}\left[\left(\dfrac{\partial}{\partial x}\right)^2 + \cdots\right]u(r) + V u(r)$ **5.** 260 MeV/c
**7. (a)** charge negative, baryon number 0, spin 0, lepton number 0 **(b)** No, $X$ could be two particles (e.g., $n \rightarrow p + e^- + \nu_e$)   **9. (a)** charge 0, baryon number 0, spin 1/2, lepton number $-1$ **(b)** yes (e.g., $\rightarrow \nu_e + \nu_e$)   **11.** 2.43 keV   **13.** $2 \times 10^{-18}$ meter   **15.** (i) fast (ii) forbidden (iii) forbidden (iv) slow (v) slow (vi) forbidden   **17.** $\gamma$ baryon number = 0; baryon number not conserved, so forbidden
**19. (a)** $uuu$, $uud$, $udd$, $ddd$; $uus$, $uds$, $dds$; $uss$, $dss$, $sss$ **(b)** 411. MeV/c$^2$;   563. MeV/c$^2$   **23.** $m = 306.$ MeV/c$^2$; $a = \left(3.84 \times 10^3 \text{ MeV/c}^2\right)^2 = 1.47 \times 10^7$ MeV$^2$/c$^4$
**25.** $m_b = 4.823$ GeV/c$^2$; 9832. Mev, $n = 1$ triplet; 9,600. and 9,692. MeV $n = 2$ $s$ and $t$

## CHAPTER 17

**1. (a)** He arranged their weights to be the same; weight depends upon gravitational mass **(b)** $T = 2\pi\sqrt{\dfrac{m_i L}{m_g g}}$;
$m_i = m_g$ to within 2 parts in 10,000.   **3.** $\Delta f/f = 2.5 \times 10^{-15}$
**5.** $f/f = 1 = gh/c^2$   **7. (a)** $-0.30$   (30% reduction)
**(b)** Transverse shift at center, $\Delta f/f = -1.3 \times 10^{-8}$ longitudinal shift at limb, $\Delta f/f = \pm 1.6 \times 10^{-4}$; both are $\ll$ gravitational shift   **9.** $15. \times 10^6$ K   **13.** $\Delta g \propto 1/M^2$, so larger if $M$ is smaller   **15.** $7.4 \times 10^{12}$ meters = 6.9 light-hours; $R \ll R_s$   **17.** $2.5 \times 10^{-16}$ s   **19.** Expect $kT = h^\alpha c^\beta (GM)^\gamma$; equate powers of units on each side to find $\alpha = 1$, $\beta = 3$, $\gamma = -1$   **21. (a)** $c(1 - R_s/r)$ **(b)** $c\sqrt{1 - \dfrac{2GM}{c^2 r}} < c$
**(c)** Supposing a Newtonian mass, force $GmM/r^2$ would "speed up" mass light as it approached $M$.

## CHAPTER 18

**5.** $v = \sqrt{\dfrac{G\left(M + 2\pi k[R^2 - R_0^2]\right)}{R}}$   **7.** $\Delta f = 10^{-3}f$ moving toward, $-10^{-3}f$ moving away   **9.** $R = \exp\left(\sqrt{8\pi G\rho/3}\, t\right)$
**11.** $^2$H $+$ $^1$H $\rightarrow$ $^3$He $+$ $\gamma$; $^3$He $+$ $^3$He $\rightarrow$ $^4$He $+$ $^1$H $+$ $^1$H

# Bibliography

Almost every topic covered in this book may be viewed as a prelude to a deeper and more intensive study of the subject. There are so many books that treat the topics herein in a more advanced manner that it would not be useful to list them. Our selective bibliography is based on two very modest goals: First, we want to lead the reader to books in which he or she may find the same material at the same level, but treated somewhat differently; and second, we want to guide the reader to places in which very specific topics that are treated only briefly in our book are described in more detail. The following list of books is designed to meet the first goal.

## Chapters 2, 3

### Special Relativity

Many textbooks cover this subject. The books that we have found most useful and stimulating are A. P. French, *Special Relativity* (New York: Norton, 1968) and J. H. Smith, *Introduction to Special Relativity* (New York: Dover Publications, 1995), for straightforward treatments of the subject in more detail that we can give here. A rich development of special relativity can be found in E. F. Taylor and J. A. Wheeler, *Spacetime Physics* (San Francisco: W. H. Freeman, 1966). The simplest treatment is N. D. Mermin, *Space and Time in Special Relativity* (New York: McGraw-Hill, 1968).

## Chapters 4, 5

### Old Quantum Theory

A good historical account of the development of the old quantum theory is M. Jammer, *The Conceptual Development of Quantum Mechanics* (New York: McGraw-Hill, 1966). Excellent discussions of some of the topics we cover may also be found in E. H. Wichmann, *Quantum Physics*, vol. 4 of the Berkeley Physics Course (New York: McGraw-Hill, 1971).

## Chapters 6, 7, 8, and 9

### Introductory Quantum Mechanics

Material at the level covered in this book, and slightly higher, can be found in R. Eisberg and R. Resnick, *Quantum Physics of Atoms, Molecules, Solids, Nuclei and Particles*, 2d ed. (New York: John Wiley, Inc., 1985). Some material can be found in more advanced texts on quantum mechanics. A long list of topics and sources is in the reference section of S. Gasiorowicz, *Quantum Physics*, 2d ed. (New York: John Wiley, Inc., 1996).

## Chapters 10, 11

### Exclusion Principle, Atoms and Molecules

In contrast to our discussion of the exclusion principle, that topic is usually introduced in connection with atomic structure. [See for example Eisberg and Resnick (*op. cit.*).] Our approach treats the properties of identical particles more broadly, and we could not find a book at our level that does the same. For the general subject of atoms and molecules, we recommend M. Karplus and R. N. Porter, *Atoms and Molecules: An Introduction for Students of Physical Chemistry* (New York: W. A. Benjamin, 1970).

## Chapters 12, 13, 14

### Statistical Physics, Radiation, and Solid-state Physics

For other treatments of these subjects, the reader may look to any number of books that have "Modern Physics" in their title or subtitle. The derivation of the fermion and boson distributions are usually done, if at all, using combinatorial methods for counting states. [See, for example, Eisberg and Resnick (*op. cit.*).] A beautiful book on solid-state physics is L. Solymar and D. Walsh, *Electrical Properties of Materials*, 6th ed. (New York: Oxford University Press, 1998).

## Chapters 15, 16

### Nuclear Physics and Particle Physics

There are many books on nuclear physics. For a modern coverage of some of the topics that we treat here, see H. Frauenfelder and E. Henley, *Subatomic Physics*, 2d ed. (Englewood Cliffs, NJ: Prentice Hall, 1991) or K. S. Krane, *Introductory Nuclear Physics* (New York: John Wiley, 1987). For quarks we refer the reader to

L. B. Okun, *Particle Physics, the Quest for Substance of Substructure* (New York: Harwood Academic, 1985).

## Chapters 17, 18
### Gravitation and Cosmology

Two excellent books that deal with these subjects, both with technical appendices that are on the level of this book, are S. Weinberg, *The First Three Minutes, a Modern View of the Origin of the Universe*, updated edition (New York: Basic Books, 1993) and J. Silk, *The Big Bang*, updated edition (San Francisco: W. H. Freeman, 1988). An appropriate textbook on this material is J. Bernstein, *An Introduction to Cosmology* (Englewood Cliffs, NJ: Prentice Hall, 1995).

For the second goal of this bibliography we have concentrated on *Scientific American* articles that have appeared in the last 20 years and that are in one way or another relevant to the topics we have covered in the text. Some of the earlier articles may no longer be completely up to date; some may be speculative, and the hopes expressed by the authors may not have come to fruition. Some have rather technical parts. Moreover, the list of articles does not cover every subject covered in this book. Current research in quantum mechanics, for example, is on more esoteric questions than we deal with. There are also no articles on special relativity, which is not currently an active area of research. In spite of these potential difficulties, the articles we have listed all have the very considerable merit of providing the reader with some idea of what is going in the various fields, and they are gorgeously illustrated. Finally, *Scientific American* is available in all college and university libraries, and these articles are easily accessible to the interested reader.

We list the articles in the order appropriate to the order of presentation of material in this book, at least where that is possible; otherwise they are by subject as listed. We have also separated out some historical articles that we found interesting, from the same *Scientific American* source.

### Historical

"Robert A. Millikan" D. J. Kevles, January 1979, p. 142.

"Lise Meitner and the Discovery of Nuclear Fission" R. L. Sime, January 1998, p. 80.

"Heike Kamerlingh Onnes' Discovery of Superconductivity" R. d B. Ouboter, March 1997, p. 98.

"The Reluctant Father of Black Holes" J. Bernstein, June 1996, p. 80.

"Heisenberg, Uncertainty and the Quantum Revolution" D. C. Cassidy, May 1992, p. 106.

"Dirac and the Beauty of Physics" R. C. Hovis and H. Kragh, May 1993, p. 104.

"Edwin Hubble and the Expanding Universe" D. E. Osterbrock, J. A. Gwinn, and R. S. Brashear, July 1993, p. 84.

"How Cosmology Became a Science" S. G. Brush, August 1992, p. 62.

### Atoms

"The Spectrum of Atomic Hydrogen" T. W. Hansch, A. L. Schawlow, and G. W. Series, March 1979, p. 94

"Highly Excited Atoms" D. Kleppner, M. G. Littman and M. L. Zimmerman, May 1981, p. 130

"The Global Positioning System" T. A. Herring, February 1996, p. 44

"Accurate Measurement of Time" W. M. Itano and N. Ramsey, July 1993, p. 56

"The Quantum Mechanics of Materials" M. L. Cohen, V. Heine, and J. C. Phillips, June 1982, p. 82

"NMR Imaging in Medicine" I. L. Pykett, May 1982, p. 78

"The Bose Einstein Condensate" E. A. Cornell and C. E. Wieman, March 1998, p. 40.

"High-Temperature Superconductors" P. C. W. Chu, September 1995, p. 162.

"Squids" J. Clarke, August 1994, p. 46.

### Lasers and their Applications

"Cooling and Trapping Atoms" W. D. Phillips and H. J. Metcalf, March 1987, p. 50

"Laser Separation of Isotopes" R. N. Zare, February 1977, p. 86

"Detecting Individual Atoms and Molecules with Lasers" V. S. Letokhov, September 1988, p. 54

"Optical Gyroscopes" D. Z. Anderson, April 1986, p. 94

"Microlasers" J. L. Jewell, J. P. Harbison, and A. Scherer, November 1991, p. 86

"Laser Trapping of Neutral Particles" S. Chu, January 1992, p. 70

"Cavity Quantum Electrodynamics" S. Haroche and J-M. Raimond, April 1993, p. 54

"Picosecond Ultrasonics" H. Maris, January 1998, p. 86

"The Single-Atom Laser" M. S. Feld and K. An, July 1998, p. 56

"Laser Scissors and Tweezers" M. W. Berns, April 1998, p. 62

"Nanolasers" P. L. Gourley, March 1998, p. 56

### Semiconductors and Microelectronics

"High Speed Silicon–Germanium Electronics" B. S. Meyerson, March 1994, p. 62

"Diminishing Dimensions" E. Corcoran, November 1990, p. 122

"Gallium Arsenide Transistors" W. R. Frensley, August 1987, p. 80

"Ballistic Electrons in Semiconductors" M. Heiblum and L. F. Eastman, February 1987, p. 102

"The Scanning Tunneling Microscope" G. Binnig and H. Rohrer, August 1985, p. 50

"The Quantum-Effect Device: Tomorrow's Transistor?" R. T. Bate, March 1988, p. 96

"Progress in Gallium Arsenide Semiconductors" M. H. Brodsky, February 1990, p. 68

"Single Electronics" K. K. Likharev and T. Claeson, June 1992, p. 80

"The Future of the Transistor" R. W. Keyes, June 1993, p. 70

"Quantum Dots" M. A. Reed, January 1993, p. 118

"Scanned Probe Microscopes" H. K. Wickramasinghe, October 1989, p. 98

## Nuclear Physics

"Vibrations of the Atomic Nucleus" G. F. Bertsch, May 1983, p. 62

"A Natural Fission Reactor" G. A. Cowan, July 1976, p. 36

"Radiocarbon Dating by Accelerator Mass Spectroscopy" R. E. M. Hedges, J. A. J. Gowlett, January 1986, p. 100.

"Progress towards a Tokamak Fusion Reactor" H. P. Furth, August 1979, p. 50.

"Fusion" H. P. Furth, September 1995, p. 174.

"Creating Superheavy Elements" P. Armbruster and G. Munzenberg, May 1989, p. 66.

"Progress in Laser Fusion" R. S. Craxton, R. L. McCrory, and J. M. Soures, August 1986, p. 68.

## Particle Physics and Accelerators

"Elementary Particles and Forces" C. Quigg, April 1985, p. 84

"Quarks with Color and Flavor" S. L. Glashow, October 1975, p. 38

"The Mass of the Photon" A. S. Goldhaber and M. M. Nieto, May 1976, p. 86

"Quarkonium" E. D. Bloom and G. J. Feldman, May 1982, p. 66

"The LEP Collider" S. Myers and E. Picasso, July 1990, p. 54

"The Discovery of the Top Quark" T. M. Liss and P. L. Tipton, September 1997, p. 54

"Tracking and Imaging Elementary Particles" H. Breuker, H. Drevermann, C. Grab, A. D. Rademakers, and H. Stone, August 1991, p. 58

"The Tevatron" L. M. Lederman, March 1991, p. 48

"The Stanford Linear Collider" J. R. Rees, October 1989, p. 58

## Gravity

"Black Holes in Galactic Centers" M. J. Rees, November 1990, p. 56.

"The Quantum Mechanics of Black Holes" S. W. Hawking, January 1977, p. 34.

"Gravitational Lenses" E. L. Turner, July 1988, p. 54.

"Gravitational Waves from an Orbiting Pulsar" J. M. Weisberg, J. H. Taylor, and L. A. Fowler, October 1981, p. 74.

## Astrophysics

"The Early Life of Stars" S. W. Stahler, July 1991, p. 48

"Dark Matter in Spiral Galaxies" V. C. Rubin, June 1983, p. 96

"Colossal Galactic Explosions" S. Veilleux, G. Cecil, and J. Bland-Hawthorn, February 1996, p. 98

"The Earth's Elements" R. P. Kirshner, October 1994, p. 58

## Cosmology

"Primordial Deuterium and the Big Bang" C. J. Hogan, December 1996, p. 68.

"The Evolution of the Universe" P. J. E. Peebles, D. N. Schramm, E. L. Turner, and R. G. Kron, October 1994, p. 52.

"The Inflationary Universe" A. H. Guth, P. J. Steinhardt, May 1984, p. 116.

"The Cosmic Background Explorer" S. Gulkis, P. M. Lubnin, S. S. Meyer, and R. F. Silverman, January 1990, p. 132.

"Surveying Space-time with Supernovae" C. J. Hogan, R. P. Kirshner, and N. B. Suntzeff. January 1999, p. 46

"Particle Acclelerators Test Cosmological Theory" D. N. Schramm and G. Steigman, June 1988, p. 66.

# Photo Credits

## Chapter 2

**2–3** U.S. Courtesy of the U.S. Naval Academy Museum
**2–8** Permission granted by The Albert Einstein Archives, The Jewish National and University Library, The Hebrew University of Jerusalem, Israel. AIP/ESVA Neils Bohr Library

## Chapter 3

**3–6** Courtesy of Lawrence Berkeley National Laboratory

## Chapter 4

**4–2** AIP Emilio Segré Visual Archives, W. F. Meggers Collection
**4–7** Courtesy of the Archives, California Institute of Technology. Albert Einstein™. Represented by The Roger Richman Agency, Inc., Beverly Hills, CA 90212    **4–13** AIP Meggers Gallery of Nobel Laureates    **4–18** Hitachi, Ltd. Advanced Research Laboratory

## Chapter 5

**5–3** The Niels Bohr Archive, Copenhagen

## Chapter 6

**6–3** Copyright © by Photo Pfaundler, A–6020 Innsbruck

## Chapter 7

**7–1** Reprinted by permission of CERN–Press Office    **7–8** AIP Emilio Segré Visual Archives, Segré Collection

## Chapter 8

**8–12** JET Joint Undertaking    **8–17** National Institute of Standards and Technology (NIST)    **8–18** IBM Corporation, Research Division, Almaden Research Center

## Chapter 9

**9–12** Yerkes Observatory Photograph    **9–19** Tony Stone Images

## Chapter 10

**10–2** AIP Emilio Segré Visual Archives, Goudsmit Collection
**10–3** Photo by Samuel A. Goudsmit, courtesy of AIP Emilio Segré Visual Archives    **10–4** AIP Emilio Segré Visual Archives

## Chapter 12

**12–16** Professor Carl E. Wieman/The JILA BEC Group
**12–18** Professor J. F. Allen

## Chapter 14

**14–29** Courtesy of Lucent Technologies Bell Labs Innovations

## Chapter 15

**15–1** Photograph by C. E. Wynn–Williams, courtesy AIP Emilio Segré Visual Archives    **15–2** AIP Emilio Segré Visual Archives, Margrethe Bohr Collection    **15–6** AIP Emilio Segré Visual Archives

## Chapter 16

**16–1** Courtesy of Lawrence Berkeley National Laboratory
**16–2b** ICRR (Institute for Cosmic Ray Research), The University of Tokyo    **16–3** AIP Emilio Segré Visual Archives, Physics Today Collection    **16–7** Rare Book and Manuscript Library, Columbia University    **16–16** AIP Meggers Gallery of Nobel Laureates
**16–21b** CERN/European Organization for Nuclear Research
**16–21c** CERN/European Organization for Nuclear Research

## Chapter 17

**17–10 (1)** Kavan Ratnatunga (Carnegie Mellon Univ.), NASA
**(2)** Kavan Ratnatunga (Carnegie Mellon Univ.), NASA    **(3)** Kavan Ratnatunga (Carnegie Mellon Univ.), NASA    **(4)** Kavan Ratnatunga (Carnegie Mellon Univ.), NASA    **17–13** NASA/Ames Research Center/Ligo Project    **17–14** Princeton Alumni Weekly Photograph Collection. University Archives, Princeton University Library    **17–15** AIP Emilio Segré Visual Archives, Physics Today Collection

## Chapter 18

**18–3** Palomar Observatory, courtesy AIP Emilio Segré Visual Archives    **18–6** AIP Emilio Segré Visual Archives    **18–7 (1)** Courtesy of Lucent Technologies Bell Labs Innovations    **(2)** Courtesy of Lucent Technologies Bell Labs Innovations    **(3)** Lucent Technologies/Bell Laboratories

## About the Authors

**(1)** Courtesy of Jeremy Bernstein    **(2)** Courtesy of Paul Fishbane
**(3)** Courtesy of Stephen Gasiorowicz

# Index